The Handbook of Plant Functional Genomics

Edited by
Günter Kahl and Khalid Meksem

Related Titles

Meksem, K., Kahl, G. (eds.)

The Handbook of Plant Genome Mapping

Genetic and Physical Mapping

2005

ISBN: 978-3-527-31116-3

Kahl, G.

The Dictionary of Gene Technology

Genomics, Transcriptomics, Proteomics

2004

ISBN: 978-3-527-30765-4

Dolezel, J., Greilhuber, J., Suda, J. (eds.)

Flow Cytometry with Plant Cells

Analysis of Genes, Chromosomes and Genomes

2007

ISBN: 978-3-527-31487-4

Ahmad, I., Pichtel, J., Hayat, S. (eds.)

Plant-Bacteria Interactions

Strategies and Techniques to Promote Plant Growth

2008

ISBN: 978-3-527-31901-5

Cullis, C. A.

Plant Genomics and Proteomics

2004

ISBN: 978-0-471-37314-8

The Handbook of Plant Functional Genomics

Concepts and Protocols

Edited by
Günter Kahl and Khalid Meksem

WILEY-VCH Verlag GmbH & Co. KGaA

The Editors

Prof. Dr. Günter Kahl
Mohrmühlgasse 3
63500 Seligenstadt
Germany

Prof. Dr. Khalid Meksem
Department of Plant, Soil and
Agricultural Systems
Southern Illinois University
Carbondale, IL 62901-4415
USA

Library of Congress Card No.: applied for

British Library Cataloguing-in-Publication Data
A catalogue record for this book is available from the British Library.

Bibliographic information published by the Deutsche Nationalbibliothek
Die Deutsche Nationalbibliothek lists this publication in the Deutsche Nationalbibliografie; detailed bibliographic data are available in the Internet at http://dnb.d-nb.de.

Typesetting Thomson Digital, Noida, India
Printing Strauss GmbH, Mörlenbach
Binding Litges & Dopf GmbH, Heppenheim
Cover Design Adam-Design, Weinheim

Printed in the Federal Republic of Germany
Printed on acid-free paper

ISBN: 978-3-527-31885-8

Dedicated to
Sigrid (Siggi) Kahl
for her life-long patience and understanding

Contents

The Handbook of Plant Functional Genomics: Concepts and Protocols.
Edited by Günter Kahl and Khalid Meksem

ISBN: 978-3-527-31885-8

Preface

More than 612 bacterial, 51 archaeal, 1283 mitochondrial, 122 plastid, 47 fungal, and 82 higher eukaryote genomes have now been fully sequenced (http://www.genomesonline.org/), and in consequence, a wealth of sequence data has become available in the various public and private data banks. Notwithstanding the massively accumulated genome sequence information, an impressive number of additional pro- and eukaryotic genomes are currently being sequenced (some 1752 bacterial, 91 archaeal, and 908 eukaryotic genomes, plus an additional 116 metagenomes). The speed of whole genome sequencing is ever-increasing and will surpass any predictions of the past, given the available series of novel second- and third-generation sequencing technologies such as picoliter pyrosequencing (454 Life Sciences, see *Nature* (2005) **437**: 376–380), the 1G Genetic Analyzer System manufactured by Solexa, the SOLiD system developed by Applied Biosystems, or the truly single-molecule sequencing platform produced by Helicos Biosciences, to name but a few. And with the increased speed of sequencing, more and more sequence data will swamp the data banks.

While the genome sequences will inform us of the precise genome size, the gross and fine genome architecture, various parameters such as e.g. the GC content, the distribution of various sequence elements (e.g. microsatellites, transposons and retrotransposons) and putative open reading frames (among many other features), the identification of regulatory sequences (promoters, enhancers, silencers) and of genes themselves still poses an extraordinary, mostly bioinformatic challenge. However, the functions of the various sequence elements still remain largely obscure. For example, of the 20–25 000 putative human genes, only about 30% encode proteins with known functions. The situation looks much worse for other animals, not to mention plants. Therefore, a great deal of analyses will necessarily be devoted to the *functions of all sequence elements* of a genome, and this enormous task will be handled by what has been coined *functional genomics*, and will be the focus of the so-called post-genomic era.

The present book *The Handbook of Plant Functional Genomics (Concepts and Protocols)* follows the unexpectedly successful and well received first book *The Handbook of Plant Genome Mapping* in the Wiley-VCH series of *Handbooks of Plant*

The Handbook of Plant Functional Genomics: Concepts and Protocols.
Edited by Günter Kahl and Khalid Meksem

ISBN: 978-3-527-31885-8

Genome Analysis. The present second handbook compiles all techniques presently developed for the various functional genomics approaches, presents informative introductions to each chapter, and robust and ready-to-go laboratory protocols written by internationally renowned experts in their research fields. Although the Handbook focuses on *plant* functional genomics, some promising techniques that have been successfully introduced into the field of medicine are also included, some of which may be new to plant biologists. Yet the potential of these technologies may well catalyze plant research. In any case, they reflect the present state-of-the-art in functional genomics generally.

The editors very much appreciate that all the authors have contributed excellent chapters, and expect that this book *The Handbook of Plant Functional Genomics (Concepts and Protocols)* will reproduce the worldwide success of the first handbook in this series.

Günter Kahl
March 2008 Frankfurt am Main (Germany)
Khalid Meksem
Carbondale (USA)

List of Contributors

Christian W. B. Bachem
University and Research Centre
Department of Plant Sciences
Laboratory of Plant Breeding
Wageningen
Droevendaalsesteeg 1
6708PB Wageningen
The Netherlands

Anke Becker
Bielefeld University
Institute for Genome Research and Systems Biology
Center for Biotechnology
33594 Bielefeld
Germany

Abdelhafid Bendahmane
DR2 INRA
INRA/CNRS-URGV
2 Rue Gaston Crémieux CP 5708
91057 EVRY Cedex
France

Thomas Berberich
Iwate Biotechnology Research Center (IBRC)
22-174-4 Narita, Kitakami
Iwate 024-0003
Japan

Steven Bernacki
North Carolina State University
Plant Biology and Genetics
851 Main Campus Drive
Raleigh, NC 27606
USA

Craita E. Bita
Radboud University Nijmegen
Department of Plant Cell Biology
Toernooiveld 1
6525 ED Nijmegen
The Netherlands

José R. Botella
ARC Centre of Excellence for Integrative Legume Research
Brisbane
4072 Queensland
Australia

Michael Braverman
454 Life Sciences
VP, Molecular Biology
20 Commercial Street
Branford, CT 06405
USA

The Handbook of Plant Functional Genomics: Concepts and Protocols.
Edited by Günter Kahl and Khalid Meksem

ISBN: 978-3-527-31885-8

Chris A. Brosnan
ARC Centre of Excellence for Integrative Legume Research
Brisbane
4072 Queensland
Australia

Thomas P. Brutnell
Cornell University
Boyce Thompson Institute for Plant Research
Tower Road
Ithaca, NY 14853-1801
USA

C. Robin Buell
The Institute for Genomic Research (TIGR)
Plant Genomics Group
9712 Medical Center Drive
Rockville, MD 20850
USA

Asun Fernandez del Carmen
Universidad Politécnica
Instituto de Biología Molecular y Celular de Plantas
Consejo Superior de Investigaciones Científicas
Avenida Tarongers s/n
46022 Valencia
Spain

Piero Carninci
RIKEN Genomic Sciences Center (GSC)
Laboratory for Genome Exploration
RIKEN Yokohama Institute
1-7-22 Suehiro-cho
Tsurumi-ku
Yokohama, Kanagawa 230-0045
Japan

Bernard J. Carroll
ARC Centre of Excellence for Integrative Legume Research
Brisbane
4072 Queensland
Australia

Toni L. Ceccardi
Applied Biosystems
Molecular Biology Division
850 Lincoln Centre Drive
Foster City, CA 94404
USA

Caifu Chen
Applied Biosystems
Molecular Biology Division
850 Lincoln Centre Drive
Foster City, CA 94404
USA

Liza Conrad
Boyce Thompson Institute for Plant Research
Tower Road
Ithaca, NY 14853
USA

Rick C. Conrad
Ambion
An Applied Biosystems Business
Austin, TX 78744
USA

Lei Du
454 Life Sciences
VP, Molecular Biology
20 Commercial Street
Branford, CT 06405
USA

Michael Egholm
454 Life Sciences
VP, Molecular Biology
20 Commercial Street
Branford, CT 06405
USA

Marianna M. Goldrick
Ambion
An Applied Biosystems Business
Austin, TX 78744
USA

Zhen Guo
GenHunter Corporation
624 Grassmere Park Drive
Nashville, TN 37211
USA

Matthias Harbers
DNAFORM Inc.
Leading Venture Plaza 2
75-1 Ono-cho, Tsurumi-ku
Yokohama
Kanagawa 230-0046
Japan

Reinhard Hehl
Technische Universität Braunschweig
Institut für Genetik
Spielmannstr. 7
38106 Braunschweig
Germany

Richard Henfrey
Applied Biosystems
Molecular & Cell Biology Division
850 Lincoln Centre Dr.
Foster City, CA 94404
USA

Aziz Jamai
Dartmouth College
Department of Biological Sciences
Hanover
NH, 03755-3576
USA

Thomas Jarvie
454 Life Sciences
20 Commercial Street
Branford, CT 06405
USA

Günter Kahl
GenXPro GmbH
Frankfurt Innovation Center
Biotechnology (FIZ)
Altenhöferallee 3
60438 Frankfurt am Main
Germany

Kazuhiro Kikuchi
Boyce Thompson Institute for Plant
Research
Tower Road
Ithaca, NY 14853
USA

Juergen Kleffe
Institut für Molekularbiologie und
Bioinformatik
Charite-Campus Benjamin Franklin
Arnimallee 22
14195 Berlin
Germany

Leonard Krall
Max Planck Institute of Molecular Plant
Physiology
Research Group Genes and Small
Molecules
Am Muehlenberg 1
14476 Potsdam-Golm
Germany

Erhard Kranz
Universität Hamburg
Biozentrum Klein Flottbek und
Botanischer Garten
Entwicklungsbiologie und
Biotechnologie
Ohnhorststr. 18
22609 Hamburg
Germany

Nandini Krishnamurthy
University of California at Berkeley
Department of Bioengineering
473 Evans Hall
Berkeley, CA 94720-1762
USA

Detlev H. Krüger
Institute of Virology
Helmut-Ruska-Haus
Charité Medical School
Campus Charité Mitte
10098 Berlin
Germany

Helge Küster
Bielefeld University
Institute for Genome Research and
Systems Biology
Center for Biotechnology
33594 Bielefeld
Germany

Yen-Ling Lee
GENOME
Genome Institute of Singapore
60 Bipolis Street
Singapore 138672
Singapore

Jim Leebens-Mack
University of Georgia
Department of Plant Biology
Athens, GA 30602-7271
USA

Joshua G. Liang
Montgomery Bell Academy
Nashville, TN 37205
USA

Julia Z. Liang
Harpeth Hall School
Nashville, TN 37215
USA

Peng Liang
Vanderbilt University
Vanderbilt-Ingram Cancer Center
School of Medicine
691 Preston Building
Nashville, TN 37232
USA

Shiming Liu
Southern Illinois University at
Carbondale
Plants and Microbes Genomics and
Genetics Laboratory
Carbondale, IL 62901-4415
USA

Xiao Hong Liu
University of Missouri-Columbia
Division of Plant Sciences
371H Life Sciences Center
Columbia, MO, 65211-7310
USA

Hideo Matsumura
Iwate Biotechnology Research Center
(IBRC)
Narita 22-174-4, Kitakami
Iwate 024-0003
Japan

Emily J. McCallum
ARC Centre of Excellence for Integrative Legume Research
Brisbane
4072 Queensland
Australia

Jonathan Meade
GenHunter Corporation
624 Grassmere Park Drive
Nashville, TN 37211
USA

Khalid Meksem
Southern Illinois University at Carbondale
Plants and Microbes Genomics and Genetics Laboratory
Carbondale, IL 62901-4415
USA

Tarik El Mellouki
Southern Illinois University at Carbondale
Plants and Microbes Genomics and Genetics Laboratory
Carbondale, IL 62901-4415
USA

Melissa Goellner Mitchum
University of Missouri-Columbia
Division of Plant Sciences
371H Life Sciences Center
Columbia, MO, 65211-7310
USA

Nooduan Muangsan
Khon Kaen University
Biology Department
123 Mittraparb Road, Muang District
Khon Kaen, 40002
Thailand

Samuel Nahashon
Tennessee State University
Institute of Agricultural and Environmental Research
Nashville, TN 37209
USA

Patrick Wei Pern Ng
GENOME
Genome Institute of Singapore
60 Biopolis Street
Singapore 138672
Singapore

Robert C. Nutter
Applied Biosystems
High Throughput Discovery Business Unit
850 Lincoln Centre Drive
Foster City, CA 94404
USA

Yoshiyuki Ogata
Kazusa DNA Research Institute
The NEDO Team of Applied Plant Genomics
2-6-7 Kazusa-Kamatari, Kisarazu
Chiba 292-0818
Japan

Giles Oldroyd
John Innes Centre
BBSRC David Phillips Fellow
Department of Disease and Stress Biology
Norwich Research Park
Colney Lane
Norwich NR4 7UH
UK

Nicholas J. Provart
University Toronto
Plant Bioinformatics
Department of Cell and Systems Biology
25 Willcocks St.
Toronto, ON, M5S 3B2
Canada

Peifeng Ren
BASF Plant Sciences L.L.C.
26 Davis Drive
Research Triangle Park, NC 27709
USA

Dominique Robertson
North Carolina State University
Plant Biology and Genetics
851 Main Campus Drive
Raleigh, NC 27606
USA

Christian Rogers
John Innes Centre
Department of Disease and Stress
Biology
Colney
Norwich NR4 7UH
UK

Yijun Ruan
GENOME
Genome Institute of Singapore
60 Bipolis Street
Singapore 138672
Singapore

Hiromasa Saitoh
Iwate Biotechnology Research Center
(IBRC)
22-174-4 Narita, Kitakami
Iwate 024-0003
Japan

Nozomu Sakurai
Kazusa DNA Research Institute
2-6-7 Kazusa-Kamatari, Kisarazu
Chiba 292-0818
Japan

Stefan Scholten
Universität Hamburg
Biozentrum Klein Flottbek und
Botanischer Garten
Entwicklungsbiologie und
Biotechnologie
Ohnhorststr. 18
22609 Hamburg
Germany

Blake Shester
GenHunter Corporation
624 Grassmere Park Drive
Nashville, TN 37211
USA

Jan Frederik Simons
454 Life Sciences
20 Commercial Street
Branford, CT 06405
USA

Kimmen Sjölander
University of California at Berkeley
Department of Bioengineering
473 Evans Hall
Berkeley, CA 94720-1762
USA

Maithreyan Srinivasan
454 Life Sciences
20 Commercial Street
Branford, CT 06405
USA

Dirk Steinhauser
Max Planck Institute of Molecular Plant Physiology
Research Group Genes and Small Molecules
Am Muehlenberg 1
14476 Potsdam-Golm
Germany

C. Neal Stewart Jr.
2431 Joe Johnson Dr.
University of Tennessee
Department of Plant Sciences
Knoxville, TN 37996-4561
USA

Bruce Taillon
454 Life Sciences
VP, Molecular Biology
20 Commercial Street
Branford, CT 06405
USA

Yoshihiro Takahashi
Iwate Biotechnology Research Center (IBRC)
22-174-4 Narita, Kitakami
Iwate 024-0003
Japan

Raimo Tanzi
Applied Biosystems
Applera Italia
Via Tiepolo 18
20052 Monza MI
Italy

Ryohei Terauchi
Iwate Biotechnology Research Center (IBRC)
Narita 22-174-4, Kitakami
Iwate 024-0003
Japan

Françoise Thibaud-Nissen
The J. Craig Venter Institute
9704 Medical Center Drive
Rockville, MD 20850
USA

John Richard (Rich) Tuttle
North Carolina State University
Plant Biology and Genetics
851 Main Campus Drive
Raleigh, NC 27606
USA

Wim H. Vriezen
Radboud University Nijmegen
Department of Plant Cell Biology
Toernooiveld 1
6525 ED Nijmegen
The Netherlands

Jamie C. Walden
GenHunter Corporation
624 Grassmere Park Drive
Nashville, TN 37211
USA

Chia-Lin Wei
GENOME
Genome Institute of Singapore
60 Bipolis Street
Singapore 138672
Singapore

Thomas Werner
Genomatix Software GmbH
Bayerstrasse 85a
80335 Munich
Germany

Jennifer Wortman
The Institute for Genomic Research (TIGR)
Informatics Department
9712 Medical Center Drive
Rockville, MD 20850
USA

Wei Zhu
The J. Craig Venter Institute
9704 Medical Center Drive
Rockville, MD 20850
USA

Suping Zhou
Tennessee State University
Institute of Agricultural and Environmental Research
Nashville, TN 37209
USA

I
Transcriptome Analysis

A
Whole Genome Expression Analysis

The Handbook of Plant Functional Genomics: Concepts and Protocols.
Edited by Günter Kahl and Khalid Meksem

ISBN: 978-3-527-31885-8

1
Single Cell Expression Profiling: Transcript and Protein Analyses in Isolated Higher Plant Gametes and Zygotes

Stefan Scholten and Erhard Kranz

Abstract

In recent years the interest in analyses of single cells has increased continually in the biological and medical sciences. Knowledge regarding the behavior of the single cell as the basic unit of an organism is important for our understanding of the whole organism. Thus, a broad and detailed knowledge of the processes of, for example, signal transduction, cellular communication, cell division, differentiation and development of cells is therefore important in basic and applied sciences. In this chapter we will describe techniques of micromanipulation, including micro-dissection of tissues, isolation and handling of individual cells, cell fusion and analyses of transcript and protein expression of single or a small number of cells. This chapter reviews these methodologies and applications, focuses in particular on gamete isolation, *in vitro* fertilization (IVF) and studies on gamete identity and early seed development in higher plants. The potential of single cell expression profiling for plant genome analysis are discussed.

1.1
Introduction

Gametes and zygotes are special cells that are worthy of study at the single cell level, and *in vitro* fertilization of such isolated cells provides a powerful system for developmental studies. Therefore, methods were developed to isolate single gametes from higher plants in order to study methods of gamete characterization, and techniques for gamete fusion and single cell culture systems for the exploration of the first steps in zygote and embryo formation. Double fertilization is the fusion of one sperm with the egg to create the embryo and the fusion of the other sperm with the central cell to form the endosperm [1]. Gamete fusion occurs *in vivo* deep within the ovule tissues in the embryo sac, and generally with the help of one of the two synergids. In contrast to animals and lower plants, *in vitro* fertilization in higher plant

The Handbook of Plant Functional Genomics: Concepts and Protocols.
Edited by Günter Kahl and Khalid Meksem

ISBN: 978-3-527-31885-8

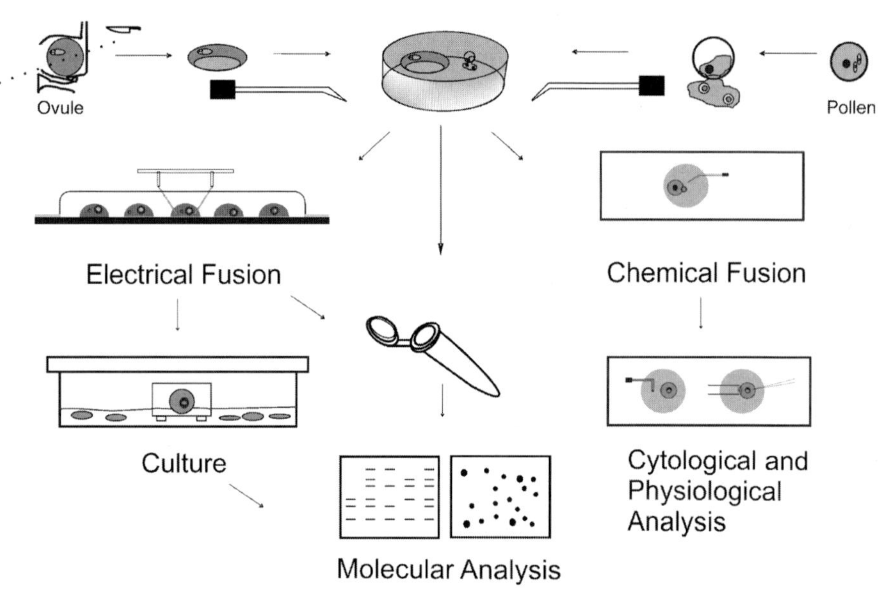

Figure 1.1 *In vitro* fertilization of single gametes of maize. Isolation: ear spikelets are cut as indicated (dotted lines, left). Cells from the embryo sac are manually isolated from nucellar tissue pieces in a plastic dish using a needle, following transfer with a capillary into microdroplets on a coverslip for gamete fusion. Subsequently, sperm cells are selected in the isolation chamber after release from pollen grains by osmotic shock (right), and a sperm is transferred into a microdroplet containing an egg cell. The figure shows manipulations in microdroplets placed on a coverslip for electrofusion (left), calcium-mediated cell fusion (right), cytological and physiological analyses, for example, microinjection and ion measurements, staining and immunochemistry on a coverslip (right, below), and culture in 'Millicell'-dishes for growth and developmental analyses (left, below) as well as molecular analyses (in the middle, below) ([46] modified).

gametes presupposes their isolation. The egg and central cells need to be isolated from an embryo sac which is generally embedded in the nucellar tissue of the ovule and normally contains two synergids and some antipodal cells. Moreover, sperm cells should be isolated from pollen grains or tubes.

With modified microtechniques, originally developed for somatic protoplast fusion [2,3], defined gamete fusion is possible (Figures 1.1 and 1.2). Because they are protoplasts, individual isolated gametes have been fused electrically, for example, [4–12] and chemically using calcium [13–15] or by polyethylene glycol [16–18]. Zygotes and primary endosperm cells can be cultured in tiny droplets of culture medium which are covered by mineral oil [5,7,10,19]. Thus, very early steps in zygote and endosperm development can be analyzed without feeder cells. Sustained growth of zygotes and endosperm can be achieved by co-cultivation with feeder cells [4,7–9,20,21]. Embryogenesis and plant formation from isolated male and female gametes fused *in vitro* have been achieved in maize and rice using electrofusion techniques

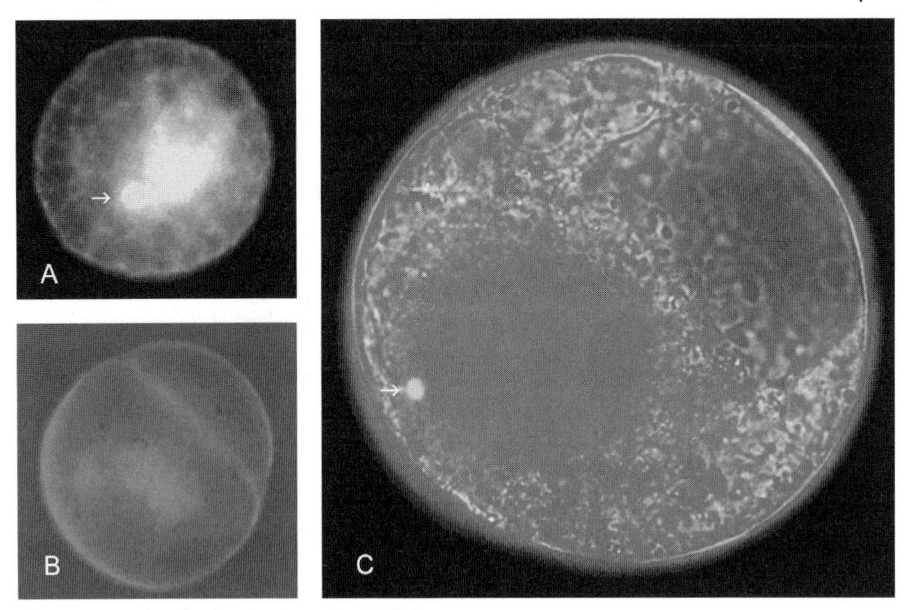

Figure 1.2 *In vitro* fertilization in maize. Epifluorescence micrographs of a fertilized egg cell, stained with DAPI (A; [7]), two-celled embryo, cell wall stained with calcofluor white (B; [7]) and fertilized central cell, stained with DAPI (C; [8]). Arrows indicate integrated sperm nuclei.

and nurse culture [4,12]. In maize, endosperm development has also been triggered after *in vitro* fusion of the sperm-central cell [8]. In these systems, zygote, embryo, plant and endosperm development take place in the absence of mother tissue, as is the case with endosperm formation without an embryo, and embryo development without endosperm.

By using single-cell micromanipulation techniques, the possibility of selection, transfer and handling of single gametes, zygotes and primary endosperm cells together with a high frequency of fusion and cell divisions enables physiological and molecular studies at the single cell level. Even using only a relatively minute amount of material, sufficient numbers of such cells can be obtained to allow the study of gene and protein expression especially of those genes involved in early events in zygote formation, early embryogenesis and endosperm development [21–26]. For example, we developed an immunocytochemical procedure to examine subcellular protein localization in isolated and cultured single cells [27]. This method is described in this chapter. Using RT-PCR (reverse transcription-polymerase chain reaction) methods, cDNA-libraries have been generated from egg cells [28] and *in vitro* zygotes [29] to isolate egg- and central cell-specific [30] and fertilization-induced genes (e.g. [24]). Further, we describe the use of micromanipulation and IVF techniques for example, to separate apical and basal cells from the two-celled embryo. Additionally, specific gene expression has been introduced into these cells in order to elucidate mechanisms of early embryonic patterning in higher plants [24,25]. Moreover, a description of an adapted method for the analyses of lysates from a few egg cells and zygotes by polyacrylamide gel electrophoresis and subsequent mass spectrometry-based proteomics technology is presented to identify major protein components expressed in these cells [23,26,31]. The techniques are described for maize (*Zea mays*) unless otherwise mentioned. IVF systems offer the potential to (1) analyze zygotes, primary endosperm cells, very young embryos and endosperm at stages characterized by an exactly defined time after fertilization, (2) study cellular events which take place immediately after fertilization and (3) produce zygotes from gametes of different cultivars. The potential of IVF to elucidate mechanisms of fertilization and early development have been reviewed, for example [32–35]. Here, we describe the application of molecular tools to studies aimed at producing a more detailed characterization of gametes and early post-fertilization events which are based upon microdissection and *in vitro* fertilization techniques.

1.2 Microdissection, Cell Isolation

In contrast to animals and lower plants, female angiosperm gametes are deeply embedded within the maternal tissues. Sperm cells need to be isolated from pollen grains or tubes, and female gametes from an embryo sac which is generally embedded in the nucellar tissue of the ovule (Figure 1.1). Double fertilization which is frequently assisted by one of the two synergids, generally occurs within the ovular tissue of the embryo sac. Sperm cells are isolated by osmotic burst, squashing or grinding of the pollen grains or tubes. Female gametes can be obtained by mechanical means using, for example, thin glass needles [6,7,20,36–38], but also by using mixtures of cell wall-degrading enzymes in combination with a manual isolation procedure. In maize, for example, treatment of the nucellar tissue with such a mixture of enzymes for a short period of time prior to the manual isolation step is often useful for softening this tissue to avoid rupture of the gametic protoplasts [5].

1.3 *In Vitro* Fertilization

Three basic microtechniques are involved in *in vitro* fertilization (IVF): (1) the isolation, handling and selection of male and female gametes, (2) the fusion of pairs of gametes, and (3) the single cell culture. The first *in vitro* fusion of isolated, single female and male angiosperm gametes was developed nearly 20 years ago as a result of the application and development of several new micromanipulation techniques. Some of these were originally developed for somatic protoplasts. Additionally, it became necessary to develop an efficient single cell culture system to ensure the sustained growth of single or small numbers of zygotes and primary endosperm cells.

Fusion of individual isolated plant gametes can be achieved by three different methods: (1) electrically (for example [5,6,8,9,11,12]), (2) chemically, using calcium [8,13–15,39–41] or (3) polyethylene glycol (for example [16–18,42]). Cell fusion using electrical pulses is a well-established and efficient method for producing sufficient numbers of zygotes for use in growth studies and molecular analyses. It is conceivable that media including calcium may be used to determine the conditions and factors which promote adhesion, *in vivo* membrane fusion, and possibly recognition events taking place during the fertilization process. However, the efficiency of this method has still to be optimized [13,15]. To date, this method has been used to study the differential contribution of cytoplasmic Ca^{2+} and Ca^{2+} influx during gamete fusion and egg activation [34,39–41]. Although early development of zygotes can be initiated after Ca^{2+}-mediated gamete fusion [13], there are no reports of sustained development of embryos or plant regeneration resulting from this type of fusion. One reason for this might be that the number of zygotes obtained by this method is insufficient for use in growth experiments.

Development after *in vitro* gamete fusion was achieved in monocots (e.g. in maize, wheat and rice), but not in dicots. The cell size of the gametes of various dicotyledonous plants (e.g. *Arabidopsis*) is small compared to that of cereals, a characteristic which hampers their manipulation and handling. In addition, a particularly critical factor for egg activation might be that the stages of the cell cycle in the isolated male and female gametes are not synchronous and therefore cannot activate the nuclear and cell division processes in the zygote [43,44]. Micromanipulation, microdissection and single gamete fusion techniques have been described in detail elsewhere [45,46].

1.4 Techniques for Molecular Analyses of Single Cell Types

1.4.1 Sampling of Single, Living Cells

The collection and storage of single, living cells is an essential step which must be carried out prior to molecular analysis. This can be achieved in various ways depending on the approach, and the number of cells required at one time. Generally, it is important to freeze or lyse the cells as fast as possible after isolation and to minimize the number of handling steps in order to avoid degradation and loss of the limited material. Mannitol is usually used to adjust the osmolality during cell isolation and does not interfere with any of the approaches described here. Long-term storage of single cells or groups of cells in mannitol solution droplets is feasible after snap-freezing in liquid nitrogen at −70 to −80 °C for periods of more than 1 year. Transient storage in this way can accumulate sufficient cells for use in experiments which require a specific number of cells. An alternative procedure is the direct lysis of cells after isolation in sodium dodecyl sulfate- (SDS) or lithium dodecyl sulfate (LiDS)-containing buffer at concentrations of at least 1%. These buffers inhibit all enzymatic activity and therefore preserve the integrity of the nucleic acids and proteins. For that reason, it is practical to collect a number of cells during the isolation procedure in lysis buffer and freeze them collectively for storage. However, the following approach should either be SDS insensitive, or it should be possible to remove the SDS quantitatively during the procedure preceding enzymatic reactions. For analysis of single cells the individual cells should be directly transferred into and stored in the tube which is to be used in subsequent steps. For single cell reverse transcription-polymerase chain reaction (RT-PCR) it is essential that the cell isolation buffer is compatible with the reverse transcriptase.

1.4.2 Analyses of Gene Expression

The analysis of gene expression at the transcript level in specific cell types always requires a step in which the message is amplified to detectable amounts. Selected combinations of techniques for transcript analyses using limited amounts of sample which were all successfully carried out with single plant gametes or small numbers of single cell types from the female or male gametophyte are described in the following sections.

1.4.2.1 Single Cell Gene-by-Gene Analysis

The simplest method of obtaining information relating to gene expression in a single cell is direct RT-PCR with the single cell as template. This method is useful if, for example, the segregation of a heterozygous locus is to be tested in gametes. Obvious disadvantages of this method are the limited number of genes which can be analyzed in one reaction, and the lack of quantitative expression data. The first disadvantage may be overcome, at least to some extent, by multiplex reactions.

Both RT and PCR reactions are carried out in the same tube in a thermocycler. Several one-tube RT-PCR kits are available from different suppliers and may be used for the analysis of single cells. However, the approach described in [22] was adapted for the analysis of plant gametes. It uses standard reagents in a two-step protocol and is modifiable to specific requirements depending on the genes of interest. The first step consists of the RT reaction. Most important is the rapid addition of the RT master-mix either before or at the time the cells are thawing, which avoids any RNA degradation. Richert *et al.* [22] did not use an initial step at 70 °C, but elevated reaction temperatures to melt secondary structures of the RNA template. Nevertheless, in some cases where the RNA templates have more stable secondary structures, higher temperatures may be necessary during some steps in the procedure. If a higher temperature step is necessary, the reaction mix protocol should be divided up. The primers alone should first be added to the cell, followed by the addition of the remainder of the RT reaction mixture in a second step preceding RT. It is reasonable to use gene-specific reverse primers for the RT reaction because this decreases the generation of unspecific products at elevated temperatures. PCR is the second step of the procedure, whereby the highest sensitivity is obtained when the PCR master mix is added to the complete RT reaction. After 40 cycles of PCR, highly abundant transcripts can be detected by agarose gel electrophoresis. If higher sensitivity is required, the gels may be blotted

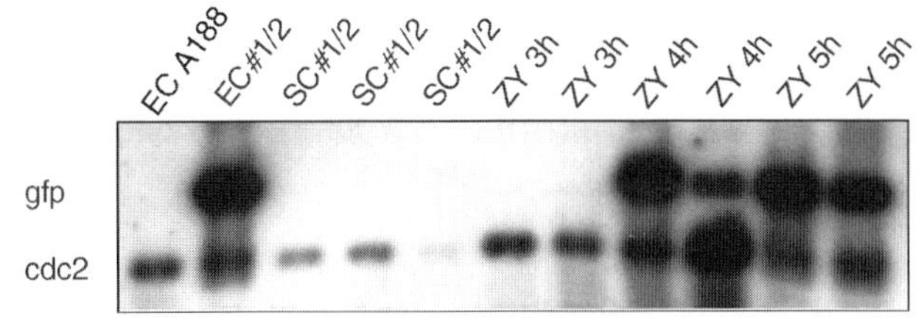

Figure 1.3 Multiplex transcript analysis in single cells. Transcripts of the GFP gene and the CDC2 gene were simultaneously detected in single egg cells (EC), groups of five sperm cells (SC) and single zygotes (ZY) at the indicated time points after fertilization. Egg cells from wild-type and transgenic plants line #1/2 were used as negative and positive control, respectively. GFP transcripts were not detectable in transgenic sperm cells. Between 3 and 4 h after fertilization of wild-type egg cells with transgenic sperm cells the paternal expression of the transgene was initiated. To enhance the signal, the PCR products were transferred to membranes and hybridized with probes against GFP and CDC2 sequences at the same time ([21], modified).

and hybridized with gene-specific probes. Further details of the procedure and some of the modifications mentioned here can be found in [21,47].

To analyze more than one gene simultaneously, a multiplex reaction can be performed. This is a simple way of reducing the time spent on isolating the cells and, importantly, provides amplification of a positive control in the same reaction, that is, by using the same cell. Designing primers for multiplex reactions follows similar principles to those used in normal PCR reactions. The primers should not fold into a hairpin, and none of the primers used in one reaction should form dimers. Primers should anneal to only one transcript sequence of the multiplex reaction. The performance of specific primer combinations should always be evaluated empirically. Figure 1.3 shows a multiplex reaction after blotting and detection by hybridization with two gene-specific probes in parallel. These experiments were performed in the context of paternal genome activation in maize zygotes and used the second gene of the multiplex RT-PCR, which is known to be constitutively expressed, as the positive control within the same cell.

1.4.2.2 **Amplification of Whole cDNA Populations**

Synthesis and amplification of cDNA populations representing all transcripts within specific cell types constitutes the basis for a number of approaches to the analysis of gene expression. Some of these approaches such as real time PCR, cDNA library construction, cDNA subtraction and microarray analyses are introduced in the following sections. Here we discuss important aspects of a cDNA synthesis and amplification protocol which has been successfully applied to a limited number of cells.

The first consideration is to decide how many cells should be used in a single cDNA synthesis reaction. Certainly, a larger number of cells will provide a better representation of low abundance transcripts. To achieve a representative cDNA population, we recommend using no less than 20 cells, because material is lost at each step in the procedure. However, some approaches have employed far fewer cells. For example, 10 cells were used to generate cDNA libraries of wheat egg cells and two-celled embryos [48], and five cells were sufficient to explore the differential gene expression in the apical and basal cell after the first zygotic division [24].

An important prerequisite for representative cDNA populations is highly efficient reverse transcription. Because reverse transcriptases are strongly inhibited by any impurities, the RNA preparations should be as clean as possible. An approved method for mRNA isolation which gives highly reproducible results in cDNA synthesis and amplification is the use of oligo $dT_{(25)}$-coated magnetic beads (Dynal, Invitrogen). The washing steps applied to the bead-bound mRNA are highly effective and result in pure mRNA. Another advantage is that the volume of the elution buffer can be reduced without the loss of any mRNA or the need to precipitate the sample. However, column-based kits or alternative protocols might also be suitable or adaptable for RNA isolation from small samples. With oligo $dT_{(25)}$-coated magnetic beads, only an analysis of transcripts with a poly(A) tail is possible. This, of course, represents a drawback if RNA species other than mRNAs are to be analyzed but offers a great advantage in transcript analyses, since it greatly reduces sample complexity and therefore the background. We always used SMART cDNA synthesis and amplification (Clontech) with our maize gamete samples. This method utilizes the ability of the Moloney murine leukemia virus reverse transcriptase (MMLV RT) to add a few non-template deoxynucleotides (mostly cytosines) to the 3′ end of a newly synthesized cDNA strand upon reaching the 5′ end of the RNA template. An oligonucleotide containing an oligo(rG) sequence at the 3′ end is added to the RT reaction together with base pairs containing the deoxycytidine stretch produced by MMLV RT. The reverse transcriptase then switches the templates, and continues replicating using the oligonucleotide as template. In this way, a unique and known sequence complementary to the added oligonucleotide is attached to the 3′ terminus of the first strand of the cDNA synthesized, which can serve as a universal forward primer-binding site to amplify the whole cDNA population. A universal reverse primer-binding site is also introduced during the RT reaction with a sequence at the 5′ end of the poly (T) primer. Although this method is less efficient than 'conventional' first and second strand synthesis followed by adapter ligation for the generation of universal primer-binding sites for the amplification of the whole cDNA population, it is simpler and quicker. The reduced handling requirements may outweigh these disadvantages, because loss of material may occur during each step. For a more detailed description and comprehensive discussion of both methods the reader is referred to [49]. Control of the amplification reaction is essential in both these cDNA synthesis methods. To avoid distortion of the cDNA samples and to preserve the original relative abundance of the transcript, the cycle number must be adjusted for each individual sample. The optimal cycle number required to produce the maximum concentration of non-distorted cDNA can easily be determined by running a test reaction and taking samples every three cycles. It is therefore important that the whole reaction mixture (usually 5 times 100 μl) is prepared immediately and cycled together for a specific number of cycles. After taking a sample for gel electrophoresis, an aliquot of the reaction mixture is cycled for another three cycles, with samples being taken between cycles. Meanwhile the remainder of the reaction is stored on ice. Once the optimal cycle number is determined, the PCR reaction is continued with

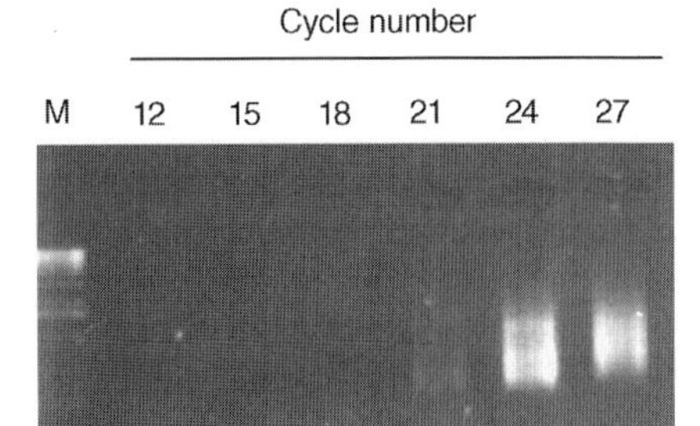

Figure 1.4 Determination of an optimal cycle number for global cDNA amplifications. Test reactions were withdrawn at the number of PCR cycles indicated, and the quantity of amplification product is shown. mRNA was isolated from 25 primary endosperm cells using SMART RT-PCR. The products first became visible after 21 PCR cycles. The concentration of PCR product leveled off between 24 and 27 cycles. Therefore, the optimal cycle number was determined to be 23 which was used for the rest of the reaction.

the appropriate cycle number. A typical cycle number optimization is shown in Figure 1.4. The cycle number for cDNA amplification should always be well below the plateau of the reaction (i.e. one or two cycles less than the cycle number after which no further increase is observed). In case of doubt we recommend a conservative approach in choosing the lower cycle number. Preservation of the original relative abundance of the transcript is the most important prerequisite for the generation of reliable data in downstream approaches using amplified cDNAs.

1.4.2.3 Quantification of Transcript Levels

Devising approaches to measure relative transcript levels quantitatively in single cells is challenging. A two step RT-PCR-based method with pre-amplification of several transcripts in a multiplex reaction followed by real-time PCR quantification of single gene transcripts in aliquots of the first reaction has been developed [50]. Interestingly, this approach revealed considerable cell-to-cell variations in an apparently homogenous T cell population. These findings highlight the individual transcriptional state of each single cell, and means that sampling a number of cells for analysis implies averaging transcriptional states. However, the number of genes which can be analyzed using this approach is limited. To obtain information about the expression levels of a large number of genes in specific cell types or to compare the expression level between two different cell types, sampling of several cells followed by cDNA synthesis and global amplification, as described above, provides a basis for generating a significant amount of expression data. By global amplification of all expressed genes within a cell type, a permanent cDNA archive can be generated and may serve as a template for hundreds of quantitative reactions. Direct comparisons of the relative levels of transcripts between first strand cDNA and amplified cDNA populations revealed that both methods produced comparable results [51]. In our hands, the comparison of quantitative RT-PCR results from amplified and non-amplified cDNA by quantification of the actin gene with two different template concentrations for each cDNA, revealed a high correlation coefficient of 0.96 [52].

1.4.2.4 Library Construction and EST Sequencing

To identify and clone new genes, the generation of cDNA libraries is a reasonable step especially if the work is being carried out with non-model species. In addition, cDNA libraries may serve as templates to produce arrays of individual, unknown cDNA fragments which can be used to characterize the expression pattern of the corresponding genes by hybridization methods. This procedure is useful for selecting genes which are potentially involved in a process of interest. To obtain information about the expression profile of single cell types and identification of new genes or transcripts within a cell, the generation of ESTs is an effective method. Several examples show that ESTs of specific cell types uncover previously unknown sequences, even in model species, where several hundred thousand ESTs already exist [48,53].

The preparation of cDNA libraries from small samples starts with the global amplification of cDNAs as described. It is important to implement restriction enzyme sites in the primer or adaptor sequences to simplify the cloning procedure and to increase its efficiency. Because the cDNA is not limited after the global amplification procedure a size fractionation step before cloning may enhance the fraction of full-length cDNAs, or at least long cDNA fragments. Various standard procedures can be followed to generate cDNA libraries with amplified cDNA populations. The SMART cDNA library construction (kit manufactured by Clontech) is a convenient method for cDNA library construction which we used successfully with samples of 25 maize central cells in combination with mRNA isolation on oligo $dT_{(25)}$-coated magnetic beads.

1.4.2.5 Targeted Approaches Using cDNA Subtraction

Specific approaches aimed at identifying differentially expressed genes between various cell types involved in plant reproduction by applying randomly amplified polymorphic DNA (RAPD) primer-driven PCR or suppression subtractive hybridization (SSH) were successful [24,54]. To identify highly abundant, differentially expressed transcripts in subtracted cDNA populations, differential screening using microarray hybridizations is highly effective [30]. The basis for all these approaches is a global cDNA amplification procedure. A disadvantage of the SSH technique is that the cDNAs must be restricted before hybridization to equalize the hybridization efficiencies of the diverse cDNAs within the population. Full-length cDNAs of interesting clones need to be reconstituted in a second step using other methods such as rapid amplification of cDNA ends (RACE). Control PCR reactions after SSH are shown in Figure 1.5. These controls are important in indicating whether the subtraction was successful before cloning and screening of the subtracted cDNA populations. Targeted approaches with gametes and fertilization products demonstrate that these types of experiments provide valuable insights into reproduction-related gene expression differences and, moreover, identify candidate genes with a high potential for important roles during plant reproduction and early seed development for further characterization.

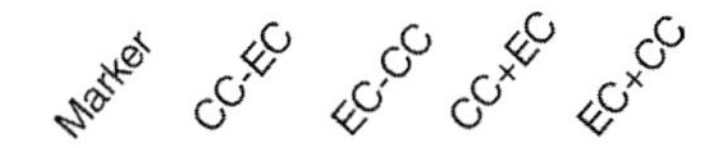

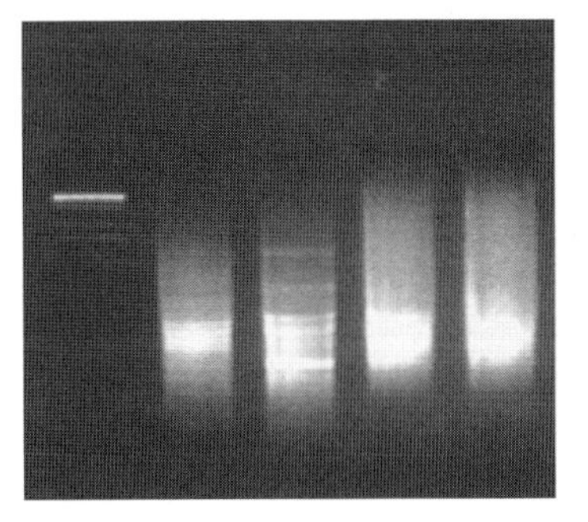

Figure 1.5 Control of suppression subtractive hybridization reactions. Subtracted cDNA populations and mixes of both cDNAs used for subtraction were amplified by PCR and separated using agarose gel electrophoresis. In this example, cDNAs of egg cells (EC) and central cells (CC) were used for subtraction in both directions. Subtracted cDNAs are indicated by '−,' corresponding control mixes are indicated by '+'. The different appearances of subtracted and non-subtracted cDNA populations indicate successful subtraction reactions. Differences between the two subtractions using the same cell types but different directions of subtraction indicate the diverse transcriptional profiles of the two cell types.

1.4.2.6 **Microarray Analyses**

Various types of amplification techniques have been developed to enable microarray gene expression analysis when the starting material is limited. The two main strategies are linear amplification, using *in vitro* transcription, and exponential amplification, based on PCR. If the reactions are well controlled, both methods preserve the relative abundance of the transcripts to a comparable extent (see, e.g. [55]).

Aspects of cDNA amplification methods have already been described in detail above. T7-based amplification of copy RNA (cRNA) by *in vitro* transcription, a method originally developed in the laboratory of James Eberwine [56], was recently evaluated in our laboratory as an alternative to produce targets for oligonucleotide microarray hybridizations. The protocol starts with mRNA isolation using oligo $dT_{(25)}$-coated magnetic beads. The mRNA of 25 egg cells and zygotic cells was then amplified using the Amino Allyl MessageAMP II aRNA Amplification kit (Ambion). After reverse transcription, second strand synthesis, cDNA purification and *in vitro* transcription for 14 hours (the maximum time period recommended), the reactions yielded around 1 μg of cRNA. This cRNA was subjected to a second round of the procedure combined with the incorporation of aminoallyl-modified nucleotides during *in vitro* transcription. After coupling fluorescent dyes to the labeled cRNAs hybridization of 70mer oligonucleotide microarrays (www.maizearray.org) high quality hybridization results were obtained. An example of these hybridizations is shown in Figure 1.6. These initial experiments demonstrated that only a few cells involved in the reproduction of higher plants are needed to obtain data on

Figure 1.6 Microarray hybridization with cRNA targets from single cell types. The hybridization of maize 57 k oligonucleotide arrays (www.maizearray.org) with cRNA generated in two rounds from 25 egg cells to 25 zygotes resulted in high-quality hybridization signals. The various red and green spots indicate highly differential gene expression.

global expression related to the fertilization event and specific tissue formation in very early seed development.

1.5 Analyses of Protein Expression

Traceable quantities of proteins can be detected in a small volume of single cells by minimizing the gel size for one- and two-dimensional polyacrylamide gel electrophoresis. Protein components can be identified by highly sensitive liquid chromatography coupled in tandem with mass spectrometry (LC-MS/MS) [23,31]. Protein patterns of differentiated cells reflect the biological function of these cells. Egg cells are such highly specialized cells which are fertilized by sperm to undergo early embryogenesis. Thus, the identification of proteins in gametes and zygotes will provide important data for understanding the mechanisms of gametogenesis, fertilization and early embryogenesis of higher plants.

Proteomics is an area of research that evaluates protein expression by resolving, identifying, quantifying, and characterizing proteins. Techniques for such studies include two-dimensional polyacrylamide gel electrophoresis, and tandem mass spectrometry and computer analysis [57]. These technologies now make it possible to identify the proteins in a relatively low concentration of cells. Such analyses were initiated to reveal which proteins are present in abundance in plant egg cells [23,31]. By minimizing the gel size in polyacrylamide gel electrophoresis, proteins of only a few egg cells can be detected. Fifteen or 45 cells respectively, are sufficient to produce detectable silver-stained protein bands or spots in SDS- or 2D-PAGEs using small sized gel molds ($50 \times 60 \times 1$ mm). Egg cell lysates from 75 to 180 cells respectively, were used for both SDS- and 2D-PAGEs for in-gel tryptic digestion and subsequent highly sensitive LC-MS/MS analyses. Three cytosolic enzymes in the glycolytic pathway, glyceraldehyde-3-phosphate dehydrogenase, 3-phosphoglycerate kinase and triosephosphate isomerase, two mitochondrial proteins, an ATPase β-subunit and adenine nucleotide transporter, and annexin p35 were identified as major proteins in maize egg cells using tandem mass spectrometric analysis and amino acid microsequencing. Thus, five of the six major egg proteins identified are thought to be involved in energy production pathways, suggesting that the egg cell has sufficient enzymes and transporters to produce and transport an energy source. The amount of protein in a maize egg cell was estimated to be 100–200 pg [23].

It is reasonable to assume, that energy-consuming serial zygotic events, such as migration of cytoplasmic organelles, the formation of a new cell wall around the zygote and nuclear division, explain why egg cells contain an abundance of energy-producing proteins.

In addition to the initial data concerning the protein composition of higher plant egg cells, these protein analyses also provide an indication of the sensitivity of and number of cells required to achieve comprehensive protein profiles of single cell types.

1.6 Prospects

These days, micromanipulation methods are routinely used to isolate gametes from higher plants and to fertilize them *in vitro*. From some higher plants, zygotes and embryos, and fertile plants and endosperm can be obtained by *in vitro* fusion of pairs of sperm and egg cells, and of pairs of sperm and central cells, respectively. This makes it possible to examine the earliest developmental processes precisely timed after fertilization. Furthermore, single zygotes, young embryos and endosperm can be isolated from *in vivo* material. Obviously only a small amount of such material can be obtained, especially if it is produced *in vitro*. In addition, micromanipulation techniques are not restricted to reproductive cells, but may be used to isolate and select various other specific cell types. The adaptation of highly sensitive molecular methods to specific cell types or even single cells, as described in this chapter, significantly expands our insight into gene expression. These methods provide a high degree of sensitivity and specificity which is necessary to understand the role of genes in differentiation, and especially in reproductive processes. With this information the genes involved in developmental processes can be defined and reverse genetic approaches to characterize their function can be initiated. Because 'whole transcriptome' arrays for various species are currently available, the exploitation of these arrays to analyze expression information in specific cell types will provide comprehensive and conclusive genetic information. Together with the emerging technologies for the analysis of proteins in the same cell type, this repertoire of methods will greatly enhance our understanding of developmental and reproductive plant biology.

References

1 Goldberg, R.B., de Paiva, G. and Yadegari, R. (1994) Plant embryogenesis: Zygote to seed. *Science*, **266**, 605–614.

2 Koop, H.-U. and Schweiger, H.-G. (1985) Regeneration of plants after electrofusion of selected pairs of protoplasts. *European Journal of Cell Biology*, **39**, 46–49.

3 Spangenberg, G. and Koop, H.-U. (1992) Low density cultures: microdroplets and single cell nurse cultures, in *Plant Tissue Culture Manual A10* (ed. K. Lindsey), Kluwer Academic Publishers, Dordrecht, pp 1–28.

4 Kranz, E. and Lörz, H. (1993) *In vitro* fertilization with isolated, single gametes results in zygotic embryogenesis and fertile maize plants. *Plant Cell*, **5**, 739–746.

5 Kranz, E., Bautor, J. and Lörz, H. (1991) *In vitro* fertilization of single, isolated gametes of maize mediated by electrofusion. *Sexual Plant Reproduction*, **4**, 12–16.

6 Kranz, E., Bautor, J. and Lörz, H. (1991) Electrofusion-mediated transmission of cytoplasmic organelles through the *in vitro* fertilization process, fusion of sperm cells with synergids and central cells, and cell reconstitution in maize. *Sexual Plant Reproduction*, **4**, 17–21.

7 Kranz, E., von Wiegen, P. and Lörz, H. (1995) Early cytological events after induction of cell division in egg cells and zygote development following *in vitro* fertilization with angiosperm gametes. *The Plant Journal*, **8**, 9–23.

8 Kranz, E., von Wiegen, P., Quader, H. and Lörz, H. (1998) Endosperm development after fusion of isolated, single maize sperm and central cells *in vitro*. *Plant Cell*, **10**, 511–524.

9 Kovács, M., Barnabás, B. and Kranz, E. (1995) Electro-fused isolated wheat (*Triticum aestivum* L.) gametes develop into multicellular structures. *Plant Cell Reports*, **15**, 178–180.

10 Faure, J.-E., Mogensen, H.L., Dumas, C., Lörz, H. and Kranz, E. (1993) Karyogamy after electrofusion of single egg and sperm cell protoplasts from maize: Cytological evidence and time course. *Plant Cell*, **5**, 747–755.

11 Uchiumi, T., Komatsu, S., Koshiba, T. and Okamoto, T. (2006) Isolation of gametes and central cells from *Oryza sativa* L. *Sexual Plant Reproduction*, **19**, 37–45.

12 Uchiumi, T., Uemura, I. and Okamoto, T. (2007) Establishment of an *in vitro* fertilization system in rice (*Oryza sativa* L.). *Planta*, 10.1007/s00425-007-0506-2.

13 Kranz, E. and Lörz, H. (1994) *In vitro* fertilisation of maize by single egg and sperm cell protoplast fusion mediated by high calcium and high pH. *Zygote*, **2**, 125–128.

14 Faure, J.-E., Digonnet, C. and Dumas, C. (1994) An *in vitro* system for adhesion and fusion of maize gametes. *Science*, **263**, 1598–1600.

15 Khalequzzaman, M. and Haq, N. (2005) Isolation and *in vitro* fusion of egg and sperm cells in *Oryza sativa*. *Plant Physiology and Biochemistry*, **43**, 69–75.

16 Sun, M.-X., Yang, H.-Y., Zhou, C. and Koop, H.-U. (1995) Single-pair fusion of various combinations between female gametoplasts and other protoplasts in *Nicotiana tabacum*. *Acta Botanica Sinica*, **37**, 1–6.

17 Sun, M.-X., Moscatelli, A., Yang, H.-Y. and Cresti, M. (2000) *In vitro* double fertilization in *Nicotina tabacum* (L.): fusion behavior and gamete interaction traced by video-enhanced microscopy. *Sexual Plant Reproduction*, **12**, 267–275.

18 Sun, M.-X., Moscatelli, A., Yang, H.-Y. and Cresti, M. (2001) *In vitro* double fertilization in *Nicotiana tabacum* (L.): the role of cell volume in cell fusion. *Sexual Plant Reproduction*, **13**, 220–225.

19 Tirlapur, U.K., Kranz, E. and Cresti, M. (1995) Characterization of isolated egg cells, *in vitro* fusion products and zygotes of *Zea mays* L. using the technique of image

analysis and confocal laser scanning microscopy. *Zygote*, **3**, 57–64.

20 Holm, P.B., Knudsen, S., Mouritzen, P., Negri, D., Olsen, F.L. and Roué, C. (1994) Regeneration of fertile barley plants from mechanically isolated protoplasts of the fertilized egg cell. *Plant Cell*, **6**, 531–543.

21 Scholten, S., Lörz, H. and Kranz, E. (2002) Paternal mRNA and protein synthesis coincides with male chromatin decondensation in maize zygotes. *The Plant Journal*, **32**, 221–231.

22 Richert, J., Kranz, E., Lörz, H. and Dresselhaus, T. (1996) A reverse transcriptase-polymerase chain reaction assay for gene expression studies at the single cell level. *Plant Science*, **114**, 93–99.

23 Okamoto, T., Higuchi, K., Shinkawa, T., Isobe, T., Lörz, H., Koshiba, T. and Kranz, E. (2004) Identification of major proteins in maize egg cells. *Plant & Cell Physiology*, **45**, 1406–1412.

24 Okamoto, T., Scholten, S., Lörz, H. and Kranz, E. (2005) Identification of genes that are up- or down-regulated in the apical or basal cell of maize two-celled embryos and monitoring their expression during zygote development by a cell manipulation- and PCR-based approach. *Plant & Cell Physiology*, **46**, 332–338.

25 Okamoto, T. and Kranz, E. (2005) *In vitro* fertilization – a tool to dissect cell specification from a zygote. *Current Science*, **89**, 1861–1869.

26 Okamoto, T. and Kranz, E. (2005) Major proteins in plant and animal eggs. *Acta Biologica Cracoviensia Series Botanica*, **47**, 17–22.

27 Hoshino, Y., Scholten, S., von Wiegen, P., Lörz, H. and Kranz, E. (2004) Fertilization-induced changes in the microtubular architecture in the maize egg cell and zygote – an immunocytochemical approach adapted to single cells. *Sexual Plant Reproduction*, **17**, 89–95.

28 Dresselhaus, T., Lörz, H. and Kranz, E. (1994) Representative cDNA libraries from few plant cells. *The Plant Journal*, **5**, 605–610.

29 Dresselhaus, T., Hagel, C., Lörz, H. and Kranz, E. (1996) Isolation of a full-length cDNA encoding calreticulin from a PCR library of *in vitro* zygotes of maize. *Plant Molecular Biology*, **31**, 23–34.

30 Lê, Q., Gutièrrez-Marcos, J., Costa, L., Meyer, S., Dickinson, H., Lörz, H., Kranz, E. and Scholten, S. (2005) Construction and screening of subtracted cDNA libraries from limited populations of plant cells: a comparative analysis of gene expression between maize egg cells and central cells. *The Plant Journal*, **44**, 167–178.

31 Uchiumi, T., Shinkawa, T., Isobe, T. and Okamoto, T. (2007) Identification of the major protein components of rice egg cells. *Journal of Plant Research*, 10.1007/s10265-007-0095-y.

32 Kranz, E. and Dresselhaus, T. (1996) *In vitro* fertilization with isolated higher plant gametes. *Trends in Plant Science*, **1**, 82–89.

33 Kranz, E. and Kumlehn, J. (1999) Angiosperm fertilization, embryo and endosperm development *in vitro*. *Plant Science*, **142**, 183–197.

34 Antoine, A.F., Dumas, C., Faure, J.-E., Feijó, J.A. and Rougier, M. (2001) Egg activation in flowering plants. *Sexual Plant Reproduction*, **14**, 21–26.

35 Wang, Y.Y., Kuang, A., Russell, S.D. and Tian, H.Q. (2006) *In vitro* fertilization as a tool for investigating sexual reproduction of angiosperms. *Sexual Plant Reproduction*, **19**, 103–115.

36 Kovács, M., Barnabás, B. and Kranz, E. (1994) The isolation of viable egg cells of wheat (*Triticum aestivum* L.). *Sexual Plant Reproduction*, **7**, 311–312.

37 Katoh, N., Lörz, H. and Kranz, E. (1997) Isolation of viable egg cells of rape (*Brassica napus* L.). *Zygote*, **5**, 31–33.

38 Kumlehn, J., Brettschneider, R., Lörz, H. and Kranz, E. (1997) Zygote implantation to cultured ovules leads to direct embryogenesis and plant regeneration of wheat. *The Plant Journal*, **12**, 1473–1479.

39 Digonnet, C., Aldon, D., Leduc, N., Dumas, C., and Rougier, M. (1997) First evidence of a calcium transient in flowering plants at fertilization. *Development*, **124**, 2867–2874.

40 Antoine, A.F., Faure, J.-E., Cordeiro, S., Dumas, C., Rougier, M. and Feijó, J.A. (2000) A calcium influx is triggered and propagates in the zygote as a wave front during *in vitro* fertilization of flowering plants. *Proceedings of the National Academy of Sciences of the United States of America*, **97**, 10643–10648.

41 Antoine, A.F., Faure, J.-E., Dumas, C. and Feijó, J.A. (2001) Differential contribution of cytoplasmic Ca^{2+} and Ca^{2+} influx to gamete fusion and egg activation in maize. *Nature Cell Biology*, **3**, 1120–1123.

42 Tian, H.Q. and Russell, S.D. (1997) Micromanipulation of male and female gametes of *Nicotiana tabacum*: II. Preliminary attempts for *in vitro* fertilization and egg cell culture. *Plant Cell Reports*, **16**, 657–661.

43 Friedman, W.E. (1999) Expression of the cell cycle in sperm of *Arabidopsis*: implications for understanding patterns of gametogenesis and fertilization in plants and other eukaryotes. *Development*, **126**, 1065–1075.

44 Tian, H.Q., Yuan, T. and Russell, S.D. (2005) Relationship between double fertilization and the cell cycle in male and female gametes of tobacco. *Sexual Plant Reproduction*, **17**, 243–252.

45 Kranz, E. (1992) *In vitro* fertilization of maize mediated by electrofusion of single gametes, in *Plant Tissue Culture Manual E1* (ed. K. Lindsey), Kluwer Academic Publishers, Dordrecht, pp 1–12.

46 Kranz, E., (1999) *In vitro* fertilization with isolated single gametes, in *Methods in Molecular Biology 111, Plant Cell Culture Protocols* (ed. R. Hall), Humana Press Inc, Totowa, NJ, pp 259–267.

47 Sauter, M., von Wiegen, P., Lörz, H. and Kranz, E. (1998) Cell cycle regulatory genes from maize are differentially controlled during fertilization and first embryonic cell division. *Sexual Plant Reproduction*, **11**, 41–48.

48 Sprunk, S., Baumann, U., Edwards, K., Langride, P. and Dresselhaus, T. (2005) The transcript composition of egg cells change significantly following fertilization in wheat (*Triticum aestivum* L.). *The Plant Journal*, **41**, 660–672.

49 Matz, M.V. (2002) Amplification of representative cDNA samples from microscopic amounts of invertebrate tissue to search for new genes. *Methods in Molecular Biology*, **183**, 13–18.

50 Peixoto, A., Monteiro, M., Rocha, B. and Veiga-Fernandes, H.D (2004) Quantification of multiple gene expression in individual cells. *Genome Research*, **14**, 1938–1947.

51 Al Taher, A., Bashein, A., Nolan, T., Hollingsworth, M. and Brady, G. (2000) Global cDNA amplification combined with real-time RT-PCR: accurate quantification of multiple human potassium channel genes at the single cell level. *Yeast*, **17**, 201–210.

52 Meyer, S., Pospisil, H. and Scholten, S. (2007) Heterosis-associated gene expression in maize embryo six days after fertilization exhibits additive, dominant and overdominant pattern. *Plant Molecular Biology*, **63**, 381–391.

53 Yang, H., Kaur, N., Kiriakopolos, S. and McCormick, S. (2006) EST generation and analyses towards identifying female gametophyte-specific genes in *Zea mays* L. *Planta*, **224**, 1004–1014.

54 Ning, J., Peng, X.-B., Qu, L.-H., Xin, H.-P., Yan, T.-T. and Sun, M.-X. (2006) Differential gene expression in egg cells and zygotes suggests that the transcriptome is restructured before the first zygotic division in tobacco. *FEBS Letters*, **580**, 1747–1752.

55 Laurell, C., Wirta, V., Nilsson, P. and Lundeberg, J. (2007) Comparative analysis of a 3′ end tag PCR and a linear RNA amplification approach for microarray analysis. *Journal of Biotechnology*, **127**, 638–646.

56 Van Gelder, R.N., von Zastrow, M.E., Yool, A., Dement, W.C., Barchas, J.D. and Eberwine, J.H. (1990) Amplified RNA synthesized from limited quantities of heterogeneous cDNA. *Proceedings of the National Academy of Sciences of the United States of America*, **87**, 1663–1667.

57 Celis, J., Ostergaard, M., Jensen, N., Gromova, I., Rasmussen, H. and Gromov, P. (1998) Human and mouse proteomic databases: novel resources in the protein. *FEBS Letters*, **430**, 64–72.

2
AFLP-Based RNA Fingerprinting: Novel Variants and Applications

Christian W.B. Bachem, Wim H. Vriezen, Craita E. Bita, and Asun Fernandez del Carmen

Abstract

Developed in the mid 1990s, the RNA fingerprinting variant of AFLP technology has nowadays become an essential tool in gene discovery. As a result, cDNA-AFLP is the method of choice for many scientists looking into biological systems with limited genome sequence information. Basically, the process consists of the (semi-) quantitative anchored PCR of cDNA restriction fragments. The use of selective restriction fragment amplification allows the systematic visualization of over 80% of the transcriptome. This method is highly adaptable to technological developments following on from the recent rapid changes in sequencing methods. Moreover, it has been applied to diverse biological systems from microbes to humans, but remains most frequently applied to plants. In this chapter we describe the applications and provide a generalized protocol.

2.1
Introduction

Plant responses to developmental and environmental stimuli result in rapid changes of gene transcription. Such alterations are generally mediated through signal transduction pathways switched on by master regulators or receptors which initiate cascades of gene induction processes and thus evoke the developmental progression and/or the responses to the environmental situation. These reactions invariably result in an adjustment of the metabolic flux pathways and the equivalent coordinated transcriptional regulation of the genes coding for the active enzymes in these pathways. Furthermore, the genes at the end-points of the pathways generally tend to be more highly activated than those at the beginning. Thus, historically, some of the first differentially transcribed (plant) genes, such as leghemoglobin [1] or induction

The Handbook of Plant Functional Genomics: Concepts and Protocols.
Edited by Günter Kahl and Khalid Meksem

ISBN: 978-3-527-31885-8

patterns such as the WUN genes [2], were isolated purely on the basis of their extreme differential expression patterns.

The conception that, obtaining information on transcriptional changes will yield information on biologically relevant metabolism pathways lies at the heart of RNA transcript profiling [3]. Thus, if a gene is specifically induced during a biological process, knowledge of its identity may reveal information about that process.

Essential to this approach is a large resource of biochemical pathway data and annotated sequence databases, though not necessarily from the system under investigation. Numerous RNA transcription profiling technologies have been developed (several reviewed in this volume). Most can be characterized methodologically into PCR- and hybridization-based systems. Moreover, these methods can be distinguished into open (without requirement of prior DNA sequence information) and closed systems (relying on a fixed set of coding regions, usually cDNA-based).

The subject of this chapter features a PCR-based open architecture system which provides high transcriptome coverage by delivering the capacity for discovery of rare messages and an increased sensitivity for detection of small changes in gene expression [4]. The method is a variation of the amplified fragment length polymorphism (AFLP) method [5] for the visualization of genomic DNA polymorphisms within a genome. The method is based on restriction enzyme digestion with two enzymes differing in the frequency of recognition sites within the given genome, followed by the ligation of anchors (also known as adaptors) on the sticky ends left by the restriction enzymes. PCR amplification, using primers corresponding to the anchors with one or more additional nucleotides extending beyond the restriction recognition site into the target fragments, is used to create the final fingerprint. Initially the method was used to construct genetic linkage maps in segregating populations, to determine genetic distances and to identify individual organisms with unknown genotypes. As a further extension of these purposes, the method has been applied to cDNA with the aim of visualizing gene expression rather than detecting DNA polymorphism (and is thus termed cDNA-AFLP). In this case, the mRNA is isolated, for example, from different tissues or various developmental stages and cDNA is synthesized. The cDNA is then processed in the same way as in genomic AFLP to prepare a template which is amplified to generate the RNA fingerprint.

The intensity of individual signals in the fingerprint is then taken as the measure of the strength of expression for the gene corresponding to the band or 'transcript derived fragment' (TDF) seen on the AFLP gel.

The method of cDNA-AFLP can be fine-tuned to suite a wide range of systems. Some minor adaptations of the procedure have been published [6] and several successful applications have been presented for a wide range of biological systems. Merging expression profiling with genetic mapping, we have applied cDNA-AFLP to RNA isolated from crossing populations and have been able to show that transcript maps and transcript bulk segregant analysis can be constructed in this way [7,8].

In this chapter, we present recent advances in the cDNA-AFLP technology, including investigation applications and combinations which particularly suit the method. We provide a state-of-the-art protocol, and present applications, possible future directions and extensions of the technology.

2.2 Methods and Protocols

2.2.1 Theoretical Considerations

The cDNA-AFLP method selectively displays transcript-derived restriction fragments as bands on a gel electrophoresis platform. Amplification is achieved by providing cDNA fragments with anchors at their termini, which serve as primer sites in the subsequent PCR amplifications. From the genes likely to be expressed in an organ, tissue or cell-type, only a fraction of the transcripts can be practically visualized in a single fingerprint by gel electrophoresis. To achieve a selective reduction, two strategies are adopted. Firstly, the restriction enzyme used to digest the cDNA can be chosen to limit the number of transcripts visualized. Secondly, by using different lengths of the so-called 'selective bases' on the primer termini, a high level of tuning in the number of targeted fragments can be achieved per amplification.

As with all such techniques, the fidelity of the RNA fingerprint in representing the expression of genes is cumulatively relative to every step of the protocol. In the first instance, it is dependent on an efficient and reliable method of RNA extraction. RNA can be isolated using any method that produces good quality and non-degraded total RNA. It is not necessary to remove traces of contaminating genomic DNA as these will be washed away during the cDNA-AFLP procedure. In general it is advisable to work as much as possible with mixes of reagents so to avoid differences in amplification between the different reactions. The synthesis for template preparation can be carried out using available protocols, the aim being to produce a high quality double-stranded cDNA. However, since the template preparation is focused on the 3′-end, the preparation of full-length cDNA is not a prerequisite for achieving good profiles.

The restriction enzyme digestion of cDNA has two main aims. The first is to provide sticky ends for the efficient ligation of anchors to the ends of restriction fragments. The second is to reduce the size of DNA fragments for separation using electrophoresis (between 50 and 800 bp). This aim can be achieved in two consecutive steps: (a) the cDNA is digested with a rare cutting enzyme such as an enzyme with a 6-nucleotide recognition sequence, and (b) the second enzyme is used to generate fragments of the desired size, for this a restriction enzyme recognizing four nucleotides is usually employed. The selection of enzymes is crucial for the optimization of results since it will affect the number of different sequences represented in the fingerprint. The correct choice of enzymes allows fragments to be produced that have sufficient sequence to allow unequivocal detection of identity and a suitable length to visualize on the chosen detection platform. By eliminating the 5′-ends of the cDNA fragments after the first digestion, as described in the current protocol, the redundancy in displayed fragments is circumvented.

Thus, using a solid support for cDNA synthesis has several advantages: firstly, it allows for the simple elimination of contaminants such as genomic DNA and other contaminants; secondly, it allows for a reduction in the complexity of the final template by washing away the 5′-ends of the cDNAs after digestion with the first restriction enzyme and the liberation of a single fragment per transcript after digestion with the second enzyme.

In cDNA-AFLP, the pre-amplification of the primary template allows the production of large amounts of working template stock. This feature not only facilitates the analysis of transcription in very small tissue samples but also delivers an almost unlimited supply of template for fingerprinting and band isolation.

The protocol described below implements radioactive labeling and visualization using an X-ray film or a phosphor-imager. However, the same protocol can equally well be used with different visualization platforms such as silver staining or fluorescent labeling systems. Hence, the technological flexibility of cDNA-AFLP is one of key advantages of this method.

2.2.2
State-of-the-Art cDNA-AFLP Protocol

This protocol can be divided in two major steps: isolation of cDNA fragments and pre-amplification of the cDNA fragments. Due to the significance of this procedure, we recommend that there is no break during the first or second step, except the O/N pause in between.

2.2.2.1 Isolation of cDNA Fragments

Equipment and Reagents

- Thermo-cycler.
- mRNA Capture Kit (Roche Diagnostics, GmbH, Germany) containing: lysis buffer, biotinylated oligo-dT (0.1 mM), streptavidin-coated PCR-tubes, ready-to-use washing buffer.
- Nuclease-free water.
- Dithiothreitol (DTT) 0.1 M.
- dNTPs 10 mM.
- SuperScript III (200 U/μl) and 5× first strand buffer (Invitrogen, Carlsbad CA, USA).
- *E. coli* ligase (10 U/μl) and 10× *E. coli* ligase buffer (Invitrogen, Carlsbad CA, USA).
- *E. coli* DNA polymerase I (10 U/μl; Fermentas, Hanover MD, USA).
- RNase H (5 U/μl; Amersham Bioscience, Uppsala, Sweden).
- 5× restriction-ligation (RL) buffer: 50 mM Tris-HAc pH 7.5, 50 mM $MgAc_2$, 250 mM KAc, 25 mM DTT.
- *Bst*YI and *Mse*I (both 10 units/μl, New England Biolabs, Beverly MA, USA).

Procedure

1. Combine 5 μg of total RNA in 10 μl nuclease-free water with 40 μl lysis buffer and 0.05 μl biotinylated oligo-dT, then transfer the mixture to the streptavidin-coated PCR-tube(s). Incubate at 37 °C for 5 min in a thermo-cycler, allow to cool to room temperature and store on ice. Discard the liquid and wash the bound mRNA by

simply adding 100 μl wash buffer, incubating the sample(s) for 1 min at room temperature and subsequently *gently* removing the liquid with a pipette. Repeat the washing step three times without touching the walls of the PCR tube(s).

2. The reverse transcription is started by adding 50 μl of the first strand cDNA synthesis mixture (31.3 μl water, 10 μl 5× first strand buffer, 5 μl 0.1 M DTT, 2.5 μl 10 mM dNTPs and 1.2 μl SuperScript III (200 U/μl – per sample) to the PCR tube(s) now containing the bound mRNA. The reaction mix is incubated for 2 h at 42 °C. The PCR tube(s) are then stored on ice and, *important*, 10 μl of the reaction volume is discarded to obtain a total volume of 40 μl.

3. For the second strand cDNA synthesis, a mixture containing 91 μl demineralized water, 16 μl 10× *E. coli* ligase buffer, 6 μl 0.1 M DTT, 3 μl 10 mM dNTPs, 1.5 μl *E. coli* ligase (15 units), 1.5 μl (15 units) *E. coli* DNA polymerase I, and 1 μl RNase H (5 units) per sample, is added to the first strand mix and incubated for 1 h at 12 °C, then subsequently for 1 h at 22 °C in the thermo-cycler. The reaction mixture is discarded (the double-stranded cDNA is still attached to the tube wall) and the PCR tube(s) *gently* washed twice with 200 μl wash buffer.

4. Digestion of the dsDNA starts by adding a mixture containing 38.8 μl water, 10 μl 5× RL buffer and 1.2 μl *Bst*YI (12 units) per sample. After incubation for 2 h at 60 °C, the PCR tubes are *gently* washed three times with 100 μl wash buffer. The second digestion is initiated by adding a mixture of 38.8 μl water, 10 μl RL-buffer and 1.2 μl *Mse*I (12 units) per sample, followed by incubation for 2 h at 37 °C.

5. Now the reactions can be stored overnight at 4 °C in a thermo-cycler/refrigerator.

2.2.2.2 Non-Selective Pre-Amplification

Equipment and Reagents

- Non-phosphorylated oligonucleotides for the anchors (*Bst*YI- and *Mse*I- anchors) are:
  - *Bst*YI-F: 5′ – CTCGTAGACTGCGTAGT – 3′ 100 μM.
  - *Bst*YI-R: 5′ – GATCACTACGCAGTCTAC – 3′ 100 μM.
  - *Mse*I-F: 5′ – GACGATGAGTCCTGAG – 3′ 100 μM.
- *Mse*I-R: 5′ – TACTCAGGACTCAT – 3′ 100 μM.
- ATP (100 mM; GE Healthcare Bio-Sciences AB, Uppsala, Sweden).
- T4 DNA ligase (5 U/μl; Fermentas, Hanover MD, USA).
- *Bst*YI (10 units/μl).
- *Mse*I (10 units/μl).
- T4 DNA ligase (5 U/μl; Fermentas, Hanover MD, USA).
- Non-selective primers for the pre-amplification:
  - *Bst*YI(T) + 0: 5′ – GACTGCGTAGTGATCT – 3′ 10 μM.
  - *Bst*YI(C) + 0: 5′ – GACTGCGTAGTGATCC – 3′ 10 μM.
  - *Mse*I + 0: 5′ – GATGAGTCCTGAGTAA – 3′ 10 μM.

- Red Hot Taq DNA polymerase (5 U/μl), 10× PCR-buffer, $MgCl_2$ (25 mM) (ABgene, Epsom, UK).
- Tris-Cl (pH 8.0) 1 mM.

Procedure

1. The reaction mixture containing the cDNA fragments liberated from the tube wall can now be transferred to clean Eppendorf PCR tubes.

2. Prepare the *Bst*YI anchor by combining, per sample, 0.05 μl of the *Bst*YI-R and 0.05 μl of the *Bst*YI-F with 0.9 μl water and the *Mse*I anchor by combining 0.5 μl *Mse*I-R with 0.5 μl *Mse*I-F, respectively. Incubate the oligo mixtures at 37 °C for 5 min and allow to cool to room temperature.

3. Prepare the anchor ligation mix (total 10 μl per sample): 1 μl *Bst*YI anchor (5 pmol), 1 μl *Mse*I anchor (50 pmol), 0.1 μl 100 mM ATP, 2 μl 5× RL-buffer, 1 μl T4 DNA ligase (5 units), 0.5 μL *Bst*YI (5 units), 0.5 μL *Mse*I (5 units) and 3.9 μl H_2O.

4. Combine 10 μl of the anchor ligation mix with the 40 μl of each sample containing the isolated *Bst*YI-*Mse*I fragments and incubate for 3 h at 37 °C.

5. Dilute the anchor ligation with 50 μl water (we use TRIS EDTA) and use 5 μl as a template in the pre-amplification procedure.

6. Prepare two primer mixes containing per PCR reaction: *either* 1.5 μl of the *Bst*YI(C) + 0-primer *or* 1.5 μl of the *Bst*YI(T) + 0 primer, 1.5 μl *Mse*I + 0-primer, 1.0 μl 10 mM dNTPs and 21.0 μl water.

7. Prepare the DNA polymerase mix (per PCR reaction): 0.2 μl Red Hot Taq DNA polymerase (1 unit), 5.0 μl 10× PCR-buffer, 5.0 μl $MgCl_2$ (25 mM) and 9.8 μl water.

8. Prepare the PCR reaction by combining 5 μl template (from step 4) with 25 μl of *either* the *Bst*YI(C) + 0-primer mix *or* the *Bst*YI(T) + 0 primer mix and 20 μl of the DNA polymerase mix. Amplify (without a hotstart) the cDNA fragments with 25 cycles of 30 s at 94 °C, 60 s at 56 °C and 60 s at 72 °C.

9. Check the pre-amplification by running 10 μl of the reaction mixture on an agarose gel with ethidium bromide. A smear of DNA fragments, ranging between 50 and 600 base pairs should be visible for all samples (Figure 2.1A).

10. Dilute a part of the pre-amplification reactions 400-fold in 1 mM Tris-Cl (pH 8.0) (to be used for the selective amplifications, store at 4 °C) and store the remainder at − 20 °C.

2.2.2.3 Selective Amplification-Reaction Using ^{33}P-Labeled Primer and Gel Analysis

Equipment and Reagents

- ^{33}P-γ-ATP (≈370 MBq/ml, MP Biomedicals, Irvine, CA, USA).
- T4-polynucleotide-kinase (10 U/μl), 10× Reaction Buffer A (Eurogentec, Seraing, B).

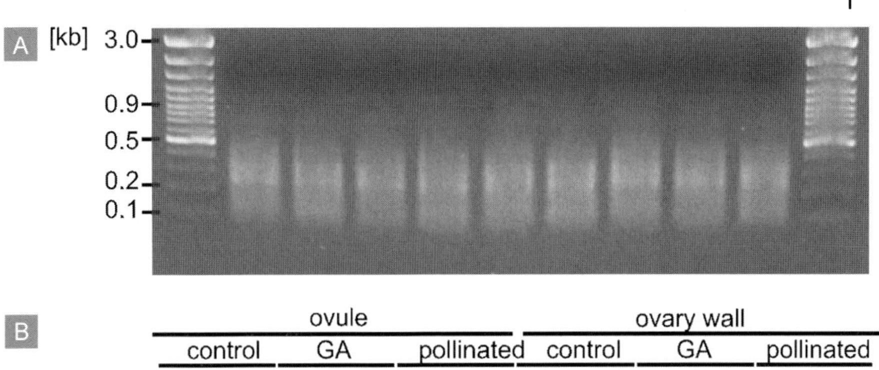

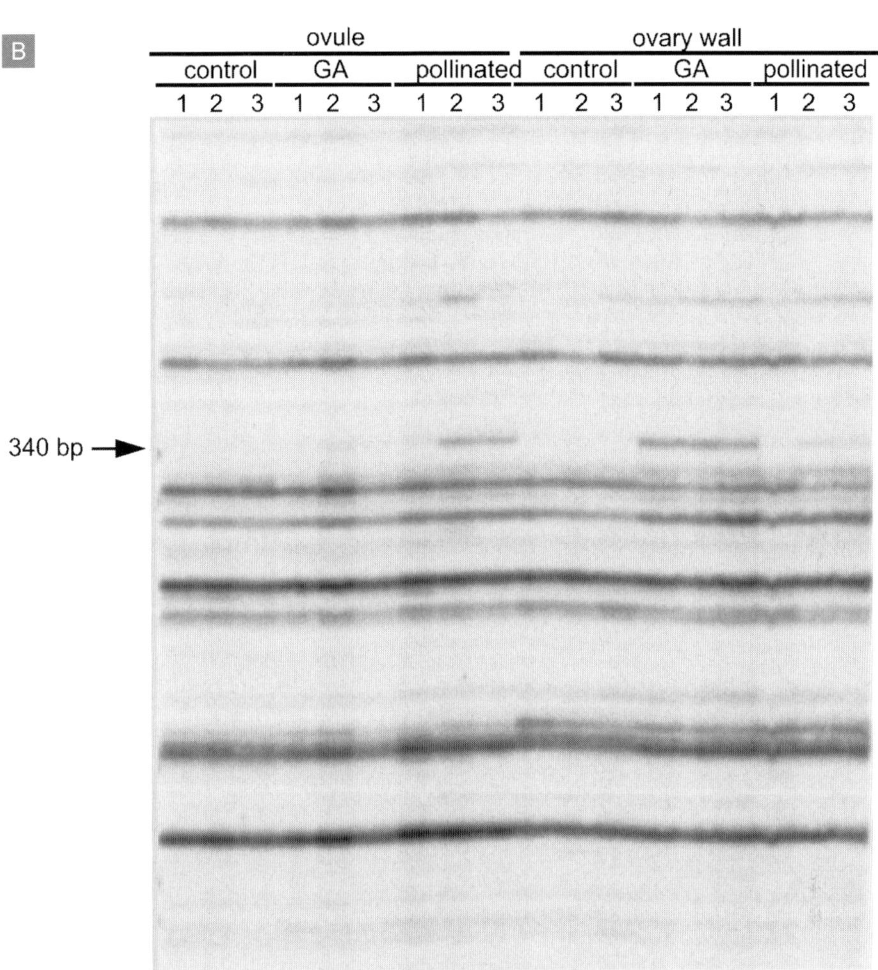

Figure 2.1 Pre-amplification and selective amplification products from cDNA of tomato mRNA. (A) Ethidium bromide-stained agarose gel loaded with cDNA obtained after a pre-amplification reaction with nonselective primers on cDNA fragments. A 100-bp DNA ladder is loaded on both sides. (B) Acrylamide gel displaying a representative result obtained after amplification of a part of the cDNA fragments using primers containing two selective nucleotides each. In this example the arrow indicates a fragment (340 bp) corresponding to a gene whose mRNA level is induced after pollination in tomato ovary and also after gibberellic acid (GA_3) treatment in the ovary wall.

- Selective primers.
  - *Bst*YI(T) + N: 5′ – GACTGCGTAGTGATCT – 3′ 10 μM.
  - *Bst*YI(C) + N: 5′ – GACTGCGTAGTGATCC – 3′ 10 μM.
  - *Mse*I + NN: 5′ – GATGAGTCCTGAGTAA – 3′ 10 μM.
- dNTPs (5 mM).
- Red Hot Taq DNA polymerase (5 U/μl), 10× PCR buffer, $MgCl_2$ (25 mM) (ABgene, Epsom, UK).
- Formamide loading dye: 2 ml EDTA 0.5 M (pH 8.0), 98 ml formamide, 0.06 g bromophenol blue.
- 10× TB: 108 g Tris base and 55 g boric acid.
- Electrophoresis buffer (TBE): 100 ml 10× TB, 4 ml 0.5 M EDTA (pH 8.0), adjust volume to 1 l with water.
- Gel mix: dissolve 450 g urea (electrophoresis grade) in 600 ml water in a 55 °C water bath. Add 112.5 ml Rotiphorese Gel 40 mix (37.5 : 1, Carl Roth GMBH), 100 ml 10× TB and 4 ml 0.5 M EDTA (pH 8.0). Adjust the volume to 1 l and filter the solution through a 0.45-μm filter.
- Sequencing gel system, e.g. Sequi-Gen GT Sequencing Cell (Bio-Rad, Hercules, CA).
- 3 MM Whatmann paper (Whatmann Int. Ltd., Maidstone, UK).
- Slab gel dryer.
- X-ray film (e.g. Kodak BioMax MR film; Kodak, USA) (see Figure 2.1B).

Procedure

1. The products of the selective PCR reactions are labeled by using *Bst*YI selective primers, previously labeled by phosphorylation using ^{33}P-γ-dATP as the phosphate donor. For 100 selective PCR reactions mix: 10 μl of *Bst*YI selective primer (10 μM), 10 μl ^{33}P-γ-ATP, 10 μl 10× Reaction Buffer A and 1 μl T4-polynucleotide kinase (10 units) and 119 μl water (150 μl total volume; the actual reaction volumes depend on the number of cDNA-AFLP amplifications that need to be performed with each *Bst*YI selective primer). Incubate for 30 min at 37 °C and, subsequently, for 5 min at 65 °C (to inactivate the kinase).
2. For each amplification reaction, 5 μl of the 400-fold diluted pre-amplification mixture is used as template. Add 15 μl of a reaction mix for each primer combination containing per sample: 1.5 μl labeled *Bst*YI + N-primer, 0.6 μl *Mse*I + NN-primer, 0.8 μl 5 mM dNTPs, 2.0 μl 10× PCR-buffer, 2.0 μl $MgCl_2$ (25 mM), 0.12 μl Red Hot *Taq* DNA polymerase (0.6 unit) and H_2O to make up the final volume to 15 μl.
3. The amplification reaction consists of 13 cycles starting with 30 s at 94 °C, 30 s at 65 °C and 60 s at 72 °C and in each subsequent cycle the annealing temperature is decreased by 0.7 °C. Then 23 additional cycles are carried out for 30 s at 94 °C, 30 s at 56 °C and 60 s at 72 °C.

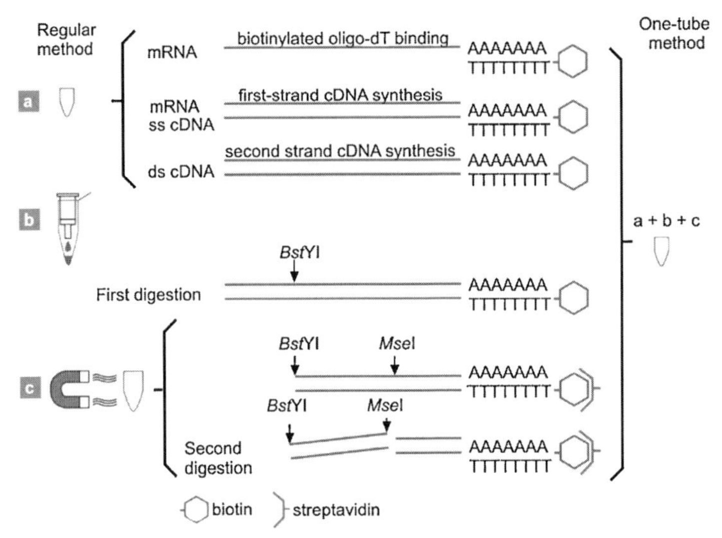

Figure 2.2 First steps in the regular cDNA-AFLP procedure to obtain gene-specific cDNA fragments. (a) Double-stranded (ds) cDNA is synthesized in a 200-μL PCR tube. (b) The cDNA obtained is purified using a column, and is then quantified. An aliquot of the cDNA is digested. (c) Subsequently, the 3′-ends of the cDNAs are captured with streptavidin-coated paramagnetic beads. Washing and capture occurs in several steps. With the one-tube method all these steps are carried out in one streptavidin-coated PCR tube.

4. Add 20 μl formamide loading dye to each sample and store the samples at $-20\,°C$ until further analysis on gel.
5. Prepare the gel by combining 100 ml gel mix, 500 μl APS (10 % (w/v)) and 100 μl TEMED. Pour the gel and let polymerization take place for at least 3 h or overnight.
6. Run the gel at constant power (at 100 W for a 38×50 cm gel). Transfer the gel to a sheet of 3 MM Whatmann paper and dry at 60 °C on a slab gel dryer until completely dry. Expose to an X-ray film and mark the orientation of the gel either by using fluorescent markers or alternatively, by stapling the film to the paper on which the gel has been dried.

A graphical summary of the procedure is presented in Figure 2.2.

2.2.2.4 **Downstream Analysis**

Fragment Isolation To isolate the DNA fragments of interest from the gel, the developed X-ray film has to be mounted to the gel using the fluorescent markers or staple holes for a perfect re-alignment. Excise the band of interest with a scalpel by cutting through the film and the paper onto which the gel is adsorbed. After excising the bands it is advisable to expose the gel again to a new X-ray film (or phosphor-image screen) to check whether the bands of interest have actually been removed from the gel. Soak the excised pieces in 100 μl of water and incubate for at least 2 h at room temperature. The water with some of the DNA can be transferred to a new tube and used as the template for re-amplification. It is not necessary to boil the gel or to use methods to precipitate the DNA in order to obtain the maximal amount of DNA from the piece of gel. In contrast, the simple method of combining a re-amplification reaction using a high quality proofreading DNA polymerase most often leads to a specific PCR product.

Re-amplification takes place in a reaction mixture containing 5 μl template, 1 μl *Bst*YI(T) + 0 (10 μM) (or the *Bst*YI(C) + 0 primer), 1 μl *Mse*I + 0 (10 μM) primer, 0.4 μl dNTPs (10 μM), 4 μl 5× buffer, 0.2 μl DNA polymerase (0.4 unit; Phusion High-Fidelity DNA Polymerase, Finnzymes Oy, Espoo, Finland), and 13.4 μl water. The PCR consisted of an initial 30 s at 98 °C, followed by 35 cycles of 10 s at 98 °C, 30 s at 52 °C and 30 s at 72 °C, and a final extension at 72 °C for 3 min.

The re-amplified product can be analyzed by running 10 μl on an ethidium bromide-stained agarose gel (2% w/v) and should be one DNA fragment of the expected size. Although it is not possible to determine the exact fragment size on an agarose gel, it is possible to compare the approximate sizes of the different fragments relative to each other, which should correspond to the relative positions on the PAGE gels. The DNA sequence can be determined either by direct sequencing or by sequencing the DNA fragments after cloning them into a vector. The first method has the advantage that only one sequence can be obtained from the re-amplification product. A disadvantage is that small amounts of contaminating DNA often lead to bad and unusable sequence data. Thus, with direct sequencing the chances of interference from contaminant DNA sequences unrelated to the DNA fragment of interest, are low but the percentage of successful sequencing reactions is often also low. Alternatively the DNA fragment can be cloned into a PCR-product cloning vector and subsequently transformed to *E. coli*. In this case it is important to isolate plasmid DNA from several *E. coli* colonies to confirm that most plasmids contain the same DNA fragment of the expected size before sequencing one or more of them.

Fragment Analysis After DNA sequencing of isolated DNA fragments the next step is to find homologous or similar sequences in order to obtain more information about the identity of the corresponding genes. This can be done by searching DNA sequence databases such as GenBank at http://www.ncbi.nlm.nih.gov/ and 'The Gene Index Project' at http://compbio.dfci.harvard.edu/tgi/ for homologous ESTs (Expressed Sequence Tags) or genes using BLAST [9].

It is necessary to confirm the expression profiles of identified genes to ensure that they correspond to the selected bands from the PAGE gel. This can be done by RNA gel blot analysis. However, cDNA-AFLP fragments that, for example correspond to genes encoding transcription factors or other low expression genes may be too low to quantify with hybridization techniques. A more sensitive method is real time PCR on cDNA. This procedure requires gene-specific primers to specifically amplify the cDNA of interest. Gene specific primers can be defined on the DNA sequence of the cDNA-AFLP fragment or on the corresponding gene sequence available in one of the databases. When working with species whose genome is not (fully) sequenced it is frequently not possible to design primers which are gene specific. In that case it will be necessary to analyze the homogeneity of the amplified DNA to be sure that this is derived from only one gene.

A further possibility is to use an AFLP fragment sequencing procedure. For this, sequential AFLP reactions are carried out where the selective extensions on the primers are increased one base at a time in sets of four [7]. This results is a successive reduction in the number of amplification products in the fingerprint until ultimately the band under investigation remains as the only band. The sequence-specific primers that are generated in this process can be used to amplify the band of interest directly and to verify the expression profile of the corresponding gene.

As mentioned, the method described above is optimized for radioactive PAGE. However, it should be noted that the cDNA-AFLP method is technologically highly versatile. In the simplest case silver staining can be used to visualize TDFs from PAGE [10]. In a semi-automated scenario fluorescent labeling can be used in combination with sequencing apparatuses [11,12].

2.3 Applications of the Technology

cDNA-AFLP has been widely applied to all biological systems. However, the primary implementation is found in the plant sciences. Here, plant development, biotic and abiotic stress is investigated in equal measure using cDNA-AFLP. Increasingly the method is now also used in combination with genetic analysis, taking advantage of the fact that cDNA-AFLP detects polymorphisms from SNPs in and around the restriction sites as well as variation in gene expression. In the following section, we highlight some examples from our own laboratories that demonstrate the flexibility of the system. It should be noted, however, that this is not intended to be a comprehensive review of the potential applications as documented in over 200 published articles which employ the method.

2.3.1 Fruit Development

Seedless fruit development (parthenocarpy) can be induced by application of auxin or gibberellin to the flower, as these plant hormones are considered to be important mediators of the signal initiating the development of fruit after pollination. To obtain a better understanding of the role these plant hormones play in the induction of tomato (*Solanum lycopersicum*) fruit growth, a transcriptome analysis was performed using two complementary approaches, cDNA-AFLP and micro-array analysis. cDNA-AFLP analysis of the ovary wall (termed pericarp) and ovules at several time-points after the induction of fruit development showed that the greatest differences in gene expression occur after 3 days and this time-point was therefore chosen for micro-array analysis. These analyses produced profiles that were partly overlapping but both suggested the same thing: in addition to auxin and gibberellin, ethylene and abscisic acid (ABA) are also involved in regulating fruit set [13]. Many of the genes identified with cDNA-AFLP were not present on the micro-array which represented 9254 known tomato transcripts. In addition, 25% of the 283 obtained sequences from cDNA-AFLP had not been previously identified as they were not present in the EST databases that contained 16 000 tomato expressed sequence tags at the time of analysis.

2.3.2
Tuber Development

Potato tuber development has been investigated using cDNA-AFLP [14]. Using a highly synchronous *in vitro* tuberization system, transcriptional changes at and around the time-point at which potato tuberization occurred were analyzed. The targeted expression analysis of a specific transcript coding for the major potato storage protein, patatin and a second transcript, coding for ADP-glucose pyrophosphorylase, a key gene in the starch biosynthetic pathway, were described. This paper confirmed that the kinetics of expression revealed by cDNA-AFLP analysis are comparable to those found by Northern analysis. Furthermore, the isolation of two tuber-specific TDFs coding for the lipoxygenase enzyme, which are differentially induced around the time-point of tuber formation has also been achieved. Analysis of the two lox TDFs demonstrates that it is possible to dissect the expression modalities of individual transcripts, which are not independently expressed in Northern analysis.

2.3.3
Transcript BSA

Quantitative trait locus (QTL) mapping represents an alternative approach in the identification of genes responsible for the naturally occurring allelic variation in complex traits. Several QTLs affecting agronomically important traits have been identified in a wide range of crop plants. However, the genes responsible for those QTLs remain difficult to track down. Map-based cloning is still the most common approach for the identification of the specific gene accounting for a QTL. Saturation of the QTL region with DNA markers tightly linked to the target gene becomes an essential step prior to positional cloning. In recent years, the benefits of using markers representing the transcribed region of the genome as opposed to anonymous genomic DNA markers, has been emphasized [15]. The combination of cDNA-AFLP and bulked segregant analysis (BSA) [16] is a particularly suitable method for saturation of QTL regions with markers that represent expressed genes. We have shown that polymorphic cDNA-AFLP fragments detected in a segregating population can be directly used as genetic markers in the construction of a linkage map [7]. In combination with BSA, the cDNA-AFLP analysis of mapping populations can lead us to the identification of those polymorphic genes linked to the trait of interest, and can potentially result in the direct identification of candidate genes [8].

A combination of cDNA-AFLP and bulked segregant analysis (BSA) was used to identify genes co-segregating with earliness of tuberization in a diploid potato population. This approach identified 37 transcript-derived fragments with a polymorphic segregation pattern between early and late tuberizing bulks. Most of the identified transcripts mapped to chromosomes 5 (19 markers) and 12 (eight markers) of the paternal map. Quantitative trait locus (QTL) mapping of tuberization time also identified earliness QTLs on these two chromosomes. A potato BAC library was screened with four of the markers linked to the main QTL. BAC contigs containing the markers showing the highest association to the trait have been identified. One of these contigs has been anchored to chromosome 5 on an ultra-dense genetic map of potato, which could be used as starting point for the map-based cloning of genes associated with earliness.

2.3.4 Domain Profiling

The possibility of using conserved domains of genes to profile a gene family using RNA fingerprinting has been demonstrated on the MADS-box gene family [17]. This method has been further developed using a cDNA-AFLP approach for the detection of functional copies of NBS-LRR resistance genes in potato. [11]. To explore the potential of NBS profiling in RGA expression analyses, RNA isolated from different tissues was used as a template for NBS-profiling [18]. Of all the fragments amplified approximately 15% showed intensity or even absent/present differences between different tissues implying tissue-specific R-gene expression. Absent/present differences between individuals were also found. In addition to being a powerful tool for generating candidate gene markers linked to R-gene loci, NBS profiling, when applied to cDNA, can be instrumental in identifying those members of an R-gene cluster that are expressed and therefore putatively functional.

2.3.5 VIDISCA

Identification of unknown pathogens using molecular biology tools is difficult because the target sequence is not known, so genome-specific PCR primers cannot be designed. To overcome this problem, we developed the VIDISCA method based on the cDNA-AFLP technique4. The advantage of VIDISCA is that prior knowledge of the sequence is not required, as the presence of restriction enzyme sites is sufficient to guarantee PCR amplification. The input sample can be either blood plasma or serum, or culture supernatant. Whereas cDNA-AFLP starts with isolated mRNA, VIDISCA begins with a treatment to selectively enrich for viral nucleic acid, including a centrifugation step to remove residual cells and mitochondria (Figure 2.1A). A DNase treatment is also used to remove interfering chromosomal and mitochondrial DNA from degraded cells (viral nucleic acid is protected within the viral particle). Finally, by choosing frequently cutting restriction enzymes, the method can be fine-tuned such that most viruses will be amplified. We were able to amplify viral nucleic acids in EDTA-treated plasma from an individual with hepatitis B viral infection, and from an individual suffering from an acute parvovirus B19 infection.

2.4 Perspectives

cDNA-AFLP remains one of the methods of choice for analyzing differential gene expression. It provides a particular advantage in biological systems where little sequence information is available and/or technical resources are limited. Amongst the open architecture technologies, cDNA-AFLP is a good alternative for the discovery of novel, low expression genes which can help towards their characterization and to link them to particular phenotypes and biochemical pathways.

The development of detection tools for specific sequences by tagging has developed rapidly along with the progress of new sequencing strategies. In principle cDNA-AFLP can be adapted to make use of any of these technologies. Clearly, both the 454 and the Solexa sequencing technologies are in the forefront of sequencing developments and both methods would be obvious candidates for application to direct sequencing of cDNA species. The possibility of using AFLP type tags on cDNA fragments derived from restriction digests in these technologies opens up the possibility of not only identifying very large numbers of expressed genes but also of retrieving SNP data in addition to the expression level based on abundance of specific TDFs. A similar approach using a 454 platform has already been demonstrated for genomic AFLP [19] and could be equally well implemented for cDNA.

Despite the rapid development of technologies in molecular biology, cDNA-AFLP is unlikely to become obsolete due to its adaptability to the new technological developments.

References

1 Baulcombe, D. and Verma, D.P. (1978) Preparation of a complementary DNA for leghaemoglobin and direct demonstration that leghaemoglobin is encoded by the soybean genome. *Nucleic Acids Research*, **5**, 4141–4155.

2 Logemann, J., Mayer, J.E., Schell, J. and Willmitzer, L. (1988) Differential expression of genes in potato tubers after wounding. *Proceedings of the National Academy of Sciences of the United States of America*, **85**, 1136–1140.

3 Bachem, C.W.B., Oomen, R.J.F.J. and Visser, R.G.F. (1998) Transcript imaging with cDNA-AFLP: a step-by-step protocol. *Plant Molecular Biology Reporter*, **16**, 157–173.

4 Bachem, C., Hoeven, R.v.d., Lucker, J., Oomen, R., Casarini, E., Jacobsen, E. and Visser, R. (2000) Functional genomic analysis of potato tuber life-cycle. *Potato Research*, **43**, 297–312.

5 Vos, P., Hogers, R., Bleeker, M., Reijans, M., van de Lee, T., Hornes, M., Frijters, A., Pot, J., Pelman, J., Kuiper, M. and Zabeau, M. (1995) AFLP: a new technique for DNA fingerprinting. *Nucleic Acids Research*, **23**, 4407–4414.

6 Vuylsteke, M., Peleman, J.D. and van Eijk, M.J. (2007) AFLP-based transcript profiling (cDNA-AFLP) for genome-wide expression analysis. *Nature Protocols*, **2**, 1399–1413.

7 Brugmans, B., van der Hulst, R.G., Visser, R.G., Lindhout, P. and van Eck, H.J. (2003) A new and versatile method for the successful conversion of AFLP markers into simple single locus markers. *Nucleic Acids Research*, **31**, e55.

8 Fernandez-del-Carmen, A., Celis-Gamboa, C., Visser, R.G. and Bachem, C.W. (2007) Targeted transcript mapping for agronomic traits in potato. *Journal of Experimental Botany*, **58**, 2761–2774.

9 Altschul, S.F., Madden, T.L., Schaffer, A.A., Zhang, J., Zhang, Z., Miller, W. and Lipman, D.J. (1997) Gapped BLAST and PSI-BLAST: a new generation of protein database search programs. *Nucleic Acids Research*, **25**, 3389–3402.

10 Guo, J.R., Schnieder, F. and Verreet, J.A. (2006) Differences between the fingerprints generated from total RNA and poly-A RNA using a modified procedure of cDNA-AFLP and silver staining. *Biotechnology Letters*, **28**, 267–270.

11 Brugmans, B. (2005) Development of tools and strategies towards marker-assisted selection and gene cloning. Ph.D. thesis in Dept. of Plant Sciences, Laboratory of Plant Breeding, WUR, Wageningen.

12 Reijans, M., Lascaris, R., Groeneger, A.O., Wittenberg, A., Wesselink, E., van Oeveren, J., de Wit, E., Boorsma, A., Voetdijk, B., van der Spek, H., Grivell, L.A. and Simons, G. (2003) Quantitative comparison of cDNA-AFLP, microarrays, and GeneChip expression data in *Saccharomyces cerevisiae. Genomics*, **82**, 606–618.

13 Vriezen, W.H., Feron, R., Maretto, F., Keijman, J. and Mariani, C. (2007) Changes in tomato ovary transcriptome demonstrate complex hormonal regulation of fruit set. New Phytologist. (in press).

14 Bachem, C.W., van der Hoeven, R.S., de Bruijn, S.M., Vreugdenhil, D., Zabeau, M. and Visser, R.G. (1996) Visualization of differential gene expression using a novel method of RNA fingerprinting based on AFLP: analysis of gene expression during potato tuber development. *The Plant Journal*, **9**, 745–753.

15 Gupta, P.K. and Rustgi, S. (2004) Molecular markers from the transcribed/ expressed region of the genome in higher plants. *Functional and Integrative Genomics*, **4**, 139–162.

16 Michelmore, R.W., Paran, I. and Kesseli, R.V. (1991) Identification of markers linked to disease-resistance genes by bulked segregant analysis: a rapid method to detect markers in specific genomic regions by using segregating populations. *Proceedings of the National Academy of Sciences of the United States of America*, **88**, 9828–9832.

17 Fischer, A., Saedler, H. and Theissen, G. (1995) Restriction fragment length polymorphism-coupled domain-directed differential display: a highly efficient technique for expression analysis of multigene families. *Proceedings of the National Academy of Sciences of the, United States of America*, **92**, 5331–5335.

18 van der Linden, C.G., Wouters, D.C., Mihalka, V., Kochieva, E.Z., Smulders, M.J. and Vosman, B. (2004) Efficient targeting of plant disease resistance loci using NBS profiling. *Theoretical and Applied Genetics*, **109**, 384–393.

19 Eijk, v.M.J.T. (2006) Complexity reduction of polymorphic sequences (CRoPS): A novel approach for high throughput polymorphism discovery. *Proceedings of the Plant & Aminal Genome Conference XIV*, San Diego, USA. (www.intl-pag.org)

3
SuperSAGE: The Most Advanced Transcriptome Technology for Functional Genomics

Ryohei Terauchi, Hideo Matsumura, Detlev H. Krüger, and Günter Kahl

Abstract

SuperSAGE is a substantially improved version of Serial Analysis of Gene Expression (SAGE), a tag-based method of gene expression profiling. Owing to its tag size (26 bp), SuperSAGE allows a secure tag-to-gene annotation by BLAST search against genomic DNA databases. For non-model organisms without DNA sequence information, the 26-bp tag sequence can be used directly as a PCR primer to carry out 3′- or 5′-RACE to recover the sequence adjacent to the tag which facilitates tag-to-gene annotation. Highly parallel sequencing is perfectly suited to sequencing SuperSAGE tags, dramatically facilitating the experimental protocol, and allowing ultra-detailed expression profiling. Furthermore, oligonucleotides corresponding to SuperSAGE tag sequences can be synthesized on a glass slide to make a high performance custom microarray (SuperSAGE-array). Thus, SuperSAGE combined with SuperSAGE-array allows a detailed, and at the same time high-throughput, expression analysis. SuperSAGE promises to be one of the most efficient available techniques for transcpriptome analysis. Here we present an overview and an up-to-date protocol of the technique. Application of SuperSAGE to plant functional genomics will be discussed with examples.

3.1
Introduction

Techniques in transcriptome analysis can be divided into two major classes. The first is based on hybridization of complementary nucleotide strands to immobilized target sequences (such as cDNAs, oligonucleotides, or PCR fragments). Microarray analysis [1] is a representative of this class. The second class involves sequencing and counting of transcripts, the prototype of which is expression sequence tag (EST) analysis [2]. In EST analysis, transcripts are converted into

The Handbook of Plant Functional Genomics: Concepts and Protocols.
Edited by Günter Kahl and Khalid Meksem

ISBN: 978-3-527-31885-8

cDNAs, and are individually cloned into a plasmid vector to generate a cDNA library. After sequencing thousands of such clones, the number of transcripts from different genes can be enumerated. The frequency of transcripts from different genes provides a gene expression profile of the biological sample. Since one DNA sequence path generates information relevant to only one transcript, throughput of EST analysis has been low. In 1995, Victor Velculescu and co-workers [3] invented a method to count transcripts in a high-throughput manner, and named it Serial Analysis of Gene Expression (SAGE). In SAGE, a short fragment of 13 bp in size (tag) is isolated from a defined position in each cDNA, and the tags are concatenated and cloned into a plasmid vector. One single DNA sequencing run of the plasmid insert can generate information of up to 50 tags, so that throughput was increased ~50-fold over EST sequencing. The key to the SAGE technique is the use of a type IIS restriction endonuclease, BsmFI, for the isolation of tag fragments. Type IIS restriction endonucleases cleave the DNA substrate outside of their specific recognition sequences in the DNA molecule [4]. BsmFI cuts 13–15 bases apart from its recognition site, which allows the isolation of tag sequences from cDNAs.

For SAGE, each mRNA is reversely transcribed to a single-stranded cDNA (sscDNA) by using a biotinylated oligo-d(T) primer. This sscDNA is then converted into a double-stranded cDNA, and digested with the four-base cutter NlaIII which recognizes and cuts the sequence 5′-CATG-3′. The most 3′-end cDNA fragments generated are collected using streptavidin-coated magnetic beads. A linker fragment is ligated to the 5′-end of the fragments collected. This linker fragment is designed to harbor a five-base sequence motif 5′-GGGAC-3′, i.e. the recognition site of BsmFI. Digestion of the linker-cDNA fragment with BsmFI therefore releases a 13-bp fragment from each cDNA, resulting in the generation of a tag fragment. Importantly, the tag sequence is always derived from the identical position in a cDNA, i.e. the most 3′-end NlaIII recognition site. Therefore, each transcript can be uniquely represented by a short tag fragment, and the tag frequency in the sample (tag count) represents the abundance of the corresponding transcript. In turn, the 13-bp tag sequence can be used as a query for a BLAST search [5] against EST databases of the species to identify the gene from which the tag sequence is derived (tag annotation). By combining the tag frequency list and tag annotation, a comprehensive and quantitative profile of gene expression can be derived. In contrast to analog datasets generated by hybridization-based methods, SAGE data are digital and easy to handle by bioinformatics approaches. SAGE facilitates the direct comparison of expression strength of different genes, and the comparison of cross-databases of gene expression. SAGE is an open-architecture method whereby the researcher can theoretically address all the expressed transcripts simply by increasing the number of tags to be analyzed. This feature is not available for microarray which is a closed-architecture method and can only be used to evaluate genes spotted on the chip. Furthermore, in contrast to microarray and another tag-based transcriptome platform, massively parallel signature sequencing (MPSS [6]), SAGE does not require sophisticated apparatus, and can easily be carried out in any laboratory equipped with basic molecular biology facilities. SAGE has been applied to several plant species including rice [7–9], barley [10], *Arabidopsis* [11–14] and cassava [15].

Although SAGE is very a useful method for gene expression profiling, there are drawbacks to the original protocol. The 13-bp tag sequence is frequently too short to uniquely identify the gene of origin by BLAST search against EST or genome DNA databases. Furthermore, in organisms for which DNA sequence databases are not available, SAGE is almost powerless, since BLAST searches cannot be undertaken without a DNA database. It is also difficult to recover experimentally the DNA sequences adjacent to the 13-bp tag sequences; SAGE tags are too short for designing PCR primers for 3′-RACE and oligonucleotide probes to screen a cDNA library, although such attempts have been reported several times [3,16,17]. To improve this situation, Saha *et al.* [18] replaced BsmFI with another type IIS enzyme, MmeI. Using this enzyme they succeeded in extending the tag size to 21 bp and called this version LongSAGE. For the application of LongSAGE to plant species, see [19].

Independently, we adopted a type III restriction endonuclease, EcoP15I [20,21], to SAGE to isolate tags as long as 26 bp, and named this version SuperSAGE [22].

Due to its 26-bp tag size, SuperSAGE has several advantageous properties for transcriptome analysis of eukaryotes:

(a) In organisms for which genomic DNA databases are available, 26-bp SuperSAGE tag sequences allow almost perfect gene annotation *in silico* by BLAST.

(b) Owing to property (a), a biological sample in which two or more eukaryotes are mixed (as e.g. in parasite–host, pathogen–host, or commensal–host interactions) can be analyzed by a single SuperSAGE experiment, and tags can be properly annotated to the genes of the corresponding species by BLAST against genomic sequences of the relevant (or related) species. This makes it possible to study an 'interaction transcriptome [23]' in order to address the biological interactions of the organisms.

(c) In the organisms for which DNA databases are not available, 26-bp SuperSAGE tag sequences can be used directly for synthesizing PCR primers to carry out 3′-RACE PCR to recover the adjacent sequences for the purpose of tag-to-gene annotation by BLASTX.

(d) Highly parallel sequencing, including the 454 sequencing technology [24], dramatically facilitates sequencing of SuperSAGE tags, and reduces the time and cost of the experiment.

(e) Oligonucleotides corresponding to the 26-bp SuperSAGE tag sequences can be synthesized on a glass slide to establish a microarray (SuperSAGE-array) in order to analyze multiple samples with high throughput [25].

In this chapter, we provide an up-to-date protocol of SuperSAGE, and present an overview of the utility of SuperSAGE in plant functional genomics.

3.2
Methods and Protocols

The original SuperSAGE protocols have been described elsewhere [26,27]. Briefly, the experiment involves the following steps: (1) mRNA extraction, (2) cDNA synthesis using a biotinylated oligo-d(T) primer and conversion of single-stranded cDNA into double-stranded cDNA, (3) digestion of the cDNAs with a four-base cutter, NlaIII, and collection of the 3′-end cDNA fragments using streptavidin-coated magnetic beads, (4) division of the collected cDNAs into two tubes, and ligation of different linker fragments to the 5′-ends of the cDNAs in each tube, (5) mixing the contents of the two tubes, and digestion of the linker-cDNA fragments with EcoP15I, and release of 'linker-tag' fragments, (6) ligation of two 'inker-tag' fragments in head-to-head orientation to generate 'linker-ditag-linker' fragments, (7) PCR amplification of 'linker-ditag-linker' fragments, (8) removal of the linker fragments by digestion with NlaIII to generate 'ditags', (9) concatenation of 'ditags', (10) cloning of 'ditag' concatemers into a plasmid vector, (11) sequencing of the plasmid insert, (12) extraction of tag sequence and compilation of the data. The latest version of the SuperSAGE protocol differs from the original in that highly parallel sequencing (reviewed in [28]) has been incorporated. In this modification, we use the 454 sequencing platform (GS-20), which allows a parallel sequencing of more than 200 000 DNA fragments at a time [24]. In sequencing with the GS-20 platform, the size of a reliable sequence read for each fragment is around 100 bp. Coincidentally, the size of the SuperSAGE 'linker-ditag-linker' fragment generated after step (7) above, is 96–98 bp (Figure 3.1), which perfectly fits the size of a single sequence path

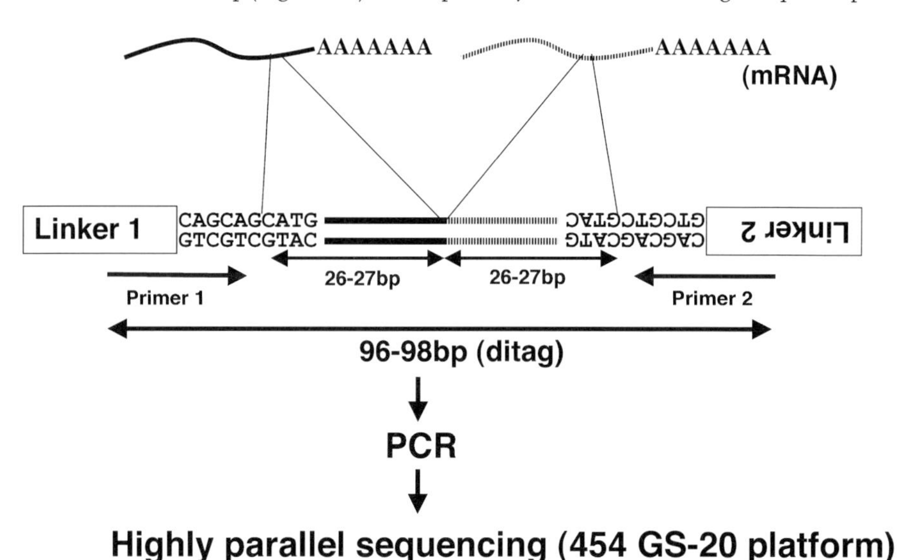

Figure 3.1 The size of a 'linker-ditag-linker' fragment after ditag PCR is 96–98 bp, and perfectly fits the size of single sequence path (~100 bp) of the 454 GS-20 highly parallel sequencing platform. This compatibility of SuperSAGE and '454 sequencing' allows a cost effective high-throughput transcriptome analysis.

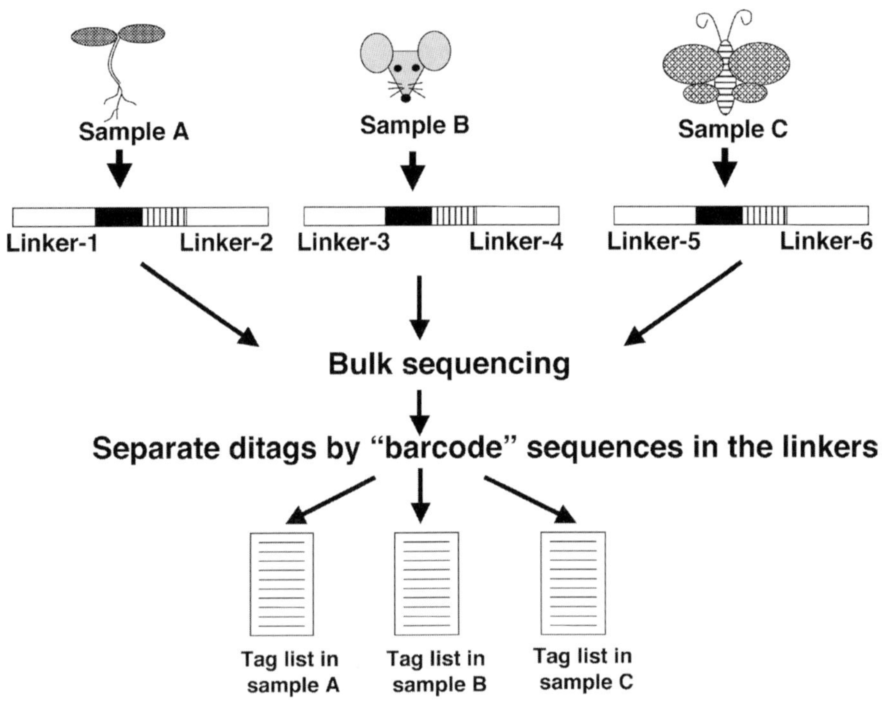

Figure 3.2 Ditag PCR products from different SuperSAGE libraries can be bulked, and sequenced together. After sequencing, each ditag is assigned to the corresponding library according to the 'barcode' sequence embedded in the linker fragments.

of GS-20 sequencing. Therefore, by using 454 sequencing after step (7), steps (8)–(11) can be omitted. Since the ditag concatenation and cloning steps (steps (9) and (10)) have been the most difficult of the entire SuperSAGE experiment, this shortcut significantly facilitates the technique. Furthermore, the sequencing of multiple SuperSAGE libraries can be achieved in a single run of 454 sequencing. For this purpose different linker fragments with unique sequences can be used to generate individual SuperSAGE libraries (Figure 3.2). These are mixed, and sequenced together. Later, the sequences can be sorted to each library according to the 'barcode' embedded in the linker sequence. This multiplexing results in a significant reduction in the cost of sequencing.

Using the improved protocol as described below, the whole process excluding DNA sequencing can be completed within a week. The amount of starting material should be in excess of 20–30 μg of total RNA.

3.2.1 Linker Preparation

Linker DNAs for SuperSAGE are prepared by annealing the two complementary oligonucleotides, as shown in Figure 3.3 (Linker-1A, 1B, 2A, 2B). Linker DNAs have

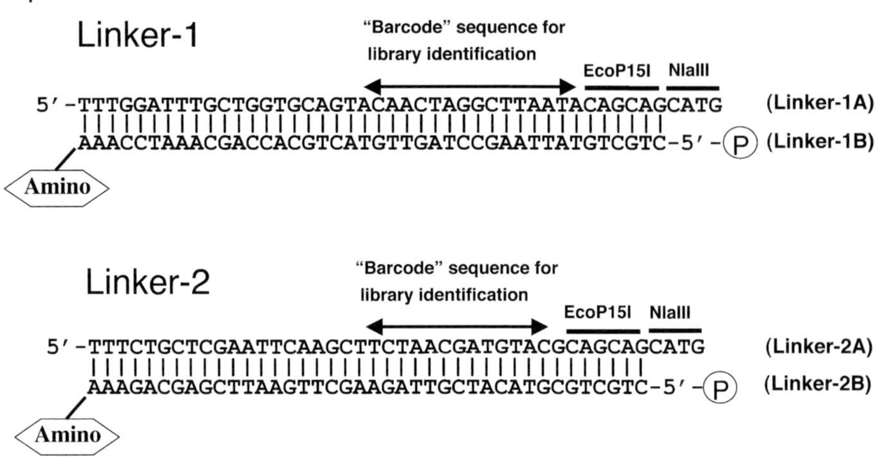

Figure 3.3 Structure of SuperSAGE linker fragments. In the upper strand, a 'barcode' sequence for the identification of the ditag library is located 5′ to the EcoP15I recognition site. An arbitrary sequence can be incorporated into this 'barcode' region for the purpose of assignment of ditags to a library.

cohesive ends, which are compatible with the end generated by NlaIII-digestion (5′-CATG-3′), and an EcoP15I-recognition site (5′-CAGCAG-3′) is present adjacent to the 5′-CATG-3′ site. The 3′-ends of the Linker-XBs should be amino-modified to prevent ligation occurring at this site. In the original protocol, only one pair of linker DNAs was used for SuperSAGE. Currently, we prepare several different pairs of linker DNAs (Linker 1, 2, 3, 4 and so on) for the creation of multiple SuperSAGE libraries. In these linkers, sequence variation of 5–6 bp is incorporated within the 10-bp region upstream of the EcoP15I recognition site (5′-CAGCAG-3′). This sequence variation serves as the 'barcode' for assigning ditags to different libraries after bulk sequencing using the 454 platform (Figure 3.2).

For preparation of Linker-1 and Linker-2 as shown in Figure 3.3, dissolve the synthesized linker oligonucleotides (Linker-1A, 1B, 2A, 2B) in LoTE buffer (3 mM Tris-HCl, pH7.5; 0.2 mM EDTA), so that their concentration is 1 μg/μl. Mix 1 μl Linker-1B (or Linker-2B), 1 μl 10× polynucleotide kinase buffer, 1 μl 10mM ATP, 7 μl H_2O and 1 μl T4 polynucleotide kinase, and incubate at 37 °C for 30 min to phosphorylate the 5′-ends. Add 1 μl Linker-1A or -2A to the 5′-phosphorylated Linker-1B or -2B solution respectively, from the previous step. After mixing, denature by incubating at 95 °C for 2 min and cool down to 20 °C for annealing. The annealed double-stranded DNAs (200 ng/μl) are designated as Linker-1 and Linker-2, respectively.

3.2.2
RNA Sample

In our experience, using 20–30 μg total RNA as the starting material leads to a successful outcome for a SuperSAGE experiment. However, the more the total RNA used (>50 μg), the higher the success rate. Before cDNA synthesis, mRNA (poly(A)$^+$ RNA) should be purified from total RNA using an oligo-dT column or other methods.

3.2.3 cDNA Synthesis

Any cDNA synthesis protocols are applicable to SuperSAGE, but biotinylated adapter-oligo dT primer harboring the EcoP15I-recognition site ('5′-CAGCAG-3′') should be used for reverse transcription. We use SuperScriptII double-strand cDNA synthesis kit (Invitrogen) following the experimental procedures given in its instruction manual. After second-strand cDNA synthesis, double-strand cDNA is purified by passing it through a column (Qiaquick PCR purification kit; Qiagen), instead of phenol/chloroform extraction and ethanol precipitation. Follow the kit instruction. This purification step can help in avoiding failure in the subsequent NlaIII digestion reaction.

3.2.4 Tag Extraction from cDNA

Purified cDNA (50 μl eluted DNA from a column) is completely digested with NlaIII, by adding 20 μl NlaIII digestion buffer (NEBuffer 4), 2 μl BSA, 123 μl LoTE, 5 μl NlaIII (10 U/μl; NEB), and incubating at 37 °C for 1.5 h. After digestion, a small aliquot of the digestion reaction (around 5 μl from 200 μl reaction solution) is loaded onto a 1% agarose gel. NlaIII digestion can be confirmed by the shift-down of cDNA sizes from 500–2000 bp to 100–300 bp. Digested cDNA solution (without purification) is divided between two tubes, tube A and tube B (each 100 μl). Tubes A and B both contain cDNA to be ligated with Linker-1 and Linker-2, respectively, as described above. Equal volume of 2× B&W buffer (10 mM Tris-HCl, pH 7.5; 1 mM EDTA; 2 M NaCl) is added to each of the tubes A and B. The contents of tubes A and B are separately added to the washed streptavidin-coated magnetic beads (Streptavidin MagneSphere Paramagnetic Particles, Promega). Biotinylated cDNA fragments are associated with streptavidin-coated magnetic beads by incubation at room temperature for 30 min. After washing the beads three times with 1× B&W buffer and once with LoTE buffer, Linker-1 and Linker-2 are ligated to the ends of cDNAs on the magnetic beads in the two tubes, respectively. For ligation, 200 ng linker DNA is usually added to a tube. However, if the amount of total RNA used as the starting material is less than 50 μg, the amount of the linker should be reduced (~100 ng in a tube). To ligate linkers to digested cDNAs bound to the magnetic beads, add 21 μl LoTE, 6 μl 5× T4 DNA ligase buffer, and either 1 μl Linker-1 or -2 solution (~100–200 ng), respectively, to the magnetic beads. The bead suspension is incubated at 50 °C for 2 min to effect the dissociation of linker dimers, and kept at room temperature for 15 min. T4 DNA ligase (10U) is then added, and the tubes are incubated at 16 °C for 2 h. After ligating the linker, the bead suspension from the two tubes is mixed. The beads are washed four times with 1× B&W buffer, followed by three washes with LoTE buffer. The resulting linker-cDNA fragments on the beads are digested with

EcoP15I to release 'linker-tag' fragments. For EcoP15I digestion, 10 μl 10 × EcoP15I digestion buffer (100 mM Tris-HCl, pH8.0; 100 mM KCl; 100 mM $MgCl_2$; 1 mM EDTA; 1 mM DTT; 50 μg/ml BSA), 2 μl 100 mM ATP, 83 μl sterile water, and 5 μl EcoP15I (2 U/μl; NEB) are added to the washed magnetic beads. Tubes are incubated at 37 °C for 2 h.

3.2.5
Purification of Linker-Tag Fragment

The DNA released from the beads after EcoP15I digestion is extracted with phenol/chloroform, and precipitated by adding 100 μl 10 M ammonium acetate, 3 μl glycogen, and 950 μl cold ethanol. The tube is maintained at −80 °C for 1 h, and the DNA precipitated by centrifugation at 15 000 × g for 40 min at 4 °C, and the resulting pellet is washed once with 70% ethanol. After drying, the pellet is dissolved in 10 μl LoTE buffer. Dissolved DNA solution is loaded onto an 8% PAGE gel, which is prepared by mixing 3.5 ml 40% acrylamide/bis solution, 13.5 ml distilled water, 350 μl 50× TAE buffer, 175 μl 10% ammonium persulfate, and 15 μl TEMED. The polyacrylamide gel is run at 75 V for 10 min, and then at 150 V for around 30 min. The gel is stained with SYBR-Green (Molecular Probes) and the DNA visualized on a UV trans-illuminator. The 'linker-tag' fragments of expected size (around 70 bp) are cut out and put into a 0.5-ml tube. Holes are made at the top and the bottom of the tube with a needle, and it is placed in a 2-ml tube. The tube is centrifuged at the maximum speed for 2–3 min. Polyacrylamide gel pieces are collected at the bottom of the 2-ml tube, and 300 μl LoTE is added to the gel pieces for resuspension. After incubation at 37 °C for 2 h, the gel suspension is transferred to a Spin-X column (Corning) and centrifuged at maximum speed for 2 min. The solution which collects at the bottom of the tube is extracted with phenol/chloroform, and precipitated as described above. After washing once with 70% ethanol, the dried linker-tag DNA is dissolved in 8 μl LoTE buffer.

3.2.6
Ditag Formation and Amplification

Purified 'linker-tag' fragments (a mixture of Linker-1-tag and Linker-2-tag fragments) are blunt-ended by fill-in reaction using the Blunting High Kit (TOYOBO). To the linker-tag solution (8 μl), 1 μl 10× blunting buffer and 1 μl KOD DNA polymerase (TOYOBO) are added. The tube is incubated at 72 °C for 2 min, and immediately transferred into ice. For ditag formation, 30 μl LoTE and 40 μl Ligation High (TOYOBO) are added to the 10 μl blunt-ended reaction. After incubation of the ligation reaction mixture at 16 °C for 4 h to overnight, a small aliquot of the ligation product is taken and diluted (1/5 and 1/10) with LoTE buffer. These two diluents are used as the template for the pilot experiment of PCR amplification of the 'linker-ditag-linker' fragments. PCR primers are designed from linker sequences, including 'barcode' sequences. For Linker-1 and

Linker-2, we use PCR primers with the sequence 5′-CAACTAGGCTTAATACAG CAGCA-3′ and 5′-CTAACGATGTACGCAGCAGCA-3′, respectively. Hot-start PCR is not always necessary for amplifying 'linker-ditag-linker' fragments. We amplify 'linker-ditag-linker' in a reaction mixture containing 5 μl 10 × PCR buffer, 5 μl 2 mM dNTP, 0.2 μl of each primer (350 ng/μl), 38.34 μl distilled water, 1 μl diluted template solution and 0.26 μl Taq polymerase (5 U/μl). Amplification of the 'linker-ditag-linker' in the reaction cycle is carried out as follows: 94 °C for 2 min, then 25 cycles each at 94 °C for 40 s, and 60 °C for 40 s. To avoid a preferential amplification of a subset of ditags, the number of PCR cycles is less than that used in the previous SuperSAGE protocol (27–29 cycles) [20]. It can be determined from the pilot PCR experiment which of the 1/5 and 1/10 template dilutions gives the better amplification of the 'linker-ditag-linker' PCR products (96–98 bp), as observed in a SYBR green-stained acrylamide gel. A bulk PCR is carried out under the same condition for 40–48 tubes each containing 50 μl, using the diluted template (either 1/5 or 1/10 dilutions) that gave the better amplification in the pilot PCR (see above). All PCR products are collected in a tube, and purified with Qiaquick PCR purification kit (Qiagen). Six to eight columns are used for the purification procedure, and eluted DNAs from all the columns are collected in a single tube. This DNA solution (180–240 μl) is loaded onto an 8% polyacrylamide gel. After running the gel and staining with SYBR Green as described above, the separated DNA fragments of the expected size (96–98 bp) are excised from the gel. DNA is eluted from the polyacrylamide gel, and purified by ethanol precipitation after phenol/chloroform extraction, as described above. From 40 to 48 PCR reaction tubes, around 1 μg of purified 'linker-ditag-linker' fragments can be obtained. This DNA is ready for sequencing in the GS20 sequencer manufactured by the 454 Life Sciences Company. We routinely order sequencing from this company.

3.2.7
Tag Extraction from Sequence Data

Some typical sequence readouts are shown in Figure 3.4. Successful sequencing will provide the experimenter with more than 200 000 such sequence reads. To extract tag sequences from the raw data, we developed a simple program pipeline known as 'SuperSAGE_tag_extract_pipe'. This pipeline sequentially removes annotation lines, sorts sequences into libraries according to the 'barcode' sequences embedded in the linker fragments (Figure 3.2), extracts ditags, removes duplicated ditags, extracts tags, and sorts tags according to their frequency. This program written in Perl can be implemented on a UNIX platform, and is freely available from RT upon request. Comparison of two or more tag frequency files can be made using 'SuperSAGE_tag_freq_comp'. In cases where there are genomic DNA databases available for particular organisms, the tag frequency list can be used for a BLAST search and the automatic tag-to-gene annotation can be implemented using the 'SuperSAGE_tag_BLAST' suite of programs. All the programs are available from RT upon request.

```
>000203_0029_0696 length=86 uaccno=D2488HV01ACUG6
CTATCGATGCACGCAGCAGCATGATTGGCAAAGAGGCAAAATTCCTAAACAGCAGAAGAT
ATGAAGTCCCATGCTGCTGTATTAAG
>000204_0204_1301 length=77 uaccno=D2488HV01AR703
CAACTAGGCTTAATACAGCAGCATGATTGAATGAGGCTCTATTGTGTTATTCACACTGGC
TGCAAATGCCCATGCTG
>000211_1090_2768 length=94 uaccno=D2488HV01CX1CY
CAACTACGCTAAATACAGCAGCATGCTGAATCAGGAGGAATAAATAAGCAAAGTAAAAGT
GAAAGAATAAACATGCTGCTGCGCACACCGTTAG
>000215_0029_0504 length=88 uaccno=D2488HV01ACUBU
CAACTAGGCTTAATACAGCAGCATGATGTGATAATTTATAACCAAATAGGATTGGAGAAG
GAGGGATGACTACATGCTGCTGCGTACA
........
```

1. Remove the annotation line (i.e. > 000203...), tidy up sequences

2. Sort sequences into different libraries by the "barcode" embedded in the linker sequence

3. Extract di-tag sequences

4. Remove duplicated ditag sequences

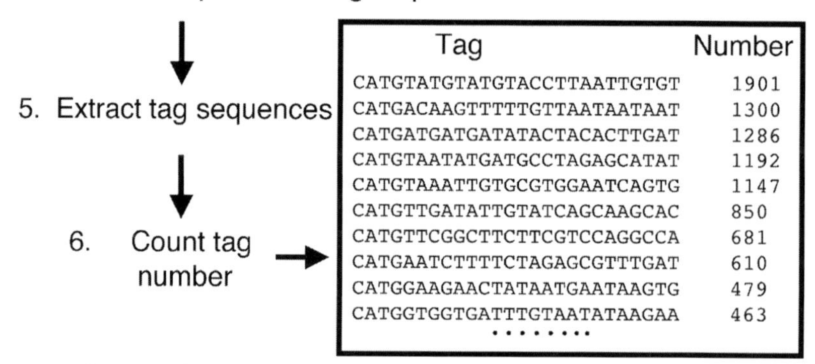

Tag	Number
CATGTATGTATGTACCTTAATTGTGT	1901
CATGACAAGTTTTTGTTAATAATAAT	1300
CATGATGATGATATACTACACTTGAT	1286
CATGTAATATGATGCCTAGAGCATAT	1192
CATGTAAATTGTGCGTGGAATCAGTG	1147
CATGTTGATATTGTATCAGCAAGCAC	850
CATGTTCGGCTTCTTCGTCCAGGCCA	681
CATGAATCTTTTCTAGAGCGTTTGAT	610
CATGGAAGAACTATAATGAATAAGTG	479
CATGGTGGTGATTTGTAATATAAGAA	463
....	

Figure 3.4 A flowchart of the isolation of SuperSAGE tag sequences from the 454 GS-20 sequence readout. Raw sequences, as shown at the top correspond to 'linker-ditag-linker' fragments (NlaIII site underlined). They are processed by a pipeline of programs to generate a list of tags and their abundances (bottom).

3.3 Applications of the Technology

3.3.1 Interaction Transcriptome

The specificity of a 26-bp SuperSAGE tag is high, so that a BLAST search against the genomic sequence of a species in most cases identifies a single position exhibiting a perfect match to the tag sequence. This high tag-to-gene annotation power of SuperSAGE allows the transcriptome study of two or more interacting eukaryotes, provided that genomic sequences are available for all the species involved. Therefore, we used SuperSAGE to study the interaction of rice (*Oryza sativa*) and the rice blast fungus (*Magnaporthe grisea*) [22]. Total RNA of blast-infected rice leaves was isolated, and subjected to SuperSAGE. A total of 12 119 tags derived from 7546 unique tags were isolated. BLAST search of these tags against rice and *Magnaporthe* genome databases showed that the majority of tags matched a corresponding location in the rice genome and only 35 unique tags in the *Magnaporthe* genome. This experiment demonstrates that SuperSAGE can be used to evaluate the 'interaction transcriptome' [29,30].

3.3.2 Application of SuperSAGE to Non-Model Organisms

SuperSAGE can be applied to organisms for which DNA databases are not available. Here we show two examples. *Nicotiana benthamiana* is a plant species frequently used for reverse genetics studies employing virus-induced gene silencing (VIGS) [31]. However, at the time we carried out the experiment, no extensive DNA database was available for the species. We have been using *N. benthamiana* for studying hypersensitive response (HR)-like cell death. A *Phytophthora infestans* eliciting INF1 is known to cause HR cell death in *N. benthamiana* [32]. To study the changes in gene expression following INF1 treatment of *N. benthamiana*, we carried out SuperSAGE [22]. *N. benthamiana* plants were infiltrated with 100 nM INF1 protein and water as control. One hour after the infiltration, leaves were harvested and total RNA extracted and subjected to SuperSAGE. A total of 5089 and 5095 tags were isolated from INF1-treated and water-treated leaves, respectively. We selected 14 tags whose frequency was drastically reduced in the INF1-treated sample as compared to the control. PCR primers containing these 14 SuperSAGE tag sequences were synthesized, and used for 3′RACE-PCR. For all of the primers, partial cDNA fragments containing polyA-tails were easily amplified. Of 14 cDNA sequences, 11 showed significant homology to known protein genes of higher plants. Many of the down-regulated genes encoded chloroplast-localized and photosynthesis-related proteins. Using the same method, *N. benthamiana* genes induced by transient overexpression of a transcription factor were also studied [33].

Coemans *et al.* [34] applied SuperSAGE to banana (*Musa acuminata*) leaves. The authors combined 3′RACE and TAIL-PCR [35] to successfully recover the coding region as well as a promoter region of a gene corresponding to a SuperSAGE tag. These experiments demonstrate that SuperSAGE is widely applicable to eukaryotic species.

3.3.3 SuperSAGE-Array

Currently the most widely used transcriptome technique is the microarray. The advantage of microarrays over tag-based techniques is that in the former a large number of biological samples can be rapidly analyzed once relevant chips are mass-fabricated. Even with the incorporation of highly parallel sequencing, SuperSAGE will not compete with microarrays in the near future, as far as the handling of multiple samples is concerned. In view of this situation, we sought to combine the advantages of both SuperSAGE and microarray to study multiple biological samples in high-throughput, which resulted in the development of the SuperSAGE-array [25]. SuperSAGE-array is a microarray consisting of oligonucleotides corresponding to 26-bp SuperSAGE tags synthesized on a glass-slide. It was shown that 26-bp SuperSAGE tag oligonucleotide probes give highly reproducible hybridization results, suitable for analyzing many samples. As an example, we first compared the gene expression profiles of leaves and suspension-cultured cells of rice (*O. sativa*) by SuperSAGE, and selected 1000 tags showing different expression patterns between the two samples. The first group of tags (78 tags) were equally represented in the two samples, the second group (438 tags) were more prevalent in leaves, and the third group (484 tags) more prevalent in cultured cells. Oligonucleotides for these 1000 tags were synthesized on a chip, and hybridized to the labeled RNAs derived from leaves and cultured cells, respectively. Notably, the SuperSAGE results were faithfully recapitulated in hybridization experiments for most of the tags: 80.4% of SuperSAGE tags more prevalent in leaves hybridized more strongly to leaf RNA, and 87% of SuperSAGE tags more prevalent in cultured cells hybridized more strongly to cultured cell RNA. If we focus on the tags that are statistically significantly differentially represented between the two samples, reproducibility was even higher: 87.7 and 89.2% of such tags showed significantly stronger hybridization to leaf and cell culture RNAs, respectively. This result indicates that oligo-arrays with immobilized SuperSAGE tag sequences can be mass-fabricated, and used to rapidly probe multiple biological samples.

We propose two major applications of SuperSAGE-array. In the first, in-depth SuperSAGE analysis is carried out for two samples that are subjected to different treatments, and the expression profiles are compared to select a subset of tags (~1000 tags) that are differentially represented between the samples. These tags are put onto the oligo-chips, and used for hybridization to RNAs from multiple samples. By this means we can address expression kinetics of selected genes over time after a distinct treatment of cells. The second application of SuperSAGE-array is simple, but has a profound potential. Since microarrays can be made for any eukaryote at minimum cost, a SuperSAGE experiment can be carried out using the species that is of particular interest to the researcher. The tag sequences (~10 000) obtained in

SuperSAGE are synthesized on a chip, and used for hybridization experiments using RNAs from relevant treatments in the target species. We do not need to know the name of gene corresponding to each tag at this stage. After hybridization experiments, it may be possible to identify tags that show interesting expression patterns. Only after that, can the annotation of the genes be considered using the technique as described above. In essence, SuperSAGE-array makes possible the fabrication of an 'easy array' for any eukaryote. In SuperSAGE-array all the procedures that are required for conventional microarray construction are omitted, including cDNA library construction, selection of unique genes, and amplification of cDNAs. Therefore, SuperSAGE-array reduces costs and time, and is widely applicable to the functional genomics of various species.

3.3.4 GMAT

Original SAGE and SuperSAGE technologies have been used to calculate the frequency of individual transcripts. However, these tag-counting techniques can also be extended to the study of genomic DNA. A genome-wide mapping technique (GMAT; [36,37]) combines chromatin immunoprecipitation (ChIP) and SAGE to study genomic sequences in samples immunoprecipitated by a specific antibody. Roh *et al.* [36] used antibodies raised against acetylated histone 3 and acetylated histone 4 to immunoprecipitate yeast chromatin. Genomic DNA in ChIPed chromatin was collected, and LongSAGE was applied. The tags of 21–22 bp were mapped on the genomic sequence of yeast to reveal the region associated with the acetylated histones. Thus, GMAT can be used to study epigenetics, and to expand the utility of tag-based methods as represented here by SAGE. Since the complexity of genomic DNA increases with increasing size of genome, the use of SuperSAGE with 26-bp tags is positively preferable in future studies to obtain a higher success rate of mapping in organisms with large genome sizes. We are currently applying SuperSAGE-GMAT to *Arabidopsis thaliana*. With the introduction of highly parallel sequencing, SuperSAGE-GMAT might become more advantageous than the microarray-based ChIP system (ChIP-chip; [38]).

3.4 Perspectives

Though SuperSAGE has convincingly shown great potential for genome-wide expression profiling in plants and animals (including humans), it has additionally given rise to the design of a series of satellite techniques such as SuperSAGE-arrays [25], and could easily be integrated into already existing technologies; the technique nevertheless is still at an early stage of development. On one hand, SuperSAGE, despite all its virtues, cannot yet solve several problems of transcriptome architecture. The technique, as is the case with all tag-based transcript profiling methods, can only give a snapshot of the transcriptome. Of course, the sampling of original RNA could be more intense along the time axis for example, during the development of an organ or the reaction towards a pathogen. However, the actual cost of carrying out many SuperSAGE transcriptome analyses during such processes is prohibitive. SuperSAGE then will doubtless remain the future method of choice for point-by-point selective transcriptome studies in eukaryotic organisms. Moreover, important parameters of the transcriptome escape detection. Foremost amongst these are the intracellular localizations of single messenger RNAs, the duration of their half-lives, and their migrations in space and time, all of which are valuable parameters of the transcriptome which remain obscure. Therefore, the dynamics of the transcriptome cannot be tackled with SuperSAGE, despite it being one of the best expression profiling technologies.

On the other hand, the technical procedure of SuperSAGE restricts the analysis to polyadenylated messenger RNAs, poly(A)$^+$-mRNAs. However, most bacterial and archaeal mRNAs are not polyadenylated, so that SuperSAGE is not yet applicable to the study of prokaryotic or archaeal transcriptomes. With its selection of only poly (A)$^+$-mRNAs, the technique also excludes analysis of the full transcriptome of a eukaryotic organism. For example, the human transcriptome, at least as estimated for 10 human chromosomes, contains about 2.2 times more unique poly(A)$^-$- than poly(A)$^+$-mRNAs, and many of the poly(A)$^-$ transcripts are encoded by intergenic genomic regions [39]. It also became clear that many of the originally polyadenylated transcripts are rapidly processed *in nucleo* to poly(A)$^-$-mRNAs, and certainly escape customary cDNA cloning procedures. This is especially true for the precursors of the hundreds, if not thousands of microRNAs; whereas the nuclear pre-miRNA transcripts are still polyadenylated, the maturation process actually starts with the removal of a substantial part of the precursor together with the poly(A)-tail by the double-strand RNA-specific ribonuclease Drosha. The resulting hairpin RNAs (precursor miRNAs, 'pre-miRNAs') are consequently non-polyadenylated.

We foresee three basic routes which will change the use of SuperSAGE for transcriptome analysis. First, the sequencing of ditags (or multiples of them) will become faster and cheaper than ever before. The advent of the emulsion-PCR-coupled pyrosequencing procedure developed by 454 Life Sciences (see Chapter 22, this volume) and the recently launched ultrahigh-throughput SOLiD sequencing technology by Applied Biosystems (see Chapter 21, this volume) have literally revolutionized tag-based transcript profiling techniques both in speed and costs. Therefore, for the first time, we here describe the sequencing of SuperSAGE tags using the 454 Life Sciences technology. Since a series of other companies are in the process of developing novel non-Sanger sequencing procedures (e.g. nanopore sequencing, capillary electrophoretic sequencing and its variant microelectrophoretic sequencing, or clonal single molecule array technology as exemplified by the reversible terminator sequencing of Solexa), we expect a further improvement in sequencing efficiency in the very near future.

A second route will be the miniaturization and automation of the SuperSAGE protocol which as yet has not been streamlined. One approach towards miniaturization is the expansion of SuperSAGE analysis to very low quantities of input material. Presently SuperSAGE still requires initial RNA concentrations of 20–30 μg, well above those found in precious needle aspirate, thin tissue section, tumor biopsy or single cell samples. Whereas laser capture microdissection, well established in the first author's laboratory, allows the sampling of single cells, the protocol has nevertheless to be adapted to expectedly minute amounts of RNA from these samples. One attractive approach, published as MicroSAGE [40], is likely to become an integral component of the procedure.

A third route of improvement will certainly be the expansion of SuperSAGE to meet requirements for transcript profiling in prokaryotes. To this end, the poly(A)$^-$-mRNAs of for example, bacteria will be captured and polyadenylated, and subsequently subjected to SuperSAGE. This 'bacterial' SuperSAGE in combination with other techniques (e.g. *in vitro* polyadenylation) will broaden its spectrum of applications significantly.

It is clear that the bioinformatics part of the SuperSAGE platform needs continuous up-dating and development. Aside from the annotation of tag-to-gene sequences in the databases a bioinformatics tool will be needed to associate tag abundances with metabolic pathways, as exemplified and illustrated by the MapMan software package [41]. All these necessary improvements will only accelerate our knowledge of the molecular biology of a cell if combined with exhaustive data from the proteome and metabolome, or, in other words, a systems biology approach that is, however, still in its earliest infancy.

Acknowledgments

This work was in part supported by the 'Program for Promotion of Basic Research Activities for Innovative Bioscience', 'Iwate University twenty first century COE Program: Establishment of Thermo-Biosystem Research Program', Deutsche Forschungsgemeinschaft (grant KR1293/4), ERA-PG grant FR/06.075B from the European Union (LegResist) to GK, and JSPS grants no. 18 310 136 and 18 688 001 to RT and HM, respectively. We thank Matt Shenton for his constructive comments during the preparation of the manuscript.

References

1 Schena, M., Shalon, D., Davis, R.W. and Brown, P.Q. (1995) Quantitative monitoring of gene expression patterns with a complementary DNA microarray. *Science*, **270**, 467–470.

2 Adams, M.D., Kelley, J.M., Gocayne, J.D., Dubnick, M., Polymeropoulos, M.H., Xiao, H., Merril, C.R., Wu, A., Olde, B., Moreno, R.F. *et al.* (1991) Complementary DNA sequencing: expressed sequence tags and human genome project. *Science*, **252**, 1651–1656.

3 Velculescu, V.E., Zhang, L., Vogelstein, B. and Kinzler, K.W. (1995) Serial analysis of gene expression. *Science*, **270**, 484–487.

4 Roberts, R.J., Belfort, M., Bestor, T., Bhagwat, A.S., Bickle, T.A., Bitinaite, J., Bluenthal, R.M., Degryarev, S.K., Dryden, D.T.F., Dybvig, K. *et al.* (2003) A nomenclature for restriction enzymes,

DNA methyltransferases, homing endonucleases and their genes. *Nucleic Acids Research*, **31**, 1805–1812.

5 Altschul, S.F., Madden, T.L., Schaffer, A.A., Zhang, J., Zhang, Z., Miller, W. and Lipman, D.J. (1997) Gapped BLAST and PSI-BLAST: a new generation of protein database search programs. *Nucleic Acids Research*, **25**, 3389–3402.

6 Brenner, S., Johnson, M., Bridgham, J., Golda, G., Lloyd, D.H., Johnson, D., Luo, S., McCurdy, S., Foy, M., Ewan, M. *et al.* (2000) Gene expression analysis by massively parallel signature sequencing (MPSS) on microbead arrays. *Nature Biotechnology*, **18**, 630–634.

7 Matsumura, H., Nirasawa, S. and Terauchi, R. (1999) Transcript profiling in rice (*Oryza sativa* L.) seedlings using serial analysis of gene expression (SAGE). *Plant Journal*, **20**, 719–726.

8 Matsumura, H., Nirasawa, S., Kiba, A., Urasaki, N., Saitoh, H., Ito, M., Kawai-Yamada, M., Uchimiya, H. and Terauchi, R. (2003) Overexpression of Bax inhibitor suppresses the fungal elicitor-induced cell death in rice (*Oryza sativa* L.) cells. *Plant Journal*, **33**, 425–434.

9 Gibbings, J.G., Cook, B.P., Dufault, M.R., Madden, S.L., Khuri, S., Turnbull, C.J. and Dunwell, M. (2003) Global transcript analysis of rice leaf and seed using SAGE technology. *Plant Biotechnology Journal*, **1**, 271–285.

10 Ibrahim, A.F., Hedley, P.E., Cardie, L., Kruger, W., Marshall, D.F., Muehlbauer, G.J. and Waugh, R. (2005) A comparative analysis of transcript abundance using SAGE and Affymetrix arrays. *Functional & Integrative Genomics*, **5**, 163–174.

11 Chakravarthy, S., Tuori, R.P., D'Ascenzo, M.D., Fobert, P.R., Despres, C. and Martin, G.B. (2003) The tomato transcription factor Pti4 regulates defense-related gene expression via GCC box and non-GCC box cis elements. *Plant Cell*, **15**, 3033–3050.

12 Ekman, D.R., Lorenz, W.W., Przybyla, A.E., Wolfe, N.L. and Dean, J.F.D. (2003) SAGE analysis of transcriptome responses in Arabidopsis roots exposed to 2,4,6-trinitrotoluene. *Plant Physiology*, **133**, 1397–1406.

13 Jung, S., Lee, J. and Lee, D. (2003) Use of SAGE technology to reveal change in gene expression in Arabidopsis leaves undergoing cold stress. *Plant Molecular Biology*, **52**, 553–567.

14 Fizames, C., Munos, S., Cazettes, C., Nacry, P., Boucherez, J., Gaymard, F., Piquemal, D., Delorme, V., Commes, T., Doumas, P. *et al.* (2004) The Arabidopsis root transcriptome by serial analysis of gene expression. Gene identification using the genome sequence. *Plant Physiology*, **134**, 67–80.

15 Fregene, M., Matsumura, H., Akano, A., Dixon, A. and Terauchi, R. (2004) Serial analysis of gene expression (SAGE) of host-plant resistance to the cassava mosaic disease (CMD). *Plant Molecular Biology*, **56**, 563–571.

16 Van den Berg, A., van der Leij, J. and Poppema, S. (1999) Serial analysis of gene expression: rapid RT-PCR analysis of unknown SAGE tags. *Nucleic Acids Research*, **27**, e17.

17 Chen, J.-J., Rowley, J.D. and Wang, S.M. (2000) Generation of longer cDNA fragments from serial analysis of gene expression tags for gene identification. *Proceedings of the National Academy of Sciences of the United States of America*, **97**, 349–353.

18 Saha, S., Sparks, A.B., Rago, C., Akmaev, V., Wang, C.J., Vogelstein, B., Kinzler, K.W. and Velculescu, V.E. (2002) Using the transcriptome to annotate the genome. *Nature Biotechnology*, **20**, 508–512.

19 Gowda, M., Jantasuriyarat, C., Dean, R.A. and Wang, G.L. (2004) Robust-LongSAGE (RL-SAGE): a substantially improved LongSAGE method for gene discovery and transcriptome analysis. *Plant Physiology*, **134**, 890–897.

20 Meisel, A., Mackeldanz, P., Bickle, T.A., Krüger, D.H. and Schroeder, C. (1995) Type III restriction endonuclease translocate DNA in a reaction driven by

recognition site-specific ATP hydrolysis. *EMBO Journal*, **14**, 2958–2966.

21 Wagenführ, K., Pieper, S., Mackeldanz, P., Linscheid, M., Krüger, D.H. and Reuter, M. (2007) Structural domains in the Type III restriction endonuclease *Eco*P15I: Characterization by limited proteolysis, mass spectrometry and insertional mutagenesis. *Journal of Molecular Biology*, **366**, 93–102.

22 Matsumura, H., Reich, S., Ito, A., Saitoh, H., Kamoun, S., Winter, P., Kahl, G., Reuter, M., Krüger, D.H. and Terauchi, R. (2003) Gene expression analysis of host–pathogen interactions by SuperSAGE. *Proceedings of the National Academy of Sciences of the United States of America*, **100**, 15718–15723.

23 Birch, P.R.J. and Kamoun, S. (2000) Studying interaction transcriptome: coordinated analyses of gene expression during plant-microorganism interactions, in *New Technologies for Life Sciences: A Trends Guide*, (supplement to Elsevier Trends Journals, December 2000) Elsevier, London, pp. 77–82.

24 Margulies, M., Egholm, M., Altman, W.E., Attiya, S., Bader, J.S., Bemben, L.A., Berka, J., Braverman, M.S., Chen, Y.-J., Chen, Z. *et al.* (2005) Genome sequencing in microfabricated high-density picolitre reactors. *Nature*, **437**, 376–380.

25 Matsumura, H., Bin Nasir, K.H., Yoshida, K., Ito, A., Kahl, G., Krüger, D.H. and Terauchi, R. (2006) SuperSAGE array: the direct use of 26-base-pair transcript tags in oligonucleotide arrays. *Nature Methods*, **3**, 469–474.

26 Matsumura, H., Reich, S., Reuter, M., Kruger, D.H., Winter, P., Kahl, G. and Terauchi, R. (2004) SuperSAGE: a potent transcriptome tool for eukaryotic organisms, in *SAGE: Current Technologies and Application* (ed. S.M. Wang), Horizon Scientific Press, Norwich, UK, pp. 77–90.

27 Matsumura, H., Reuter, M., Kruger, D.H., Winter, P., Kahl, G. and Terauchi, R. (2007) SuperSAGE, in Serial Analysis of Gene Expression, Methods and Protocols (ed. N. Kåtre Lehmann), Humana Press, Totowa, NJ, USA.

28 Fan, J.-B., Chee, M.S. and Gunderson, K.L. (2006) Highly parallel genomic assays. *Nature Reviews Genetics*, **7**, 632–644.

29 Matsumura, H., Ito, A., Saitoh, H., Winter, P., Kahl, G., Reuter, M., Kruger, D.H. and Terauchi, R. (2004) SuperSAGE. *Cellular Microbiology*, **7**, 11–18.

30 Terauchi, R., Matsumura, H., Ito, A., Fujisawa, S., Bin Nasir, K.H., Saitoh, H., Kamoun, S., Winter, P., Kahl, G., Reuter, M. and Kruger, D. (2006) SuperSAGE, a potent tool to dissect plant–microbe interactions, in *Biology of Plant–Microbe Interactions*, Vol. 5 (eds F. Sanchez, C. Quinto, I.M. Lopez-Lara and O. Geiger), International Society for Molecular Plant-Microbe Interactions, St. Paul, Minnesota, pp. 569–575.

31 Baulcombe, D.C. (1999) Fast forward genetics based on virus-induced gene silencing. *Current Opinion in Plant Biology*, **2**, 109–113.

32 Kamoun, S., van West, P., Vleeshouwers, V.G., de Groot, K.E. and Govers, F. (1998) Resistance of *Nicotiana benthamiana* to *Phytophthora infestans* is mediated by the recognition of the elicitor protein I NF1. *Plant Cell*, **10**, 1413–1426.

33 Bin Nasir, K.H., Takahashi, Y., Ito, A., Saitoh, H., Matsumura, H., Kanzaki, H., Shimizu, T., Ito, M., Sharma, P.C., Ohme-Takagi, M. *et al.* (2005) High-throughput in plant expression screening identifies a class II ethylene-responsive element binding factor-like protein that regulates plant cell death and non-host resistance. *Plant Journal*, **43**, 491–505.

34 Coemans, B., Matsumura, H., Terauchi, R., Remy, S., Swennen, R. and Sagi, L. (2005) SuperSAGE combined with PCR walking allows global gene expression profiling of banana (*Musa acuminata*), a non-model organism. *Theoretical and Applied Genetics*, **111**, 1118–1126.

35 Liu, Y. and Whittier, R.F. (1995) Thermal asymmetric interlaced PCR: automatable amplification and sequencing of insert end fragments from P1 and YAC clones for chromosome walking. *Genomics*, **25**, 674–681.

36 Roh, T.Y., Ngau, W.C., Cui, K., Landsman, D. and Zhao, K. (2004) High-resolution genome-wide mapping of histone modifications. *Nature Biotechnology*, **22**, 1013–1016.

37 Roh, T.Y., Cuddapah, S. and Zhao, K. (2005) Active chromatin domains are defined by acetylation islands revealed by genome-wide mapping. *Genes & Development*, **19**, 542–552.

38 Horak, C.E. and Snyder, M. (2002) ChIP-chip: a genomic approach for identifying transcription factor binding sites. *Methods in Enzymology*, **350**, 469–483.

39 Cheng, J., Kapranov, P., Drenkow, J., Dike, S., Brubaker, S., Patel, S., Long, J., Stern, D., Tammana, H., Helt, G. *et al.* (2005) Transcriptional maps of 10 human chromosomes at 5-nucleotide resolution. *Science*, **308**, 1149–1154.

40 Datson, N.A., van der Perk-de Jong, J., van der Berg, M.P., de Kloet, E.R. and Vreugdenhil, E. (1999) MicroSAGE: a modified procedure for serial analysis of gene expression in limited amounts of tissues. *Nucleic Acids Research*, **27**, 1300–1307.

41 Thimm, O., Blaesing, O., Gibon, Y., Nagel, A., Meyers, S., Krüger, P., Selbig, J., Müller, L.A., Rhee, S.Y. and Stitt, M. (2004) MAPMAN: a user-driven tool to display genomics data sets onto diagrams of metabolic pathways and other biological processes. *Plant Journal*, **37**, 914–939.

4
From CAGE to DeepCAGE: High-Throughput Transcription Start Site and Promoter Identification for Gene Network Analysis

Matthias Harbers, Thomas Werner, and Piero Carninci

Abstract

The availability of whole genome sequences has opened up a range of entirely new possibilities in genomic research. However, our understanding of the regulatory elements within genomes and the utilization of genomic information in a biological context are only partially understood. With the discovery of new classes of RNA and many rare transcripts, genome annotations remain a challenge that requires new approaches for the identification of transcripts and the regulatory elements controlling their expression. In this chapter we describe Cap Analysis Gene Expression (CAGE) and its extension to DeepCAGE, which are methods of obtaining sequencing tags from the true 5′-ends of mRNA at a very high throughput. Computational analysis of CAGE tags in combination with genomic sequence information allows for genome-wide mapping of transcription starting sites along with promoter and transcript identification. The mapping of promoter features along with expression profiles provides important information relating to genome annotations which is required to produce or analyze the results of studies in functional genomics.

4.1
From Genomes to Transcriptomes

The ability to sequence entire genomes and their annotation by various computational and empirical approaches has been a major breakthrough in genome research. However, the currently available computational methods, although very powerful, cannot reliably predict genes [1] and their regulatory regions [2] in plants and other genomes. Therefore experimental approaches in gene discovery are essential to provide the necessary information to generate information-driven computational methods, and to provide genomic resources for hypothesis-driven research in functional genomics. These needs are further underlined by the fact that one genome contained by all cells within an organism effectively gives rise to very different

The Handbook of Plant Functional Genomics: Concepts and Protocols.
Edited by Günter Kahl and Khalid Meksem

ISBN: 978-3-527-31885-8

transcriptomes depending on the cell type, developmental stage or various other parameters such as for instance, environmental stimuli.

From the plant perspective, genomic research has been pioneered by studies on the premiere model plant *Arabidopsis thaliana*, which was the first plant genome to be sequenced [3], followed by large-scale efforts in rice genome research (*Oryza sativa*) [4,5]. Along with new data from additional plant genomes such as that of the black cottonwood tree (*Populus trichocarpa*) [6], it was realized that plants, surprisingly, have a larger number of gene than mammalian genomes. Computational predictions originally estimated some 25 000 genes in *Arabidopsis* [3], some 32 000–50 000 [5] or 46 022–55 615 [4] in rice, and even over 45 000 for the black cottonwood tree [6]. Many of these transcripts are still awaiting experimental validation, while at the same time the discovery of new classes of RNA, such as short non-coding RNAs (ncRNAs), added a new level of complexity to the picture [7]. In *Arabidopsis* and rice, tag-based approaches as exemplified by Serial Analysis of Gene Expression (SAGE) [8] and Massively Parallel Signature Sequencing (MPSS) [9–11] as well as whole-genome tiling arrays [12,13] have been successfully used to confirm the expression of many genes at different stages. A first boost to promoter identification in plants was provided by full-length cDNA sequencing, in particular in *Arabidopsis* [14] and rice [15,16], where promoter regions could be identified around the 5′-ends of full-length cDNA sequences. Using different approaches, research has moved on further to the characterization of putative regulatory elements including transcription factor binding sites [17,18], scaffold/matrix attachment regions [19], and DNA methylation sites [20] paving the way for mapping not only transcripted regions but also putative regulatory elements within promoters. These studies must rely on multiple approaches applied to many different samples to uncover all transcripts and the regulatory networks in control of their expression (refer to the NIH ENCODE Project (ENCyclopedia Of DNA Elements) as an example of further annotation of regions within the human genome [21]). Even taking into account the great achievements of the past, we are still at a very early stage in our understanding of how these networks function in the utilization of genomic information.

4.2
Addressing the Complexity of Transcriptomes

Reliable and detailed mapping of all transcripts and regulatory elements within genomes will only be possible by integrating different experimental approaches in combination with computational methods. New developments in empirical methods have focused on whole-genome tiling arrays [22] and tag-based approaches [23]. Whole-genome tiling arrays was first successfully used for the annotation of human chromosomes [24], but have lately been used in studies on other organisms including plants [12,25]. Their applications rank from whole transcriptome annotation, chromatin-immunoprecipitation-chip (ChIP) studies, analysis of alternative splicing, DNA foot printing, and polymorphism discovery to genome re-sequencing. Although very powerful with their instant genome-wide view, tiling arrays do not allow for the identification of individual transcripts and their accurate borders. Therefore alternative approaches making use of partial cDNA sequences or sequencing tags are gaining increasing importance in transcriptome analysis. This trend will certainly benefit from the additional momentum provided by new developments in high-throughput sequencing technologies [26–28]. In combination with alignments to genomic sequences, sequence tags can achieve a single-base pair resolution presently unmatched by any other high-throughput approach. In this chapter, we focus on a tag-based method known as Cap Analysis of Gene Expression (CAGE) for genome-wide mapping of Transcriptional Start Sites (TSS) and their link to expression profiles.

4.3 The Shift From CAGE to DeepCAGE

Initiated by large-scale sequencing of Expressed Sequence Tags (ESTs) from cDNA libraries [29], it was assumed that short sequences obtained in large numbers were sufficient to identify individual transcripts. Limited in throughput by the high sequencing cost, subtraction technologies have recently been used to give deeper sequencing coverage after removal of known transcripts from cDNA libraries prior to deep sequencing [30]. However, although they are a powerful tool in gene discovery, subtraction approaches do not allow for EST quantification and the removal of isoforms transcribed from different promoters leads to biased transcript coverage. Therefore to further improve the throughput of random sequencing, SAGE pioneered the use of even shorter fragments (14–25 bp) that can be concatenated for obtaining many sequence tags with a single sequencing read (about 15–20 tags/read; commonly up to 50 000 tags can be obtained from a SAGE library) [31]. SAGE paved the way for the development of many tag-based approaches focusing on different regions within RNA molecules or on the identification of genomic regions [23]. Although SAGE is now becoming an obsolete technique, with the introduction of MPSS about 1 000 000 tags from the most 3′-ends of RNA can be obtained in a single experiment [32].

CAGE was the first approach capable of isolating tags from the true 5′-ends of mRNAs [33] to link transcripts to promoters and TSS identification. CAGE and similar approaches such as 5′-SAGE [34] make use of the 5′-end specific cap structure which identifies transcripts derived from RNA polymerase II-driven transcription. In the classical CAGE protocol, concatemers of some 800 bp were cloned into a vector for sequencing [35]. However, the throughput of capillary sequencing is a limiting factor in the analysis of classical CAGE libraries, and with the development of the new 454 sequencing method the protocol was modified for the preparation of shorter concatemers (about 100–200 bp) that can be directly sequenced on G20 or GS FLX sequencers. In analogy to the DeepSAGE method [36], we named the new high-throughput method DeepCAGE. DeepCAGE libraries have been successfully prepared at RIKEN from mouse and human samples, and up to 1 900 000 tags have been sequenced from a single DeepCAGE library in two consecutive 454 sequencing runs demonstrating the strength of combining CAGE and 454 sequencing. Using the power of DeepCAGE RIKEN has obtained over 70 million human CAGE tags at the time of writing. Similarly, in the future modified DeepCAGE protocols will be developed for use with other high-throughput sequencers such as the recently released Solexa 1 G Genome Analyzer or ABI SOLiD system, and other high-throughput analyzers presently under development [28]. It has been estimated that Solexa sequencing can yield 20 million tags per run, whereas the SOLiD system may even provide 250 million tags per run.

4.4
Applications of CAGE and DeepCAGE Libraries

CAGE and DeepCAGE can be used in two principally distinct experimental designs: (1) genome-wide mapping of TSS for genome and promoter annotation; (2) linking promoter regions to expression profiles in a biological context to construct gene networks.

For studies targeting genome-wide mapping of TSS and promoter regions, it is preferable to obtain the highest possible number of tags from one well-defined sample. This sample will define one specific tissue/stage/time-point for a given organism or cell line. Depending on the objectives of the study, many such libraries may be required to obtain an overview of TSS and promoter usage in different tissues or at different developmental stages for the same organism. Since this sample is viewed as 'representative', special precautions need to be taken when collecting RNA for the preparation of CAGE and DeepCAGE libraries. Due to the inherited variations between individual animals, plants or even cell lines, an appropriate number of samples per library should be taken (e.g. by pooling RNAs or tissue from six individual animals or plants having the same biological background) to ensure that the library is indeed 'representative' of the particular tissue/stage/time-point under study. Unfortunately this may not be possible for many valuable samples.

Using CAGE and DeepCAGE for the analysis of a biological system under different conditions requires an experimental design that targets the production of multiple CAGE or DeepCAGE libraries per data point from repeated experiments and their suitable controls. Each experiment/data point should be repeated at least three times, which may lead to a rather large number of CAGE or DeepCAGE libraries. Therefore, depending on the expression levels of the target genes within the study, pooled CAGE or DeepCAGE libraries could be prepared to keep the overall sequencing cost down even where rare transcripts may be missed. Protocols for preparing pooled CAGE libraries have been developed, in which the origin of each tag within the library can be identified by a short barcode sequence (commonly 3–5 bp) introduced during the first linker ligation step [35]. Moreover, pooled CAGE libraries can be a very good way of making the best use of the very high throughput of new sequencing instruments providing many millions of tags per run.

4.5
Preparation of a DeepCAGE Library

A very detailed protocol for the preparation of CAGE libraries has been published [35], and therefore, in addition to identifying the key steps in library preparation, we provide here only additional experimental advice in respect of those steps in the CAGE protocol which have been altered specifically for the preparation of a DeepCAGE library (refer to Box 4.1 for more experimental details). The worksheet for the production of a DeepCAGE library is outlined in Figure 4.1.

Preparation of DeepCAGE libraries commonly starts from 50 μg of total RNA, but as little as 25 μg has been successfully used. The use of purified mRNA fractionations is not recommended as it leads to a loss of nonpolyadenylated mRNA that can no longer be covered by the DeepCAGE library. Further interesting aspects can arise from RNAs derived from cell fractionation as already used in tiling array studies to elucidate ncRNA precursors and link some of them to promoter regions, for instance [37]. cDNA synthesis is primed by a set of random primers to assure priming from non-polyadenylated mRNA and to reach the 5′-end of very long transcripts. A key step for CAGE library preparation is the selection of true 5′-ends by the so-called Cap-Trapper method [38]. This involves the chemical biotinylation on streptavidin-coated beads of the cap structure for enrichment of RNA/cDNA hybrids comprising 5′-ends. Other RNA molecules including tRNAs and rRNAs are not selected because they lack a biotinylated cap structure. The Cap-Trapper approach has provided more complex libraries as compared to another cap selection method, the so-called oligo capping [34] which uses a number of enzymatic reactions to modify the 5′-ends of mRNA (unpublished data). In the first linker ligation step, a double-stranded linker harboring a recognition site for the class II restriction endonuclease *MmeI* is ligated adjacent to the 3′-end of the cDNA fragments (equal to the 5′-end of the parental mRNA). This linker also contains a recognition site for a second restriction endonuclease (in this case *XmaJI* is used for concatenation) and a biotin group for recovery of the DeepCAGE tags. For the preparation of pooled CAGE libraries, the linker should also contain a DNA barcode sequence of 3–5 bp between the two recognition sites that will be carried over into the concatemer to mark the origin of different tags. This can be used to maximize the reliability of CAGE experiments when unequal PCR amplification of the CAGE tags at later stages is a concern (see below). Therefore, particularly when CAGE is applied to the analysis of biological systems at different time-points, subtle measuring differences between similar samples can be enhanced by pooling CAGE libraries. *MmeI* cuts the DNA at about 20 bp from its recognition site yielding short 20-bp tags (denoted as CAGE tags) that can be further manipulated for sequencing. An alternative protocol is under development using *Eco*P15I to obtain 27-bp long CAGE tags. After ligation of a second linker to the positions opened up by *MmeI* digestion, DNA fragments comprising the 5′-end specific tags are captured on streptavidin-coated beads using the biotin group introduced during the first linker ligation step. Due to the demands of the 454 sequencing process, PCR conditions must be optimized to provide at least 15 μg of double-strand PCR product. The number of cycles per PCR reaction should

Box 4.1
DeepCAGE-454 Library Preparation Protocol

For the preparation of a DeepCAGE library refer to the CAGE protocol published by Kodzius *et al.* [35]. This protocol describes in detail all the steps required for the preparation of CAGE libraries including the option of preparing pooled CAGE libraries. This protocol should be followed until Step 61 has been successfully completed. For the concatenation step 1 μg of *Xma*JI digested and purified tag DNA should be prepared. In addition, double-stranded 454 adaptors A and B [39] with the following sequences should also be prepared:

A-up	CCATCTCATCCCTGCGTGTCCCATCTGTTCCCTCCCTGTCTCAG
A-down	(phosphate)-CTAGCTGAGACAGGGAGGGAACAGATGGGACACGCAGGGATGAGATGG
B-up	BioTEG-CCTATCCCCTGTGTGCCTTGCCTATCCCCTGTTGCGTGTCTCAG
B-down	(phosphate)-CTAGCTGAGACACGCAACAGGGGATAGGCAAGGCACACAGGGGATAGG

Double-stranded linkers are prepared by mixing equal amounts of the 'up' and 'down' oligonucleotides before use. Annealing is achieved by heating the oligonucleotides to 65 °C for 10 min in 50 mM NaCl and then slowly cooling the mixture down to room temperature.

In the concatenation reaction the ratio of tag DNA to adaptor A and adaptor B must be 20 : 1 : 1. Briefly, a ligation reaction is set up with 1 μg of tag DNA (the amounts of adaptor should be scaled down for lower concentrations of tag DNA; it is desirable to have at least 500 ng of tag DNA per ligation reaction) and 50 ng of each annealed adaptor A and adaptor B:

• Tag DNA sample, 1 μg	X μl
• Adaptor A (50 ng/μl)	1 μl
• Adaptor B (50 ng/μl)	1 μl
• 10× T4 Ligation Buffer (500 mM Tris-HCl, 100 mM $MgCl_2$, 10 mM ATP, 100 mM Dithiothreitol, 250 μg/ml BSA, pH 7.5)	1 μl
• T4 Ligase (2000 u/μl)	1 μl
• Water	to 10 μl
• Total	10 μl

The ligation reaction should be incubated overnight at 16 °C. At the end of the reaction, the concatenation product is purified using a GFX PCR DNA and Gel Band Purification Kit (www.gelifesciences.co.jp/tech_support/manual/pdf/dnapcr/74003977aa.pdf) to eliminate shorter reaction products. Alternative purification kits may also be used, however, it is important to ensure that the kit purifies the concatenation products while removing short DNA fragments.

The purified concatenation product can be directly subjected to 454 sequencing. The efficiency of the concatenation reaction can be confirmed by chromatography using an Agilent 2100 bioanalyzer microfluidics-based platform or by agarose electrophoresis using small aliquots of the ligation reactions.

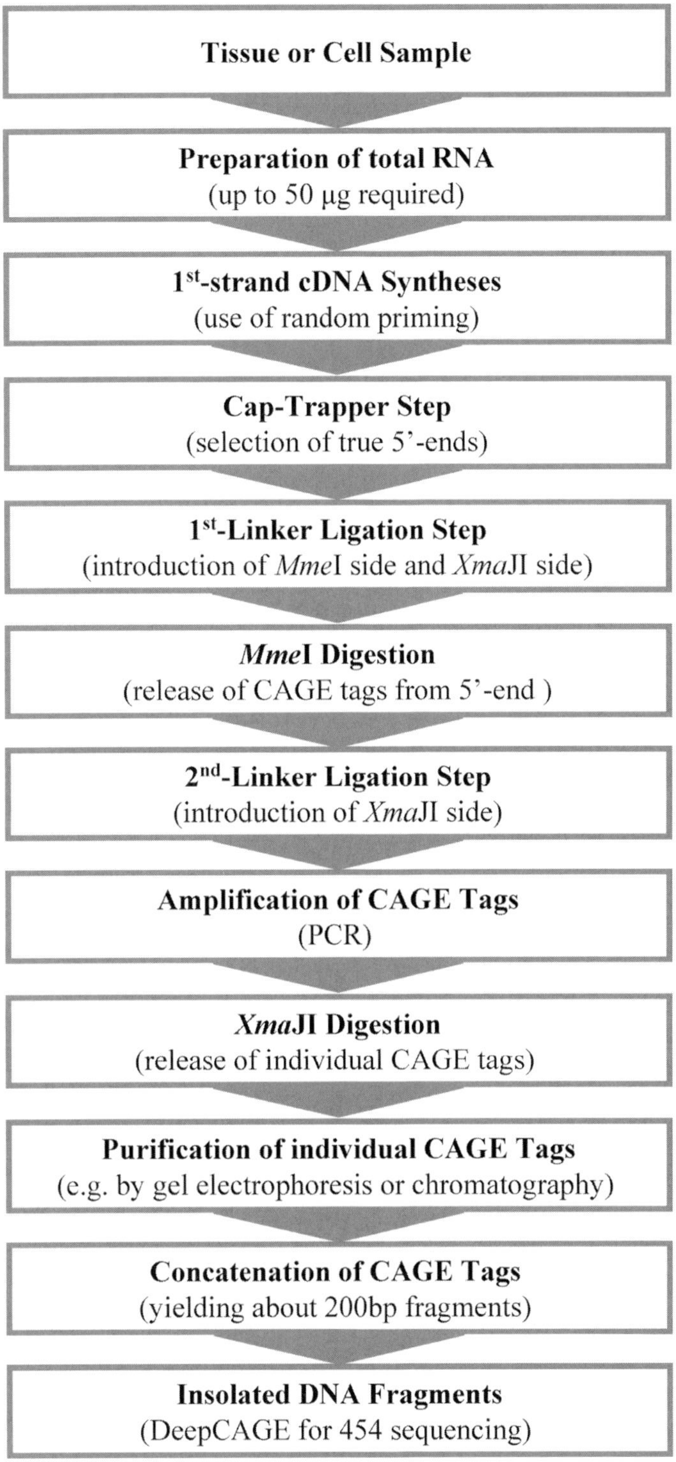

Figure 4.1 Schematic workflow for the preparation of a DeepCAGE library. Only key steps are indicated.

be kept to a minimum, and it is preferable to carry out many PCR reactions in parallel (in most cases up to 20 PCR reactions are required for a DeepCAGE library). The PCR products are digested with *Xma*JI, and the isolation of about 1 μg of DNA containing the tags is necessary for the concatenation step; with experience, as little as 200 ng of tag DNA may be used in the concatenation step. It is recommended that the DNA fragments are well purified prior to concatenation, for example by chromatography or gel electrophoresis, to remove all remaining linker fragments and PCR primers. For the preparation of DeepCAGE libraries, individual tags are ligated into short concatemers, where the concatenation reaction is carried out in the presence of two different terminators denoted as A and B adaptors in the 454 process [39]. These adaptors comprise sequences needed for the 454 sequencing process and have *Xma*JI compatible ends. In the ligation process, the open *Xma*JI site at one end of these A and B adaptors ligates to other DNA fragments; however the other ends of the adaptors cannot be ligated as they are blocked, thus preventing any further extension of the concatemers. The ratio of tag DNA to A adaptor and B adaptor is 20 : 1 : 1, which ensures the formation of relatively large concatamers that are ideal for the 454 emulsion PCR, as the background is composed up of only a few empty A-B adaptor dimers and short concatamers. The concatenation reaction products are purified using a GFX DNA purification kit (GE Healthcare Life Sciences) to further eliminate any short ligation products which would reduce the sequencing yield. Alternatively, separation may use a spin column to remove shorter fragments, but in any case careful selection of the exclusion size is necessary to avoid losing the desired longer concatemers. Concatemers obtained in this way can be used directly in a 454 sequencing reaction according to the maker's directions or can be sent to a 454 sequencing service provider. These concatemers commonly yield about three CAGE tags per read using a G20 Sequencer. However, 454 Life Sciences has already launched an improved GS FLX sequencer that will increase the output to 5–6 CAGE tags per run while enabling a much larger number of parallel reads. Further improvements in the data output of new sequencing methods are to be expected in the coming years, leading to the production of millions of CAGE tags from a single sequencing run.

4.6 CAGE Data Analysis and Genome Mapping Approaches

The recent discovery of very large numbers of new RNAs, often expressed at very low levels, has initiated a discussion concerning the nature of 'transcriptional noise' and which RNAs are 'functional transcripts' Although it is beyond the scope of this review, we would like to make some brief remarks here due to the general importance of these concerns. For example, the ratio of transcriptional noise has been estimated in yeast based on the basic features of biological processes and expression levels [40]. In this study, 'noise' is used in the sense of specific, low-level transcription with unknown function and not in the sense of spontaneous random transcription. Accordingly, in yeast almost 90% of the initial transcripts were considered to be 'transcriptional noise', including transcription initiated outside of 'classical' promoters and not associated with any known transcripts as well as individual transcripts of very low abundance. In contrast, although 'meaningful transcripts' represent only about 10% of the total transcripts for the majority of RNAs in yeast, their individual abundances are 10 000 times greater than those of 'noise' RNAs. Some groups are challenging the assumed phenomenon of 'transcriptional noise' by using various computational methods such as comparison with orthologs and identification of conservation, to establish evidence to support the notion that although new transcripts may be expressed at low levels they do in fact have a function [41]. Many studies have now identified large numbers of intercalating transcripts that span the majority of the human or mouse genomes. A common feature of protein-coding transcripts is that they are often linked to other transcripts of unknown biological function, which suggests a new means of co-regulation. Although we do not yet understand the underlying principles behind these new transcripts, the discovery of so many rare transcripts cannot be dismissed. Only functional studies on these new transcripts will elucidate their potential functions or characterize them as the products of what may seem to be useless or error-prone transcription. With the growing number of newly identified transcripts such future studies will be labour intensive. However to progress this area of research we would suggest that all CAGE tags and related TSS should be reported and stored in public databases even in the absence of evidence to suggest that they possess a particular function. Dismissal of what may currently appear to be 'negative data' may have a potentially negative effect on future research and may also make the search for evidence to support the potential function of rare transcripts more difficult. On the other hand, in cases where data derived from CAGE libraries are used for the selection of targets in experimental studies, we strongly advise a careful and more conservative approach which should include experimental validation. Clearly, the throughput of experiments will be a limiting factor for the number of targets, and studies in functional genomics including knock-out or knock-down experiments should be based on high quality targets for which there is sufficient evidence to support their proposed function. Using bioinformatics to analyze conservation patterns and other features of novel TSS candidates is one method of obtaining the required supporting evidence.

Since the linker sequences in concatemers are known, the sequences of individual CAGE tags can easily be retrieved from the original reads by computational analysis which should also include a review of the sequence quality. The essential first step in CAGE tag analysis is thus the mapping to genomic sequences followed by connecting the CAGE tags to genome annotations. Up to 65% (in some cases up to 70%) of the CAGE tags can be mapped to unique genomic regions, and these are at first utilized to identify known transcripts based on genome annotations. Many gene annotations are based on computational predictions, and therefore it is preferable to confirm whether an annotated locus is supported by full-length cDNA data. Mapping to the genome is also essential in assigning CAGE tags to tag clusters, which group all the CAGE tags that overlap on the same genomic location (see below). For many positions, only a single CAGE tag can be found (a so-called 'singleton'), or CAGE tags having multiple mapping positions.

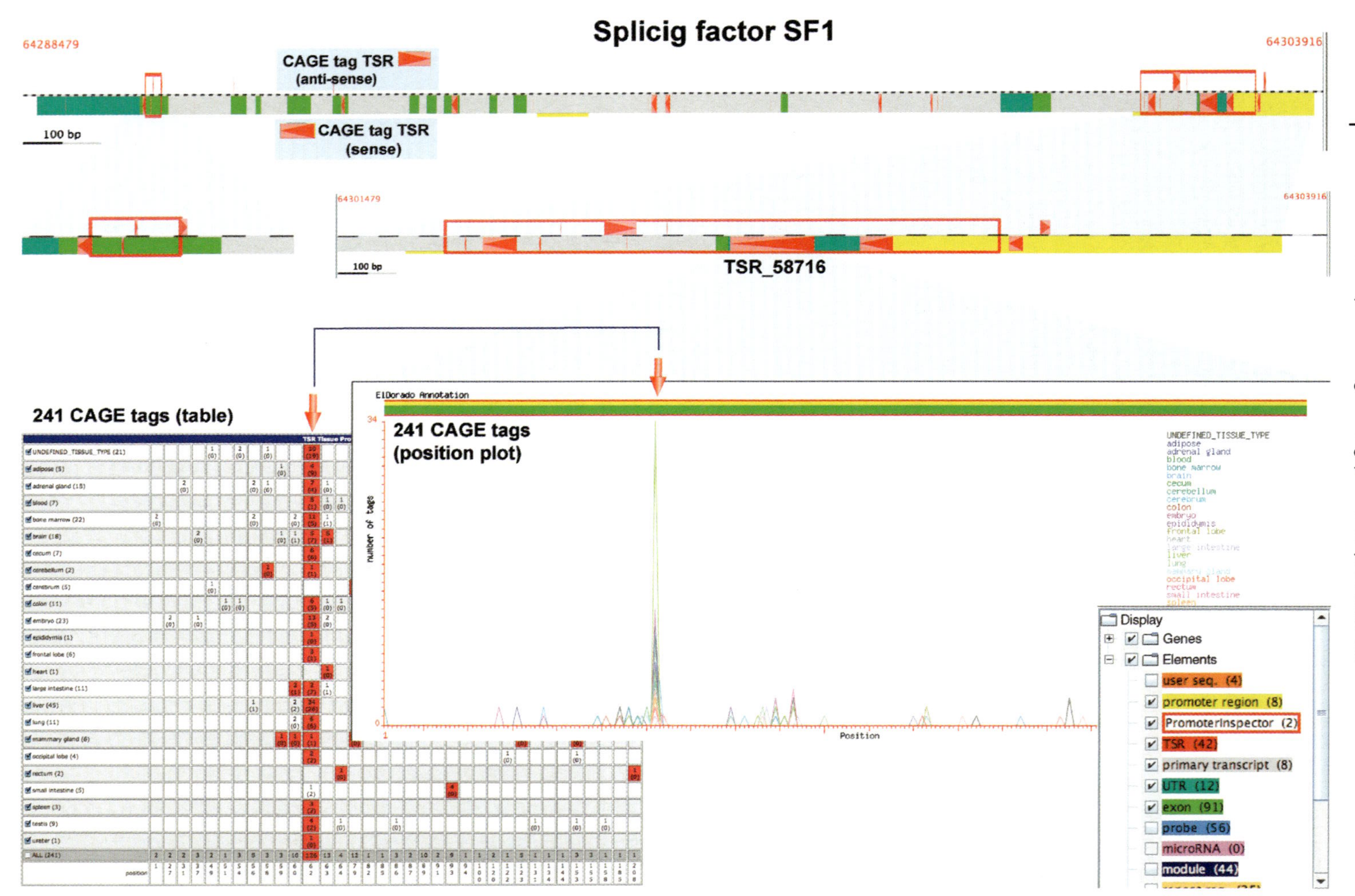

Splicig factor SF1
64288479
64303916
CAGE tag TSR (anti-sense)
CAGE tag TSR (sense)
100 bp
64301479
64303916
100 bp
TSR_58716
241 CAGE tags (table)
241 CAGE tags (position plot)
ElDorado Annotation
number of tags
Position
Display
Genes
Elements
user seq. (4)
promoter region (8)
PromoterInspector (2)
TSR (42)
primary transcript (8)
UTR (12)
exon (91)
probe (56)
microRNA (0)
module (44)

Multiple mapping positions may represent the expression of repeat elements used as promoters for stage-specific expression [42], which may in the future reveal new methods of transcriptional regulation (unpublished data). Singletons require additional experimental validation where for example RACE (Rapid Amplification of cDNA Ends) experiments have demonstrated that over 90% of the singletons could be related to true initiation sites [43].

However, experimental validation of all singletons is not feasible for large-scale CAGE analysis and therefore in an alternative approach to tag annotation and target selection only tags that appear in duplicate either on the same nucleotide or within a window of up to 40 bp around initial mapping positions in the genome forming the CAGE tag clusters, would be considered. These simple measures reduce the number of CAGE tags to be considered for target selection by almost threefold. Genomic mapping of this reduced set of 'reliable' CAGE tags from human samples revealed that even in such a reduced set CAGE tag clusters appear that not only occur within known promoters (62%) but also in thus far unannotated regions. CAGE tag clusters that map on unannotated regions are likely to identify novel promoters, which depending on the number of CAGE tags within a cluster, are likely to include TSS for rare transcripts. Moreover, it has been noted that many CAGE tags not only map to the 5′-ends of mRNA, but also map within existing transcripts and often within exons (so-called 'exon painting' or 'exonic TSS' [44]). Although the function of these unconventional promoters has not been demonstrated, bioinformatics evidence may suggest that in many cases they may have regulatory functions. Hence, more conservative data analysis also confirms the existence of unexpected initiation sites, and in these cases it is noteworthy that 'exon painting events' are associated with the TATA-box driven transcripts rather than CpG-rich promoters. Just to complete the list of difficulties caused by such transcriptome complexity, there are CAGE tags which point in both directions including exons at 3′-ends [44], thus indicating a high level of antisense transcription. This is demonstrated in Figure 4.2 which shows an example for the human Splicing factor 1 (SF1, D11S636, ZFM1, ZNF162) gene. In mouse, the presence of CAGE tags producing antisense, mapping to known mRNAs, indicated that up to 72% of the transcriptional units may have antisense counterparts [45].

Figure 4.2 The human Splicing factor 1 or zinc finger protein 162 (Synonyms are: D11S636, SF1, ZFM1, ZNF162, Gene ID 7536) gene located on chromosome 11q13 is shown on the lower strand and is transcribed from right to left. The graphic was taken from the ElDorado database. TSR, Transcriptional Start Region. A TSR is defined by a cluster of at least two CAGE tags with 0 or < 40 bp distance between them. Top: complete view of the SF1 region. On the top strand a number of antisense TSRs are visible throughout the SF1 major transcript. Next line: light blue areas indicate the zoomed regions, comprising the major promoter region of SF1 (right side) and the 3′ exon showing 'exon painting' by four distinct TSRs (left side). Bottom: the most prominent TSR containing 241 individual CAGE tags is shown both in positional plot as well as in table view. Both views differentiate the tissues where the CAGE tags were found. The red arrows joined by a blue line indicate the major previously unknown transcription start site for which 34 CAGE tags were detected in various tissues.

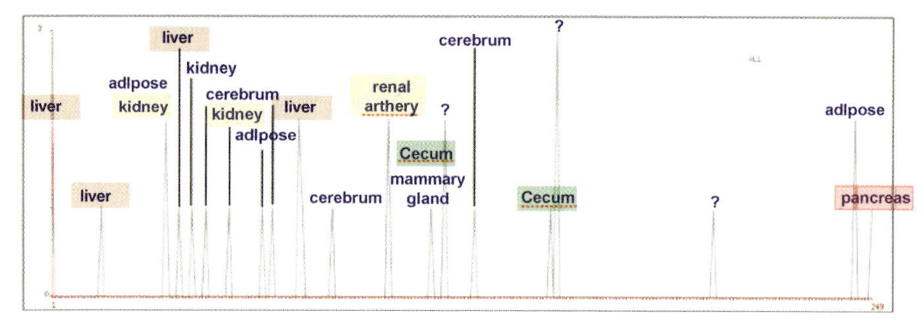

Figure 4.3 Tissue-associated CAGE tags within the major promoter region of the human glucocorticoid receptor gene (The human Nuclear receptor subfamily 3, group C, member 1, Synonyms are: GCCR, GCR, GR, GRL, NR3C1, Gene ID 2908) located on chromosome 5q31.3. The graphic was taken from the ElDorado database and the tissues were labeled manually. The complete region encompasses 249 bp (TSR281068) and the horizontal scale is in nucleotides. The vertical scale indicates the number of CAGE tag clusters located at the same nucleotide (maximum, three). Question marks indicate unknown tissue associated with the corresponding CAGE clusters.

If CAGE tags from different tissues are mapped to the same genome another very interesting feature is revealed: very often CAGE tags that cluster to one promoter region originate from different tissues. As shown in Figure 4.3 for the major promoter region of the human glucocorticoid receptor gene (NR3C1), CAGE tags derived from different tissues tend to cluster locally within the promoter region suggesting that there is tissue-specific use of alternative promoters. Such features only become apparent after mapping of CAGE tags to a genomic template sequence which leads to the identification of the corresponding genes, promoters and transcripts. As can be seen from Figure 4.3, meaningful statistical analysis of the mapping results requires an exceptionally large number of CAGE tags in order to render the data approximately quantitative. If an average coverage of about 100 tag clusters per TSS can be achieved, mapping will reveal a realistic picture of the relative abundance of the various transcripts, and at the same time facilitate differentiation between rare transcriptional noise and relatively strong signals which represent unknown but probably meaningful transcripts. In addition, such quantitative data allows the results of the promoter analysis to be assigned to the 'correct promoters', because most of the uncertain promoter sequences that would otherwise have contaminated the comparative promoter analysis, have been removed. This situation is analogous to that which occurs with microarray data where statistical analysis is carried out prior to the selection of genes and/or promoters by promoter analysis [46]. These difficulties can be overcome by using DeepCAGE rather than CAGE experiments which have a lower sequencing depth.

There is another very important aspect of mapping CAGE tag clusters to the genome. The initial mapping and selection process is completely independent of the existing genomic annotation and relies exclusively on the genomic DNA sequence. While this ensures that promoters and transcripts are associated with CAGE tags in an unbiased manner, it also guarantees that the locations of CAGE tags are totally unbiased towards any annotation. This is important as a recent study using chromatin immunoprecipitation experiments to identify TATA-box binding protein (TBP) locations on genomic DNA have revealed that as many as 50% of human promoters remain unannotated in conventional genome databases [47]. With regard to potentially new promoters, this observation relates to CAGE tag clusters found in unannotated regions which may also represent novel promoters, especially if CAGE tags are abundant within these clusters. Complementary approaches such as the TBP-ChIP or genomic tiling arrays can be used to verify such new promoters and their transcripts.

The basic premise is that any type of statistical analysis of CAGE data can only yield results as good as the underlying genomic annotation used for mapping. Therefore, genomic mapping is the first and currently most important procedure in CAGE tag analysis as every subsequent study by whatever means is limited by the quality and completeness of the initial mapping and the underlying genome. Genomes and their annotations vary depending on the extent to which the information has been experimentally verified. To this end, full-length cDNAs remain an essential element in the annotation of genomes and transcripts [13,14] together with curated databases of promoter and genomic information.

Examples of mouse and human promoters can be queried in the CAGE Analysis and the Genomic Element Viewers that are freely available at http://fantom.gsc.riken.go.jp. Genes, transcripts and their promoter usage can be searched using the CAGE Analysis Viewer, while the Genomic Element Viewer will provide information on CAGE tags and other elements of genome sequences [48]. Another example of a highly curated and continuously updated database is the commercial ElDorado platform at http://www.genomatix.de/ providing extensive promoter annotations including information on CAGE tags presently in the public domain.

4.7 Expression Profiling: Putting CAGE Tags into a Biological Context

As detailed above, in-depth coverage of CAGE libraries as provided by DeepCAGE in particular will reveal both individual promoters which are active under specific conditions and will also facilitate the quantification of transcripts. However, as CAGE tags only identify the 5′-end of transcripts they cannot reveal any events of alternative splicing. Commonly RACE experiments are carried out to confirm CAGE tags [43] or to link certain exons to 5′-ends [49]. In addition, conventional microarray-based expression analysis can be related to CAGE data [50], where exon-arrays in particular allow at least the partial resolution of single transcripts and alternative splicing. Other approaches can be used to confirm TSS including promoter-tiling arrays as analytical endpoints of ChIP experiments (ChIP-on-chip) yielding further complementary information about the actual binding of transcription factors (TFs) to promoters. For example, CAGE data in combination with other approaches have been used to characterize functional elements in genomic DNA as part of the NIH ENCODE Project [21].

Taken together, these experimental data provide an excellent basis for condition-associated promoter analysis aiming at the elucidation of underlying regulatory networks. Even in the absence of detailed ChIP-on-chip data concerning the actual TF-binding to promoters, analysis of the presence and organization of TF-binding sites (TFBSs) can reveal a great deal about transcriptional mechanisms and regulatory networks [51–53]. In brief, comparative analysis of co-regulated genes, that is, genes co-expressed under the same conditions by the same underlying mechanism, will allow the elucidation of the organizational patterns of particular TFBSs. These organizational patterns are known as frameworks and are associated with the molecular mechanism of transcriptional control. Such frameworks can then be used to scan whole genomes for other promoters that may be regulated by the same mechanism [46]. Thus a set of genes/transcripts is revealed that can now be analyzed for additional connections to known networks/pathways, this will then lead to the identification of sub-groups forming potential regulatory networks. Once this has been achieved experimental results on this subset of genes can be used to support or negate the proposed regulatory network connection [52]. The beauty of this approach is that available information (literature, pathway analysis) is combined with experimental data (CAGE tags and microarrays, among others) and genomic molecular evidence (TFBS-frameworks) to anchor the results firmly to the genome sequence, which will provide the definitive conclusion.

4.8 Perspectives

Genome and transcriptome research has recently been advanced by the development of large-scale EST sequencing and cDNA cloning projects and by the remarkable discovery of new transcripts and RNA classes that had previously been overlooked by computational genome annotations. In addition to the dearth of knowledge concerning the structural features of new RNA classes which could not be cloned by classical cDNA library approaches, we came to realize that more highly sensitive approaches are needed to identify rare transcripts. Most of the standard expression profiling methods used these days are limited by their sensitivity or do not allow for *de novo* gene discovery. As an example, Czechowski *et al.* [54] have carried out profiling studies on 1465 *Arabidopsis* transcription factors using real-time RT-PCR and Affymetrix microarrays. In this study, RT-PCR was able to detect 83% of the target genes in a range from 0.001 to 100 copies per cell. In contrast, only 55% of the target transcripts were detected in the same sample by microarray analysis and both the sensitivity and reproducibility were much lower using this technique. In addition, RT-PCR was able to confirm the expression of target transcripts which were not found in *Arabidopsis* EST and MPSS databases. The sensitivity of microarray experiments is presently in the range of 1–10 copies per RNA molecule per cell, and repetitive experiments are needed to accurately identify and analyze rare transcripts [55]. For high-throughput sequencing approaches expression levels are commonly given in Transcripts Per Million (TPM) values. In MPSS experiments, most transcripts were found in the range of 1 to 100 TPM; statistical analysis of the detection range of MPSS has been described [32]. Similarly, deep sequencing of a DeepCAGE library yields high coverage of most sequence tags, and the detection range can be improved even further by increasing the number of sequencing reads per library. Therefore a single DeepCAGE experiment can provide a range of evidence for a TSS/transcript, where the sensitivity of the experiment directly correlates with the total number of tags obtained from the library. In this regard, sequencing-based approaches are distinct from microarray experiments, where in each experiment one hybridization reaction per location defines the signal strength. Even where microarrays use more than one probe per transcript on the same array, the number of hybridization reactions per transcript remains lower than the number of tags commonly obtained for most transcripts in DeepCAGE libraries. This is an important argument for the use of DeepCAGE in the identification of rare transcripts, where the increasing sequencing depth can provide a spectrum of evidence for weak TSSs and their transcripts. However, even in deeply sequenced DeepCAGE libraries singletons are retrieved at the present sequencing depth of about 1 900 000 tags per library. Although singletons have an equal chance of being either functional or non-functional, further analysis using CAGE and SAGE libraries has shown that the functionality of the majority of singletons can be confirmed and that they are not the consequence of experimental errors [43,56]. These observations emphasize the high reliability of sequencing-based approaches in gene discovery and genome annotation.

CAGE and DeepCAGE libraries have already been successfully used for gene discovery [43] and genome-wide mapping of TSS [44] in mouse and human. Studies on the application of DeepCAGE to the characterization of model biological systems and gene networks are also underway. These studies have revealed interesting aspects of the structure of TSS [44] and have identified a number of single-exon genes. For future studies in functional genomics complete maps on all TSS and their tissue-specific activities will become a meaningful tool to enable phenotypes to be linked their genotypes. In addition, DeepCAGE analysis of model biological systems will provide basic information regarding promoter activities and transcription factor availability so that models of regulatory networks for gene expression can be constructed. Thus the integration of DeepCAGE-derived information with genome annotations will provide important platforms for hypothesis-driven studies and data interpretation. In particular common promoter features will be used to associate co-regulated genes with particular biological conditions in an attempt to elucidate their functions. Therefore we believe that DeepCAGE will become a standard methodology in plant science contributing greatly to genome and promoter annotation as well as to gene network studies.

Acknowledgments

MH thanks the present and past members of the DNAFORM CAGE Library Team, S. Ishikawa, C. Kato, F. Kobayashi, and M. Suzuki as well as Y. Hayashizaki at RIKEN for our close collaboration on the developments in CAGE technology.

TW thanks the Genomatix team and especially Matthias Scherf and Andreas Klingenhoff for data preparation, advice and discussions. PC thanks all of the members of the RIKEN GSC-GREG-Laboratory, GSL-Laboratory and the Fantom-3 Consortium members for data preparation, analysis, advice and discussions, and Y. Hayashizaki for support. This work (PC) was supported by a Research Grant for National Project on Protein Structural and Functional Analysis from MEXT, a Research Grant for the RIKEN Genome Exploration Research Project from the Ministry of Education, Culture, Sports, Science and Technology of the Japanese Government and a grant for the Genome Network Project from the Ministry of Education, Culture, Sports, Science and Technology (Japan). Other parts of this work (TW) were also supported by grant 0 313 724A from Biochance-PLUS-3 (Germany).

References

1 Do, J.H. and Choi, D.K. (2006) Computational approaches to gene prediction. *Journal of Microbiology*, **44**, 137–144.

2 Shahmuradov, I.A., Solovyev, V.V. and Gammerman, A.J. (2005) Plant promoter prediction with confidence estimation. *Nucleic Acids Research*, **33**, 1069–1076.

3 The Arabidopsis Genome Initiative (2000) Analysis of the genome sequence of the flowering plant *Arabidopsis thaliana*. *Nature*, **408**, 796–815.

4 Yu, J., Hu, S., Wang, J., Wong, G.K., Li, S., Liu, B., Deng, Y., Dai, L., Zhou, Y., Zhang, X., Cao, M., Liu, J., Sun, J., Tang, J., Chen, Y., Huang, X., Lin, W., Ye, C., Tong, W., Cong, L., Geng, J., Han, Y., Li, L., Li, W., Hu, G., Huang, X., Li, W., Li, J., Liu, Z., Li, L., Liu, J., Qi, Q., Liu, J., Li, L., Li, T., Wang, X., Lu, H., Wu, T., Zhu, M., Ni, P., Han, H., Dong, W., Ren, X., Feng, X., Cui, P., Li, X., Wang, H., Xu, X., Zhai, W., Xu, Z., Zhang, J., He, S., Zhang, J., Xu, J., Zhang, K., Zheng, X., Dong, J., Zeng, W., Tao, L., Ye, J., Tan, J., Ren, X., Chen, X., He, J., Liu, D., Tian, W., Tian, C., Xia, H., Bao, Q., Li, G., Gao, H., Cao, T., Wang, J., Zhao, W., Li, P., Chen, W., Wang, X., Zhang, Y., Hu, J., Wang, J., Liu, S., Yang, J., Zhang, G., Xiong, Y., Li, Z., Mao, L., Zhou, C., Zhu, Z., Chen, R., Hao, B., Zheng, W., Chen, S., Guo, W., Li, G., Liu, S., Tao, M., Wang, J., Zhu, L., Yuan, L. and Yang, H. (2002) A draft sequence of the rice genome (*Oryza sativa* L. ssp. indica). *Science*, **296**, 79–92.

5 Goff, S.A., Ricke, D., Lan, T.H., Presting, G., Wang, R., Dunn, M., Glazebrook, J., Sessions, A., Oeller, P., Varma, H., Hadley, D., Hutchison, D., Martin, C., Katagiri, F., Lange, B.M., Moughamer, T., Xia, Y., Budworth, P., Zhong, J., Miguel, T., Paszkowski, U., Zhang, S., Colbert, M., Sun, W.L., Chen, L., Cooper, B., Park, S., Wood, T.C., Mao, L., Quail, P., Wing, R., Dean, R., Yu, Y., Zharkikh, A., Shen, R., Sahasrabudhe, S., Thomas, A., Cannings, R., Gutin, A., Pruss, D., Reid, J., Tavtigian, S., Mitchell, J., Eldredge, G., Scholl, T., Miller, R.M., Bhatnagar, S., Adey, N., Rubano, T., Tusneem, N., Robinson, R., Feldhaus, J., Macalma, T., Oliphant, A. and Briggs, S. (2002) A draft sequence of the rice genome (*Oryza sativa* L. ssp. *japonica*). *Science*, **296**, 92–100.

6 Tuskan, G.A., Difazio, S., Jansson, S., Bohlmann, J., Grigoriev, I., Hellsten, U., Putnam, N., Ralph, S., Rombauts, S., Salamov, A., Schein, J., Sterck, L., Aerts, A., Bhalerao, R.R., Bhalerao, R.P., Blaudez, D., Boerjan, W., Brun, A., Brunner, A., Busov, V., Campbell, M., Carlson, J., Chalot, M., Chapman, J., Chen, G.L., Cooper, D.,

Coutinho, P.M., Couturier, J., Covert, S., Cronk, Q., Cunningham, R., Davis, J., Degroeve, S., Dejardin, A., Depamphilis, C., Detter, J., Dirks, B., Dubchak, I., Duplessis, S., Ehlting, J., Ellis, B., Gendler, K., Goodstein, D., Gribskov, M., Grimwood, J., Groover, A., Gunter, L., Hamberger, B., Heinze, B., Helariutta, Y., Henrissat, B., Holligan, D., Holt, R., Huang, W., Islam-Faridi, N., Jones, S., Jones-Rhoades, M., Jorgensen, R., Joshi, C., Kangasjarvi, J., Karlsson, J., Kelleher, C., Kirkpatrick, R., Kirst, M., Kohler, A., Kalluri, U., Larimer, F., Leebens-Mack, J., Leple, J.C., Locascio, P., Lou, Y., Lucas, S., Martin, F., Montanini, B., Napoli, C., Nelson, D.R., Nelson, C., Nieminen, K., Nilsson, O., Pereda, V., Peter, G., Philippe, R., Pilate, G., Poliakov, A., Razumovskaya, J., Richardson, P., Rinaldi, C., Ritland, K., Rouze, P., Ryaboy, D., Schmutz, J., Schrader, J., Segerman, B., Shin, H., Siddiqui, A., Sterky, F., Terry, A., Tsai, C.J., Uberbacher, E., Unneberg, P. *et al.* (2006) The genome of black cottonwood, *Populus trichocarpa* (Torr. & Gray). *Science*, **313**, 1596–1604.

7 Fahlgren, N., Howell, M.D., Kasschau, K.D., Chapman, E.J., Sullivan, C.M., Cumbie, J.S., Givan, S.A., Law, T.F., Grant, S.R., Dangl, J.L. and Carrington, J.C. (2007) High-throughput sequencing of *Arabidopsis* microRNAs: Evidence for frequent birth and death of MIRNA genes. *PLoS ONE*, **2**, e219.

8 Robinson, S.J., Cram, D.J., Lewis, C.T. and Parkin, I.A. (2004) Maximizing the efficacy of SAGE analysis identifies novel transcripts in *Arabidopsis. Plant Physiology*, **136**, 3223–3233.

9 Meyers, B.C., Lee, D.K., Vu, T.H., Tej, S.S., Edberg, S.B., Matvienko, M. and Tindell, L.D. (2004) Arabidopsis MPSS. An online resource for quantitative expression analysis. *Plant Physiology*, **135**, 801–813.

10 Meyers, B.C., Tej, S.S., Vu, T.H., Haudenschild, C.D., Agrawal, V., Edberg, S.B., Ghazal, H. and Decola, S. (2004) The use of MPSS for whole-genome transcriptional analysis in *Arabidopsis. Genome Research*, **14**, 1641–1653.

11 Nakano, M., Nobuta, K., Vemaraju, K., Tej, S.S., Skogen, J.W. and Meyers, B.C. (2006) Plant MPSS databases: signature-based transcriptional resources for analyses of mRNA and small RNA. *Nucleic Acids Research*, **34**, D731–D735.

12 Li, L., Wang, X., Stolc, V., Li, X., Zhang, D., Su, N., Tongprasit, W., Li, S., Cheng, Z., Wang, J. and Deng, X.W. (2006) Genome-wide transcription analyses in rice using tiling microarrays. *Nature Genetics*, **38**, 124–129.

13 Yamada, K., Lim, J., Dale, J.M., Chen, H., Shinn, P., Palm, C.J., Southwick, A.M., Wu, H.C., Kim, C., Nguyen, M., Pham, P., Cheuk, R., Karlin-Newmann, G., Liu, S.X., Lam, B., Sakano, H., Wu, T., Yu, G., Miranda, M., Quach, H.L., Tripp, M., Chang, C.H., Lee, J.M., Toriumi, M., Chan, M.M., Tang, C.C., Onodera, C.S., Deng, J.M., Akiyama, K., Ansari, Y., Arakawa, T., Banh, J., Banno, F., Bowser, L., Brooks, S., Carninci, P., Chao, Q., Choy, N., Enju, A., Goldsmith, A.D., Gurjal, M., Hansen, N.F., Hayashizaki, Y., Johnson-Hopson, C., Hsuan, V.W., Iida, K., Karnes, M., Khan, S., Koesema, E., Ishida, J., Jiang, P.X., Jones, T., Kawai, J., Kamiya, A., Meyers, C., Nakajima, M., Narusaka, M., Seki, M., Sakurai, T., Satou, M., Tamse, R., Vaysberg, M., Wallender, E.K., Wong, C., Yamamura, Y., Yuan, S., Shinozaki, K., Davis, R.W., Theologis, A. and Ecker, J.R. (2003) Empirical analysis of transcriptional activity in the *Arabidopsis* genome. *Science*, **302**, 842–846.

14 Seki, M., Narusaka, M., Kamiya, A., Ishida, J., Satou, M., Sakurai, T., Nakajima, M., Enju, A., Akiyama, K., Oono, Y., Muramatsu, M., Hayashizaki, Y., Kawai, J., Carninci, P., Itoh, M., Ishii, Y., Arakawa, T., Shibata, K., Shinagawa, A. and Shinozaki, K. (2002) Functional annotation of a full-length *Arabidopsis* cDNA collection. *Science*, **296**, 141–145.

15 Kikuchi, S., Satoh, K., Nagata, T., Kawagashira, N., Doi, K., Kishimoto, N.,

Yazaki, J., Ishikawa, M., Yamada, H., Ooka, H., Hotta, I., Kojima, K., Namiki, T., Ohneda, E., Yahagi, W., Suzuki, K., Li, C.J., Ohtsuki, K., Shishiki, T., Otomo, Y., Murakami, K., Iida, Y., Sugano, S., Fujimura, T., Suzuki, Y., Tsunoda, Y., Kurosaki, T., Kodama, T., Masuda, H., Kobayashi, M., Xie, Q., Lu, M., Narikawa, R., Sugiyama, A., Mizuno, K., Yokomizo, S., Niikura, J., Ikeda, R., Ishibiki, J., Kawamata, M., Yoshimura, A., Miura, J., Kusumegi, T., Oka, M., Ryu, R., Ueda, M., Matsubara, K., Kawai, J., Carninci, P., Adachi, J., Aizawa, K., Arakawa, T., Fukuda, S., Hara, A., Hashizume, W., Hayatsu, N., Imotani, K., Ishii, Y., Itoh, M., Kagawa, I., Kondo, S., Konno, H., Miyazaki, A., Osato, N., Ota, Y., Saito, R., Sasaki, D., Sato, K., Shibata, K., Shinagawa, A., Shiraki, T., Yoshino, M., Hayashizaki, Y. and Yasunishi, A. (2003) Collection, mapping, and annotation of over 28,000 cDNA clones from japonica rice. *Science*, **301**, 376–379.

16 Kitagawa, N., Washio, T., Kosugi, S., Yamashita, T., Higashi, K., Yanagawa, H., Higo, K., Satoh, K., Ohtomo, Y., Sunako, T., Murakami, K., Matsubara, K., Kawai, J., Carninci, P., Hayashizaki, Y., Kikuchi, S. and Tomita, M. (2005) Computational analysis suggests that alternative first exons are involved in tissue-specific transcription in rice (*Oryza sativa*). *Bioinformatics*, **21**, 1758–1763.

17 Bulow, L., Steffens, N.O., Galuschka, C., Schindler, M. and Hehl, R. (2006) AthaMap: from *in silico* data to real transcription factor binding sites. *In Silico Biology*, **6**, 243–252.

18 Galuschka, C., Schindler, M., Bulow, L. and Hehl, R. (2007) AthaMap web tools for the analysis and identification of co-regulated genes. *Nucleic Acids Research*, **35**, D857–D862.

19 Rudd, S., Frisch, M., Grote, K., Meyers, B.C., Mayer, K. and Werner, T. (2004) Genome-wide *in silico* mapping of scaffold/matrix attachment regions in *Arabidopsis* suggests correlation of intragenic scaffold/matrix attachment regions with gene expression. *Plant Physiology*, **135**, 715–722.

20 Zhang, X., Yazaki, J., Sundaresan, A., Cokus, S., Chan, S.W., Chen, H., Henderson, I.R., Shinn, P., Pellegrini, M., Jacobsen, S.E. and Ecker, J.R. (2006) Genome-wide high-resolution mapping and functional analysis of DNA methylation in *Arabidopsis*. *Cell*, **126**, 1189–1201.

21 The ENCODE Project Consortium (2007) Identification and analysis of functional elements in 1% of the human genome by the ENCODE pilot project. *Nature*, **447**, 799–816.

22 Mockler, T.C., Chan, S., Sundaresan, A., Chen, H., Jacobsen, S.E. and Ecker, J.R. (2005) Applications of DNA tiling arrays for whole-genome analysis. *Genomics*, **85**, 1–15.

23 Harbers, M. and Carninci, P. (2005) Tag-based approaches for transcriptome research and genome annotation. *Nature Methods*, **2**, 495–502.

24 Cheng, J., Kapranov, P., Drenkow, J., Dike, S., Brubaker, S., Patel, S., Long, J., Stern, D., Tammana, H., Helt, G., Sementchenko, V., Piccolboni, A., Bekiranov, S., Bailey, D.K., Ganesh, M., Ghosh, S., Bell, I., Gerhard, D.S. and Gingeras, T.R. (2005) Transcriptional maps of 10 human chromosomes at 5-nucleotide resolution. *Science*, **308**, 1149–1154.

25 Thibaud-Nissen, F., Wu, H., Richmond, T., Redman, J.C., Johnson, C., Green, R., Arias, J. and Town, C.D. (2006) Development of *Arabidopsis* whole-genome microarrays and their application to the discovery of binding sites for the TGA2 transcription factor in salicylic acid-treated plants. *Plant Journal*, **47**, 152–162.

26 Metzker, M.L. (2005) Emerging technologies in DNA sequencing. *Genome Research*, **15**, 1767–1776.

27 Shendure, J., Mitra, R.D., Varma, C. and Church, G.M. (2004) Advanced sequencing technologies: methods and goals. *Nature Reviews Genetics*, **5**, 335–344.

28 Hall, N. (2007) Advanced sequencing technologies and their wider impact in microbiology. *Journal of Experimental Biology*, **210**, 1518–1525.

29 Adams, M.D., Kelley, J.M., Gocayne, J.D., Dubnick, M., Polymeropoulos, M.H., Xiao, H., Merril, C.R., Wu, A., Olde, B., Moreno, R.F.*et al.* (1991) Complementary DNA sequencing: expressed sequence tags and human genome project. *Science*, **252**, 1651–1656.

30 Carninci, P., Waki, K., Shiraki, T., Konno, H., Shibata, K., Itoh, M., Aizawa, K., Arakawa, T., Ishii, Y., Sasaki, D., Bono, H., Kondo, S., Sugahara, Y., Saito, R., Osato, N., Fukuda, S., Sato, K., Watahiki, A., Hirozane-Kishikawa, T., Nakamura, M., Shibata, Y., Yasunishi, A., Kikuchi, N., Yoshiki, A., Kusakabe, M., Gustincich, S., Beisel, K., Pavan, W., Aidinis, V., Nakagawara, A., Held, W.A., Iwata, H., Kono, T., Nakauchi, H., Lyons, P., Wells, C., Hume, D.A., Fagiolini, M., Hensch, T.K., Brinkmeier, M., Camper, S., Hirota, J., Mombaerts, P., Muramatsu, M., Okazaki, Y., Kawai, J. and Hayashizaki, Y. (2003) Targeting a complex transcriptome: the construction of the mouse full-length cDNA encyclopedia. *Genome Research*, **13**, 1273–1289.

31 Velculescu, V.E., Zhang, L., Vogelstein, B. and Kinzler, K.W. (1995) Serial analysis of gene expression. *Science*, **270**, 484–487.

32 Reinartz, J., Bruyns, E., Lin, J.Z., Burcham, T., Brenner, S., Bowen, B., Kramer, M. and Woychik, R. (2002) Massively parallel signature sequencing (MPSS) as a tool for in-depth quantitative gene expression profiling in all organisms. *Brief Functional Genomics and Proteomics*, **1**, 95–104.

33 Shiraki, T., Kondo, S., Katayama, S., Waki, K., Kasukawa, T., Kawaji, H., Kodzius, R., Watahiki, A., Nakamura, M., Arakawa, T., Fukuda, S., Sasaki, D., Podhajska, A., Harbers, M., Kawai, J., Carninci, P. and Hayashizaki, Y. (2003) Cap analysis gene expression for high-throughput analysis of transcriptional starting point and identification of promoter usage. *Proceedings of the National Academy of Sciences of the United States of America*, **100**, 15776–15781.

34 Hashimoto, S., Suzuki, Y., Kasai, Y., Morohoshi, K., Yamada, T., Sese, J., Morishita, S., Sugano, S. and Matsushima, K. (2004) 5′-end SAGE for the analysis of transcriptional start sites. *Nature Biotechnology*, **22**, 1146–1149.

35 Kodzius, R., Kojima, M., Nishiyori, H., Nakamura, M., Fukuda, S., Tagami, M., Sasaki, D., Imamura, K., Kai, C., Harbers, M., Hayashizaki, Y. and Carninci, P. (2006) CAGE: cap analysis of gene expression. *Nature Methods*, **3**, 211–222.

36 Nielsen, K.L., Hogh, A.L. and Emmersen, J. (2006) DeepSAGE–digital transcriptomics with high sensitivity, simple experimental protocol and multiplexing of samples. *Nucleic Acids Research*, **34**, e133.

37 Kapranov, P., Cheng, J., Dike, S., Nix, D.A., Duttagupta, R., Willingham, A.T., Stadler, P.F., Hertel, J., Hackermueller, J., Hofacker, I.L., Bell, I., Cheung, E., Drenkow, J., Dumais, E., Patel, S., Helt, G., Ganesh, M., Ghosh, S., Piccolboni, A., Sementchenko, V., Tammana, H. and Gingeras, T.R. (2007) RNA maps reveal new RNA classes and a possible function for pervasive transcription. *Science*, **316**, 1484–1488.

38 Carninci, P., Kvam, C., Kitamura, A., Ohsumi, T., Okazaki, Y., Itoh, M., Kamiya, M., Shibata, K., Sasaki, N., Izawa, M., Muramatsu, M., Hayashizaki, Y. and Schneider, C. (1996) High-efficiency full-length cDNA cloning by biotinylated CAP trapper. *Genomics*, **37**, 327–336.

39 Margulies, M., Egholm, M., Altman, W.E., Attiya, S., Bader, J.S., Bemben, L.A., Berka, J., Braverman, M.S., Chen, Y.J., Chen, Z., Dewell, S.B., Du, L., Fierro, J.M., Gomes, X.V., Godwin, B.C., He, W., Helgesen, S., Ho, C.H., Irzyk, G.P., Jando, S.C., Alenquer, M.L., Jarvie, T.P., Jirage, K.B.,

Kim, J.B., Knight, J.R., Lanza, J.R., Leamon, J.H., Lefkowitz, S.M., Lei, M., Li, J., Lohman, K.L., Lu, H., Makhijani, V.B., McDade, K.E., McKenna, M.P., Myers, E.W., Nickerson, E., Nobile, J.R., Plant, R., Puc, B.P., Ronan, M.T., Roth, G.T., Sarkis, G.J., Simons, J.F., Simpson, J.W., Srinivasan, M., Tartaro, K.R., Tomasz, A., Vogt, K.A., Volkmer, G.A., Wang, S.H., Wang, Y., Weiner, M.P., Yu, P., Begley, R.F. and Rothberg, J.M. (2005) Genome sequencing in microfabricated high-density picolitre reactors. *Nature*, **437**, 376–380.

40 Struhl, K. (2007) Transcriptional noise and the fidelity of initiation by RNA polymerase II. *Nature Structural & Molecular Biology*, **14**, 103–105.

41 Ponjavic, J., Ponting, C.P. and Lunter, G. (2007) Functionality or transcriptional noise? Evidence for selection within long noncoding RNAs. *Genome Research*, **17**, 556–565.

42 Peaston, A.E., Evsikov, A.V., Graber, J.H., de Vries, W.N., Holbrook, A.E., Solter, D. and Knowles, B.B. (2004) Retrotransposons regulate host genes in mouse oocytes and preimplantation embryos. *Developmental Cell*, **7**, 597–606.

43 Carninci, P., Kasukawa, T., Katayama, S., Gough, J., Frith, M.C., Maeda, N., Oyama, R., Ravasi, T., Lenhard, B., Wells, C., Kodzius, R., Shimokawa, K., Bajic, V.B., Brenner, S.E., Batalov, S., Forrest, A.R., Zavolan, M., Davis, M.J., Wilming, L.G., Aidinis, V., Allen, J.E., Ambesi-Impiombato, A., Apweiler, R., Aturaliya, R.N., Bailey, T.L., Bansal, M., Baxter, L., Beisel, K.W., Bersano, T., Bono, H., Chalk, A.M., Chiu, K.P., Choudhary, V., Christoffels, A., Clutterbuck, D.R., Crowe, M.L., Dalla, E., Dalrymple, B.P., de Bono, B., Della Gatta, G., di Bernardo, D., Down, T., Engstrom, P., Fagiolini, M., Faulkner, G., Fletcher, C.F., Fukushima, T., Furuno, M., Futaki, S., Gariboldi, M., Georgii-Hemming, P., Gingeras, T.R., Gojobori, T., Green, R.E., Gustincich, S., Harbers, M., Hayashi, Y., Hensch, T.K., Hirokawa, N., Hill, D., Huminiecki, L., Iacono, M., Ikeo, K., Iwama, A., Ishikawa, T., Jakt, M., Kanapin, A., Katoh, M., Kawasawa, Y., Kelso, J., Kitamura, H., Kitano, H., Kollias, G., Krishnan, S.P., Kruger, A., Kummerfeld, S.K., Kurochkin, I.V., Lareau, L.F., Lazarevic, D., Lipovich, L., Liu, J., Liuni, S., McWilliam, S., Madan Babu, M., Madera, M., Marchionni, L., Matsuda, H., Matsuzawa, S., Miki, H., Mignone, F., Miyake, S., Morris, K., Mottagui-Tabar, S., Mulder, N., Nakano, N., Nakauchi, H., Ng, P., Nilsson, R., Nishiguchi, S., Nishikawa, S.*et al.* (2005) The transcriptional landscape of the mammalian genome. *Science*, **309**, 1559–1563.

44 Carninci, P., Sandelin, A., Lenhard, B., Katayama, S., Shimokawa, K., Ponjavic, J., Semple, C.A., Taylor, M.S., Engstrom, P.G., Frith, M.C., Forrest, A.R., Alkema, W.B., Tan, S.L., Plessy, C., Kodzius, R., Ravasi, T., Kasukawa, T., Fukuda, S., Kanamori-Katayama, M., Kitazume, Y., Kawaji, H., Kai, C., Nakamura, M., Konno, H., Nakano, K., Mottagui-Tabar, S., Arner, P., Chesi, A., Gustincich, S., Persichetti, F., Suzuki, H., Grimmond, S.M., Wells, C.A., Orlando, V., Wahlestedt, C., Liu, E.T., Harbers, M., Kawai, J., Bajic, V.B., Hume, D.A. and Hayashizaki, Y. (2006) Genome-wide analysis of mammalian promoter architecture and evolution. *Nature Genetics*, **38**, 626–635.

45 Katayama, S., Tomaru, Y., Kasukawa, T., Waki, K., Nakanishi, M., Nakamura, M., Nishida, H., Yap, C.C., Suzuki, M., Kawai, J., Suzuki, H., Carninci, P., Hayashizaki, Y., Wells, C., Frith, M., Ravasi, T., Pang, K.C., Hallinan, J., Mattick, J., Hume, D.A., Lipovich, L., Batalov, S., Engstrom, P.G., Mizuno, Y., Faghihi, M.A., Sandelin, A., Chalk, A.M., Mottagui-Tabar, S., Liang, Z., Lenhard, B. and Wahlestedt, C. (2005) Antisense transcription in the mammalian transcriptome. *Science*, **309**, 1564–1566.

46 Seifert, M., Scherf, M., Epple, A. and Werner, T. (2005) Multievidence microarray mining. *Trends in Genetics*, **21**, 553–558.

47 Denissov, S., van Driel, M., Voit, R., Hekkelman, M., Hulsen, T., Hernandez, N., Grummt, I., Wehrens, R. and Stunnenberg, H. (2007) Identification of novel functional TBP-binding sites and general factor repertoires. *EMBO Journal*, **26**, 944–954.

48 Kawaji, H., Kasukawa, T., Fukuda, S., Katayama, S., Kai, C., Kawai, J., Carninci, P. and Hayashizaki, Y. (2006) CAGE Basic/Analysis Databases: the CAGE resource for comprehensive promoter analysis. *Nucleic Acids Research*, **34**, D632–D636.

49 Kapranov, P., Drenkow, J., Cheng, J., Long, J., Helt, G., Dike, S. and Gingeras, T.R. (2005) Examples of the complex architecture of the human transcriptome revealed by RACE and high-density tiling arrays. *Genome Research*, **15**, 987–997.

50 Kodzius, R., Matsumura, Y., Kasukawa, T., Shimokawa, K., Fukuda, S., Shiraki, T., Nakamura, M., Arakawa, T., Sasaki, D., Kawai, J., Harbers, M., Carninci, P. and Hayashizaki, Y. (2004) Absolute expression values for mouse transcripts: re-annotation of the READ expression database by the use of CAGE and EST sequence tags. *FEBS Letters*, **559**, 22–26.

51 Cohen, C.D., Klingenhoff, A., Boucherot, A., Nitsche, A., Henger, A., Brunner, B., Schmid, H., Merkle, M., Saleem, M.A., Koller, K.P., Werner, T., Grone, H.J., Nelson, P.J. and Kretzler, M. (2006) Comparative promoter analysis allows *de novo* identification of specialized cell junction-associated proteins. *Proceedings of the National Academy of Sciences of the United States of America*, **103**, 5682–5687.

52 Werner, T. (2007) Regulatory networks: linking microarray data to systems biology. *Mechanisms of Ageing and Development*, **128**, 168–172.

53 Werner, T., Fessele, S., Maier, H. and Nelson, P.J. (2003) Computer modeling of promoter organization as a tool to study transcriptional coregulation. *FASEB Journal*, **17**, 1228–1237.

54 Czechowski, T., Bari, R.P., Stitt, M., Scheible, W.R. and Udvardi, M.K. (2004) Real-time RT-PCR profiling of over 1400 *Arabidopsis* transcription factors: unprecedented sensitivity reveals novel root- and shoot-specific genes. *Plant Journal*, **38**, 366–379.

55 Draghici, S., Khatri, P., Eklund, A.C. and Szallasi, Z. (2006) Reliability and reproducibility issues in DNA microarray measurements. *Trends in Genetics*, **22**, 101–109.

56 Khattra, J., Delaney, A.D., Zhao, Y., Siddiqui, A., Asano, J., McDonald, H., Pandoh, P., Dhalla, N., Prabhu, A.L., Ma, K., Lee, S., Ally, A., Tam, A., Sa, D., Rogers, S., Charest, D., Stott, J., Zuyderduyn, S., Varhol, R., Eaves, C., Jones, S., Holt, R., Hirst, M., Hoodless, P.A. and Marra, M.A. (2007) Large-scale production of SAGE libraries from microdissected tissues, flow-sorted cells, and cell lines. *Genome Research*, **17**, 108–116.

5
Gene Identification Signature-Paired End diTagging (GIS-PET): A Technology for Transcriptome Characterization

Patrick Ng, Yen-Ling Lee, Chia-Lin Wei, and Yijun Ruan

Abstract

GIS-PET is an application of the paired-end ditagging (PET) concept that we developed for high-throughput transcriptome characterization. It combines the accuracy of full-length cDNA (flcDNA) sequencing for identifying transcription start and stop sites with the efficiency provided by DNA-tagging, to enable the quantitation and precise localization of all transcripts on a reference genome. The procedure described here can be carried out in any standard molecular biology laboratory. Because a high-quality flcDNA library is constructed as an intermediate step in the procedure, any transcripts of interest that are identified during data analysis can be easily recovered for further functional studies by PCR. We have previously validated the GIS-PET procedure by analyzing the transcriptome of the E14 murine embryonic stem cell line, and were able to identify splicing variants, verify predicted genes, identify novel genes, and discover unusual transcripts including sense/antisense pairs and transcripts apparently derived from *trans*- and intergenic splicing. Using Sanger sequencing for data collection, GIS-PET is estimated to be 30-fold more efficient than flcDNA sequencing for mapping transcript ends. The GIS-PET procedure is easily adapted for use with the new 454-sequencing technology, which in the context of the GS20 sequencer, increases annotation efficiency a further 100-fold. We present in this chapter a working protocol for GIS-PET, and a discussion of its characteristics relative to other methods used for transcriptome analysis.

5.1
Introduction

The completion of the draft sequence of the human genome in 2001 [1] heralded the beginning of the 'postgenomic era'. With more than 500 genomes already published, and many more in the pipeline (http://www.genomesonline.org), the

The Handbook of Plant Functional Genomics: Concepts and Protocols.
Edited by Günter Kahl and Khalid Meksem

ISBN: 978-3-527-31885-8

challenge has now become that of rapidly and accurately extracting the information contained within each genome sequence. This information includes the location and structure of every gene, and the mechanisms by which their expression is regulated.

The total complement of all expressed genes in a particular cell, under a specific set of conditions, constitutes its transcriptome. Recent studies have made us realize that many of the transcripts in each transcriptome are in fact non-coding RNAs (ncRNAs; reviewed in [2]) that, as the name implies, do not code for proteins. Furthermore, many of these ncRNAs appear to regulate the expression of other genes, for instance in the form of sense/antisense pairs ([3]; reviewed in [4]). While many transcripts are polyadenylated at the 3′-end, about 20% appear to not to be [5], which has a bearing on the experimental method used for transcriptome characterization. Additional transcriptome complexity is provided by the familiar phenomena of alternative splicing, alternative transcriptional start-sites (TSS) and alternative transcriptional termination-sites (TTS), and also by the less-familiar phenomena of tandem chimerism (also called intergenic splicing, transcription-induced chimerism, or cotranscription) [6,7] and *trans*-splicing (reviewed in [8]). As a result of the modifications exercised on primary transcripts, even mature mRNAs expressed from the same gene can be structurally different from each other, and therefore transcripts and their consequent translated proteins may differ in terms of stability, subcellular localization, post-translational modifications and of course function (reviewed in [9]).

Where the expression of transcripts is concerned, regulatory control appears to be exerted not only by classical TATA-box-associated promoters, but also by less positionally well-defined CpG-island-associated promoters that do not contain TATA-boxes [10]. It now appears that regulatory elements can be found in both intergenic as well as within gene-coding regions; they can also interact across long linear distances to control the expression of genes that are initially well-separated linearly, or even located on different chromosomes, but are apparently brought together in close spatial proximity by the interaction of DNA-associated proteins (reviewed in [11]). Some promoters are bidirectional, and can regulate the expression of proximal genes present in opposite strands of the same chromosome [12]. Finally, gene expression can be controlled by a number of epigenetic processes, including methylation, acetylation and phosphorylation (reviewed in [13]).

Transcriptome characterization therefore can be defined as the elucidation of the transcriptome of a particular system, including the structure and location of all coding and non-coding transcripts (known, as well as novel), and a quantitative measure of their individual expression levels. The information obtained from transcriptome analysis can therefore be invaluable for annotating the genome. For the purposes of this chapter, we will not extend the definition of transcriptome characterization to include promoter identification. However, it should be appreciated that a complete picture of any genome of interest would need to include both transcriptome characterization and genome annotation.

We will now briefly review the experimental techniques used to characterize transcriptomes. Low-throughput (or gene-by-gene) methods such as Northern blots, RNase-protection assays, RT-PCR, and methods that highlight differentially-expressed genes (such as differential display PCR, or suppression subtractive hybridization) will not be considered here, because these are more suited to small-scale studies, or to experimental validations. The medium- to high-throughput technologies used for transcriptome analysis include:

5.1.1
Microarray Analysis

Microarrays are perhaps the platform that immediately comes to mind when transcriptome profiling is mentioned. Microarrays are spotted with probes (also called 'features') that can be either double-stranded cDNA (usually fewer than 40 000 probes per array or chip), or short, single-stranded oligonucleotides that may or may not overlap in sequence coverage (currently up to 6.5 million features per array). Additionally, oligonucleotide arrays can either be expression arrays that represent only the coding regions of the genome, or 'whole genome tiling arrays' (WGTA) where the entire genome (usually with repeat-regions masked) is represented on one or more chips. Finally, the probes of an oligonucleotide array can be derived from either chromosomal strand, allowing a distinction to be made between sense and antisense transcription (Tiling arrays are reviewed in [14]). The advantages of using microarrays for transcriptome analysis are the ease with which the hybridization and data-collection procedures can be carried out, and the well-established software for analysis of the results. Disadvantages include experimental variability leading to non-portability of results, an inability to provide absolute quantitation of gene expression, cross-hybridization, and in the case of arrays other than WGTA, the closed-architecture of the system that only permits the study of targets corresponding to the probes which have been spotted. Where WGTAs are concerned, there are currently no commercially-available standard array-sets that cover the entire genome of more complex organisms at the same single-base resolution provided by DNA sequencing methods, and hence the identification of TSS, TTS and exon/intron junctions cannot be as precise. Furthermore, WGTAs are unable to quantitate the expression of individual splice variants amongst multiple variants expressed from the same multi-exonic gene. Nevertheless, microarray-based transcriptome analyses continue to reveal findings that continually shed new light on transcriptome complexity (see reviews [15,16]), for example, studies suggesting that a far greater than expected proportion of the genome is transcribed [5,17], and that at least as many ncRNAs are expressed as coding mRNAs. Also, only about half of all transcripts expressed are apparently found in the cytoplasm [5]. Microarray studies enabled the estimation of alternative-splicing frequency in the human genome, at between 35–60% per gene [18].

5.1.2
cDNA Sequencing, Including EST- and flcDNA-Sequencing

Here, transcripts are captured in the form of ds cDNAs, and sequenced to reveal their content. In traditional EST-sequencing approaches, a pool of transcripts is reverse-transcribed *en masse*, and single-pass sequenced from either end. Due to the nature of both the reverse transcriptases and the high CG-content in the 5′-ends of transcripts, such EST libraries are often 3′-biased, with 5′-EST sequences that do not accurately reflect the authentic TSS. Simple EST sequences are therefore of limited value in complete transcriptome characterization. In contrast, flcDNA sequencing is considered to be the 'gold standard' for transcriptome analysis: here, additional manipulations ensure that effectively all the ds cDNAs present are full-length. Each flcDNA is then sequenced bidirectionally from end-to-end. This ensures the most detailed view of the internal structure of every transcript, but is obviously tedious, very expensive, and the throughput generally too low for all but the largest genome centers to use for large-scale projects. Nonetheless, large scale transcriptome projects such as FANTOM have used flcDNA sequencing to provide valuable insights into mammalian transcriptomes [19,20].

5.1.3
DNA-Tagging Methods

Short DNA sequences, of usually between 10 and 20 bp, can be extracted from reverse-transcribed transcripts, and characterized far more efficiently than the entire transcripts themselves. Although EST-sequencing could be described as a form of tagging technique, the approach is usually thought of as being exemplified by MPSS (Massively Parallel Signature Sequencing) [21], SAGE (Serial Analysis of Gene Expression) [22] and its derivative, LongSAGE [23]. In these techniques, short DNA tags are enzymatically extracted from the 3′-most 'anchoring enzyme' (AE) site (e.g. Nla III) using a 'tagging enzyme', and concatenated prior to large-scale sequencing to increase sequencing efficiency. The problem of non-specific mapping (particularly to complex mammalian genomes) displayed by the original SAGE technique were somewhat alleviated by the use of longer Mme I-derived tags in the LongSAGE modification, and subsequently improved even further in SuperSAGE [24]). The chief advantages of SAGE and other DNA-tagging-based methods are, firstly, the portability of results, since SAGE tags provide a straightforward measure of gene-expression quantitation by simple tag-counting followed by the comparison of normalized tag-counts between libraries, and, secondly, the open-architecture of the system, that enables its application to gene discovery. However, the technique suffers from several disadvantages: transcripts lacking the appropriate AE site will be missed, incomplete AE digestion results in uncertainty of tag identity, little or no information about internal gene structure can be obtained, and retrieval of transcripts of interest for further study is difficult. Because cDNA libraries have to be constructed as part of the SAGE procedure, it is therefore also technically more challenging than microarray hybridization.

To overcome at least some of these problems, several groups have attempted a SAGE-like modification that extracts positionally-defined tags. In the case of 5′LongSAGE [25], 5′-End SAGE [26], 5′-RATE [27] and CAGE [28], tags are extracted from the 5′-terminal end, while in 3′LongSAGE [25], tags are derived from the 3′-terminal, just before the poly-A tail. 5′-Terminal tags have proven useful for promoter studies and for quantitation of gene expression [10]. The counting and mapping of both 5′- and 3′-terminal tags to the genome allows, in theory, both efficient gene quantitation and gene annotation, but missing tags can make it difficult to annotate genes correctly.

In 2005, we improved the 5′/3′-LongSAGE procedure, so that instead of extracting terminal tags in two separate sets from the transcriptome, we cloned flcDNA in specially-designed plasmid vectors, from which we could simultaneously extract both the terminal tags from any flcDNA, and link them together in the form of Paired-End diTags (PETs). These PETs could then be concatenated for efficient Sanger-based sequencing. In this way, we sought to combine both the accuracy provided by flcDNA sequencing with the efficiency of tag-based methods. The technique was initially called 'Gene Identification Signature' (GIS) analysis [29], but is now known as 'Gene Identification Signature- Paired-End diTagging' (GIS-PET), and was the first of several PET-based applications that were subsequently developed for genome characterization. Some of these will be described later in the chapter.

In GIS-PET, ditags extracted from raw sequences are mapped onto a reference genome assembly, thereby precisely demarcating the TSS and TTS of every transcript, including novel genes. The expression of each spliceform is provided by tag-counting. Finally, because an flcDNA library is constructed as part of the procedure, full-length clones for further study can be easily retrieved by PCR.

5.1.4
Advanced DNA Sequencing Technologies

DNA-tagging methods for transcriptome analysis rely on DNA sequencing for data collection, but improvements in sequencing technology may soon make it feasible to directly sequence every transcript.

Sanger dideoxy-sequencing has been the mainstay of DNA sequencing technology for decades, but the past 2 years has seen major innovations in the field, and next-generation sequencers running on a range of different technologies can be expected to succeed Sanger-based sequencers in the very near future (except perhaps for *de novo* genome sequencing). The first of these units to be marketed was the 454-sequencing system [30] based on pyrosequencing, in the form of the GS20 machine marketed by Roche. Its throughput of 25 million bases in a single 4-h run was unprecedented, but it suffered from short read-lengths per template (~100 bp), and the inability to provide paired-end information critical for accurate transcript demarcation and other genome applications. Using a minor modification of the protocol described in this chapter, we were able to combine the advantages of GIS-PET with the throughput provided by 454-sequencing, thereby enabling the extraction of nearly 0.5 million PETs in a single GS20 run. This technique was named 'Multiplex Sequencing of Paired-End ditags' (MS-PET) [31].

Several other next-generation systems, now with the ability to provide paired-end reads, are currently being introduced by Illumina and ABI. These are mentioned in Section 5.5.

This chapter will focus on providing a working protocol for the GIS-PET technology developed in our laboratory. Because of the estimated 30-fold greater efficiency in transcript characterization compared to flcDNA sequencing (calculations based on Sanger sequencing), fewer than 70 000 sequencing reads would be needed to completely characterize a transcriptome comprising a total of 1 million individual transcripts. With access to a GS20 sequencer, a simple modification to the basic procedure would enable the collection of 0.5 million PETs (corresponding to the same number of transcripts) in a single 4-h machine run.

More importantly, the basic concept of end-terminal DNA-tagging used in GIS-PET is general enough to be applied to any number of transcriptome and genome analysis methodologies, and we look forward to more of these being developed in the near future by interested readers.

5.2 Protocol

The GIS-PET procedure (see Figure 5.1) can be divided into three parts. In the first part, which should take about 9 days, a GIS-PET flcDNA library is constructed in the cloning vector pGIS4a using a modified Cap-Trapper approach [32]. In the second part, tagging is achieved with Mme I digestion, and self-circularization results in an intermediate Single-PET library. In the third and last part, PETs are released, then either concatenated and cloned in the pZErO-1 vector to make a GIS-PET sequencing library for efficient Sanger sequencing, or, alternatively, purified PETs may be dimerized into diPETs for 454-sequencing. From the raw sequences, PETs are

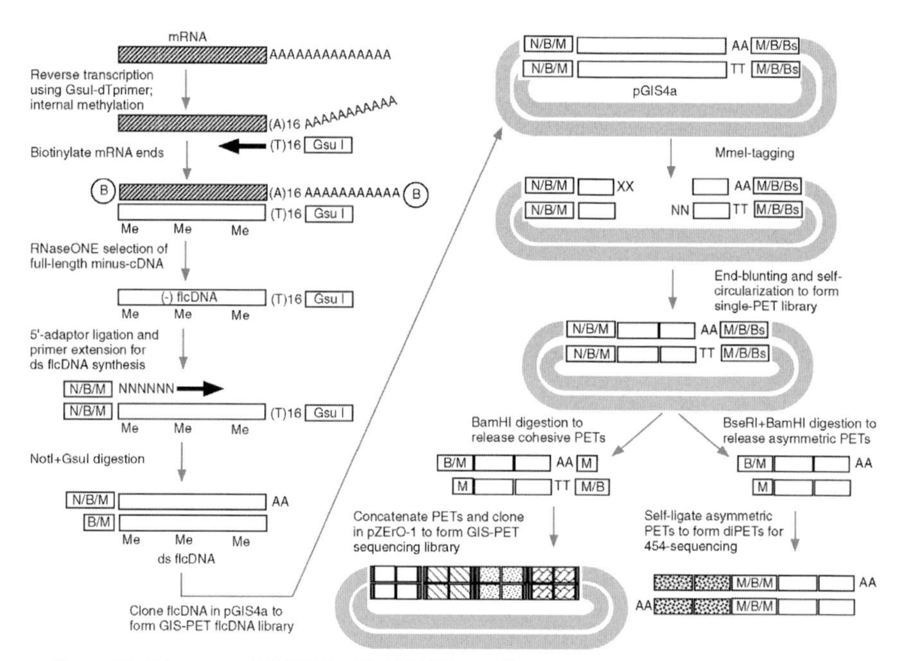

Figure 5.1 Schematic of GIS-PET, with MS-PET modification.

extracted and mapped onto the appropriate reference genome assembly using the PET-Tool software suite developed in-house [33]. In this protocol, the terms 'PET' and 'ditag' are used interchangeably.

(a) The cloning vector pGIS4a (see Figure 5.2) is an improved version of the originally-published pGIS1 vector [29]. With pGIS4a, the 3′ adaptor ligation step is now unnecessary. pGIS4a is freely available from the authors, and researchers

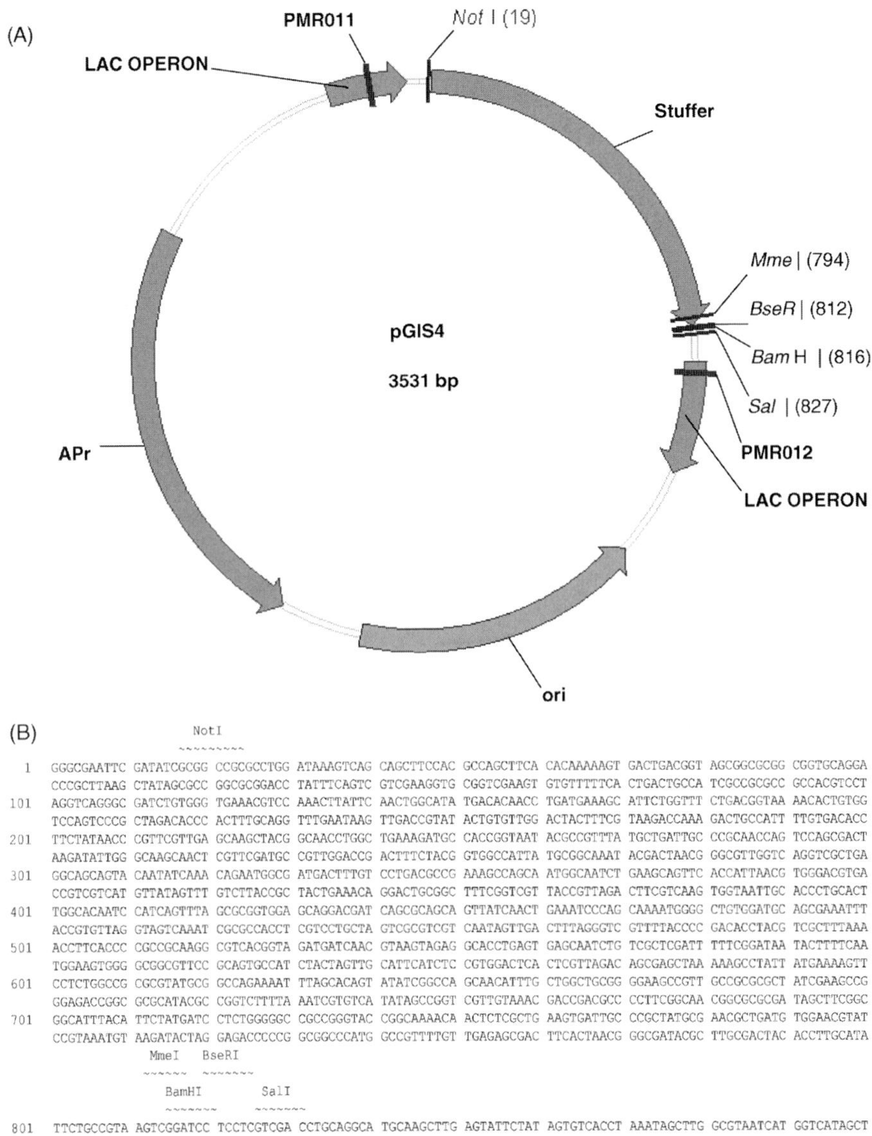

Figure 5.2 (A) The pGIS4a vector and (B) details of the cloning site.

should contact either Patrick Ng at ngwp@gis.a-star.edu.sg or Yijun Ruan at ruanyj@gis.a-star.edu.sg for details.

(b) The PET-Tool ditag extraction and mapping software is freely available for academic users by downloading from http://www.gis.a-star.edu.sg/PET_Tool/. Other users should contact ruanyj@gis.a-star.edu.sg for access details.

5.2.1 Construction of a GIS-PET flcDNA Library

All steps prior to alkaline hydrolysis of mRNA should be performed using RNase-free reagents and consumables. In the Cap-Trapper procedure, the diol structures present in the cap and 3′ terminal of the mRNA (polyA-RNA) are oxidized, then chemically biotinylated. RNase I digestion enriches for full-length (capped) mRNA, and its associated minus-strand cDNA. The resulting enriched (−) flcDNA/mRNA heteroduplexes containing biotinylated cap structures are then captured by binding to streptavidin-coated beads. After hydrolytic degradation of mRNA, the released (−) flcDNA is converted to double-stranded (ds) cDNA by 5′-adapter ligation followed by primer extension. Gsu I is used to remove the residual polyA tail, leaving an AA dinucleotide residue useful for orienting the final PETs on the genome. This is followed by Not I digestion to form the 5′ cohesive site, resulting in ds flcDNA that can be directionally inserted into the prepared pGIS4a vector.

5.2.1.1 Reverse-Transcription of mRNA (polyA-RNA) Sample

1. Mix the following in a 0.2-ml thin-walled PCR tube:

- 2 μg/μl GsuI-oligo dT primer — 3.5 μl
  (see Figure 5.3 for oligonucleotide sequences)
- 40 U/μl RNasin-PLUS inhibitor (Promega) — 1 μl
- PolyA RNA — 10 μg
- Nuclease-free water (Ambion) — to 20 μl

In a PCR machine, heat the reaction mixture to 65 °C for 10 min and cool to 37 °C for 1 min to allow annealing of oligo-dT primer. Hold at 42 °C while preparing the other components.

2. Prepare the reverse transcriptase (RT) mix in a separate 0.2-ml thin-walled PCR tube on ice as follows:

- 2× GC-I buffer (Takara) — 75 μl
  The 2× GC-I buffer is part of the LA-PCR kit from Takara
- RNasin-PLUS inhibitor — 1 μl
- Modified 10 mM dNTP (with 5-Me-dCTP instead of dCTP) — 4 μl
  (see Reagents for details of modified dNTP)
- 4.9 M sorbitol (Sigma) — 26 μl
- 200 U/μl Superscript II RT (Invitrogen) — 15 μl

GsuI-oligo dT primer
5′-GAGCTAGTTCTGGAGTTTTTTTTTTTTTTTTVN-3′

GIS-(N)6 adapter
5′-CTAAACTCGAGGCGGCCGCGGATCCGACNNNNNN-3′
3′-TTTGAGCTCCGCCGGCGCCTAGGCTG-p-5′

GIS-(N)5 adapter
5′-CTAAACTCGAGGCGGCCGCGGATCCGACGNNNNN-3′
3′-TTTGAGCTCCGCCGGCGCCTAGGCTG-p-5′

PMR011 (M13(+)
5′-GATGTGCTGCAAGGCGATTAAG-3′

PMR012 (M13(-))
5′-AGCGGATAACAATTTCACACAGG-3′

pGIS7/4VERIF-TOP
5'-GGCCGCGTGAGCAAGAAGAAGCAGCAGAGAAGACCCTAGGATCCAACTCGAGAA-3'

pGIS7/4VERIF-BOT
5'-CTCGAGTTGGATCCTAGGGTCTTCTCTGCTGCTTCTTCTTGCTCACGC-3'

Structure of a generic ~ 50bp BamHI-cohesive PET
5′-GATCCGACXXXXXXXXXXXXXXXXXXNNNNNNNNNNNNNNNNAAGTCG-3′
3′-GCTGXXXXXXXXXXXXXXXXXXNNNNNNNNNNNNNNNNTTCAGCCTAG -5′
Where X and N may be any of A, C, G or T.

Structure of a generic diPET
dλλλλλλλλλλλλλλλλ 81ZZZZZZZZZZZZZZZZZZ ƃʇɔƃ**GATCCGAC** XXXXXXXXXXXXXXXXXX$_{18}$NNNNNNNNNNNNNNNN$_{16}$ **AA**
∀∀ 91λλλλλλλλλλλλλλλλ 81ZZZZZZZZZZZZZZZZZZ ɔ∀ƃɔɔʇ∀ƃ**GCTG** XXXXXXXXXXXXXXXXXX$_{18}$NNNNNNNNNNNNNNNNp

Figure 5.3 Oligonucleotides and adaptors used in this protocol.

3. Place 10 μl saturated trehalose (see Reagents) into yet another 0.2-ml thin-walled PCR tube, and leave warming at 42 °C in a thermal cycler.

4. When the oligo-dT/mRNA annealing step is complete, place the RT mix into the 42 °C-thermal cycler for 2 min to preheat. Add the pre-warmed trehalose to the warm RT mix (total volume now 131 μl), and quickly transfer the entire reaction mix into the tube containing the annealed primer/mRNA (final volume 151 μl). Immediately incubate as follows:

   - 40 min at 42 °C, then
   - 20 min at 50 °C, then
   - 20 min at 55 °C.

5. Add 2 μl of 20 mg/ml proteinase K (Ambion) to degrade all enzymes and incubate for 15 min at 45 °C.

6. Extract the reaction with (25 : 24 : 1) phenol/chloroform/IAA (Ambion; equilibrated at pH 6.6) to remove proteins, and then re-extract with 150 μl nuclease-free water.

Note that, unlike subsequent phenol-chloroform extraction steps in this protocol, Phase-Lock gel cannot be used here, due to the very high density of the solution.

Precipitate the aqueous layer containing the (−) cDNA/RNA heteroduplex with ethanol as follows:

- (−) cDNA/RNA heteroduplex 300 μl
- 3 M sodium acetate, pH 5.2 30 μl
- Absolute ethanol 825 μl
- Do *not* add glycogen

Maintain at −80 °C for 30 min, then microcentrifuge for 30 min at maximum speed (∼20 000 g), and 4 °C. Wash with 70% ethanol and resuspend the sizeable white pellet in 44.5 μl nuclease-free water.

5.2.1.2 **Oxidation**

1. Prepare the following stocks fresh in water, using 1.7-ml microcentrifuge tubes:
   - 10 mM long-arm biotin hydrazide (Vector Laboratories)
   - 100 mM sodium periodate ($NaIO_4$) (Sigma).
2. Combine the following in a 1.5-ml siliconized or low-binding microcentrifuge tube:
   - (−) cDNA/RNA heteroduplex (from step 6 of Section 2.1.1) 44.5 μl
   - 1.1 M sodium acetate, pH 4.5 3 μl
   - Freshly-prepared 100 mM $NaIO_4$ 2.5 μl

Incubate on ice, for 45 min in the dark.

3. Add the following to the 50-μl reaction to precipitate the (−) cDNA/RNA heteroduplex:
   - 10% SDS 0.5 μl
   - 5 M NaCl 11 μl
   - Isopropanol 61 μl

Maintain at −80 °C for 30 min, then microcentrifuge for 30 min at maximum speed, and 4 °C. Wash with 70% ethanol and resuspend the small white pellet in 50 μl nuclease-free water.

5.2.1.3 **Biotinylation of RNA Ends**

1. To the 50 μl oxidized (−) cDNA/RNA, add the following:
   - 1 M sodium acetate, pH 6.1 5 μl
   - 10% SDS 5 μl
   - 10 mM fresh biotin hydrazide 150 μl

Incubate overnight (12–16 h) at room temperature (∼25 °C), in the dark.

5.2.1.4 **RNaseONE Selection for Full-Length (−) cDNA/RNA Heteroduplex**

1. Precipitate the biotinylated (−) cDNA/RNA heteroduplex (210 μl) by adding:

- 5 M NaCl 5 µl
- 1 M sodium acetate, pH 6.1 7 µl
- Absolute ethanol 750 µl

Maintain at −80 °C for 30 min, then microcentrifuge for 30 min at maximum speed, and 4 °C. Wash once with 70% ethanol. Resuspend the small white pellet in 170 µl nuclease-free water.

2. Carry out RNaseONE digestion to select for protected, biotinylated, full-length (−) cDNA/RNA hybrid. Use approximately 5 U RNaseONE per µg of mRNA sample used at the start of the protocol:

- Biotinylated (−) cDNA/RNA (from step 1, above) 170 µl
- 10× RNaseONE buffer 0 µl
- 10 U/µl RNaseONE (Promega) 4.5 µl
- Nuclease-free water 5.5 µl

Incubate for 30 min at 37 °C.

3. Quench the reaction by adding:

- 10 mg/ml yeast tRNA 4 µl
- 5 M NaCl 50 µl

5.2.1.5 Binding Biotinylated (−) cDNA/RNA Heteroduplex to Streptavidin Beads

The user should be familiar with the use of the magnetic stand for handling M280 beads. All procedures involving M280 beads should be performed in siliconized or low-binding microcentrifuge tubes.

1. During the RNaseONE selection procedure, prepare M-280 streptavidin Dynabeads (200 µl bead suspension per RNA sample) for use as follows:

Wash three times with 200 µl of 1× binding buffer (see Reagents) at room temperature, then pre-block by adding 200 µl of 1× binding buffer plus 0.25 µg/µl yeast tRNA, and incubate for 30 min at 4 °C on a benchtop hot/cold shaking incubator at 800 rpm. Remove supernatant.

Wash three times with 200 µl of 1× binding buffer, at room temperature, leaving the beads in the final wash until use.

2. Remove supernatant from beads and add the ~254 µl RNaseONE-treated (−) cDNA/RNA heteroduplex from step 3 of Section 2.1.4). Rotate for 30 min at room temperature on a benchtop rotator to enable binding to occur.

3. Wash heteroduplex-bound beads at room temperature as follows:

- Twice with 200 µl of 1× binding buffer
- Once with 200 µl of 1× blocking buffer (see Reagents)
- Once with 200 µl of 1× wash buffer (see Reagents)
- Once with 200 µl of 50 µg/ml yeast tRNA.

5.2.1.6 Alkaline Hydrolysis to Release (−) Strand flcDNA

1. Remove the supernatant produced in step 3 of Section 2.1.5 from the beads, and add 50 μl freshly-made alkaline hydrolysis buffer (see Reagents). Shake for 10 min at 65 °C at 1400 rpm in a benchtop hot/cold shaking incubator. At this stage, the minus-strand flcDNA will be released into the supernatant.

2. Collect (do not discard!) the supernatant containing the (−) flcDNA into a tube containing 150 μl of 1 M Tris-HCl, pH 7.5, to neutralize the hydrolysis buffer.

3. Repeat the hydrolysis and collection (steps 1 and 2, above) twice more, collecting all fractions in the same tube, to give a final volume of 300 μl.

4. Perform an extraction with an equal volume (300 μl) of (25 : 24 : 1) phenol/chloroform/IAA, pH 7.9 (Ambion), then precipitate with ethanol to recover the (−) flcDNA as follows:

   - (−) flcDNA 300 μl
   - 3 M sodium acetate, pH 5.2 30 μl
   - 15 mg/ml GlycoBlue (Ambion) 4 μl
   - Absolute ethanol 800 μl

Maintain at −80 °C for 30 min, then microcentrifuge for 30 min at maximum speed, and 4 °C.

Note that in this and subsequent phenol–chloroform extractions the use of Phase Lock Gel tubes (Eppendorf) greatly facilitates the extraction procedure.

5. Wash with 70% ethanol. Resuspend the (barely visible) pellet in 5 μl LoTE buffer (see Reagents) or 10 mM Tris-HCl, pH 8.5 (identical to Qiagen EB buffer).

5.2.1.7 Double-Stranded cDNA (ds cDNA) Synthesis

1. Carry out single-stranded linker (SSL) ligation of the mixed 5′ adaptors (see Figure 5.3 for oligonucleotide sequences) by combining the following, on ice, in a 1.7-ml microcentrifuge tube:

   - (−)flcDNA (from step 5 of Section 2.1.6) 5 μl
   - 0.4 μg/μl GIS-(N)5 adaptor 4 μl
   - 0.4 μg/μl GIS-(N)6 adaptor 1 μl
   - Takara solution II 10 μl
   - Takara solution I (ligase) 20 μl

   (Takara Solution I and II are from the Takara Ligation Kit version 2)

Incubate overnight (12–16 h) at 16 °C.

2. Set up the following primer extension reaction, on ice, in a 0.2-ml thin-walled PCR tube:

   - Overnight ligation reaction (step 1, above) 40 μl
   - Deionized water 20 μl

- 10× ExTaq buffer with Mg^{2+} (Takara) 8 µl
- 2.5 mM dNTP mix 8 ul
- 5 U/µl ExTaq DNA polymerase 4 µl

Note: do not use the Hot Start ExTaq DNA polymerase from Takara.

3. Transfer the tube directly from ice to a 65 °C-preheated thermal cycler, and incubate as follows:

   - 5 min at 65 °C
   - 30 min at 68 °C
   - 10 min at 72 °C
   - Hold at 4 °C.

4. Add 2 µl of 20 mg/ml proteinase K and incubate for 15 min at 45 °C to degrade any remaining DNA polymerase.

5. Adjust the volume to 200 µl with deionized water, extract with perform phenol-chloroform (pH 7.9), and then precipitate with ethanol as follows:

   - ds flcDNA 200 µl
   - 3 M sodium acetate, pH 5.2 20 µl
   - 15 mg/ml GlycoBlue 2 µl
   - Absolute ethanol 600 µl

Maintain at −80 °C for 30 min, then microcentrifuge for 30 min at maximum speed, and 4 °C.

6. Wash with 70% ethanol and resuspend the ds cDNA pellet in 70 µl water. Set aside 5 µl to run on an agarose gel to determine whether the ds cDNA synthesis was successful (a faint smear should be visible).

5.2.1.8 **Further Processing of ds flcDNA**

1. To remove the polyA tail, set up the following reaction in a 1.7-ml microcentrifuge tube:

   - ds flcDNA (from step 6 of Section 2.1.7) 65 µl
   - 10× Tango buffer (Fermentas) 8.6 µl
   - 10× SAM (optional, but stimulates GsuI activity) (NEB) 8.6 µl

The commercially available 32 mM SAM is diluted in water to a 500 uM working solution, which is referred to as 10× SAM.

   - 10 mg/ml BSA (NEB) 1 µl
   - 5 U/µl Gsu I (Fermentas) 2 µl

Note that the isoschizomer Bpm I should not be used, as it is insensitive to methylation and may therefore cut within the ds flcDNA.

   - Deionized water 0.8 µl

Digest for 4 h to overnight at 30 °C (not 37 °C).

2. To form the Not I-cohesive 5′ terminal site, add the following to the reaction (final volume 100 μl):

   - 10× Tango buffer 11.4 μl
   - 10 U/μl Not I (NEB) 2.6 μl

Incubate for 4 h at 37 °C. Heat-inactivate the enzymes by incubating for 15 min at 65 °C and then placing the tube in ice.

5.2.1.9 cDNA Size Fractionation

Fractionation of the ds cDNA using commercial gel-filtration mini-columns is carried out to remove all excess adapters, enzymes, and small digestion products prior to insertion into the selected cloning vector.

1. Prepare one cDNA size fractionation column (Invitrogen) as follows:

Allow column to equilibrate to room temperature. Remove the top cap first then the bottom, and allow column to drain. Add 0.8 ml of 1× TEN buffer (see Reagents) and allow to drain completely. Repeat three times (this process will take ~1 h).

2. Label 20 1.7-ml microcentrifuge tubes to be used for fraction collection.
3. Add 100 μl of digestion reaction (step 3 of Section 2.1.8; adjust volume with 1× TEN buffer) to the prepared column. Collect the entire flowthrough in collection tube 1.
4. Add 100 μl 1× TEN buffer and collect the entire flowthrough in tube 2.
5. Add another 100 μl 1× TEN buffer and start collecting single drops, 1 drop (~35 μl) per tube, for subsequent tubes 3–20. Allow column to drain completely before adding another 100 μl 1× TEN buffer.
6. Measure the absorbance at 260 nm of each fraction, preferably using a Nanodrop instrument, which requires only 1 μl of sample, or by PicoGreen fluorimetry (Invitrogen).
7. Run 5 μl of each fraction on an agarose gel to assist in determining which fractions are suitable for cloning (usually fractions 7 through 10).
8. Pool desired fractions, and precipitate with ethanol to concentrate the ds flcDNA for ligation to the vector. Wash pellet with 70% ethanol and resuspend pellet in 6 μl EB buffer (Qiagen).

5.2.1.10 Cloning of flcDNA in pGIS4a Vector

1. The pGIS4a vector (see Figure 5.2) must be prepared and validated prior to use in library construction. It contains an ~800-bp stuffer that must be excised by Not I and Bse RI digestion, which simultaneously creates a Not I-cohesive site to receive the 5′-adaptor terminal region of the cDNA, and a TToverhang ready to receive the corresponding AA overhang at the 3′-end of the cDNA insert. Set up the digestion as follows:

• Purified pGIS4a plasmid DNA	10 μg
• 10× NEBuffer 3 (NEB)	20 μl
• 10 U/μl Not I (i.e. fourfold excess of enzyme)	4 μl
• 4 U/ul Bse RI (i.e., fivefold excess of enzyme)	12.5 μl
• 10 mg/ml BSA	2 μl
• Deionized water	to 200 μl

Incubate for 3 h (maximum!) at 37 °C, no longer because Bse RI exhibits some non-specific activity.

2. Purify the Not I/Bse RI-digested pGIS4a by agarose gel extraction using a Qiagen Gel Extraction Kit or similar. Quantitate using a Nanodrop or other spectrophotometer, and resuspend at 40 ng/μl in EB buffer.

3. Validate the prepared vector by performing the following ligations:

   (a) Not I/Bse RI-digested pGIS4a vector only, no-ligase control (this tests for the presence of contaminating uncut vector).
   (b) Not I/Bse RI-digested pGIS4a vector self-ligation (self-ligation background control).
   (c) Not I/Bse RI-digested pGIS4a with test-insert (positive control).

An appropriate test-insert would contain a Not I-cohesive site at one end, and an AA overhang at the other. We use a synthetic adaptor made by annealing the oligonucleotides pGIS7/4VERIF-TOP and pGIS7/4VERIF-BOT (see Figure 5.3). Incubate overnight (12–16 h) at 16 °C, then heat-inactivate at 65 °C for 10 min.

4. Adjust volume of ligation reaction to 200 μl with deionized water, extract with phenol-chloroform (pH 7.9) and precipitate with ethanol. Wash the pellet at least twice with 70% ethanol to remove salt, and resuspend in 20 μl EB buffer.

5. Transform 25 μl of electrocompetent cells with 1 μl of each purified ligation mix, recover in 1 ml of SOC medium (see Reagents) and plate 20–50 μl (out of 1 ml) of the culture on Lennox LB-amp agar plates (see Reagents). There should be zero or very few colonies for ligations (a) and (b) above, and many for the positive control (c).

6. Using prepared, validated pGIS4a vector, set up the following ligation reaction, on ice, in a 1.7-ml microcentrifuge tube as follows:

• 40 ng/μl Not I/Bse RI-cut pGIS4a	1 μl
• ds flcDNA fraction(s) (100 ng minimum; step 8 of Section 2.1.9)	6 μl
• 5× ligase buffer with PEG (Invitrogen)	2 μl
• 5 U/μl T4 DNA ligase (Invitrogen)	1 μl

Also set up a vector self-ligation control. Incubate overnight (12–16 h) at 16 °C, then heat-inactivate at 65 °C for 10 min.

7. Adjust volume to 200 μl with deionized water, extract with phenol–chloroform (pH 7.9) and precipitate with ethanol. Wash the pellet at least twice with 70% ethanol to remove salt, and resuspend in 20 μl EB buffer.

8. Transform competent cells (we use OneShot electrocompetent TOP10 *E. coli* cells from Invitrogen) with 1 µl of the 20 µl purified ligation by electroporation. Recover with 1 ml of SOC medium held at room temperature (see Reagents) by shaking at 200 rpm for 1 h at 37 °C using 15-ml Falcon tubes.
9. Plate 20–50 µl (out of 1 ml) on Lennox LB-amp agar plates for quality control (QC) screening and library efficiency calculations. Incubate overnight at 37 °C.

5.2.1.11 Perform QC on flcDNA Library

1. Count the numbers of colonies and determine the library efficiency taking into consideration the self-ligation background. Pick colonies (24–48 colonies are sufficient and convenient) for screening by PCR using primers PMR011 and PMR012 (see Figure 5.3). Analyze PCR products by agarose gel electro-phoresis.

If PCR shows a satisfactory range of insert sizes (typically a range of products from 200 to 5000 bp; there should not be a predominance of a single-sized band), pick one to four 96-well plates of colonies for DNA sequencing to determine full-length efficiency (by BLAST alignment of sequences against the GenBank nr database). The quality of the flcDNA library can also be assayed using a PCR-based commercial cDNA Integrity Kit (KPL).

2. Store the library in the form of the purified ligation mix (step 2 of Section 2.1.10) frozen indefinitely at −80 °C) until ready to proceed to the construction of the Single-PET library.

5.2.2 Construction of a Single-PET Library

In this part of the protocol, the flcDNA library is expanded by growing on Lennox LB-amp agar instead of liquid culture, to minimize competition and thus preserve the representation of gene expression. Mme I digestion of plasmid DNA results in the retention of 5′- and 3′-terminal tags and the removal of most of the intervening DNA from each flcDNA insert in pGIS4a. The plasmids are then re-circularized and transformed to give the single-PET library.

5.2.2.1 Plasmid DNA Preparation

1. Transform all of the remaining purified ligation mix (from step 2 of Section 2.1.10) and amplify once by plating an appropriate number of clones (70 000–100 000 cfu/tray to allow growth of each colony without excessive overcrowding) on large (22 × 22 cm) plates (Genetix Q-trays) containing Lennox LB-amp agar. Use a maximum of 700 µl of culture per Q-tray, and plate 10–15 trays. Incubate overnight at 37 °C.

The number of colonies required is determined by the estimated transcriptome size; we assume here that 1 million flcDNA clones provides sufficient coverage. Although only a small fraction of the plasmid DNA obtained after solid-phase amplification is subsequently used for Mme I digestion, it is still critical for the sake of proper representation and complexity to first obtain the benchmark 1 million cfu.

2. Harvest the resulting bacterial colonies by manually scraping into Lennox LB medium (20–30 ml per Q-tray) using disposable plastic 'hockey-puck' spreaders (e.g. Lazy-L; Sigma), and transfer to 500-ml plastic centrifuge bottles. Centrifuge the cells for 20 min at 4000 g, and 4 °C, in a floor-standing ultracentrifuge.

3. Prepare plasmid DNA using the HiSpeed Plasmid Maxi kit (Qiagen) or any other preferred method, and quantify the amount of DNA recovered by spectrometry.

Ten Q-trays of scraped bacteria usually produce ~1 mg of plasmid DNA. Therefore, in theory, two Qiagen Maxi tips (per library) should be sufficient. However, to avoid clogging, we usually use three to four tips per library. Suggested volumes of buffers to use with bacteria from 10 Q-trays are 20–50 ml each of P1, P2, and P3 buffers, instead of the standard manufacturer's conditions.

5.2.2.2 **Tagging by Mme I Digestion**

1. Digest ~10 μg plasmid DNA using Mme I. It is important to ensure that the enzyme is always present in less than fourfold excess to prevent methylation-induced inhibition of digestion. Suggested reaction conditions are as follows:

   - Approximately 10 μg plasmid DNA (from step 3 of Section 2.2.1) 100 μl
   - 10× NEBuffer 4 (NEB) 20 μl
   - 10× SAM (500 μM) 20 μl
   - 2 U/μl MmeI 12 μl
   - Deionized water 48 μl

Incubate for 4 h to overnight at 37 °C, and run an aliquot on an agarose gel to determine the efficiency of the restriction digestion. A strong band of ~2800 bp is the desired linear single-PET plasmid DNA.

2. Purify the entire digestion reaction on a 0.7% agarose gel, loading the digestion products in as few lanes as possible to facilitate excision. Run controls on the same gel, comprising uncut as well as linearized pGIS4a cloning vector, to ensure that the correct band is excised. It is critical to excise only the band corresponding to the linear single-PET DNA, and to avoid contamination with uncut flcDNA plasmid DNA.

3. Excise the ~2800 bp linear single-PET plasmid DNA band and purify using a Qiaquick gel extraction kit (Qiagen) or similar. Quantify the amount of DNA recovered from the gel by spectrometry.

5.2.2.3 **Intramolecular Circularization to Create Single-PET Plasmids**

1. The two-base 3′-overhangs created by Mme I digestion must be blunted as follows:

- Approximately 0.5–2.0 μg DNA 50 μl
- 10× Tango buffer 6 μl
- 0.1 M DTT 0.3 μl
- T4 DNA polymerase use 5 U/μg DNA
- 10 mM dNTP mix 0.6 μl
- Deionized water to 60 μl

Incubate for 5 min at 37 °C, then inactivate for 10 min at 75 °C.
Alternatively, the End-It blunting kit (Epicentre) can be used.

2. Adjust volume to 200 μl with deionized water, extract with phenol–chloroform (pH 7.9) and precipitate with ethanol. Wash the pellet with 70% ethanol, and resuspend in an appropriate volume of EB buffer so that the final DNA concentration is ~2 ng/μl. The exact concentration is not important, it is only important that it is dilute enough to favor intramolecular ligation.

3. Set up the self-ligation reaction on ice as follows:

- 100 ng DNA 50 μl
- 5× ligation buffer with PEG (Invitrogen) 20 μl
- 5 U/μl T4 DNA ligase 1 μl
- The concentration of DNA in the ligation reaction should be 1 ng/μl or less.

Incubate overnight (12–16 h) at 16 °C, then heat-inactivate at 65 °C for 10 min.

4. Adjust volume to 200 μl with deionized water, extract with phenol–chloroform (pH 7.9) and precipitate with ethanol. Wash the pellet at least twice with 70% ethanol to remove salt, and resuspend in 20 μl EB buffer.

5.2.2.4 **Transform Cells**

1. Transform competent cells with 1 μl of the purified ligation reaction and plate as in steps 3–4 of Section 2.1.10. Due to the (usually) much higher library titers, it may be necessary to plate the transformed cells at higher dilutions to facilitate subsequent counting.

5.2.2.5 **Perform QC on GIS Single-PET Library**

1. Count the numbers of colonies and determine library efficiency taking into consideration the self-ligation background. Pick colonies (24–48 colonies are sufficient and convenient) for screening by PCR using primers PMR011 and PMR012 (see Figure 5.3). Analyze PCR products by agarose gel electrophoresis.

It is important that the titer of the single-PET library is high (typically, even a 1 : 1000 plating is nearly confluent) as this indicates successful self-circularization and that the resulting library will be of sufficient complexity. PCR should show a single band of ~300 bp in size in >90% of the samples, indicating the presence of a single PET insert in each plasmid. A vector-only PCR control will give a band of about 250 bp.

2. Store the library in the form of the purified ligation mix (from step 4 of Section 2.2.3), frozen indefinitely at −80 °C).

At this stage, if Sanger sequencing is to be used to obtain ditag data, proceed to Section 2.3 'Construction of a GIS-PET sequencing library for Sanger sequencing of ditags'. If the ditags are to be sequenced using a GS20 sequencer, proceed to Section 2.4 'Construction of diPETs for 454-sequencing'.

5.2.3 Construction of a GIS-PET Sequencing Library for Sanger Sequencing of Ditags

5.2.3.1 Single-PET Plasmid DNA Preparation

1. For large-scale plating, based on the number of colonies observed in step 1 of Section 2.2.5, spread enough of the remaining transformed bacterial culture onto an appropriate number of Lennox LB-amp Q-trays to obtain at least the same number (or greater) of colonies as in the original flcDNA library (i.e. usually ~1 million colonies). This ensures that the library remains representative of the original sample. Incubate overnight at 37 °C.

5.2.3.2 Bam HI-Digestion to Release Single-PETs

1. Digest 500 μg or more plasmid DNA using Bam HI. Suggested reaction conditions are:

• 500 μg plasmid DNA	x μl
• 10× BamHI buffer (NEB)	100 μl
• 10 mg/ml BSA	10 μl
• 20 U/μl Bam HI (NEB)	50 μl
• Deionized water	to 1 ml.

Dispense in 100-μl aliquots for enhanced digestion efficiency. Incubate overnight (12–16 h) at 37 °C.

There is no upper limit to the amount of DNA that can be cut here. It is advisable to cut as much plasmid DNA as possible to obtain a large quantity of PETs.

2. After the digestion, pool all the aliquots, then re-dispense in 200-ul aliquots for convenience, and extract with phenol–chloroform (pH 7.9). Then precipitate with ethanol as follows:

• Bam HI-digested single-PET plasmid DNA	200-μl aliquots
• 3 M sodium acetate, pH 5.2	20 μl
• 1 M $MgCl_2$	4.5 μl

The addition of $MgCl_2$ enhances the precipitation of the short PET DNA fragments.

• 15 mg/ml GlycoBlue	2 μl
• Absolute ethanol	600 μl

Maintain at −80 °C for 30 min, then microcentrifuge for 30 min at maximum speed, and 4 °C. Wash with 70% ethanol, and resuspend all the pellets in a combined total of 500 μl EB buffer, which is suitable for purification using PAGE.

5.2.3.3 PAGE-Purification of 50-bp BamHI-Cohesive Single PETs

Although 2% agarose gel extraction can be used for PET purification, PAGE is preferred as the higher resolving power results in fewer impurities. We use a Hoefer vertical electrophoresis system with 15 × 15 cm gels.

1. Cast a 15 × 15 cm, 1.5-mm thick, 15 well, 10% polyacrylamide gel according to the following recipe:

   - 40% acrylamide/bis (29 : 1) solution (Bio-Rad) 10 ml
   - 5× TBE buffer 8 ml
   - TEMED 16 μl
   - Deionized water 21.6 ml

When ready to pour, add 0.4 ml of freshly-prepared 10% APS solution. Cast the polyacrylamide gel, and allow it to set at room temperature for at least 1 h. If desired, the gel can be kept at 4 °C for several days for later use.

2. Load a maximum of 20 μg DNA per well (with 6× bromophenol blue loading dye; see Reagents), as excessive DNA results in fluorescence quenching that interferes with DNA excision. Any remaining DNA can be loaded in a separate gel, or frozen for later PET purification. Also load appropriate DNA ladders (such as the 25-bp DNA ladder from Invitrogen, or the Wider Range DNA Ladder from Takara) in separate wells.

3. Electrophorese at 200 V (constant) until the bromophenol blue band has almost reached the bottom of the gel. Stain the gel with SYBR Green I for 30 min and visualize on a Dark Reader blue-light transilluminator (Clare Chemical).

It is preferable to use the Dark Reader transilluminator for visualization, as exposure to UV light (especially short-wavelength UV) will damage DNA. If this is not possible, at least ensure that long-wavelength (365 nm) UV light is used.

4. Excise the 50-bp Bam HI-cohesive PETs and collect gel fragments in 0.6-ml microcentrifuge tubes that have each been pierced at the bottom with a 21-G needle. Use DNA from two lanes per pierced-tube. Place the pierced tubes inside standard 1.7-ml microcentrifuge tubes and microcentrifuge for 5 min at maximum speed, and 4 °C.

The gel pieces are conveniently shredded and collected at the bottom of each 1.7-ml microcentrifuge tube.

5. Add 300 μl of 5 : 1 (v/v) LoTE buffer/7.5 M ammonium acetate to each tube. Elute the PETs from the gel by first heating at 65 °C for 30 min, then leaving each tube overnight (12–16 h) at 4 °C and finally re-heating for 30 min at 65 °C.

6. Separate the supernatant (containing eluted 50-bp PETs) from the gel pieces with the aid of microspin plastic centrifuge tube filter units (e.g. SpinX (Costar) or Mermaid (Bio 101)). Using a 1-ml pipet tip, aspirate the liquid and gel from the tubes in step 5 into the filter units (contents of two tubes into each filter column), then microcentrifuge for 10 min at maximum speed, and 4 °C.

The procedure from steps 4–6 is known as the 'gel-crush method' of DNA purification.

7. Pool the collected supernatants, extract with phenol–chloroform and precipitate with ethanol as in step 2 of Section 2.3.2. Resuspend the pellets from all the tubes in a combined total of 20 μl EB buffer, and quantify by PicoGreen fluorimetry or, if available, by using an Agilent BioAnalyzer with a DNA 1000 kit.

The presence of GlycoBlue (or glycogen) at this stage precludes spectrophotometric quantitation. It is advisable to examine the quality of the eluted PETs by running a small portion on a polyacrylamide minigel before proceeding to concatenation.

5.2.3.4 **PET Concatenation**

1. Set up the following ligation reaction:

   - 50 bp cohesive PETs — 200–1000 ng
   - 10× ligase buffer with spermidine (see Reagents) — 1 μl
     Addition of spermidine is important for enhancing ligase activity and favors linear concatenation.
   - 5 U/μl T4 DNA ligase — 1 μl
   - Deionized water — to 10 μl

Incubate for 30 min to overnight at 16 °C. Heat-inactivate at 65 °C for 10 min.

The ligation time must be optimized empirically by running an aliquot on a polyacrylamide minigel. Over-concatenation will result in excessively high-molecular-weight DNA that is difficult to clone. Aim for a smear of DNA averaging around 1 kb.

5.2.3.5 **Purification of Concatenated PETs**

1. Purify the ligation reaction using the Qiaquick PCR purification kit. Elute DNA with 50 μl EB buffer. Quantify DNA by Nanodrop or other spectrophotometer.

2. Concatenated PETs are now partially re-digested with Bam HI to ensure the presence of cohesive termini suitable for insertion into the Bam HI-digested pZErO-1 vector. This step is critical to the success of concatemer cloning. This Bam HI re-digestion is performed only on BamHI cohesive-ended concatemers. Do not exceed the 30-min digestion time as this will destroy the concatemers. It is helpful initially to visualize the digestions at several time points on a polyacrylamide minigel.

Perform a partial BamHI re-digest as follows:

• Spin-purified concatemer DNA	50 μl
• 10× Bam HI buffer	6 μl
• Diluted Bam HI enzyme	Use 1–3 U/μg DNA
• 10 mg/ml BSA	1 μl
• Deionized water	to 60 μl

Incubate for 30 min at 37 °C, and rapidly add 12 μl of 6× bromophenol blue loading dye, heat for 15 min at 65 °C, and then chill on ice.

3. Load the 72 μl of concatemer DNA (containing loading dye) into as few wells of a 10% polyacrylamide minigel as possible. Electrophorese for ~1 h at 200 V (constant) or until the bromophenol blue tracking dye is at the bottom of the gel. Stain for 15–30 min in SYBR Green I and visualize on a Dark Reader transilluminator for band excision.

4. Excise the concatenated DNA in three separate fractions by molecular weight: low (400–1000 bp), medium (1000–2000 bp), and high (>2000 bp). Avoid collecting the DNA trapped within the wells.

5. Extract DNA from each gel slice by the gel-crush method (step 4 of Section 2.3.3): place the gel slice of each excised size fraction into a pierced 0.6-ml microcentrifuge tube and proceed as described previously.

6. Separate the supernatant (containing eluted concatenated PETs) from the gel pieces with the aid of microspin filter units.

7. Extract each eluted size fraction with phenol–chloroform and precipitate with ethanol. Wash each pellet with 70% ethanol and resuspend each in 6 μl EB or LoTE buffer.

If there is more than one pellet per size fraction, pool pellets from the same size fraction and resuspend in a total combined volume of 6 μul LoTE or EB buffer.

5.2.3.6 **Cloning Concatenated PETs in pZErO-1 Vector**

Any general-purpose cloning vector that permits insertion within a Bam HI site can be used for cloning and sequencing the concatenated PETs. The reason pZErO-1 was chosen is because it positively selects for plasmids with inserts: bacterial cells containing empty vectors are killed (refer to Invitrogen for more details).

1. The pZErO-1 vector (Invitrogen) must be prepared prior to use in library construction. Set up the Bam HI digestion as follows:

• pZErO-1 plasmid DNA	2 vg
• 10× Bam HI buffer	5 μl
• 20 U/μl Bam HI	0.5 μl
• 10 mg/ml BSA	0.5 μl
• Deionized water	to 50 μl

Incubate for 2 h at 37 °C. Extract with phenol–chloroform and precipitate with ethanol, wash the pellet with 70% ethanol, and resuspend in EB or LoTE buffer at a concentration of 33 ng/μl. It is not necessary to further purify the vector preparation. Ideally, the preparation should be validated before use by carrying out a vector self-ligation step, as well as inserting a suitable test fragment, to determine the cloning efficiency and background.

The 1-kb-plus DNA ladder from Invitrogen comprises Bam HI-cohesive fragments, and is convenient for use as a test-insert.

2. Set up the ligation as follows:

• Concatemer DNA fraction (from step 7 of Section 2.3.5)	6 μl
• BamHI-digested pZErO-1 DNA	1 μl
• 5× ligase buffer with PEG	2 μl
• 5 U/μl T4 DNA ligase	1 μl

Also set up a vector self-ligation in parallel as a control. Incubate overnight (12–16 h) at 16 °C, then heat-inactivate at 65 °C for 10 min.

3. Adjust volume to 200 μl with deionized water, extract with phenol–chloroform (pH 7.9) and precipitate with ethanol. Wash the pellet at least twice with 70% ethanol to remove salt, and resuspend in 20 μl EB buffer.

5.2.3.7 **Transform Cells**

1. Transform 25 μl electrocompetent cells (we use OneShot electrocompetent TOP10) with 1 μl of the purified ligation reaction. In contrast to the previous transformations in this protocol, recovery of the transformed cells is accomplished using 1 ml of Lennox LB medium, not SOC, because we have found that higher titers are achieved at this step.

2. Plate 20–50 μl (out of 1 ml) on a small Lennox LB agar plus Zeocin (25 μg/ml final Zeocin concentration) (see Reagents) plate. Incubate overnight at 37 °C.

It is convenient to use pre-mixed powdered media, such as imMedia Zeo agar (Invitrogen).

5.2.3.8 **Carry out QC on GIS-PET Sequencing Library**

1. Count the numbers of colonies and determine library efficiency taking into consideration the self-ligation background. Pick colonies (24–48 colonies are sufficient and convenient) for screening by PCR using primers PMR011 and PMR012. Analyze PCR products by agarose gel electrophoresis

If PCR shows a satisfactory range of insert sizes, pick one to four 96-well plates of colonies for overnight culture (in Lennox LB medium plus 25 μg/ml Zeocin) and sequencing to determine the integrity and average number of PETs per insert.

2. Store the library in the form of a purified ligation mix frozen at either −20 or −80 °C until required for large-scale transformations, plasmid extractions, and sequencing of GIS-PETs for complete transcriptome characterization. See Section 5.3 of this protocol for details of data analysis.

5.2.4 Construction of diPETs for 454-Sequencing

The procedures in this section should only be undertaken if the user intends to obtain ditag data using 454-sequencing. Here, asymmetric PETs with only one cohesive site each are extracted from the single-PET library. Self-ligation of these asymmetric PETs results in the formation of dimerized PETs (diPETs) that fit conveniently within the GS20 sequencing read-length of ~100 bp.

5.2.4.1 Single-PET Plasmid DNA Preparation

1. For large-scale plating, based on the number of colonies observed in step 1 of Section 2.2.5, spread enough of the remaining transformed bacterial culture on an appropriate number of Lennox LB-amp Q-trays to obtain at least the same number (or greater) of colonies as in the original flcDNA library (i.e. usually ~1 million colonies). This ensures that the library remains representative of the original sample. Incubate overnight at 37 °C.

5.2.4.2 Bse RI Linearization of Single-PET Plasmid DNA

1. Set up the following digestion reaction:

• Single-PET plasmid DNA (from step 1 of Section 2.4.1)	1000 μg
• 10× NEBuffer 2 (NEB)	400 μl
• 10 mg/ml BSA	40 μl
• 4 U/μl Bse RI (i.e. fourfold excess of enzyme)	1 ml
• Deionized water	to 4 ml

Incubate for 3 h (maximum!) at 37 °C. For efficient enzymatic digestion, we dispense 100-μl aliquots of the reaction mix into individual tubes. Do not incubate for >3 h, as this results in non-specific bands. Run a 1% agarose gel to check the quality and completeness of the digestion, with uncut single-PET plasmid DNA as a sizing control.

2. Extract the entire 4 ml digestion reaction with an equal volume of (25:24:1) phenol/chloroform/IAA, pH 7.9, then collect the upper aqueous layer and precipitate 500-μl aliquots (for convenience) with isopropanol as follows:

• Bse RI-digested, linearized single-PET plasmid DNA	500 μl
• 3 M sodium acetate, pH 5.2	50 μl
• 15 mg/ml GlycoBlue	5 μl
• Isopropanol	600 μl

Maintain at −80 °C for 30 min, then microcentrifuge for 30 min at maximum speed, and 4 °C. Wash with 70% ethanol, and resuspend all the DNA pellets in EB buffer, to a combined total volume of 1.5 ml.

5.2.4.3 BamHI Digestion to Release Asymmetric PETs

Because Bse RI digestion of pGIS4a-derived plasmid DNA results in the formation of non-complementary 3′-AA overhangs that cannot ligate to each other (in the context of Bam HI-released ditags), there is no need to dephosphorylate the Bse RI-released ends.

1. Digest the BseRI-linearized single-PET plasmid DNA using BamHI as follows:

• BseRI-linearized single-PET plasmid DNA (~1000 µg)	500 µl
• 10× Bam HI buffer	100 µl
• 10 mg/ml BSA	10 µl
• 20 U/µl BamHI (i.e. twofold excess of enzyme)	100 µl
• Deionized water	290 µl

Incubate overnight (12–16 h) at 37 °C. For efficient enzymatic digestion, we dispense the reaction mix into tubes each containing a 100-µl aliquot.

2. To facilitate gel loading, the volume of the digestion mixture needs to be reduced by isopropanol precipitation as in step 2 of Section 2.4.2. Wash with 70% ethanol, and resuspend all the DNA pellets in EB buffer, to a total combined volume of 350 µl.

3. Run an aliquot of the reaction mixture on a 2% agarose gel to check the quality and completeness of the digestion, and observe the presence of released asymmetric PETs. If the digestion products are as expected, proceed to the gel-purification of the asymmetric PETs.

5.2.4.4 Recovery and Quantitation of Purified Asymmetric PETs

The purification of asymmetric PETs using agarose gel is described here. While agarose gel purification works well, some users have experienced subsequent difficulty in PET dimerization, which was resolved when PAGE purification was used instead of agarose. This we attribute to the greater purity of PAGE-purified DNA. If desired, refer to Section 2.3.3 for details. Note that if PAGE-purification is preferred, the PETs in step 2 of Section 2.4.3 should be resuspended in a smaller volume of buffer to facilitate gel-loading.

1. Run the entire 350 µl BamHI digestion reaction (from step 2 of Section 2.4.3) on a 2% agarose gel, and carefully excise the 40–50-bp asymmetric PET DNA from the gel. The DNA is eluted from the gel slices by an agarose gel-crush method (similar to that described for polyacrylamide in step 4 of Section 2.3.3) as follows:

Place the excised gel fragments into 0.6-ml microcentrifuge tubes (1 gel slice per pierced-tube) that have been pierced at the bottom with a 21-G needle. Use one gel slice per tube. Place the pierced tubes inside 1.7-ml microcentrifuge tubes and microcentrifuge for 5 min at maximum speed, and 4 °C. The gel pieces are conveniently shredded and collected at the bottom of each 1.7-ml microcentrifuge tube.

2. Add 300 μl of 5 : 1 (v/v) LoTE buffer/7.5 M ammonium acetate to each tube, then freeze at −80 °C (or in a dry-ice/ethanol bath) for 1–2 h. Thaw the tubes at room temperature, then allow the DNA to elute overnight (12–16 h) at 4 °C followed by 30 min to 2 h at 37 °C.

3. Separate the supernatant (containing eluted asymmetric PETs) from the gel pieces using microspin filter units, extract with perform phenol–chloroform and precipitate with ethanol as described in steps.6 and 7 of Section 2.3.3. Resuspend the ethanol-precipitated asymmetric PET DNA pellets in a total volume of 12 μl using LoTE or EB buffer.

4. Quantify the purified PET DNA either by PicoGreen fluorimetry or, ideally, using an Agilent BioAnalyzer with a DNA 1000 kit. A rough estimation of the quantity of PETs recovered is necessary before proceeding to PET dimerization (diPET formation) to ensure that sufficient diPET DNA is produced for GS20 454-sequencing.

5.2.4.5 **Formation of diPETs**

1. The asymmetric PETs have only one Bam HI cohesive site each, and can be dimerized as follows (volumes are illustrative):

• Purified asymmetric single-PET DNA	~2–5 μg
• 10× ligase buffer with spermidine	1 μl
• 5 U/μl T4 DNA ligase	2 μl
• Deionized water	to 10 μl (or 20 μl if PETs are too dilute)

Incubate for 16–30 h at 16 °C, then heat-inactivate for 10 min at 65 °C. Some users have found that extending the ligation time to 30 h or longer results in a more complete dimerization.

2. Electrophorese an aliquot (5% is convenient) of the ligation reaction on a 4–20% polyacrylamide minigel (or, preferably, in an Agilent Bioanalyzer with a DNA 1000 kit) to determine dimerization efficiency. If the dimerization appears mostly complete (the benchmark we set is 5% or less of unligated single-PETs), proceed to step 3. However, if dimerization was poor, it will be necessary to purify the desired diPETs by PAGE (as in Section 2.3.3).

3. Adjust volume of ligation reaction to 200 μl with deionized water, extract with phenol–chloroform and precipitate with ethanol. Resuspend the purified diPET DNA pellet in 20 μl of EB buffer, and end-blunt as in step 1 of Section 2.2.3, or using the End-It blunting kit from Epicentre. Repeat phenol–chloroform extraction and ethanol precipitation. The diPET DNA is now ready to be processed for 454-sequencing according to the manufacturer's Library Preparation Protocol (not described here).

5.3 Data Analysis

For the purposes of extracting PETs from raw sequence data and mapping them onto the appropriate genome assembly, we developed the PET-Tool software suite [33] that comprises four modules: the Extractor module for PET extraction; the Examiner module for analytic evaluation of PET sequence quality; the Mapper module for locating PETsequences in the genome sequences; and the ProjectManager module for data organization. The salient feature of this software is the mapping algorithm based on Compressed-Suffix Array that was ~60× faster than BLAST for mapping ditags to the genome (unpublished results). For details on ditag mapping criteria, and the subsequent visualization of the mapped PETs on the UCSC genome browser, please refer to the Supplementary Information accompanying the GIS-PET publication [29]. Figure 5.4 shows an example of the mapping of PETs to a reference genome assembly.

PET-Tool is freely available for academic users and can be downloaded from http://www.gis.a-star.edu.sg/PET_Tool/.

5.4 Discussion

A complete and rigorous transcriptome analysis requires that every transcript in the sample of interest be characterized. This includes obtaining information about the expression level of each splice variant, distinguishing the structure of each splice

(A)

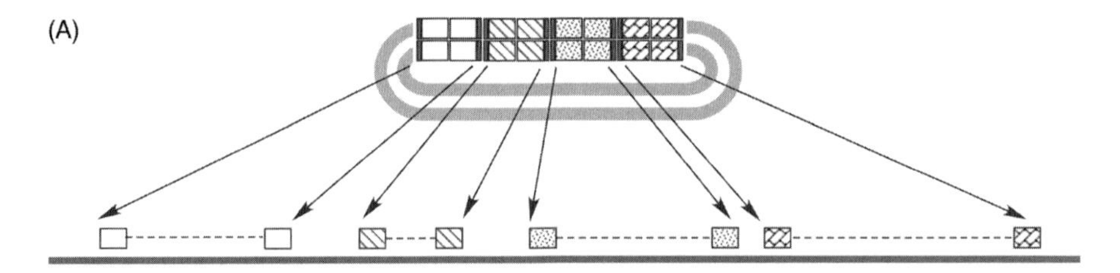

(B)

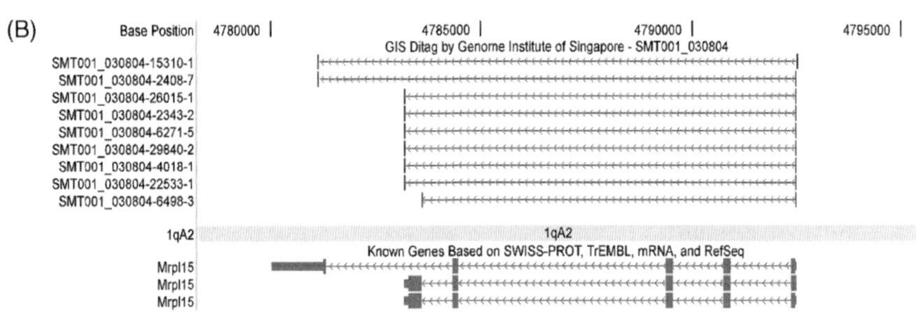

Figure 5.4 Mapping of PETs on a reference genome assembly. (A) Schematic of the mapping process. (B) An example of nine PETs from GIS-PET library SMT001 mapping to various known splicing variants of *Mrpl15*.

variant, precisely locating the TSS, TTS and exon/intron junctions of each transcript on the genome, and identifying novel transcripts or transcripts displaying unusual properties. These include ncRNAs, non-polyadenylated RNAs and rare transcripts such as those derived from intergenic- and *trans*-splicing.

We previously validated the GIS-PET procedure by using it to study the transcriptome of the mouse E14 embryonic stem cell-line [29]. We found that >70% of the 63 467 unique PETs that were identified were immediately mapped to the mouse genome assembly (UCSC mm3; http://genome.ucsc.edu/), and this figure rose to >80% when we allowed for single-base polymorphisms. We feel that this may still be an underestimate of the true accuracy of GIS-PET, since the E14 cell-line was derived from a different mouse strain than the one used in the UCSC mm3 reference genome. The mapped PETs spanned known genes, predicted genes, previously unknown transcripts, and genes for which only EST information was available in the public domain. Of the PET-identified transcripts from all categories, 94% could be verified by PCR and sequencing. Interestingly, GIS-PET enabled the identification and verification of what we called 'unconventional fusion transcripts', such as one which appeared to be a product of *trans*-splicing between the genes *Ppp2r4* and *Set*, resulting in a hitherto unidentified TU which coded for a novel fusion ORF containing elements of both genes. Where the quantitation of gene expression is concerned, a comparison of tag-counts obtained from the same system by EST analysis and by GIS-PET revealed that the results were largely similar ($r = 0.75$), though not identical. This was to be expected due to unavoidable bias during the multi-step bacterial cloning process, but precautions such as solid-phase library amplification minimized any inherent bias.

As with other methods which allow the extraction of positionally-defined tags, GIS-PET relies entirely on the quality of the flcDNA library that is constructed. We had selected the Cap-Trapper [32] flcDNA procedure because it enabled us to routinely obtain >90% full-length clones, but this method is technically demanding and lengthy, and requires a large amount of starting mRNA (~10 ug). If ease of use is preferred, and/or sample quantities are limited, there are many other methods available that can easily be adapted for use in GIS-PET with some sacrifice in full-length quality. However, it should be kept in mind that a good flcDNA library facilitates subsequent functional validation of any transcripts of interest, as full-length clones can be easily recovered from the library by simple PCR.

Another potential limitation of the GIS-PET procedure is the use of a plasmid cloning vector (pGIS4a) that may result in the under-representation of long transcripts. Again, if desired, this can be resolved by using an alternative cloning system with some modification on the part of the user.

If the MS-PET option is adopted, the user should be aware that the homopolymer errors that are a characteristic of 454-sequencing will result in a first-pass mapping rate that is lower than that obtained from Sanger-sequenced ditag data [31].

Finally, although GIS-PET is superior by far to single-tag methods in defining the exact boundaries of TUs, like any other tagging method it is still unable to present information on internal gene structure, so splicing variants differing in internal exons cannot be distinguished from full-length transcripts. This remains the province of flcDNA cloning and sequencing. However, for the purposes of identifying novel transcripts, particularly unconventional transcripts produced by rare biological events such as *trans*- and intergenic splicing, GIS-PET offers the ideal platform.

In terms of efficiency, a single read of 700–1000 bases using a Sanger-based sequencer should reveal 15 PETs (equivalent to 15 transcripts). By contrast, to obtain information from the same number of transcripts by flcDNA sequencing, each of the 15 transcripts would need to be sequenced from each end, that is, 30 reads of 700–1000 bases each. Hence, GIS-PET is 30-fold more efficient than flcDNA sequencing for demarcating TUs. If the user has access to a GS20 sequencer, the MS-PET modification [31] can be employed. Taking into account the throughput of the GS20 machine relative to an ABI 3730xl, MS-PET results in a further 100-fold increase in efficiency compared to GIS-PET. Expressed in another way, the profiling of 1 million transcripts would require 2 million Sanger sequencing reads using the flcDNA approach, about 66 000 Sanger sequencing reads using GIS-PET, and only two runs using MS-PET. At an estimated cost of US$1 per Sanger sequencing read, compared to US$5000 for an entire GS20 run (capital costs ignored), the savings in terms of cost and time per project are significant.

Proposed improvements to the existing GIS-PET procedure include incremental enhancements such as replacing the tedious and potentially biased bacterial amplification steps with rolling-circle amplification; replacing the manual flcDNA fractionation step with automated HPLC, and developing a RecA-based procedure to rapidly recover full-length clones of interest from the existing plasmid preparation without resorting to PCR. In the longer term, it should be possible to replace the entire GIS-PET protocol with a much simpler, *in vitro* cloning-based procedure that will eliminate any bias caused by bacterial cloning.

5.5 Perspectives

The techniques employed for transcriptome characterization can, broadly speaking, be divided into two categories: sequencing-based approaches, and those relying on array hybridization.

All current tag-based methods including GIS-PET rely on DNA sequencing for data collection. With some technical expertise, any standard molecular biology laboratory can apply the protocol described in this chapter to characterize any transcriptome for which a reference genome assembly exists. However, it is only one instance of the general usefulness of the paired-end-ditagging (PET) concept. Our group has since developed other PET-based applications, such as ChIP-PET for regulome [34] and methylome analysis, and MS-PET for ultra-high-throughput transcriptome and genome analysis. We are currently in the final stages of developing two other PET applications wherein *in vitro* cloning is used for genome assembly, and for the elucidation of long-distance regulatory interactions. Other PET applications that are in the pipeline include methods for identifying polymorphisms, mutations and genomic rearrangements.

It is increasingly obvious that a combination of paired-end read capability, long read-length, high throughput, multiplex-run capability, low cost and ease of template preparation are all attributes of the ideal sequencing system, and already we are seeing commercially available systems that possess many, if not all, of these attributes. Only a few years ago, it would have been unthinkably expensive to consider characterizing a transcriptome by directly sequencing every transcript, but this may change with the advent of the new generation of sequencers.

The Roche/454 GS20 system is now being superseded by the GS FLX, which promises a fourfold improvement in throughput (100 million bases per run) coupled with a twofold enhancement of read-length, to an average of 200–250 bases per template. To accelerate the throughput of this technology, modifications enabling multiple samples to be mixed in one run are the subject of ongoing work, both in our laboratory and by other end-users [35]. The current drawbacks are the high capital outlay, the technically-challenging sample preparation procedure that is prone to cross-contamination, and the lack of a protocol to fully exploit the enhanced read-length of the GS FLX to increase PET-mapping specificity and/or efficiency: the current paired-end kit is limited to extracting 20-bp tags from 2.5-kb DNA targets.

The 1G Genome Analyzer from Solexa (now Illumina) can perform direct paired-end reads on 600 bp-long templates with a throughput of 1 billion bases per 3-day run, which, while somewhat slower than the GS FLX, is far easier to use in terms of sample preparation. There is apparently a modified procedure for longer targets [36]. Competition between all four nucleotides at each addition step reduces the homopolymer errors found in pyrosequencing-based technologies. The current read-length is still poor at 40 bp per read.

ABIs approach to paired-end sequencing used in its SOLiD (Supported Oligonucleotide Ligation and Detection) sequencing technology is straightforward and also based on pyrosequencing: any DNA fragment (up to 8 kb) can be circularized with an adaptor, and subsequent enzymatic manipulations extract 26-bp tags from each terminal for sequencing. The reported throughput is 2–3 million bases per run using two full slides, although no information is available on the time needed per run. Read-length is the poorest among the three next-generation sequencers, only 25 bases per read. However base-calling accuracy is apparently higher due to a two-base encoding system [37].

The various platforms mentioned above are already well-suited to DNA-tag sequencing, and as read-lengths improve and prices fall, direct end-to-end transcript sequencing should become feasible.

Where microarray hybridization is concerned, it can be expected that high-resolution true WGTAs with one-base resolution will eventually become practical to manufacture commercially, the main question being that of demand. Currently, chips containing about 6.5 million oligonucleotide features per array can be purchased, but for a one-direction, one-base resolution human WGTA, approximately 3 billion features would have to be contended with, which works out to about 200 chips per set even if repeat regions are omitted, far too unwieldy and expensive for practical use. The actual number of arrays will of course vary depending on specific chip characteristics and user requirements, but clearly, assuming the number of features per chip doubles annually, it should become feasible to perform experiments on such an array set within 2 to 3 years' time. The point to note though is that, for many purposes except detailed identification of exon/intron junctions or SNPs, lower-resolution arrays already provide sufficiently detailed data. More importantly, for purposes involving the identification of non-contiguous phenomena such as chromosomal rearrangements and long-distance interactions, microarray hybridization will never be the most practical approach, and alternative experimental strategies should be selected.

From the above, it is clear that each approach has its inherent strengths and weaknesses, and no single method currently fulfills all the requirements for complete transcriptome profiling. Until such time as each transcript (including those unconventional transcripts produced by rare events) in a transcriptome can be analyzed individually, a combined approach using two or more technologies would appear to be ideal, within the restrictions of available resources.

Reagents

Modified 10mM dNTP (with 5-Me-dCTP instead of dCTP) (RNase-free)

• 100mM stock solution of dATP (final 10mM)	100 µl
• 100mM stock solution of dTTP (final 10mM)	100 µl
• 100mM stock solution of dGTP (final 10mM)	100 µl
• 100mM stock solution of 5-Me-dCTP (final 5mM)	50 µl
• 1 *M* Tris-HCl, pH 8.0 (final 10mM)	10 µl
• Nuclease-free water	640 µl
• Total volume	1 ml

Saturated trehalose (RNase free).
Place a magnetic stir bar into a small beaker and heat 10 ml water to 42 °C. Slowly add 8 g D-(+)-trehalose dihydrate powder (Sigma) and allow it to dissolve by stirring. While carefully maintaining the temperature at 42 °C, continue adding trehalose powder until saturation is reached. Cool solution to room temperature, allowing trehalose crystals to form. At this stage, add enough diethylpyrocarbonate (DEPC) to give a final concentration of 0.1%. Shake vigorously for 5–10 min (or stir overnight), then autoclave the saturated trehalose solution to deactivate the DEPC. Store saturated trehalose in aliquots at −20 °C.

Because the solubility of trehalose increases with temperature, it is important to maintain the temperature at not more than 42 °C when making this solution. This is the temperature at which the trehalose is used in the protocol.

1× Binding buffer (RNase free)

- 2 M NaCl
- 50 mM EDTA, pH 8.0.

1× Blocking buffer (RNase free)

- 0.4% SDS
- 50 μg/ml yeast tRNA.

1× Wash buffer (RNase free)

- 10 mM Tris-HCl, pH 7.5
- 0.2 mM EDTA, pH 8.0
- 10 mM NaCl
- 20% glycerol
- 40 μg/ml yeast tRNA (Sigma).

Alkaline hydrolysis buffer

- 50 mM NaOH
- 5 mM EDTA, pH 8.0
- Prepare fresh before use.

LoTE buffer

- 3 mM Tris-HCl, pH 7.5
- 0.2 mM EDTA.

1× TEN buffer

- 10 mM Tris-HCl, pH 8.0
- 0.1 mM EDTA, pH 8.0
- 25 mM NaCl.

SOC medium

- 0.5% yeast extract
- 2% tryptone
- 10 mM NaCl
- 2.5 mM KCl
- 10 mM $MgCl_2$
- 10 mM $MgSO_4$
- 20 mM glucose

Lennox LB medium

- 10 g/l Tryptone
- 5 g/l Yeast extract
- 5 g/l NaCl

Lennox LB-amp agar

- LB medium (Lennox)
- 15 g/l Agar-B
- 100 μg/ml ampicillin

Lennox LB agar plus Zeocin

- LB medium (Lennox)
- 15 g/l Agar-B
- 25 μg/ml ampicillin

6× Bromophenol blue loading dye

- 10 mM Tris-HCl pH 8.0
- 0.03% bromophenol blue
- 60% glycerol
- 60 mM EDTA pH 8.0.

10× Ligase buffer with spermidine

- 60 mM Tris-HCl, pH 7.5
- 60 mM $MgCl_2$
- 50 mM NaCl
- 1 mg/ml BSA
- 70 mM 2-mercaptoethanol
- 5 mM ATP
- 20 mM DTT
- 10 mM spermidine (Sigma)

Acknowledgments

The authors thank all contributors to the GIS-PET project. In particular, Chiu Kuo-Ping and Wing-Kin Sung who developed the PET-Tool software, while How Choon Yong, Azmi Ridwan and Atif Shahad developed the T2G browser for mapped PET visualization. Chee Hong Wong and Leonard Lipovich provided additional bioinformatics support.

References

1 Lander, E.S., Linton, L.M., Birren, B. *et al.* (2001) Initial sequencing and analysis of the human genome. *Nature*, **409** (6822), 860–921.

2 Mattick, J.S. and Makunin, I.V. (2006) Non-coding RNA. *Human Molecular Genetics*, **15**, (Spec No 1), R17–R19.

3 Katayama, S., Tomaru, Y., Kasukawa, T. *et al.* (2005) Antisense transcription in the mammalian transcriptome. *Science*, **309** (5740), 1564–1566.

4 Lapidot, M. and Pilpel, Y. (2006) Genome-wide natural antisense transcription: coupling its regulation to its different regulatory mechanisms. *EMBO Reports*, **7** (12), 1216–1222.

5 Cheng, J., Kapranov, P., Drenkow, J. *et al.* (2005) Transcriptional maps of 10 human chromosomes at 5-nucleotide resolution. *Science*, **308** (5725), 1149–1154.

6 Akiva, P., Toporik, A., Edelheit, S. *et al.* (2006) Transcription-mediated gene fusion in the human genome. *Genome Research*, **16** (1), 30–36.

7 Parra, G., Reymond, A., Dabbouseh, N. *et al.* (2006) Tandem chimerism as a means

to increase protein complexity in the human genome. *Genome Research*, **16** (1), 37–44.

8 Horiuchi, T. and Aigaki, T. (2006) Alternative trans-splicing: a novel mode of pre-mRNA processing. *Biology of the Cell*, **98** (2), 135–140.

9 Stamm, S., Ben-Ari, S., Rafalska, I. *et al.* (2005) Function of alternative splicing. *Genetics*, **344**, 1–20.

10 Carninci, P., Sandelin, A., Lenhard, B. *et al.* (2006) Genome-wide analysis of mammalian promoter architecture and evolution. *Nature Genetics*, **38** (6), 626–635

11 Dean, A. (2006) On a chromosome far, far away: LCRs and gene expression. *Trends in Genetics*, **22** (1), 38–45.

12 Trinklein, N.D., Aldred, S.F., Hartman, S.J. *et al.* (2004) An abundance of bidirectional promoters in the human genome. *Genome Research*, **14** (1), 62–66.

13 Bernstein, B.E., Meissner, A. and Lander, E.S. (2007) The mammalian epigenome. *Cell*, **128** (4), 669–681.

14 Mockler, T.C., Chan, S., Sundaresan, A. *et al.* (2005) Applications of DNA tiling arrays for whole-genome analysis. *Genomics*, **85** (1), 1–15.

15 Frith, M.C., Pheasant, M. and Mattick, J.S. (2005) The amazing complexity of the human transcriptome. *European Journal of Human Genetics*, **13** (8), 894–897.

16 Johnson, J.M., Edwards, S., Shoemaker, D. *et al.* (2005) Dark matter in the genome: evidence of widespread transcription detected by microarray tiling experiments. *Trends in Genetics*, **21** (2), 93–102.

17 Kapranov, P., Drenkow, J., Cheng, J. *et al.* (2005) Examples of the complex architecture of the human transcriptome revealed by RACE and high-density tiling arrays. *Genome Research*, **15** (7), 987–997.

18 Johnson, J.M., Castle, J., Garrett-Engele, P. *et al.* (2003) Genome-wide survey of human alternative pre-mRNA splicing with exon junction microarrays. *Science*, **302** (5653), 2141–2144.

19 Okazaki, Y., Furuno, M., Kasukawa, T. *et al.* (2002) Analysis of the mouse transcriptome based on functional annotation of 60,770 full-length cDNAs. *Nature*, **420** (6915), 563–573.

20 Kawai, J., Shinagawa, A., Shibata, K. *et al.* (2001) Functional annotation of a full-length mouse cDNA collection. *Nature*, **409** (6821), 685–690.

21 Brenner, S., Johnson, M., Bridgham, J. *et al.* (2000) Gene expression analysis by massively parallel signature sequencing (MPSS) on microbead arrays. *Nature Biotechnology*, **18** (6), 630–634.

22 Velculescu, V.E., Zhang, L., Vogelstein, B. *et al.* (1995) Serial analysis of gene expression. *Science*, **270** (5235), 484–487.

23 Saha, S., Sparks, A.B., Rago, C. *et al.* (2002) Using the transcriptome to annotate the genome. *Nature Biotechnology*, **20** (5), 508–512.

24 Matsumura, H., Reich, S., Ito, A. *et al.* (2003) Gene expression analysis of plant host–pathogen interactions by SuperSAGE. *Proceedings of the National Academy of Sciences of the United States of America*, **100** (26), 15718–15723.

25 Wei, C.-L., Ng, P., Chiu, K.P. *et al.* (2004) 5′ Long serial analysis of gene expression (LongSAGE) and 3′ (LongSAGE) for transcriptome characterization and genome annotation. *Proceedings of the National Academy of Sciences of the United States of America*, **101** (32), 11701–11706.

26 Hashimoto, S., Suzuki, Y., Kasai, Y. *et al.* (2004) 5′-end SAGE for the analysis of transcriptional start sites. *Nature Biotechnology*, **22** (9), 1146–1149.

27 Gowda, M., Li, H., Alessi, J. *et al.* (2006) Robust analysis of 5′-transcript ends (5′-RATE): a novel technique for transcriptome analysis and genome annotation. *Nucleic Acids Research*, **34e**, 126.

28 Shiraki, T., Kondo, S., Katayama, S. *et al.* (2003) Cap analysis gene expression for high-throughput analysis of transcriptional starting point and identification of promoter usage. *Proceedings of the National Academy of

Sciences of the United States of America, **100** (26), 15776–15781.

29 Ng, P., Wei, C.L., Sung, W.K. *et al.* (2005) Gene identification signature (GIS) analysis for transcriptome characterization and genome annotation. *Nature Methods*, **2** (2), 105–111.

30 Margulies, M., Egholm, M., Altman, W.E. *et al.* (2005) Genome sequencing in microfabricated high-density picolitre reactors. *Nature*, **437** (7057), 376–380.

31 Ng, P., Tan, J.J., Ooi, H.S. *et al.* (2006) Multiplex sequencing of paired-end ditags (MS-PET): a strategy for the ultra-high-throughput analysis of transcriptomes and genomes. *Nucleic Acids Research*, **34e**, 84.

32 Carninci, P. and Hayashizaki, Y. (1999) High-efficiency full-length cDNA cloning. *Methods in Enzymology*, **303**, 19–44.

33 Chiu, K.P., Wong, C.H., Chen, Q. *et al.* (2006) PET-Tool: a software suite for comprehensive processing and managing of Paired-End diTag (PET) sequence data. *BMC Bioinformatics*, **7**, 390.

34 Wei, C.L., Wu, Q., Vega, V.B. *et al.* (2006) A global map of p53 transcription-factor binding sites in the human genome. *Cell*, **124** (1), 207–219.

35 Binladen, J., Gilbert, M.T., Bollback, J.P. *et al.* (2007) The use of coded PCR primers enables high-throughput sequencing of multiple homolog amplification products by 454 parallel sequencing. *PLoS ONE*, **2e**, 197.

36 Karow, J. (2007) As users demand paired-end sequencing, 454, Illumina, and ABI work on new kit. *In Sequence* 1 (9). http://www.in-sequence.com/issues/1_9/features/138789-1.html.

37 Karow, J. (2006) A 'Solid' Debut: ABI Sheds Light on Agencourt's Sequencing Technology. *GenomeWeb Daily News*. http://www.genomeweb.com/issues/news/135186-1.html.

6
High-Throughput Functional Screening of Genes *In Planta*

Thomas Berberich, Yoshihiro Takahashi, Hiromasa Saitoh, and Ryohei Terauchi

Abstract

The function of genes (except for genes encoding structural or regulatory RNAs) is determined by the activities of their encoded proteins. Alterations in the expression and activity of the proteins cause phenotypic changes in a plant during development as well as in response to environmental stimuli. Functional screening for phenotypic changes *in planta*, by randomly expressing a population of cDNAs, can identify novel factors involved in the specific processes under survey.

Transformation of plants by infection with *Agrobacterium tumefaciens* and virus-based expression of the transfected cDNAs have been combined in a method that allows high-throughput functional screening *in planta*. Several studies have demonstrated that high-throughput functional screening with the *Agrobacterium*-mediated virus gene expression system is a powerful tool. A step-by-step protocol is provided starting from cDNA library construction, through to *in planta* expression and observation of phenotypes.

6.1
Introduction

In plants developmental and metabolic changes as well as responses to alterations in the environment are often initiated by differential gene expression, reflected by variations in the transcriptome. Miscellaneous methods are available to examine changes in transcriptomes as described in Part I of this handbook. The function of genes (except for genes encoding structural or regulatory RNAs) is determined by the activities of their-encoded proteins, which then produce phenotypic changes. A survey based on the expression of proteins encoded by a population of cloned cDNAs *in vivo* followed by screening for phenotypic changes has frequently been applied to identify the function of genes in bacteria, yeast and animal cells [1,2]. Such functional

The Handbook of Plant Functional Genomics: Concepts and Protocols.
Edited by Günter Kahl and Khalid Meksem

ISBN: 978-3-527-31885-8

screening in a high-throughput format for *in planta* analysis of randomly expressed cDNAs is described in this chapter.

The application of several findings and inventions in plant biology and biotechnology are the basis of the method described here. Initially, a plant transformation and *in planta* gene expression system is required. Here, the *Agrobacterium tumefaciens*-based transformation system is utilized. *A. tumefaciens* is a gram-negative, non-sporeforming, rod-shaped bacterium which, at the beginning of the 20th century, was identified as the causative agent of crown gall disease [3]. Later it was found that the symptoms of the disease are based on the ability of *A. tumefaciens* to transfer a particular DNA segment (T-DNA) of the tumor-inducing plasmid (Ti-plasmid) into the nucleus of infected cells where it is integrated into the host genome and transcribed. This observation recommended *A. tumefaciens* as a tool for gene transfer into plant cells [4–6]. Some features are especially important for the use of *A. tumefaciens* for plant transformation: the infection process results from transfer and integration of T-DNA and the subsequent expression of T-DNA genes; the T-DNA genes are transcribed only in plant cells and play no role in the transfer process; and any foreign DNA placed between the T-DNA borders can be transferred to plant cells, regardless of the source from which it was derived [7,8]. Consequently vectors and bacterial plant transformation systems were developed and further improved by employing a modified Ti plasmid that had been 'disarmed' by deletion of the tumor-inducing genes [9,10]. With this method stable transformation of plants with *A. tumefaciens* was established. However, for a high-throughput screening approach the traditional method of stable transformation of plants is unsuitable since it is overly time-consuming and laborious. In contrast, transient transformation methods have the advantage of being rapid procedures thus making them appropriate for high-throughput application. A drawback of the commonly used transient transformation systems such as electroporation [11] or particle bombardment [12] however, is that they require extra equipment and that only protoplasts or single cells can be transformed. Consequently it is difficult, if not impossible, to observe phenotypic changes. Fortunately, the *A. tumefaciens*-mediated transformation system has been further developed into an efficient method for transient expression of transgenes by agroinfiltration [13]. Whole intact leaves can easily be transformed by infiltration with a suspension of *A. tumefaciens* cells carrying T-DNA harboring the transgene of interest. Using agroinfiltration a large number of clones can be tested in a screening approach. Throughput is still limited however, since a relatively large leaf sector has to be infiltrated to produce clear phenotypes. This means that a large number of plants and extensive space is required to carry out the screening. Only the combination of agroinfiltration with a virus-based transgene expression system finally led to the development of an *in planta* functional screening method in high-throughput format [14]. The advantage of this method is the use of virus-derived vectors for *in planta* transgene expression, first described for cauliflower mosaic virus (CaMV) [15] and later applied to other viruses such as tobacco mosaic virus (TMV) [16,17] and potato virus X (PVX) [18,19]. The recombinant viruses are able to multiply and disperse in the tissue which leads to a high level of gene expression [17] and restriction of the inoculation site to a small area. Application of the viral transgene expression method *per se* for high-throughput screening is disadvantageous because for each individual cDNA to be tested, infectious particles or transcripts of the recombinant virus have to be created. The two systems, agroinfiltration and virus-based transgene expression have been combined by construction of binary plasmids harboring the viral vector in which cDNAs are placed under the control of a promoter for viral coat protein genes. These binary plasmid vectors can be transferred into plant cells by *A. tumefaciens* where transcription of the infectious recombinant viral RNA is then driven by the 35S CaMV promoter. The transcripts initiate the formation of virus particles that infect the surrounding tissue and express the recombinant cDNAs [14,20]. In one of the first applications of high-throughput functional cloning, cDNAs from the plant pathogen *Cladosporium fulvum* were screened for eliciting hypersensitive response (HR) cell death in tomato plants [14]. From the cDNA library 9600 of the *A. tumefaciens* colonies were individually inoculated with toothpicks onto leaves of tomato plants that were resistant to *C. fulvum* and subsequently four cDNAs were identified whose expression induced formation of necrotic lesions around the inoculation site. One of the cDNAs coded for the known avirulence factor protein, AVR4 which elicits HR in tomato carrying the *Cf4* resistance gene. These results showed that the method could indeed identify cell death-inducing factors by screening for the corresponding phenotype. In another approach using the *Agrobacterium*-mediated virus gene expression system, 16 unique cDNAs predicted to encode secreted proteins from *Phytophthora sojae* were expressed in *N. benthamiana* leaves, resulting in identification of the necrosis-inducing factor PsoNIP [21]. Furthermore, expression in *N. benthamiana* of 63 cDNAs coding for putative extracellular proteins from *Phytophthora infestans* identified two novel necrosis-inducing cDNAs, *crn1* and *crn2* [22]. In a large-scale screening for cell death-causing factors in *N. benthamiana* 40 000 individual *Agrobacterium* colonies were inoculated with toothpicks onto leaf blades of *N. benthamiana* [23]. This screen identified 30 clones which elicited cell death *in planta* including one encoding an ethylene-responsive element binding factor (ERF) [23] and another coding for a mitogen activated protein kinase kinase (MAPKK) [24] both of which are involved in the response of plants to pathogens.

The above-mentioned studies have demonstrated that high-throughput functional screening with the *Agrobacterium*-mediated virus gene expression system is a powerful tool for the identification of novel factors involved in the phenotypic changes that are under investigation.

6.2 Methods and Protocols

Before starting a high-throughput *in planta* screening two questions should be answered: first, how can the phenotypic change under investigation be examined in the testing system? Second, what is the best source of mRNA for the construction of a cDNA library? The most efficient testing systems are those in which the phenotypic changes are visible as in the examples discussed later in Section 6.3, where a grayish area or yellow coloration around inoculation sites indicates cell death and senescence or chlorophyll degradation respectively. Secondary treatment of transformed leaves with drugs, stains, pathogens or elicitors can also be used to detect the phenotypic changes which are under investigation. The source material for construction of a cDNA library should originate from the organism, using either treated or untreated tissue which probably contains the mRNAs which code for the proteins involved in the particular phenotypic changes which are being studied.

As in other methods which employ cDNA, the quality of the mRNA from the tissue of choice used as the starting material is a crucial factor, particularly since only cDNAs comprising full coding regions are able to produce reliable phenotypes after the encoded proteins have been expressed *in planta*. There are many different methods available for the extraction and purification of RNA from plant and other tissues, some of which are described in Part I of this book. In our laboratory we routinely follow the protocols for TRI Reagent (Sigma) or TRIzol (Invitrogen) and the RNAeasy Plant Kit (Qiagen) to isolate RNA from plant and fungal tissues. These methods give high quality total RNA from which poly(A)$^+$-RNA can be further purified on the oligo-dT matrix contained in mRNA purification kits from several manufacturers. The mRNA is then converted into double stranded cDNA using commercially available synthesis and cloning kits which produce a population of cDNAs bearing the 5′- and 3′-overhangs necessary for direct ligation into potato virus X (PVX)-based binary plasmid vectors. The cDNA library is finally transformed into *Agrobacterium* cells which are then plated on agar containing appropriate antibiotics for selection of transformants. Single colonies are transferred to 96-well microtiter plates containing growth medium. These bacterial cultures are inoculated onto leaves with toothpicks. The clones that display a phenotype are subsequently used for agroinfiltration of half a leaf to confirm the positive result. The plasmids from bacterial clones which show the selected phenotype after *in planta* expression are recovered and the nucleotide sequences of the cDNA inserts determined; the corresponding genes are then identified by BLAST search. The final stage is to design an analytical method to determine the function of the gene in bringing about the observed phenotypic changes. The design of this final step is however dependent on the type of protein which is encoded by the gene. The steps involved in this method are illustrated in Figure 6.1. The whole procedure starting from the extraction of total RNA and including the toothpick inoculation of leaves will take approximately 2 weeks.

6.2.1
Extraction of Total RNA

Researchers who are not used to working with RNA should create a RNase-free working environment to avoid contamination and all materials, solutions and reagents should be sterilized (for details see [25,26]).

The procedure for extraction of total RNA described here is based on the single-step method [27] which uses guanidinium thiocyanate, phenol and chloroform for homogenization of samples and separation of RNA, DNA and proteins.

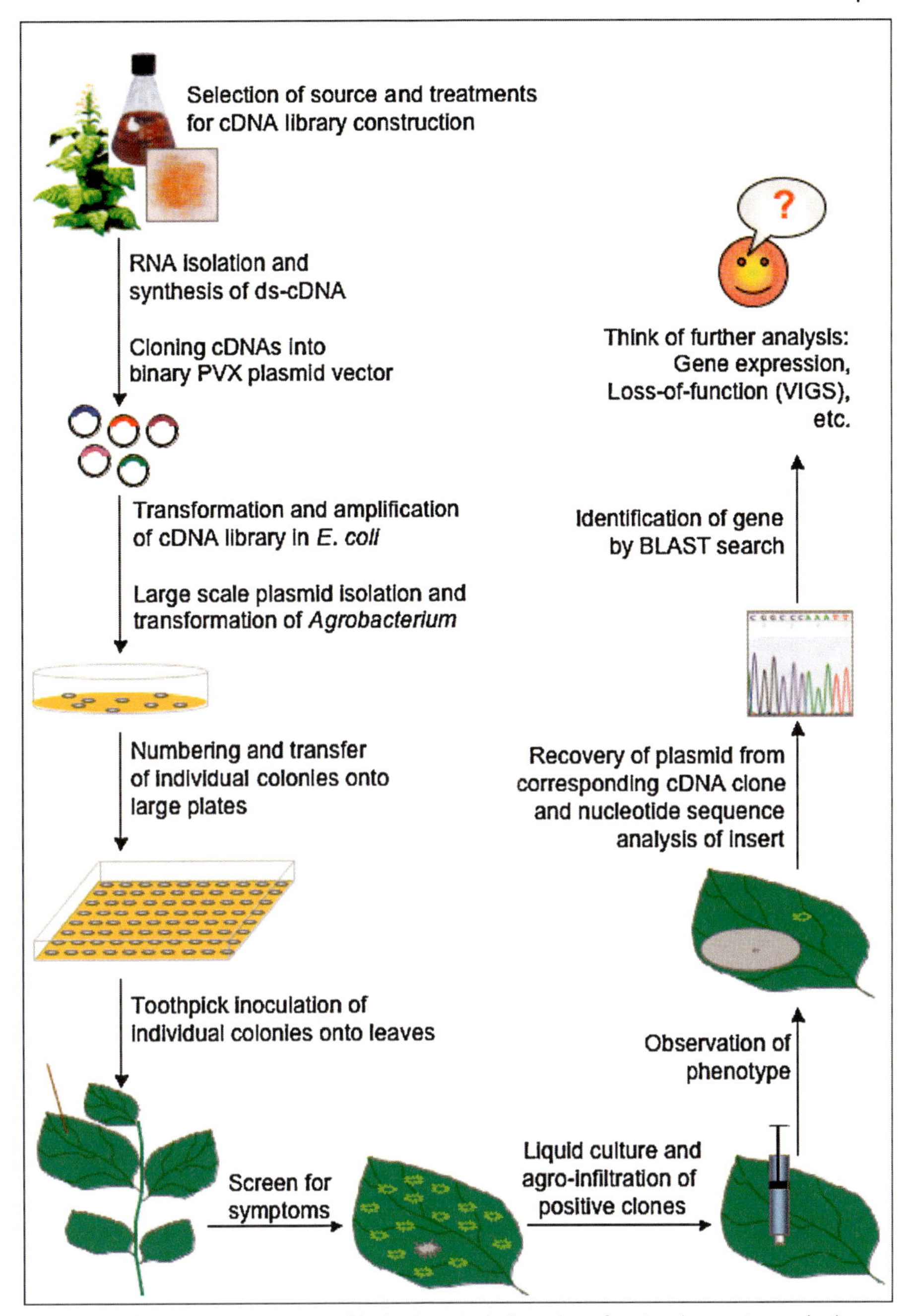

Figure 6.1 Schematic representation of the *in planta* high-throughput functional screening method.

Reagents and Equipment

- Extraction buffer: TRI Reagent (Sigma product no. T 9424) or TRIzol (Invitrogen cat. no. 15596-018)
- Chloroform
- Isopropanol
- 75% (v/v) ethanol in RNase-free water
- RNase-free water
- Liquid nitrogen
- Sterile, RNase-free polypropylene centrifuge tubes
- Mortar and pestle
- Cooled centrifuge

Procedure

1. Grind the tissue under liquid nitrogen to a fine powder using a pestle and mortar and transfer the powder into an appropriate, pre-chilled polypropylene tube (e.g. 50–100 mg per 1.5 ml microfuge tube). The powder must not thaw before coming into contact with the extraction buffer. Add 1 ml extraction buffer per 50–100 mg powder and mix well. Alternatively, tissue can be directly homogenized in extraction buffer (1 ml per 50–100 mg tissue) in a Polytron or other appropriate homogenizer. Let the samples stand for 5 min at room temperature.

2. Add 0.2 ml chloroform per ml of extraction buffer used and shake vigorously for 15 s, let samples stand at room temperature for 2–15 min.

3. Centrifuge samples at $12\,000 \times g$ for 15 min at 4 °C. After centrifugation the colorless upper aqueous phase contains RNA (the red organic lower phase contains proteins, and the interphase contains DNA)[1].

4. Transfer the upper phase to a fresh tube without carrying over traces of the interphase or lower phase.

5. Add 0.5 ml isopropanol per ml of extraction buffer used in step 1, mix and allow to stand for 5–10 min at room temperature.

6. Centrifuge samples at $12\,000 \times g$ for 10 min at 4 °C.

7. Remove supernatant and wash the sediment by vortexing with 1 ml 75% (v/v) ethanol per ml of extraction buffer used in step 1.

8. Centrifuge samples at $7500 \times g$ for 5 min at 4 °C.

9. Remove supernatant and allow the sedimented RNA to dry briefly. Do not let the RNA dry completely as this will reduce its solubility.

1) If required, DNA and/or proteins of the same sample can be extracted from the interphase or organic phase, respectively, following the manuals of TRI reagent or TRIzol.

10. Dissolve the RNA in 20–40 μl per 100 mg powder (step 1) of RNase-free water or elution buffer supplied with a poly(A)$^+$ purification kit (see Section 6.2.2). The RNA can be stored at −80 °C.

11. Determine concentration using a spectrophotometer at 260 nm. Absorbance of a diluted RNA sample is measured in a quartz microcuvette and the concentration can be calculated using the following equation.
[RNA] = Absorbance 260 nm × 0.04 × dilution of sample [mg/ml]

6.2.2 Purification of Poly(A)$^+$-mRNA

The purification of mRNA of eukaryotic origin is based on its main characteristic, the polyadenylation at the 3′-end. The poly-(A)-tail binds to oligo(dT) which is fixed to a solid support, mostly cellulose, and can be eluted after washing off the non-polyadenylated RNA. Only 1–5% of total RNA is poly(A)$^+$-RNA whereas 80–85% is rRNA and 15–20% low-molecular weight RNA such as tRNA. Taking into account that approximately 1–5 μg of mRNA is needed for the optimal construction of a cDNA library, the input for mRNA purification should be 100–500 μg of total RNA. However, if the source of RNA is limited smaller amounts can be used for cDNA synthesis even without mRNA purification (see Section 6.2.3). Purification of poly(A)$^+$-RNA using the Micro-FastTrack 2.0 Kit (Invitrogen) is described below. Many other purification systems are available which also give similar results for the purification of mRNA from prokaryotic systems (Epicentre Biotechnologies).

Reagents and Equipment

Micro-FastTrack 2.0 Kit (Invitrogen) containing:

- Binding buffer (500 mM NaCl, 10 mM Tris-HCl, pH 7.5)
- Low salt wash buffer (250 mM NaCl, 10 mM Tris-HCl, pH 7.5)
- Elution buffer (10 mM Tris-HCl, pH 7.5)
- 5 M NaCl
- 2 M Na acetate
- 2 mg/ml glycogen carrier
- Oligo(dT) cellulose powder, 25 mg per vial
- Spin-columns
- Ethanol: 75% (v/v) ethanol in RNase-free water
- Sterile, RNase-free polypropylene microcentrifuge tubes
- Heating block set to 65 °C
- Cooled centrifuge

Procedure

1. A solution containing 300–500 μg of total RNA (Section 6.2.1) is adjusted to 1 ml final volume and 500 mM NaCl final concentration by adding the appropriate

volume of 5 M NaCl. Alternatively, or if the starting RNA solution is very dilute, RNA can be precipitated in a centrifuge tube by adding 0.15 volumes of 2 M Na acetate and 2.5 volumes of ice-cold ethanol to the RNA solution. The mixture is maintained at −20 °C for 1 h and centrifuged at 12 000 × g for 15 min at 4 °C. The sedimented RNA is washed once with 1 ml 75% (v/v) ethanol in RNase-free water (see Section 6.2.1), dried briefly and resuspended in 10 μl of Elution buffer. This solution is then added to 1 ml of Binding buffer.

2. Heat the RNA sample to 65 °C for 5 min, then immediately place sample on ice for exactly 1 min.
3. Add the sample to a vial of oligo(dT) cellulose and allow it to swell for 2 min.
4. Rock or rotate the sample at room temperature for 30 min.
5. Centrifuge vial at 4000 × g for 5 min at room temperature.
6. Remove supernatant carefully without disturbing the cellulose sediment. Resuspend the sediment in 1.3 ml of Binding buffer.
7. Centrifuge vial at 4000 × g for 5 min at room temperature and remove supernatant. Repeat the washing step twice using 1.3 ml of Binding buffer for each wash.
8. Resuspend the final washed sediment in 300 μl Binding buffer and transfer the sample to a spin-column plugged into a microcentrifuge tube. Centrifuge at 4000 × g for 10 s at room temperature. Repeat this step as many times as necessary until all the oligo(dT) cellulose has been transferred to the column. Remove flow-through from the tube and replace column.
9. Add 500 μl of Binding buffer to the column and centrifuge at 4000 × g for 10 s at room temperature. Repeat this step a further three times using 500 μl Binding buffer each time.
10. Add 200 μl of Low salt wash buffer to the column and gently resuspend the cellulose with a sterile pipette tip. Take care not to damage the membrane in the column. Centrifuge at 4000 × g for 10 s at room temperature. Repeat this step once with 200 μl of Low salt wash buffer.
11. Place the spin-column in a new RNase-free microcentrifuge tube, add 100 μl of Elution buffer and resuspend cellulose with a pipette tip as in step 10. Centrifuge at 4000 × g for 10 s at room temperature. The liquid in the tube is the mRNA sample, DO NOT discard! Add another 100 μl of Elution buffer to the column, resuspend cellulose and centrifuge again in the same tube. The final volume of the mRNA sample should be 200 μl.
12. Quantitate the yield of RNA as described in Section 6.2.1, step 11. RNA can be concentrated by precipitation from the 200-μl sample by adding 10 μl glycogen carrier, 30 μl 2 M Na acetate and 600 μl ethanol. Freeze the sample in dry ice until solid or maintain at −20 °C for a minimum of 1 h, then centrifuge at maximum speed (≥15 000 × g) for 20 min at 4 °C. Remove supernatant, briefly centrifuge

again and remove residual ethanol. Dry the sedimented RNA and resuspend in 2–10 μl of Elution buffer. The RNA can be stored at −80 °C.

6.2.3 Synthesis of cDNA and Ligation to Binary, PVX-Based Expression Vectors

Dependent on the plasmid vector used for cloning, suitable overhangs at the 5′- and 3′-ends of the double stranded cDNAs (ds-cDNAs) need to be created for directional cloning into the sites produced by the cut of the vector with specific restriction enzymes. Oligonucleotides used for reverse transcription of the mRNA define the 3′-overhang of the cDNA. In the case of poly(A)$^+$-RNA as the template it was an oligo-dT primer to which the overhang sequence was added. The 5′-overhang is defined by the oligonucleotide that is used for initiation of second strand synthesis. The synthesis of ds-cDNA with asymmetric *Sfi*I-overhangs for directional cloning into the pSfinx vector is described below starting from poly(A)$^+$-mRNA using the Creator SMART cDNA Library Construction Kit (Clontech). Other cDNA synthesis kits that produce cDNAs with 5′- and 3′-ends compatible to other cloning sites of appropriate vectors can be used similarly.

Reagents and Equipment

Creator SMART cDNA Library Construction Kit (Clontech) containing:

- 10 μM CDS III/3′ primer (for oligo(dT) priming)
- 10 μM SMART IV oligonucleotide
- Reverse transcriptase (PowerScript)
- 5× first-strand buffer (250 mM Tris-HCl, pH 8.3, 30 mM $MgCl_2$, 375 mM KCl)
- 20 mM dithiothreitol (DTT)
- 10 μM 5′ PCR primer
- 20 μg/μl proteinase K
- 20 units/μl *Sfi*I restriction enzyme
- 10× *Sfi*I buffer
- 100× bovine serum albumin (BSA)
- 10 mM dNTP mix (dATP, dCTP, dGTP, dTTP)
- 25 mM Na hydroxide
- 10× Advantage 2 PCR buffer
- 50× Advantage 2 polymerase mix
- 3 M Na acetate, pH 4.8
- 20 μg/ml glycogen carrier
- 1% xylene cyanol dye
- Column buffer
- 10× ligation buffer
- 10 mM ATP
- T4 DNA ligase
- phenol/chloroform/isoamyl alcohol (25 : 24 : 1)

- chloroform/isoamyl alcohol (24 : 1)
- 80% ethanol
- Deionized, sterile water
- 0.5-ml microcentrifuge tubes
- Air incubator or a thermal cycler with heatable lid set to 42 °C
- Heating block or a thermal cycler set to 65 °C
- Cooled microcentrifuge
- Agarose and other reagents needed for horizontal analytical gel

Procedure

1. For first-strand cDNA synthesis combine the following reagents in a 0.5-ml microcentrifuge tube and make up to a total volume of 5 μl with deionized water: 1–3 μl polyA + RNA (1.0 μg), 1 μl SMART IV oligonucleotide, 1 μl CDS III/3′ oligo(dT) primer. Mix and centrifuge briefly.

2. Incubate at 72 °C for 2 min, then cool on ice for 2 min.

3. Centrifuge the tube briefly and then add 2 μl 5× first-strand buffer, 1 μl DTT, 1 μl dNTP mix and 1 μl of reverse transcriptase. The total volume is now 10 μl. Mix and briefly centrifuge. Incubate at 42 °C for 1 h in an air incubator or a thermal cycler with a heated lid to prevent evaporation.

4. Place the tube on ice, add 1 μl Na hydroxide to the mixture and incubate at 65 °C for 30 min.

5. Place the tube back on ice, use immediately for second-strand cDNA synthesis or store at −20 °C.

6. For second-strand cDNA synthesis the following components are added to the 11 μl of first-strand cDNA synthesis from step 5: 71 μl deionized water, 10 μl 10× PCR buffer, 2 μl dNTP mix, 2 μl 5′ PCR primer, 2 μl CDS III/3′ primer and 2 μl of 50× Advantage 2 polymerase mix. Mix gently and centrifuge briefly. Put the tube into a preheated (95 °C) thermal cycler with heated lid and carry out primer extension as follows: 72 °C for 10 min then 95 °C for 1 min followed by three cycles of 95 °C for 10 s and 68 °C for 8 min.

7. After the reaction is complete, use 5 μl to run on a 1% agarose gel alongside a DNA size marker. In the ethidium bromide stained gel the cDNA should appear as a smear between 0.1–9 kbp. The ds-cDNA can be stored at −20 °C.

8. For inactivation of the DNA polymerases 50 μl of the reaction mixture (step 6) is transferred into a 0.5-ml microcentrifuge tube and 2 μl of proteinase K are added. Mix and briefly centrifuge. After incubation at 45 °C for 20 min add 50 μl of deionized water followed by 100 μl phenol/chloroform/isoamyl alcohol. Mix by gentle inversion for 2 min, then centrifuge at 14 000 × g for 5 min to achieve phase separation.

9. Transfer the upper aqueous phase into a new 0.5-ml microcentrifuge tube, add 100 μl chloroform/isoamyl alcohol, mix by gentle inversion for 2 min, then centrifuge at 14 000 × g for 5 min at room temperature to achieve phase separation.

10. Transfer the upper aqueous phase into a new 0.5-ml microcentrifuge tube, add 10 μl 3 M Na acetate, 1.3 μl glycogen carrier and 260 μl ethanol. Mix and immediately centrifuge at 14 000 × g for 5 min at room temperature.

11. Remove the supernatant carefully with a pipette without disturbing the sediment. Add 80% ethanol to wash the sediment, centrifuge at 14 000 × g for 5 min at room temperature, remove the supernatant carefully with a pipette and let the sediment air-dry for 10 min.

12. To produce the *Sfi*I-digested 5′- and 3′-ends, dissolve the sediment from step 11 in 79 μl deionized water, then add 10 μl 10× *Sfi*I buffer, 10 μl *Sfi*I enzyme and 1 μl 100× BSA. Mix well and incubate at 50 °C for 2 h.

13. Size fractionation of the ds-cDNA is achieved by adding 2 μl of 1% xylene cyanol dye to the restriction reaction of step 12. The *Sfi*I-digested ds-cDNA is then loaded onto a CHROMA SPIN-400 column that has been washed once with 700 μl of column buffer. After the sample is fully absorbed rinse the tube that contained the cDNA sample with 100 μl of column buffer and apply to the column. After the buffer has stopped dripping out of the column carefully apply 600 μl of column buffer and immediately collect single drop fractions into microcentrifuge tubes that have been labeled #1–#16 and store them on ice.

14. Separate 3 μl of adjacent fractions together with a DNA size marker on a 1% agarose gel containing 0.1 μg/ml ethidium bromide. Under UV light identify the first three fractions which contain ds-cDNA. These fractions are pooled in a 1.5-ml microcentrifuge tube.

15. Precipitate the ds-cDNA by adding 1/10 volume 3 M Na acetate, 1.3 μl glycogen and 2.5 volumes ethanol. Mix well and incubate at −20 °C overnight.

16. Centrifuge at 14 000 × g for 20 min at room temperature, remove the supernatant carefully with a pipette, centrifuge again briefly and remove remaining liquid. Let the sediment air-dry for about 10 min, then resuspend in 7 μl deionized water. This is the *Sfi*I-cut ds-cDNA which is ready for ligation into the pSfinx plasmid vector.

17. The pSfinx plasmid DNA is cut by the *Sfi*I restriction enzyme followed by dephosphorylation with calf intestine phosphorylase (CIP) according to basic protocols [25,26] and is adjusted to a concentration of 0.2 μg/ml.

18. Ligation reactions with three different ratios of cDNA to plasmid vector are prepared. Into three 0.5-ml microcentrifuge tubes labeled A, B and C and stored on ice, pipette 1 μl of *Sfi*I-cut pSfinx plasmid, 0.5 μl 10× ligation buffer, 0.5 μl 10 mM ATP and 0.5 μl T4 DNA ligase. To tube A add 0.5 μl of *Sfi*I-cut ds-cDNA (from step 16) and 2 μl deionized water; to tube B add 1 μl of *Sfi*I-cut ds-cDNA and

1.5 μl deionized water; and to tube C add 1.5 μl of *Sfi*I-cut ds-cDNA and 1 μl deionized water. Mix the reagents gently, centrifuge briefly and incubate at 16 °C overnight.

19. To each of the ligation reactions add 95 μl of deionized water, 1.5 μl glycogen carrier and mix well, then add 280 μl ice-cold ethanol. Mix gently and cool to −70 °C for at least 4 h.

20. Centrifuge at maximum speed in a microcentrifuge for 20 min at room temperature. Remove the supernatant carefully without disturbing the precipitated DNA. After the sediments (A, B, and C) are air dried, resuspend each in 5 μl deionized water. The cDNAs ligated to pSfinx vector are now ready for transformation into *E. coli* cells for amplification of the library (Section 6.2.4) or can be directly used for transformation of electrocompetent Agrobacteria (Section 6.2.5) (Figure 6.2).

6.2.4
Amplification of the cDNA Library in *E. coli*

In some cases it is advisable to amplify the plasmid cDNA library prior to transformation of Agrobacteria to enhance the number of transformants. The ligation reactions from step 20 in Section 6.2.3 are transformed with high efficiency into

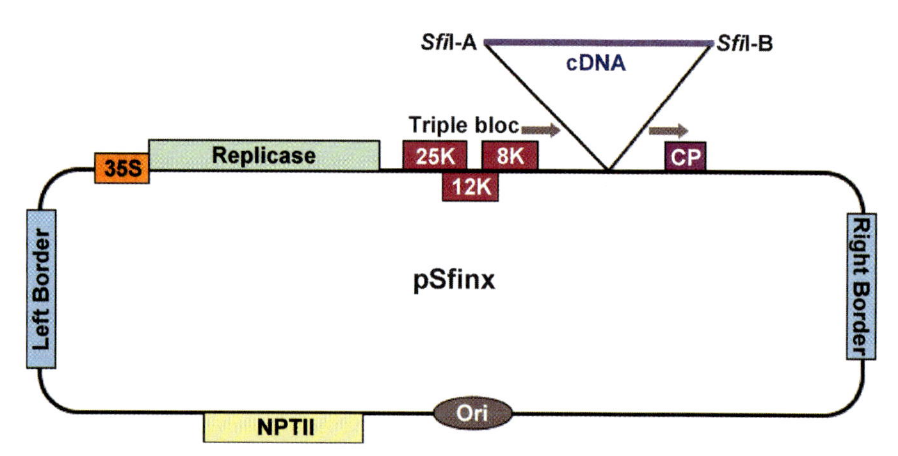

Figure 6.2 Map of the binary PVX-based expression vector pSfinx. The vector is derived from pGR106 [28] by inserting additional restriction sites between the *Cla*I and *Asc*I sites resulting in the cloning region (5′-*Cla*I-*Sfi*I-*Sma*I-*Eco*RV-*Sfi*I-*Asc*I-*Not*I-*Sal*I-3′) [14]. The vectors pGR106 and pGR107 are similar to pSfinx except for the cloning sites 5′-*Cla*I-*Asc*I-*Not*I-*Sal*I-3′ and 5′-*Cla*I-*Sma*I-*Sal*I-3′, respectively [28]. The plasmids contain left and right borders of the T-DNA, origin of replication (Ori) and the antibiotic resistance gene for neomycin-phospho- transferase II (NPTII). The expression of the PVX sequence based on the cDNA of PVX strain UK3, comprising the replicase, triple block and coat protein (CP) genes, is driven by the cauliflower mosaic virus 35S promoter. The two asymmetric *Sfi*I restriction sites, *Sfi*I-A and *Sfi*I-B, are used for directional cloning of the cDNAs downstream of a duplicated coat protein promoter (→). Genebank accession number for pGR106 is AY297843.

E. coli DH5α cells for amplification of the plasmids. Either of the two standard methods, electroporation of electrocompetent cells or transformation of chemically-competent cells can be used for transformation if the transformation efficiency is higher than 1×10^8 colony forming units per μg (cfu/μg) plasmid [25,26]. Such highly competent *E. coli* DH5α cells are commercially available. The procedure of plasmid isolation from *E. coli* cells in midi or maxi format is not described in detail here because this is a common basic technique and plasmid isolation kits including detailed manuals are freely available.

Reagents and Equipment

- Equipment for bacteriological work [25,26]
- Highly competent *E. coli* DH5α cells
- LB medium (10 g/l tryptone, 5 g/l yeast extract, 5 g/l NaCl. 1 ml/l 1 N NaOH)
- LB agar plates (90 mm diameter) containing 50 μg/ml kanamycin(Kan)
- Solutions or kits for plasmid isolation from *E. coli* cells in midi or maxi format

Procedure

1. Transform the three ligation reactions separately into *E. coli* cells and grow for 1 h at 37 °C after adding the recommended quantity of growth medium (usually SOC or LB). Remove 1 μl of the bacterial suspension and dilute in 50 μl of LB medium in a fresh tube for each transformation reaction and plate onto 90-mm LB-agar plates containing the appropriate antibiotic, which is kanamycin in the case of the pSfinx vector. Incubate at 37 °C overnight. Store the remaining transformation mixtures at 4 °C.

2. Examine the bacterial growth on the plates. The transformations of at least two of the three ligation reactions should produce a substantial number of colonies. From the number of colonies and the volume plated (1 μl) calculate the volume required to obtain confluent or nearly confluent growth of bacteria.

3. Plate the rest of the appropriate transformation mixtures (stored at 4 °C) onto as many agar plates as necessary to produce confluent or nearly confluent growth and incubate at 37 °C for 18–20 h.

4. Collect and pool all colonies from the agar plates by pipetting LB medium onto the agar and scraping the colonies into the liquid with a spreader or equivalent sterile tool. Distribute approximately 5–6 ml each of the dense bacterial suspension into 50-ml polypropylene centrifuge tubes and centrifuge at $6000 \times g$ for 5 min. Discard the supernatants and determine the wet weight of each of the bacterial sediments. These can be directly used for bulk plasmid isolation or can be stored at −20 °C.

5. Methods of plasmid isolation can be found in laboratory manuals and handbooks [25,26] or in the manuals accompanying plasmid isolation kits such as those from Qiagen or Promega. The final plasmid solution should be adjusted to a concentration of 0.5–1.0 μg/μl.

6. The quality of the cDNA library can be tested by picking 10 single bacterial colonies from the agar plates (step 2) and growing each in 2 ml LB liquid medium to produce plasmid mini preparations. Aliquots of the isolated plasmids are cut with *Sfi*I restriction enzyme and analyzed using agarose gel electrophoresis. The plasmids should contain various cDNA inserts of sizes greater than 500 base pairs.

6.2.5
Transformation of cDNA Library into *Agrobacterium tumefaciens* Cells

- YEP medium (10 g/l yeast extract, 10 g/l peptone, 5 g/l NaCl, pH 7.0)
- 2× 250 ml YEP medium + 0.5% glucose in 1-l flasks
- LB medium (see Section 6.2.4)
- 50 mg/ml Rifampicin (Rif) in dimethylformamide (DMSO)
- 50 mg/ml kanamycin (Kan)
- 10% (v/v) glycerol in water (sterile)
- LB agar plates (90 mm diameter) containing 50 μg/ml Kan
- 384-well plates containing LB agar and 50 μg/ml Kan or 96-well plates containing LB agar and 50 μg/ml Kan (As an alternative to the 96-well plates standard square LB agar plates (10 × 14 cm) containing 50 μg/ml Kan can be used. Each square plate should be numbered and a grid comprising eight lines parallel to the long edges (line A–H) and 12 lines parallel to the short edges (lines 1–12) should be drawn on the underside of the plate. This will give a pattern of squares with 96 intersections and each square can be numbered individually. For example 4-B-10 marks the point were line B crosses line 10 on plate number 4).
- 150 mM NaCl
- 20 mM $CaCl_2$
- 250-ml centrifuge tubes with appropriate rotor and centrifuge
- 50-ml centrifuge tubes with appropriate rotor and centrifuge
- Heat block or water bath set to 37 °C
- Incubator set to 28 °C

6.2.5.1 Preparation of Competent *Agrobacterium tumefaciens* Cells

The *A. tumefaciens* strains MOG101, GV3101 and LBA4404 carry chromosomal resistance to rifampicin.

Preparation of Chemically-competent *Agrobacterium tumefaciens* Cells

1. Grow *Agrobacterium* strain in 5 ml YEP containing 20 μg/ml Rif overnight at 28 °C with vigorous shaking.

2. Use 2.5 ml each of this culture to inoculate 250 ml YEP + 0.5% glucose and grow bacteria at 28 °C with vigorous shaking to a density of 0.5–0.6 OD_{600} which takes about 4–5 h.

3. Harvest bacteria by centrifugation at 4000 × g for 5 min at 4 °C. Discard supernatant, resuspend each pellet in 25 ml ice-cold 150 mM NaCl, pool in a 50-ml centrifuge tube and allow to cool on ice for 15 min.

4. Centrifuge at 4000 × g for 5 min at 4 °C. Discard supernatant and resuspend the sediment in 5 ml ice-cold 20 mM $CaCl_2$ and store on ice.

5. Aliquot 100 µl of the bacterial suspension into 1.5-ml microcentrifuge tubes on ice and quick-freeze in liquid nitrogen. Store the competent Agrobacteria at −80 °C.

Transformation of Chemically-competent *Agrobacterium tumefaciens* Cells

1. Take 10 tubes of competent Agrobacteria from the −80 °C freezer and add approximately 1 µg plasmid (Section 6.2.4 step 5) in a maximum volume of 10 µl to each of the frozen bacterial samples and immediately incubate at 37 °C for 5 min. After 1 min mix the contents by flicking the tubes briefly.

2. Add 1 ml of YEB medium (without antibiotics) to each tube and shake at 28 °C for 2–4 h.

3. Centrifuge the tubes at 4000 × g for 10 min.

4. Discard supernatants and add 100 µl YEB medium to each tube and resuspend the bacterial pellets by pipetting up and down.

5. Plate the bacteria from each tube onto a 90-mm LB agar plate containing Kan.

6. Incubate the plates upside down at 28 °C in the dark for 2–3 days.

7. With sterile tooth picks transfer individual colonies from the plates of step 6 to numbered 384-well or 96-well plates filled with LB agar containing Kan or to the alternative square plates (see Section 6.2.5). Incubate the plates upside down at 28 °C in the dark for 2–3 days. These are individually numbered *Agrobacterium* clones from your cDNA library which will be used for tooth pick inoculation of leaves.

Preparation of Electrocompetent *Agrobacterium tumefaciens* Cells

1. Grow *Agrobacterium* strain in 5 ml YEP containing 20 µg/ml Rif overnight at 28 °C with vigorous shaking.

2. Use 2.5 ml each of this culture to inoculate 250 ml LB medium and grow bacteria at 28 °C with vigorous shaking to an optical density of 0.5–0.8 OD_{600}.

3. Harvest bacteria by centrifugation at 4000 × g for 5 min at 4 °C. Discard supernatant, completely resuspend the sediment in 50 ml of ice-cold deionized water and transfer to a 50-ml centrifuge tube. Centrifuge again and wash the sediment a further three times with 50 ml ice-cold deionized water.

4. The final bacterial sediment of each 250-ml culture is resuspended in 1.25 ml of ice-cold 10% (v/v) glycerol (0.5% of the original volume).

5. Aliquot 50 μl of the bacterial suspension into 1.5-ml microcentrifuge tubes on ice and quick-freeze in liquid nitrogen. Store the electrocompetent Agrobacteria at −80 °C. For reference see [29].

Electroporation of *Agrobacterium tumefaciens* Cells

1. Thaw competent cells on ice (50 μl per transformation, see paragraph "Preparation of Electrocompetent *Agrobacterium tumefaciens* Cells" above).

2. Add plasmid DNA (2.5 μl cDNA each from Section 6.2.3 step 20 or 1 μl each from Section 6.2.4 step 5) to the cells, and mix on ice.

3. Transfer the mixture to a pre-chilled electroporation cuvette. Carry out electroporation as recommended by the manufacturer of the chosen electroporator. For example, for the GenePulser (Bio-Rad) electroporator with a 1-mm cuvette, use the following conditions: capacitance: 25 μF, voltage: 2.4 kV, resistance: 200 Ω.

4. Immediately after electroporation, add 1 ml of YEB medium (without antibiotics) to the cuvette, and transfer the bacterial suspension to a 1.5-ml microcentrifuge tube. Incubate for 4 h at 28 °C with gentle agitation.

5. Follow steps 3–7 from paragraph "Transformation of Chemically-competent *Agrobacterium tumefaciens* Cells" above.

6.2.6
Toothpick Inoculation of Leaves

1. Many plant species can be screened and transformed using this method, but the most commonly employed species are *Nicotiana benthamiana*, *Nicotiana tabacum* and tomato. In this example *Nicotiana benthamiana* plants are used which have been grown in soil in separate pots. Fully expanded leaves (four to five per plant) of 6–8-week-old plants are used for toothpick inoculation.

2. Before starting inoculation, label the plant or pot and the leaves with a waterproof marker pen and/or adhesive tape so that each inoculation site can be attributed to the bacterial colony used. A scheme for labeling the leaves is given in Figure 6.3.

3. With sterile wooden toothpicks, carefully pick individual Agrobacteria colonies from the agar plates (Section 6.2.5.1, paragraph "Transformation of Chemically-competent *Agrobacterium tumefaciens* Cells", step 7) and pierce the leaves.

4. After completing inoculation incubate the agar plates again at 28 °C overnight to ensure re-growth of the colonies. Seal the plates with parafilm and store at 4 °C.

5. Place the inoculated plants into a greenhouse or growing chamber regulated to a temperature of 20–25 °C.

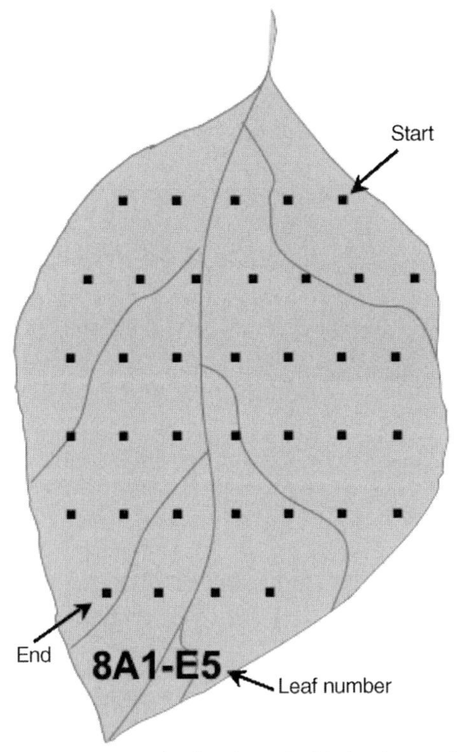

Figure 6.3 Example of a scheme for labeling leaves after toothpick inoculation. In the leaf number 8A1-E5, the digit 8 stands for the well number in the 96-well plate and A1 for the colony number from which the inoculum originated and E5 for the last of 37 colonies that were used on this leaf. Thus, next to A1 colony number B1 from plate number 8 was inoculated followed by C1 to H1 then A2, B2 and so forth.

6. Depending on the phenotype that is under investigation, observe the leaves visually and/or use them for the appropriate assays.

6.2.7 Agroinfiltration

For confirmation of the phenotype produced by individual *Agrobacterium* clones in the screening procedure, the clones are cultured in liquid medium and used for infiltration of larger areas of leaves. In contrast to the tooth pick inoculation procedure leaves are not wounded in the agroinfiltration process. To ensure that infection with Agrobacteria is effective, acetosyringone is added prior to infiltration. Phenolic compounds, such as acetosyringone are released as signals from wounded plants to activate the *VirA* gene of *Agrobacterium* encoding a transmembrane dimeric sensor which then triggers the plant–bacterium interaction [30,31].

- 1-ml plastic syringes without needles.
- 100 mM acetosyringone in DMSO (stock solution, store at 4 °C).
- Infiltration buffer (10 mM MES-KOH, pH 5.6, 10 mM $MgCl_2$, 150 μM acetosyringone).

1. Individual *Agrobacterium* clones that gave a positive phenotype in the toothpick inoculation site are selected. The corresponding colony on the agar plate is used for inoculation of a 4-ml LB liquid culture including the appropriate antibiotic.
2. Grow the bacteria at 28 °C to saturation (1–2 days) with vigorous shaking.
3. Collect the bacteria by centrifugation at 4000 × g for 5 min.
4. Discard supernatant and resuspend the bacteria in infiltration buffer to an optical density of 0.8–1.0 OD_{600}.
5. Incubate the *Agrobacterium* suspension at room temperature for 2–4 h in the dark.
6. Aspirate about 0.5 ml of the suspension using a 1-ml syringe.
7. Press the tip of the syringe against the underside of a fully expanded leaf while simultaneously applying gentle counter-pressure to the other side of the leaf with a finger of the other hand. The *Agrobacterium* solution is then injected into the airspaces inside the leaf through stomata, which can be monitored by the darkening of the infiltrated leaf area.
8. Usually only one half of the leaf is infiltrated, leaving the other half as a negative control.
9. Place the infiltrated plants into a greenhouse or growing chamber maintained at a temperature of 20–25 °C.
10. Depending on the phenotype that is under investigation, observe the leaves visually and/or use them for the appropriate assays.

6.2.8
Recovery of the cDNA Fragments

Once positive clones have been identified, the cDNA fragment is recovered and the nucleotide sequence is analyzed. The simplest way to recover the cDNA insert is by colony-PCR amplification using specific oligonucleotide primers corresponding to vector sequences left and right of the cloning site. Also, the recombinant plasmids can be isolated from an *Agrobacterium* liquid culture and the inserts recovered by digestion with restriction enzyme(s), *Sfi*I in case of pSfinx. Both, the colony-PCR fragments and isolated plasmids can be used for nucleotide sequence analysis.

Standard protocols for colony-PCR and plasmid isolation for *E. coli* can also be used for *Agrobacterium* and are not described here.

The sequences of oligonucleotide primers for amplification of inserts and nucleotide sequence analysis from pSfinx (pGR106, pGR107) are based on the nucleotide sequence of plasmid vector pGR106 (Genebank accession number AY297843):

- Forward primer: 5′-CAATCACAGTGTTGGCTTGC-3′.
- Reverse primer: 5′-GACCCTACGGGCTGTGTTG-3′.

6.3 Application of the Technology

Until now high-throughput *in planta* screening has mostly been applied to the identification of cell death-inducing factors in host–pathogen interactions [14,21–24]. Programmed cell death (PCD) causes a cell to commit suicide and helps organisms to contain sites of infections and eliminate old or surplus cells. In these cases the changes in phenotype in leaves are rather easy to detect as a grayish color develops around the toothpick inoculation site when cell death occurs (Figure 6.4a). In our own approaches to identify such cell death-causing factors we first used a cDNA library derived from messenger RNA of leaves that had been infiltrated with the elicitor INF1 [32] from *Phytophtora infestans* for several periods of time. From the cDNA library of more than 100 000 clones 40 000 *Agrobacterium* clones were inoculated to leaves of *Nicotiana benthamiana* plants. This screening led to the identification of 30 candidate genes encoding proteins that are involved in the onset of cell death which was confirmed by the severe phenotypes displayed after agroinfiltration (Figure 6.4b). From these NbCD1 encoding an ethylene-responsive element binding factor (ERF) and NbMKK1 coding for a mitogen activated protein kinase kinase (MAPKK) were analyzed in detail [23,24]. Since in cell death, especially in the hypersensitive response in plants, reactive oxygen species (ROS) play a pivotal role [33], we used a cDNA library from *N. benthamiana* leaves under chemically induced oxidative stress for a second screening approach. More than 30 000 clones from this cDNA library have been screened on *N. benthamiana* leaves and 232 clones that caused cell death after less than

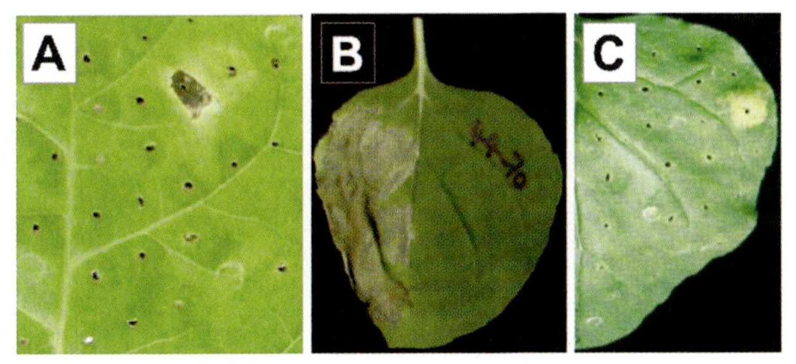

Figure 6.4 (A) Typical cell death-like phenotype observed in toothpick inoculation screening. (B) Cell death-like phenotype after agroinfiltration of the same clone as in A. (C) Typical yellowing phenotype indicating expression of a senescence-associated cDNA after toothpick inoculation.

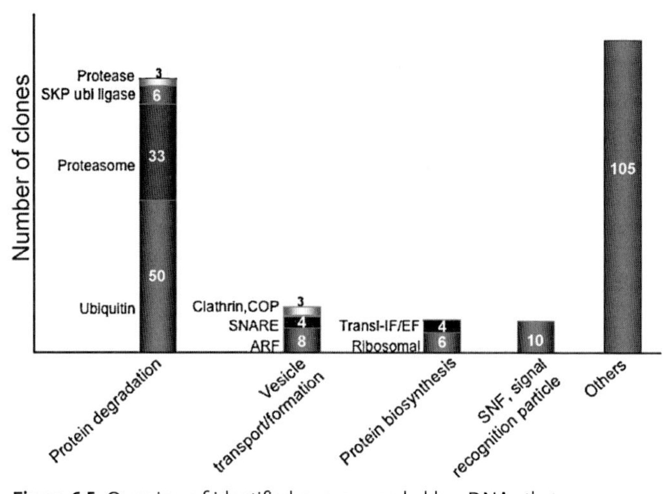

Figure 6.5 Overview of identified genes encoded by cDNAs that produced cell death-like phenotypes in a screening with a cDNA library constructed from leaves under chemically induced oxidative stress. Where possible genes have been grouped according to their cellular function.

20 days after inoculation were further analyzed. BLAST [34] searches with the determined nucleotide sequences identified the encoded proteins allowing the positive clones to be grouped according to their function. A summary of this grouping is given in Figure 6.5. For example the largest group of 92 members is represented by cDNAs coding for proteins involved in protein degradation. Another group comprises proteins which play a role in vesicle formation and transport systems. It has been shown that such proteins indeed are involved in pathogen defense [35]. Although all the clones caused cell death after overexpression in leaves, further investigation of each individual clone is necessary to clarify the function of the encoded protein in the observed phenotypic change. Further experiments include analysis of gene expression by RNA blot hybridization or RT-PCR. Inducible expression of a tagged version of the encoded proteins *in planta* can be used to monitor the time course of phenotype development dependent on expression of the protein detected by immunoblotting. Analysis of the cellular localization of the proteins may provide further insight into their function, as may the analysis of their interactions with other proteins or factors, for example by yeast-2-hybrid analysis. Finally a loss-of-function analysis by virus induced gene silencing (VIGS) should be incorporated into the characterization of the cDNAs. Methods of further analyses are being developed for the clones which have been identified and are shown in Figure 6.5.

As an example of the use of the high-throughput *in planta* screening method to identify factors other than those involved in programmed cell death, we initiated a screen for cDNAs encoding proteins that are involved in the process of leaf senescence. cDNA libraries were constructed in the pSfinx vector from *N. benthamiana* leaves at different stages of senescence and from young green leaves. In a first test of toothpick inoculation with the senescent leaf-derived cDNA library typical senescence-like phenotypes were observed showing yellowing around the inoculation site as a result of chlorophyll degradation (Figure 6.4c). This result is encouraging with regard to the further expansion of this screening approach. Furthermore, screening for inhibitors of leaf senescence is possible using the cDNA library derived from young green leaves. After toothpick inoculation leaves are placed in the dark to induce senescence and examined for phenotypes characterized by 'staying green' around inoculation sites.

Using specific testing systems for phenotypic changes in leaves, the method described here can be used to screen for factors involved in the response to a variety of stressors or in response to drug application, for example. Modifier screens represent a particularly powerful technique as the expression of cDNA clones after toothpick inoculation can be used to identify genes and pathways that can either enhance or suppress a given phenotype of interest. For example, inoculated leaves can be challenged with an elicitor or other effectors and the speed of cell death compared to control inoculations (empty vector) can be monitored. With such an approach clones can be identified which encode proteins that accelerate or inhibit the response to the effector, thus indicating probable involvement in the underlying signaling pathways.

The clear advantage of the method described in this chapter is the real high-throughput format achieved by using toothpick inoculation which allows the screening of more than 100 000 clones in a relatively short time-frame with the use of relatively little greenhouse and laboratory space. Because expression of the proteins encoded by the cDNAs is responsible for the phenotypes under investigation, novel factors may be identified which might otherwise not be obvious from transcriptome analysis. Another advantage, especially for research on plant–microbe interactions, is that any cDNAs from organisms other than plants, for example from bacterial or fungal pathogens, can be expressed within the leaves to test for the effect of the encoded proteins. However, a portion of proteins will escape detection if their roles in bringing about a phenotypic change are dependent on a network of interactions with other factors that are not present in the inoculated leaves.

6.4 Perspectives

As described in Section 6.3 the high-throughput *in planta* screening method can be employed for searching proteins involved in any imaginable process as long as a well-defined detection system for the particular phenotypic change is available. Researchers planning to make use of *in planta* screening will need to develop detection systems to adapt the method for their individual needs. The basis of the screening described here is the expression of cDNA clones *in planta* which is a gain-of-function strategy. Complementary to this method, a loss-of-function strategy based on post-transcriptional gene silencing (PTGS) which was first described in plants as co-suppression [36,37] and is now also described as RNA interference

(RNAi) in other systems [38], has been adopted as a screening method with a high-throughput format for plants [39,40]. PTGS is based on the sequence specific degradation of endogenous mRNAs by homologous double-stranded RNA (dsRNA). In brief, dsRNA is cleaved into smaller molecules known as short interfering RNA (siRNA) which then guide the degradation of homologous target transcripts by association with the siRNA antisense strand in the RNAi silencing complex (RISC) [41]. The use of recombinant viruses for specific silencing of endogenous genes by PTGS has been developed and is termed 'virus-induced gene silencing' (VIGS) [42]. Tobacco mosaic virus (TMV), potato virus X (PVX) and vectors derived from other viral systems are now routinely used for gene silencing in plants. With these tools for VIGS it is possible to conduct high-throughput *in planta* screenings using a loss-of-function approach similar to the gain-of-function approach described here.

Acknowledgments

We thank Mattieu Joosten, Wageningen University, for providing the pSfinx vector. Excellent technical assistance in our screening projects from Akiko Hirabuchi, Akiko Ito and Hiroe Utsushi is very much appreciated. We thank all members of the Rice Research Group at IBRC for their contributions, especially Matthew Shenton for careful scrutiny of the text. The work in our laboratory is supported by the 'Program for Promotion of Basic Research Activities for Innovative Bioscience' (Japan).

References

1 Rine, J. (1991) Gene overexpression in studies of *Saccharomyces cerevisiae*. *Methods in Enzymology*, **194**, 239–251.

2 Grimm, S. (2004) The art and design of genetic screens: mammalian culture cells. *Nature Reviews. Genetics*, **5**, 179–189.

3 Smith, E.F. and Townsend, C.O. (1907) A plant tumor of bacterial origin. *Science*, **25**, 671–673.

4 Schell, J. and Van Montagu, M. (1977) The Ti-plasmid of *Agrobacterium tumefaciens*, a natural vector for the introduction of nif genes in plants? *Basic Life Sciences*, **9**, 159–179.

5 Nester, E.W., Gordon, M.P., Amasino, R.M. and Yanofsky, M.F. (1984) Crown gall: a molecular and physiological analysis. *Annual Review of Plant Physiology and Plant Molecular Biology*, **35**, 387–413.

6 Binns, A.N. and Thomashow, M.F. (1988) Cell biology of *Agrobacterium* infection and transformation of plants. *Annual Review of Microbiology*, **42**, 575–606.

7 Hooykaas, P.J.J. and Shilperoort, R.A. (1992) *Agrobacterium* and plant genetic engineering. *Plant Molecular Biology*, **19**, 15–38.

8 Zupan, J.R. and Zambryski, P.C. (1995) Transfer of T-DNA from *Agrobacterium* to the plant cell. *Plant Physiology*, **107**, 1041–1047.

9 Herrera-Estrella, L., Depicker, A., Van Montagu, M. and Schell, J. (1983) Expression of chimaeric genes transferred into plant-cells using a Ti-plasmid-derived vector. *Nature*, **303**, 209–213.

10 Birch, RG. (1997) Plant transformation: problems and strategies for practical

application. *Annual Review of Plant Physiology and Plant Molecular Biology*, **48**, 297–326.

11 Lindsey, K. and Jones, M.G.K. (1987) Transient gene expression in electroporated protoplasts and intact cells of sugar beet. *Plant Molecular Biology*, **10**, 43–52.

12 Klein, T.M., Wolf, E.D., Wu, R. and Sanford, J.C. (1987) High-velocity microprojectiles for delivery of nucleic acids into living cells. *Nature*, **327**, 70–73.

13 Kapila, J., De Rycke, R., Van Montagu, M. and Angenon, G. (1997) An *Agrobacterium*-mediated transient gene expression system for intact leaves. *Plant Science*, **122**, 101–108.

14 Takken, F.L.W., Luderer, R., Gabriels, S.J.E.J., Westerink, N., Lu, R., de Wit, P.J.G.M. and Joosten, M.H.A.J. (2000) A functional cloning strategy, based on a binary PVX-expression vector, to isolate HR inducing cDNAs of plant pathogens. *Plant Journal*, **24**, 275–283.

15 Brisson, N., Paszkowski, J., Penswick, J.R., Gronenborn, B., Potrykus, I. and Hohn, T. (1984) Expression of a bacterial gene in plants by using a viral vector. *Nature*, **310**, 510–514.

16 Takamatsu, N., Ishikawa, M., Meshi, T. and Okada, Y. (1987) Expression of bacterial chloramphenicol acetyltransferase gene in tobacco plants mediated by TMV-RNA. *EMBO Journal*, **6**, 307–311.

17 Kumagai, M.H., Turpen, T.H., Weinzettl, N., Della-Cioppa, G., Turpen, A.M., Donson, J., Hilf, M.E., Grantham, G.L., Dawson, W.O., Chow, T.P., Piatak, M. and Grill, L.K. (1993) Rapid, high-level expression of biologically active a-trichosanthin in transfected plants by an RNA viral vector. *Proceedings of the National Academy of Sciences of the United States of America*, **90**, 427–430.

18 Chapman, S., Kavanagh, T. and Baulcombe, D. (1992) Potato virus X as a vector for gene expression in plants. *Plant Journal*, **2**, 549–557.

19 Baulcombe, D.C., Chapman, S. and Santa Cruz, S. (1995). Jellyfish green fluorescent protein as a reporter for virus infections. *Plant Journal*, **7**, 1045–1053.

20 Turpen, T.H., Turpen, A.M., Weinzettl, N., Kumagai, M.H. and Dawson, W.O. (1993) Transfection of whole plants from wounds inoculated with *Agrobacterium tumefaciens* containing cDNA of tobacco mosaic virus. *Journal of Virological Methods*, **42**, 227–239.

21 Qutob, D., Kamoun, S. and Gijzen, M. (2002) Expression of a *Phytophthora sojae* necrosis-inducing protein occurs during transition from biotrophy to necrotrophy. *Plant Journal*, **32**, 361–373.

22 Torto, T.A., Li, S., Styer, A., Huitema, E., Testa, A., Gow, N.A.R., van West, P. and Kamoun, S. (2003) EST mining and functional expression assays identify extracellular effector proteins from the plant pathogen Phytophthora. *Genome Research*, **13**, 1675–1685.

23 Nasir, K.H.B., Takahashi, Y., Ito, A., Saitoh, H., Matsumura, H., Kanzaki, H., Shimizu, T., Ito, M., Fujisawa, S., Sharma, P.C., Ohme-Takagi, M., Kamoun, S. and Terauchi, R., (2005) High-throughput *in planta* expression screening identifies a class II ethylene-responsive element binding factor-like protein that regulates plant cell death and non-host resistance. *Plant Journal*, **43**, 491–505.

24 Takahashi, Y., Nasir, K.H.B., Ito, A., Kanzaki, H., Matsumura, H., Saitoh, H., Fujisawa, S., Kamoun, S. and Terauchi, R. (2007) High-throughput screen of cell-death-inducing factors in *Nicotiana benthamiana* identifies a novel MAPKK that mediates INF1-induced cell death signaling and non-host resistance to *Pseudomonas cichorii*. *Plant Journal*, **49**, 1030–1040.

25 Ausubel, F.M., Brent, R., Kingston, R.E., Moore, D.D., Seidman, J.G., Smith, J.A. and Struhl, K. (1994) *Current Protocols in Molecular Biology*, Green Publishing Associates and Wiley-Interscience, New York (USA).

26 Sambrook, J., Fritsch, E.F. and Maniatis, T. (1989) *Molecular Cloning: A Laboratory Manual*, 2nd edn., Cold Spring Harbor Laboratory Press, Cold Spring Harbor, USA.

27 Chomczynski, P. and Sacchi, N. (1987) Single-step method of RNA isolation by acidic guanidinium thiocyanate-phenol-chloroform extraction. *Analytical Biochemistry*, **162**, 156–159.

28 Jones, L., Hamilton, A.J., Voinnet, O., Thomas, C.L., Maule, A.J. and Baulcombe, D.C. (1999) RNA–DNA interactions and DNA methylation in posttranscriptional gene silencing. *Plant Cell*, **11**, 2291–2301.

29 Mersereau, M., Pazour, G.J. and Das, A. (1990) Efficient transformation of *Agrobacterium tumefaciens* by electroporation. *Gene*, **90**, 149–151.

30 Winans, S.C. (1992) Two-way chemical signalling in *Agrobacterium*–plant interactions. *Microbiological Reviews*, **56**, 12–31.

31 Pan, S.Q., Charles, T., Jin, S., Wu, Z.L. and Nester, E.W. (1993) Pre-formed dimeric state of the sensor protein VirA is involved in plant–*Agrobacterium* signal transduction. *Proceedings of the National Academy of Sciences of the United States of America*, **90**, 9939–9943.

32 Kamoun, S., van West, P., Vleeshouwers, V.G., de Groot, K.E. and Govers, F. (1998) Resistance of *Nicotiana benthamiana* to *Phytophtora infestans* is mediated by the recognition of the elicitor I NF1. *Plant Cell*, **10**, 1414–1426.

33 Van Breusegem, F. and Dat, J.F. (2006) Reactive oxygen species in plant cell death. *Plant Physiology*, **141**, 384–390.

34 Altschul, S.F., Gish, W., Miller, W., Myers, E.W. and Lipman, D.J. (1990) Basic local alignment search tool. *Journal of Molecular Biology*, **215**, 403–410.

35 Lee, W.Y., Hong, J.K., Kim, C.Y., Chun, H.J., Park, H.C., Kim, J.C., Yun, D.-J., Chung, W.S., Lee, S.-H., Lee, S.Y., Cho, M.J. and Lim, C.O. (2003) Over-expressed rice ADP-ribosylation factor 1 (RARF1) induces pathogenesis-related genes and pathogen resistance in tobacco plants. *Physiologia Plantarum*, **119**, 573–581.

36 van der Krol, A.R., Mur, L.A., Beld, M., Mol, J.N. and Stuuitje, A.R. (1990) Flavonoid genes in petunia: addition of a limited number of gene copies may lead to a suppression of gene expression. *Plant Cell*, **2**, 291–299.

37 Napoli, C., Lemieux, C. and Jorgensen, R. (1990) Introduction of a chimeric chalcone synthase gene into petunia results in reversible co-suppression of homologous genes *in trans*. *Plant Cell*, **2**, 279–289.

38 Cogoni, C., Irelan, J.T., Schumacher, M., Schmidhauser, T.J., Selker, E.U. and Macino, G. (1996) Transgene silencing of the al-1 gene in vegetative calls of *Neurospora* is mediated by a cytoplasmic effector and does not depend on DNA-DNA interactions or DNA methylation. *EMBO Journal*, **15**, 3153–3163.

39 Lu, R., Malcuit, I., Moffett, P., Ruiz, M.T., Peart, J., Wu, A.J., Rathjen, J.P., Bendahmane, A., Day, L. and Baulcombe, D.C. (2003) High throughput virus-induced gene silencing implicates heat shock protein 90 in plant disease resistance. *EMBO Journal*, **22**, 5690–5699.

40 Burch-Smith, T.M., Anderson, J.C., Martin, G.B. and Dinesh-Kumar, S.P. (2004) Application and advantages of virus-induced gene silencing for gene function analysis in plants. *Plant Journal*, **39**, 734–746.

41 Bartel, D.P. (2004) MicroRNAs: genomics, biogenesis, mechanism, and function. *Cell*, **116**, 281–297.

42 Rui, L., Martin-Hernandez, A.M., Peart, J.R., Malcuit, I. and Baulcombe, D.C. (2003) Virus-induced gene silencing in plants. *Methods*, **30**, 296–303.

7
Microarrays as Tools to Decipher Transcriptomes in Symbiotic Interactions

Helge Küster and Anke Becker

Abstract

In the past it has been found that microarrays constructed from longmer oligonucleotides provide efficient tools to derive snapshots of gene expression in different species. In order to profile gene expression during the symbiotic interaction of the model legume *Medicago truncatula* with beneficial arbuscular mycorrhizal fungi and nitrogen-fixing rhizobial prokaryotes, we developed 70mer oligonucleotide microarrays designated Mt16kOLI1Plus. These expression profiling tools are based on 16 470 oligonucleotide probes representing more than 14 000 genes of the model legume *Medicago truncatula*, an estimated 35% of the gene space. To derive transcription profiles from symbiotic tissue samples, robust target labeling, microarray hybridization, and data evaluation protocols were established. Our microarrays were applied to the identification of genes up-regulated in arbuscular mycorrhiza as well as genes activated both during nodulation and mycorrhization. Our transcriptome profiling experiments not only identified a range of genes associated with different cellular functions required for the formation of efficient root endosymbioses, such as the facilitation of transport processes across perisymbiotic membranes, but also specified putative signaling components as well as transcriptional regulators.

7.1
Introduction

In the era of genomics, high-throughput experiments can be performed to specify the expression of all genes of a given organism in anzy condition of interest. Usually, such strategies are referred to as 'omics' approaches, integrating the level of gene transcription ('transcriptomics'), messenger RNA translation and protein modification ('proteomics'), as well as the synthesis of metabolites ('metabolomics'). Together, these experiments can be regarded as untargeted, since in contrast to classical studies in molecular genetics, their common concept is to measure the expression of as many genes as possible in a single experiment without introducing a bias in the

The Handbook of Plant Functional Genomics: Concepts and Protocols.
Edited by Günter Kahl and Khalid Meksem

ISBN: 978-3-527-31885-8

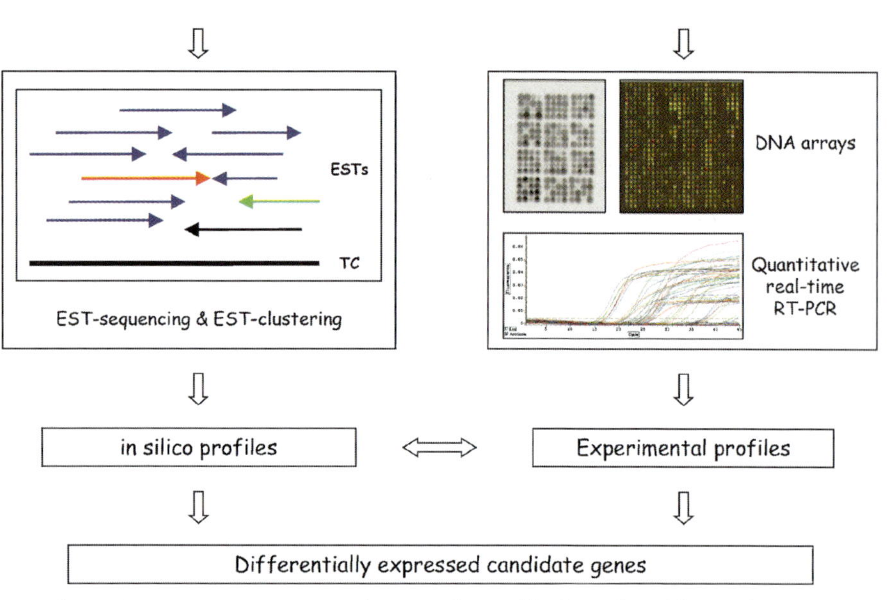

Figure 7.1 Discovery science: a combination of *in silico* and experimental transcriptome profiling approaches supports the identification of candidate genes.Left panel: deep EST sequencing leads to the identification of tentative consensus sequences (TC) by EST clustering. The distribution of ESTs in different cDNA libraries (indicated by different colours) can be traced back to infer a differential expression of the corresponding gene. Right panel: macro- and microarray hybridizations in addition to qRT-PCR experiments can be used to obtain or verify snapshots of gene expression. Both the *in silico* and experimental expression profiles can be related to identify differentially expressed candidate genes.

pre-selection of the genes analyzed. It should be stressed that in the first instance high-throughput expression profiling experiments are not hypothesis-, but rather discovery-driven. Nevertheless, due to their genome-wide scope, they ultimately allow the derivation of novel hypotheses on the conditions investigated [1,2]. Thus, global gene expression studies have the potential to generate biological information that could not have been derived from targeted analyses undertaken on the basis of a limited number of genes being pre-selected according to particular criteria.

With respect to transcriptomics, *in silico* and experimental expression profiling approaches (Figure 7.1) need to be differentiated [3,4]. Typically, *in silico* studies build on the calculation and statistical validation of the frequencies of Expressed Sequence Tags (ESTs) in clustered transcript sequences (referred to as EST-clusters or Tentative Consensus sequences, TCs). The prerequisite for these studies is the existence of comprehensive collections of ESTs derived from the deep terminal sequencing of random and usually non-normalized cDNA libraries that have been constructed from as many different tissues or conditions as possible [5]. In contrast, experimental expression profiling relies on the measurement of gene activities by quantitative reverse transcription PCR experiments (qRT-PCR, [6]), by Serial Analysis of Gene Expression

(SAGE, [7,8]) or by hybridization-based methods using DNA arrays [9]. Whereas real-time RT-PCR and DNA arrays require knowledge of EST or gene sequences and thus constitute 'closed' platforms, the SAGE technology is an 'open' platform that does not require any sequence knowledge prior to an expression profiling experiment [10].

Although real-time RT-PCR can nowadays be scaled-up to allow the simultaneous profiling of several hundred genes [11], only SAGE and DNA array hybridizations offer the possibility of deriving genome-wide snapshots of gene expression in a single experiment [12]. It should be mentioned here that with the emergence of ultra fast sequencing technologies, for example 454 sequencing [13], the application of high-throughput EST-sequencing has recently gained considerable momentum in expression profiling, and it can be expected that these technologies will become a major component of future transcriptomics approaches, in particular for those species where complete genome sequences are not yet available [10].

Various model plants have been selected to study biological conditions of interest using genomic approaches. Examples include *Arabidopsis thaliana* as a general plant model but also as a model for crucifers [14], *Oryza sativa* as a model for grasses [15], and poplar as a model for woody species [16]. In legume plants, two annual species proved to be excellent genomic models: *Medicago truncatula* (barrel medic, [17]) and *Lotus japonicus* (bird's foot trefoil, [18]). In addition to well-advanced genome sequencing projects and the existence of comprehensive mutant collections [19–21], high-throughput EST-sequencing was used to profile gene expression during symbiotic and pathogenic interactions in particular, as well as during seed development [22–24]. In the case of *M. truncatula*, use of these techniques resulted in the deposition of ~225 000 ESTs in three publicly accessible databases: the DFCI *M. truncatula* Gene Index (formerly hosted by TIGR, [5]), the *Medicago* EST Navigation System MENS [25] and the *M. truncatula* DataBase MtDB [26]. In addition to storing EST and TC annotations, these EST databases also provide *in silico* profiling tools [25,27].

Although useful for identifying candidate genes, there is an obvious need to complement and validate *in silico* analyses by experimental expression profiling and gene identification techniques, for example, by carrying out DNA array hybridizations [3,4]. Depending on their mode of manufacture, DNA arrays can be differentiated into macro- and microarrays as well as DNA chips [28,9]. Whereas macro- and microarrays are constructed by spotting defined probes, usually 50–70mer oligonucleotides or PCR fragments covering cDNA in addition to genomic sequences, with the aid of robotic arrayers, DNA chips are obtained by the *in situ* synthesis of either 50–70mer long or 25mer short oligonucleotides [29–31]. In contrast to spotted macro- and microarrays that are usually designed and constructed in the frame of collaborative research projects, chip production is a largely commercial activity, with prominent representatives being Agilent [29], Nimblegen [30], and Affymetrix [31] platforms. Nowadays, microarray or chip expression profiling tools are available for the major model and crop plants studied [32]. Recently, an Affymetrix *Medicago* GeneChip was developed that carries 51 k non-redundant *M. truncatula* probes [33], following the earlier construction of a more targeted 11-k Symbiosis GeneChip [34]. The 51-k Affymetrix *Medicago* GeneChip was designed on the basis of expressed sequence tags from the DFCI *Medicago truncatula* Gene Index [5] and all gene models from the *M. truncatula* genome project available by

July 2005; probably representing more than 80% of all *M. truncatula* genes [35]. Due to the high costs associated with GeneChip hybridizations, these tools have not been widely used in the *Medicago* community up to now.

In the field of spotted arrays, two principle versions can be distinguished: (1) macroarrays that comprise nylon membranes carrying PCR-products at a density of usually less than 100 probes/cm^2, and (2) microarrays where PCR- or 50–70mer oligonucleotide probes are spotted at a density of up to 50 000 probes/cm^2 on surface-modified glass slides [28]. In contrast to macroarrays which are hybridized successively using radioactively labeled nucleic acids (designated "target") representing the transcriptome of one biological condition, microarrays are hybridized simultaneously with combined fluorescently-labeled targets from an experimental and a reference condition [4]. Here, the reference condition serves as an internal standard to correct for inevitable array-to-array variations in spot content (Figure 7.2). That way, microarray hybridizations deliver expression ratios in each experiment [36]. In contrast to spotted arrays, DNA chips display a lower chip-to-chip variation [37], allowing usage of single-label hybridizations and the comparison of expression profiles across different chips without reference hybridizations for every chip. Using appropriate algorithms that apply unsupervised techniques, for example, the generation of self-organizing maps, hierarchical clustering or principal component analyses [38,2], the vast amount of data obtained via microarray or chip experiments can be efficiently mined to identify networks of co-regulated genes as well as marker genes for specific developmental stages [39,40].

In the frame of different international projects, a range of expression profiling tools was developed for the model legume *Medicago truncatula* [4]. As for other plant species, the field moved from the community-driven construction of PCR-product based cDNA-macroarrays, cDNA-microarrays, and 70mer oligonucleotide microarrays to

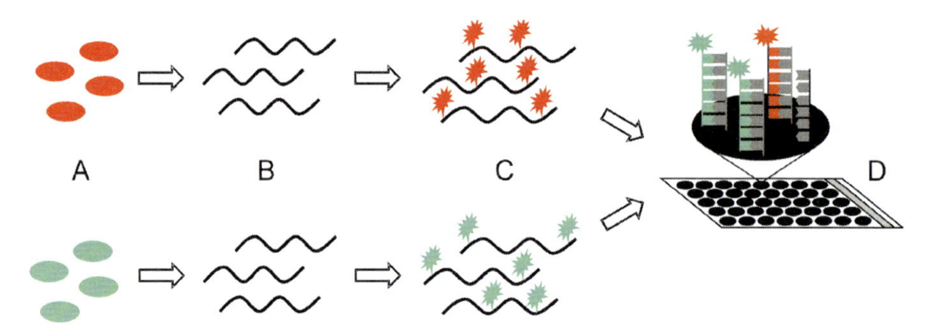

Figure 7.2 Concept of a microarray hybridization using two combined fluorescently labeled targets representing an experimental and a reference condition. (A) Biological samples in an experimental (represented in red) and a reference (represented in green) condition. (B) Pools of total RNA are isolated from the two samples. (C) Labeled targets are synthesized from the two RNA populations by reverse transcription and subsequent Cy-coupling. Two fluorescent dyes with different spectral properties are used, in this example Cy5 and Cy3. (D) Hybridizing the combined labeled targets to a microarray carrying gene-specific probes (black circles) allows the determination of the ratio of Cy-labeled experimental and reference targets. This ratio is proportional to the differences in distribution of the corresponding mRNAs in the original samples.

Table 7.1 Expression profiling tools available for the model legume *Medicago truncatula*.

Array	Platform	Probes	Reference
Mt1k	cDNA microarray	Probes representing a 1-k unigene set of *M. truncatula*	[42]
Mt2.5k	cDNA macroarray	Probes representing 2.5-k EST clusters from arbuscular mycorrhizal roots of *M. truncatula*	[23]
Mt6k	cDNA microarray	Probes representing a 6-k unigene set of *M. truncatula*	[42]
Mt6kRIT	cDNA macroarray	Probes representing 6 k EST-clusters from *M. truncatula* root nodules, AM roots, and uninfected roots	[41]
Mt6kRIT	cDNA microarray	Probes representing 6-k EST-clusters from *M. truncatula* root nodules, AM roots, and uninfected roots	[41]
Mt8k	cDNA microarray	Mt6kRIT probe set plus 2-k probes representing EST clusters from *M. truncatula* flowers and pods	[24]
Mt16kOLI1	70mer oligonucleotide microarray	Probes representing 16-k EST clusters of the TIGR *M. truncatula* Gene Index	[43]
Mt16kOLI1Plus	70mer oligonucleotide microarray	Mt16kOLI1 probe set plus 384 probes representing regulators	[48]
Medicago Symbiosis Chip	Affymetrix GeneChip	10-k probes representing *M. truncatula* EST clusters plus probes representing the *S. meliloti* genome	[34]
Medicago GeneChip	Affymetrix GeneChip	51-k probes representing *M. truncatula* ESTs, EST clusters, and genome sequences plus probes representing the *S. meliloti* genome	[33], [35]

commercially available Affymetrix GeneChips (Table 7.1). With respect to cDNA macroarrays, a 2.5-k array representing genes expressed in arbuscular mycorrhizal roots [23] and a 6-k macroarray from different symbiotic root interactions (Mt6k-RIT, [41]) were developed. The Mt6k-RIT cDNA collection in addition to curated 1- and 6-k unigene sets of *Medicago truncatula* [42], was subsequently used to establish microarray tools. Concomitantly, the Mt6k-RIT microarray was extended to Mt8k versions by the addition of probes from developing flowers and pods [24].

During the first International Conference on Legume Genomics and Genetics (St. Paul, Minnesota, USA, 2002) the *Medicago* community decided to commission a 16-k 70mer oligonucleotide collection representing all publicly available TCs from Operon Biotechnologies (Köln, Germany). The 16-k collection of probes was recently extended by 384 probes targeted against transcription factors and other regulators [4]. In the frame of the EU Integrated Project GRAIN LEGUMES and the DFG mycorrhiza network MolMyk, this probe collection was used to construct 70mer oligonucleotide microarray tools referred to as Mt16kOLI1 and Mt16kOLI1Plus, respectively. In total, it can be estimated that these arrays represent more than 14 000 *M. truncatula* genes, an estimated 35% of the gene space. It should be mentioned that the 16 470 Mt16kOLI1-Plus oligonucleotide probes were derived from a clustered EST collection (TIGR Medicago truncatula Gene Index; [5]) representing more than 50 random cDNA libraries from a range of organs, tissues, and growth conditions. Thus, the Mt16kO-

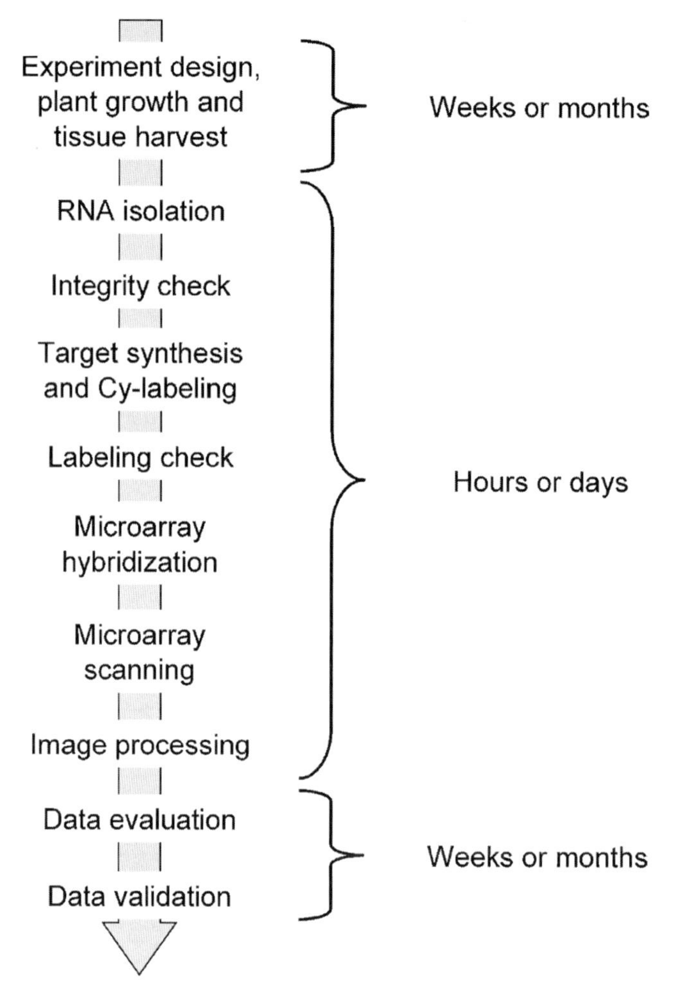

Figure 7.3 Workflow of a typical microarray experiment. The diagram shows the key steps in a microarray experiment. Whereas technically demanding experimental steps associated with target isolation, labeling and hybridization are usually a matter of several hours or a few days for medium-sized experiments, it should be taken into account that the input (i.e. harvesting material grown under defined growth conditions) and data evaluation in general are much more time consuming. Prior to any microarray experiments, care must be taken to select an appropriate set-up for the experiment, for example, to plan for a sufficient number of biological replicates for a subsequent statistical assessment of the data. Following data evaluation, validation experiments are usually required, for example, the confirmation of selected expression patterns by quantitative real-time RT-PCR.

LI1/Mt16kOLI1Plus tools represent a whole-plant collection of *M. truncatula* genes, with no particular bias towards a particular organ or growth condition.

Due to their extensive use in the *M. truncatula* community during the last 2 years [4,43–48], the Mt16kOLI1/Mt16kOLI1Plus 70mer oligonucleotide microarray tools will be the focus of the 'Methods and Protocols' section. It has to be emphasized here that the successful use of these protocols and hence the outcome of a microarray experiment depends on a proper experimental design. In this respect, it is important that a sufficient number of biological replicates are studied to allow a statistical assessment of the data. No fixed number can be given here, but the use of three independent biological replicates can be regarded as a minimum requirement for current standards. Also, appropriate reference conditions need to be identified, possibly by pre-checking the expression of selected marker genes. Since a certain proportion of false-positive or false-negative results is inevitable in global transcriptomics studies, the expression profiles of a subset of differentially expressed genes is usually validated by independent methods, for example, Northern blot hybridizations or qRT-PCR experiments [6]. Figure 7.3 illustrates the key steps of a microarray experiment, starting from the experimental design, and relates these steps to their approximate duration. The most relevant sections of this experimental workflow are detailed in the subsequent 'Methods and Protocols' section.

7.2 Methods and Protocols

7.2.1 Spotting and Storage of Mt16kOLI1/Mt16kOLI1Plus 70mer Oligonucleotide Microarrays

A range of microarray spotters and several microarray slide surfaces are commercially available, the latter requiring surface-specific storage, blocking, and prehybridization steps. Here, we can only briefly summarize the conditions relevant for Mt16kOLI1/Mt16kOLI1Plus microarrays.

All 70mer oligonucleotide probes were designed and synthesized by Operon Biotechnologies (Köln, Germany). Lyophilized oligonucleotides were delivered in 384-well microplates and were subsequently dissolved in 15 μl of a sterile-filtered and autoclaved solution of 3× SSC/1.5 M betaine (20× SSC is a solution of 3 M NaCl and 0.3 M Na_3-citrate at pH 7.4) to yield a 40 μM solution. In order to completely dissolve the 70mer oligonucleotides, the microplates were incubated for 48 h at 4 °C, and shaken for 10 min at 400 rpm every 10–12 h. The microplates were sealed with sealing foil (Greiner bio-one, Essen, Germany) and stored at −20 °C until use. Immediately before microarray printing, the plates were thawed to room temperature and spun down for 5 min at 2000 rpm.

Using a MicroGrid II (Zinsser Analytics GmbH, Frankfurt, Germany) according to the operation manual, 70mer oligonucleotide probes were arrayed on 'Nexterion E' epoxy-modified microarray slides (Peqlab Biotechnologie GmbH, Erlangen,

Germany). The layout and design of the Mt16kOLI1/Mt16kOLI1Plus microarrays is documented in array definition files A-MEXP-85 and A-MEXP-138 from the ArrayExpress database. Immediately after printing, the probes were cross-linked by baking the slides for 2 h at 80 °C. Cross-linked slides were sealed in plastic bags together with fresh desiccation packs and stored at 18–20 °C in a humidor.

7.2.2
Synthesis of Targets by Indirect Reverse Transcription Cy-Labeling

This section describes the synthesis of Cy-labeled targets by reverse transcription of total RNA, using a mixture of double anchored oligo-dT primers and random hexamers in conjunction with indirect aminoallyl-coupling of fluorescent dyes. This procedure requires 10–20 μg of total RNA, an amount that can usually be obtained from pooled tissue samples. Since RNA integrity is important in order to derive meaningful expression profiles, we recommend using the commercial column-based RNA purification systems available from various suppliers (e.g. Qiagen, Hilden, Germany). In addition, it is important to check RNA integrity, using for example an Agilent 2100 Bioanalyzer (Agilent Technologies, Waldbronn, Germany). In cases where the amount of total RNA available is less than 10 μg, for instance where samples have been obtained by microdissection, target amplification protocols must be applied. These can be either based on exponential PCR-amplification or on linear T7 amplification. A range of commercial kits is available for these procedures (BD Biosciences, Heidelberg, Germany; Biozym Scientific GmbH, Hessisch Oldendorf, Germany), and the Cy-labeled targets obtained can be used with the hybridization protocols described below.

7.2.2.1 Components Stored at −20 °C

- 5× Reaction buffer (Bioline GmbH, Luckenwalde, Germany).
- BioScript Reverse Transcriptase (200 U/μl; Bioline GmbH, Luckenwalde, Germany).
- RNAse inhibitor (40 U/μl; Invitrogen GmbH, Karlsruhe, Germany).
- 0.2 M NaOH and 0.2 M HCl (Merck KGaA, Darmstadt, Germany).
- 4 M hydroxylamine (Sigma-Aldrich Chemie GmbH, München, Germany; dissolve in MilliQ water).
- 1 M sodium bicarbonate pH 9.0 (Sigma-Aldrich Chemie GmbH, München, Germany; dissolve in MilliQ water and adjust pH).
- 25 × dNTP (4 : 1 aa-dUTP/dTTP mix) stock as follows, store at −20 °C in aliquots.

– 100 mM dATP	31.25 μl (final concentration 12.5 mM)
– 100 mM dCTP	31.25 μl (final concentration 12.5 mM)
– 100 mM dGTP	31.25 μl (final concentration 12.5 mM)
– 100 mM dTTP	6.25 μl (final concentration 2.5 mM)
– 50 mM aa-dUTP	50.00 μl (Fermentas GmbH, St. Leon-Rot, Germany; final concentration 10.0 mM)
– RNAse-free water	100 00 μl

- Unmodified or amino-modified random hexamer primers (dissolved in DEPC-water).
- Double anchored oligo-dT_{15}VN primers (dissolved in DEPC-water).

7.2.2.2 Components Stored at 4 °C (−20 °C after Aliquoting in 1/10 Volumes)

- Cy3-NHS or Alexa555/Alexa532/Alexa546-NHS ester (GE Healthcare, Freiburg, Germany; Invitrogen GmbH, Karlsruhe, Germany).
- Cy5-NHS or Alexa647-NHS ester (GE Healthcare, Freiburg, Germany; Invitrogen GmbH, Karlsruhe, Germany).
- In each case, 1/6th of one aliquot of the monoreactive dye is used for one labeling.
- To prepare aliquots, dissolve NHS esters in 10 µl of water-free DMSO; it is essential to avoid any contact of the dyes with water prior to labeling. Immediately re-seal DMSO with fresh desiccation packs, aliquot 1.5 µl of NHS esters into brown Eppendorf tubes, and speed-vac in the dark for 45 min. Seal dried NHS esters in plastic bags together with desiccation packs, and store them at −20 °C.

7.2.2.3 Components Stored at Room Temperature

- CyScribe GFX Purification Kit (GE Healthcare, Freiburg, Germany)
- RNAse-free Eppendorf tubes and filter tips
- Autoclaved MilliQ water
- DEPC-treated water
- 80% (v/v) ethanol (diluted from absolute ethanol with DEPC-treated water).

7.2.2.4 Reverse Transcription of Total RNA to obtain Aminoallyl-Labeled First-Strand cDNA

1. Preheat 42 and 70 °C heating blocks for 30 min before starting, prepare an ice bucket.
2. Wear gloves, use filter tips, autoclavable pipettmen, and RNAse-free Eppendorf tubes.
3. Thaw DEPC H_2O, 5 × Reaction Buffer and primers.
4. Combine up to 18.8 µl (10–20 µg) of total RNA, 1.0 µl of oligo-dT_{15}VN primers (2.5 µg/µl), and 1.0 µl of 5′-aminomodified random hexamers (5 µg/µl); if necessary add DEPC-treated H_2O to obtain a final volume of 20.8 µl.
5. Mix by flicking and quickly spin down.
6. Incubate at 70 °C for 10 min in a heating block.
7. Incubate for 5 min on ice to effect primer annealing, quickly spin down.
8. During the incubation on ice, prepare a master mix in an RNAse-free Eppendorf tube, adding an extra volume of 10% if multiple labeling is required.

RNAse Inhibitor, BioScript and 25 × dNTP should be added immediately before use.

• 5 × Reaction Buffer	6.0 μl
• RNase inhibitor (40 U/μl)	0.5 μl
• BioScript reverse transcriptase (200 U/μl)	1.5 μl
• 25 × dNTP stock solution including aa-dUTP	1.2 μl
• Mix by flicking, spin down, and leave at RT until use (do NOT store this mix on ice)	

9. At RT, add 9.2 μl of the master mix to each annealing reaction, mix by flicking, spin down.
10. Incubate at 42 °C for 90 min in a heating block.
11. Allow 0.2 N NaOH, 0.2 N HCl, 1 M sodium bicarbonate and 4 M hydroxylamine to stand at RT. From now on, RNAse-free conditions are not required.
12. Wearing gloves, place CyScribe GFX columns (one per labeling) in collection tubes and prepare one empty 1.5-ml tube per labeling.
13. Thaw and vortex 1 M sodium bicarbonate, pH 9.0 to dissolve white precipitates.
14. Prepare 0.1 M sodium bicarbonate (pH 9.0) by diluting the 1 M stock solution in MilliQ water. Note that 60 μl 0.1 M sodium bicarbonate (pH 9.0) is required per labeling.
15. Prepare 1.8 ml 80% (v/v) ethanol per labeling.

7.2.2.5 Hydrolysis of RNA

1. Add 15 μl of commercial 0.2 M NaOH with an arrested pipettman. Observe exact flow out.
2. Mix by flicking and spin down.
3. Incubate at 70 °C for 10 min in a heating block.
4. Add 15 μl of commercial 0.2 M HCl using the arrested pipettman. Ensure that an exact flow out is observed, mix immediately by pipetting up and down to avoid precipitation.
5. Immediately after each target is neutralized, quickly proceed with CyScribe GFX column purification. In these and the following steps, do NOT use Tris-containing buffers instead of water, since the amino groups will interfere with the subsequent Cy-coupling.

7.2.2.6 Clean-Up of Aminoallyl-Labeled First-Strand cDNA

1. Directly after neutralization of one labeling reaction add 450 μl capture buffer to the reaction and mix by pipetting up and down. Proceed with this step until all

labeling reactions have been neutralized and mixed with capture buffer. Note that samples should not stay in capture buffer longer than 5 min.

2. Add the complete neutralized mix to a CyScribe GFX column.
3. Spin at 13 000 rpm for 30 s at 20 °C in a microcentrifuge and discard the flow-through.
4. Add 600 μl of 80% (v/v) ethanol.
5. Spin at 13 000 rpm for 30 s at 20 °C in a microcentrifuge and discard the flow-through.
6. Repeat this washing step twice.
7. Spin at 13 000 rpm for 10 s at 20 °C in a microcentrifuge and place column in a new 1.5-ml tube.
8. Add 60 μl 0.1 M sodium bicarbonate (pH 9.0).
9. Incubate for 5 min at RT.
10. Spin at 13 000 rpm for 1 min at 20 °C in a microcentrifuge.
11. Immediately proceed with Cy-coupling or store purified first strand cDNA at −20 °C.

7.2.2.7 Coupling of Fluorescent Dyes to Aminoallyl-Labeled First-Strand cDNA

1. Protect samples from light by using brown Eppendorf tubes from now on. Avoid exposure to room light and direct sunlight.
2. Dissolve Cy3- or Cy5-NHS esters (aliquoted in brown Eppendorf tubes) in the complete aa-containing first strand cDNA solution by pipetting up and down several times until the dye is completely dissolved.
3. Do NOT spin down, just tap down drops from the side of the Eppendorf tubes.
4. Incubate for 1–2 h at RT in the dark.

7.2.2.8 Quenching of all Remaining NHS Esters

1. Add 4.5 μl of 4 M hydroxylamine
2. Mix by flicking, do NOT spin down
3. Leave for 15 min at RT in the dark

7.2.2.9 Clean-up of Fluorescently Labeled Targets

1. Work quickly to protect labeled targets from light, Cy5 bleaches quickly and is particularly sensitive to high ozone concentrations.
2. Add 600 μl capture buffer (CyScribe GFX Purification Kit) to the Cy5-labeled sample and mix by pipetting up and down, then add the Cy3-labeled sample to this solution and mix by pipetting up and down.

3. Apply all to a GFX column in a collection tube (CyScribe GFX Purification Kit); do not leave the sample in capture buffer for more than 5 min.
4. Spin at full speed (~10 000–13 000 rpm) for 30 s and discard the flow-through.
5. Add 600 µl washing buffer (CyScribe GFX Purification Kit).
6. Spin at full speed (~10 000–13 000 rpm) for 30 s and discard the flow-through.
7. Add 600 µl washing buffer (CyScribe GFX Purification Kit).
8. Spin at full speed (~10 000–13 000 rpm) for 30 s and discard the flow-through.
9. Add 600 µl washing buffer (CyScribe GFX Purification Kit).
10. Spin at full speed (~10 000–13 000 rpm) for 30 s and discard the flow-through.
11. Spin at full speed (~10 000–13 000 rpm) for 10 s.
12. Transfer the dried GFX column to a fresh brown Eppendorf tube.
13. Add 60 µl elution buffer (CyScribe GFX Purification Kit) to the center of the filter.
14. Leave for 5 min at RT, then spin at full speed (~10 000–13 000 rpm) for 1 min.
15. The resulting 60 µl of combined Cy3/Cy5-labeled targets is transferred to a fresh brown Eppendorf tube with a screw cap. Remove 2 µl into a normal Eppendorf tube for checking the quality of targets.
16. Freeze Cy-labeled targets at −20 °C until use.

7.2.2.10 Quality Control of Fluorescently Labeled Targets

1. Combine 1 µl of labeled targets with 4 µl 80% (v/v) glycerol, and run a 0.8% (w/v) agarose gel in TA buffer (40 mM Tris-Cl, 10 mM sodium acetate, 1 mM EDTA; adjust to pH 7.8 with pure acetic acid) for 20 min at 80 V.
2. Scan the gel first in the Cy5 and then the Cy3 channel on a Typhoon phosphoimager (GE Healthcare, Freiburg, Germany) for example.
3. Check for the presence of a fluorescent smear indicating the size of labeled targets.
4. Analyze 1 µl in the ND-1000 Spectrophotometer (Peqlab Biotechnologie GmbH, Erlangen, Germany). Typical values for reverse-transcription labeling from 10 to 20 µg total RNA are in the range of 80–150 ng/µl cDNA with 0.03–0.06 pg Cy-dye per ng cDNA.

7.2.3 Pre-Processing, Hybridization and Scanning of Mt16kOLI1/Mt16kOLI1Plus Microarrays

Mt16kOLI1/Mt16kOLI1Plus oligonucleotide microarrays are spotted on Nexterion E (Peqlab Biotechnologie GmbH, Erlangen, Germany) slides that must be processed prior to hybridizations to block free epoxy groups using the following solutions.

- Rinsing solution 1: Mix 250 ml MilliQ H_2O and 250 µl Triton X100, dissolve at 80 °C for 5 min, cool down to room temperature.
- Rinsing solution 2: Mix 500 ml MilliQ H_2O and 50 µl 32% (v/v) HCl
- Rinsing solution 3: Mix 225 ml MilliQ H_2O and 25 ml 1 M KCl

- Blocking solution: Mix 150 ml MilliQ H_2O with 47 µl 32% (v/v) HCl and 50 ml 4× blocking solution as follows. Pre-warm the MilliQ/HCl mix to 50 °C, add the 4× blocking solution 5 min before use and pre-warm this solution to 50 °C for at least 5 min. The temperature of the blocking solution must be 50 °C at the beginning of the 15-min blocking step. Please note that the blocking solution is unstable and must not be stored longer!

1. Remove a sealed slide package from the humidor.
2. Wearing gloves, take out the desired slides (only touch them in the area of the code no., the DNA side faces up when the number can be read). Seal the remaining slides of the box together with a new desiccation pack and return the sealed package to the humidor maintained at 18–20 °C.
3. Place the slides in a plastic rack and carry out processing by transferring the racks from one container to another, occasionally lift the rack up and down during washing.
4. Wash slides for 5 min at room temperature in 250 ml of rinsing solution 1.
5. Wash slides twice for 2 min at room temperature in 250 ml of rinsing solution 2.
6. Wash slides for 10 min at room temperature in 250 ml of rinsing solution 3.
7. Wash slides for 1 min at room temperature in 250 ml of MilliQ H_2O.
8. Incubate slides for 15 min at 50 °C in 200 ml prewarmed blocking solution in a glass container, shake at least every 5 min or apply constant shaking. Use a flat bottomed glass container to process one to two slides (20 ml of blocking solution) and a multiple glass container to process multiple slides (200 ml blocking solution).
9. Wash slides for 1 min at room temperature in 250 ml of MilliQ H_2O
10. Place rack on an approximately 12 × 8-cm plastic microplate cover containing two Kim-wipes and immediately spin in the microplate centrifuge at 1200 rpm for 3 min. Use a stack of three used glass slides at every side of the plastic dish to elevate the rack containing the slides, this avoids the occurrence of precipitation artefacts on the side of the slide.

Several commercial systems can be used to hybridize Mt16kOLI1/Mt16kOLI1Plus microarrays using Cy-labeled targets, for example the ASP station (GE Healthcare, Freiburg, Germany) or the HS4800 (Tecan Deutschland GmbH, Crailsheim, Germany). Follow the instructions given in the manufacturer's manual. In our hands, manual washing is superior to washing in hybridization stations, and essential steps in manual washing of Mt16kOLI1/Mt16kOLI1Plus microarrays are as follows.

1. Prepare 500 ml of each washing buffer in demineralized water from appropriate stocks (20× SSC, 10% (w/v) SDS; 20× SSC is a solution of 3 M NaCl and 0.3 M Na_3-citrate at pH 7.4). Preheat the 2× SSC, 0.2% (w/v) SDS washing buffer to 42 °C and adjust the temperature of the 0.05× SSC washing buffer to exactly 21 °C.

2. Immediately before the hybridization program terminates, pour 250 ml 2× SSC, 0.2% (w/v) SDS washing buffer prewarmed to 42 °C into a black plastic box (app. 12 × 8 × 5 cm; Carl Roth GmbH, Karlsruhe, Germany).

3. Remove the slides from the hybridization machine wearing gloves and only touching the edges and place them into a plastic slide rack (Carl Roth GmbH, Karlsruhe, Germany) This rack should be immersed in the prewarmed 2× SSC, 0.2%(w/v) SDS washing buffer (poured into a black plastic box as mentioned above) to prevent the hybridized slides drying out.

4. Move up and down several times immediately and shake for 1 min on a horizontal shaker at 50–100 rpm after removal of the last slide. From now on move slide racks up and down every 15 s to avoid the formation of air bubbles.

5. Transfer to 0.2× SSC, 0.1% (w/v) SDS (RT) in a plastic slide rack, shake for 1 min.

6. Transfer to 0.2× SSC, 0.1% (w/v) SDS (RT) in a plastic slide rack, shake for 1 min.

7. Transfer to 0.2× SSC (RT) in a plastic slide rack and shake for 1 min.

8. Transfer to 0.2× SSC (RT) in a plastic slide rack and shake for 1 min.

9. Transfer to 0.05× SSC (held at exactly 21 °C) in a plastic slide rack, shake for 1 min.

10. Place rack on an approximately 12 × 8-cm plastic microplate cover containing two Kim-wipes and immediately spin at 1200 rpm for 3–5 min. Use a stack of three used glass slides at every side of the plastic dish to elevate the rack containing the slides, this avoids the occurrence of precipitation artefacts on the side of the slide. If necessary, dry corners of the slide with a Kim-wipe.

11. Place dried slides in a slide box in the dark until required for scanning to avoid bleaching of the Cy5 dyes. Work quickly at all times, avoid direct light and exposure to high ozone concentrations, since this strongly enhances Cy5-bleaching.

12. Proceed with microarray scanning, using an instrument that provides automatic gain control functionality to adjust photomultiplier gain (e.g. LS Reloaded; Tecan Deutschland GmbH, Karlsruhe, Germany), and follow the manufacturer's instructions.

7.2.4
Handling and Evaluation of Microarray Data

Microarray scanners usually scan hybridized slides in steps of 1–10 μm, recording pixels of (1–10 μm) × (1–10 μm). As a rule of thumb, spot diameters should be broken down into approximately 10 pixels. Since typical microarray spots are approximately 80–100 μm in diameter, recording 10 × 10 μm pixels is sufficient to provide enough independently measured pixels per spot.

Scanning Mt16kOLI1/Mt16kOLI1Plus microarrays at a $10 \times 10\,\mu m$ resolution yields two primary data files of approximately 25 Mb, one for each Cy-channel. These primary files are stored in 16-bit TIFF format that encodes the fluorescence intensity per pixel on a 16-step $\log_2$ scale with a non-logarithmic signal intensity range up to an absolute value of 65 535. Thus, microarrays offer a dynamic range of five orders of magnitude for measuring signal intensities.

Subsequent to scanning, the TIFF files obtained should be analyzed using image processing software to identify spots by placing grids on the overlaid Cy5- and Cy3-TIFF files to separate signal pixels from local background pixels via image segmentation, and to record pixel intensities for signal and background pixels, respectively. Usually, the output of image processing software such as ImaGene (BioDiscovery, El Segundo, CA, USA) or GenePix (Molecular Devices Corporation, Sunnyvale, CA, USA) are tab-deliminated text files specifying at least the arithmetic mean of signal and background pixels for each spot, together with information on spot quality that is usually encoded by flag numbers.

Following image processing, the data files obtained can be imported into a range of commercial microarray evaluation software packages such as GeneSight (BioDiscovery, El Segundo, CA, USA) and GenePix (Molecular Devices Corporation, Sunnyvale, CA, USA), or to shareware software [49]. All software tools mentioned are capable of running different statistical analyses which specify lists of differentially expressed genes and offer higher-order unsupervised techniques, for example hierarchical clustering or principal component analysis [9,38]. For further details of the specific properties of microarray evaluation software, the experimenter should consult the detailed manuals supplied with the respective software.

7.3 Applications of the Technology

7.3.1 Microarray-Based Identification of *Medicago truncatula* Genes Induced during Different Arbuscular Mycorrhizal Interactions

Some 80% of all terrestrial plants enter an arbuscular mycorrhiza (AM) symbiosis with *Glomeromycota* fungi [50]. During AM, fungal hyphae penetrate the root epidermis via appressoria, pass through the outer cortical cells, and proliferate in the inner cortex. Here, they form highly branched intracellular structures known as arbuscules. In addition to the intraradical hyphae, arbuscules are the sites of nutrient exchange between the macro- and microsymbiont, with plant hexoses being exchanged for phosphate and other mineral nutrients [51,52]. Ultimately, AM formation leads to a substantial reprogramming of root physiology, with approximately 20% of all plant photosynthates being allocated to AM roots to support growth under nutrient-limiting conditions [53].

For a long time molecular research on AM was hindered by the development of asynchronous symbiosis leading to the concomitant presence of different AM stages and by the obligate biotrophy of AM fungi. As a result, only a few AM-induced plant genes were reported some years ago [54,55]. With the availability of global transcriptomics platforms for the model legume *M. truncatula*, identification of comprehensive collections of genes activated during AM became possible. To specify the common genetic program induced by the two widely studied fungal microsymbionts *Glomus mosseae* and *Glomus intraradices*, we applied our 70mer oligonucleotide microarrays [4,43] in conjunction with the target labeling protocols detailed above. This global transcriptome profiling approach was based on pooled tissue samples harvested from AM roots 4 weeks after inoculation with either AM fungus. Before conducting microarray experiments, all AM root samples were checked by histological staining for the comparable presence of fungal structures and by real-time RT-PCR [43] to assure a similar induction of the phosphate transporter gene MtPt4, a marker for efficient AM formation [56]. Using the ImaGene software (BioDiscovery, El Segundo, CA, USA) for image processing of primary hybridization data and the EMMA software [49] for data normalization and statistical analyses [57], a total of 201 *M. truncatula* genes were found to be significantly induced at least twofold in colonized roots of both AM interactions [43,44,47]. The complete dataset on *M. truncatula* AM-related gene expression can be retrieved from the ArrayExpress database using accession number E-MEXP-218.

The 201 AM-induced genes comprised two AM marker genes previously identified by targeted molecular studies, thus validating our global transcriptomics dataset. Amongst those marker genes were the AM-specific phosphate transporter MtPt4 [58], the germin-like protein MtGlp1 [59], the glutathione S-transferase MtGst1 [60], the serine carboxypeptidase MtScp1 [23], the hexose transporter MtSt1 [61], the 1-deoxy-D-xylulose 5-phosphate synthase MtDXS2 [62], and a multifunctional aquaporin [63]. Most of these genes are specifically expressed in arbuscule-containing cells (Figure 7.4), suggesting that other co-regulated AM-induced genes are also expressed in these specialized symbiotic cells.

Amongst the 201 co-induced genes, we identified more than 150 genes that had not previously been described as AM-related. These genes specified previously unknown nitrate, manganese and sugar transporters with possible relevance to symbiotic nutrient exchange, a range of enzymes involved in secondary and hormone metabolism, different Kunitz-type proteases, and several protease inhibitors; the latter two categories being known characteristics of AM roots [52,64,65]. With respect to symbiotic signaling [66], several co-induced genes encoded LRR-type receptor kinases and other putative receptor kinases. In addition, novel AM-induced transcriptional regulators including different Myb and bZIP transcription factors were identified [23,43]. Apart from more than 200 co-induced genes, several hundred genes were specifically up-regulated during the *Glomus mosseae* or the *Glomus intraradices* AM, implying that the plant genetic program activated in AM roots to some extent depends on the colonizing microsymbiont.

Subsets of genes identified by global transcriptome profiling studies were subjected to real-time RT-PCR to verify their symbiosis-induced expression [43]. These experiments, together with similar results from other AM transcriptome profiling experiments using cDNA arrays [23,67], confirmed the differential expression of the

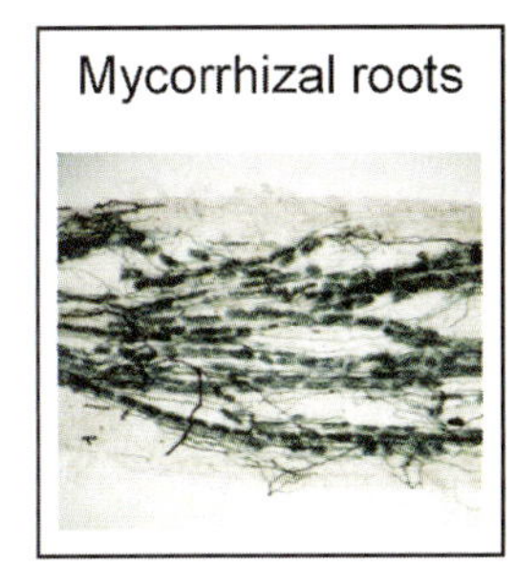

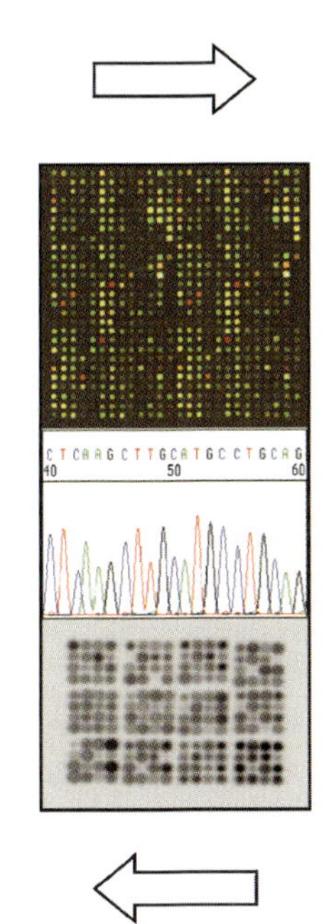

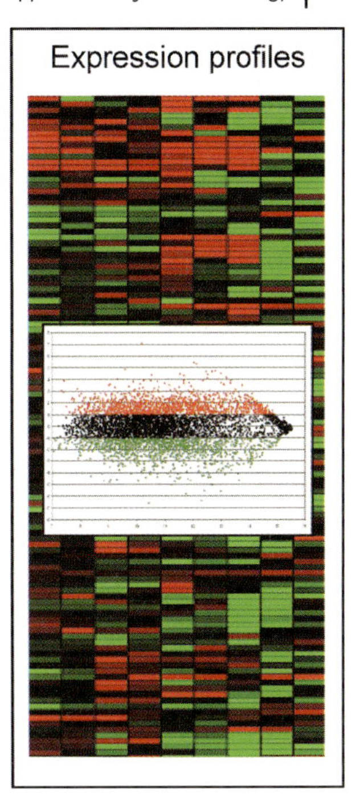

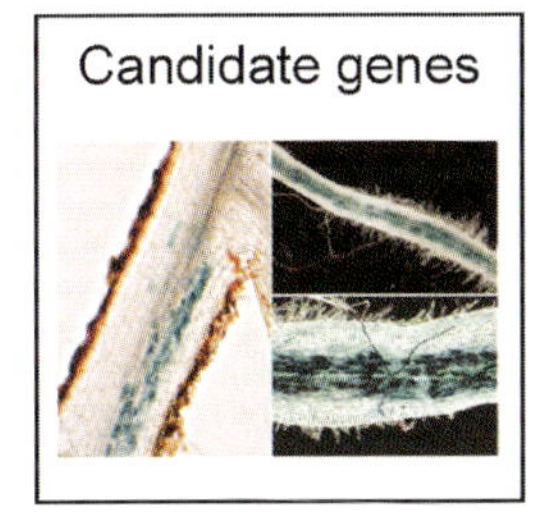

Figure 7.4 Transcriptomics-based identification of genes activated in arbuscular mycorrhizal roots. Microarray hybridizations and *in silico* studies (central panel) were undertaken to identify the genetic program common to different arbuscular mycorrhizal interactions. The data obtained were evaluated using plotting tools and hierarchical clustering algorithms (right panel) to identify genes specifically induced during AM. A subset of genes was checked by expressing promoter-reporter gene fusions in transgenic roots (lower left panel) to specify an arbuscule-related expression.

vast majority of genes studied [43,47]. In addition, the real-time RT-PCR studies led to the conclusion that microarray-based expression data from pooled tissue samples in general are more reliable than *in silico* predictions relying on EST distributions [67]; an observation corroborating the theoretical considerations documented in [2]. In particular TCs that are only represented by a limited number of ESTs cannot be reproducibly detected across the different biological replicates used for microarray hybridization and real-time RT-PCR experiments. In such cases, either the *in silico* profiles suffer from a lack of biological repetition or the microarray hybridizations fail to detect locally expressed genes that can conversely, be detected by high-throughput EST sequencing. On the other hand, for TCs represented by many ESTs from deeply sequenced cDNA libraries, there appears to be a reasonable overlap between *in silico* and experimental transcriptome profiles [22,44].

These observations support the complementarities of 'closed' (e.g. microarray-based) and 'open' (e.g. SAGE- or 454-technology-based) transcriptome profiling approaches [10]. It can be expected that the future application of 'open' transcriptome profiling technologies – in particular when coupled to cellular expression profiling – will generate substantial information concerning gene expression during AM interactions.

7.3.2 Microarray-Based Identification of *Medicago truncatula* Genes Activated during Nodulation and Mycorrhization

In addition to the formation of AM symbioses with mycorrhizal fungi, legumes have the unique capacity to enter a nitrogen-fixing root nodule symbiosis with soil prokaryotes from different genera. In the case of *M. truncatula*, the nitrogen-fixing microsymbiont *Sinorhizobium meliloti* has been sequenced [68] and constitutes an excellent model system to study symbiotic nitrogen fixation [69,70]. It should be emphasized that although the different symbiotic microbes colonize root tissues intracellularly during nodulation and mycorrhization, they are separated from the plant cytoplasm by highly specialized perisymbiotic membranes [71,72]. Considering the apparent analogies in the infection processes [73,74], an overlap in gene expression was proposed [75]. Common gene activation is particularly evident for the signaling cascades that initiate both symbioses [76–78]. Thus, it is tempting to speculate that the root nodule symbiosis had adopted ancient signaling pathways leading to AM formation that had already been established 400 million years ago [79].

In the past few years, expression profiling strategies have been pursued to identify symbiotically-induced (symbiosin) genes co-activated during nodulation and mycorrhization. These strategies combined deep EST sequencing, analysis of suppressive subtractive (SSH) cDNA libraries, *in silico* and microarray-based profiling of symbiosis-related gene expression. Together, these approaches have identified several genes that were co-activated in the two legume symbioses [25,41,43,63,65,67,80,81].

To relate the specific gene expression profiles from AM roots described above to those from nitrogen-fixing root nodules, we applied our 70mer oligonucleotide microarrays to obtain gene expression profiles from different stages of root nodule development (Figure 7.5, [43,45]). For these experiments, *M. truncatula* plants were grown under defined conditions in aeroponic caissons [82], and root nodules were collected between 2 and 3 weeks post-inoculation. The complete nodulation-related dataset can be retrieved from the ArrayExpress database using accession number E-MEXP-238.

Interestingly, our microarray experiments revealed only a limited overlap between the transcription profiles of mycorrhizal roots and root nodules, with approximately 12% of the genes that were identified as AM-induced also being activated at least twofold in nitrogen-fixing root nodules [67,43]. Although delivering novel genes with a symbiosis-related expression profile, this overlap was somewhat lower than expected [75]. Dilution effects probably masked the detection of locally expressed genes in addition to problems related to the lack of resolution of cellular expression due to the use of pooled tissue samples [47].

In general, the genes co-activated during nodulation and mycorrhization (symbiosin genes) were correlated to later stages of the symbioses, where the encoded proteins facilitated transport processes across perisymbiotic membranes, formation of symbiotic membrane structures, and specific modification of extracellular matrices [67]. Amongst the symbiotically induced genes encoding membrane proteins, the

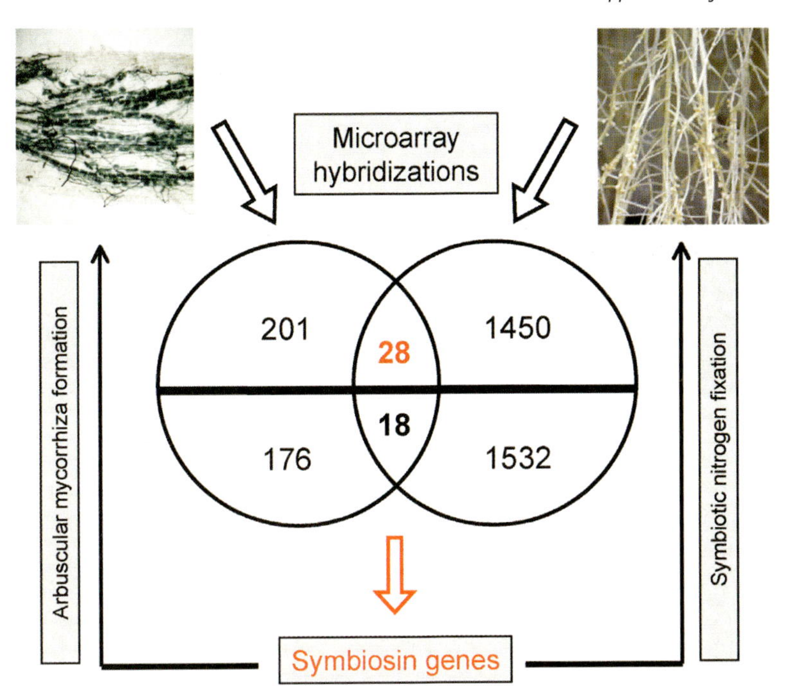

Figure 7.5 Identification of symbiotically activated genes by transcriptomics. Microarray hybridizations were performed to identify the genetic program common to different arbuscular mycorrhizal interactions (left side) and nitrogen-fixing root nodules (right side) of *Medicago truncatula*. Venn diagrams are used to visualize the number genes found to be differentially expressed in either symbiosis, with the overlap of symbiotically co-induced genes shown in red. These symbiosin genes can be related to different cellular functions required for the formation of efficient root endosymbioses.

MtHa1 gene encoding a plasma membrane H^+-ATPase localized in periarbuscular membranes [83] deserved attention. The induction of *MtHa1* also during nodulation illustrates the common requirement for acidification of the perisymbiotic space in both root endosymbioses [67], most probably to facilitate energy-dependent nutrient transport across perisymbiotic membranes. Additional symbiosin genes encoded the multifunctional aquaporin gene MtNip1 [63] and a membrane nodulin of unknown function originally identified in soybean symbiosome membranes [84]. The activation of different genes encoding predicted symbiosome membrane proteins supports the hypothesis that peribacteroid and periarbuscular membranes share common structural properties to support symbiotic metabolite exchange.

With respect to membrane formation, the symbiotically induced *MtAnn2* gene encoding an annexin might play a role in the calcium-dependent reorganization of membranes during the colonization of plant tissues by microbial structures, a function proposed for other plant annexins [67]. Since the invading microsymbionts remain surrounded by membranes of plant origin during all stages of infection and intracellular colonization of root tissues, such processes are obviously relevant for both root nodule and AM symbiosis. An *in situ* localization of *MtAnn2* promoter activity showed an induction of the gene in the nodule primordium and in arbuscule-containing cells [67], an observation consistent with a function during initiation or establishment of membrane-surrounded endosymbiotic structures.

Finally, the co-induction of polygalacturonase and endo-1,3-1,4-β-D-glucanase genes indicates the recruitment of similar cell wall-modifying enzymes in root nodules and AM, possibly related to the modification of extracellular matrices surrounding symbiotic structures. Interestingly, genes encoding enzymes involved in protein processing, for example, the serine carboxypeptidase gene *MtScp1* [23], were also activated in both symbioses.

To obtain transcriptional snapshots from earlier stages of nodulation, we initiated microarray experiments studying the response of *M. truncatula* roots to secreted rhizobial nodulation factors which trigger nodulation (Andreas Niebel, LIPM, INRA Toulouse, and Helge Küster, unpublished data). In addition, we started laser-capture-microdissection experiments to obtain cellular snapshots of gene expression in mature root nodules, that way differentiating gene expression in the infected, symbiotic cells and gene expression in the infection area of the nodule (Erik Limpens, Wageningen University, and Helge Küster, unpublished data). Due to an increased temporal and spatial resolution, it can be expected that these experiments will further advance our understanding of the symbiosis-specific gene expression in legume plants in response to colonizing micro-organisms.

7.4 Perspectives

One limitation of 70mer-based microarrays is the cross-hybridization of probes that covers parts of the coding regions, thus abolishing the differentiation of closely related members of gene families. This problem can only in part be alleviated by the placement of probes in 3′ untranslated regions, since for species without a complete genome sequence probes are not available for a number of genes. It can be expected that such problems will be overcome, once the 70mer oligonucleotide microarrays presented here are replaced by the recently released Affymetrix *Medicago* GeneChips which rely on shorter and more gene-specific oligonucleotide probes.

A major drawback of current symbiosis research is the use of pooled tissue samples, where different cell types and different stages of development are mixed, thus obscuring the detection of genes differentially expressed only in specific cell types. To solve this problem, the development of robust single cell expression profiling technologies for plant tissues has become the current focus of research [10,85]. Once global transcriptomics experiments advance to the cellular level, an integrated view of the symbiotic *M. truncatula* transcriptome based on specific cell types can be obtained, for example, infected root nodule cells as well as arbuscule-containing cells of AM roots. Combined with mutant analyses, it can be expected that the resulting cellular picture of symbiotic interactions will provide molecular information regarding how *M. truncatula* root cells are reprogrammed to accommodate beneficial micro-organisms.

References

1 Provart, N.J. and McCourt, P. (2004) Systems approaches to understanding cell signaling and gene regulation. *Current Opinion in Plant Biology*, **7**, 605–609.

2 Allison, D.B., Cui, X., Page, G.P. and Sabripour, M. (2006) Microarray data analysis: from disarray to consolidation and consensus. *Nature Reviews. Genetics*, **7**, 55–65.

3 Alba, R., Zhangjun, F., Payton, P., Liu, Y., Moore, S.L., Debbie, P., Cohn, J., D'Ascenzo, M., Gordon, J.S., Rose, J.K., Martin, G., Tanksley, S.D., Bouzayen, M., Jahn, M.M. and Giovannoni, J. (2004) ESTs, cDNA microarrays, and gene expression profiling: tools for dissecting plant physiology and development. *Plant Journal*, **39**, 697–714.

4 Küster, H., Becker, A., Firnhaber, C., Hohnjec, N., Manthey, K., Perlick, A.M., Bekel, T., Dondrup, M., Henckel, K., Goesmann, A., Meyer, F., Wipf, D., Requena, N., Hildebrandt, U., Hampp, R., Nehls, U., Krajinski, F., Franken, P. and Pühler, A. (2007) Development of bioinformatic tools to support EST-sequencing, *in silico*- and microarray-based transcriptome profiling in mycorrhizal symbioses. *Phytochemistry*, **68**, 19–32.

5 Lee, Y., Tsai, J., Sunkara, S., Karamycheva, S., Pertea, G., Sultana, R., Antonescu, V., Chan, A., Cheung, F. and Quackenbush, J. (2005) The TIGR Gene Indices: clustering and assembling EST and known genes and integration with eukaryotic genomes. *Nucleic Acids Research*, **33**, D71–D74.

6 Bustin, S.A. (2000) Absolute quantification of mRNA using real-time reverse transcription polymerase chain reaction. *Journal of Molecular Endocrinology*, **25**, 169–193.

7 Matsumura, H., Reich, S., Ito, A., Saitoh, H., Kamoun, S., Winter, P., Kahl, G., Reuter, M., Krüger, D.H. and Terauchi, R. (2005) Gene expression analysis of plant host–pathogen interactions by SuperSAGE. *Proceedings of the National Academy of Sciences of the United States of America*, **100**, 15718–15723.

8 Matsumura, H., Ito, A., Saitoh, H., Winter, P., Kahl, G., Reuter, M., Krüger, D.H. and Terauchi, R. (2005) SUPERSAGE. *Cellular Microbiology*, **7**, 11–18.

9 Galbraith, D.W. (2006) Links DNA microarray analyses in higher plants. *OMICS*, **10**, 455–473.

10 Ohtsu, K., Takahashi, H., Schnable, P.S. and Nakazono, M. (2007) Cell type-specific gene expression profiling in plants by using a combination of laser microdissection and high-throughput technologies. *Plant & Cell Physiology*, **48**, 3–7.

11 Czechowski, T., Bari, R.P., Stitt, M., Scheible, W.R. and Udvardi, M.K. (2004) Real-time RT-PCR profiling of over 1400 Arabidopsis transcription factors: unprecedented sensitivity reveals novel root- and shoot-specific genes. *Plant Journal*, **38**, 366–379.

12 Richmond, T. and Somerville, S. (2000) Chasing the dream: plant EST microarrays. *Current Opinion in Plant Biology*, **3**, 108–116.

13 Margulies, M., Egholm, M., Altman, W.E., Attiya, S., Bader, J.S., Bemben, L.A., Berka, J., Braverman, M.S., Chen, Y.J., Chen, Z., Dewell, S.B., Du, L., Fierro, J.M., Gomes, X.V., Godwin, B.C., He, W., Helgesen, S., Ho, C.H., Irzyk, G.P., Jando, S.C., Alenquer, M.L., Jarvie, T.P., Jirage, K.B., Kim, J.B., Knight, J.R., Lanza, J.R., Leamon, J.H., Lefkowitz, S.M., Lei, M., Li, J., Lohman, K.L., Lu, H., Makhijani, V.B., McDade, K.E., McKenna, M.P., Myers, E.W., Nickerson, E., Nobile, J.R., Plant, R., Puc, B.P., Ronan, M.T., Roth, G.T., Sarkis, G.J., Simons, J.F., Simpson, J.W., Srinivasan, M., Tartaro, K.R., Tomasz, A., Vogt, K.A., Volkmer, G.A., Wang, S.H., Wang, Y., Weiner, M.P., Yu, P., Begley, R.F. and Rothberg, J.M. (2005) Genome

sequencing in microfabricated high-density picolitre reactors. *Nature*, **437**, 376–380.

14 The *Arabidopsis* Genome Initiative. (2000) Analysis of the genome sequence of the flowering plant *Arabidopsis thaliana*. *Nature*, **408**, 796–815.

15 Goff, S.A., Ricke, D., Lan, T.H., Presting, G., Wang, R., Dunn, M., Glazebrook, J., Sessions, A., Oeller, P., Varma, H., Hadley, D., Hutchison, D., Martin, C., Katagiri, F., Lange, B.M., Moughamer, T., Xia, Y., Budworth, P., Zhong, J., Miguel, T., Paszkowski, U., Zhang, S., Colbert, M., Sun, W.L., Chen, L., Cooper, B., Park, S., Wood, T.C., Mao, L., Quail, P., Wing, R., Dean, R., Yu, Y., Zharkikh, A., Shen, R., Sahasrabudhe, S., Thomas, A., Cannings, R., Gutin, A., Pruss, D., Reid, J., Tavtigian, S., Mitchell, J., Eldredge, G., Scholl, T., Miller, R.M., Bhatnagar, S., Adey, N., Rubano, T., Tusneem, N., Robinson, R., Feldhaus, J., Macalma, T., Oliphant, A. and Briggs, S. (2002) A draft sequence of the rice genome (*Oryza sativa* L. ssp. japonica). *Science*, **296**, 92–100.

16 Jansson, S. and Douglas, C.J. (2007) Populus: A Model System for Plant Biology. Annual Review of Plant Biology, **58**, 435–458.

17 Barker, D.G., Bianchi, S., Blondon, F., Dattée, Y., Duc, G., Essad, S., Flament, P., Gallusci, P., Génier, G., Guy, P., Muel, X., Tourneur, J., Dénarié, J. and Huguet, T. (1990) *Medicago truncatula*, a model plant for studying the molecular genetics of the *Rhizobium*–legume symbiosis. *Plant Molecular Biology Reporter*, **8**, 40–49.

18 Handberg, K. and Stougaard, J. (1992) *Lotus japonicus*, an autogamous, diploid legume species for classical and molecular genetics. *Plant Journal*, **2**, 487–496.

19 Tadege, M., Ratet, P. and Mysore, K.S. (2005) Insertional mutagenesis: a Swiss Army knife for functional genomics of *Medicago truncatula*. *Trends in Plant Science*, **10**, 229–235.

20 Udvardi, M.K., Tabata, S., Parniske, M. and Stougaard, J. (2005) *Lotus japonicus*: legume research in the fast lane. *Trends in Plant Science*, **10**, 222–228.

21 Town, C.D. (2006) Annotating the genome of *Medicago truncatula*. *Current Opinion in Plant Biology*, **9**, 122–127.

22 Fedorova, M., van de Mortel, J., Matsumoto, P.A., Cho, J., Town, C.D., VandenBosch, K.A., Gantt, J.S. and Vance, C.P. (2002) Genome-wide identification of nodule-specific transcripts in the model legume *Medicago truncatula*. *Plant Physiology*, **130**, 519–537.

23 Liu, J., Blaylock, L.A., Endre, G., Cho, J., Town, C.D., VandenBosch, K.A. and Harrison, M.J. (2003) Transcript profiling coupled with spatial expression analyses reveals genes involved in distinct developmental stages of an arbuscular mycorrhizal symbiosis. *Plant Cell*, **15**, 2106–2123.

24 Firnhaber, C., Pühler, A. and Küster, H. (2005) EST sequencing and time course microarray hybridizations identify more than 700 *Medicago truncatula* genes with developmental expression regulation in flowers and pods. *Planta*, **222**, 269–283.

25 Journet, E.P., van Tuinen, D., Gouzy, J., Crespeau, H., Carreau, V., Farmer, M.J., Niebel, A., Schiex, T., Jaillon, O., Chatagnier, O., Godiard, L., Micheli, F., Kahn, D., Gianinazzi-Pearson, V. and Gamas, P. (2002) Exploring root symbiotic programs in the model legume *Medicago truncatula* using EST analysis. *Nucleic Acids Research*, **30**, 5579–5592.

26 Lamblin, A.F., Crow, J.A., Johnson, J.E., Silverstein, K.A., Kunau, T.M., Kilian, A., Benz, D., Stromvik, M., Endre, G., VandenBosch, K.A., Cook, D.R., Young, N.D. and Retzel, E.F. (2003) MtDB: a database for personalized data mining of the model legume *Medicago truncatula* transcriptome. *Nucleic Acids Research*, **31**, 196–201.

27 Stekel, D.J., Git, Y. and Falciani, F. (2000) The comparison of gene expression from multiple cDNA libraries. *Genome Research*, **10**, 2055–2061.

28 Becker, A. (2004) Design of microarrays for genome-wide expression profiling, in *Molecular Microbial Ecology Manual*, 2nd edn. (eds A. Akkermans, F.J. de Bruijn, G. Kowaltchuk and J. van Elsas), Kluwer Academic Publishers, Dordrecht, The Netherlands.

29 Wolber, P.K., Collins, P.J., Lucas, A.B., De Witte, A. and Shannon, K.W. (2006) The Agilent *in situ*-synthesized microarray platform. *Methods in Enzymology*, **410**, 28–57.

30 Nuwaysir, E.F., Huang, W., Albert, T.J., Singh, J., Nuwaysir, K., Pitas, A., Richmond, T., Gorski, T., Berg, J.P., Ballin, J., McCormick, M., Norton, J., Pollock, T., Sumwalt, T., Butcher, L., Porter, D., Molla, M., Hall, C., Blattner, F., Sussman, M.R., Wallace, R.L., Cerrina, F. and Green, R.D. (2002) Gene expression analysis using oligonucleotide arrays produced by maskless photolithography. *Genome Research*, **12**, 1749–1755.

31 Dalma-Weiszhausz, D.D., Warrington, J., Tanimoto, E.Y. and Miyada, C.G. (2006) The Affymetrix GeneChip platform: an overview. *Methods in Enzymology*, **410**, 3–28.

32 Rensink, W.A. and Buell, C.R. (2005) Microarray expression profiling resources for plant genomics. *Trends in Plant Science*, **10**, 603–609.

33 Tesfaye, M., Silverstein, K.A.T., Bucciarelli, B., Samac, D.A. and Vance, C.P. (2006) The Affymetrix *Medicago* GeneChip® array is applicable for transcript analysis of alfalfa (*Medicago sativa*). *Functional Plant Biology*, **33**, 783–788.

34 Barnett, M.J., Toman, C.J., Fisher, R.F. and Long, S.R. (2004) A dual-genome Symbiosis Chip for coordinate study of signal exchange and development in a prokaryote–host interaction. *Proceedings of the National Academy of Sciences of the United States of America*, **101**, 16636–16641.

35 Benedito, V.A., Dai, X., He, J., Zhao, P.X. and Udvardi, M.K. (2006) Functional genomics of plant transporters in legume nodules. *Functional Plant Biology*, **33**, 731–736.

36 Quackenbush, J. (2002) Microarray data normalization and transformation. *Nature Genetics*, **32**, S496–S501

37 MAQC Consortium, (2006) The MicroArray Quality Control (MAQC) project shows inter- and intraplatform reproducibility of gene expression measurements. *Nature Biotechnology*, **24**, 1151–1161.

38 Rhee, S.Y., Dickerson, J. and Xu, D. (2006) Bioinformatics and its applications in plant biology. *Annual Review of Plant Biology*, **57**, 335–360.

39 Slonim, D.K. (2002) From patterns to pathways: gene expression data analysis comes of age. *Nature Genetics*, **32**, S502–S508.

40 Zhu, T. (2003) Global analysis of gene expression using GeneChip microarrays. *Current Opinion in Plant Biology*, **6**, 418–425.

41 Küster, H., Hohnjec, N., Krajinski, F., El Yahyaoui, F., Manthey, K., Gouzy, J., Dondrup, M., Meyer, F., Kalinowski, J., Brechenmacher, L., van Tuinen, D., Gianinazzi-Pearson, V., Pühler, A., Gamas, P. and Becker, A. (2004) Construction and validation of cDNA-based Mt6k-RIT macro- and microarrays to explore root endosymbioses in the model legume *Medicago truncatula*. *Journal of Biotechnology*, **108**, 95–113.

42 Lohar, D.P., Sharopova, N., Endre, S., Peñuela, S., Samac, D., Town, C., Silverstein, K.A.T. and VandenBosch, K.A. (2005) Transcript analysis of early nodulation events in *Medicago truncatula*. *Plant Physiology*, **140**, 221–234.

43 Hohnjec, N., Vieweg, M.F., Pühler, A., Becker, A. and Küster, H. (2005) Overlaps in the transcriptional profiles of *Medicago truncatula* roots inoculated with two different *Glomus* fungi provide insights into the genetic program activated during arbuscular mycorrhiza. *Plant Physiology*, **137**, 1283–1301.

44 Hohnjec, N., Henckel, K., Bekel, T., Gouzy, J., Dondrup, M., Goesmann, A. and Küster, H. (2006) Transcriptional snapshots provide insights into the molecular basis of arbuscular mycorrhiza in the model legume *Medicago truncatula*. *Functional Plant Biology*, **33**, 737–748.

45 Barsch, A., Tellström, V., Patschkowski, T., Küster, H. and Niehaus, K. (2006) Metabolite profiles of nodulated alfalfa plants indicate that distinct stages of nodule organogenesis are accompanied by global physiological adaptations. *Molecular Plant–Microbe Interactions: MPMI*, **19**, 998–1013.

46 Buitink, J., Leger, J.J., Guisle, I., Ly Vu, B., Wuillème, S., Lamirault, G., Le Bars, A., Le Meur, N., Becker, A., Küster, H. and Leprince, O. (2006) Transcriptome profiling uncovers metabolic and regulatory processes occurring during the transition from desiccation-sensitive to desiccation-tolerant stages in *Medicago truncatula* seeds. *Plant Journal*, **47**, 735–750.

47 Küster, H., Vieweg, M.F., Manthey, K., Baier, M.C., Hohnjec, N. and Perlick, A.M. (2007) Identification and expression regulation of symbiotically activated legume genes. *Phytochemistry*, **68**, 1–18.

48 Tellström, V., Usadel, B., Thimm, O., Stitt, M., Küster, H. and Niehaus, K. (2007) The lipopolysaccharide of *Sinorhizobium meliloti* suppresses defense-associated gene expression in cell cultures of the host plant *Medicago truncatula*. *Plant Physiology*, **143**, 825–837.

49 Dondrup, M., Goesmann, A., Bartels, D., Kalinowski, J., Krause, L., Linke, B., Rupp, O., Szyrba, A., Pühler, A. and Meyer, F. (2003) EMMA: a platform for consistent storage and efficient analysis of microarray data. *Journal of Biotechnology*, **106**, 135–146.

50 Schüssler, A., Schwarzott, D. and Walker, C. (2001) A new fungal phylum, the Glomeromycota: phylogeny and evolution. *Mycological Research*, **105**, 1413–1421.

51 Smith, S.E. and Read, D.J. (1997) *Mycorrhizal Symbiosis*, Academic Press, London.

52 Balestrini, R. and Lanfranco, L. (2006) Fungal and plant gene expression in arbuscular mycorrhizal symbiosis. *Mycorrhiza*, **16**, 509–524.

53 Harrison, M.J. (2005) Signaling in the arbuscular mycorrhizal symbiosis. *Annual Review of Microbiology*, **59**, 19–42.

54 Franken, P. and Requena, N. (2001) Analysis of gene expression in arbuscular mycorrhiza: new approaches and challenges. *The New Phytologist*, **150**, 431–439.

55 Gianinazzi-Pearson, V. and Brechenmacher, L. (2004) Functional genomics of arbuscular mycorrhiza: decoding the symbiotic cell programme. *Canadian Journal of Botany*, **82**, 1228–1234.

56 Isayenkov, S., Fester, T. and Hause, B. (2004) Rapid determination of fungal colonization and arbuscule formation in roots of *Medicago truncatula* using real-time (RT) PCR. *Journal of Plant Physiology*, **161**, 1379–1383.

57 Dudoit, S., Yang, Y.H., Callow, M.J. and Speed, T.P. (2002) Statistical methods for identifying differentially expressed genes in replicated cDNA microarray experiments. *Statistica Sinica*, **12**, 111–139.

58 Harrison, M.J., Dewbre, G.R. and Liu, J. (2002) A phosphate transporter from *Medicago truncatula* involved in the acquisition of phosphate released by arbuscular mycorrhizal fungi. *Plant Cell*, **14**, 2413–2429.

59 Doll, J., Hause, B., Demchenko, K., Pawlowski, K. and Krajinski, F. (2003) A member of the germin-like protein family is a highly conserved mycorrhiza-specific induced gene. *Plant & Cell Physiology*, **44**, 1208–1214.

60 Wulf, A., Manthey, K., Doll, J., Perlick, A.M., Linke, B., Bekel, T., Meyer, F., Franken, P., Küster, H. and Krajinski, F. (2003) Transcriptional changes in response to arbuscular mycorrhiza development in the model plant *Medicago truncatula*. *Molecular Plant–Microbe Interactions: MPMI*, **16**, 306–314.

61 Harrison, M.J. (1996) A sugar transporter from *Medicago truncatula*: altered expression pattern in roots during vesicular–arbuscular (VA) mycorrhizal associations. *Plant Journal*, **9**, 491–503.

62 Walter, M.H., Hans, J. and Strack, D. (2002) Two distantly related genes encoding 1-deoxy-D-xylulose 5-phosphate synthases: differential regulation in shoots and apocarotenoid-accumulating mycorrhizal roots. *Plant Journal*, **31**, 243–254.

63 Brechenmacher, L., Weidmann, S., van Tuinen, D., Chatagnier, O., Gianinazzi, S., Franken, P. and Gianinazzi-Pearson, V. (2004) Expression profiling of up-regulated plant and fungal genes in early and late stages of *Medicago truncatula–Glomus mosseae* interactions. *Mycorrhiza*, **14**, 253–262.

64 Grunwald, U., Nyamsuren, O., Tamasloukht, M., Lapopin, L., Becker, A., Mann, P., Gianinazzi-Pearson, V., Krajinski, F. and Franken, P. (2004) Identification of mycorrhiza-regulated genes with arbuscule development-related expression profile. *Plant Molecular Biology*, **55**, 553–566.

65 Frenzel, A., Manthey, K., Perlick, A.M., Meyer, F., Pühler, A., Krajinski, F. and Küster, H. (2005) Combined transcriptome profiling reveals a novel family of arbuscular mycorrhizal-specific *Medicago truncatula* lectin genes. *Molecular Plant–Microbe Interactions: MPMI*, **18**, 771–782.

66 Paszkowski, U. (2006) A journey through signaling in arbuscular mycorrhizal symbioses. *The New Phytologist*, **172**, 35–46.

67 Manthey, K., Krajinski, F., Hohnjec, N., Firnhaber, C., Pühler, A., Perlick, A.M. and Küster, H. (2004) Transcriptome profiling in root nodules and arbuscular mycorrhiza identifies a collection of novel genes induced during *Medicago truncatula* root endosymbioses. *Molecular Plant–Microbe Interactions: MPMI*, **17**,1063–1077.

68 Galibert, F., Finan, T.M., Long, S.R., Puhler, A., Abola, P., Ampe, F., Barloy-Hubler, F., Barnett, M.J., Becker, A., Boistard, P., Bothe, G., Boutry, M., Bowser, L., Buhrmester, J., Cadieu, E., Capela, D., Chain, P., Cowie, A., Davis, R.W., Dreano, S., Federspiel, N.A., Fisher, R.F., Gloux, S., Godrie, T., Goffeau, A., Golding, B., Gouzy, J., Gurjal, M., Hernandez-Lucas, I., Hong, A., Huizar, L., Hyman, R.W., Jones, T., Kahn, D., Kahn, M.L., Kalman, S., Keating, D.H., Kiss, E., Komp, C., Lelaure, V., Masuy, D., Palm, C., Peck, M.C., Pohl, T.M., Portetelle, D., Purnelle, B., Ramsperger, U., Surzycki, R., Thebault, P., Vandenbol, M., Vorhölter, F.J., Weidner, S., Wells, D.H., Wong, K., Yeh, K.C. and Batut, J. (2001) The composite genome of the legume symbiont *Sinorhizobium meliloti*. *Science*, **293**, 668–672.

69 Pühler, A., Arlat, M., Becker, A., Gottfert, M., Morrissey, J.P. and O'Gara, F. (2004) What can bacterial genome research teach us about bacteria–plant interactions? *Current Opinion in Plant Biology*, **7**, 137–147.

70 Pobigaylo, N., Wetter, D., Szymczak, S., Schiller, U., Kurtz, S., Meyer, F., Nattkemper, T.W. and Becker, A. (2006) Construction of a large signature-tagged mini-Tn5 transposon library and its application to mutagenesis of *Sinorhizobium meliloti*. *Applied and Environmental Microbiology*, **72**, 4329–4337.

71 Day, D.A., Kaiser, B.N., Thomson, R., Udvardi, M.K., Moreau, S. and Puppo, A. (2001) Nutrient transport across symbiotic membranes from legume nodules. *Australian Journal of Plant Physiology*, **28**, 667–674.

72 Provorov, N.A., Borisov, A.Y. and Tikhonovich, I.A. (2002) Developmental genetics and evolution of symbiotic structures in nitrogen-fixing nodules and arbuscular mycorrhiza. *Journal of Theoretical Biology*, **214**, 215–232.

73 Parniske, M. (2000) Intracellular accommodation of microbes by plants: a common developmental program for symbiosis and disease? *Current Opinion in Plant Biology*, **3**, 320–328.

74 Genre, A., Chabaud, M., Timmers, T., Bonfante, P. and Barker, D.G. (2005) Arbuscular mycorrhizal fungi elicit a novel intracellular apparatus in *Medicago* *truncatula* root epidermal cells before infection. *Plant Cell*, **17**, 3489–3499.

75 Lum, M.R. and Hirsch, A.M. (2002) Roots and their symbiotic microbes: strategies to obtain nitrogen and phosphorus in a nutrient-limiting environment. *Journal of Plant Growth Regulation*, **21**, 368–382.

76 Parniske, M. (2004) Molecular genetics of the arbuscular mycorrhizal symbiosis. *Current Opinion in Plant Biology*, **7**, 414–421.

77 Geurts, R., Fedorova, E. and Bisseling, T. (2005) Nod factor signaling genes and their function in the early stages of *Rhizobium* infection. *Current Opinion in Plant Biology*, **8**, 346–352.

78 Stacey, G., Libault, M., Brechenmacher, L., Wan, J. and May, G.D. (2006) Genetics and functional genomics of legume nodulation. *Current Opinion in Plant Biology*, **9**, 110–121.

79 Kistner, C. and Parniske, M. (2002) Evolution of signal transduction intracellular symbiosis. *Trends in Plant Science*, **7**, 511–518.

80 Weidmann, S., Sanchez, L., Descombin, J., Chatagnier, O., Gianinazzi, S. and Gianinazzi-Pearson, V. (2004) Fungal elicitation of signal transduction-related plant genes precedes mycorrhiza establishment and requires the dmi3 gene in *Medicago truncatula*. *Molecular Plant–Microbe Interactions: MPMI*, **17**, 1385–1393.

81 El Yahyaoui, F., Küster, H., Ben Amor, B., Hohnjec, N., Pühler, A., Becker, A., Gouzy, J., Vernié, T., Gough, C., Niebel, A., Godiard, L. and Gamas, P. (2004) Expression profiling in *Medicago truncatula* identifies more than 750 genes differentially expressed during nodulation, including many potential regulators of the symbiotic program. *Plant Physiology*, **136**, 3159–3176.

82 Journet, E.P., El-Gachtouli, N., Vernoud, V., de Billy, F., Pichon, M., Dedieu, A., Arnould, C., Morandi, D., Barker, D.G. and Gianinazzi-Pearson, V. (2001) *Medicago truncatula* ENOD11: a novel RPRP-encoding early nodulin gene expressed during mycorrhization in arbuscule-containing cells. *Molecular Plant–Microbe Interactions: MPMI*, **14**, 737–748.

83 Valot, B., Negroni, L., Zivy, M., Gianinazzi, S. and Dumas-Gaudot, E. (2006) A mass spectrometric approach to identify arbuscular mycorrhiza-related proteins in root plasma membrane fractions. *Proteomics*, **6**, S145–S155.

84 Winzer, T., Bairl, A., Linder, M., Linder, D., Werner, D. and Müller, P. (1999) A novel 53-kDa nodulin of the symbiosome membrane of soybean nodules, controlled by *Bradyrhizobium japonicum*. *Molecular Plant–Microbe Interactions: MPMI*, **12**, 218–226.

85 Kehr, J. (2003) Single cell technology. *Current Opinion in Plant Biology*, **6**, 617–621.

Links

DFCI *Medicago truncatula* Gene Index http://compbio.dfci.harvard.edu/.

Medicago EST Navigation System (MENS) http://medicago.toulouse.inra.fr/Mt/EST/.

Medicago truncatula DataBase (MtDB) http://www.medicago.org/.

Operon Biotechnologies array database http://www.operon.com/arrays/omad.php/.

EU Integrated Project: 'GRAIN LEGUMES' http://www.eugrainlegumes.org/.

DFG Mycorrhiza Network 'MolMyk' http://www.genetik.uni-bielefeld.de/MolMyk/.

EBI ArrayExpress database http://www.ebi.ac.uk/arrayexpress/.

B
Gene-by-Gene Analysis

The Handbook of Plant Functional Genomics: Concepts and Protocols.
Edited by Günter Kahl and Khalid Meksem

ISBN: 978-3-527-31885-8

8
Genome-Wide Analysis of mRNA Expression by Fluorescent Differential Display

Suping Zhou, Jonathan D. Meade, Samuel Nahashon, Blake R. Shester, Jamie C. Walden, Zhen Guo, Julia Z. Liang, Joshua G. Liang, and Peng Liang

Abstract

Fluorescent differential display (FDD) was developed from conventional radioactive DD using modified and fluorescently labeled anchor primers in optimized polymerase chain reaction (PCR) mixtures. It is a highly sensitive technique for mRNA fingerprinting to isolate rare gene transcripts, detect minute variations in transcript levels, and to identify both increased and decreased mRNAs. The high reproducibility, high throughput, and operation safety makes this technique suitable for rapid and large-scale screening of differentially expressed genes. Here, we present a detailed description of the procedure for carrying out FDD analysis on plant tissues. Its applications and future technical improvements are also discussed.

8.1
Introduction

The angiosperms of the plant kingdom contain 250 000 species with varied floral forms and different developmental patterns. Each individual plant undergoes a typical life cycle starting from seed germination, through vegetative and reproductive stages, and ultimately concluding with the production of seeds for future generations. In general, all of these steps are fine-tuned to environmental changes; for instance, floral initiation of many plants such as winter annuals, biennials and perennials requires a cold temperature (vernalization) and certain day length (photoperiodism). Plants are constantly exposed to endogenous and exogenous stresses, and thereby have developed various adaptive mechanisms. Modern plant scientists have been seeking to understand how the genetic codes are programmed to allow plants to sustain immense organic evolution and mechanical adjustment. The completed and ongoing genome projects on model species of *Arabidopsis thaliana*, tomato, rice, and so

The Handbook of Plant Functional Genomics: Concepts and Protocols.
Edited by Günter Kahl and Khalid Meksem

ISBN: 978-3-527-31885-8

on, have revealed that only a small fraction of the genes embedded in the plant genome are transcribed into mRNAs for functional protein synthesis. Essentially, interpretation of the genomic instructions in the post-genome era will have to rely, at least in large part, on tools which allow us to determine when and where a gene, or a group of genes, will be turned on or off during a biological process.

Differential display (DD) is a powerful tool for studying differential gene expression [1] in any eukaryotic species, and has been successfully adapted to plant tissues [2,3]. Fluorescent differential display (FDD), which employs fluorescently labeled anchor primers alongside a fluorescent DNA imaging system has high reproducibility, high throughput, and is safe to carry out [4]. Using this technique, it is possible to undertake rapid and large-scale screening for differentially expressed genes.

FDD consists of four steps: RT conversion of mRNA into single strand cDNA, fluorescent labeling cDNA fragment via FDD-PCR, DNA sequencing gel electrophoresis, and cDNA cloning and characterization (Figure 8.1). Like conventional DD, it begins with total RNA being harvested from the cells/tissues being compared. The messenger RNAs (mRNAs) within the total RNA populations are then converted into single stranded cDNA by reverse transcription. The current methodology makes use of 3′ 'anchored' oligo-dT primers that target the poly-adenylation site of eukaryotic mRNA and have the form H-T_{11}M, where H is a *Hind* III restriction site (AAGCTT), T_{11} is a string of 11 Ts (though the first two Ts come from the *Hind* III site), and M is G, C, or A [5,6]. They are referred to as 'anchor' primers because the non-T base after the string of 11 Ts enables the primer to be anchored to the same spot for each round of amplification, in contrast to standard oligo-dT primers that only contain a string of Ts and will anneal in multiple spots, creating a smear. The *Hind* III restriction site is usually incorporated into the anchor primer to elongate the primer and to make it more efficient in annealing to the targeted poly-A site, as well as to improve its downstream applications such as cDNA cloning. Using the current anchor primer design, the cDNA populations are subsequently divided into three subpopulations that represent one-third of the potential mRNA expressed in the cell at any given time.

The next step in FDD is the PCR-amplification of the cDNA subpopulations utilizing a combination of fluorochrome-labeled anchor primers (known generically as FH-T_{11}M) and with a set of 'arbitrary' primers that are random and short in length. The design of these arbitrary 13-mers (H-AP primers) utilized in DD technology also includes a *Hind* III restriction site (AAGCTT) and a 7-base-pair backbone of random base combinations. The *Hind* III restriction site is included in both the anchor and arbitrary primers for more efficient primer annealing and easier downstream manipulation of the cDNA [5]. The primers used in DD represent a random selection from over 16 000 (4^7) base-pair combinations. Additionally, the length of an arbitrary primer is so designed that by probability each will recognize 50–100 mRNAs under a given PCR condition [7]. As a result, mRNA 3′ termini defined by any given pair of anchored-primer and arbitrary primer are amplified and displayed by denaturing polyacrylamide gel electropho-

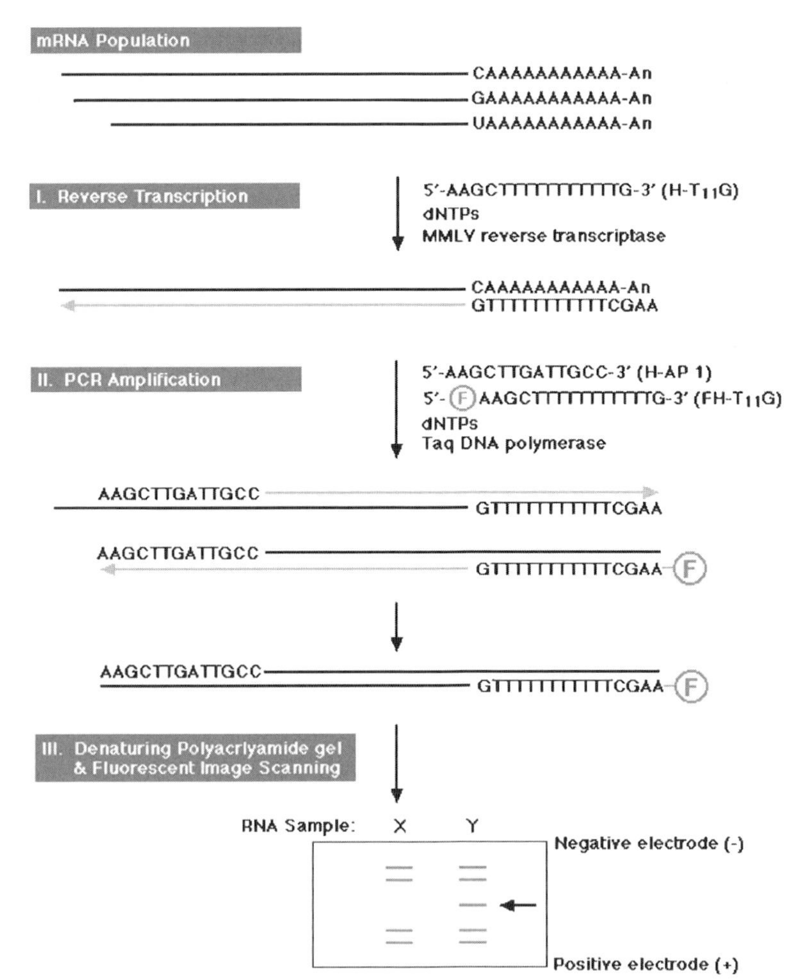

Figure 8.1 Schematic representation of fluorescent mRNA differential display (FDD). Three fluorescently-labeled one-base anchored oligo-dT primers with 5′ *Hind* III sites are used in combination with a series of arbitrary 13-mers (also containing 5′ *Hind* III sites) to reverse transcribe and amplify the mRNAs from a cell.

resis. A mathematical model of estimated gene coverage utilizing various combinations of anchor and arbitrary primers was developed shortly after the advent of differential display technology [7]. This mathematical model indicated that approximately 240 primer combinations (three anchor primers with 80 arbitrary primers) were needed to approach the level of estimated genome-wide screening for eukaryotes (~95%). A newer mathematical model [6] predicts that more primer combinations are required to give that level of coverage; using 480 primer

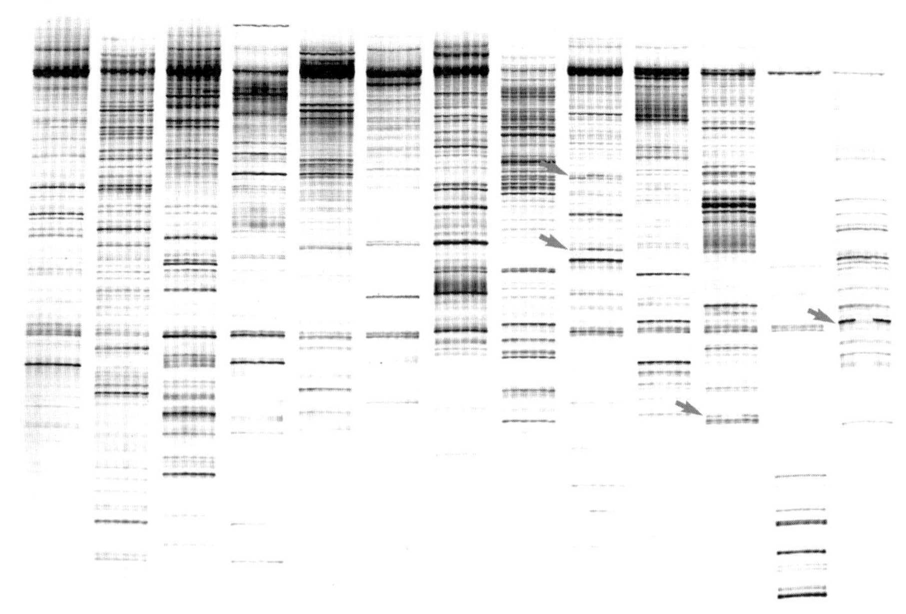

Figure 8.2 Automated FDD of plant RNA reveals flower color-related genes. Three RNA samples (white, yellow, and red petals from the *Mirabilis jalapa* (4 o'clock) flower) were compared in-duplicate with three anchor primers in combination with 80 arbitrary primers using automation in liquid-handling, a 132-lane electrophoresis unit, and digital acquisition of the gel images. This image shows 13 of the 240 primer combinations used in our experiment; each set of six lanes represents one primer combination with the three samples in-duplicate. Arrows indicate reproducible differences in mRNA expression linked to a specific floral color.

combinations (three anchor primers with 160 arbitrary primers) would provide approx 93% coverage.

After PCR amplification, gel electrophoresis is carried out to separate the resulting PCR products by size. Reactions are run side-by-side so that the samples being compared are next to one another for each primer combination. Comparison of the cDNA patterns between or among relevant RNA samples reveals differences in the gene expression profile for each sample (see Figure 8.2). Electrophoresis can be performed with denaturing polyacrylamide sequencing gels [1,8], non-denaturing polyacrylamide gels [9], or with agarose gels [10]. Sequencing gels offer the best band resolution and allow for easy and efficient recovery of cDNA, and accommodate a large number of reactions on each gel and are therefore the most commonly used. Because the resulting cDNAs are fluorescently labeled, the digital cDNA profiles can be acquired on a fluorescent imager scanner such as the FMBIO laser imager series (MiraiBio, Alameda, CA), the Typhoon (GE Healthcare, Piscataway, NJ) or FLA-5000 (FUJIFILM Medical Systems, Stamford, CT).

Upon completion of the gene expression profiling by gel electrophoresis, the next step is to begin characterization of the genes of interest. Bands are excised from the gel matrix and reamplified with the same primer combinations as the original FDD-PCR and under the same reaction conditions. Generally, a PCR-product cloning step is recommended before differential gene confirmation and sequencing, but this is determined by the preferences of the researcher. The PCR-TRAP Cloning System (GenHunter Corporation, Nashville, TN) is recommended because it is designed specifically for cloning the DD bands and employs highly efficient positive-selection cloning. More than one colony (three to five) should be screened for the correct size to allow for the possibility that more than one distinct cDNA is contained within an excised band. Furthermore, if the screening results indicate that more than one cDNA is present in the colony population, each of the different fragments should then be further characterized.

Characterization of each potential gene includes sequencing of the cloned cDNAs of interest, with the results giving an indication of whether the cDNA is a known or unknown sequence. As with any differential gene expression technology, it should be ensured that the characterized sequences are actually differentially regulated, that is a 'real difference', and not a false positive. A variety of confirmation techniques, including Northern blot analysis, reverse Northern blot analysis, and quantitative RT-PCR (qRT-PCR) can be used. Quantitative RT-PCR is a sensitive and reliable method for determining gene copy number; however it requires very stringent primer design. Some of the short cDNA fragments of unknown genes may not contain such sequences. Instead, Northern blot analysis is by far the most accepted tool to validate both high- and low-level mRNA expression, despite being labor-intensive, time-consuming, and requiring a significant amount of RNA.

8.2 Methods and Protocols

8.2.1 Materials

8.2.1.1 Total RNA Isolation and Removal of Genomic DNA from Total RNA

- RNA isolation reagent: a phenol–guanidinum monophasic solution such as RNApure (GenHunter Corporation, Nashville, TN, Cat. Nos. P501 to P503) is recommended.
- Polytron homogenizer for RNA extraction from tissue (Biospec Products Inc., Bartlesville, OK), or mortar and pestle (pre-baked at over 100 °C overnight).
- Diethyl pyrocarbonate-(DEPC)-treated water (GenHunter, Cat. No. R105).
- 1.5-ml DNase- and RNase-free microcentrifuge tubes. RNase-free DNase I. The MessageClean DNA Removal Kit (GenHunter, Cat. No. M601) is recommended which contains RNase-free DNase I (10 units/μl), 10× reaction buffer, 3 M sodium acetate (pH 5.5), DEPC-treated water, and RNA Loading Mix.
- Agarose, ultraPure (Invitrogen, Carlsbad, CA).

- Phenol/chloroform (3 : 1) solution, Tris saturated: 30 ml melted crystalline phenol, 10 ml chloroform, 10 ml Tris-HCl, pH 7.0.
- 10× MOPS buffer: 0.2 M MOPS, 0.05 M sodium acetate, 0.01 M ethylenediamine tetraacetic acid (EDTA), pH 6.5.
- 12.3 M (37%) formaldehyde, pH > 4.0.

8.2.1.2 Single-Strand cDNA Synthesis by Reverse Transcription

- RNAspectra Fluorescent Differential Display Kit (GenHunter, Cat. Nos. R501-R510 and F501-F510) including distilled water, 5× RT buffer, dNTP mix (FDD), oligo-dT anchor primers (H-T_{11}M), and MMLV Reverse Transcriptase.
- 0.2-mL thin-walled PCR tube, RNase-free (GenHunter, Catalog No. T101).
- Thermal cycler. Eppendorf* Mastercycler* Thermal Cyclers or The GeneAmp PCR System 9600 (Applied Biosystems, Foster City, CA).

8.2.1.3 Fluorescent Differential Display-PCR (FDD-PCR)

- RNAspectra Fluorescent Differential Display Kit (GenHunter, Cat. Nos. R501-R510 and F501-F510) including distilled water, 10× PCR buffer (100 mM Tris-Cl, pH 8.4, 500 mM KCl, 15 mM $MgCl_2$, and 0.01% gelatin), FDD dNTP mix, fluorescent anchor primers (R-H-T_{11}M or F-H-T_{11}M), and arbitrary primers (H-AP).
- *Taq* DNA polymerase (Qiagen, Valencia, CA, Catalog No. 201207).
- 0.2-ml thin-walled PCR tube, RNase-free (GenHunter) or 96-well PCR plates (Thermo-Fast 96 Detection Plate, ABgene Inc., Rochester, NY, Catalog No. AB-1100).
- Liquid-handling Robot. GenHunter uses the Biomek 2000 (Beckman Coulter Inc., Fullerton, CA).

8.2.1.4 Gel Electrophoresis

- Gel apparatus with low-fluorescent (borosilicate) glass plates such as Horizontal or Vertical FDD Electrophoresis Systems (GenHunter, Catalog #s SA101 or SA201).
- Sigmacote (Sigma, St. Louis, MO) or similar product.
- 6% denaturing gel solution such as Sequagel 6 Ready-To-Use 6% Sequencing Gel (National Diagnostics, Atlanta, GA, Cat. No. EC-836) or 5.5–6.5% LI-COR KBplus Gel Matrix (LI-COR Biosciences, Lincoln, NE).
- 10× TBE: 0.89 M Tris-borate, pH 8.3; 20 mM disodium ethylenediamine tetraaceticacid (Na_2EDTA).
- 10% ammonium persulfate (APS).
- *N,N,N′,N′*-Tetramethylethylenediamine (TEMED) (if using LI-COR gel system).
- FDD Loading Dye from RNAspectra Kit (GenHunter, Catalog No. F201).
- Fluorescent Laser Scanner. The FMBIO II or III Series (MiraiBio, Alameda, CA) is recommended.
- FDD locator dye (GenHunter, Catalog # F202 and R202).

8.2.1.5 **Reamplification of Selected Differentially Expressed Bands**

- Glycogen, 10 mg/ml (Sigma, GenHunter).
- 2.10× Agarose DNA loading dye (40% sucrose, 0.1% bromophenol blue, 0.1% xylene cyanole FF, 2.5 mM in distilled water).

8.2.1.6 **Cloning of Reamplified PCR Products**

- PCR-TRAP Cloning System (GenHunter, Catalog # P404) including insert-ready PCR-TRAP cloning vector, T4 DNA ligase, distilled water, 10× ligase buffer, Lgh/Rgh primers (2 μM), Colony Lysis Buffer, 10× PCR buffer, dNTP 250 μM, tetracycline, and GH competent cells.

- LB media. Make 1 l LB with 10 g Bacto-tryptone, 5 g Bacto-yeast extract, 10 g NaCl, and make up to 1 l with dH_2O.

- LB-Agar-TET plates. Make 1 l LB-Agar-TET plates with LB and 15 g Bacto-agar. After autoclaving at 121 °C for 20 min, add 1 ml tetracycline (20 mg/ml) when liquid cools to approximately 50 °C. Or microwave until the agar is melted, and then add tetracycline as above. Pour into bacterial polystyrene Petri dishes.

- QIAEX II Gel Extraction Kit (Qiagen, Catalog # 20021).

8.2.1.7 **Verification of Cloned PCR Products**

- AidSeq Primer Set C (GenHunter, Catalog # P203): includes Lseq and Rseq primers.
- Bigdye-terminator v3.1 cycle sequencing kit (Applied Biosystem), DyeEx 2.0 Spin Kit, or DyeEx 96 kit (Qiagen).

8.2.1.8 **Confirmation of Differential Gene Expression by Northern Blot**

- HotPrime DNA Labeling Kit (GenHunter, Catalog # H501) including Klenow DNA polymerase, 10× labeling buffer, dNTP (-dATP) or dNTP (-dCTP) (500 μM), stop buffer, and distilled water.
- Lock-top microfuge (USA Scientific, Ocala, FL, Catalog # 1415–5100).
- Alpha-[^{32}P] dATP (3000 curies/mmole) (PerkinElmer Life Sciences, Boston, MA, Catalog # BLU512H).
- Sephadex G50 column (Roche Applied Science, Indianapolis, IN, Catalog # 1814419). Salmon Sperm DNA (10 mg/ml) (GenHunter, Catalog # ML2).
- Nylon Membrane: Nytran SuperCharge Nylon Transfer Membrane (Schleicher and Schuell, Keene, NH, Catalog # 10 416 216).
- Single emulsion scientific imaging film. Kodak Biomax MS (Kodak-Eastman, Rochester, NY, Catalog # 8715187) is recommended.
- 20× Saline-Sodium Citrate (SSC): 3 M NaCl, 0.3 M trisodium citrate $\cdot 2H_2O$. Adjust pH to 7.0 with 1 M HCl.
- Formamide prehybridization/hybridization solution (GenHunter, Catalog # ML1)

If preparing in the laboratory, use the following protocol for 500 ml:

• 20× Saline-Sodium Phosphate-EDTA (SSPE)[a]	125 ml
• 50× Denhardt's Solution[b]	50 ml
• 20% Sodium Dodecyl Sulfate (SDS)	2.5 ml
• Formamide	250 ml
• Distilled water	Up to 500 ml

[a]To make 20× SSPE: 3 M NaCl, 0.1 M NaH_2PO_4 (dibasic), 0.01 M EDTA
[b]To make Denhardt's solution, 50 ml:

• Ficoll	0.5 g
• Polyvinylpyrrolidone	0.5 g
• BSA (Pentax Fraction V)	0.5 g
• Distilled water	Up to 50 ml

Mix well, aliquot into smaller volumes, and store at −20 °C until use.

8.2.2 Methods

8.2.2.1 Total RNA Isolation and Removal of Genomic DNA

To screen a 240-primer combination using FDD, 12–15 μg of DNA-free total RNA is required. Approximately 50 μg of total RNA should be sufficient for the FDD and a subsequent confirmation step. An average yield of 10–15 μg of total RNA can be obtained from 100 mg of fresh plant tissues (either leaf, stem, root or callus tissue) using a reagent based on the standard phenol/guanidine thiocyanate technique such as RNApure. The contaminating genomic DNA must be removed from the total RNA because any primers with matching sequence to the contaminating DNA will anneal during the FDD-PCR reactions, thereby causing amplification of DNA sequences and leading to a higher false-positive rate. Generally, 50–80% of the starting amount of total RNA can be retrieved after DNase I digestion.

1. Freeze 100–200 mg fresh weight (FW) of fresh tissues in liquid nitrogen and grind into fine powder, and transfer into a 1.5-ml microcentrifuge tube.
2. Add 1 ml of RNApure RNA isolation reagent. Ideally, the volume ratio of RNA isolation reagent to tissue should range from 5 : 1 to 10 : 1. Young leaf, root and callus tissues require a 5 : 1 ratio, and mature tissues with a high content of polysaccharides and phenolic compounds should be used in the upper range.
3. After resting on ice for 10 min, add 150 μl of chloroform to each tube. Vortex for 10 s. The protocol can be stopped here by cooling the mixtures to −20 °C or to −80 °C overnight.
4. Centrifuge the tubes at 14 000 rpm and 4 °C for 10 min.
5. Carefully remove the upper phase into a clean, labeled 1.5-ml centrifuge tube.
6. To each 500 μl of the supernatant, add 400 μl of phenol/chloroform (3 : 1, pH 7.0), vortex to mix and repeat steps 4 and 5.

7. Add an equal volume of isopropanol. Mix and allow to stand on ice for 10 min. The protocol can be stopped here by cooling the mixture to −20 °C or to −80 °C overnight.

8. Centrifuge for 10 min at 4 °C at maximum speed.

9. Rinse the RNA pellet with 1 ml of cold 70% ethanol (in DEPC-treated water). Centrifuge for 2 min at 4 °C at maximum speed.

10. Remove the ethanol. Spin briefly and remove the residual wash solution with a pipette; air-dry the pellet for 10–15 min.

11. Resuspend the RNA in DEPC-treated water to a concentration of above 1 μg/μl. Do not use SDS in the resuspension if using RNA for any PCR application.

12. Measure the concentration on a Nanodrop-spectrometer; or read at 260 nm in a UV-spectrometer after dilution in 1 ml of water (a 1 : 1000 dilution). $1\ OD_{260} = 40\ \mu g$.

13. Move on to step 14 for DNase digestion and store RNA that has not been 'cleaned' in aliquots at −80 °C until next use.

14. If necessary, dilute desired amount of RNA to be digested (maximum of 50 μg) with DEPC-treated water to a volume of 50 μl.

15. In a 1.5-ml centrifuge tube, add the following *in the order shown* (to a total reaction volume of 56.7 μl):

    - Total RNA (10–50 μg) 50 μl
    - 10× Reaction buffer 5.7 μl
    - RNase-free DNase I (10 units/μl) 1.0 μl
    - Mix gently and incubate at 37 °C for 30 min.

16. Add 40 μl of phenol/chloroform (3 : 1, pH 7.0) solution to each DNase I reaction and vortex for 30 s.

17. Place on ice for 10 min.

18. Centrifuge at maximum speed (14 000 rpm) for 5 min at 4 °C.

19. Collect upper phase and place in a clean, labeled 1.5-mL microfuge tube.

20. Add 5 μl 3 M sodium acetate and 200 μl 100% ethanol. Mix well.

21. Store for at least 1 h at −80 °C. Overnight to a few days at −80 °C is acceptable.

22. Centrifuge at 4 °C for 10 min at maximum speed to pellet the RNA.

23. Carefully remove the supernatant and rinse the RNA pellet with 0.5 ml of 70% ethanol (in DEPC-treated water). Do not disturb the pellet.

24. Centrifuge for 5 min at maximum speed at 4 °C and remove supernatant. Centrifuge again briefly, removing the residual liquid without disturbing the RNA pellet.

25. Air-dry the pellet for 10–15 min at room temperature and resuspend the RNA in 10–20 μl of DEPC-treated water.

26. Quantify the RNA as described in step 12.

27. Check the integrity of RNA on 7% denaturing formaldehyde agarose gel with MOPS and formaldehyde.

8.2.2.2 Gel Preparation

1. Add the following to a microwave-safe container:

   - 10× MOPS 10 ml
   - Agarose 1–1.5 g
   - Distilled water 83 ml

2. Microwave for approximately 3 min or until agarose is melted.
3. Let agarose cool to at least 50 °C (just within the limits of skin tolerance to heat).
4. Add 7 ml of a 12.3 M (37%) formaldehyde solution. Gently mix.
5. Pour into prepared gel casting plate and add gel comb.
6. Running buffer (1 l) is prepared by diluting 100 ml of 10× MOPS with 900 ml of distilled water to a 1× concentration. Cover agarose gel with running buffer.

8.2.2.3 RNA Loading Sample Preparation

1. Add 1–10 μl (2–3 μg) of RNA to 20 μl RNA loading mix in a labeled 1.5-ml microfuge tube. Mix well and incubate at 65 °C for 10 min; centrifuge sample briefly to collect condensate; place samples on ice for 5 min.
2. Load entire amount onto RNA gel.
3. Run at 50–60 V for approximately 45 ×min or until resolution of the ribosomal subunits is achieved.

8.2.2.4 Single-Strand cDNA Synthesis by Reverse Transcription

Generally, two RT reactions are preferred per sample (known as 'in-duplicate') to ensure reproducibility and as a method of reducing any false positives. If 240 primer combinations are to be carried out, it is recommended that separate RT core mixes for each individual H-T_{11}M are set up in 200-μl volume RT reactions. The total volume of core mix should be adjusted according to the number of primer combinations.

1. Dilute 40 μl of each RNA sample to a final concentration of 0.1 μg/μl with DEPC-treated water and mix thoroughly. Place on ice.

2. For an RT core mix with two samples in-duplicate for one H-T_{11}M primer (H-T_{11}G is shown here), add the following:
   - 376 μl distilled water
   - 160 μl 5× RT buffer

- 64 µl FDD dNTP mix
- 80 µl H-T_{11}G primer
- 680 µl total volume
- Mix well.

3. Divide the above 680 µl evenly into four tubes labeled with the sample name (e.g. RTG-1a, RTG-1b, RTG-2a, RTG-2b), and aliquot 170 µl into each tube.
4. Add 20 µl of the corresponding total RNA (0.1 µg/µl, freshly diluted) to each tube.
5. Program the thermal cycler as follows: (65 °C for 5 min, ≥37 °C for 60 min, ≥75 °C for 5 min, ≥4 °C).
6. Place tubes on thermal cycler and begin program.
7. After 10 min at 37 °C, pause the thermal cycler and add 10 µL of MMLV reverse transcriptase to each tube. Quickly mix well by finger-tipping or pipetting up and down before continuing the incubation program.
8. At the end of the reverse transcription, briefly spin the tube at maximum speed to collect condensate. Place the tubes on ice or store at −20 °C for later use; and
9. Repeat steps 1–8 for H-T_{11}A and H-T_{11}C primers.

8.2.2.5 **Fluorescent Differential Display-PCR**

This protocol is designed for 240 primer combinations in-duplicate per sample using three fluorescent dye-labeled anchor primers (FH-T_{11}M) and 80 upstream arbitrary primers (H-AP). A separate FDD-PCR core mix for each individual FH-T_{11}M primer must be prepared. A core mix for all 80 H-AP primers for FH-T_{11}G primer is shown here. This will be called the 'FDD Core Mix G'.

1. FDD Core Mix G
   - 4080 µl distilled water
   - 800 µl 10× PCR buffer
   - 640 µl dNTP mix (FDD)
   - 800 µl FH-T_{11}G primer
   - 6320 µl total volume
   - Mix well
2. Aliquot 1896 µl of FDD Core Mix G into three separate tubes labeled 'FDD Core Mix G'. Aliquot the remaining amount into a fourth tube labeled 'FDD Core Mix G-remainder' (approximately 632 µl).
3. To one of the tubes labeled 'FDD Core Mix G', add 24 µl *Taq* DNA polymerase. Mix well. Freeze the other three aliquots (see step 2 above) at −80 °C for later PCR reactions.
4. Aliquot 480 µl of 'FDD Core Mix G/Taq' mixture into four separate tubes labeled to identify the corresponding RT reaction.

5. Add 60 μl of the corresponding cDNA from RT to each of the four tubes. Mix well.
6. Using either a robot or by hand, add 2 μl of H-AP primers 1–24 to the corresponding wells of a 96-well plate.
7. Using either a robot or by hand, add 18 μl of the FDD Core Mixes to the corresponding wells of a 96-well plate.
8. The total reaction volume will be 20 μl. Add 25 μl of mineral oil if required.
9. Program the thermal cycler to:
   - 94 °C for 15 s
   - 40 °C for 2 min
   - 72 °C for 60 s
   - for 40 cycles
   - ≥72 °C for 5 min
   - ≥4 °C soak.
10. Put the 96-well plate on the thermal cycler and initiate program. Once completed, store reaction mixtures at −20° C in the dark; and
11. Repeat steps 3–10 for the other primers.

8.2.2.6 Gel Electrophoresis

The Horizontal FDD Electrophoresis System has 132 lanes and a Microtrough System with grooved glass plates. This apparatus allows an entire 96-well plate to be loaded onto one gel. The Microtrough System makes it very easy to load the gels using standard 10-μl pipet tips instead of the more difficult to manipulate flat gel-loading tips that are employed in standard sequencing apparatuses. A multi-channel pipetter, such as the 8-channel Matrix Equalizer 384 with a volume range of 0.5–12.5 μl (Matrix Technologies, Hudson, NH), also works fairly well for gel loading. Under all circumstances, it is necessary to make sure that the tip space of the pipetter matches the distance between the grooves of the Microtrough System and that the PCR reaction set-up is configured accordingly.

For the experiments such as that described above which comprise 960 PCR reactions carried out using 10 96-well plates, it is recommended that 10 separate gels are run, each derived from one 96-well plate. One to two gels can generally be run per day, requiring 5–10 days to complete all the electrophoreses. The Sequagel 6 Ready-To-Use 6% Sequencing Gel (National Diagnostics) or the 5.5–6.5% Gel Matrix (LI-COR) are recommended for denaturing gel electrophoresis. A general protocol is given here for the 6% denaturing polyacrylamide gel which is recommended for the resolution of cDNA profiles.

Thoroughly clean both sides of the glass plates to be used with warm water and soap, ensuring that there is no previous gel debris or streaks. Be sure to rinse thoroughly afterward as any soap residue may cause problems. KOH can be used occasionally for this purpose to strip off hard-to-clean residue.

The glass plates should be cleaned again by wiping with a 50% ethanol (EtOH) solution, or by spraying with isopropanol. Make sure plates are completely dry.

1. Coat the interior surface of one of the plates (usually the notched plate) with 500 µl Sigmacote or similar product using a Kim-Wipe to achieve an even spread across the surface. Allow to dry for 1 min. This coating step ensures that the gel adheres preferentially to the non-coated plate during separation of plates for excision of bands after the gel has been run.
   - Use 60 ml of the gel mixture for a 45 × 28 × 0.04 cm gel.
   - Add 0.5 ml of 10% APS solution and mix thoroughly.
   - Pour gel into sequencing gel cast and allow it to polymerize for 1–2 h or overnight. Cover the gel assembly with a damp paper towel and wrap in plastic film or Saran Wrap to prevent the gel from cracking due to loss of moisture.
   - After polymerization, load the glass plates into the sequencing apparatus and add 1× TBE buffer to upper and lower buffer chambers.
   - Flush the urea from the gel wells and pre-run the sequencing gel in 1× TBE buffer for 30 min.
   - Add 3.5 µl of each FDD-PCR reaction to 2 µl of FDD loading dye. Alternatively, an appropriate ratio of loading dye (8 µl for 20 µl PCR reactions) can be added directly to the PCR reaction if it is only going to be used for running gels. Incubate at 80 °C for 2 min immediately before loading onto the gel to denature the cDNA samples, and then cool on ice for 1–2 min.
2. Load equal amounts of sample (usually 3–4 µl) into each well. It is crucial that all the urea is flushed out of the wells before loading samples with a syringe. For best results, load four to six lanes and then stop briefly to re-flush the unloaded wells. Load in appropriate groups, usually by primer combination.
3. Electrophoresis should be carried out for 1.5 to 3 h at 60 W constant power (voltage not to exceed 2000 V) until the xylene cyanole dye (the slower moving dye) reaches the bottom of the gel. In a 6% gel, the xylene cyanole will co-migrate with DNA of approximately 106 bp as a reference point. The gel should be kept in the dark while running to prevent photo-bleaching of samples either by using a dark room, turning off the light, or covering the gel apparatus with a cardboard box.
4. Turn off power supply and remove the plates from the gel apparatus. Tear off the gel tape and remove spacers and comb. Clean the outside of the glass plates with warm water and 50% ethanol to remove any residue left by the gel or tape. Thorough cleaning is required to reduce background signals produced by gel particles sticking on the plate and fingerprints; and
5. Scan the gel on a fluorescence imager with an appropriate filter, following the manufacturer's instructions based on the particular fluorophore being used.

8.2.2.7 **Reamplification of Selected Differentially Expressed cDNA Bands**

cDNA bands that show reproducible differences between the samples being compared should be excised from the gel and reamplified using the same anchor–arbitrary primer combinations and reaction conditions as used in the initial FDD-PCR reactions.

1. Separate the glass plates by taking off the notched/smaller glass plate (coated plates) leaving the gel attached to the un-notched/larger plate.
2. Place a layer of UV-transparent plastic wrap (Saran Wrap) on top of the gel. This prevents contamination of the gel as well as making gel cutting easier.
3. Spot 0.5 μl of FDD Locator Dye at the upper and lower corners of the gel to facilitate orientation of the gel pattern. The FDD locator dye, with its combination of fluorescent and visible dyes, can be used to easily align the gel with the printed template for band excision.
4. Re-scan the gel with the gel facing upwards.
5. Print a real-size image on appropriately sized paper (11 × 17 inch) using an appropriate printer. This printed image will be used as the template for cutting out the bands.
6. Choose and label the bands to be excised. A band ID should contain RN-G-1A (RN = researcher name; G = FH-T_{11}G anchor primer; 1 = H-AP1 arbitrary primer; A = top differentially expressed band in lane).
7. Place the printout on the table-top and lay the glass plate on top of it. Orient the plate so that the locator dye spots on the printout align with those on the gel.
8. Excise each band with a razor blade and place it into a 1.5-mL microfuge tube labeled with the corresponding band name.
9. Add 100 μl of distilled water to the tube containing the gel slice; soak for 10 min at room temperature; boil the tightly-sealed tube (using parafilm or a lock-top tube), or incubate on a hot plate at100 °C, for 15 min to elute the cDNA from the gel slice.
10. Spin for 2 min at maximum speed to collect condensate and pellet the gel.
11. Transfer the supernatant to a clean 1.5-ml microfuge tube labeled with the same ID, and add 10 μl of 3 M sodium acetate, 5 μl of glycogen (10 mg/ml) and 450 μl of 100% ethanol per tube. Allow to stand for at least 30 min on dry ice or in a −80 °C freezer.
12. Centrifuge at 13 000 rpm for 10 min at 4 °C to pellet the DNA. Remove the supernatant and rinse the pellet with 200 μl of ice-cold 85% ethanol. Spin briefly and remove the residual ethanol.
13. Air-dry the pellet and dissolve in 10 μl of dH_2O.
14. Calculate the number of cDNA bands obtained from the same anchor primer, and
15. Prepare an Anchor Primer Re-amplification Core Mix. Calculate the number of cDNA bands produced from each arbitrary primer, and divide the Anchor Primer Re-amplification Core Mix accordingly.
16. Increasing the volume by an extra 10% is recommended to ensure that there is sufficient liquid to aliquot. A Standard Reamplification Reaction will contain:

- Distilled water 23.3 μl
- 10× PCR buffer 4.0 μl
- dNTP Mix (FDD) 0.3 μl
- H-AP primer (2 μM)* 4.0 μl
- H-T_{11}M (2 μM) 4.0 μl
- cDNA template* 4.0 μl
- *Taq* DNA polymerase 0.4 μl
- Total volume 40.0 μl

17. After core mixes have been prepared, aliquot 32 μl into 0.2-ml tubes (individually, as strip tubes, or in a 96-well plate) labeled with band ID.
18. Add 4 μl of the corresponding cDNA template from step 15.
19. Place the reamplification reactions in the thermal cycler and carry out a PCR reaction using the same conditions as those in the FDD-PCR.
20. repare a 1.5% agarose gel with ethidium bromide by adding 1.5 g of agarose to 100 ml of 1× TAE. When the agarose/1× TAE mix cools to approximately 50 °C (just within the limits of skin tolerance to heat), add 3 μl of ethidium bromide, swirl to mix, and pour the solution into a plastic agarose-casting tray.
21. Add 30 μl of the reamplification reaction to 5 μl of agarose DNA loading dye in a 0.5-ml microfuge tube. Load the 35 μl volume onto the 1.5% agarose gel. Store the remaining 10 μl of the PCR samples at −20 °C for future cloning.
22. Carry out electrophoresis at 70 V for approximately 45–60 min; and
23. Confirm correct cDNA reamplification by visualizing gel using a UV transilluminator. The reamplified band should be approximately the same size as the band that was excised from the original FDD gel.

8.2.2.8 **Cloning of Reamplified PCR Products**

It is highly recommended that the amplified cDNAs be cloned directly into the PCR-TRAP cloning vector. The ligation is conducted at 16 °C overnight after making the following additions in the order shown:

- dH_2O 10 μl
- 10× ligase buffer 2 μl
- PCR-TRAP Vector 2 μl
- PCR product 5 μl
- T4 DNA ligase (add last!) 1 μl
- Total volume 20 μl

The ligation products are used immediately for transformation or stored at −20 °C. For transformation, add 10 μl of each ligation mix to freshly thawed GH-competent cells and mix well by finger-tipping and incubate on ice for 45 min. Heat shock the cells for 2 min at 42 °C and then replace the tubes on ice for 2 min. Add 0.4 ml of LB medium without tetracycline and incubate the cells at 37 °C for 1 h. It is important to ensure that there is no tetracycline in the LB during this step because the bacteria with recombinant plasmids need time to express the tetracycline-resistance gene. After vortexing briefly, plate 200 μl of cells on a pre-warmed LB-Tet plate (containing 20 μg/ml of tetracycline) for 1 h. Store the remaining cells at 4 °C if they are to be replated within 1 week. Once the plate surface is dry, incubate the plate upside-down overnight at 37 °C. Score the Tet^R colonies and store the plate upside-down at 4 °C for further analysis.

8.2.2.9 **Verification of the Cloned Inserts**

The DNA insert into the plasmid is verified by the colony-PCR method using primers that flank the cloning site of the PCR-TRAP Vector.

Colony Lysis

1. Mark each Tet^R colony on Petri dishes, and label a corresponding microfuge tube containing an aliquot of 50 μl of colony lysis buffer.
2. Pick each colony with a clean pipet tip (try not to pick too much of the colony; a tiny amount that can be seen by the naked eye is usually more than enough) and transfer the cells into the colony lysis buffer in the labeled tube.
3. Incubate the tubes in boiling H_2O, or on hot plate at 100 °C, for 10 min.
4. Spin at room temperature for 2 min to pellet the cell debris, and transfer the supernatant into a clean tube; and
5. Immediately use the lysate for PCR analysis or store at −20 °C for future amplification.

PCR Reaction

1. For each colony lysate add:

• dH_2O	20.4 μl
• 10× PCR buffer	4.0 μl
• dNTPs (250 μM)	3.2 μl
• Lgh primer	4.0 μl
• Rgh primer	4.0 μl
• Colony lysate	4.0 μl
• *Taq* DNA Polymerase	0.4 μl
• Total volume	40.0 μl
• Mix well and add 30 ml mineral oil if required for the thermal cycler.	

2. PCR parameters are as follows: 30 cycles of 94 °C for 30 s, ≥52 °C for 40 s, ≥72 °C for 1 min, with a final extension at 72 °C for 5 min, and 4 °C holding temperature.

For confirmation of cDNA > 700 bp, increase the elongation time at 72 °C from 1 to 2 min.

3. Analyze 20 µl of the PCR product on a 1.5% agarose gel with ethidium bromide staining, while saving the remainder for sequencing. Plasmids with an insert should produce an easily visible band. Verify the insert size by comparing the molecular weight of the PCR product before and after cloning. The PCR product after colony-PCR should be 120 bp larger than the original PCR insert before cloning due to the flanking vector sequence being amplified.
4. The bands should then be purified from the agarose gel using a QIAEX II kit and saved for Northern blot probe generation using GenHunter's HotPrime DNA Labeling Kit; and
5. After a plasmid has been determined to contain an insert of interest, the corresponding Tet^R colony should be re-streaked to produce single colonies on a new LB-Tet plate:
   (a) Locate the colony marked with the number on the original plate, and streak the cells onto a new LB-Tet plate.
   (b) Change to another clean tip, rotate the plate through 90°, and streak a second time in order to obtain single colonies.
   (c) Incubate the plate overnight at 37 °C.
   (d) Inoculate a single Tet^R colony into 5 ml of LB culture (without tetracycline, and use 3 ml for plasmid miniprep. Save the remainder in glycerol (50%) as cell stock at −70 °C.

Sequencing of Cloned PCR Products

If using the PCR-TRAP Cloning System, sequencing can be conducted utilizing vector-specific primers such as Lseq/Rseq or Lgh/Rgh. If using a cloning vector other than the one recommended, consult the manufacturer's guidelines for sequencing instructions.

1. For bands of the correct size, purify the remaining 20 µl of the retained colony PCR reaction using the QIAquick PCR Purification Kit and continue with direct sequencing; and
2. For the cloned inserts, sequence the plasmids using the Bigdye-terminator sequencing kit.

8.2.2.10 Confirmation of Differential Gene Expression by Northern Blot

The Northern blot technique is technically simple and straight-forward in approach, requiring no manipulation of the RNA sequences from which differential gene expression has been detected. The HotPrime DNA Labeling Kit, a random decamer prime labeling kit which incorporates the anchored oligo-dT primers ($H\text{-}T_{11}M$) into the labeling buffer to ensure full-length anti-sense cDNA probe labeling and uses radioactive dATP to take advantage of the AT-rich nature of DD bands, is specifically designed to efficiently label DNA probes isolated from differential display for Northern blot analysis.

1. For Northern blot using the QIAEX II kit purified PCR products and HotPrime DNA Labeling Kit, set up the following reaction in a 1.5-ml microfuge tube with a locking cap (so the cap will not loosen during boiling):

   - Distilled water 11 μl
   - 10× Labeling buffer 3 μl
   - DNA template to be labeled (10–50 ng) 7 μl

2. Incubate the mixture in a boiling water bath for 10 min; rapidly chill the tubes on ice. Spin the tube briefly to collect the condensate.

3. To the reaction, add the following in the order shown:

   - dNTP (-dATP) (500 μM)[a] 3 μl
   - Alpha-[^{32}P] dATP (3000 Ci/millimole)[a] 5 μl
   - Klenow DNA polymerase 1 μl

   [a] If using alpha-[^{32}P] dCTP instead of alpha-[^{32}P] dATP, substitute dNTP (-dCTP) for dNTP (-dATP).

4. Incubate for 20 min at room temperature, followed by incubation at 37 °C for an additional 10 min.

5. Add 6 μl of the Stop buffer and mix well.

6. Purify the labeled probe on a Sephadex G50 column. Collect the purified probe in a 1.5-ml microfuge tube with a lock-on cap. Count 1 μl of labeled probe in a scintillation counter. A total of 10 million or more CPM can be obtained for most of the labeled DNA probes.

7. RNA gel and transference to nitrocellulose or nylon membrane using the standard procedures.

8. If the prehybridization buffer has been stored at −20 °C, thaw at 37 °C for 20 min.

9. Denature the salmon sperm DNA by incubating for 10 min in a boiling water bath.

10. Add salmon sperm DNA (to a final concentration of 100–200 μg/ml) in the prehybridization solution. Mix well.

11. Use 5 ml of prehybridization solution or enough to cover the membrane.

12. Prehybridize at 42 °C for at least 4 h.

13. Denature the purified probe in a 1.5-m-microfuge tube with a lock-on cap (otherwise the cap may loosen) by boiling for 10 min in a water bath.

14. Chill on ice for 2 min.

15. Spin down the condensate and add the probe directly to the prehybridization solution.
16. Hybridize overnight.
17. Carefully decant the radioactive hybridization solution and dispose of it in an appropriate container for radioactive waste.
18. Wash with 1× SSC containing 0.1% SDS twice at room temperature, each time disposing of the wash solution in an appropriate container.
19. Wash for 15–20 min with 0.25× SSC containing 0.1% SDS prewarmed to the final washing temperature of 50–55 °C.
20. Blot the membrane dry with paper towels and cover using UV-transparent plastic wrap; and
21. Expose blot to single emulsion film with an intensifying screen at −70 °C overnight for optimum signal detection.

The bands that are confirmed by replications, are considered 'real' differences, will warrant further study and downstream functional characterizations.

8.3 Applications of the Technology

Fluorescent differential display (FDD) is a highly sensitive mRNA fingerprinting technique to isolate rare transcripts, detect minute variations in transcript levels, and identify both increased and decreased mRNAs [3]. The technique has been used for the isolation of genes encoding transcription factors, membrane proteins and rare enzymes, and in profiling of various genes that are involved in physiological events, stress responses, signal transduction, secondary metabolism, and hormonal regulation [11–13]. Novel genes associated with photomorphogenesis, photoperiod control and circadian pathways have also been characterized in various plant species using the FDD technique [14–17]. Zhou *et al.* [18–20] conducted a genome-wide scan of the Japanese spurge, which has a high tolerance to cold temperatures and had not been the subject of previous relevant genetic studies. This genome-wide screening using all 240 primer combinations revealed several hundred cold-inducible genes.

FDD is also used to identify genes expressed at certain genetic backgrounds, locate the genes to specific chromosomes, and improve efficiency of functional genome studies. As genome projects on various plant species are completed, more complete gene sequences will become available and accessible to the public and this will make it easier to define gene identity based on short DNA fragments. The FDD cDNA sequence can be used for reverse genetic analysis of the genes of interest and of closely related sequences [21]. This technique will have wider application in the isolation of the clue-sequences which can be applied in the functional characterization of economically important genes.

FDD, with optimized procedures and gel apparatus, is a highly efficient technique for large-scale gene expression studies. However, the cDNA cloning and the sequence analysis of these fragments as well as the downstream verification process can be time consuming, especially when a large number of gene fragments appear to be different.

8.4 Perspectives

5This automated FDD platform has been shown to be accurate and high throughput [22–26] for large-scale screenings. Data analysis tools such as spectra overlay, which allows digital data presentation and quantification, have been developed by Hitachi Genetics Systems for use with the FMBIO series of fluorescent scanners.

Another option for visualization of fluorescent labeled PCR reactions is to run samples on an automated sequencer. The capillary array-based automated DNA sequencers, such as Applied Biosystems ABI3100, can detect FDD bands with several different fluorophores. The results of FDD are seen as a series of spectral peaks for each lane, which can be compared to show differences in a very sensitive and reproducible way. The use of this Capillary Electrophoresis (CE) can dramatically cut down on the time and labor required for large-scale FDD screenings. However, the major drawback and bottleneck for using this technology with FDD is that, at this time, there is no way to retrieve bands from the CE results.

Other steps in the FDD process have been analyzed and targeted for further streamlining and optimization. Direct sequencing of differentially expressed cDNAs of interest without subcloning using the corresponding H-AP primer is one area of improvement [24]. Furthermore, computer programs have been developed to automatically allow positive band identification from an FDD image [25–27]. The most sophisticated attempt in FDD downstream automation could have been the development of a prototype computer-controlled system for positive-band identification and retrieval by Hitachi [28]. This approach employed capillary array gel electrophoresis coupled with fraction collection using sheath flow technology. Automation in PCR set-up and fluorescent data analysis for TOGO and GeneCalling has also been described [29,30]. Elimination of the manual reaction set-up, through the use of a robotic liquid dispenser, not only ensures reproducibility by reducing pipetting errors, but, in combination with the elimination of conventional DD autoradiography, also increases the efficiency of differential gene expression screening.

FDD, together with cDNA microarray, are the two major platforms that can accommodate gene expression analysis on a genome-wide scale. The microarray procedure can only be used for gene expression studies of a few model plant species with completely, or partially sequenced genomes, such as Arabidopsis, rice, and tomatoes, and so on. Although many of the features in model plant species are shared among a wide range of related taxa, the differences, even though they may be a small part of the whole genome, appear to be the most important in determination of distinct traits for individual species, or genotypes. In contrast to DNA microarrays, DD is an 'open' system which is not dependent on any prior knowledge of the genes to be analyzed. As such, novel genes can be discovered using DD which is readily applicable to any biological system where no microarray 'chips' are available. Furthermore, DD can compare more than two RNA samples side-by-side without the need for data normalization. Through further refinement and automation, DD will undoubtedly continue to play a key role in gene discovery research in the post-genome era.

References

1 Liang, P. and Pardee, A.B. (1992) Differential display of eukaryotic messenger RNA by means of the polymerase chain reaction. *Science*, **257**, 967–971.

2 Yamazaki, M. and Saito, K. (2002) Differential display analysis of gene expression in plants. *Cellular and Molecular Life Sciences: CMLS*, **59**, 1246–1255.

3 Kuno, N., Muramatsu, T., Hamazato, F. and Furuya, M. (2000) Identification of large-scale screening of phytochrome-regulated genes in etiolated seedlings of *Arabidopsis* using a fluorescent differential display technique. *Plant Physiology*, **122**, 15–24.

4 Ito, T., Kito, K., Adati, N., Mitsui, Y., Hagiwara, H. and Sakaki, Y. (1994) Fluorescent differential display: arbitrarily primed RT- PCR fingerprinting on an automated DNA sequencer. *FEBS Letters*, **351**, 231–236.

5 Liang, P., Averboukh, L. and Pardee, A.B. (1994) Method of differential display, in *Methods in Molecular Genetics* (ed. K.W. Adolph), Academic Press, San Diego, CA, pp. 3–16.

6 Yang, S. and Liang, P. (2005) Global analysis of gene expression by differential display, in *Differential Display Methods and Protocols* (eds P. Liang, J.D. Meade and A.B. Pardee), Vol. 317, Humana Press, Totowa, New Jersey, USA, pp. 3–21.

7 Liang, P., Averboukh, L. and Pardee, A.B. (1993) Distribution and cloning of eukaryotic mRNAs by means of differential display: Refinements and optimization. *Nucleic Acids Research*, **21**, 3269–3275.

8 Hsu, D.K., Donohue, P.J., Alberts, G.F. and Winkles, J.A. (1993) Fibroblast growth factor-1 induces phosphofructokinase, fatty acid synthase and Ca^{2+}-ATPase mRNA expression in NIH 3T3 cells. *Biochemical and Biophysical Research Communications*, **197**, 1483–1491.

9 Liang, P., Bauer, D., Averboukh, L., Warthoe, P., Rohrwild, M., Muller, H., Strauss, M. and Pardee, A.B. (1995) Analysis of altered gene expression by differential display. *Methods in Enzymology*, **254**, 304–321.

10 Sokolov, B.P. and Prockop, D.J. (1994) A rapid and simple PCR-based method for isolation of cDNAs from differentially expressed genes. *Nucleic Acids Research*, **22**, 4009–4015.

11 Chen, W.J. and Zhu, T. (2004) Networks of transcription factors with roles in environmental stress response. *Trends in Plant Science*, **9**, 591–596.

12 Friedrichsen, F.M., Nemhauser, J., Muramitsud, T., Maloofa, J.N., Alonsoa, J., Eckera, J.R., Furuyad, M. and Chory, J. (2002) Three redundant brassinosteroid early response genes encode putative bHLH transcription factors required for normal growth. *Genetics*, **162**, 1445–1456.

13 Chaban, C., Waller, F., Furuya, M. and Nick, P. (2003) Auxin responsiveness of a novel cytochrome P450 in rice coleoptiles. *Plant Physiology*, **133**, 2000–2009.

14 Hayama, R., Izawa, T. and Shimamoto, K. (2002) Isolation of rice genes possibly

involved in the photoperiodic control of flowering by a fluorescent differential display method. *Plant and Cell Physiology*, **43**, 494–504.

15 Loyall, L., Uchida, K., Braun, S., Furuya, M. and Frohnmeyer, H. (2000) Glutathione and a UV-induced glutathione S-transferase are involved in signaling to chalcone synthase in cell cultures. *The Plant Cell*, **12**, 1939–1950.

16 Kuno, N., Møller, S.G., Shinomura, T., Xu, X.M., Chua, N.-H. and Furuya, M. (2003) The novel MYB protein EARLY-PHYTOCHROME-RESPONSIVE1 is a component of a slave circadian oscillator in *Arabidopsis*. *Plant Cell*, **15**, 2476–2488.

17 Higuchi, Y., Sage-Ono, K., Kamada, H. and Ono, M. (2007) Isolation and characterization of novel genes controlled by short-day treatment in *Pharbitis nil*. *Plant Biotechnology*, **24**, 201–207.

18 Zhou, S., Sauve, R. and Abudullah, A. (2005) Identification of genes regulated by low temperature in *Pachysandra terminalis* Sieb.et Zucc using cDNA differential display. *Horticultural Science*, **40**, 1995–1997.

19 Zhou, S., Chen, F.-C., Nahashon, S. and Chen, T.T. (2006) Cloning and characterization of glycolate peroxidase and NADH-dependent hydropyruvate reductase genes in *Pachysandra terminals*. *Horticultural Science*, **41**, 1226–1230.

20 Zhou, S., Sauve, R. and Chen, F.-C. (2007) Structure and temperature regulated expression of a cysteine protease gene in *Pachysdandra terminalis* Sieb & Zucc. *Journal of the American Society for Horticultural Science*, **13**, 97–101.

21 Scutt, C.P., Vinauger-Douard, M., Fourquin, C., Ailhas, J., Kuno, N., Uchida, K., Gaude, T., Furuya, M. and Dumas, C. (2003) The identification of candidate genes for a reverse genetic analysis of development and function in the *Arabidopsis* gynoecium. *Plant Physiology*, **132**, 653–665.

22 Cho, Y.-j., Meade, J.D., Walden, J.C., Chen, X., Guo, Z. and Liang, P. (2001) Multicolor fluorescent differential display. *Biotechniques*, **30**, 562–572.

23 Liang, P. (2000) Gene discovery using differential display. *Genetic Engineering News*, **20**, 37.

24 Buess, M., Moroni, C. and Hirsch, H.H. (1997) Direct identification of differentially expressed genes by cycle sequencing and cycle labeling using the differential display PCR primers. *Nucleic Acids Research*, **25**, 2233–2235.

25 Aittokallio, T., Ojala, P., Nevalainen, T.J. and Nevalainen, O. (2000) Analysis of similarity of electrophoretic patterns in mRNA differential display. *Electrophoresis*, **21**, 2947–2956.

26 Aittokallio, T., Ojala, P., Nevalainen, T.J. and Nevalainen, O. (2001) Automated detection of differentially expressed fragments in mRNA differential display. *Electrophoresis*, **22**, 1935–1945.

27 Qin, L., Prins, P., Jones, J.T., Popeijus, H., Smant, G., Bakker, J. and Helder, J. (2001) GenEST, a powerful bidirectional link between cDNA sequence data and gene expression profiles generated by cDNA-AFLP. *Nucleic Acids Research*, **29**, 1616–1622.

28 Irie, T., Oshida, T., Hasegawa, H., Matsuoka, Y., Li, T., Oya, Y., Tanaka, T., Tsujimoto, G. and Kambara, H. (2000) Automated DNA fragment collection by capillary array gel electrophoresis in search of differentially expressed genes. *Electrophoresis*, **21**, 367–374.

29 Green, C.D., Simons, J.F., Taillon, B.E. and Lewin, D.A. (2001) Open systems: panoramic views of gene expression. *Journal of Immunological Methods*, **250**, 67–79.

30 Lo, D., Hilbush, B. and Sutcliffe, J.G. (2001) TOGA analysis of gene expression to accelerate target development. *European Journal of Pharmaceutical Sciences*, **14**, 191–196.

9
Real-Time Quantitation of MicroRNAs by TaqMan MicroRNA Assays

Toni L. Ceccardi, Marianna M. Goldrick, Peifeng Ren, Rick C. Conrad, and Caifu Chen

Abstract

MicroRNAs (miRNAs) are powerful regulators of gene expression that work through binding to complementary regions in target mRNAs. In plants, the major effect of miRNAs upon their targets is in the form of down-regulation by cleavage. Many plant miRNAs have been shown to regulate transcription factors, and therefore the roles of specific miRNAs are often related to plant growth, reproduction and development. With such an important regulatory role, the reliable and accurate quantitation of specific miRNAs is desirable. Short sequences and high homology within large miRNA gene families make them difficult and unreliable targets for traditional hybridization-based detection methods. Real-time PCR methods provide greater specificity and sensitivity to quantitation of miRNAs, with a TaqMan assay approach being the most sensitive and specific in its ability to distinguish between closely related family members. We have developed and validated a set of *Arabidopsis* TaqMan MicroRNA Assays, many of which, due to sequence conservation within the plant kingdom, will work with other plant species. The TaqMan MicroRNA Assays enable researchers to order an off-the-shelf assay and generate accurate quantitative results within 3 h of obtaining purified RNA. MicroRNA quantitation can be used for a myriad of downstream applications, including validation of predicted miRNAs, absolute quantitation, tissue expression profiling, tandem quantitation of miRNA with mRNA, and biomarker discovery.

9.1 Introduction

9.1.1 What are microRNAs?

MicroRNAs (miRNAs) are short single-stranded noncoding RNAs of approximately 22 bases in length, whose biological function is to down-regulate expression of

The Handbook of Plant Functional Genomics: Concepts and Protocols.
Edited by Günter Kahl and Khalid Meksem

ISBN: 978-3-527-31885-8

protein-coding genes. MicroRNAs can exert their effects through several different mechanisms, including translational repression, mRNA cleavage/destabilization, and epigenetic effects mediated through binding to nascent transcripts. Primary miRNA transcripts are sequentially processed by RNase-III-family enzymes in a two-step process. In plants both cleavage steps are thought to occur in the nucleus, and in rapid succession, suggested by the fact that pre-miRNAs are rarely detected [1]. Following processing, mature miRNAs are transported into the cytoplasm. The effects of miRNAs on mRNA are mediated by proteins of the Argonaute family, which interact with miRNAs to form a functional complex known as RISC (RNA-induced silencing complex). The miRNA component guides the RISC to target mRNAs through base-pairing interactions. The extent of complementarity between the miRNA and its target site in the mRNA can vary from perfect to more limited pairing restricted to the seed region in positions 2–7 of miRNAs at the 5′ end. In plants, miRNAs tend to interact with mRNAs through almost perfect base-pairing, leading to cleavage of the target mRNA. Plant miRNAs also tend to target the coding region of their target mRNAs, in contrast to the 3′ untranslated regions (UTRs) targeted by most animal miRNAs. Binding in the coding regions is hypothesized to be a trigger for cleavage of the mRNA target [2]. In some cases, plant miRNAs can also participate in translational repression [3].

Most plant miRNAs are found to belong to large gene families. These families are often conserved among plant species, suggesting that they play important roles as regulators of plant growth and development. In fact, the majority of plant miRNAs appear to down-regulate genes for transcription factors, placing miRNAs in a key role of establishing organ patterns and fertility in the developing plant [1]. Although animal miRNAs also frequently regulate transcription factors, there is little or no overlap in primary sequences of plant and animal miRNAs. The biological processes that control transcription of miRNAs are poorly understood, but RNA Pol II is thought to be primarily responsible [1]. The expression of miRNAs in plants and animals can vary from being broadly expressed, to being restricted to specific tissues and developmental stages.

9.1.2
Why are Researchers Interested?

In the past, plant biologists have focused on gene regulation in plants by protein-coding genes such as transcription factors. Discovery of plant miRNAs has provided new insight into how gene expression in plants is fine-tuned via miRNAs at post-transcriptional levels. Researchers are interested in addressing many questions: How did plant miRNAs evolve and how many are there? When, where and to what extent, are plant miRNAs expressed? What roles do miRNAs take in plant development, biotic and abiotic stress-response and other physiological responses? How do plant miRNAs function individually and as a family? How do miRNAs regulate target genes and how are miRNAs themselves regulated? How would miRNA and miRNA-based technology benefit agriculture and plant biotechnology? Lastly, what other small RNAs play regulatory roles in plants?

9.1.3 Current Technologies for miRNA Quantitation

The abundance of miRNAs in animals has been shown to range from only a few to 100 000 or more copies per cell, making some miRNAs among the most abundantly expressed RNAs known. The expression levels in plants have not been similarly quantitated, but knowing that certain miRNAs have been cloned hundreds of times, they are clearly among the more highly abundant RNAs expressed in plants [1].

The miRNA quantitation methods most commonly used are hybridization-based, with Northern blots being a common method both for validation of expression and quantitation [4]. Microarray technologies can likewise be used to screen and roughly quantify many miRNA species at once [5,6]. However, these technologies are limited in both sensitivity and specificity. Other hybridization-based assays include RNase protection [7], a signal-amplifying ribozyme method [8], and primer extension [9]. When using a hybridization-based method, it is difficult to distinguish miRNA precursor transcripts (the so-called pri- and pre-miRNAs) from mature miRNAs, except by size exclusion. And there is difficulty in obtaining the specificity needed to distinguish between closely related miRNA family members, which may differ by only one or a few bases. Ribonuclease protection assays and Northern blots typically require the use of radiolabeled probes, which may be undesirable. The dynamic range of detection of hybridization-based methods is limited, and very low expressers may not be detectable. Finally, Northern analyses or microarrays may take 2 to 3 days to produce results.

Bead-based methods have been adapted for miRNA profiling [10]. In this approach, a set of capture oligonucleotides complementary to the miRNA targets are coupled to a collection of polystyrene microspheres, each of which is differentially labeled with a mixture of fluorescent dyes to create a color signature corresponding to each miRNA. Small RNAs are ligated with adaptors, and the adaptor sequences used to amplify the miRNA population via RT-PCR, using a common biotinylated primer. The sample is then hybridized to the capture oligos, stained with streptaviden–phycoerythrin, and analyzed via flow cytometry. Bead-based analysis shows better sensitivity compared to Northern blot detection, and improved specificity for distinguishing related miRNAs compared to glass microarray methods.

The Invader assay from Third Wave has also been adapted for quantitative detection of miRNAs. The Invader assay uses a novel 5′ nuclease enzyme that cleaves a structure created by hybridization of two synthetic oligonucleotides designed to partially overlap the target miRNA [11]. A series of cleavages release a fluorescent label to create a signal from a single miRNA target. This method may be able to discriminate closely related miRNAs in crude cell lysates without the purification of small RNA. Using a synthetic let-7 miRNA, the assay was shown to have a 3-log dynamic range and ability to detect as few as 20 000 copies of the target.

Real-time (or quantitative) RT-PCR methods have been adapted for quantitation of miRNAs. These PCR-based methods have improved sensitivity over hybridization-based methods. Depending on the design, they can also be more specific, allowing differentiation of precursor miRNAs from mature miRNAs, and also discrimination of related miRNAs within homologous miRNA families. SYBR Green I dye detection assays may require a melting curve analysis to help improve specificity. Several recent publications describe a miRNA detection method based on adding poly(A) tails to the 3′ ends of the small RNA samples. Next, reverse transcription with oligo dT-primers incorporated universal reverse primer binding sites for subsequent real-time PCR amplification and detection with SYBR Green. This method was used to detect several miRNAs in *Arabidopsis* using as little as 100 pg of total RNA, and was able to discriminate related miRNA targets differing by as little as a single base [12]. All real-time PCR methods are convenient in that results can typically be obtained in less than a day.

Applied Biosystems offers pre-designed TaqMan MicroRNA Assays that have the ability to distinguish between closely related miRNAs with as little as one nucleotide difference. Using a stem-looped gene-specific reverse transcription (RT) primer, the assays can distinguish between precursor miRNAs and mature miRNAs. TaqMan miRNA assays have improved sensitivity over all methods with the ability to detect down to 10 copies of a single miRNA. Finally they have up to a 7-log dynamic range, confidently measuring a full range of miRNA expression levels.

9.2 Methods and Protocols

9.2.1 Bioinformatic Tools for miRNA Discovery

The first approach to discovering new miRNAs, and still the most common, was to clone and sequence individual small RNAs. Most miRNAs were identified by this approach [13–17]. Even though this molecular cloning method identified hundreds of new miRNAs, it was limited in its ability to detect rare or cell type-specific miRNAs.

The availability of full genome databases of several organisms enabled the development of an informatics approaches for the identification of new miRNAs. This was inspired by the fact that most known miRNAs are conserved among related species [13,16,18]. Many predicted plant miRNAs have relied on the degree of conservation between the eudicot *Arabidopsis thaliana* and the monocot *Oryza sativa* [19]. Aside from looking for homology between diverse organisms, another criteria relies on the observation that many miRNAs occur multiple times within a genome [20]. Thus, intragenomic matching is an alternative homology-based filtering technique. Approaches based on sequence conservation have been useful in revealing many of the large miRNA gene families.

To date, several computational methods have been developed for predicting new miRNA genes for both animals and plants [21–24]. The presence of stem-loop structures and intergenic location of miRNAs are additional criteria used in computational methods. Two examples of widely used programs are miRseeker and MiRscan.

Lai *et al.* [25] developed the miRseeker program and identified 48 miRNA candidates in Drosophila, 24 of which were validated. The other program, MiRscan has been applied to vertebrate and nematode genomes to identify new miRNA genes [18,23]. Later, similar approaches specifically designed to detect plant miRNAs identified several new candidates [24,26]. Folding predictions have been relatively less useful for plant miRNAs than for animals, because plant primary miRNA transcripts are longer and show more heterogeneity in their stem-loop structures. One filtering criterion that is being applied is the degree of pairing in a predicted stem-loop structure which resembles that of conserved and functionally validated miRNAs [1].

Finally, in plants there is high utility of algorithms that search for complementary target sites in mRNAs to computationally predict plant miRNAs. Due to the extensive base-pairing observed for plant miRNAs and their target mRNAs, the algorithms typically rely on a stringency of zero to three mismatches allowed between the predicted miRNA/mRNA pairing.

Useful on-line tools for bioinformatics-based miRNA analysis include: (1) The Sanger miRNA database, miRBase (http://microrna.sanger.ac.uk/); a searchable database of published miRNA sequences and annotation, supporting searches for both miRNAs and their target mRNAs [27,28]; (2) The *Arabidopsis* Small RNA Project Database (ASRP, http://asrp.cgrb.oregonstate.edu); contains *Arabidopsis* and rice small RNA sequences, with tools for miRNA and siRNA identification and analysis [29]; (3) Cereal Small RNA Database (CSRDB, http://sundarlab.ucdavis.edu/smrnas/); contains small RNA sequences from maize and rice derived by 454 Sequencing Technology [30]; (4) MPSS (Massively Parallel Signature Sequencing, Solexa, Inc, http://mpss.udel.edu/rice) database for rice mRNAs and small RNAs [31]; (5) miRU (http://bioinfo3.noble.org/miRU.htm), a web server for plant target prediction [32].

9.2.2
MicroRNA Isolation from Plants

Isolation of RNA from plants is routinely characterized by two major problems. The first is the robustness of the cell walls in some tissues. The second is the presence of polyphenolic compounds and other problematic biomolecules, especially polysaccharides. The former can be addressed by freezing the samples in liquid nitrogen or dry ice, then crushing them in a cold mortar. Transferring frozen samples to RNA*later* ICE (Ambion, Austin, TX) is an alternative method which alleviates some of the drudgery of this task. More robust samples are perfused with RNA*later* ICE overnight at −20 °C and then immediately processed in the lysis solution (see below) with a rotor-stator homogenizer. This provides protection from RNases that can be released during the thawing process and leads to more efficient homogenization of the sample. But RNA*later* ICE is not recommended for TriReagent extractions. For diminishing the effects of polyphenolic compounds, the Ambion Plant RNA Isolation Aid can be used to eliminate many of these in a pre-extraction spin step. The procedure given here uses the Plant RNA Isolation Aid in combination with the *mir*Vana miRNA Isolation Kit (Ambion).

9.2.2.1 Extraction from Plant Tissue

The following procedure isolates total RNA including miRNA. An alternative procedure using the same kit can also be carried out to split the RNA into larger- and smaller-species-enriched fractions. For some downstream protocols, such as microarrays, it is beneficial to begin with an enriched small RNA fraction. However, for the TaqMan MicroRNA Assays this subfractionation is unnecessary.

Initial Sample Treatment

Weigh *Arabidopsis thaliana* rosettes, flash freeze and grind under liquid nitrogen in a pre-frozen mortar and pestle. Alternatively, use fresh *Arabidopsis* leaf, weigh first (we used 0.1 g starting material), or store the tissue first in RNA*later* ICE at −20 °C until needed. Blot dry before weighing and continue with the protocol below.

The powder, fresh tissue or soaked cold tissue should be placed in at least 10 vol (ml per gram of tissue) of the *mir*Vana Lysis/Binding Solution and 1 vol of the Ambion Plant RNA Isolation Aid. Homogenize for approximately 1 min, for example at setting 5 in a rotor/stator mechanical homogenizer such as the PRO250 homogenizer (Pro Scientific, Inc. Oxford, CT) with a 10-mm probe.

Organic Extraction

Acidify the lysate (0.4 ml) with the addition of the miRNA Homogenate Additive (0.1 vol) and mix. Incubate on ice for 10 min prior to the addition of 0.4 ml of the acid phenol–chloroform mix that comes with the kit. Vortex vigorously for 1 min, then centrifuge for 5 min to separate the organic and aqueous phases (Note: All centrifugation steps were performed at maximum speed in a microfuge at room temperature).

Isolation of Total RNA (with miRNA)

1. *Adsorbing to Glass Matrix:* Remove 350 μl of the upper phase, being careful to ensure that there is no contamination from the interface. Add 438 μl (1.25 vol) of ethanol, mix thoroughly, and pass the solution through the provided glass fiber filter (GFF) by centrifugation for 30 s. The GFF is contained in a plastic 'basket' which is placed into a 2-ml collection tube before use. After passing the preparation through the GFF, remove the filter basket, discard the filtrate, and replace the basket into the collection tube.

2. *Washing the GFF:* Pass washes 1 to 3 through the filter in subsequent centrifugations for 30 s (700 μl for wash 1, 500 μl for washes 2 and 3), discarding the filtrate after each spin.

3. *Eluting the Sample:* Spin the filter thoroughly to ensure dryness (30 s at maximum speed), and transfer the filter basket to a fresh collection tube. Add 100 μl of hot (95 °C before pipetting) Elution Solution to the GFF and centrifuge for 30 s to elute the RNA.

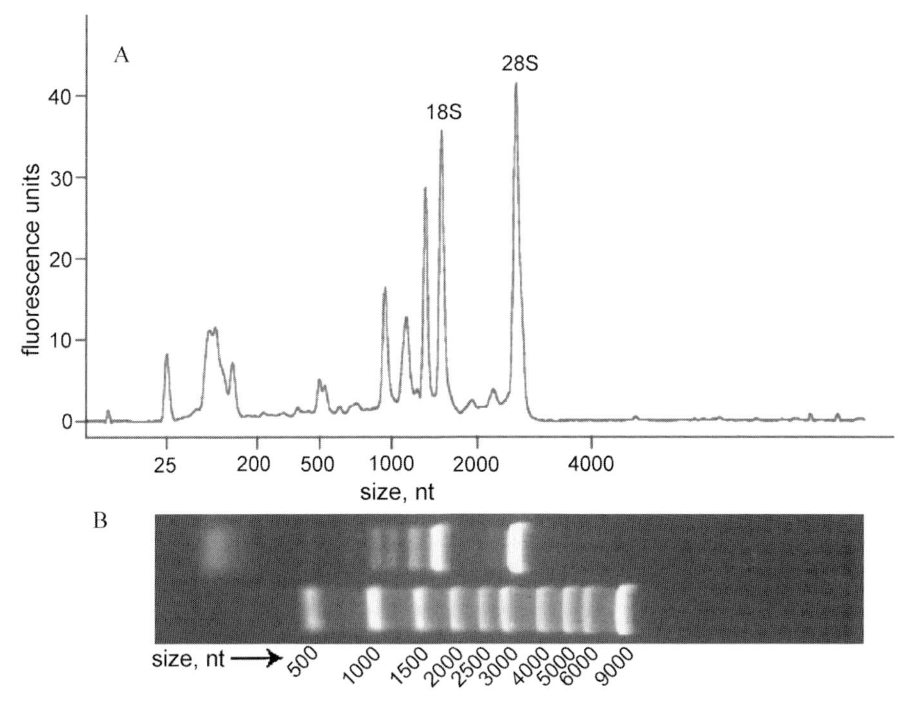

Figure 9.1 *mir*Vana isolated *Arabidopsis thaliana* RNA. RNA was isolated using the *mir*Vana RNA Isolation Kit following the standard protocol (Total), (A) 150 ng loaded onto a capillary electrophoresis system (Agilent Bioanalyzer 2100); (B) One microgram of RNA loaded onto a 1% denaturing agarose gel (Ambion NorthernMax Kit). Note the additional bands (besides the major ribosomal RNA species) in the sample are indicative of plant samples and represent chloroplast and mitochondrial rRNA.

Quantitation of RNA

The amount of total RNA can be calculated from the A_{260} using standard methods (1 OD_{260} unit = 40 μg/ml; Figure 9.1).

9.2.2.2 **PCR Directly from Cells in Culture**

Plant cells grown in suspended cultures can be used directly in the quantitative RT-PCR (qRT-PCR, or TaqMan) assays. For the process presented here, Black Mexican Sweet (BMS) maize cells were used. Propagate cells in MS2D medium [33] with 2 mg 2,4-D/l in Erlenmeyer flasks at 25 °C in the dark on a rotary shaker. Collect the cells from log-phase culture by centrifugation (8000 rpm, 5 min) then wash briefly with DEPC-treated water. After counting under a microscope, add 2× lysis buffer (AB P/N: 4305895) at a ratio of 1 μl per 20–40 cells. Incubate (5 min at room temperature) the cells and dilute at least 10 times with 0.1× TE buffer to make a diluted lysate that serves as an RNA stock solution (each microliter represents total RNA from two to four cells). The RT reaction can be set up with a series of dilutions of this RNA stock.

9.2.3 Description of TaqMan MicroRNA Assays

9.2.3.1 Principle of TaqMan MicroRNA Assays

The TaqMan MicroRNA Assays are quantitative RT-PCR assays and are designed to detect and accurately quantify mature miRNAs using Applied Biosystems real-time PCR instruments and reagents.

The principle of the TaqMan MicroRNA Assays is similar to conventional TaqMan RT-PCR. A major difference is the use of a novel stem-loop primer during the reverse transcriptase (RT) reaction (Figure 9.2A and B). Stem-loop RT primers have several advantages. First, by annealing a short RT priming sequence to the 3′ end of the miRNA, better specificity is achieved for discriminating similar miRNAs. Secondly, its double-stranded stem structure inhibits hybridization of the RT primer to miRNA precursors and other long RNAs. Thirdly, the base stacking of the stem enhances the stability of miRNA and DNA hetero-duplexes, improving the RT efficiency for relatively short RT primers (the

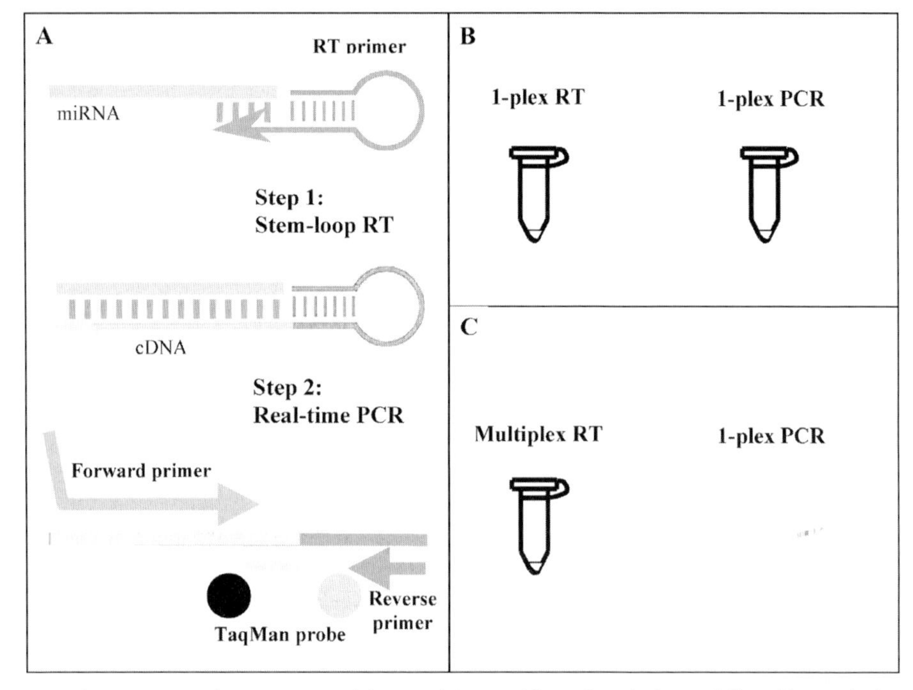

Figure 9.2 (A) The structure and design of the miRNA stem-loop RT primer is shown as it first binds to the specific miRNA target, and is extended during the RT reaction (step 1). In step 2, the stem-loop opens up during annealing and extension in the TaqMan PCR reaction. (B) The workflow of singleplex RT followed by singleplex PCR, as available for plant biologists. (C) The workflow of multiplex RT followed by singleplex PCR, currently available for animal biologists, or by application note for plant biologists wishing to adopt a more convenient workflow.

portion bound to the 3′ end of miRNAs). Finally, the stem-loop structure, when unfolded, adds sequence downstream of the miRNA after reverse transcription (Figure 9.2A). The resulting longer RT product presents a template which is more amenable to real-time TaqMan assay design with great sensitivity and specificity that are largely contributed by specific PCR primers and the TaqMan probe.

9.2.3.2 Performing the TaqMan MicroRNA Assay

Step 1 – RT Reaction

The RT reactions can be conducted using either a single miRNA-specific stem-loop RT primer (Figure 9.2B) or a mixture of multiple RT primers (Figure 9.2C). There are several advantages to using multiplex RT, which may: (1) improve workflow, (2) reduce reagent cost, (3) minimize sample input, (4) allow normalization to control gene(s) detected in the same reaction as the miRNAs, and (5) allow compatibility with the current single tube assays and TaqMan Low Density Array platform.

Prior to performing the individual TaqMan miRNA Assays, the RNA should be extracted using the methods described above or another PCR-compatible method. Then, 1 to 10 ng of total RNA is combined in the 15 μl (total reaction volume) RT reaction with: 1.5 μl 10 × RT-PCR buffer, 1 μl of 50 U/μl MultiScribe RT enzyme, 0.15 μl 100 × dNTP mix, 0.19 μl 20 U/μl RNase-inhibitor, and 3 μl 5 × specific RT-primer (the remainder of the reaction is made up with RNA and nuclease-free water). All components (except RNA and water) can be found in the TaqMan MicroRNA Reverse Transcription Kit (Applied Biosystems, Foster City, CA, part numbers 4366596 or 4366597), which is optimized for use with the Applied Biosystems TaqMan MicroRNA Assays. Incubate the RT reaction(s) on a GeneAmp PCR System 9700 (Applied Biosystems, or other compatible) thermocycler: 16 °C/30 min, 42 °C/30 min, 85 °C/5 min, 4 °C/hold. Abundant miRNAs can be detected and accurately quantitated using reduced total RNA input (Figure 9.3).

Step 2 – Real-Time PCR Reaction

Each individual RT reaction has a unique set of PCR primers and a TaqMan probe designed for it. Once the RT reactions are complete, the PCR reactions can be assembled. For a 10-μL reaction (recommended if using 384-well plates), combine 5 μl 2× Universal Master Mix without UNG (Applied Biosystems), 0.66 μl of the RT reaction, 0.5 μl 20× TaqMan Assay, and 3.84 μl nuclease-free water. If using 96-well plates, the recommended reaction size is 20 μl, so each component should be doubled in volume.

Optimized running conditions on the 7900 HT Real-Time PCR System (Applied Biosystems): 95 °C/10 min, (95 °C/15 s, 60 °C/60 s) × 40 cycles. It is recommended that the real-time PCR reactions should be run in triplicate or quadruplicate and the averages should be used, discarding any outlier (>2 standard deviations), for use in subsequent analyses.

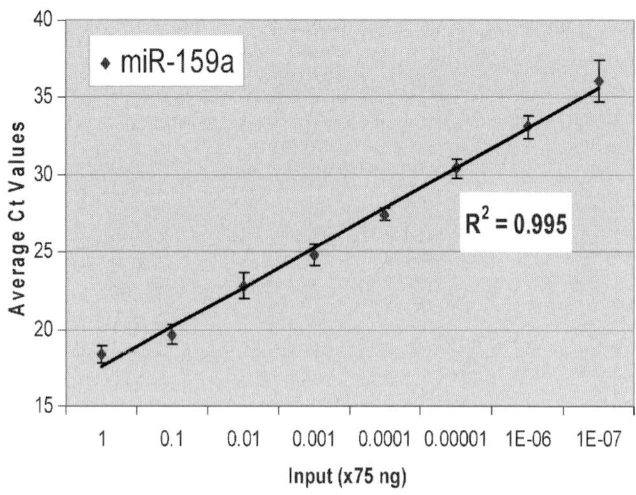

Figure 9.3 Ath-miR159a TaqMan MicroRNA Assay run against total RNA (input based on OD) extracted from *A. thaliana* seedling, serially diluted over 7-logs. The input amount (75 ng to 7.5 fg) represents the total input into the RT step.

The current catalog of TaqMan miRNA assays are maintained in synch with the current miRBase release, and can be downloaded from the Applied Biosystems website. Sixty-five pre-validated assays have been designed for *A. thaliana*, many of which are homologous to other plant species.

9.2.4
Data Normalization

Careful selection of an appropriate control or set of controls is extremely important as significant variations have been observed between samples even for the most commonly used housekeeping genes, including ACTB (β-Actin) and GAPDH [34]. An ideal endogenous control generally demonstrates gene expression that is relatively constant and abundant across different tissues or cell types. However, the chosen endogenous control or set of controls must still be validated for the target cell, tissue, or treatment [35], since no given gene can serve as a universal endogenous control for all experimental conditions.

When considering endogenous controls suitable for use with the TaqMan MicroRNA Assays, it is important that they share similar properties to miRNAs, such as RNA stability and size, and are amenable to the miRNA assay design. A number of reports indicate that other classes of small non-coding RNAs (ncRNAs) are expressed both abundantly and stably making them good candidates for use as endogenous controls [36,37]. For this purpose, we have carried out a systematic study of a set of 32 *Arabidopsis* ncRNA species ranging in size from 24 to 82 nucleotides, including small nuclear RNA (snRNA), small non-messenger RNA (snmRNA), and small nucleolar RNA (snoRNA) [38], across a variety of plant tissues to determine their suitability as endogenous controls for miRNA expression normalization. Our results showed stable expression of several ncRNA genes including snoR41Y, snoR65, snoR66, snoR85, and U60 which are recommended for data normalization.

In addition to snoRNA controls, we recommend the use of: (1) endogenous miRNA genes; (2) structural RNAs (18S rRNA, U6 snRNA); and (3) house-keeping genes (GAPDH, actin, β-tubulin). It is recommended that several control genes are tested first and then the two best normalization controls are chosen for the particular experiment being undertaken.

9.3 Applications of the Technology

9.3.1 Quantitation of miRNAs

TaqMan MicroRNA Assays offer several distinct advantages over conventional miRNA detection methods such as Northern blots and microarrays. They include: (1) high-quality quantitative data – the assays can detect and quantify miRNA up to 7 logs of dynamic range (Figure 9.3); (2) better sensitivity – the assays can detect miRNAs in as little as 10 pg of total plant RNA, allowing the conservation of limited samples (Figure 9.3); (3) high specificity – the assays detect only mature miRNA, not its precursor, with single-base discrimination (Figure 9.3); (4) reproducibility – TaqMan MicroRNA Assays yield highly reproducible results reflecting the high accuracy with which miRNAs can be measured; and (5) rapid and simple methodo-logy. The two-step protocol takes less than 4 h and can be used with any Applied Biosystems Real-Time PCR instrument [39].

Over 90 different miRNA families have been identified in plants. In each family, miRNAs members often differ from each other by only one or two nucleotides. The specificity of TaqMan MicroRNA Assays can be illustrated in a test case of differentiating between two members of the miR159 family that differ from each other only in the last nucleotide: miR159a (5′ UUUGGAUUGAAGGGAGCUCU<u>A</u> 3′) and miR159b (5′ UUUGGAUUGAAGGGAGCUCU<u>U</u> 3′). The assay was conducted with synthetic RNA oligos, either miR159a or miR159b, as the input. As shown in Figure 9.4, the miR159a assay (i.e. miR159a-specific RT and TaqMan assay) predominantly detected miR159a with extremely low cross-reaction with miR159b. Conversely, the miR159b assay (i.e. miR159b-specific RT and TaqMan assay) detected miR159b with very low cross-reaction with miR159a.

In a challenging test case, the well-known let-7 miRNA family was used to demonstrate TaqMan miRNA Assay specificity. Each assay was tested with its intended miRNA target as well as with each member of the let-7 family. The C_T difference between perfectly matched (to its template) and mismatched assays was used to calculate the percent relative detection. Although let-7a–let-7c and let-7b–let-7c pairs differ from each other by only a single nucleotide, the relative detection was two orders of magnitude higher for the matched assays than the mismatched assays [39].

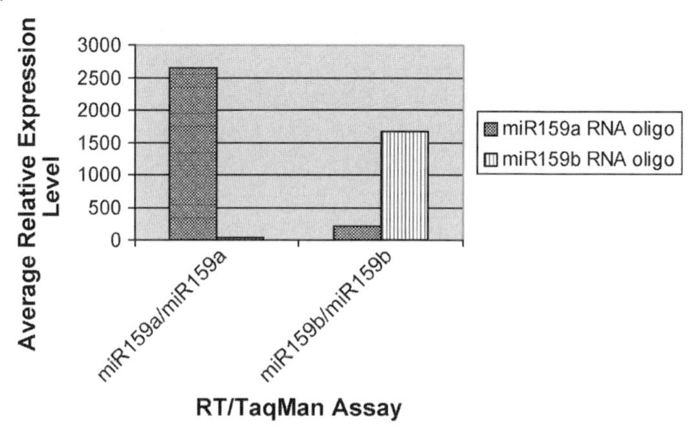

Figure 9.4 Specificity of TaqMan miRNA Assays demonstrated for two miRNA sequences that differ by only the 3′-most base, miR159a and miR159b. Each assay was paired with an artificial RNA template for either its intended miRNA target, or for the closely related but unintended miRNA target.

9.3.2
Absolute Quantitation of miRNAs

Absolute quantitation is necessary if researchers are interested in determining the miRNA copy number per cell. It requires serially diluted synthetic miRNA standards of known concentrations to generate a standard curve, which is used to determine the concentration of unknowns based on their C_T values. This method assumes that all standards and samples have approximately equal amplification efficiencies. Furthermore, the concentration of serial dilutions should stay within the range of accurately quantifiable and detectable levels. Each miRNA may require its own standard curve for better estimates of quantity. TaqMan miRNA assay results from a synthetic template can also serve as a positive control for normalizing plate-to-plate or day-to-day variations.

Synthetic RNA should be quantified based on the A_{260} and diluted over several orders of magnitude ranging from 10^2 to 10^8 copies as template in RT reactions. The unknown RNA sample can be compared to the standard curve to calculate the absolute concentration, or copy number, of the miRNA of interest. Ideally, a new standard curve must be generated each time an absolute quantitation reaction is conducted.

9.3.3
Expression Profiling of miRNAs

There are a number of methods available for constructing an expression heat-map based on real-time PCR results. These methods include: (1) use of normalized C_T to one or several endogenous controls; or (2) use of ΔC_T where ΔC_T is calculated from the formula: C_T of averaged control(s) − C_T of a miRNA gene. ΔC_T represents the relative expression changes of miRNAs over endogenous control(s). Agglomerative hierarchical clustering can be performed to construct an expression profile using commercially or publicly available software such as the CLUSTER program [40]. No further data normalization such as log transformation, gene or sample normalization, and data centering is needed to ensure a true view of relative expression changes among miRNAs and samples.

TaqMan miRNA assay data can be applied to miRNA profiling to reveal spatial and temporal expression of plant miRNAs. Figure 9.5 shows an example of 50 TaqMan miRNA assays run on *Arabidopsis* seedling total RNA. Because many plant miRNAs are conserved among different plant species, assays designed for such conserved miRNAs should be readily used in multiple species. For example, *Arabidopsis* miRNA assays were used to detect miRNA expression in maize. While many miRNAs were expressed at relatively low levels, some miRNAs were expressed at relatively high levels in maize leaf tissue (data not shown). One of the more abundantly expressed miRNAs was miR159, which has been shown to be involved in leaf morphogenesis in *Arabidopsis* [41].

9.3.4 Verification of Predicted Novel miRNAs

Computational prediction of miRNA genes avoids the cloning bias of detecting the more abundant species. However, validating expression of bioinformatically predicted miRNAs presents significant technical challenges. MicroRNAs are tiny molecules of only ~22 nucleotides in length. Many miRNAs are expressed at a very low level, and in some cases are highly similar in sequence to other miRNAs. Therefore, traditional gene expression techniques are not always suitable, or sufficiently sensitive and specific, for identifying bioinformatically predicted miRNAs. However, significant progress has been made in recent years in the development, implementation and refinement of several validation approaches, which include: (1) miRNA microarrays; (2) Northern hybridization; (3) primer extension [9]; (4) *in situ* hybridization [42–44]; and (5) quantitative RT-PCR. In general, cloning and sequencing provide the highest level of validation for predicted miRNAs. Even though the microarray is the mostly commonly used method, highly sensitive and specific RT-PCR methods have become increasingly popular for the verification of novel miRNA candidates.

Future extensions to the product line of the Applied Biosystems TaqMan miRNA Assays include a validated design pipeline, where either a 'virtual' assay will be designed for every miRNA available in public databases, or researchers will be able to enter a miRNA sequence of interest and have a TaqMan miRNA assay designed for them. These customized assays in combination with TaqMan Low Density Arrays will offer specificity, sensitivity, and convenient mid-throughput.

9.3.5 MicroRNAs in Plant Growth and Development

The role of miRNAs in plant growth and development can be categorized in two different ways. First, mutations in the key genes involved in miRNA biogenesis lead to pleiotropic developmental defects [3]. The effects can range from embryonic lethality

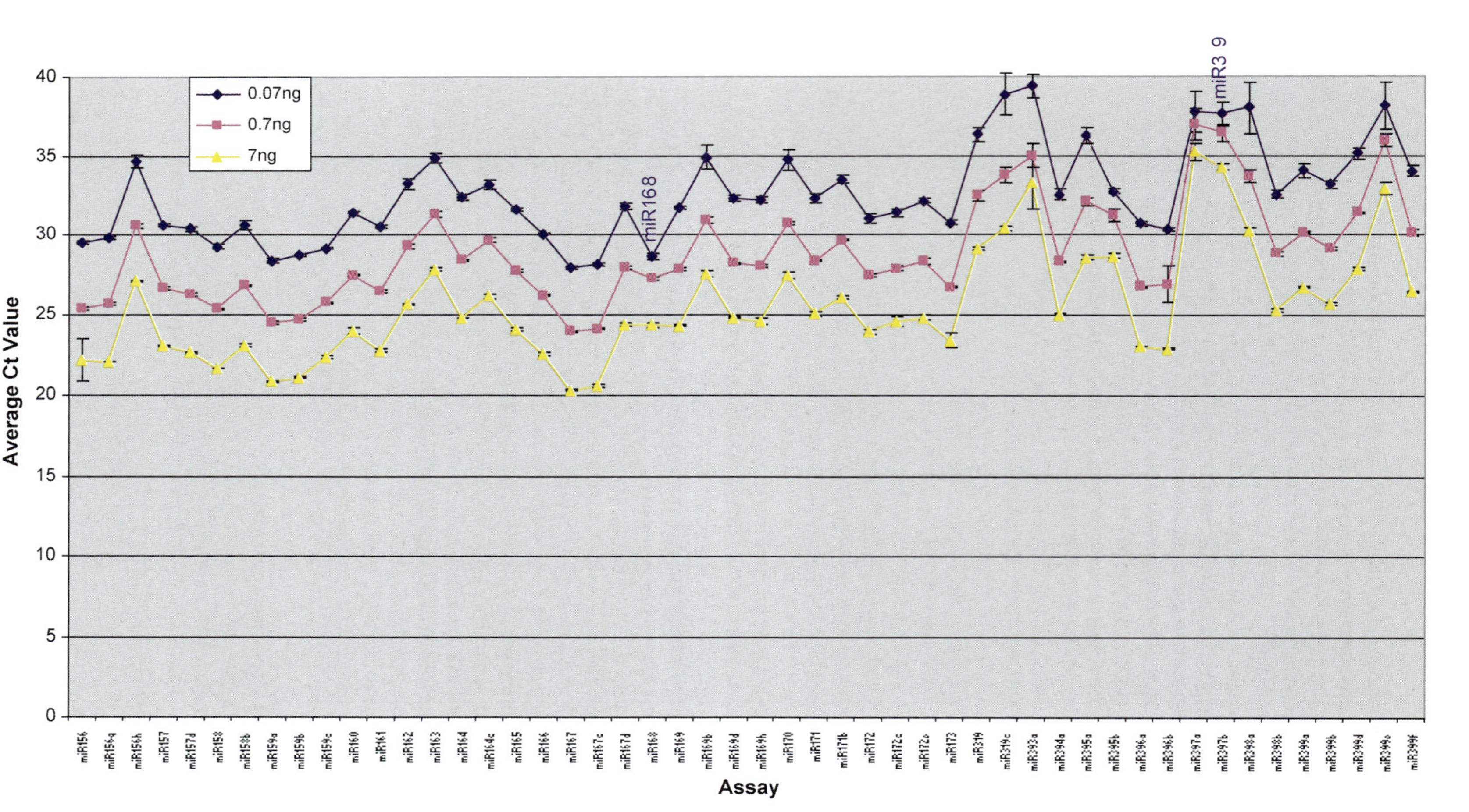

Figure 9.5 Fifty TaqMan MicroRNA Assays were run on a sample of RNA from *A. thaliana* seedlings. This experiment demonstrates the range of expression levels seen among different miRNAs. Three different inputs of total RNA ranging from 7 to 0.07 ng show the sensitivity of the assays, while the reproducibility of the assays is demonstrated by the similarity in the three graphs.

to dysmorphic growth. Secondly, specific mutations that disrupt the binding between miRNA and the target mRNA cause specific developmental defects to occur. Using techniques to block the expression of miRNAs, or the binding sites between miRNAs and their target mRNAs, researchers have determined the roles of many miRNAs. Among the specific functions found to be associated with miRNAs are the signaling of leaf polarity, reproductive development [45], and gravitropic responses in roots [1].

Likewise, over-expression of a miRNA itself can be used to characterize its biological function. Most studies so far have been conducted on the model plant *Arabidopsis*. Over-expression of miR156 targeting the SPL transcription factor increased leaf initiation and biomass, and decreased apical dominance [46]. Over-expression of miR159 and its targeted MYB transcription factors caused male sterility and delayed flowering time [47]. MicroRNAs found to play a role in the development of economically important crops include maize Corngrass 1 which encodes two tandem miR156 genes that appear to have contributed to the evolution of maize from grass-like ancestors [48].

The importance of miRNAs in plant development is also revealed in the redundancy and specialization among the members of the miRNA family and their role in developmental robustness [49]. The ability of TaqMan miRNA assays to differentiate closely related miRNA family members should facilitate research in this area.

9.3.6
Discovery of miRNA Biomarkers

It has been well documented that miRNAs can serve as biomarkers in animal systems. In plants, a few potential miRNA biomarkers have been identified in an effort to discover novel plant miRNAs, especially those regulated by abiotic stress. Environmental triggers can result in up- or down-regulation of specific miRNAs, and whether a miRNA is up- or down-regulated in response to a stress is hypothesized to depend on what transcripts or proteins need to be called into play during stress responses. For instance, miR393 is strongly up-regulated by cold, dehydration, NaCl and treatment with the stress hormone ABA, whereas miR319c is up-regulated by cold but not by other environmental stimuli [2]. MiR393 targets TIR1, a positive regulator of auxin signaling. The up-regulation of miR393 therefore signals the down-regulation of auxin signaling and seedling growth. Thus the stress response which turns on miR393 will inhibit plant growth during stress [2].

9.3.7
Discovery and Validation of Plant miRNA Targets

The high degree of sequence complementarity between a plant miRNA and its target facilitates target discovery via computational methods which search for regions of complementarity of known miRNAs to annotated mRNAs. Furthermore, the predominant mode of action for plant miRNAs is miRNA-mediated mRNA cleavage. This leads to an experimental approach to the discovery of miRNA targets using global microarray analysis to identify mRNAs whose levels are decreased in response to a particular miRNA. For example, mRNA expression levels are compared between wild-type plants and their counterparts lacking expression of a particular miRNA due to mutation. This approach was successfully used to identify mRNA targets of the JAW-D miRNA in *Arabidopsis*, by hybridization of labeled cRNA derived from wild-type and JAW-mutant plants to Affymetrix arrays containing probes for >24 000 annotated genes [41]. Target genes identified by their higher levels of expression in JAW mutants included several members of the TCP gene family, which encode transcription factors that control leaf morphogenesis.

Further validation of plant miRNA targets has been carried out by the identification of miRNA-mediated cleavage products. Validation can be achieved *in vitro* by using wheat-germ lysate or in a transient assay with *Agrobacterium* infiltration to observe miRNA-mediated cleavage of target mRNA [50,51]. However, the most informative approach is to directly detect *in vivo* products of miRNA-mediated cleavage by using 5′ RACE (Rapid Amplification of 5′ cDNA Ends; [24,52]). In this method, an adaptor is added to the 5′ end of the predicted miRNA target using T4 RNA ligase, followed by RT-PCR with adaptor-specific and miRNA-target gene-specific primers. Following nested PCR using the 5′ adaptor and 3′ primer sites, the product is cloned and sequenced. For a validated target, the 5′ end of the sequence of the cloned PCR product must match the predicted cleavage site of targeted mRNA (i.e. between positions 10 and 11 of miRNA; [53]).

9.4
Perspectives

Initially, the methods used in miRNA research involved ligation and cloning of size-fractionated RNA followed by sequencing of concatenated clones to identify candidate miRNAs. Relatively insensitive Northern blot assays were then carried out to provide experimental verification that the candidate sequences were actually expressed in a tissue-specific and developmental stage-specific manner, as would be expected if this newly defined class of noncoding RNA had regulatory functions. These early efforts led to the discovery of abundant miRNAs in different species, first in animals and then in plants. Subsequently, biochemical approaches were successfully used to elucidate many of the details of miRNA biogenesis, and functional studies were carried out aimed at defining the biological role of miRNAs. Bioinformatic tools were developed and used in conjunction with experimental techniques to further define the extent of the 'miRNA world'. Limitations to the use of standard qRT-PCR were then overcome to allow this powerful technique to be used for quantitative analysis of miRNA. The use of qRT-PCR opened the door to detection of very low-abundance miRNAs, some of which are only expressed during narrow developmental windows. Compared to microarray-based miRNA analysis, qRT-PCR is more sensitive, and therefore permits miRNA detection in total RNA preparations, avoiding the need to size-fractionate the small RNA species prior to analysis and enabling miRNA analysis in microdissected samples. The additional specificity associated with the internal hybridization probes used for TaqMan-based qRT-PCR is especially advantageous for miRNA analysis, since the very short amplicons and very limited sequence space within which to design amplification primers can be problematic for SYBR-Green-based detection. The role of miRNA-regulated gene expression in plant biology is just beginning to be appreciated, and many ongoing studies are directed towards understanding the scope and mechanism of miRNA-mediated effects in plants. With the availability of straightforward methods for extraction of total RNA (including the small RNA fraction) from plants and sensitive qRT-PCR assays that allow analysis of miRNA in minute amounts of total RNA, the stage is set for rapid progress that promises to yield important insights into this fascinating area of botanical science.

References

1 Jones-Rhoades, M.W., Bartel, D.P. and Bartel, B. (2006) MicroRNAs and their regulatory roles in plants. *Annual Review of Plant Biology*, **57**, 19–53.
2 Sunkar, R. and Zhu, J.K. (2004) Novel and stress-regulated microRNAs and other small RNAs from *Arabidopsis*. *Plant Cell*, **16**, 2001–2019.
3 Mallory, A.C. and Vaucheret, H. (2006) Functions of microRNAs and related small RNAs in plants. *Nature Genetics*, **38**, (Suppl), S31–S36.
4 Sempere, L.F. *et al.* (2004) Expression profiling of mammalian microRNAs uncovers a subset of brain-expressed microRNAs with possible roles in murine and human neuronal differentiation. *Genome Biology*, **5**, R13.
5 Liu, C.G. *et al.* (2004) An oligonucleotide microchip for genome-wide microRNA profiling in human and mouse tissues. *Proceedings of the National Academy of Sciences of the United States of America*, **101**, 9740–9744.
6 Barad, O. *et al.* (2004) MicroRNA expression detected by oligonucleotide microarrays: system establishment and expression profiling in human tissues. *Genome Research*, **14**, 2486–2494.
7 Lee, Y. *et al.* (2002) MicroRNA maturation: stepwise processing and subcellular localization. *The EMBO Journal*, **21**, 4663–4670.
8 Hartig, J.S. *et al.* (2004) Sequence-specific detection of MicroRNAs by signal-amplifying ribozymes. *Journal of the American Chemical Society*, **126**, 722–723.
9 Altuvia, Y. *et al.* (2005) Clustering and conservation patterns of human microRNAs. *Nucleic Acids Research*, **33**, 2697–2706.
10 Lu, J. *et al.* (2005) MicroRNA expression profiles classify human cancers. *Nature*, **435**, 834–838.
11 Allawi, H.T. *et al.* (2004) Quantitation of microRNAs using a modified Invader assay. *RNA*, **10**, 1153–1161.
12 Shi, R. and Chiang, V.L. (2005) Facile means for quantifying microRNA expression by real-time PCR. *Biotechniques*, **39**, 519–525.
13 Lagos-Quintana, M. *et al.* (2001) Identification of novel genes coding for small expressed RNAs. *Science*, **294**, 853–858.
14 Lagos-Quintana, M. *et al.* (2003) New microRNAs from mouse and human. *RNA*, **9**, 175–179.
15 Lagos-Quintana, M. *et al.* (2002) Identification of tissue-specific microRNAs from mouse. *Current Biology*, **12**, 735–739.
16 Lau, N.C. *et al.* (2001) An abundant class of tiny RNAs with probable regulatory roles in *Caenorhabditis elegans*. *Science* **294**, 858–862.
17 Lee, R.C. and Ambros, V. (2001) An extensive class of small RNAs in *Caenorhabditis elegans*. *Science*, **294**, 862–864.
18 Lim, L.P. *et al.* (2003) Vertebrate microRNA genes. *Science*, **299**, 1540.

19 Bonnet, E. *et al.* (2004) Detection of 91 potential conserved plant microRNAs in *Arabidopsis thaliana* and *Oryza sativa* identifies important target genes. *Proceedings of the National Academy of Sciences of the United States of America*, **101**, 11511–11516.

20 Lindow, M. and Gorodkin, J. (2007) Principles and limitations of computational microRNA gene and target finding. *DNA and Cell Biology*, **26**, 339–351.

21 Chen, H.M., Li, Y.H. and Wu, S.H. (2007) Bioinformatic prediction and experimental validation of a microRNA-directed tandem trans-acting siRNA cascade in *Arabidopsis. Proceedings of the National Academy of Sciences of the United States of America*, **104**, 3318–3323.

22 Grad, Y. *et al.* (2003) Computational and experimental identification of *C. elegans* microRNAs. *Molecules and Cells*, **11**, 1253–1263.

23 Lim, L.P. *et al.* (2003) The microRNAs of *Caenorhabditis elegans. Genes & Development*, **17**, 991–1008.

24 Wang, X.J. *et al.* (2004) Prediction and identification of *Arabidopsis thaliana* microRNAs and their mRNA targets. *Genome Biology*, **5**, R65.

25 Lai, E.C. *et al.* (2003) Computational identification of *Drosophila* microRNA genes. *Genome Biology*, **4**, R42.

26 Jones-Rhoades, M.W. and Bartel, D.P. (2004) Computational identification of plant microRNAs and their targets, including a stress-induced miRNA. *Molecules and Cells*, **14**, 787–799.

27 Griffiths-Jones, S. (2006) miRBase: the microRNA sequence database. *Methods in Molecular Biology*, **342**, 129–138.

28 Griffiths-Jones, S. *et al.* (2006) miRBase: microRNA sequences, targets and gene nomenclature. *Nucleic Acids Research*, **34**, D140–144.

29 Gustafson, A.M. *et al.* (2005) ASRP: the *Arabidopsis* Small RNA Project Database. *Nucleic Acids Research*, **33**, D637–640.

30 Johnson, C. *et al.* (2007) CSRDB: a small RNA integrated database and browser resource for cereals. *Nucleic Acids Research*, **35**, D829–833.

31 Nakano, M. *et al.* (2006) Plant MPSS databases: signature-based transcriptional resources for analyses of mRNA and small RNA. *Nucleic Acids Research*, **34**, D731–735.

32 Zhang, Y. (2005) miRU: an automated plant miRNA target prediction server. *Nucleic Acids Research*, **33**, W701–704.

33 Murashige, T. and Skoog, F. (1962) A revised medium for rapid growth and bioassays with tobacco tissue cultures. *Physiologia Plantarum*, **15**, 473–497.

34 de Kok, J.B. *et al.* (2005) Normalization of gene expression measurements in tumor tissues: comparison of 13 endogenous control genes. *Laboratory Investigation*, **85**, 154–159.

35 Nicot, N. *et al.* (2005) Housekeeping gene selection for real-time RT-PCR normalization in potato during biotic and abiotic stress. *Journal of Experimental Botany*, **56**, 2907–2914.

36 Eddy, S.R. (2001) Non-coding RNA genes and the modern RNA world. *Nature Reviews. Genetics*, **2**, 919–929.

37 Fedorov, A. *et al.* (2005) Computer identification of snoRNA genes using a Mammalian Orthologous Intron Database. *Nucleic Acids Research*, **33**, 4578–4583.

38 Marker, C. *et al.* (2002) Experimental RNomics: identification of 140 candidates for small non-messenger RNAs in the plant *Arabidopsis thaliana. Current Biology*, **12**, 2002–2013.

39 Chen, C. *et al.* (2005) Real-time quantification of microRNAs by stem-loop RT-PCR. *Nucleic Acids Research*, **33**, e179.

40 Eisen, M.B. *et al.* (1998) Cluster analysis and display of genome-wide expression patterns. *Proceedings of the National Academy of Sciences of the United States of America*, **95**, 14863–14868.

41 Palatnik, J.F. *et al.* (2003) Control of leaf morphogenesis by microRNAs. *Nature* **425**, 257–263.

42 Kloosterman, W.P. *et al.* (2006) *In situ* detection of miRNAs in animal embryos using LNA-modified oligonucleotide probes. *Nature Methods*, **3**, 27–29.

43 Wienholds, E. *et al.* (2005) MicroRNA expression in zebrafish embryonic development. *Science*, **309**, 310–311.

44 Juarez, M.T. *et al.* (2004) MicroRNA-mediated repression of rolled leaf1 specifies maize leaf polarity. *Nature*, **428**, 84–88.

45 Chen, X. (2005) MicroRNA biogenesis and function in plants. *FEBS Letters*, **579**, 5923–5931.

46 Schwab, R. *et al.* (2005) Specific effects of microRNAs on the plant transcriptome. *Developmental Cell*, **8**, 517–527.

47 Achard, P. *et al.* (2004) Modulation of floral development by a gibberellin-regulated microRNA. *Development*, **131**, 3357–3365.

48 Chuck, G. *et al.* (2007) The heterochromic maize mutant Corngrass1 results from overexpression of a tandem microRNA. *Nature Genetics*, **39**, 544–549.

49 Sieber, P. *et al.* (2007) Redundancy and specialization among plant microRNAs: role of the MIR164 family in developmental robustness. *Development*, **134**, 1051–1060.

50 Kasschau, K.D. *et al.* (2003) P1/HC-Pro, a viral suppressor of RNA silencing, interferes with *Arabidopsis* development and miRNA function. *Developmental Cell*, **4**, 205–217.

51 Llave, C. *et al.* (2002) Cleavage of Scarecrow-like mRNA targets directed by a class of *Arabidopsis* miRNA. *Science*, **297**, 2053–2056.

52 Talmor-Neiman, M. *et al.* (2006) Identification of trans-acting siRNAs in moss and an RNA-dependent RNA polymerase required for their biogenesis. *Plant Journal*, **48**, 511–521.

53 Lu, D.P. *et al.* (2005) PCR-based expression analysis and identification of microRNAs. *Journal of RNAi and Gene Silencing*, **1**, 44–49.

II
Gene Silencing, Mutation Analysis and Functional Genomics

The Handbook of Plant Functional Genomics: Concepts and Protocols.
Edited by Günter Kahl and Khalid Meksem

ISBN: 978-3-527-31885-8

10
RNA Interference

Chris A. Brosnan, Emily J. McCallum, José R. Botella, and Bernard J. Carroll

Abstract

RNA interference (RNAi) occurs when double-stranded RNA (dsRNA) is processed into small interfering RNAs (siRNAs) and micro RNAs (miRNA), 21–24 nucleotides (nt) in length, by DICER or DICER-like (DCL) enzymes [1,2]. These siRNAs guide ARGONAUTE-like (AGO) proteins to target and destroy homologous RNA sequences, or to down-regulate transcription of homologous DNA [3–5]. As an effective tool for functional genomics, transgenes can be engineered to express dsRNA to down-regulate homologous mRNAs [6–8]. This chapter will review the subject of RNAi in plants, based largely on research findings in *Arabidopsis*, and provide some simple examples of how RNAi can be used as a tool for functional genomics in plants.

10.1
Introduction

RNA silencing was first discovered in plants where it was known as post-transcriptional gene silencing (PTGS) [9–11]. The central trigger for RNA silencing is the production of double-stranded RNA (dsRNA) and its processing by DICER or DICER-like (DCL) proteins into small interfering RNAs (siRNAs) or micro RNAs (miRNAs) [3–5]. In 1998, when Fire and Mello [12] demonstrated that dsRNA induced RNA silencing in *Caenorhabditis elegans*, it was called RNA interference (RNAi), and this term has since been adopted to describe the phenomenon in all classes of eukaryotic organisms. In the same year, it was also shown that dsRNA induced gene silencing in plants [13], and in the following year, Hamilton and Baulcombe [14] working with tomato and *Nicotiana benthamiana*, discovered that small RNAs were associated with RNA silencing. RNAi has subsequently been demonstrated to be a highly conserved process across eukaryotic organisms, playing key roles in processes such as viral resistance, developmental gene regulation, and transposon silencing via DNA methylation and chromatin remodeling [3–5].

The Handbook of Plant Functional Genomics: Concepts and Protocols.
Edited by Günter Kahl and Khalid Meksem

ISBN: 978-3-527-31885-8

In plants, RNAi can be initiated by sense, antisense [10,13,15] or dsRNA [6] expressing transgenes, or by viruses [16,17] and all of these can produce dsRNA. When a sense or antisense transgene is expressed above a threshold level, it is somehow recognized as a foreign or aberrant RNA (abRNA) [16,18], and amplified into dsRNA [19]. However, expression of dsRNA from a transgene containing inverse repeat DNA sequences (hairpin transgenes) is more efficient than sense or antisense transgenes for inducing gene silencing or producing artificial virus resistance [6,13,19].

An innate form of viral resistance can be induced in plants upon virus infection, and this defense is dependent on the RNAi machinery [20–26]. Numerous plant viruses replicate via a dsRNA intermediate that is produced by a viral-encoded RNA-dependent RNA polymerase, and this can induce RNAi against the virus. Similarly, virus-induced gene silencing (VIGS) can be initiated by viruses containing a portion of the gene to be silenced (either an endogenous gene or a transgene) [17]. In some cases of VIGS, the virus is eliminated by RNA silencing [17], but silencing of the target gene persists [17,27]. This indicates that there is a maintenance phase, as distinct from an initiation phase of RNA silencing. The use of VIGS as a tool for functional genomics will be described in detail in this book (*see* chapter 11 by Steven Bernacki and colleagues).

The first plant genes to be identified as being involved in RNAi were required for sense-induced transgene silencing. These genes were initially discovered by mutagenesis studies in *Arabidopsis* and include *RNA-dependent RNA polymerase 6* (*RDR6*) [22,23] (Figure 10.1), *SGS3/SDE2* (encoding a plant-specific coiled-coil protein) [23] and *SDE3* (encoding an RNA helicase) [24]. mRNA is characterized by a 7-methyl G cap at the 5′ end, and it is now known that decapped RNA serves as a substrate for amplification of dsRNA by RDR6 [28]. Thus, decapped RNA produced from sense and antisense transgenes is a likely inducer of RNAi in plants. Subsequent studies demonstrated the involvement of *ARGONAUTE1* (*AGO1*) [29], *NRPD1a/SDE4* (encoding the plant-specific RNA polymerase IVa) and *RDR2* in transgene silencing [30]. Despite the importance of DCLs in RNAi, no *dcl* mutants were recovered by these early genetic screens due to the redundant nature of the four *DCL* genes in *Arabidopsis* [31]. Construction of double and triple *dcl* mutants was required to define the complementary and hierarchical roles for these proteins in processing dsRNA into small RNAs [25,26,32].

One of the most intriguing features of transgene silencing is that once initiated, silencing can move cell-to-cell, and over longer distances to tissues distinct from the initiating source. In plants, the cell-to-cell (local silencing) movement is through plasmodesmata [33], and longer distance (systemic silencing) movement is most likely through the phloem, in a way similar to viral movement in plants [34–36]. Recently, there have been some significant advances in the understanding of both cell-to-cell spread [37–39] and long-distance transmission [40] of transgene silencing. Regarding long-distance transmission, movement of transgene silencing up the plant to the shoot apex is much more efficient than downward transmission [41]. So far, long-distance transmission of gene silencing has only been widely demonstrated for transgenic traits.

Another feature of transgene silencing in plants is the spreading of siRNA production along the length of the transcript, a process referred to as transitivity [42].

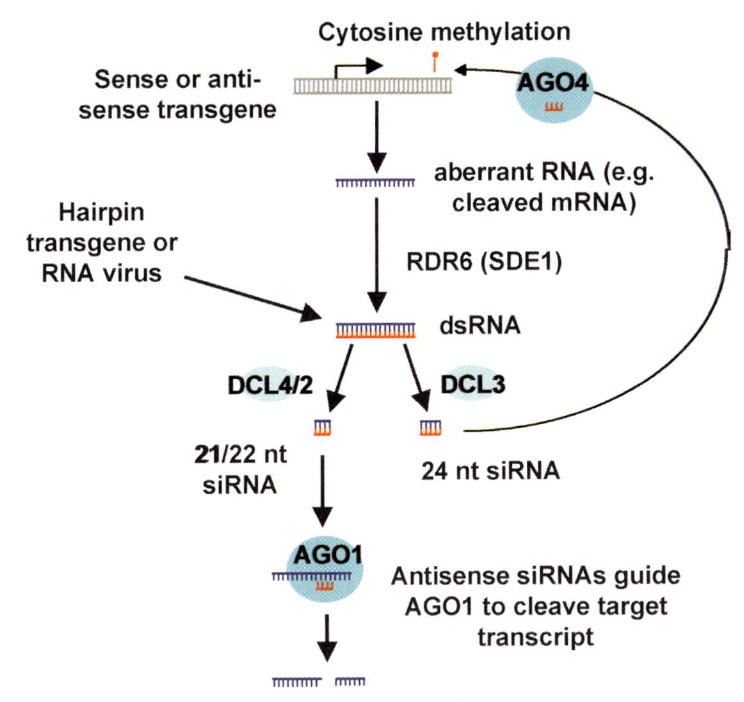

Figure 10.1 Mechanism of siRNA silencing of target genes in *Arabidopsis*. RNAi can be initiated by sense, antisense or dsRNA expressing transgenes, or by viruses. Sense and antisense-induced RNAi requires RNA-dependent RNA polymerase 6 (RDR6), and hairpin transgenes or virus induced gene silencing (VIGS) are more efficient tools for functional genomics in plants. VIGS is considered separately in the following chapter. DCL4, DCL2 and DCL3 process dsRNA into 21, 22, and 24-nucleotide (nt) siRNAs, respectively. 21 and 22-nt siRNAs guide AGO1 and/or perhaps other AGOs to cleave homologous mRNA, and 24-nt siRNAs guide AGO4 to direct DNA methylation and transcriptional silencing [25,26,32]. For further details see text.

When a portion of a transgene is targeted for silencing by a hairpin or sense transgene, siRNA production can extend to other parts of the transcript via the action of RDR6 [42]. Surprisingly, transitivity does not occur when endogenous genes are targeted [43], but irrespective of this lack of transitivity, effective knockouts of endogenous transcripts can be achieved using hairpin transgenes [6,7].

In plants, small RNA produced by DCLs are double-stranded, typically 21–24 bp in length, and have a characteristic two or three nucleotide (nt) 3′ overhang. HEN1 recognizes 21–24 nt small RNA duplexes and attaches a methyl group onto the 2′ hydroxyl group of the 3′ terminal nt [44]. This modification is thought to stabilize siRNAs and miRNAs [44,45]. The small RNA duplexes are then loaded into an RNA-induced Silencing Complex (RISC), where an AGO protein cleaves and removes the passenger strand, leaving the antisense strand small RNA to guide sequence-specific degradation of target transcripts [46–50].

siRNAs and miRNAs have vastly different functions in eukaryotic organisms. Besides guiding defense against foreign RNA, siRNA pathways play a crucial role in transcriptional silencing of transposons. In *Drosophila* embryo extracts [51] and in *C. elegans* [52], siRNAs can also act as primers to re-amplify dsRNA from the target transcript via the action of an RDR. This function is also assumed to occur in plants, although primer independent amplification can also occur [53,54]. In contrast to the siRNA pathways, the complex network of miRNAs in plants and animals plays a crucial role in developmental regulation of endogenous gene expression [55]. The first miRNA was discovered in *C. elegans*, and was shown to be necessary for the timing of larval development [56]. This miRNA, *lin-4*, is a negative regulator of the developmental gene *Lin-14* [56]. Since the discovery of *lin-4*, the fundamental role that miRNAs play in endogenous gene regulation and development of multicellular eukaryotes has been frequently reported [55].

Besides having different biological functions, siRNAs and miRNAs are also generated in different ways [4]. siRNAs are produced from perfectly complementary dsRNA that is expressed either from endogenous repetitive loci, or from an introduced transgene or virus (Figure 10.1). siRNAs are widely recognized as a cellular or genomic immune system, either affecting transcription or mRNA stability of transposons and foreign elements. In *Arabidopsis*, DCL2, DCL3 and DCL4 are responsible for producing siRNAs [25,26,32].

In contrast, miRNAs are expressed from endogenous non-coding loci, and when transcribed produce a fold-back, stem-loop RNA structure, with the stem usually being an imperfect inverted repeat sequence [3]. These primary miRNA transcripts are processed by DCL1 in *Arabidopsis* to produce mature miRNAs, which are then loaded into AGO1 within RISC to guide post-transcriptional gene silencing [3]. The endogenous genes regulated by a miRNA are distinct from the pre-miRNA loci itself (as opposed to siRNAs, which are usually generated from the loci which they target, such as a transgene or repetitive DNA). But despite the distinction between siRNAs and miRNAs in origin and function, they do however share similar modes of action via association with similar biochemical complexes.

The mechanism by which a miRNA post-transcriptionally down-regulates gene expression depends on the level of complementarities it has with the target mRNAs. The 3′ end of miRNA incorporates into the RISC complex via a groove in the PAZ domain of AGO1 [3]. The RISC complex is then aligned with the target RNA, which is thought to be held in the groove of the PIWI domain of AGO1 [57]. As opposed to mammals, where most miRNAs contain several mismatches to their targets and are involved in translational repression [3], the known plant miRNAs have almost perfect complementarity with their targets and usually cleave the target mRNA [58]. There are only single reports of miRNAs regulating gene expression at the transcriptional [59] and translational [60] level in plants.

Small RNA-mediated mRNA cleavage in plants usually occurs about half way along the region of complementarity [49,50]. Recent work with the targets of miR171 by Parizotto *et al.* [61], and also by Mallory *et al.* [62], has shown that complementarity of the 5′ end of the miRNA with the mRNA is required for target cleavage, and that the 3′ end is less important. Upon cleavage, the exosome may degrade the 5′ portion of the cleaved mRNA [63]. The de-capped 3′ end of the *Scarecrow* mRNA is still readily detectable in the case of miRNA171-mediated cleavage of the transcript [64]. However, there are other fates for de-capped 3′ ends of cleaved mRNA from transgenes, including degradation by 5′–3′ exonucleases (e.g. EXORIBONUCLEASE4; XNR4) or amplification into dsRNA by RDR6 and the consequent induction of RNAi [28,40].

Approximately half of the known miRNA targets are transcription factors, the correct regulation of which is vital for eukaryotic development. Mutations in both *AGO1* and *DCL1*, two of the main components of the miRNA biogenesis pathway, produce severe developmental defects. The *ago1* null mutant is seedling lethal [65,66], and the *dcl1* null mutant is a more severe embryo lethal [67,68]. Various leaky *ago1* mutants have radial organs, are sterile and have no meristem function [66,69]. Mutations in *PINHEAD/ZWILLE* (*AGO10*) lead to developmental defects that overlap with those seen in *ago1* mutants [69]. Leaky *dcl1* mutants also have phenotypes similar to *ago1* [70]. Mutations in two of the other miRNA biogenesis proteins, HYL1 and HEN1, produce phenotypes resembling weak *dcl1* mutants such as those with altered leaf development, lack of apical dominance and enhanced hormone sensitivity [45,71].

In 2004, there were 118 potential miRNA genes identified in *Arabidopsis* and their targets could be divided into 42 families [72]. Currently, there are about 180 *Arabidopsis* miRNAs registered in the Sanger miRBase (http://microrna.sanger.ac.uk/cgi-bin/sequences/mirna_summary.pl?org=ath). Many of these miRNAs are likely to be highly conserved between most plant species [73]. The miRNA branch of RNA silencing therefore represents an ancient, evolutionarily conserved form of gene regulation of fundamental importance to both plant and animal development.

10.2 Methods and Protocols

The protocol described here involves using the vector pHANNIBAL [7], which expresses an intron-splicible dsRNA under the control of the constitutive Cauliflower mosaic virus *35S* promoter (Figure 10.2). Targeted knockouts of desired genes are based on the homology of the initiating RNAi transgene and the target transcript. The first step in designing an effective RNAi transgene is identifying the sequences within the target mRNA that are unique, such that the specificity of the desired transcript is maximized and the chances of off-target gene regulation by RNAi are minimized [74]. The first step is therefore to conduct a sequence homology search with the full-length coding sequence of the gene of interest using a standard BLASTN search (http://www.ncbi.nlm.nih.gov/BLAST). If the genome of the species being used has not been completely sequenced, searches of EST databases for that species, in combination with searches of fully sequenced plant genomes should be conducted. The length of sequence required for effective RNAi can be as low as 100 nt, although the optimal length is 300–600 nt [7]. Targeting of untranslated regions (UTRs) may also be used if the target gene is a member of a highly conserved family of genes, as 3′-UTRs are less conserved than coding sequences.

Once an effective length of unique target sequence has been identified in the mRNA of interest, the next step is to look for potential enzyme sites that are not present within the target sequence but are present within either of the two multiple

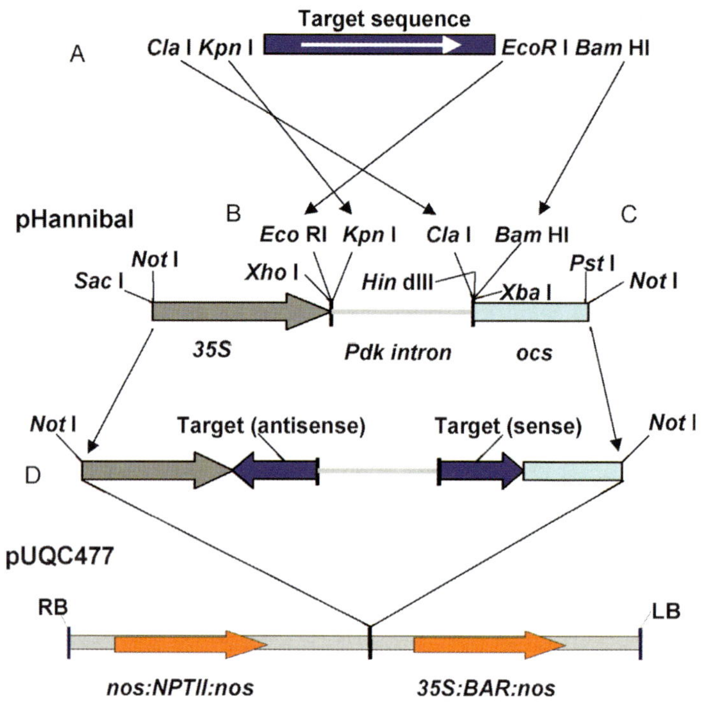

Figure 10.2 Construction of pHANNIBAL-derived RNAi transgenes and *Agrobacterium* binary vectors for plant functional genomics. Unique restriction enzyme sites present in the pHANNIBAL multiple cloning sites [7] are incorporated at the 5′ end of oligonucleotides designed to amplify the target region from the cDNA of interest. The single PCR product (A) can be used in consecutive ligations to clone the antisense (B) and then the sense (C) arms of the hairpin transgene into pHANNIBAL. In the example shown in the figure, the antisense arm is cloned into the first multiple cloning site of pHANNIBAL using *Eco*RI and *Kpn*I (B), and then the sense arm is cloned into the second multiple cloning site using *Cla*I and *Bam*HI (C). The RNAi transgene cassette is then cloned as a *Not*I fragment into an appropriate binary vector such as pUQC477 (D) [40] for transformation into plants. The binary vector pUQC477 has dual plant transformation markers, *nos:NPTII* and *35S:BAR*, which confer resistance to the antibiotic kanamycin and the herbicide BASTA, respectively. For further details, see text.

cloning site of the pHANNIBAL vector (Figure 10.2). Unique restriction enzyme sites are incorporated at the 5′ end of oligonucleotides designed to amplify the target region from the cDNA of interest. In order to reduce costs and working time, two sites may be incorporated at the ends of each oligonucleotide primer (Figure 10.2). In this case, a single PCR product can be used for cloning the sense and antisense arms of the hairpin transgene into pHANNIBAL. Once the target PCR product has been amplified, it is cloned into pGEM-T Easy (Promega, USA), and sequenced to confirm authenticity. For the examples described below, *Kpn*I and *Eco*RI were used to insert the antisense arm into the first multiple cloning site of pHANNIBAL, and *Cla*I and *Bam*HI were used to insert the sense arm into the second multiple cloning site (Figure 10.2). The antisense arm of the hairpin is first excised from pGEM-T clone with *Kpn*I and *Eco*RI (New England Biolabs) using 3–5 U of enzyme per μg of plasmid

DNA. The pHANNIBAL vector [7] is also cut with the same enzymes. The insert and vector fragments are gel-purified using QIAGEN's gel extraction kit (QIAGEN, Valencia, California, USA), according to the manufacturer's instructions. After overnight ligation using T4 DNA ligase (New England Biolabs), DH5α *E. coli* cells are transformed with the ligated products, plated on LB containing 100 μg/ml ampicillin and selected overnight at 37 °C. The resulting colonies are screened by PCR to confirm the presence of the insert, using oligonucleotides corresponding to the *35S* promoter (35S-F; 5′-CACTATGTCGACCAAGACCCTTCCTCTATATAAG-3′) and *PDK* intron (PDK5′; 5′- TCGAACATGAATAAACAAGG-3′) of pHANNIBAL (Figure 10.2). An empty vector will result in a product size of ~100 bp, whereas successful cloning of the first arm will result in a product of 100 bp plus the size of the insert. Positive colonies are then grown overnight at 37 °C in 5 ml LB under ampicillin selection. Following plasmid miniprep purification, diagnostic digests are carried out to confirm the presence of the insert. Although the insert should only be in one direction based on the restriction enzymes used, diagnostic digests can be used to confirm the antisense orientation of the insert using enzyme sites contained within the insert in combination with another restriction site in the vector.

The sense arm of the hairpin transgene is then excised and gel-purified from the pGEM-T Easy clone of interest using *Bam*HI and *Eco*RI (Figure 10.2). The pHANNIBAL-derived vector containing the first antisense arm of the hairpin is also digested with these enzymes and gel-purified. The vector and insert fragments are then ligated and transformed into *E. coli* DH5α, as described above for the first cloning step into pHANNIBAL. The resulting colonies are screened by PCR for the sense insert using the oligonucleotides PDK3′ (5′-TTCGTCTTACACATCACTTG -3′) and OCS-R (5′-GGTAAGGATCTGAGCTACACATGCTCAGG-3′). These primers have annealing sites in the *Pdk* intron and *ocs* terminator, respectively (Figure 10.2). The PCR product will be ~200 bp in length for an empty vector, and 200 bp plus the size of the insert for those containing the sense arm of the RNAi transgene. Positive colonies are cultured, and miniprep-purified plasmids are digested to confirm the presence of the sense-orientated insert. Sequencing of both arms of the transgene can be achieved using both the 35S-F and OCS-R primers listed above. Although the examples described here have the antisense arm upstream of the sense arm, effective RNAi transgenes have been reported containing the sense arm first followed by the antisense [7].

At this stage, if tissue-specific RNAi is desired, the constitutive *35S* promoter can be replaced with one that drives tissue-specific expression. The final RNAi transgene cassette is then purified as a *Not*I fragment and cloned into an appropriate binary vector such as pUQC477 (Figure 10.2) [40] for plant transformation. The binary vector pUQC477 has dual selection markers, *nos:NPTII* and *35S:BAR*, which provide resistance to the antibiotic kanamycin and the herbicide BASTA, respectively (Figure 10.2). The presence and orientation of the insert within the final binary vector can be determined by diagnostic restriction enzyme digest. Binary vectors are transferred into the *Agrobacterium* strain LBA4404 or GV3101 by a tri-parental mating method or electroporation. Tri-parental mating is routinely used to transfer binaries into the *Agrobacterium* strain LBA4404, for example, [75]. This involves growing LBA4404 in a 10-ml LB culture containing 50 μg/ml rifampicin for 36 h at

28 °C. Ten-milliliter cultures of both a helper *E. coli* strain (pRK2013) and *E. coli* DH5α containing the binary vector of interest are also grown for 12 h at 37 °C under appropriate selection. Each culture is spun down at 3000 rpm for 10 min and the pellet resuspended in 1 ml of LB. On an LB plate free of selection, 30 μl of each suspension are combined together and incubated at 28 °C for 16 h. A streak from this plate is then grown for 48 h at 28 °C, on a plate containing 50 μg/ml rifampicin (selection for *Agrobacterium*) and additional selection for the binary vector (50 μg/ml kanamycin in the case of pUQC477). A single colony from this plate is then selected and streaked on a plate containing the same selection agents. The integrity of the transgene in *Agrobacterium* can be confirmed with a diagnostic PCR test. For the examples described here, *Agrobacterium* strain GV3101 was also used for plant transformation but was transformed with the binary vector using electroporation, and plated on 50 μg/ml gentamycin to select for the vector. *Arabidopsis* can be efficiently transformed using *Agrobacterium* via the floral dip method as described by Clough and Bent [76], particularly with strain GV3101 [77]. For all other plant species, transformation involves biolistic bombardment with transgenes or *Agrobacterium*-mediated gene transfer to cell or tissue cultures, followed by plant regeneration.

Once putative transgenic RNAi plants have been generated, transformation and down-regulation of the target transcript is confirmed. Southern detection of the transgene [78] and more importantly, Northern detection of siRNAs homologous to the target [79] should be used to confirm transformation with the RNAi transgene. In our experience, the number of copies of the RNAi transgene has little effect on the efficiency of silencing and it is therefore not normally necessary to determine the transgene copy number in independent transgenic lines. Demonstration of down-regulation of the target transcript is achieved using either Northern blot analysis [78] or reverse transcriptase PCR (RT-PCR), for example, [40]. For genes expressed at very low levels, RT-PCR would be the preference as transcripts even from non-transformed plants may be difficult to detect using Northern blots. If RT-PCR is used to screen for RNA silencing, it is important to include 'reverse transcriptase minus' controls in the analysis to exclude the possibility of genomic or plasmid DNA contamination. It is also important to design oligonucleotides outside the region of sequence used for the RNAi transgene otherwise there will be a significant amount of background signal derived from the hairpin transgene.

10.3 Applications of the Technology

10.3.1 Targeting Transgenes in *Arabidopsis* using RNAi

The effectiveness of the intron-splicible RNAi was tested by targeting two different transgenes. The first example involved targeting the *BAR* gene which confers resistance to the herbicide BASTA. An RNAi transgene targeting the first 350 nt of the *BAR* coding sequence was inserted into pUQC477 (Figure 10.2) to produce pUQC1081 [40]. Thus,

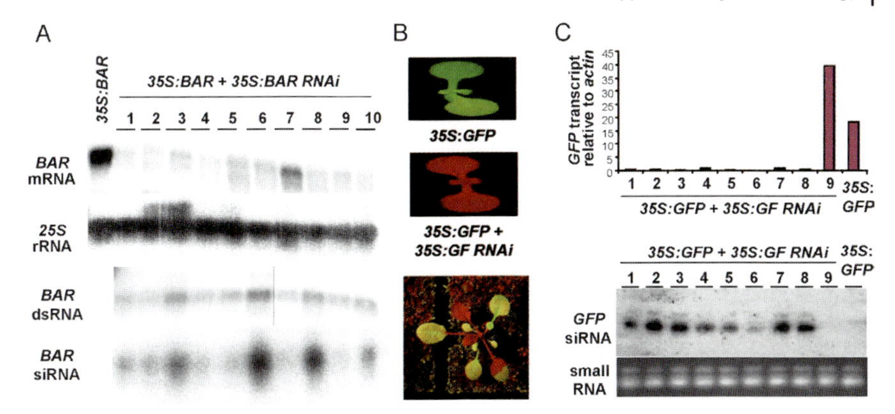

Figure 10.3 RNAi knockout of *BAR* and *GFP* in *Arabidopsis*. (A) RNAi silencing of the *BAR* transcript. Lines 1–10 represent 10 independent transgenic lines transformed with pUQC1081 (an intact *35S:BAR* transgene linked to a *BAR* RNAi transgene, i.e. *35S:BAR* + *35S:BAR RNAi*) [40], compared to a *35S:BAR* control (expressing high levels of the *BAR* transcript). Top two panels represent Northern analysis of the full length *BAR* transcript and the *25S* loading control. Third and fourth panels represent Northern analysis of dsRNA (total RNA was treated with single strand ribonuclease RNaseONE before analysis; [40]) and *BAR*-specific siRNAs, respectively. Despite the significant decrease in transcript levels, all 10 independent pUQC1081 (*35S:BAR* + *35S:BAR RNAi*) transformants were completely resistant to BASTA (data not shown). (B) Phenotype of pUQC214 (*35S:GFP*) and pUQC218 (*35S:GFP* + *35S:GF RNAi*) seedlings [40] photographed under blue light. Red color represents chlorophyll autofluorescence but no GFP fluorescence under blue light. The lower panel is the phenotype of a pUQC218 (*35S:GFP* + *35S:GF RNAi*) rootstock grafted onto a pUQC214 (*35S:GFP*) scion, demonstrating the spread of RNAi from the root to the apex (red areas) of the shoot [40]. (C) Top: *GFP* transcript level in independent pUQC218 (*35S:GFP* + *35S:GF RNAi*) transgenics and a non-silenced pUQC214 (*35S:GFP*) expressing line [40], as determined by real-time RT-PCR. 1–9 represent independent transgenic lines of pUQC218 (*35S:GFP* + *35S:GF RNAi*). Bottom: Northern blot analysis of *GFP*-specific siRNAs on the same lines depicted in the upper panel. Small RNA represents an ethidium bromide-stained agarose gel of low MW RNAs that serves as a loading control.

the *BAR* RNAi transgene is linked to an intact target *35S:BAR* transgene in pUQC1081. Transformation of *Arabidopsis* with this binary vector resulted in a significant decrease in the level of intact *BAR* transcript, as well as the presence of *BAR*-specific dsRNA and siRNAs (Figure 10.3A). However, despite the significant decrease in transcript levels, all independent transformants of pUQC1081 were completely resistant to BASTA (data not shown). Thus, in this example, the *BAR* transcript was almost completely silenced but sufficient BAR protein was still produced to confer resistance to the herbicide. The lack of BASTA sensitivity in these transgenic lines could be due to the BAR protein being particularly stable. Alternatively, resistance to BASTA conferred by the *BAR* transgene is a non-cell autonomous trait [80] and expression of the BAR protein in a small proportion of cells (where RNAi may not be effective) could be sufficient to confer herbicide resistance to the whole plant.

In contrast to BAR, Green Fluorescent Protein (GFP) confers a cell autonomous phenotype, and it has been used extensively as a reporter to study RNAi and its systemic transmission in plants [22,35,40]. In this example, an RNAi transgene targeting the first 400 nt of the GFP was linked to an intact *35S:GFP* transgene in the binary vector pUQC218 [40]. In contrast to the BAR example, complete phenotypic silencing of GFP was observed (Figure 10.3B, middle panel). Eight out of nine pUQC218 T1 transgenics displayed efficient silencing of GFP as characterized by the lack of GFP fluorescence under blue light (Figure 10.3B, middle panel). Silenced plants had up to a 100-fold decrease in *GFP* transcript levels and contained high levels of GFP-specific siRNAs (Figure 10.3C). The pUQC218 line in Figure 10.3C that failed to silence was a result of an incomplete T-DNA insertion resulting in the absence of the RNAi transgene (as determined by Southern analysis; data not shown).

By adapting an *Arabidopsis* grafting approach using a pUQC218 RNAi line as a silenced rootstock and a pUQC214 (*35S:GFP* alone) expressing line as the scion, the cell-autonomous nature of GFP can be demonstrated [40]. The long-distance silencing phenotype of GFP is shown in Figure 10.3B (lower panel); a mobile signal from the rootstock is transmitted to the shoot apex to induce RNAi against GFP. Recently, this grafting and GFP reporter system was used together with T-DNA insertion mutants, to gain insights into the mechanism, and the genes involved in long-distance reception of mRNA silencing in newly formed shoot tissue of *Arabidopsis* [40]. Surprisingly, it was shown that nuclear gene silencing plays a key role in the reception of mRNA silencing in the shoot apex [40].

10.3.2
Tissue-Specific RNA Silencing of an Endogenous Gene in Tobacco

The use of RNAi transgenes driven by a constitutive promoter can provide an efficient tool for plant functional genomics. However, tissue-specific expression of the RNAi transgene can be useful when targeting genes that are essential for plant development or viability. In order to achieve altered levels of terpene biosynthesis specifically in flowers, a floral-specific *chalcone synthase* (*CHS*) promoter was used to drive expression of an RNAi transgene targeting the 1-deoxy-D-xylulose-5-phosphate reductoisomerase (*DXR*) coding sequence. The *CHS* promoter was from snapdragon (*Antirrhinum majus*) and drives pigment biogenesis largely in petals and in the seed coat [81]. The promoter was cloned upstream of an intron-splicible inverted repeat to target 695 nt of the *DXR* mRNA (Figure 10.4). The DXR enzyme catalyzes one of the initial and rate-limiting steps of terpene and chlorophyll biosysnthesis [81–83]. Transformation of *Nicotiana tobaccum* with the *CHS:DXR* RNAi transgene resulted in ~25% of independent tranformants displaying a predominantly floral-specific bleaching phenotype (Figure 10.4), a result of the knockout of chlorophyll biosynthesis in this organ. This bleaching phenotype is similar to that seen in phytoene desaturase knockout, an enzyme that is also required for chlorophyll biosynthesis [32], but in our case with the tissue-specific promoter driving expression of the RNAi transgene, the bleaching occurs specifically in the floral organs. The reduction in the number of independent transformants displaying the knockout phenotype when compared to the *GFP* example described above, and to other published examples [6,7], could be due to a number of factors. These include the lower strength

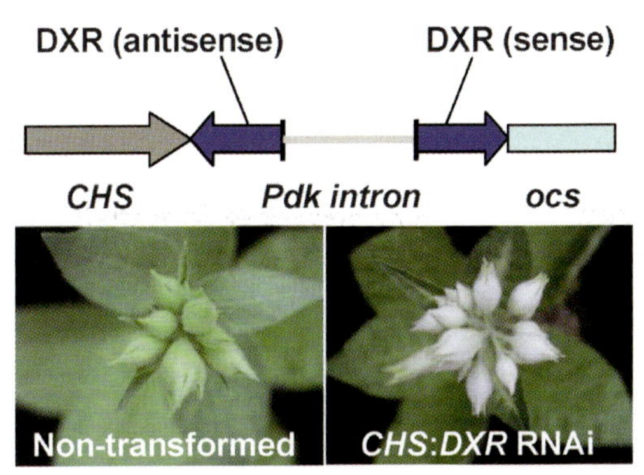

Figure 10.4 Tissue-specific RNAi silencing of *DXR* in *N. tobaccum*. The *DXR* hairpin is driven by the floral-specific *chalcone synthase* (*CHS*) promoter from snapdragon (*A. majus*). Also shown are the phenotypes of non-transformed tobacco (left) and plants transformed with the *CHS:DXR RNAi* transgene (right).

of the tissue-specific *CHS* promoter in comparison to the constitutive *35S* promoter, the efficiency of silencing the *DXR* mRNA, and the stability of the DXR protein.

Numerous other examples of the use of pHANNIBAL or related derivatives to down-regulate expression of endogenous genes have been reported in the literature, for example, [6,7,32].

10.3.3 Advantages of Using RNAi in Plant Functional Genomics

In recent years, T-DNA tagging approaches have become the primary method of attaining functional knockouts in the model plant species, *Arabidopsis thaliana* [84,85]. However, T-DNA knockouts may be lethal if the gene of interest is essential for plant development. In addition, the identification of an effective T-DNA line can be time consuming and occasionally not successful. For these reasons, the use of RNAi knockouts in *Arabidopsis* is an extremely useful option to complement T-DNA insertion mutants. In all other plant species, however, genome-saturating insertion lines are not available, and the use of RNAi is an essential tool for functional genomics. The improvements in RNAi vectors for stable transformation of plants, including tissue specificity and inducible expression [86], provides advantages over other methods of RNA-based knockout systems such as Virus Induced Gene Silencing (VIGS) [87–89] and *Agrobacterium* infiltration [35].

10.4 Perspectives

We have described some simple examples of the use of the pHANNIBAL RNAi system [6,7] to silence genes of interest, and there are numerous other examples in the literature [7,32]. The pHANNIBAL system involves using the constitutive and powerful *35S* promoters to drive expression of hairpin RNA [7] (also see Figure 10.2), but the option exists for using other constitutive promoters, and tissue-specific (e.g. Figure 10.4) or inducible [86] promoters.

Waterhouse and colleagues [7,90] have improved on their pHANNIBAL system and developed high-throughput hairpin RNA vectors (known as the pHELLGATE series) that utilize the Gateway unidirectional *in vitro* cloning system (Invitrogen, Carlsbad, CA). This more advanced system has been described in detail, and allows high throughput cloning and the possibility of targeting large numbers of plant genes for silencing [90].

Highly specific gene silencing has also been recently achieved using artificial miRNAs (amiRNAs) in *Arabidopsis*, and the specificity of amiRNAs in silencing genes is similar to natural plant miRNAs [8]. AmiRNAs can also be expressed from constitutive, tissue-specific and inducible promoters [8]. As a single amiRNA is produced for each target transcript, compared to a population of siRNAs from a perfectly complementary inverted repeat transgene, this approach could further minimize the chances of off-target gene regulation by RNAi [8,74].

RNA interference of gene expression is often incomplete [6,8] (Figure 10.3), and this can be a disadvantage compared to insertion mutagenesis. However, while T-DNA insertion mutants representing complete gene knockouts are an option for most genes in *Arabidopsis*, in all other plant species, we are largely dependent on RNAi as a tool for functional genomics.

References

1 Zamore, P.D., Tuschl, T., Sharp, P.A. and Bartel, D.P. (2000) RNAi: Double-stranded RNA directs the ATP-dependent cleavage of mRNA at 21 to 23 nucleotide intervals. *Cell*, **101**, 25–33.

2 Bernstein, E., Caudy, A.A., Hammond, S.M. and Hannon, G.J. (2001) Role for a bidentate ribonuclease in the initiation step of RNA interference. *Nature*, **409**, 363–366.

3 Bartel, D.P. (2004) MicroRNAs: Genomics, biogenesis, mechanism, and function. *Cell*, **116**, 281–297.

4 Baulcombe, D. (2004) RNA silencing in plants. *Nature*, **431**, 356–363.

5 Lippman, Z. and Martienssen, R. (2004) The role of RNA interference in heterochromatic silencing. *Nature*, **431**, 364–370.

6 Smith, N.A., Singh, S.P., Wang, M.B., Stoutjesdijk, P.A., Green, A.G. and Waterhouse, P.M. (2000) Total silencing by intron-spliced hairpin RNAs. *Nature*, **407**, 319–320.

7 Wesley, S.V., Helliwell, C.A., Smith, N.A., Wang, M.B., Rouse, D.T., Liu, Q., Gooding, P.S., Singh, S.P., Abbott, D., Stoutjesdijk, P.A. *et al.* (2001) Construct design for efficient, effective and high-throughput gene silencing in plants. *The Plant Journal*, **27**, 581–590.

8 Schwab, R., Ossowski, S., Riester, M., Warthmann, N. and Weigel, D. (2006) Highly specific gene silencing by artificial microRNAs in *Arabidopsis*. *Plant Cell*, **18**, 1121–1133.

9 Smith, H.A., Swaney, S.L., Parks, T.D., Wernsman, E.A. and Dougherty, W.G. (1994) Transgenic plant virus resistance mediated by untranslatable sense RNAs: Expression, regulation and fate of

nonessential RNAs. *Plant Cell*, **6**, 1441–1453.

10 Jorgensen, R.A., Cluster, P.D., English, J., Que, Q. and Napoli, C.A. (1996) Chalcone synthase cosuppression phenotypes in petunia flowers: Comparison of sense vs. antisense constructs, and single-copy vs. complex T-DNA sequences. *Plant Molecular Biology*, **31**, 957–973.

11 Matzke, M.A. and Jorgensen, R.A. (1996) From plants to mammals. *Science*, **271**, 1347–1348.

12 Fire, A., Xu, S., Montgomery, M.K., Kostas, S.A., Driver, S.E. and Mello, C.C. (1998) Potent and specific genetic interference by double-stranded RNA in *Caenorhabditis elegans*. *Nature*, **391**, 806–811.

13 Waterhouse, P.M., Graham, M.W. and Wang, M.B. (1998) Virus resistance and gene silencing in plants can be induced by simultaneous expression of sense and antisense RNA. *Proceedings of the National Academy of Sciences of the United States of America*, **95**, 13959–13964.

14 Hamilton, A.J. and Baulcombe, D.C. (1999) A species of small antisense RNA in posttranscriptional gene silencing in plants. *Science*, **286**, 950–952.

15 Napoli, C., Lemieux, C. and Jorgensen, R. (1990) Introduction of a chimeric chalcone synthase gene into petunia results in reversible co-suppression of homologous genes *in trans*. *Plant Cell*, **2**, 279–289.

16 Lindbo, J.A., Silva-Rosales, L., Proebsting, W.M. and Dougherty, W.G. (1993) Induction of a highly specific antiviral state in transgenic plants: implications for regulation of gene expression and virus resistance. *Plant Cell*, **5**, 1749–1759.

17 Ruiz, M.T., Voinnet, O. and Baulcombe, D.C. (1998) Initiation and maintenance of virus-induced gene silencing. *Plant Cell*, **10**, 937–946.

18 Baulcombe, D.C. (1999) Gene silencing: RNA makes RNA makes no protein. *Current Biology*, **9**, R599–R601.

19 Beclin, C., Boutet, S., Waterhouse, P. and Vaucheret, H. (2002) A branched pathway for transgene-induced RNA silencing in plants. *Current Biology*, **12**, 684–688.

20 Al-Kaff, N.S., Covey, S.N., Kreike, M.M., Page, A.M., Pinder, R. and Dale, P.J. (1998) Transcriptional and posttranscriptional plant gene silencing in response to a pathogen. *Science*, **279**, 2113–2115.

21 Ratcliff, F.G., MacFarlane, S.A. and Baulcombe, D.C. (1999) Gene silencing without DNA. RNA-mediated cross-protection between viruses. *Plant Cell*, **11**, 1207–1216.

22 Dalmay, T., Hamilton, A., Rudd, S., Angell, S. and Baulcombe, D.C. (2000) An RNA-dependent RNA polymerase gene in *Arabidopsis* is required for posttranscriptional gene silencing mediated by a transgene but not by a virus. *Cell*, **101**, 543–553.

23 Mourrain, P., Beclin, C., Elmayan, T., Feuerbach, F., Godon, C., Morel, J.B., Jouette, D., Lacombe, A.M., Nikic, S., Picault, N., Remoue, K., Sanial, M., Vo, T.A. and Vaucheret, H. (2000) *Arabidopsis SGS2* and *SGS3* genes are required for posttranscriptional gene silencing and natural virus resistance. *Cell*, **101**, 533–542.

24 Dalmay, T., Horsefield, R., Braunstein, T.H. and Baulcombe, D.C. (2001) *SDE3* encodes an RNA helicase required for post-transcriptional gene silencing in *Arabidopsis*. *EMBO Journal*, **20**, 2069–2078.

25 Deleris, A., Gallego-Bartolome, J., Bao, J., Kasschau, K.D., Carrington, J.C. and Voinnet, O. (2006) Hierarchical action and inhibition of plant Dicer-like proteins in antiviral defense. *Science*, **313**, 68–71.

26 Bouche, N., Lauressergues, D., Gasciolli, V. and Vaucheret, H. (2006) An antagonistic function for *Arabidopsis* DCL2 in development and a new function for dcl4 in generating viral siRNAs. *EMBO Journal*, **25**, 3347–3356.

27 Jones, L., Hamilton, A.J., Voinnet, O., Thomas, C.L., Maule, A.J. and Baulcombe, D.C. (1999) RNA–DNA interactions and DNA methylation in post-transcriptional gene silencing. *Plant Cell*, **11**, 2291–2301.

28 Gazzani, S., Lawrenson, T., Woodward, C., Headon, D. and Sablowski, R. (2004) A link between mRNA turnover and RNA interference in *Arabidopsis*. *Science*, **306**, 1046–1048.

29 Morel, J.B., Godon, C., Mourrain, P., Beclin, C., Boutet, S., Feuerbach, F., Proux, F. and Vaucheret, H. (2002) Fertile hypomorphic *ARGONAUTE (ago1)* mutants impaired in post-transcriptional gene silencing and virus resistance. *Plant Cell*, **14**, 629–639.

30 Herr, A.J., Jensen, M.B., Dalmay, T. and Baulcombe, D.C. (2005) RNA polymerase IV directs silencing of endogenous DNA. *Science*, **308**, 118–120.

31 Gasciolli, V., Mallory, A.C., Bartel, D.P. and Vaucheret, H. (2005) Partially redundant functions of *Arabidopsis* DICER-like enzymes and a role for DCL4 in producing trans-acting siRNAs. *Current Biology*, **15**, 1494–1500.

32 Fusaro, A.F., Matthew, L., Smith, N.A., Curtin, S.J., Dedic-Hagan, J., Ellacott, G.A., Watson, J.M., Wang, M.B., Brosnan, C., Carroll, B.J. and Waterhouse, P.M. (2006) RNA interference-inducing hairpin RNAs in plants act through the viral defence pathway. *EMBO Reports*, **7**, 1168–1175.

33 Voinnet, O. (2005) Non-cell autonomous RNA silencing. *FEBS Letters*, **579**, 5858–5871.

34 Palauqui, J.C., Elmayan, T., Pollien, J.M. and Vaucheret, H. (1997) Systemic acquired silencing: transgene-specific post-transcriptional silencing is transmitted by grafting from silenced stocks to non-silenced scions. *EMBO Journal*, **16**, 4738–4745.

35 Voinnet, O. and Baulcombe, D.C. (1997) Systemic signalling in gene silencing. *Nature*, **389**, 553.

36 Voinnet, O., Vain, P., Angell, S. and Baulcombe, D.C. (1998) Systemic spread of sequence-specific transgene RNA degradation in plants is initiated by localized introduction of ectopic promoterless DNA. *Cell*, **95**, 177–187.

37 Dunoyer, P., Himber, C. and Voinnet, O. (2005) Dicer-like 4 is required for RNA interference and produces the 21-nucleotide small interfering RNA component of the plant cell-to-cell silencing signal. *Nature Genetics*, **37**, 1356–1360.

38 Smith, L.M., Pontes, O., Searle, I., Yelina, N., Yousafzai, F.K., Herr, A.J., Pikaard, C.S. and Baulcombe, D.C. (2007) An SNF2 protein associated with nuclear RNA silencing and the spread of a silencing signal between cells in *Arabidopsis*. *Plant Cell*, **19**, 1507–1521.

39 Dunoyer, P., Himber, C., Ruiz-Ferrer, V., Alioua, A. and Voinnet, O. (2007) Intra- and intercellular RNA interference in *Arabidopsis thaliana* requires components of the microRNA and heterochromatic silencing pathways. *Nature Genetics*, **39**, 848–856.

40 Brosnan, C.A., Mitter, N., Christie, M., Waterhouse, P.M. and Carroll, B.J. (2007) Nuclear gene silencing pathway directs the reception of long-distance mRNA silencing in *Arabidopsis*. *Proceedings of the National Academy of Sciences of the United States of America*, **104**, 14741–14746.

41 Tournier, B., Tabler, M. and Kalantidis, K. (2006) Phloem flow strongly influences the systemic spread of silencing in GFP *Nicotiana benthamiana* plants. *The Plant Journal*, **47**, 383–394.

42 Himber, C., Dunoyer, P., Moissiard, G., Ritzenthaler, C. and Voinnet, O. (2003) Transitivity-dependent and -independent cell-to-cell movement of RNA silencing. *EMBO Journal*, **22**, 4523–4533.

43 Vaistij, F.E., Jones, L. and Baulcombe, D.C. (2002) Spreading of RNA targeting and DNA methylation in RNA silencing requires transcription of the target gene and a putative RNA-dependent RNA polymerase. *Plant Cell*, **14**, 857–867.

44 Yang, Z., Ebright, Y.W., Yu, B. and Chen, X. (2006) HEN1 recognizes 21–24 nt small RNA duplexes and deposits a methyl group onto the 2′ OH of the 3′ terminal nucleotide. *Nucleic Acids Research*, **34**, 667–675.

45 Boutet, S., Vazquez, F., Liu, J., Beclin, C., Fagard, M., Gratias, A., Morel, J.B., Crete, P., Chen, X. and Vaucheret, H. (2003) *Arabidopsis* HEN1: A genetic link between endogenous miRNA controlling development and siRNA controlling transgene silencing and virus resistance. *Current Biology*, **13**, 843–848.

46 Hammond, S.M., Caudy, A.A. and Hannon, G.J. (2001) Post-transcriptional gene silencing by double-stranded RNA. *Nature Reviews. Genetics*, **2**, 110–119.

47 Elbashir, S.M., Harborth, J., Lendeckel, W., Yalcin, A., Weber, K. and Tuschl, T. (2001) Duplexes of 21-nucleotide RNAs mediate RNA interference in cultured mammalian cells. *Nature*, **411**, 494–498.

48 Nykanen, A., Haley, B. and Zamore, P.D. (2001) ATP requirements and small interfering RNA structure in the RNA interference pathway. *Cell*, **107**, 309–321.

49 Qi, Y. and Hannon, G.J. (2005) Uncovering RNAi mechanisms in plants: Biochemistry enters the foray. *FEBS Letters*, **579**, 5899–5903.

50 Baumberger, N. and Baulcombe, D.C. (2005) *Arabidopsis* ARGONAUTE1 is an RNA slicer that selectively recruits microRNAs and short interfering RNAs. *Proceedings of the National Academy of Sciences of the United States of America*, **102**, 11928–11933.

51 Lipardi, C., Wei, Q. and Paterson, B.M. (2001) RNAi as random degradative PCR: siRNA primers convert mRNA into dsRNAs that are degraded to generate new siRNAs. *Cell*, **107**, 297–307.

52 Sijen, T., Fleenor, J., Simmer, F., Thijssen, K.L., Parrish, S., Timmons, L., Plasterk, R.H. and Fire, A. (2001) On the role of RNA amplification in dsRNA-triggered gene silencing. *Cell*, **107**, 465–476.

53 Brodersen, P. and Voinnet, O. (2006) The diversity of RNA silencing pathways in plants. *Trends in Genetics*, **22**, 268–280.

54 Moissiard, G., Parizotto, E.A., Himber, C. and Voinnet, O. (2007) Transitivity in *Arabidopsis* can be primed, requires the redundant action of the antiviral Dicer-like 4 and Dicer-like 2, and is compromised by viral-encoded suppressor proteins. *RNA*, **13**, 1268–1278.

55 Carrington, J.C. and Ambros, V. (2003) Role of microRNAs in plant and animal development. *Science*, **301**, 336–338.

56 Lee, R.C., Feinbaum, R.L. and Ambros, V. (1993) The *C. elegans* heterochromic gene *lin-4* encodes small RNAs with antisense complementarity to *lin-14*. *Cell*, **75**, 843–854.

57 Jones-Rhoades, M.W., Bartel, D.P. and Bartel, B. (2006) MicroRNAs and their regulatory roles in plants. *Annual Review of Plant Biology*, **57**, 19–53.

58 Rhoades, M.W., Reinhart, B.J., Lim, L.P., Burge, C.B., Bartel, B. and Bartel, D.P. (2002) Prediction of plant microRNA targets. *Cell*, **110**, 513–520.

59 Bao, N., Lye, K.W. and Barton, M.K. (2004) MicroRNA binding sites in *Arabidopsis* class III HD-ZIP mRNAs are required for methylation of the template chromosome. *Developmental Cell*, **7**, 653–662.

60 Chen, X. (2004) A microRNA as a translational repressor of *APETALA2* in *Arabidopsis* flower development. *Science*, **303**, 2022–2025.

61 Parizotto, E.A., Dunoyer, P., Rahm, N., Himber, C. and Voinnet, O. (2004) *In vivo* investigation of the transcription, processing, endonucleolytic activity, and functional relevance of the spatial distribution of a plant miRNA. *Genes & Development*, **18**, 2237–2242.

62 Mallory, A.C., Reinhart, B.J., Jones-Rhoades, M.W., Tang, G., Zamore, P.D., Barton, M.K. and Bartel, D.P. (2004) MicroRNA control of *PHABULOSA* in leaf development: importance of pairing to the microRNA 5′ region. *EMBO Journal*, **23**, 3356–3364.

63 van Hoof, A. and Parker, R. (1999) The exosome: a proteasome for RNA? *Cell*, **99**, 347–350.

64 Llave, C., Xie, Z.X., Kasschau, K.D. and Carrington, J.C. (2002) Cleavage of *Scarecrow-like* mRNA targets directed by a class of *Arabidopsis* miRNA. *Science*, **297**, 2053–2056.

65 Vaucheret, H., Vazquez, F., Crete, P. and Bartel, D.P. (2004) The action of *ARGONAUTE1* in the miRNA pathway and its regulation by the miRNA pathway are crucial for plant development. *Genes & Development*, **18**, 1187–1197.

66 Kidner, C.A. and Martienssen, R.A. (2004) Spatially restricted microRNA directs leaf polarity through argonaute1. *Nature*, **428**, 81–84.

67 Schauer, S.E., Jacobsen, S.E., Meinke, D.W. and Ray, A. (2002) *DICER-LIKE1*: Blind men and elephants in *Arabidopsis* development. *Trends in Plant Science*, **7**, 487–491.

68 Williams, L., Grigg, S.P., Xie, M., Christensen, S. and Fletcher, J.C. (2005) Regulation of *Arabidopsis* shoot apical meristem and lateral organ formation by microRNA miR166g and its AtHD-ZIP target genes. *Development*, **132**, 3657–3668.

69 Kidner, C.A. and Martienssen, R.A. (2005) The role of *ARGONAUTE1 (AGO1)* in meristem formation and identity. *Developmental Biology*, **280**, 504–517.

70 Jacobsen, S.E., Running, M.P. and Meyerowitz, E.M. (1999) Disruption of an RNA helicase/RNAse III gene in *Arabidopsis* causes unregulated cell division in floral meristems. *Development*, **126**, 5231–5243.

71 Vazquez, F., Gasciolli, V., Crete, P. and Vaucheret, H. (2004) The nuclear dsRNA binding protein HYL1 is required for microRNA accumulation and plant development, but not posttranscriptional transgene silencing. *Current Biology*, **14**, 346–351.

72 Griffiths-Jones, S. (2004) The microRNA registry. *Nucleic Acids Research*, **32**, D109–111.

73 Floyd, S.K. and Bowman, J.L. (2004) Gene regulation: ancient microRNA target sequences in plants. *Nature*, **428**, 485–486.

74 Jackson, A.L., Bartz, S.R., Schelter, J., Kobayashi, S.V., Burchard, J., Mao, M., Li, B., Cavet, G. and Linsley, P.S. (2003) Expression profiling reveals off-target gene regulation by RNAi. *Nature Biotechnology*, **21**, 635–637.

75 Carroll, B.J., Klimyuk, V.I., Thomas, C.M., Bishop, G.J., Harrison, K., Scofield, S.R. and Jones, J.D.G. (1995) Germinal transpositions of the maize element *Dissociation* from T-DNA loci in tomato. *Genetics*, **139**, 407–420.

76 Clough, S.J. and Bent, A.F. (1998) Floral dip: a simplified method for *Agrobacterium*-mediated transformation of *Arabidopsis thaliana*. *The Plant Journal*, **16**, 735–743.

77 McGinnis, K., Chandler, V., Cone, K., Kaeppler, H., Kaeppler, S., Kerschen, A., Pikaard, C., Richards, E., Sidorenko, L., Smith, T. *et al.* (2005) Transgene-induced RNA interference as a tool for plant functional genomics. *Methods in Enzymology*, **392**, 1–24.

78 Sambrook, J., Fritsch, E.F. and Maniatis, T. (1989) *Molecular Cloning: A Laboratory Manual*, Cold Springs Harbor, Laboratory Press, New York.

79 Mitter, N., Sulistyowati, E. and Dietzgen, R.G. (2003) Cucumber mosaic virus infection transiently breaks dsRNA-induced transgenic immunity to Potato Virus Y in tobacco. *Molecular Plant–Microbe Interactions: MPMI*, **16**, 936–944.

80 Jones, J.D.G., Jones, D.A., Bishop, G.J., Harrison, K., Carroll, B.J. and Scofield, S.R. (1993) Use of the maize transposons *Activator* and *Dissociation* to show that phosphinothricin and spectinomycin resistance genes act non-cell-autonomously in tobacco and tomato seedlings. *Transgenic Research*, **2**, 63–78.

81 Fritze, K., Staiger, D., Czaja, I., Walden, R., Schell, J. and Wing, D. (1991) Developmental and UV light regulation of the snapdragon chalcone synthase promoter. *Plant Cell*, **3**, 893–905.

82 Estevez, J.M., Cantero, A., Reindl, A., Reichler, S. and Leon, P. (2001) 1-Deoxy-d-xylulose-5-phosphate synthase, a limiting enzyme for plastidic isoprenoid biosynthesis in plants. *The Journal of Biological Chemistry*, **276**, 22901–22909.

83 Mahmoud, S.S. and Croteau, R.B. (2002) Strategies for transgenic manipulation of monoterpene biosynthesis in plants. *Trends in Plant Science*, **7**, 366–373.

84 Alonso, J.M., Stepanova, A.N., Leisse, T.J., Kim, C.J., Chen, H., Shinn, P., Stevenson, D.K., Zimmerman, J., Barajas, P., Cheuk, R. *et al.* (2003) Genome-wide insertional mutagenesis of *Arabidopsis thaliana*. *Science*, **301**, 653–657.

85 Rosso, M.G., Li, Y., Strizhov, N., Reiss, B., Dekker, K. and Weisshaar, B. (2003) An *Arabidopsis thaliana* T-DNA mutagenized population (GABI-Kat) for flanking sequence tag-based reverse genetics. *Plant Molecular Biology*, **53**, 247–259.

86 Guo, H.S., Fei, J.F., Xie, Q. and Chua, N.H. (2003) A chemical-regulated inducible RNAi system in plants. *The Plant Journal*, **34**, 383–392.

87 Dalmay, T., Hamilton, A., Mueller, E. and Baulcombe, D.C. (2000) Potato Virus X amplicons in *Arabidopsis* mediate genetic and epigenetic gene silencing. *Plant Cell*, **12**, 369–379.

88 Ratcliff, F., Martin-Hernandez, A.M. and Baulcombe, D.C. (2001) Technical advance. Tobacco rattle virus as a vector for analysis of gene function by silencing. *The Plant Journal*, **25**, 237–245.

89 Turnage, M.A., Muangsan, N., Peele, C.G. and Robertson, D. (2002) Geminivirus-based vectors for gene silencing in *Arabidopsis*. *The Plant Journal*, **30**, 107–114.

90 Helliwell, C.A. and Waterhouse, P.M. (2005) Constructs and methods for hairpin RNA-mediated gene silencing in plants. *Methods in Enzymology*, **392**, 24–35.

11
Extending Functional Genomics: VIGS for Model and Crop Plants

Steven Bernacki, John Richard Tuttle, Nooduan Muangsan, and Dominique Robertson

Abstract

Bioinformatics can be used to identify small sets of genes whose expression changes in response to a stimulus, but whether the changes are correlative or functional remains unclear until the expression of each gene can be modulated individually or in tandem with a second gene. Virus induced gene silencing (VIGS) can modulate the expression of individual or combinations of plant genes, providing a glimpse into what they can do. The attractiveness of VIGS is its speed; the function of genes producing visible phenotypes, such as *PDS* (*Phytoene Desaturase*) or *ChlI*, (*Magnesium Chelatase subunit I*) can be seen in as little as 21 days after deployment of the vector. Because a reduction of messenger RNA levels for most genes does not usually produce visible phenotypes, it is essential to have a predefined goal before initiating a VIGS experiment. Some of the factors to consider in setting up a VIGS screen include availability of a suitable VIGS vector for the target plant, whether the tissue is susceptible to VIGS, and how to optimize VIGS conditions for different developmental stages of the plant. VIGS vectors are ideal for the direct cloning of PCR-based suppressive, subtractive hybridization (SSH) libraries because they accept fragments from about 100–800 kb and the whole gene sequence is not needed for silencing. VIGS vectors are also useful for testing gene function in crop plants, which are often difficult to transform and lack adequate sequence information or testable mutants. We provide an overview of experiments that have been reported in the literature using different viruses, genes, and host plants. We also describe a VIGS experiment and the controls that are necessary for understanding the results.

11.1
Introduction

Viral-Induced Gene Silencing (VIGS) is a practical application resulting from brilliant experiments demonstrating that plants purposely silence genes to defend themselves against viruses [1]. The first documentation of a successful VIGS vector occurred in 1995

The Handbook of Plant Functional Genomics: Concepts and Protocols.
Edited by Günter Kahl and Khalid Meksem

ISBN: 978-3-527-31885-8

and used *Tobbaco Mosaic Virus* (TMV) to silence *Phytoene Desaturase* (*PDS*) in *Nicotiana benthamiana* by incorporating a homologous sequence to the gene of interest into the viral genome [2]. In 1998, *Potato Virus X* (PVX) and a ssDNA virus, *Tomato Golden Mosaic Virus* (TGMV), were used to silence both endogenous genes and transgenes in *N. benthamiana* [3,4]. *Tobacco Rattle Virus* (TRV), reported in 2001, became the vector of choice for down-regulating genes in *N. benthamiana* [5]. TRV is a bipartite RNA virus that produces minimal symptoms, has a broad host range, and can be easily inoculated using *Agrobacterium* vectors (Figure 11.1). Although TRV is seed transmitted, VIGS of an endogenous gene post meiosis has never been reported. It was not until 2002 that a silencing vector was developed for *Arabidopsis thaliana* [6]. This vector was derived from *Cabbage Leaf Curl Virus* (CaLCuV), a geminivirus in the same genus as TGMV. In 2005, TRV was also adapted for use with *Arabidopsis* and made available to the scientific community [7,8]. Today, several different viruses have been adapted for VIGS and the range of host plants includes both model and crop plants (Table 11.1).

The attraction of VIGS is the rapid characterization of gene function. No matter how good the information from proteomics, expression arrays, and other large-scale analyses, at some point it is necessary to prove cause and effect for a single gene or group of genes. Modulation of gene expression has to date been the only real way to do this. Gene knockouts, over-expression experiments, or mutagenesis can provide as close to a controlled experiment for this as is possible by asking: what happens when gene X changes and nothing else does? VIGS introduces variables due to virus–host interactions but including an empty VIGS vector control can mitigate these effects. Ideally, VIGS should only be used as part of an experimental protocol to determine gene function and stable transformation with RNAi vectors (inducible if necessary) should be used to derive conclusive evidence for gene function. In practice, this is either not necessary, bulging cell walls could clearly be attributed to loss of *CesA* expression [9], or impractical (for plants that are difficult to transform). One advantage of VIGS over transformation is that it uses wild-type plants, which show normal development until the time of inoculation.

Although *Tobacco Rattle Virus* (TRV) has become the vector of choice for down-regulating genes in *N. benthamiana* [5], TGMV is an important alternative. TGMV and other geminivirus vectors are inoculated directly from isolated DNA (Figure 11.1), allowing the assessment of gene function in mature leaves [10] in the absence of bacteria. TRV is inoculated by infiltration of leaf mesophyll with *Agrobacterium* suspensions (a mixture of TRV RNAs 1 and 2) (Figure 11.1), needle inoculation of suspensions into stem tissue (Figure 11.1) or vacuum infiltration of whole seedlings in suspensions of *Agrobacterium*. Analyses of gene function are always conducted in upper leaves, away from the *Agrobacteria*. Because all viruses move systemically through the phloem from source tissues to sink, silencing occurs primarily in tissues that are still developing and require photosynthate for growth, such as young leaves and roots, not mature tissues. However because TGMV inoculation does not require *Agrobacterium*, silencing can be analyzed in inoculated, mature tissue. A second advantage of TGMV is that it is excluded from the meristem, while TRV is seed-transmitted. The absence of virus from the meristem simplifies interpretation of results.

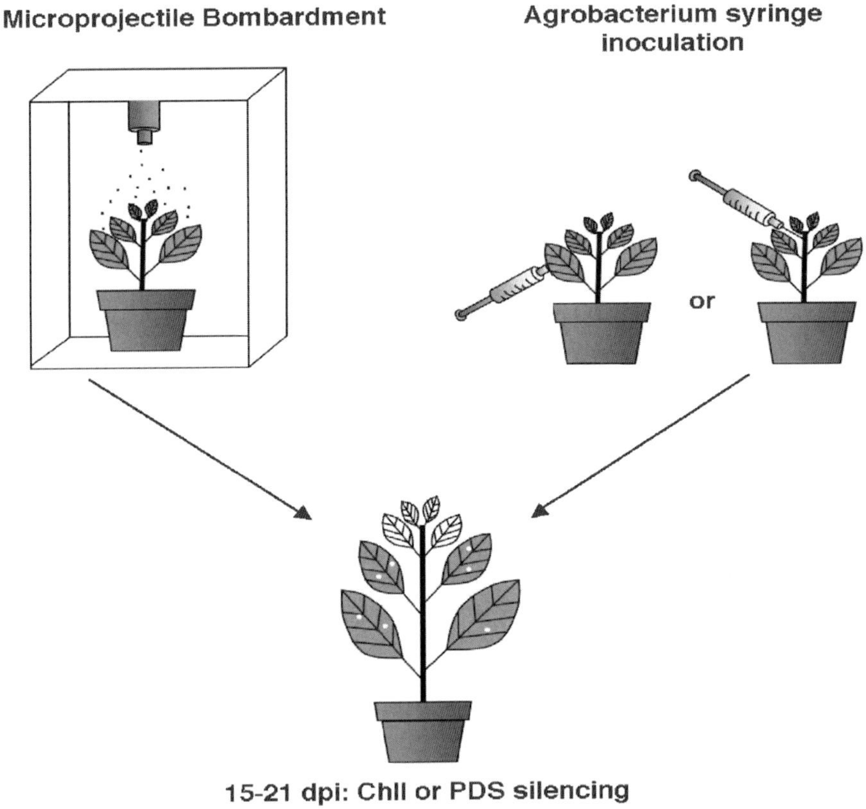

Microprojectile Bombardment

Agrobacterium syringe inoculation

or

15-21 dpi: ChlI or PDS silencing

Figure 11.1 Overview of Virus-Induced Gene Silencing. Plants at the top are wild-type host plants receiving a VIGS vector containing a gene fragment homologous to the host plant target gene, in this case *ChlI* or *PDS*. For microprojectile bombardment (left), plants are placed in a chamber and DNA-coated gold or tungsten particles are delivered from a gene gun. For inoculation with *Agrobacteria*, different methods are used depending on the anatomy of the leaf and the growth habits of the plant. *N. benthamiana* can be inoculated with a needle-less syringe on the underside of the leaf. Other plants are inoculated using a syringe needle inserted into stem tissue. Still another method uses vacuum infiltration of *Agrobacteria* into seedlings. After 15–25 days, silencing is seen in the new growth. Microprojectile bombardment of TGMV:ChlI into *N. benthamiana* produces spots on inoculated leaves where silencing radiates from the bombarded cell. Spots are seen after 5–10 days and remain for the duration of the leaf's lifespan. Silencing of *ChlI* in *N. benthamaina* and *Arabidopsis* continues throughout the plant life cycle and extends into flowers and fruits. Silencing has not been reported in progeny plants with the notable exception of the transgene, GFP.

Immunolocalization of proliferating cell nuclear antigen (PCNA), which is essential for DNA replication, in TGMV:PCNA-silenced plants demonstrated that TGMV-mediated silencing was effective throughout the meristem [11]. Similar results have not been reported for TRV, which is seed-transmitted and is thought to invade the meristem [5]. Drawbacks of geminivirus vectors include an increased level of symptoms compared to TRV and the requirement of a microprojectile delivery system, which can be expensive and time consuming.

Table 11.1 Summary of VIGS Experiments.

Parent virus	Target species	Genes	Comments	Reference
Tobacco Mosaic Virus, genus *Tobamovirus*	*N. benthamiana*	*PDS, ChlH*	Single RNA component-genome, duplicated subgenomic promoters transcribe coat, silencing sequence	[2,24]
Brome Mosaic Virus, family *Bromoviridae*	Rice, barley, one cultivar of maize (Va35)	Actin, *PDS*	Four-part RNA genome	[40]
Barley Streak Mosaic Virus; genus *Hordeivirus*	Barley, hexaploid wheat	PDS, Lr21, RAR1, SGT1, and HSP90	Tripartite RNA virus, removed coat protein to make the vector.	[41–43,3,9]
Potato Virus X; family *Flexiviridae*	*N. benthamiana*, potato	*RbcS, NtCesA, NtCDPK2, FtsH, WIPK, SIPK, NbrbohA, NbrbohB, Nb14-3-3a, Nb14-3-3b, IP-L, AtCDC5*	Monopartite positive strand virus, about 6.5 kb. Duplicated subgenomic promoter for silencing sequence	[44–50]
Poplar mosaic virus	*N. benthamiana*	GFP	GFP cloned in place of movement and coat proteins; replicates but does move	[51]
Tobacco Rattle Virus, genus *Tobravirus*	*N. benthamiana*, Poppy, *Aquilegia*, tomato, potato, transgenic 35S:N gene *N. benthamiana*, *Arabidopsis*, Petunia, deadly nightshade, two wild *Solanum* relatives, *Capsicum*, *N. attenuata*	*PDS, R1, Rx, RB, RBR, EDS1, Pto, CTR1, CTR2, CHS, ACO1, ACO4, RbcS, Rar1, NPR1/NIM1*- like genes, *NRG1, MEK1, MEK2, NTF6, WIPK, NPR1, RAR, COI1, TGA1, TGA2.2, PP2Ac, NPK1, NbNAP1, NbPHB1, NbPHB2, NbDEK, NbAXS1, Sgt1-1, Sgt1-2, Hsp90, APR134, SMO1, SMO2, NbPPS3, HXK1, NbERS, NbSRS, AtCDC5, NbDEF, MEK1 MAPKK, NTF6 MAPK, WRKY/MYB, COI1, NaGLP*, Aconitase, *NbECR, 3betaHSD/D, CITRX. ACIK1, PMT, TI, IRT1, TTG1, RHL1, RML1*, and *PB7.*	Bipartite positive strand RNA virus. RNA 2 (about 2–3 kb) used for cloning. Both genome components cloned into binary vectors for agroinoculation. Also available as Gateway vector	[5,52,53,8, 54–60,37, 31,61–63, 7,64–85]

Pea Early Browning Virus, genus *Tobravirus*	*Pisum sativum*	*PDS, LFY, KOR1*	Bipartite genome, coat protein promoter used, nematode transmission genes deleted	[86]
Bean pod mottle virus, family *Comoviridae*	Soybean	PDS	Bipartite, positive strand virus	[87]
Tomato Golden Mosaic Virus, family *Geminiviridae*	*N. benthamiana*	*ChlI, PDS, PCNA, RBR*	Insertion into B component, maximum 150 bp	[4,11]
Cabbage Leaf Curl Virus, family *Geminiviridae*	*Arabidopsis*	ChlI	Coat protein replacement, maximum 800 bp	[6]
African Cassava Mosaic Virus, family *Geminiviridae*	Cassava	*ChlI, CYP79D*	Coat protein replacement, maximum 800 bp	[19]
Beta satellite of *Tomato Yellow Leaf Curl China Virus*, family *Geminiviridae*	*N. benthamiana, N. glutinosa*, tomato	*PDS, ChlI, PCNA*	Not found in the New World; used primarily in Asia	[88,89]
Cotton Leaf Crumple Virus, family *Geminiviridae*	Cotton	ChlI	Coat protein replacement, maximum 800 bp	[90]
Pepper huasteco yellow vein virus, family *Geminiviridae*	Tobacco tomato, pepper	*NbChlI, Comt, pAmt, Kas*	Coat protein replacement, maximum 800 bp	[91]

PDS, phytoene desaturase; *R1, Rx, RB*, three defense genes; *RBR, EDS1, Pto, CTR1, CTR2*, constitutive triple response (ethylene response); *CHS*, chalcone synthase; *RbcS*, Ribulose bisphosphate carboxylase small subunit; *Rar1, NPR1/NIM1-?* like genes; *NRG1, MEK1, MEK2, NTF6, NPR1, COI1, TGA1, TGA2.2, PP2Ac*, protein phosphatase 2; *NPK1, NbAXS1; Sgt*, suppressor of G-two allele of *Skp1; DEK*, calpain; *PHB*, prohibitin; *AXS1*, UDP-D-apiose/UDP-D-xylose synthases; *At(Nb)NAP1* is a plastidic SufB protein involved in Fe-S cluster assembly; *ACO*, 1-aminocyclopropane-1-carboxylate oxidase; *APR134*, Calmodulin-related protein; *UNI*, UNIFOLIATA; *KOR1*, KORRIGAN 1; *SMO*, sterol 4alpha-methyl oxidase; *SIPK*, salicylic acid-induced protein kinase; *WIPK*, wound-induced protein kinase; *MAPK*, mitogen-activated protein kinase; *PPSs*, proteins phosphorylated by *StMPK1; Hxk1*, hexokinase; *ERS*, glutamyl-tRNA synthetase; *SRS*, seryl-tRNA synthetase; *IP-L*, ToMV CP-interacting protein-L; *CDC5*, Myb-related protein, cell division cycle; *NbDEF*, DEFICIENS; *NaGLP*, germin-like protein; *NbECR*, enoyl-CoA reductase; *3betaHSD/D*, 3beta-hydroxysteroid dehydrogenase/C-4 decarboxylases; *CITRX*, Cf-9-interacting thioredoxin; *ACIK1*, *Avr9/Cf-9* induced kinase 1; *PMT*, putrescine *N*-methyltransferase; *TI*, trypsin inhibitor; *Lr21*, leaf rust resistance gene 21; *TTG1*, transparent testa glabra; *RHL1*, root hairless; *RML1*, root meristemless1; *PB7*, 20S proteasome subunit, and nematode resistance (Mi); *Comt*, caffeic acid *O*-methyltransferase gene; *CYP79D1* [2], genes involved in linamarin synthesis; *Kas*, keto-acyl ACP synthase gene; *Amt*, possible aminotransferase gene.

It should be pointed out that *Agrobacterium* vectors have been made for geminivirus inoculation, but experiments in our laboratory have always used bombardment of DNA to avoid potential pathogen-related effects due to *Agrobacterium*.

Despite the widespread use of VIGS, it is still not straightforward to set up a reliable VIGS experiment or screen without previous experience. Here, we describe a typical VIGS experiment using microprojectile bombardment of the geminivirus, TGMV. Methods for isolating target gene inserts, collecting and analyzing data, and designing proper controls are discussed. Most of this information applies to both TRV and TGMV vectors. Websites describing how to make a gene gun could be consulted unless access to a gene gun is available. Because of the conserved nature of geminivirus genome organization, and the fact that many horticultural and crop plants are infected by these viruses, the methods presented here could be adapted for use in other virus–host plant combinations. Some suggestions for making novel VIGS vectors are provided in Step 1, along with a brief description of geminivirus gene function. It will be helpful to consult one or more excellent reviews on this topic [12–15] before starting a VIGS experiment.

11.2 Methods and Protocols

11.2.1 Constructing Geminivirus VIGS Vectors

General features of the *Begomovirus* genus is given here but for more information on different geminiviruses and their hosts, The Geminivirus Detective website can be consulted (http://gemini.biosci.arizona.edu/). Figure 11.2 shows the general structure of geminivirus genomes and describes the proteins needed for replication, movement, and encapsidation. It is important to understand the biology of the virus used for vector construction because unexpected silencing, such as lack of virus in new growth due to silencing of a required host factor, can occur. In this respect, the availability of both RNA and DNA viral vectors is an advantage because RNA viruses encode their own polymerase and replicate in the cytoplasm, not the nucleus.

Geminiviruses are unique in having to move in and out of nuclei as well as in and out of cells. They replicate by inducing cellular DNA replication machinery, much like mammalian DNA viruses. Replication occurs by rolling circle amplification or recombination-dependent replication [16]. Vectors derived from geminiviruses have two common regions, containing the origin of replication. Unit-length viral genomes are reconstituted *in planta* after the viral replication-associated proteins AL1 and AL3 are expressed [17]. Recircularized episomes move cell-to-cell and through the phloem but actually infect fewer than 10% of the cells in the plant. Because silencing is systemic, the other 90% of the cells are not impacted by cell-autonomous changes caused by virusal gene expression, although the plant response to infection may include systemic changes.

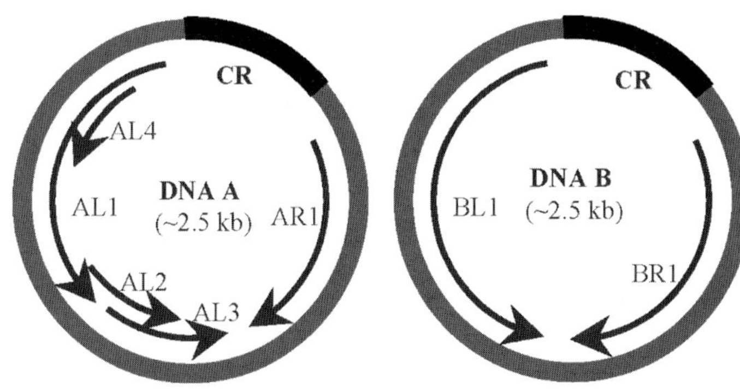

Figure 11.2 Structure of a geminivirus VIGS vector. Bipartite geminiviruses have two components, A and B, which are structurally similar. The common region is conserved between the components and contains the origin of replication. Bidirectional promoters produce rightward and leftward transcripts with putative polyadenylation sites located before the opposing open reading frame ends. The double-stranded genome of the typical *Begomovirus* A component has five open reading frames (arrows, left circle). The AL1 and AL3 proteins are required for viral replication, inducing cell cycle-related proteins necessary for DNA replication [92]. AL2 is necessary for transcriptional activation of the *AR1* and *BR1* genes [93]. AL4 has anti-silencing activity in at least some geminiviruses [94]. The coat protein is translated from *AR1* and is required for encapsidation of single-stranded viral DNA, insect transmission, and long distance movement [95,96]. Depending on the host, *AR1* is dispensable for movement allowing foreign DNA to be cloned in its place [96,97]. The B genome contains two genes (right circle). BR1 is a nuclear shuttle protein necessary for export of viral DNA from the nucleus [98] while BL1 is needed for cell-to-cell movement [99]. Limited amounts of foreign DNA can be cloned downstream of the *BR1* promoter and before the putative polyadenylation site for rightward transcription.

Some geminiviruses, such as *Cotton Leaf Crumple Virus*, are phloem-limited and do not move outside of the vascular tissue while others (TGMV and CaLCuV) can be found in mesophyll, cortical, and epidermal cells. Each of these geminiviruses has been modified for VIGS by removal of the 800-bp coat protein gene (*AR1*), resulting in an A component of 1.7 kb that can still be efficiently trafficked. Insertion of foreign DNA for silencing can therefore comprise up to 800 bp in the coat protein replacement vectors. The ability to move outside of the vascular tissue is not a requirement for the efficient spread of silencing suggesting that any geminivirus could be used as a vector, providing that symptoms are minimal. It should be noted that removing the coat protein attenuates symptoms and adding foreign DNA can further reduce symptoms. In addition, the later in development that a plant is bombarded, the more attenuated the symptoms.

11.2.1.1 **Structure of Geminivirus Plasmids**

The majority of viruses in the genus *Begomovirus*, family *Geminiviridae*, have bipartite genomes consisting of two circular molecules, the A and B components (Figure 11.2). Each component is approximately 2.5–3 kilobases and there is an upper limit to size, perhaps because viral DNA must be bound and trafficked by the cell-to-cell movement protein BL1 (also known as BC1) or due to plasmodesmatal constraints [18]. VIGS vectors have been made from several bipartite *Begomoviruses* including TGMV,

Cabbage Leaf Curl Virus (CaLCuV), *African Cassava Mosaic Virus* (ACMV), and *Cotton Leaf Crumple Virus* (CLCrV) [4,6,19,20] and these references provide useful information for vector construction. Each of these vectors have been cloned as 1.3–1.5 tandem direct repeats in an *E. coli* plasmid so that the inserted viral sequence has a single copy of the genes flanked by duplicated common regions.

If a geminivirus clone is obtained that has a single copy of the virus, it will be necessary to clone the common region and insert it so that it produces a tandem direct repeat. This will allow a unit length viral genome component to be initiated in one common region and completed at the origin of replication of the second common region. During replication, the AL1 protein nicks double-stranded DNA at the origin and then ligates the ends of the newly replicated molecule together to make a single-stranded circular DNA molecule. Host enzymes are thought to render it into a double-stranded DNA molecule capable of being transcribed, replicated, and perhaps moved throughout the plant.

11.2.1.2 Construction of an *AR1* Replacement Vector

Two sites of integration of foreign DNA have been tested in geminiviruses [6,11]. One site is downstream of the *AR1* coat protein gene promoter and before the polyadenylation site with the intent of producing a transcript consisting of the target gene fragment. The second site is downstream of the *BR1* open reading frame but before the polyadenylation site, which results in translation of *BR1* as well as silencing of the target gene. In other words, the A-component vector is an 800-bp gene replacement vector while the B-component vector is an insertion vector, and can accept up to approximately 150–160 bp of target gene sequence.

Figure 11.3 outlines the procedure for making an A-component silencing vector. The *AR1* coat protein gene is the best candidate for making geminivirus VIGS vectors because the vectors are not transmissible without the coat protein. Before replacing the coat protein gene, it should be mutated and tested for infectivity because in some *Begomovirus*–host plant combinations, *AR1* is required for movement [21]. This is easily done by creating a frameshift mutation in the coat protein gene to produce a premature stop codon or by excising a portion of the coat protein gene.

To replace the *AR1* gene with a multiple cloning site, two sets of overlapping PCR products are used such that one product ends near the *AR1* start codon and the second product begins at the stop codon and both products have 5′ extensions containing the multiple cloning site (MCS) sequences. It is convenient to make the primers so that restriction enzymes at the non-MCS end of the fragment can be used to clone the fragments back into the virus. The first set of primers should have sequence for unique restriction sites in 5′ end of the reverse primer near the *AR1* start codon. The second set should have the same restriction sites at the 5′ end of the forward primer (with respect to the *AR1* start codon) and amplify a region near the *AR1* stop codon. Once the PCR products are obtained and cut with the appropriate restriction enzymes, incubate the PCR products and viral vector (cut with the same two restriction enzymes) at 95 °C, cool to allow the overlapping MCS sequences to anneal, and add ligase at room temperature. Transform *E. coli* and screen for the presence of the insert using a restriction enzyme with a site found only in the MCS, or

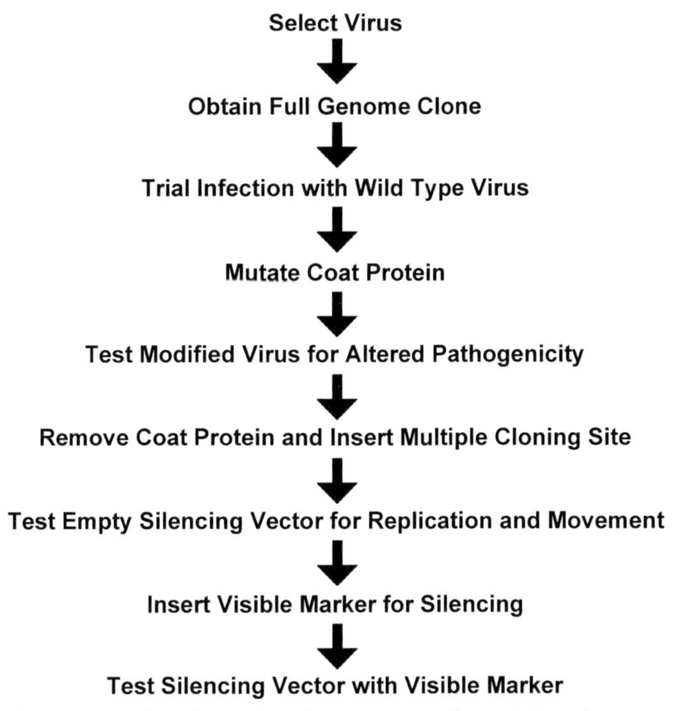

Figure 11.3 Outline of procedures for construction of a geminivirus A component silencing vector.

any enzyme to linearize the plasmid. The size differences between plasmids with and without the overlapped PCR products should be sufficient to distinguish plasmids that have the correct insert. It will be necessary to sequence at least three transformants for possible mutations in the PCR products.

11.2.1.3 Construction of an Insertion Vector

In the event that modification of the A component fails to give a vector that shows systemic movement, or if the spread of silencing is limited, the B component can be used as a vector. B-component vectors have a multiple cloning site so that a silencing fragment can be inserted downstream of a B-component gene [6,11]. Similar methods using overlapping PCR products can be used to engineer a single unique restriction site downstream of the *BR1* open reading frame and before the putative polyadenlyation site (often the sequence AATAAA). We found that a multiple cloning site takes up too much room in this type of vector. Because it is an insertion vector, like many RNA virus vectors, it retains the coat protein gene and is infectious and can be transmitted by whiteflies, the only known method of transmitting *Begomoviruses*.

Warning: Extreme care must be used to prevent escape of any infectious virus and for geminiviruses, yellow sticky tape should be used to monitor for the presence of whiteflies. The person providing viral plasmids should know about the distribution of the virus but it is up to the investigator to make sure that all regulatory procedures have been followed and clearance for working with the virus has been obtained. It is everyone's responsibility to autoclave each infected plant at the end of an experiment and to keep all working areas clean and free of any type of insect. Old World viruses, especially beta components, should NOT be used in the New World. New World viruses should be considered for use only if they are not infectious (they lack a coat protein gene) or the weather is too cold for whiteflies. The ability of these viruses to exist in mixed populations and undergo recombination should not be underestimated. They are emerging pathogens, and their effects are becoming more severe each year [22].

11.2.2 Silencing an Endogenous Gene

11.2.2.1 Visible Markers for Testing and Optimizing VIGS

VIGS experiments can be used to test individual genes of known sequence or libraries of unknown sequence. We will start with a target gene of known sequence. *PDS* (required for carotenoid synthesis) and *ChlI* and *ChlH* (different subunits of magnesium chelatase, a chloroplast enzyme required for inserting Mg into the protoporphyrin ring for chlorophyll synthesis) have all been used as visible markers for the extent of silencing [4,23,24]. It is strongly suggested that a visible marker be included in every experiment as a positive control for possible changes in the timing and extent of silencing due to environmental or other variables.

We have successfully silenced the magnesium chelatase I subunit gene, *ChlI*, (also known as *Sulfur* or *Su* in tobacco and *Chlorata42* in *Arabidopsis*) using only 92 bp of homologous sequence downstream of the *BR1* gene [11]. When a 56-bp fragment was used in the B component, silencing was initiated but the extent of silencing was limited [11]. The minimal size for silencing may depend on the size of flanking RNA in the transcript, RNA secondary structure, and other variables and these sizes are given as examples only [25,26].

In some cases, it may be desirable to clone a target gene fragment into a geminivirus vector already containing a fragment of *ChlI* or *PDS*. If tissue is to be analyzed for biochemical changes, simply excising chlorotic tissue will ensure that target gene silencing has occurred. Using a transgene fragment, such as the gene for green fluorescent protein (*GFP*), requires a host plant transgenic for *GFP*. GFP co-silencing can be used to identify silenced plants but generally should not be used as a marker for systemically silenced tissue because the spread of silencing is much greater for transgenes than for endogenous genes [11,27]. The following example uses pMTCaLCuVA:ChlI, which contains a 360-bp *ChlI* fragment from *Arabidopsis*. Fragments of between 100 and 440 bp of homologous DNA can be cloned into this vector for tracking the silencing of an unknown gene in chlorotic tissue. If pCPCaLCuVA.007 was used instead, the fragments could be up to 800 bp and secondary effects due to chlorosis would not be a problem, although determining the exact area of silencing would be more difficult. Detailed methods for using this vector have previously been published [15].

11.2.2.2 Cloning a Target Gene Fragment(s) into the CaLCuVA:ChlI Vector

1. Digest the pMTCaLCuVA:ChlI plasmid (i.e. 2.5 μg) with *Acc*65I according to the manufacturer's directions. It may be useful to run a gel to determine if the plasmid is completely linearized.

2. Following electrophoresis in agarose, purify the fragment using a gel extraction kit and quantify the DNA. Alternatively, a spin column (e.g. from Qiagen) can be used to purify the DNA.

3. Isolation of the insert fragments can be achieved by standard cloning techniques or by reverse transcriptase polymerase chain reaction (RT-PCR) using RNA prepared from the target (host) plant [28]. Primers for RT-PCR can contain an embedded restriction site for *Acc*651 at the 5′ ends, allowing a 3-bp overhang to ensure complete digestion [29]. Alternatively, a cDNA library can be constructed using RNA from the tissue of interest. A better strategy uses suppression subtractive hybridization (SSH) [30] to isolate unique RNA fragments that correspond to the genetic differences between two tissues or conditions (see http://www.evrogen.com/s5.shtml for description). In this case, cDNA fragments should be cloned directly into pCPCaLCuVA.007 lacking *ChlI* so that fragments between 100 and 800 bp can be tested.

4. After quantification of insert DNA, ligate the vector and insert in a 1 : 1 and 1 : 3 molar ratio using T4 DNA ligase and buffer according to the manufacturer's directions.

5. Transform the ligation product into competent *E. coli* cells and select transformants on Luria Broth plates containing ampicillin (100 μg/l) or carbenicillin (50 μg/l) [28].

6. Miniprep DNA from putative transformants and test for the presence of insert DNA by PCR or restriction site analyses [28]. Obtaining sequence information can be used to determine the orientation of the fragment, but both sense and antisense fragments are effective for silencing.

7. Carry out a large-scale plasmid DNA isolation. Resuspend plasmid DNA in TE (10 mM Tris pH 8.0–1 mM EDTA) and quantify. Adjust the concentration to 1 μg/μl with TE.

8. Conduct large-scale DNA isolations of pCPCaLCuVB.02 for co-bombardment with the A component. Also prepare DNA from pMTCaLCuVA:ChlI as a positive control for bombardment.

11.2.2.3 Plant Preparation

Arabidopsis seeds can be germinated (after a 3-day pretreatment at 4 °C) either in soil or on sterile medium (i.e. Murashige Skoog (MS) medium + 1% sucrose). We use short day conditions (8 h light, 16 h dark) to promote vegetative growth and slow down flowering. Temperature and, to a lesser degree, humidity have dramatic effects on symptoms and silencing in some ecotypes of *Arabidopsis* (Flores and Robertson, unpublished data). For the Col-0 ecotype, we found that symptoms are attenuated at 25/23 °C compared to 22/20 °C while silencing is still extensive. We now use 25/23 °C instead of the 22/20 °C reported earlier [6]. It is important to keep the temperature controlled in order to produce reproducible results.

After 3–4 weeks, seedlings germinated in soil should be transferred to 3 × 3 inch pots with four seedlings per pot. Bombardment is then carried out 1–2 weeks later, at the six- to eight-leaf stage. Seedlings germinated on sterile MS medium can be bombarded in plates and then transferred to soil 3 days later. It can be useful to have a second set of plants ready for bombardment 1–3 weeks later as a back-up. After the controls have served their purpose, they can be discarded.

11.2.2.4 **Microprojectile Bombardment**

Microprojectile bombardment uses plasmid DNA coated onto 1–1.5-micron particles (the microprojectiles) to transform cells and can be used on intact plants in soil or on plates. If plants in soil are to be used, special precautions must be observed to prevent contamination of other user's experiments if they require sterile conditions. For the BioRad PDS1000, manufacturer's instructions can be followed for preparation of gold or tungsten particles. When coating the particles with DNA, 5 μg of each DNA component, the A component and the B component (10 μg DNA total), should be added to the 60-μl aliquot preparation of gold or tungsten particles instead of the suggested 5 μg DNA. A 60-μl aliquot preparation provides enough particles for five bombardments (20 plants). We typically use 10 plants per test construct, and it is necessary to bombard more plants than will be used to guard against bombardment damage. One operator can carry out about 30–50 bombardments per session; coating particles in the morning and conducting bombardments in the afternoon.

It is very important to clean the chamber before a new construct is to be bombarded. We use a spray bottle with water to dissolve DNA and then wipe the upper part of the chamber and macroprojectile-holder carefully, followed by 95% ethanol. To determine if this is a problem, plants can be inoculated with the B component only to test for carryover of A-component DNA. The B component will not replicate in the absence of the A component and plants will become infected only if carryover DNA is present. Since the A component contains the test silencing sequence, it is important to ensure that each bombardment is clean.

We recently began using a home-made particle inflow gun that resembles the gun described by the Nonet laboratory, Washington University, St. Louis, designed for silencing genes in *C. elegans* (http://neuroscience.wustl.edu/nonetlab/ResourcesF/genegun/Genegun.htm). This gun is much less expensive to operate and the results of silencing experiments have been similar (if not better) using this gun compared to the PDS 1000. DNA is precipitated onto gold using the same procedure as for the PDS 1000 but Swinnex filters (Millipore) are used instead of macrocarriers to hold the particles, and rupture discs and screens are not needed. The Swinnex filters can be autoclaved and re-used. A vacuum pump and Helium tank are still required and control of the velocity of the microprojectiles is less precise than with the high tolerance BioRad rupture discs. It should also be mentioned that a hand-held Bio-Rad

PDS is available that does not require a vacuum. Very good results can be obtained with this gun (Vicki Vance, personal communication; SC, USA) although two bombardments are needed for each plant or set of plants.

11.2.2.5 Assessment of VIGS

Plants need to be assessed visually for the development of disease symptoms or silencing effects as the experiment progresses. Although general guidelines for collecting data and annotating results are presented, the details will vary according to the goal of the experiment and the expected qualities of the target gene (or screen). The time devoted to developing methods for collecting and analyzing data before the experiment begins will be well spent because it is difficult to change parameters once the experiment is underway. Methods for analysis must include labeling and tracking individual plants and photodocumentation of whole plants. Other methods for the assessment of putative silencing target areas include macrophotography, tissue sampling and fixation, and tissue sampling to be kept at −80 °C for further analyses such as the quantification of gene silencing, and assessment of viral DNA levels. Data is often collected at a specified time point, for example 21 days. The experiment (if successful) must be repeated to confirm the results. We repeated the bombardment of TGMV carrying an *RBR* silencing fragment into *N. benthamiana* over 100 times and used photography at weekly intervals, RT-PCR, qRT-PCR, and viral DNA PCR at different time points, toluidine blue-staining of mid-vein cross-sections cut with a vibratome to see vascular anatomy, trypan blue staining followed by choral hydrate clearing to determine the extent of cell death, DAPI staining and fluorescence microscopy for nuclear structure and vascular anatomy, dissecting and regular microscopy of trichomes, pavement cells, and stomata, and counting and photography of curled/straight flowers to analyze the mutant *RBR* phenotype in *N. benthamiana* [10]. Proteomics or microarrays could theoretically have been used to investigate genes regulated by *RBR*, with the empty vector-infected plants as control. Less expensive assays include RT-PCR of genes expected to be regulated by RBR and fluorescence-activated cell sorting (FACS) to look at endoreduplication, assays that were used on TRV:RBR-silenced tissues by another group [31].

It will be useful to develop a database for storing and analyzing VIGS results. The open source database MySQL is adaptable but has a steep learning curve while Filemaker Pro is expensive but web-friendly. Consultation with a bioinformatics or computer programmer is recommended before making a commitment to either platform. Once a templated database has been constructed, adding records for each plant is relatively straightforward.

Verification of Target Gene Silencing RT-PCR experiments should be performed at 3 weeks post inoculation to verify that the target gene is down-regulated. Primers should be designed to amplify a region of the gene outside of the silencing fragment because viral transcription of the fragment causes an increase in RNA levels. Additionally, it is very difficult to remove viral DNA from the cDNA and at least a 3-h incubation with RNAase-free DNAse is required for RT-PCR. It is also important to include a no-reverse-transcriptase control PCR reaction using the same reagents to guard against DNA contamination. RT-PCR of an endogenous gene, such as GAPDH or actin should be used as a control for cDNA construction and to normalize results.

If less than a twofold reduction in target mRNA is achieved, a second set of primers should be tested, or a different target gene fragment should be used for VIGS, or the type of tissue collected for sampling should be re-evaluated. The level of mRNA down-regulation needed to achieve a change in phenotype will vary for different genes depending in part on the rate of protein turnover, whether the protein is an enzyme or has a structural role, whether the protein is part of a larger complex, and other factors. Information obtained from bioinformatics will help in deciding whether the reduction in mRNA level is sufficient for an evaluation of possible phenotypic changes.

Photodocumentation Photodocumentation is essential for enabling a comparison of phenotypes over time and between experiments. A dedicated area for photography should be set up with a tripod, lighting system and black background, and method for obtaining reliable, standardized portrait and overhead photographs. Digital photographs must be saved using an appropriate title including date, VIGS construct, and plant number, or labels containing such information should be included in the picture. Photographs taken with a dissecting microscope should have similar information and magnification level. Observations are especially helpful in these experiments and should be recorded in a laboratory notebook or typed directly into the database.

Changes in leaf mid-vein anatomy can be profound without a concomitant change in overall leaf morphology. Fixation of leaf tissue followed by vibratome or paraffin sectioning, staining, and *in situ* or immunolocalization can provide very useful information about gene function. Does the gene product affect cell size, shape, development, or differentiation? Are pavement cells altered in shape or size? A very useful reference for monitoring phenotypic changes was produced by researchers at Paradigm Genetics for Arabidopsis [32,33]. Figure 11.4 shows plants at two different stages of the experiment and shows that dramatic changes can occur. Silencing of *ChlI* does not kill the plant because the lower leaves remain green. The height of the plant, size of the leaves, and time to flower are all changed. We use this plant as a control for meristem structure and other types of analyses and have compared it to uninfected *N. benthamiana* and plants infected with wild-type TGMV to visualize the extremes of possible virus-associated changes to structure and function [10,11].

One point should be kept in mind when interpreting a VIGS phenotype. VIGS is almost never complete and there may be inherent differences between genes in terms of susceptibility to silencing [34]. It is always possible that a silencing phenotype inherent to a gene will not be observed. This could be due to the target gene function being supported by a residual low level of mRNA in the infected plants. If the target gene is a member of a multiple family it is necessary to target conserved and non-conserved regions to determine whether the silencing

Figure 11.4 *N. benthamiana* inoculated with TGMV:ChlI 1 week post-bombardment (left) and 2–3 weeks post-bombardment. The pot is visible on the left and the second leaf above the edge of the pot has bombardment damage at the edge of the leaf. Note the circular yellow spots on different leaves showing where microprojectiles successfully delivered a silencing-competent virus.

phenotypes are due to one or several members of the family. VIGS can target two genes simultaneously, which can be helpful for obtaining a phenotype for genes that are part of a family (see below).

11.3 Applications of the Technology

A review of the literature shows that VIGS has been used to analyze gene function primarily in *N. benthamiana*, primarily at 21 day post infection (dpi), and primarily using TRV (Table 11.1). Applications include finding new genes using an EST library cloned into TRV; VIGS of a small number of genes to determine which affect disease resistance, or extensive analysis of one or a few genes using VIGS as one of many tools. The extensive use of *N. benthamiana* is due in part to the effectiveness of TRV-induced gene silencing and the ease of agro-inoculation in this species. Agroinoculation of other species may require vacuum infiltration of seedlings rather syringe inoculation without a needle, or using a needle to puncture the vascular tissue of the stem. *N. benthamiana* is also useful for screens because of its small size and rapid cycling time.

Of the viral vectors described to date, TRV is exceptional for its lack of symptoms, and TGMV is exceptional for the continuity of silencing over time in *N. benthamiana*. This was recently demonstrated using the retinoblastoma-related gene (*RBR*). TRV-mediated silencing of RBR was optimal at 21 days, after which time plants appeared to grow out of the phenotype or lose the silencing [31]. TGMV-mediated silencing produced similar results in new growth but also showed a novel phenotype in mature tissue – cell death after 21 days [10]. This could not be quantified in TRV-mediated silencing because of the presence of *Agrobacterium* in the mature leaf and the movement of TRV directly to meristematic tissues. Use of a phloem-limited TGMV silencing vector could demonstrate that cell death was not due to viral symptoms as cell death occurred in mesophyll and epidermal cells as well as vascular tissue. RT-PCR demonstrated that the *RBR* message was reduced compared to mock, TGMV:SU, and wild-type TGMV-inoculated plants at 21 and 28 days in the same tissues.

There are caveats to using VIGS as genetic tool to study gene function. Virus infection has to change host gene expression in order replicate and spread. Fortunately, the number of cells actually supporting virus replication is much smaller than the number of cells receiving the diffusible silencing signal. The 'empty vector' control can be used to help screen out virus-induced changes, but alterations in viral gene expression or spread can also occur due to the nature of the particular silencing target. Knowledge of the viral vector will help to determine whether the silenced target gene will compromise or augment virus–host interactions. In some cases, attenuating symptoms may prove to be the most difficult task for a new vector, and will be one of the major differences between VIGS vectors that are truly useful and VIGS vectors that never quite realize their potential. In part, this is due to our limited knowledge of virus–host interactions, such as plant responses to putative or demonstrated viral anti-silencing proteins.

A real strength of VIGS is the ability to study down-regulation of embryo-lethal or other essential genes. The fact that small amounts of transcript may remain is actually an advantage of the VIGS method. In the case of essential genes, a complete knockout results in an embryo lethal phenotype, which is uninformative. By using VIGS, the plant can be allowed to mature under conditions that allow wild-type protein expression, and then inoculated to cause transcript reduction in the gene of interest. This has allowed the study of essential genes that have previously been impossible to knockout, such as *PCNA* and *RBR* [10,11,31].

Another advantage to VIGS is the ability to silence in tandem, or even silencing entire gene families. Silencing in tandem is possible simply by inserting silencing fragments from two different genes into the vector. The geminivirus-based VIGS system allows for up to 800 bp of insert DNA, and generally only 100–200 bp is needed for efficient silencing. Therefore, more than one completely non-homologous gene can be silenced using a single vector. Another tactic that can be utilized is silencing multiple homologs or even an entire gene family by designing the silencing fragment using highly conserved regions of the gene family. Conversely, the fragment could be designed to a divergent portion of the gene in order to silence only a single member of the family.

11.4 Perspectives

Tools to study gene function in many crop plants are urgently needed. Most crop plants have not been sequenced, and only limited EST data may be available. Due to the fact that VIGS does not require stable transformation, it could be used in many crop plants assuming that suitable viruses are available. Also, limited EST data can still be used to create fragments for silencing genes, and other genes of interest may be silenced using conserved sequences from closely related species.

Two recent examples highlight the use of VIGS in forward genetics screens for identifying genes involved in a biological process. One used subtractive, suppressive hybridization (SSH) PCR to identify a subset of genes involved in tobacco-blue mold interactions and VIGS to test them [35]. A second used *Agrobacterium*-delivered VIGS vectors to identify genes with altered crown gall phenotypes on leaf discs from *N. benthamiana* [36]. Other VIGS screens have identified novel genes in disease resistance pathways, increasing fundamental knowledge about disease processes [37,38]. Although these screens have not used agriculturally important crop plants, construction of SSH libraries in crop plants is straightforward: because PCR methods are used the resulting DNA fragments are of small size (100–400 nt or up to 900 nt, in our experience), and the cDNA fragments can be cloned directly into a viral vector. Sequence information is not needed for testing such libraries by VIGS if a predetermined screen is used (such as herbicide targets). Plants with a desired VIGS phenotype can be used for DNA or RNA isolation (depending on the virus) and the insert DNA amplified by PCR using vector sequences for primers. As suggested in 1999, VIGS may yet fulfill its promise as a method for fast forward genetics [39].

References

1 Lindbo, J.A. and Dougherty, W.G. (2005) Plant pathology and RNAi: a brief history. *Annual Review of Phytopathology*, **43**, 191–204.

2 Kumagai, M.H., Donson, J., della-Cioppa, G., Harvey, D., Hanley, K. and Grill, L.K. (1995) Cytoplasmic inhibition of carotenoid biosynthesis with virus-derived RNA. *Proceedings of the National Academy of Sciences of the United States of America*, **92**, 1679–1683.

3 Ruiz, M.T., Voinnet, O. and Baulcombe, D.C. (1998) Initiation and maintenance of virus-induced gene silencing. *Plant Cell*, **10**, 937–946.

4 Kjemtrup, S., Sampson, K., Peele, C., Nguyen, L.V., Conkling, M.A., Thompson, W.F. and Robertson, D. (1998) Gene silencing from plant DNA carried by a Geminivirus. *Plant Journal*, **14**, 91–100.

5 Ratcliff, F., Martin-Hernandez, A.M. and Baulcombe, D.C. (2001) Technical Advance. Tobacco rattle virus as a vector for analysis of gene function by silencing. *Plant Journal*, **25**, 237–245.

6 Turnage, M.A., Muangsan, N., Peele, C.G. and Robertson, D. (2002) Geminivirus-based vectors for gene silencing in *Arabidopsis. Plant Journal*, **30**, 107–114.

7 Cai, X.Z., Xu, Q.F., Wang, C.C. and Zheng, Z. (2006) Development of a virus-induced gene-silencing system for functional analysis of the RPS2-dependent resistance signalling pathways in *Arabidopsis. Plant Molecular Biology*, **62**, 223–232.

8 Burch-Smith, T.M., Schiff, M., Liu, Y. and Dinesh-Kumar, S.P. (2006) Efficient virus-induced gene silencing in *Arabidopsis. Plant Physiology*, **142**, 21–27.

9 Burton, R.A., Gibeaut, D.M., Bacic, A., Findlay, K., Roberts, K., Hamilton, A., Baulcombe, D.C. and Fincher, G.B. (2000) Virus-induced silencing of a plant

cellulose synthase gene. *Plant Cell*, **12**, 691–706.

10 Jordan, C.V., Shen, W., Hanley-Bowdoin, L. and Robertson, D. (2007) Geminivirus-induced gene silencing of the tobacco retinoblastoma-related gene results in cell death and altered development. *Plant Molecular Biology*, 10.1007/s11103-007-9206-3.

11 Peele, C., Jordan, C.V., Muangsan, N., Turnage, M., Egelkrout, E., Eagle, P., Hanley-Bowdoin, L. and Robertson, D. (2001) Silencing of a meristematic gene using geminivirus-derived vectors. *Plant Journal*, **27**, 357–366.

12 Burch-Smith, T.M., Anderson, J.C., Martin, G.B. and Dinesh-Kumar, S.P. (2004) Applications and advantages of virus-induced gene silencing for gene function studies in plants. *Plant Journal*, **39**, 734–746.

13 Benedito, V.A., Visser, P.B., Angenent, G.C. and Krens, F.A. (2004) The potential of virus-induced gene silencing for speeding up functional characterization of plant genes. *Genetics and Molecular Research*, **3**, 323–341.

14 Robertson, D. (2004) VIGS vectors for gene silencing: many targets, many tools. *Annual Review of Plant Physiology and Plant Molecular Biology*, **55**, 495–519.

15 Muangsan, N. and Robertson, D. (2004) Geminivirus vectors for transient gene silencing in plants. *Methods in Molecular Biology*, **265**, 101–116.

16 Alberter, B., Ali Rezaian, M. and Jeske, H. (2005) Replicative intermediates of tomato leaf curl virus and its satellite DNAs. *Virology*, **331**, 441–448.

17 Elmer, J.S., Sunter, G., Gardiner, W.E., Brand, L., Browning, C.K., Bisaro, D.M. and Rogers, S.G. (1988) *Agrobacterium*-mediated inoculation of plants with tomato golden mosaic virus DNAs. *Plant Molecular Biology*, **10**, 225–234.

18 Gilbertson, R.L., Sudarshana, M., Jiang, H., Rojas, M.R. and Lucas, W.J. (2003) Limitations on geminivirus genome size imposed by plasmodesmata and virus-encoded movement protein: insights into DNA trafficking. *Plant Cell*, **15**, 2578–2591.

19 Fofana, I.B., Sangare, A., Collier, R., Taylor, C. and Fauquet, C.M. (2004) A geminivirus-induced gene silencing system for gene function validation in cassava. *Plant Molecular Biology*, **56**, 613–624.

20 Tuttle, J.R., Haigler, C.H., Shah, I., Brown, J. and Robertson, D. (2007) Using virus-induced gene silencing and gene expression as tools to understand virus infections in the cotton plant. (in preparation).

21 Pooma, W., Gillette, W.K., Jeffrey, J.L. and Petty, I.T. (1996) Host and viral factors determine the dispensability of coat protein for bipartite geminivirus systemic movement. *Virology*, **218**, 264–268.

22 Mansoor, S., Briddon, R.W., Zafar, Y. and Stanley, J. (2003) Geminivirus disease complexes: an emerging threat. *Trends in Plant Science*, **8**, 128–134.

23 Kumagai, M.H., Donson, J., Dellacioppa, G., Harvey, D., Hanley, K. and Grill, L.K. (1995) Cytoplasmic inhibition of carotenoid biosynthesis with virus-derived RNA. *Proceedings of the National Academy of Sciences of the United States of America*, **92**, 1679–1683.

24 Hiriart, J.B., Aro, E.M. and Lehto, K. (2003) Dynamics of the VIGS-mediated chimeric silencing of the *Nicotiana benthamiana* ChlH gene and of the tobacco mosaic virus vector. *Molecular Plant–Microbe Interactions: MPMI*, **16**, 99–106.

25 Thomas, C.L., Jones, L., Baulcombe, D.C. and Maule, A.J. (2001) Size constraints for targeting post-transcriptional gene silencing and for RNA-directed methylation in *Nicotiana benthamiana* using a potato virus X vector. *Plant Journal*, **25**, 417–425.

26 Pang, S.Z., Jan, F.J. and Gonsalves, D. (1997) Nontarget DNA sequences reduce the transgene length necessary for RNA-mediated tospovirus resistance in transgenic plants. *Proceedings of the National Academy of Sciences of the United States of America*, **94**, 8261–8266.

27 Himber, C., Dunoyer, P., Moissiard, G., Ritzenthaler, C. and Voinnet, O. (2003) Transitivity-dependent and -independent cell-to-cell movement of RNA silencing. *EMBO Journal*, **22**, 4523–4533.

28 Sambrook, J., Fritsch, E.F. and Maniatis, T. (2001) *Molecular Cloning*, Cold Spring Harbor Press, New York.

29 NEB. (2007) New England Biolabs: Cleavage close to the end of DNA fragments (oligonucleotides). http://www.neb.com/nebecomm/tech_reference/restriction_enzymes/cleavage_olignucleotides.asp

30 Diatchenko, L., Lua, Y.F., Campbell, A.P., Chenchik, A., Moqadam, F., Huang, B., Lukyanov, S., Lukyanov, K., Gurskaya, N., Sverdlov, E.D., and Siebert, P.D. (1996) Suppression subtractive hybridization: a method for generating differentially regulated or tissue-specific cDNA probes and libraries. *Proceedings of the National Academy of Sciences of the United States of America*, **93**, 6025–6030.

31 Park, J.A., Ahn, J.W., Kim, Y.K., Kim, S.J., Kim, J.K., Kim, W.T. and Pai, H.S. (2005) Retinoblastoma protein regulates cell proliferation, differentiation, and endoreduplication in plants. *Plant Journal*, **42**, 153–163.

32 Boyes, D., Zayed, A., Ascenzi, R., McCaskill, A., Hoffman, N., Davis, K. and Gorlach, J. (2001) Growth stage-based phenotypic analysis of *Arabidopsis*: A model for high throughput functional genomics in plants. *The Plant Cell*, **13**, 1499–1510.

33 Kjemtrup, S., Boyes, D.C., Christensen, C., McCaskill, A.J., Hylton, M. and Davis, K. (2003) Growth stage-based phenotypic profiling of plants, in *Plant Functional Genomics Methods and Protocols* (ed. E. Grotewold), Humana Press, p. 500, Totowa, NJ (USA).

34 McGinnis, K., Chandler, V., Cone, K., Kaeppler, H., Kaeppler, S., Kerschen, A., Pikaard, C., Richards, E., Sidorenko, L., Smith, T., Springer, N. and Wulan, T. (2005) Transgene-induced RNA interference as a tool for plant functional genomics. *Methods in Enzymology*, **392**, 1–24.

35 Borras-Hidalgo, O., Thomma, B.P., Collazo, C., Chacon, O., Borroto, C.J., Ayra, C., Portieles, R., Lopez, Y. and Pujol, M. (2006) EIL2 transcription factor and glutathione synthetase are required for defense of tobacco against tobacco blue mold. *Molecular Plant–Microbe Interactions: MPMI*, **19**, 399–406.

36 Anand, A., Vaghchhipawala, Z., Ryu, C.M., Kang, L., Wang, K., del-Pozo, O., Martin, G.B. and Mysore, K.S. (2007) Identification and characterization of plant genes involved in *Agrobacterium*-mediated plant transformation by virus-induced gene silencing. *Molecular Plant–Microbe Interactions: MPMI*, **20**, 41–52.

37 Peart, J.R., Mestre, P., Lu, R., Malcuit, I. and Baulcombe, D.C. (2005) NRG1, a CC-NB-LRR protein, together with N, a TIR-NB-LRR protein, mediates resistance against tobacco mosaic virus. *Current Biology*, **15**, 968–973.

38 Liu, Y., Schiff, M., Czymmek, K., Talloczy, Z., Levine, B. and Dinesh-Kumar, S.P. (2005) Autophagy regulates programmed cell death during the plant innate immune response. *Cell*, **121**, 567–577.

39 Baulcombe, D.C. (1999) Fast forward genetics based on virus-induced gene silencing. *Current Opinion in Plant Biology*, **2**, 109–113.

40 Ding, X.S., Schneider, W.L., Chaluvadi, S.R., Mian, M.A. and Nelson, R.S. (2006) Characterization of a Brome mosaic virus strain and its use as a vector for gene silencing in monocotyledonous hosts. *Molecular Plant–Microbe Interactions: MPMI*, **19**, 1229–1239.

41 Holzberg, S., Brosio, P., Gross, C. and Pogue, G.P. (2002) Barley stripe mosaic virus-induced gene silencing in a monocot plant. *Plant Journal*, **30**, 315–327.

42 Hein, I., Barciszewska-Pacak, M., Hrubikova, K., Williamson, S., Dinesen, M., Soenderby, I.E., Sundar, S., Jarmolowski, A., Shirasu, K. and Lacomme, C. (2005) Virus-induced gene silencing-based functional characterization of genes associated with powdery mildew resistance in barley. *Plant Physiology*, **138**, 2155–2164.

43 Scofield, S.R., Huang, L., Brandt, A.S. and Gill, B.S. (2005) Development of a virus-induced gene-silencing system for hexaploid wheat and its use in functional analysis of the Lr21-mediated leaf rust resistance pathway. *Plant Physiology*, **138**, 2165–2173.

44 Romeis, T., Ludwig, A.A., Martin, R. and Jones, J.D. (2001) Calcium-dependent protein kinases play an essential role in a plant defence response. *EMBO Journal*, **20**, 5556–5567.

45 Saitoh, H. and Terauchi, R. (2002) Virus-induced silencing of FtsH gene in *Nicotiana benthmiana* causes a striking bleached leaf phenotype. *Genes & Genetic Systems*, **77**, 335–340.

46 Yoshioka, H., Numata, N., Nakajima, K., Katou, S., Kawakita, K., Rowland, O., Jones, J.D. and Doke, N. (2003) *Nicotiana benthamiana* gp91phox homologs NbrbohA and NbrbohB participate in H_2O_2 accumulation and resistance to *Phytophthora infestans*. *Plant Cell*, **15**, 706–718.

47 Sharma, P.C., Ito, A., Shimizu, T., Terauchi, R., Kamoun, S. and Saitoh, H. (2003) Virus-induced silencing of WIPK and SIPK genes reduces resistance to a bacterial pathogen, but has no effect on the INF1-induced hypersensitive response (HR) in *Nicotiana benthamiana*. *Molecular Genetics and Genomics*, **269**, 583–591.

48 Faivre-Rampant, O., Gilroy, E.M., Hrubikova, K., Hein, I., Millam, S., Loake, G.J., Birch, P., Taylor, M. and Lacomme, C. (2004) Potato virus X-induced gene silencing in leaves and tubers of potato. *Plant Physiology*, **134**, 1308–1316.

49 Hirano, T., Ito, A., Berberich, T., Terauchi, R. and Saitoh, H. (2008) Virus-induced gene silencing of 14-3-3 genes abrogates dark repression of nitrate reductase activity in *Nicotiana benthamiana*. *Molecular Genetics and Genomics* (submitted).

50 Li, Y., Wu, M.Y., Song, H.H., Hu, X. and Qiu, B.S. (2005) Identification of a tobacco protein interacting with tomato mosaic virus coat protein and facilitating long-distance movement of virus. *Archives of Virology*, **150**, 1993–2008.

51 Naylor, M., Reeves, J., Cooper, J.I., Edwards, M.L. and Wang, H. (2005) Construction and properties of a gene-silencing vector based on Poplar mosaic virus (genus Carlavirus). *Journal of Virological Methods*, **124**, 27–36.

52 Peart, J.R., Cook, G., Feys, B.J., Parker, J.E. and Baulcombe, D.C. (2002) An EDS1 orthologue is required for N-mediated resistance against tobacco mosaic virus. *Plant Journal*, **29**, 569–579.

53 Liu, Y., Schiff, M. and Dinesh-Kumar, S.P. (2002) Virus-induced gene silencing in tomato. *Plant Journal*, **31**, 777–786.

54 Ekengren, S.K., Liu, Y., Schiff, M., Dinesh-Kumar, S.P. and Martin, G.B. (2003) Two MAPK cascades, NPR1, and TGA transcription factors play a role in Pto-mediated disease resistance in tomato. *Plant Journal*, **36**, 905–917.

55 He, X., Anderson, J.C., del Pozo, O., Gu, Y.Q., Tang, X. and Martin, G.B. (2004) Silencing of subfamily I of protein phosphatase 2A catalytic subunits results in activation of plant defense responses and localized cell death. *Plant Journal*, **38**, 563–577.

56 Chen, J.C., Jiang, C.Z., Gookin, T.E., Hunter, D.A., Clark, D.G. and Reid, M.S.

(2004) Chalcone synthase as a reporter in virus-induced gene silencing studies of flower senescence. *Plant Molecular Biology*, **55**, 521–530.

57 Chen, J.C., Jiang, C.Z. and Reid, M.S. (2005) Silencing a prohibitin alters plant development and senescence. *Plant Journal*, **44**, 16–24.

58 Hileman, L.C., Drea, S., Martino, G., Litt, A. and Irish, V.F. (2005) Virus-induced gene silencing is an effective tool for assaying gene function in the basal eudicot species *Papaver somniferum* (opium poppy). *Plant Journal*, **44**, 334–341.

59 Jin, H., Axtell, M.J., Dahlbeck, D., Ekwenna, O., Zhang, S., Staskawicz, B. and Baker, B. (2002) NPK1, an MEKK1-like mitogen-activated protein kinase kinase kinase, regulates innate immunity and development in plants. *Developmental Cell*, **3**, 291–297.

60 Brigneti, G., Martin-Hernandez, A.M., Jin, H., Chen, J., Baulcombe, D.C., Baker, B. and Jones, J.D. (2004) Virus-induced gene silencing in *Solanum* species. *Plant Journal*, **39**, 264–272.

61 Fu, D.Q., Zhu, B.Z., Zhu, H.L., Zhang, H.X., Xie, Y.H., Jiang, W.B., Zhao, X.D. and Luo, K.B. (2006) Enhancement of virus-induced gene silencing in tomato by low temperature and low humidity. *Molecules and Cells*, **21**, 153–160.

62 Fu, D.Q., Zhu, B.Z., Zhu, H.L., Jiang, W.B. and Luo, Y.B. (2005) Virus-induced gene silencing in tomato fruit. *Plant Journal*, **43**, 299–308.

63 Chung, E., Seong, E., Kim, Y.C., Chung, E.J., Oh, S.K., Lee, S., Park, J.M., Joung, Y.H. and Choi, D. (2004) A method of high frequency virus-induced gene silencing in chili pepper (*Capsicum annuum* L. cv Bukang). *Molecules and Cells*, **17**, 377–380.

64 Ahn, C.S., Lee, J.H. and Pai, H.S. (2005) Silencing of NbNAP1 encoding a plastidic SufB-like protein affects chloroplast development in *Nicotiana benthamiana*. *Molecules and Cells*, **20**, 112–118.

65 Ahn, C.S., Lee, J.H., Reum Hwang, A., Kim, W.T. and Pai, H.S. (2006) Prohibitin is involved in mitochondrial biogenesis in plants. *Plant Journal*, **46**, 658–667.

66 Ahn, J.W., Kim, M., Lim, J.H., Kim, G.T. and Pai, H.S. (2004) Phytocalpain controls the proliferation and differentiation fates of cells in plant organ development. *Plant Journal*, **38**, 969–981.

67 Ahn, J.W., Verma, R., Kim, M., Lee, J.Y., Kim, Y.K., Bang, J.W., Reiter, W.D. and Pai, H.S. (2006) Depletion of UDP-D-apiose/UDP-D-xylose synthases results in rhamnogalacturonan-II deficiency, cell wall thickening, and cell death in higher plants. *The Journal of Biological Chemistry*, **281**, 13708–13716.

68 Bhattarai, K.K., Li, Q., Liu, Y., Dinesh-Kumar, S.P. and Kaloshian, I. (2007) The MI-1-mediated pest resistance requires hsp90 and sgt1. *Plant Physiology*, **144**, 312–323.

69 Chiasson, D., Ekengren, S.K., Martin, G.B., Dobney, S.L. and Snedden, W.A. (2005) Calmodulin-like proteins from *Arabidopsis* and tomato are involved in host defense against *Pseudomonas syringae* pv. *tomato*. *Plant Molecular Biology*, **58**, 887–897.

70 Darnet, S. and Rahier, A. (2004) Plant sterol biosynthesis: identification of two distinct families of sterol 4alpha-methyl oxidases. *The Biochemical Journal*, **378**, 889–898.

71 Kim, M., Lim, J.H., Ahn, C.S., Park, K., Kim, G.T., Kim, W.T. and Pai, H.S. (2006) Mitochondria-associated hexokinases play a role in the control of programmed cell death in *Nicotiana benthamiana*. *Plant Cell*, **18**, 2341–2355.

72 Gould, B. and Kramer, E.M. (2007) Virus-induced gene silencing as a tool for functional analyses in the emerging model plant *Aquilegia* (columbine Ranunculaceae). *Plant Methods*, **3**, 6.

73 Katou, S., Yoshioka, H., Kawakita, K., Rowland, O., Jones, J.D., Mori, H. and Doke, N. (2005) Involvement of PPS3 phosphorylated by elicitor-responsive mitogen-activated protein kinases in the regulation of plant cell death. *Plant Physiology*, **139**, 1914–1926.

74 Kim, Y.K., Lee, J.Y., Cho, H.S., Lee, S.S., Ha, H.J., Kim, S., Choi, D. and Pai, H.S. (2005) Inactivation of organellar glutamyl- and seryl-tRNA synthetases leads to developmental arrest of chloroplasts and mitochondria in higher plants. *The Journal of Biological Chemistry*, **280**, 37098–37106.

75 Lin, Z., Yin, K., Wang, X., Liu, M., Chen, Z., Gu, H. and Qu, L.J. (2007) Virus induced gene silencing of AtCDC5 results in accelerated cell death in *Arabidopsis* leaves. *Plant Physiology and Biochemistry*, **45**, 87–94.

76 Liu, Y., Nakayama, N., Schiff, M., Litt, A., Irish, V.F. and Dinesh-Kumar, S.P. (2004) Virus induced gene silencing of a DEFICIENS ortholog in *Nicotiana benthamiana*. *Plant Molecular Biology*, **54**, 701–711.

77 Liu, Y., Schiff, M. and Dinesh-Kumar, S.P. (2004) Involvement of MEK1 MAPKK, NTF6 MAPK, WRKY/MYB transcription factors COI1 and CTR1 in N-mediated resistance to tobacco mosaic virus. *Plant Journal*, **38**, 800–809.

78 Lou, Y. and Baldwin, I.T. (2006) Silencing of a germin-like gene in *Nicotiana attenuata* improves performance of native herbivores. *Plant Physiology*, **140**, 1126–1136.

79 Park, J.A., Kim, T.W., Kim, S.K., Kim, W.T. and Pai, H.S. (2005) Silencing of NbECR encoding a putative enoyl-CoA reductase results in disorganized membrane structures and epidermal cell ablation in *Nicotiana benthamiana*. *FEBS Letters*, **579**, 4459–4464.

80 Rahier, A., Darnet, S., Bouvier, F., Camara, B. and Bard, M. (2006) Molecular and enzymatic characterizations of novel bifunctional 3beta-hydroxysteroid dehydrogenases/C-4 decarboxylases from *Arabidopsis thaliana*. *The Journal of Biological Chemistry*, **281**, 27264–27277.

81 Rivas, S., Rougon-Cardoso, A., Smoker, M., Schauser, L., Yoshioka, H. and Jones, J.D. (2004) CITRX thioredoxin interacts with the tomato Cf-9 resistance protein and negatively regulates defence. *EMBO Journal*, **23**, 2156–2165.

82 Rowland, O., Ludwig, A.A., Merrick, C.J., Baillieul, F., Tracy, F.E., Durrant, W.E., Fritz-Laylin, L., Nekrasov, V., Sjolander, K., Yoshioka, H. and Jones, J.D. (2005) Functional analysis of Avr9/Cf-9 rapidly elicited genes identifies a protein kinase, ACIK1, that is essential for full Cf-9-dependent disease resistance in tomato. *Plant Cell*, **17**, 295–310.

83 Saedler, R. and Baldwin, I.T. (2004) Virus-induced gene silencing of jasmonate-induced direct defences, nicotine and trypsin proteinase-inhibitors in *Nicotiana attenuata*. *Journal of Experimental Botany*, **55**, 151–157.

84 Valentine, T., Shaw, J., Blok, V.C., Phillips, M.S., Oparka, K.J. and Lacomme, C. (2004) Efficient virus-induced gene silencing in roots using a modified tobacco rattle virus vector. *Plant Physiology*, **136**, 3999–4009.

85 Ryu, C.M., Anand, A., Kang, L. and Mysore, K.S. (2004) Agrodrench: a novel and effective agroinoculation method for virus-induced gene silencing in roots and diverse *Solanaceous* species. *Plant Journal*, **40**, 322–331.

86 Constantin, G.D., Krath, B.N., MacFarlane, S.A., Nicolaisen, M., Johansen, I.E. and Lund, O.S. (2004) Virus-induced gene silencing as a tool for functional genomics in a legume species. *Plant Journal*, **40**, 622–631.

87 Zhang, C. and Ghabrial, S.A. (2006) Development of Bean pod mottle virus-based vectors for stable protein expression and sequence-specific virus-induced gene silencing in soybean. *Virology*, **344**, 401–411.

88 Tao, X. and Zhou, X. (2004) A modified viral satellite DNA that suppresses gene expression in plants. *Plant Journal*, **38**, 850–860.

89 Cai, X., Wang, C., Xu, Y., Xu, Q., Zheng, Z. and Zhou, X. (2007) Efficient gene silencing induction in tomato by a viral satellite DNA vector. *Virus Research*, **125**, 169–175.

90 Tuttle, J.R., Haigler, C., Idris, A.M., Brown, J.K. and Robertson, D. Disarming cotton *leaf crumble virus* for virus induced gene silencing in *Gossypium hirsutum*. Plant Physiology (submitted).

91 Carrillo-Tripp, J., Shimada-Beltran, H. and Rivera-Bustamante, R. (2006) Use of geminiviral vectors for functional genomics. *Current Opinion in Plant Biology*, **9**, 209–215.

92 Hanley-Bowdoin, L., Settlage, S. and Robertson, D. (2004) Reprogramming plant gene expression – a prerequisite to geminivirus DNA replication. *Molecular Plant Pathology*, **5**, 149–156.

93 Sunter, G. and Bisaro, D.M. (1992) Transactivation of geminivirus-AR1 and geminivirus-BR1 gene expression by the viral-AL2 gene product occurs at the level of transcription. *Plant Cell*, **4**, 1321–1331.

94 Vanitharani, R., Chellappan, P., Pita, J.S. and Fauquet, C.M. (2004) Differential roles of AC2 and AC4 of cassava geminiviruses in mediating synergism and suppression of posttranscriptional gene silencing. *Journal of Virology*, **78**, 9487–9498.

95 Briddon, R.W., Pinner, M.S., Stanley, J. and Markham, P.G. (1990) Geminivirus coat protein gene replacement alters insect specificity. *Virology*, **177**, 85–94.

96 Pooma, W., Gillette, W.K., Jeffrey, J.L. and Petty, I.T.D. (1996) Host and viral factors determine the dispensability of coat protein for bipartite geminivirus systemic movement. *Virology*, **218**, 264–268.

97 Qin, S., Ward, B.M. and Lazarowitz, S.G. (1998) The bipartite geminivirus coat protein aids BR1 function in viral movement by affecting the accumulation of viral single-stranded DNA. *Journal of Virology*, **72**, 9247–9256.

98 Sanderfoot, A.A., Ingham, D.J. and Lazarowitz, S.G. (1996) A viral movement protein as a nuclear shuttle. The geminivirus BR1 movement protein contains domains essential for interaction with BL1 and nuclear localization. *Plant Physiology*, **110**, 23–33.

99 Noueiry, A.O., Lucas, W.J. and Gilbertson, R.L. (1994) Two proteins of a plant DNA virus coordinate nuclear and plasmodesmal transport. *Cell*, **76**, 925–932.

12
TILLING: A Reverse Genetics and a Functional Genomics Tool in Soybean

Khalid Meksem, Shiming Liu, Xiao Hong Liu, Aziz Jamai, Melissa Goellner Mitchum, Abdelhafid Bendahmane, and Tarik El-Mellouki

Abstract

The need to provide the missing links between DNA sequences and phenotype is becoming increasingly urgent, as more genes are identified through DNA sequencing. Therefore, cost-effective and time-saving technologies are needed to validate gene function. TILLING (Targeting Induced Local Lesions IN Genomes) is a reverse genetics tool used for the identification of chemical-based mutations. TILLING was developed first in plants using *Arabidopsis* [1], the system is based on (1) the production of mutations throughout the genome using chemical mutagenesis such as ethylmethanesulfonate (EMS); and (2) the screening of the mutant plant collections developed for rapid systematic identification of mutations in target sequences using mismatch detection enzymes. The majority of the available and committed soybean genomic tools are being developed from two cultivars, 'Williams82' and 'Forrest', both cultivars are to the soybean community what 'Col' and 'Ler' are to the *Arabidopsis thaliana* community or what 'Mo17' and 'B73' are to the maize community. Therefore, we used TILLING as a reverse genetics tool for functional analysis of soybean genes using two platforms, one from Forrest and the other from Williams 82. In this chapter, we will review the technology and its applications to soybean.

12.1
Introduction

The availability of genomic data emerging from genome and EST sequencing projects is increasing; however, relating gene sequences to a phenotype is a limiting step in the functional annotation of DNA sequences derived from transcriptome analysis, positional cloning and genome sequencing projects. The sequence expression pattern, and redundancy of genes in the genome are factors that must be taken

The Handbook of Plant Functional Genomics: Concepts and Protocols.
Edited by Günter Kahl and Khalid Meksem

ISBN: 978-3-527-31885-8

into account in future genetic analysis. It should be a priority to develop tools for functional analysis in crop species that take into account these points in addition to the elucidation of gene function in the context of the whole organism.

TILLING, according to [1–6] has been adopted as a high-throughput functional genomics tool that allows for the study of a gene function in its cellular context by relating the gene sequence to a specific phenotype. This system is based on (1) the production of mutations in the plant genome using ethylmethanesulfonate (EMS), and (2) the rapid systematic screening of the EMS plant collections produced for identification of mutations in target sequences. TILLING has several advantages over other reverse genetic approaches [7–9]: the high-throughput potential; the reliability and near irreversibility of chemical mutagenesis; the provision of an allelic series of mutations in a gene of interest; the ability to target specific regions of interest; the efficiency in relation to small genes (<1 kb); the applicability to both essential and non-essential genes; and the suitability for detection of natural variation (SNP, Indel). The fact that genetic transformation is not required makes this a viable functional analysis system for plants recalcitrant to genetic transformation.

The elucidation of gene function in the context of a whole organism still requires the identification of mutant alleles to study the function of plant genes in a cellular context, and to relate the gene sequence to a phenotype. Moreover, conducting reverse genetics in model plants is limited, since important agronomic traits are difficult, if not impossible, to characterize and extrapolate to crop species. Therefore, an efficient technology for crop plants is required to be able to assign functions to genes with a high degree of confidence.

The ongoing soybean genetics and genomics projects are identifying sequences, genetic locations, and expression patterns of most or all genes. In this chapter, we will describe TILLING as a reverse genetics tool for functional analysis of soybean genes using EMS mutagenesis.

12.2 Methods and Protocols

12.2.1 Production of Suitable Mutant Population for TILLING

The size of the population needed for an efficient TILLING project is linked to its mutation load and mutations density; therefore, the first task in TILLING is to establish protocols that can link the concentration of the chemical mutagen used to the mutation density. Since there are no standard protocols, different titration treatments of soybean seeds with EMS were tested. The germination rate of the treated seeds was used as a first selection factor toward the development of a suitable TILLING population. In a suitable population the mutation load and screening cost can be balanced to provide for an efficient and cost-effective reverse genetics and functional analysis tool.

A sample of 200 soybean seeds were treated with different concentrations of EMS each varying from 0 to 120 mM (0, 20, 40, 60, 80, 100, and 120 mM EMS solutions) overnight (16–20 h). Next day seed were washed thoroughly in water and the EMS was neutralized using a solution of 10% (w/v) sodium thiosulfate. From each treatment, 100 seeds were sown and their germination rates were scored. The germination rate was calculated as a percentage of cotyledons emerging from the soil compared to the negative control (0 mM EMS). The seedlings were then grown in a greenhouse under conditions of a 12 h light/8 h dark photoperiod and a temperature of 28–30 °C. The EMS treatment that produced 50–70% germination rate was later used to process 2000 seeds each from the same seed stock to produce three M1 populations (50, 60, and 70% germination rates). After EMS treatment, the seeds were sown in the greenhouse under conditions of a 16-h photoperiod and a temperature of 28–30 °C and were grown to produce M2 seeds through self-pollination during the fall and winter. During the spring and summer, the seedlings of the M1 plants were transferred once they had developed the second trifoliate leaves (after 2–3 weeks) from the greenhouse to the field where they were grown to produce M2 seeds. M2 seeds were harvested from the same M1 plants and were collected together in a group known as M2 family. M2 plants were used for DNA extraction. The M3 seeds were harvested for long-term storage of the mutagenized populations.

12.2.2
DNA Extraction and DNA Construction of Pools

High quality DNA is essential to ensure both successful TILLING screening and stability during long-term storage. The amount and quality of DNA is tissue-, age- and quantity-dependent. Therefore, when extracting DNA from a mutagenized population, the amount and quality should be optimized. When starting with the same amount of tissue, the DNA extraction protocol described below yielded more or less the same quantity of DNA. That saved time and money otherwise needed to calibrate the M2 DNA concentrations before pooling.

DNA was extracted using a 96-well plate format kit called the MagAttract 96 DNA Plant Core Kit (QIAGEN, Valencia, CA). High-throughput disruption of plant material was carried out using the TissueLyser System (QIAGEN, Valencia, CA). In each well of the 96-well array, a tungsten bead was added to a known weight of plant material (following the kit's manufacturer) and the plates were then shaken well for 2 min to ensure that the tissue was disrupted. The quantities of DNA were estimated on a 1% (w/v) agarose gel (in 1× TBE) stained with ethidium bromide.

Following the DNA extraction, aliquots from each well plate were diluted 20 times to normalize the DNA concentration. To increase the screening throughput, diluted normalized DNAs arrayed from each 96-well plate were pooled vertically at an eightfold concentration in one row of the 96-well pool-plate. Each well of a row pool plate, contains eight individuals from the same column of the source plate (Figure 12.1). Pool-plates are then used for TILLING PCR amplifications. Each pool-plate corresponds to 768 individual M2 families.

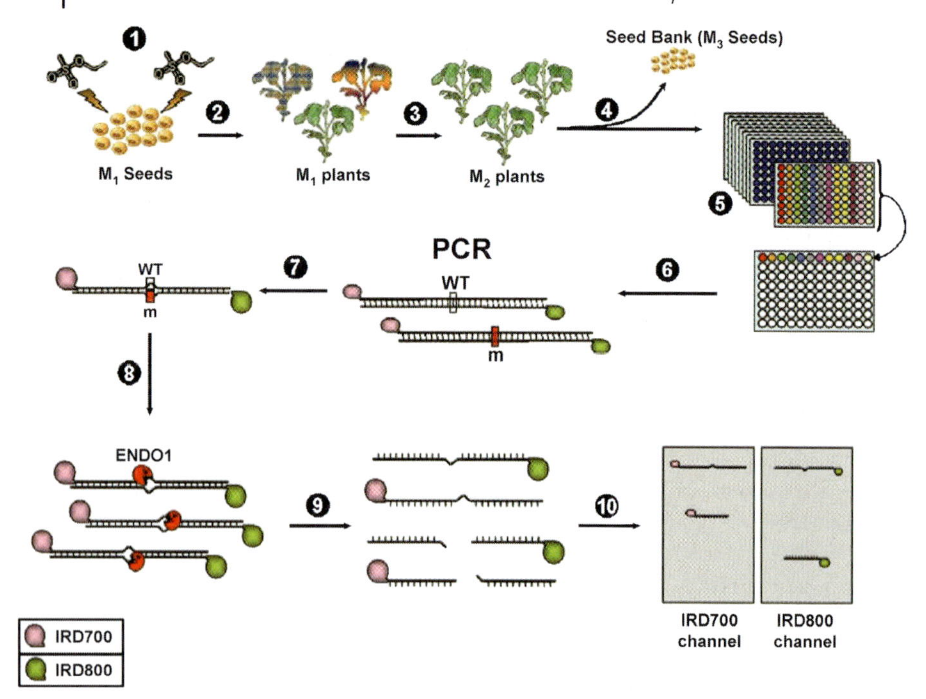

Figure 12.1 (1) Soybean TILLING Strategy: Soybean seeds are mutagenized with EMS to produce M_1 seeds, (2) M1 seeds are grown to produce M_1 plants (genetic chimeras). (3) The M_1 plants are advanced to the M_2 generation by self-pollination. (4) Leaf material is harvested for DNA extraction, M_3 seeds are harvested and stored at the seed-bank so they can be used later. The extracted DNA is normalized and arrayed in 96-well plates. (5) The DNAs are pooled up to eightfold to increase TILLING screening throughput. (6) The screening starts with a PCR amplification using fluorescently-labeled gene-specific primers. The amplified products contain sequences with no base pair changes (WT) and others carrying mutations (m). (7) Amplicons are heated to denaturation to separate the complementary strands, followed by a slow cooling step, in which PCR products re-anneal forming heteroduplexes. (8) Heteroduplexes are cut with an S1-type nuclease known as ENDO1 that cleaves the DNA at miss-pairing sites; (9) the samples are purified through a Sephadex G50 spin-column plate, denatured and loaded onto a 100-tooth absorption-membrane comb and separated by electrophoresis. (10) The gel images are analyzed for the presence of excised products representing mutations in the pooled DNAs.

12.2.3
TILLING Screening for Mutations

12.2.3.1 Gene-Specific Primers for TILLING

Gene-specific primers were designed using a web-based program called Codons Optimized to Deliver Deleterious Lesions (CODDLe), which is available through the http://www.proweb.org/input/ web-link. From an entry sequence the program generates a gene model with defined intron/exon positions; it also provides a protein conservation model using the Blocks Databases [10]. The Sorting Intolerant From Tolerant (SIFT) program can be used to provide additional blocks ([11]; http://blocks.fhcrc.org/sift/SIFT.html). The gene and protein model are than used by CODDLE to generate a graphical output where predicted induced base changes that affect the protein function are shown [12]. A window of the 1500-bp region 'Amplicon' is usually selected, and the Primer3 program [13] identifies the optimum combinations of forward and reverse primers for the TILLING assay.

To increase the stringency of the PCR reaction, the intron sequences were carefully selected so as to be certain of the specificity of the primers being used to amplify the targeted gene sequences. EST-based sequences may limit the size of the available target sequence, and its exonic composition. EST and cDNA sequences available through GenBank can be used to generate primers to amplify the corresponding genes. The amplified product can then be sequenced to generate a better template for specific primers for TILLING. To ensure the appropriate specificity of the primers being used to amplify the targeted gene sequences for TILLING in plant genomes of variable ploidy, the intron sequences should be reviewed carefully. To avoid non-specific amplifications, primers should be designed to produce a unique amplicon. For soybean the 2007 genomic WGS trace file can be searched for duplicated regions as can the BAC end sequence collections at NCBI. However, primers must still be pre-tested to determine whether they amplify a single product. This development allowed the research community to start using the soybean TILLING resources prior to the elucidation of the sequence of the whole soybean genome.

12.2.3.2 **PCR Amplification and Heteroduplex Formation**

The PCR reactions were carried out in a 10-μl total volume containing about 2 ng of DNA, 1 × ExTaq buffer (Mg^{2+} plus) (TAKARA BIO INC., Madison, WI), 0.5 Units of Ex Taq polymerase (TAKARA BIO INC., Madison, WI), 0.2 mM dNTPs, 0.2 μM primers of each the forward and reverse primers (the forward primers contained 0.08 μM of non-labeled oligonucleotides and 0.12 μM of the IRDye700-labeled oligonucleotides, and the reverse primers contained 0.04 μM of non-labeled oligonucleotides and 0.16 μM of the IRDye800-labeled oligonucleotides) and distilled sterilized water up to 10 μl.

The TILLING PCR procedure consisted of the following: an initial denaturation step at 95 °C for 2 min (to ensure the opening of the double-stranded DNA) followed by seven cycles of touchdown PCR (94 °C for 20 s, an annealing step starting initially at 73 °C for 30 s and decreasing by 1 °C per cycle, a temperature ramp increasing 0.5 °C per second to 72 °C, and 72 °C for 1 min); then 44 cycles of PCR (94 °C for 20 s, 65 °C for 30 s, a ramp of 0.5 °C per second up to 72 °C, 72 °C for 1 min); and finally an extension step at 72 °C for 5 min.

To ensure the formation of heteroduplexes, the denaturation and re-annealing step was included in the same program as follows: 99 °C for 10 min; followed by 69 cycles of 70 °C for 20 s with 0.3 °C decrease in temperature per cycle.

12.2.3.3 **M13-Tailed PCR Amplification for TILLING**

A modification of the standard TILLING protocol to optimize the GmClavata1A ortholog screen for mutants in soybean is presented in this section. The protocol is based on two PCR reactions, where the first reaction will serve to prepare a template for the second reaction (Figure 12.2). The first PCR profile was obtained as follows: an initial denaturation step at 94 °C for 5 min, followed by 35 cycles of 94 °C for 30 s,

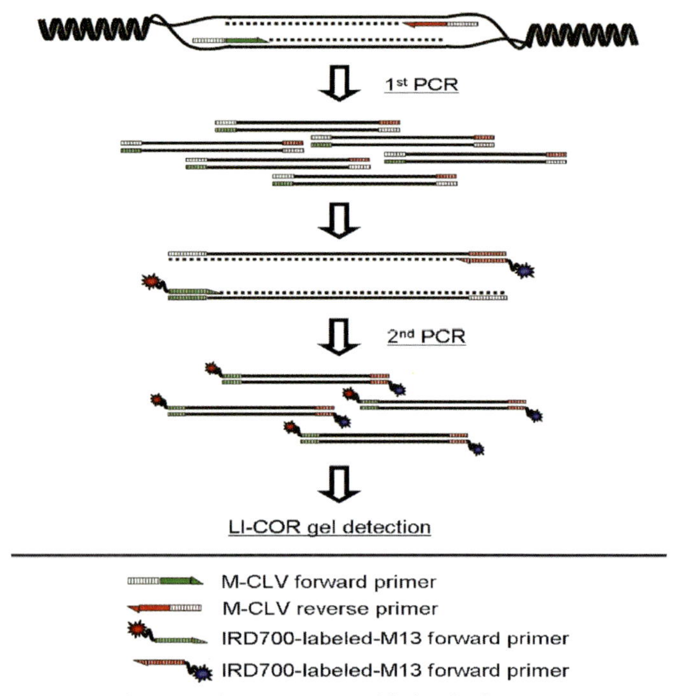

Figure 12.2 The targeted sequence is amplified in the first PCR reaction using M13-adapter-tailed primers (M-CLV forward and reverse), and serves as template in a second PCR using M13 forward and reverse universal primers that are labeled with IRD700 and IRD800 respectively. The final PCR product is subsequently heteroduplexed and cut using the ENDO1 enzyme and subjected to gel electrophoresis analysis.

T_M°C for 30 s, 72 °C for 1.5 min (time depends on fragment length, 1 min/kbp), and an extension time of 72 °C for 10 min. The forward and reverse gene-specific primers were 5′-tailed with an adaptor sequence complementary to the universal M13 forward and reverse primers (see Table 12.1 for primer sequences, note that a four-base sequence 'CAGT' was added to the 5′ end of each tailed oligo to facilitate later cloning and manipulation of the amplified DNA product).

The second PCR reaction was performed as described above for TILLING using 0.2 μM M13 forward and reverse universal primers labeled with the IRD700 and 800 fluorescent labels respectively, 2 μl DNA template from the first PCR product, 1× ExTaq buffer (Mg^{2+} plus), 0.5 Units of Ex Taq polymerase (TAKARA BIO INC., Madison, WI) and 0.2 mM dNTPs in a total reaction volume of 20 μl. The PCR product was subjected to the heteroduplex formation steps before gel analysis.

12.2.3.4 **Endonuclease Digestion and Purification of the Amplified DNA**

Following PCR amplification and heteroduplex formation, samples were digested with the mismatch repair enzyme ENDO1 ([14]; Serial Genetics, Every, France).

Table 12.1 Oligonucleotide sequences used in M13-tailed PCR amplifications for TILLING.

Gene name	Primer name	Primer sequence (5′–3′)	Size (bp)	T_m (°C)	Amplicon size (bp)
M13 primers	M-13 F	CACGACGTTGTAAAACGAC	19	49	—
	M-13 R	GGATAACAATTTCACACAGG	20	48	
GmClavata1A Primers	CLV-F	GCAGTTCCGTCAGGGATTTTCAAG	24	57	1485
	CLV-R	TACTGCTGCATCCGACGGCTGAGA	24	61	
M13-tailed amplicons	M-CLV-F	CAGTCACGACGTTGTAAAACGACATG CAGTTCCGTCAGGGATTTTCAAG	49	70	
	M-CLV-R	CAGTGGATAACAATTTCACACAGGATTA CTGCTGCATCCGACGGCTGAGA	50	71	

Exactly 5 μl of the PCR product was cut with 1 unit of the ENDO1 enzyme and incubated in a tube containing 1× ENDO1 Buffer (1× Reaction buffer: 0.01 M HEPES pH 7.5, 0.01 M $MgSO_4$, 0.002% (v/v) Triton X-100, 0.2 μg/ml BSA and 0.01 M KCl) in a total volume of 30 μl for 25 min at 42 °C. The digestion reactions were stopped by adding 5 μl of 75 mM EDTA to each tube. The DNAs were purified using a Sephadex G50 separation column (Amersham Biosciences AB, Uppsala, Sweden) as follows: using a multi-channel pipette, the total volume of each sample was deposited on top of a water-swelled Sephadex column set in a Multiscreen-HV 0.45 μm Durapore plate (Millipore Corporation, Bedford, MA). The eluants were collected in a 96-well collector plate (Millipore Corporation, Bedford, MA) containing 5 μl of formamide loading dye after 2 min centrifugation at 1200 rpm (290 × g) in an 5810/5810R centrifuge (Eppendorf, Netheler-Hinz, Germany). The eluants were concentrated to about 4 to 5 μl using a CentriVap Concentrator (LABCONCO Corporation, Kansas City, MO) for 40–60 min at 65 °C.

12.2.3.5 Gel Electrophoresis and Image Analysis

Prior to gel electrophoresis, the samples were heated at 95 °C for 5–10 min and placed on ice for 15 min. The samples were loaded by capillarity absorption onto 100-tooth membrane paper comb (The Gel Company, San Francisco, CA) using a membrane comb-loading tray (The Gel Company, San Francisco, CA).

The 100-tooth comb containing samples was inserted between the gel-containing plates until contact with the polyacrylamide gel was achieved. Sample separation was conducted by electrophoresis using a 6.5% (w/v) polyacrylamide denaturing gel in 1× TBE buffer. Electrophoresis conditions were as follows: voltage 1500 V, current 35 mA, power 35 W and temperature 50 °C; for 3–5 h using a Li-Cor sequencing machine 4200 or 4300S (LI-COR, Lincoln, NE). Images were collected in TIFF format and were analyzed visually using Adobe Photoshop software (Adobe Systems Inc., San Jose, CA) (Figure 12.3). Semi-automated programs such as the GelBuddy have been developed by several laboratories involved in TILLING and Eco-TILLING to assist with gel image analysis [15]. However, manual editing is always required for

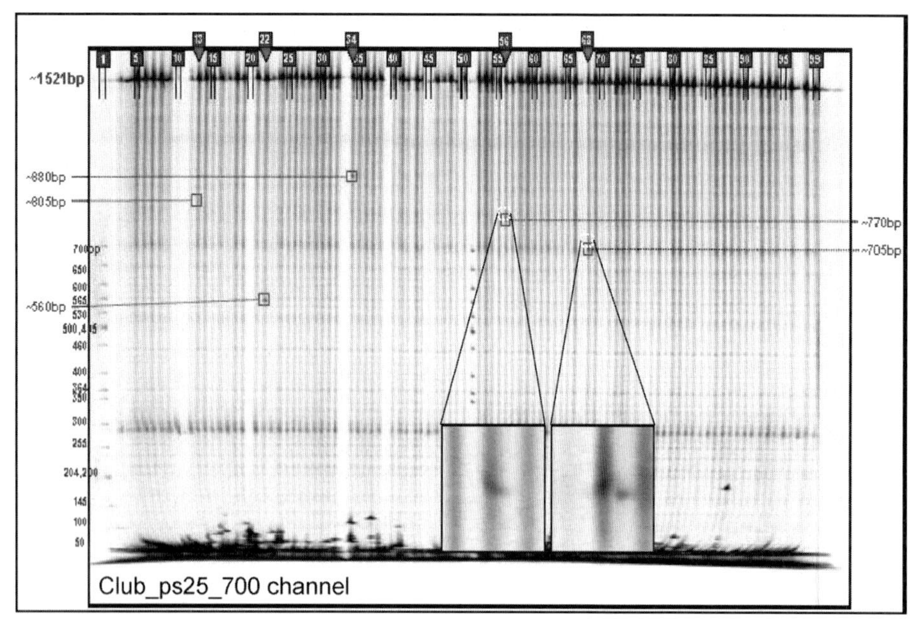

Figure 12.3 The mutagenized populations were screened for mutations induced by EMS using TILLING in eightfold-pooled DNA (each lane contains DNA from eight different individuals). Sample separation was achieved by electrophoresis using a 6.5% (w/v) polyacrylamide denaturing gel in 1 × TBE buffer. Electrophoresis run conditions were; voltage 1500 V, current 35 mA, power 35 W and temperature 50 °C; for 3–5 h using a Li-Cor sequencing machine 4300S (LI-COR, Lincoln, NE). Images were collected in the TIFF format and were analyzed visually using Adobe Photoshop software (Adobe Systems Inc.). The IR Dye 700 gel image from a 96-lane TILLING assay for mutations in a 1521-bp amplicon of the soybean GmClavata1a gene is shown. Bands corresponding to five mutations detected during this screen are boxed in red; two sections are magnified at the bottom of the image. The molecular weight of each mutant band is shown in offsets (left and right).

accurate identification of mutations and follow-up analysis. Once a candidate mutation is identified in the pool, wild-type DNA was added to each member of the pool separately and assayed by TILLING to identify the individual carrying the mutation, this step is known as 'deconvolution'.

12.3 Applications of TILLING to Soybean

12.3.1 Mutation Discovery, Density and Distribution in two Mutagenized Soybean Populations

The total number of identified mutations divided by the total number of base pairs screened provided an estimate of the mutation rate in each soybean population assayed using TILLING. The average mutation frequency of the mutagenized Forrest

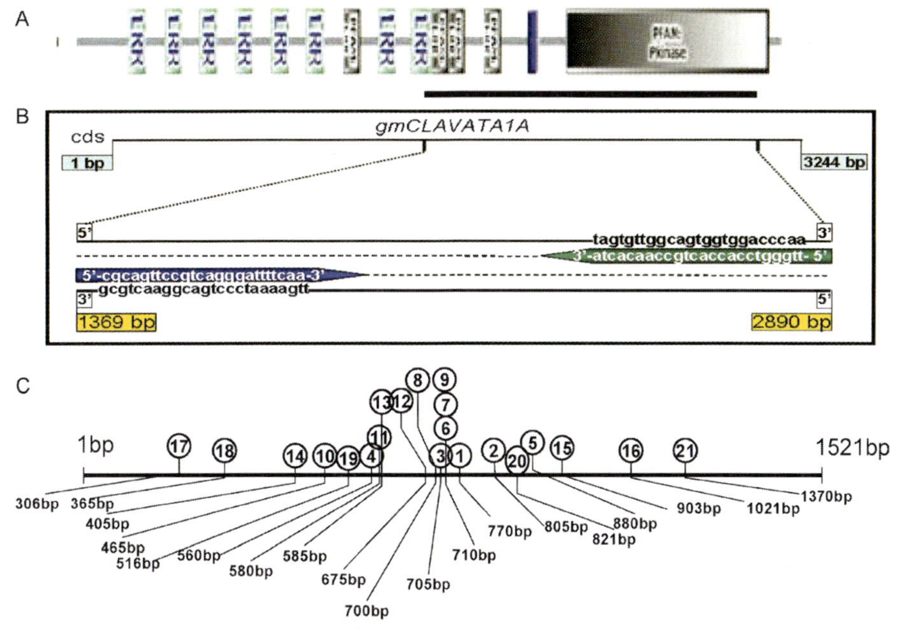

Figure 12.4 A graphic map representing all mutations detected during a TILLING assay for mutations in the 1521-bp amplicon of the soybean *GmClavata1a* gene; a total of 21 mutations were identified in 1536 soybean M2 families, the mutations were randomly distributed over the 1.5-kb amplicon, the mutations are mapped on the tested amplicon to illustrate the distribution of mutations (C). Similar graphics could be obtained using the automated program PARSESNP where triangles pointing at the sequence indicate the location of each mutation. A protein structure prediction model is presented at the top (A) of the image to show the targeted amplicon, the primers used during the screen are shown in the mid-section of the image (B).

population was estimated to be one mutation per 150 kbp, while the average mutation rate for the Williams82 population was estimated to be one mutation per 210 kbp. Although 90% of the changes induced by EMS were G/C to A/T transitions, these changes had more or less the same mutational effect, with 5% truncation, 60% missense and 35% silent mutations.

EMS treatment was expected to induce random changes throughout the genome. To assess the effect of EMS on soybean DNA, the mutation distribution was surveyed within each tested amplicon in the soybean TILLING project. EMS-induced mutations in soybean were found to be randomly distributed and covered the whole length of each amplicon tested. Figure 12.4 shows an example of mutation distribution analysis using an amplicon of the *GmClavata1A* gene in soybean. The figure shows that mutations in the first hundred base pairs neighboring the TILLING primers site were under-represented. The interpretation of this phenomenon is not linked to the nature of the mutagen but to the limitations of the detection method used. Small fragments are difficult to detect in the TILLING gels in the area covering small-sized fragments at the bottom of the gel. PCR artifacts from random mispriming which appear at the same positions in both channels could be mistaken for mutations at the bottom of the gel. Therefore, our cut-off for gel analysis was about 150 bp from the bottom of the gel. Another explanation may be linked to the amount of priming DNA sequences required by the ENDO1 mismatch enzymes to bind and cut the mismatch DNA sequences.

12.3.2
Confirmation and Segregation Patterns of TILLING Mutations in Soybean

A web-based resource called PARSESNP, which automatically analyzes mutations discovered after TILLING was used for mutation analysis. The graphical and tabular output from PARSESNP allows the user to visualize the identified mutation and its possible effect on the gene product. Missense changes are provided by PARSESNP, in which mutations predicted to be damaging to the protein have positive scores based on a scoring matrix generated from the protein conservation model. The restriction sites either gained or lost because of the induced polymorphism are listed. These sites can serve as a tool for downstream genotyping applications. Once the sequence information is completed, the identified mutant is isolated from the population for phenotypic analysis.

It is a step forward to be able to identify EMS-induced mutations in soybean; however, it is very important to confirm the presence of the identified mutation in the seed stock and analyze the segregation pattern of the mutation in the progeny in order to determine the zygosity of each plant. To do this, dCAPs primers [16] flanking the mutation sites can be used to amplify the DNA of the M3 and M4 plants of identified mutant seed stocks. PCR products are then digested with the appropriate restriction enzyme and analyzed by gel electrophoresis (Figures 12.5 and 12.6). Using this approach, the presence of a silent mutation Q263= and a nonsense mutation Q263* in a soybean leucine-rich repeat receptor-like kinase (LRR-RLK) gene (a candidate gene for *GmRhg4* = *Glyine max* resistance to *Heterodera glycines*, the soybean cyst nematode) was confirmed by genotyping individual plants from the seed stock to follow the segregation pattern of the mutation [17].

While in the majority of cases it is easy to follow the segregation pattern of an identified mutation, identical or near identical copies of genes known as homeologs and paralogs found mostly in polyploid genomes may pose an extra challenge. This was the case for the *LRR-RLK* candidate gene sequence isolated during positional cloning of the SCN resistance gene, *GmRhg4* [17]. Within the SCN resistance locus on soybean linkage group A2, the analysis of the recombinant NILs and the genomic sequence data indicated an *LRR-RLK* gene sequence as a candidate gene for resistance. To assess gene function in SCN resistance, the Forrest (SCN-resistant cultivar) TILLING population was screened to identify mutations within the *LRR-RLK* gene sequence [17]. The most interesting mutation identified was the GmRhg4 Q263* nonsense mutation which should result in a truncated protein lacking part of the leucine-rich repeat domain, the transmembrane domain, and the predicted kinase domain. An individual plant homozygous for this mutation should not make a functional LRR-RLK protein. The M2 plant for the Q263* mutation was determined

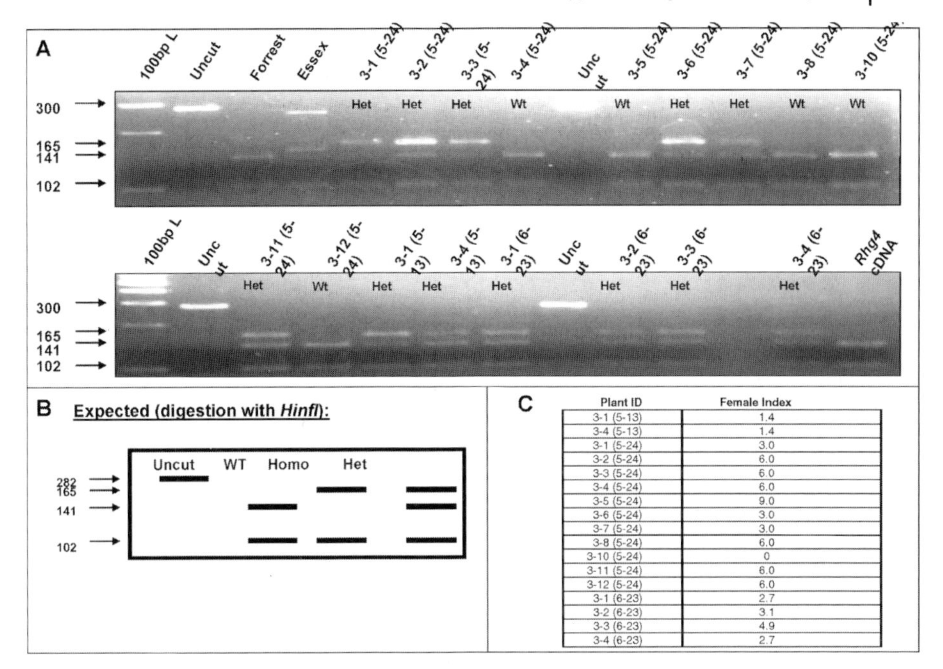

Plant ID	Female Index
3-1 (5-13)	1.4
3-4 (5-13)	1.4
3-1 (5-24)	3.0
3-2 (5-24)	6.0
3-3 (5-24)	6.0
3-4 (5-24)	6.0
3-5 (5-24)	9.0
3-6 (5-24)	3.0
3-7 (5-24)	3.0
3-8 (5-24)	6.0
3-10 (5-24)	0
3-11 (5-24)	6.0
3-12 (5-24)	6.0
3-1 (6-23)	2.7
3-2 (6-23)	3.1
3-3 (6-23)	4.9
3-4 (6-23)	2.7

Figure 12.5 Genotyping of LRR-RLK TILLING mutant Q263* using primer set M2. (A) *HinfI* restriction digestion pattern of a 282-bp product amplified using dCAPs primers (primer set M2) from M3 soybean plants segregating for the Q263* nonsense mutation in the *GmRhg4 LRR-RLK* candidate gene for SCN resistance. (B) Expected restriction digestion pattern with *HinfI*. (C) Soybean cyst nematode female indexes of Q263* M3 mutant plants. Plants with female indexes <10% are resistant to soybean cyst nematode.

to be heterozygous, thus segregation of the mutation in the M3 plants would be expected. Of 17 plants phenotyped, all of them had female indexes below 10%, similar to wild-type Forrest (Figure 12.5C). Initial genotyping of the 17 individual plants with dCAPs primer set M2 identified 12 plants heterozygous, five plants wild-type, and no plants homozygous for the mutation. This was not the expected 1 : 2 : 1 (Wt : Het : Homo) segregation pattern (Figure 12.5A). Either the homozygous mutants displayed a lethal phenotype, or the dCAPs primers were amplifying another gene that was then masking the homozygotes. To test the second possibility these same lines were re-genotyped using the dCAPs primer set 423 designed for the Q263= mutant, to identify homozygotes (data not shown). Figure 12.6A shows that using the different set of primers, five homozygotes were uncovered in the M3 Q236* plants. After sequencing the PCR fragment generated using the first set of primers, it was determined to be a mixed amplicon containing sequences differing by 12 nucleotides that resulted in two amino acid differences within a 282-bp stretch of sequence. Interestingly, this suggested that there was at least one other copy of a gene that shows similarity to this *LRR-RLK* gene. Genome analysis by hybridization to BACs and annotation of the WGS sequence suggests there are as many as nine paralogs of the RLK at *Rhg4* and three exist in closely syntenic regions with similar neighboring

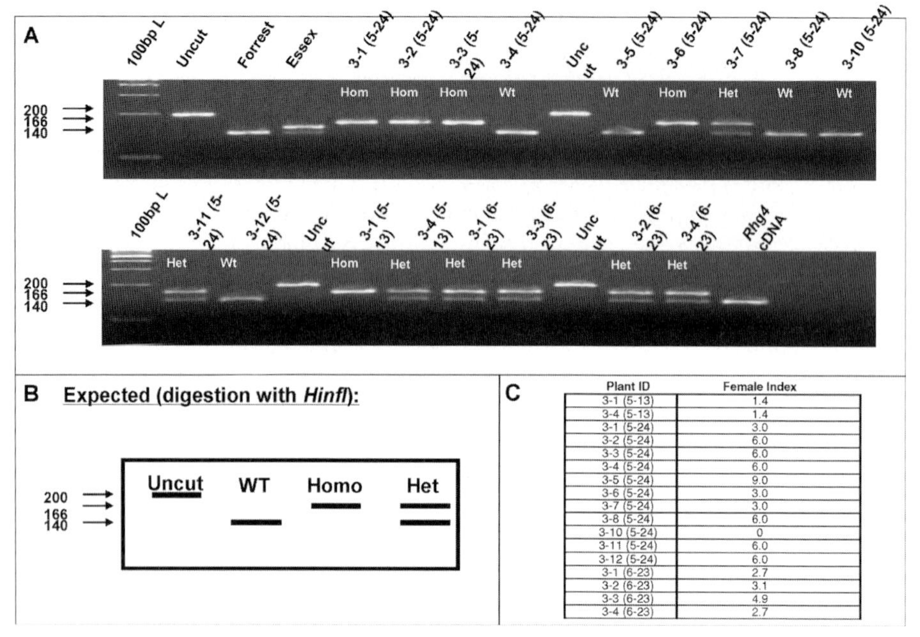

Plant ID	Female Index
3-1 (5-13)	1.4
3-4 (5-13)	1.4
3-1 (5-24)	3.0
3-2 (5-24)	6.0
3-3 (5-24)	6.0
3-4 (5-24)	6.0
3-5 (5-24)	9.0
3-6 (5-24)	3.0
3-7 (5-24)	3.0
3-8 (5-24)	6.0
3-10 (5-24)	0
3-11 (5-24)	6.0
3-12 (5-24)	6.0
3-1 (6-23)	2.7
3-2 (6-23)	3.1
3-3 (6-23)	4.9
3-4 (6-23)	2.7

Figure 12.6 Genotyping of LRR-RLK TILLING mutant Q263* using primer set 423. (A) *Hinfl* restriction digestion pattern of a 200-bp product amplified using dCAPs primers (primer set 423) from M3 soybean plants segregating for the Q263* mutation in the *GmRhg4 LRR-RLK* candidate gene for SCN resistance. (B) Expected restriction digestion pattern with *Hinfl*. (C) Soybean cyst nematode female indexes of Q263* M3 mutant plants. Plants with female indexes <10% are resistant to soybean cyst nematode.

genes. The possibility exists that these genes may be functionally redundant, so by knocking out one copy (e.g. Q263*) the other copies can still function as an SCN resistance gene. This may explain why no change in the resistance phenotype in this mutant was detected. This possibility and several others are under further investigation. This example highlights the importance of having some knowledge of gene copy number and/or gene family composition when using TILLING in soybean for functional analysis, therefore, an integrated approach to functional gene analysis is key to gene annotation in soybean and plants with more complex genomes.

12.4 Discussion and Perspectives

Gene silencing using double-stranded RNA (dsRNA)-mediated interference (known as RNA interference (RNAi; [18]) was adopted as a high throughput approach for functional analysis of plant genes [19]. Virus-based vectors known as VIGS (for Virus-Induced Gene Silencing) have been widely used to knock out the function of an endogenous gene in a transient manner. When a virus carries a sequence from a plant gene, the transcripts of both the viral and the homologous endogenous gene are degraded by a Post-Transcriptional Gene Silencing (PTGS) mechanism [20–24] and this forms the basis of VIGS. Although VIGS is easy to use and very suitable for high throughput, the limited host range and the induction of viral symptoms restrict its usefulness. Stable transgenesis has also been used to produce dsRNA in plants [25]. However, plant transformation is limited in its efficiency of producing large numbers of transgenics which reduces the attractiveness of this technology.

In the past decade, genomic resources such as insertion libraries were developed to bridge the gap between a DNA sequence and its function. Transposon and T-DNA insertions were widely used for functional annotation of plant genes [26]; insertion mutant populations were developed and the function of a gene could be assigned based on the analysis of the mutant(s) phenotype [27,28]. Although powerful in delivering valuable knockout mutants for gene function analysis, insertional mutagenesis carries the burden of being sequence target-biased; it requires a large number of mutant lines and sometimes fails to target the gene of interest.

Factors limiting the adoption of some gene knockout techniques in a major economically important crop such as the soybean include: the prevalence of gene and genome duplications; the ubiquity of gene functional homologs, orthologs and paralogs; and the low efficiency of soybean transformation. Therefore, TILLING was developed as a high throughput functional analysis tool in soybean. Although it has been useful in identifying informative mutants in plants [29–33,14], it does present several challenges that must be given serious considerations. The technology is based on random mutagenesis, which made it impossible to guarantee the desired mutation. Seed storage and viability issues are among the first issues to be solved to ensure the availability of the identified mutation for further analysis. The low to medium throughput of the technology is a challenge that several laboratories are trying to resolve by using alternative detection systems to the commonly used gel-based system.

Background mutations that are inherent to chemical mutagenesis is another major issue; although the link between the induced lesion(s) identified and their phenotype can be established by comparing the phenotypes of the identified allelic series without the need to wait until after the introgression is completed, and outcrossing is not a prerequisite for analysis [2]. It is a wise practice to backcross the identified lesions to the wild-type parent for several cycles before agronomic use.

Acknowledgments

Many thanks to the following granting agencies for supporting the Meksem laboratory and the soybean TILLING endeavors: the Illinois Missouri Biotechnology Alliance-USDA, the USDA Special Research Grants: Illinois Biotechnology: TILLING: Alternative alleles for soybean biotechnology, the United Soybean Board: SCN Biotechnology project and the USDA-NRI plant genome program: Project 2006-03573, TILLING: A Community Oriented Reverse Genetics Tool in Soybean.

References

1 Colbert, T., Till, B.J., Tompa, R., Reynolds, S., Steine, M.N., Yeung, A.T., McCallum, C.M., Comai, L. and Henikoff, S. (2001) High-throughput screening for induced point mutations. *Plant Physiology*, **126**, 480–484.

2 Henikoff, S. and Comai, L. (2003) Single-nucleotide mutations for plant functional genomics. *Annual Review of Plant Biology*, **54**, 375–401.

3 Henikoff, S., Till, B.J. and Comai, L. (2004) TILLING. Traditional mutagenesis meets functional genomics. *Plant Physiology*, **135** (2), 630–636.

4 Perry, J.A., Wang, T.L., Welham, T.J., Gardner, S., Pike, J.M., Yoshida, S. and Parniske, M. (2003) A TILLING reverse genetics tool and a web-accessible collection of mutants of the legume *Lotus japonicus*. *Plant Physiology*, **131** (3), 866–871.

5 Stemple, D.L. (2004) TILLING – a high-throughput harvest for functional genomics. *Nature Reviews. Genetics*, **5** (2), 145–150.

6 Till, B.J., Zerr, T., Comai, L. and Henikoff, S. (2006) A protocol for TILLING and Ecotilling in plants and animals. *Nature of Protocol*, **1** (5), 2465–2477.

7 Greene, E.A., Codomo, C.A., Taylor, N.E., Henikoff, J.G., Till, B.J., Reynolds, S.H., Enns, L.C., Burtner, C., Johnson, J.E., Odden, A.R., Comai, L. and Henikoff, S. (2003) Spectrum of chemically induced mutations from a large-scale reverse-genetic screen in *Arabidopsis*. *Genetics*, **164** (2), 731–740.

8 Till, B.J., Reynolds, S.H., Greene, E.A., Codomo, C.A. and Enns, L.C. (2003) Large-scale discovery of induced point mutations with high throughput TILLING. *Genome Research*, **13**, 524–530.

9 Alonso, J.M. and Ecker, J.R. (2006) Moving forward in reverse: genetic technologies to enable genome-wide phenomic screens in *Arabidopsis*. *Nature Reviews. Genetics*, **7** (7), 524–536.

10 Henikoff, J.G., Greene, E.A., Pietrokovski, S. and Henikoff, S. (2000) Increased coverage of protein families with the blocks database servers. *Nucleic Acids Research*, **28**, 228–230.

11 Ng, P.C. and Henikoff, S. (2001) Predicting deleterious amino acid substitutions. *Genome Research*, **11**, 863–874.

12 McCallum, C.M., Comai, L., Greene, E.A. and Henikoff, S. (2000) Targeted screening for induced mutations. *Nature Biotechnology*, **18**, 455–457.

13 Rozen, S. and Skaletsky, H. (2000) Primer3 on the WWW for the general users and for biologist programmers. *Methods in Molecular Biology*, **132**, 365–386.

14 Triques, K., Sturbois, B., Gallais, S., Dalmais, M., Chauvin, S., Clepet, C., Aubourg, S., Rameau, C., Caboche, M. and Bendahmane, A. (2007) Characterization of *Arabidopsis thaliana* mismatch specific endonucleases: application to mutation discovery by TILLING in pea. *Plant Journal*, **51** (6), 1116–1125.

15 Zerr, T. and Henikoff, S. (2005) Automated band mapping in electrophoretic gel images using background information. *Nucleic Acids Research*, **33** (9), 2806–2812.

16 Neff, M.M., Turk, E. and Kalishman, M. 2002 Web-based primer design for single nucleotide polymorphism analysis. *Trends in Genetics*, **18**, 613–615.

17 Liu, S., Liu, X.H., Jamai, A. Mitchum, M.G., Lightfoot, D.A. and Meksem, K. (2007) Elucidating the molecular mechanisms of soybean resistance to soybean cyst nematode (in preparation).

18 Hamilton, A., Voinnet, O., Chappell, L. and Baulcombe, D.C. (2002) Two classes of short interfering RNA in RNA silencing. *EMBO Journal*, **21**, 4671–4679.

19 Fagard, M. and Vaucheret, H. (2000) Systemic silencing signal(s). *Plant Molecular Biology*, **43** (2–3), 285–293.

20 Kumagi, M.H., Donson, J., Della-Cioppa, G., Harvey, D., Hanley, K. and Grill, L.K. (1995) Cytoplasmic inhibition of carotenoid biosynthesis with virus-derived RNA. *Proceedings of the National Academy of Sciences of the United States of America*, **92**, 1679–1683.

21 Kjemtrup, S., Sampson, K.S., Peele, C.G., Nguyen, L.V. and Conkling, M.A. (1998) Gene silencing from plant DNA carried by a geminivirus. *Plant Journal*, **14**, 91–100.

22 Ruiz, M.T., Voinnet, O. and Baulcombe, D.C. (1998) Initiation and maintenance of virus-induced gene silencing. *Plant Cell*, **10**, 937–946.

23 Burton, R.A., Gibeaut, D.M., Bacic, A., Findlay, K., Roberts, K., Hamilton, A., Baulcombe, D.C. and Fincher, G.B. (2000) Virus-induced silencing of a plant cellulose synthase gene. *Journal of Plant Cell*, **12**, 691–705.

24 Ratcliff, F., Martin-Hernandez, A.M. and Baulcombe, D.C. (2001) Technical Advance. Tobacco rattle virus as a vector for analysis of gene function by silencing. *Plant Journal*, **25**, 237–245.

25 Chuang, C.-F. and Meyerowitz, E.M. (2000) Specific and heritable genetic interference by double-stranded RNA in *Arabidopsis thaliana. Proceedings of the National Academy of Sciences of the United States of America*, **96**, 4985–4990.

26 Krysan, P.J., Young, J.C., Tax, F. and Sussman, M.R. (1996) Identification of transferred DNA insertions within Arabidopsis genes involved in signal transduction and ion transport. *Proceedings of the National Academy of Sciences of the United States of America*, **93**, 8145–8150.

27 Martienssen, R.A. (1998) Functional genomics: probing plant gene function and expression with transposons. *Proceedings of the National Academy of Sciences of the United States of America*, **95**, 2021–2026.

28 Winkler, R.G., Frank, M.R., Galbraith, D.W., Feyereisen, R. and Feldmann, K.A. (1998) Systematic reverse genetics of transfer-DNA-tagged lines of *Arabidopsis*. Isolation of mutations in the cytochrome P450 gene superfamily. *Plant Physiology*, **118**, 743–750.

29 Slade, A.J., Fuerstenberg, S.I., Loeffler, D., Steine, M.N. and Facciotti, D. (2005) A reverse genetic, nontransgenic approach to wheat crop improvement by TILLING. *Nature Biotechnology*, **23** (1), 75–81.

30 Till, B.J., Reynolds, S.H., Weil, C., Springer, N., Burtner, C., Young, K., Bowers, E., Codomo, C.A., Enns, L.C., Odden, A.R., Greene, E.A., Comai, L. and Henikoff, S. (2004) Discovery of induced point mutations in maize genes by TILLING. *BMC Plant Biology*, **4** (1), 12.

31 Till, B.J., Cooper, J., Tai, T.H., Colowit, P., Greene, E.A., Henikoff, S. and Comai, L. (2007) Discovery of chemically induced mutations in rice by TILLING. *BMC Plant Biology*, **7**, 19.

32 Mizoi, J., Nakamura, M. and Nishida, I. (2006) Defects in CTP:phosphorylethanolamine cytidyltransferase affect embryonic and postembryonic development in *Arabidopsis. Plant Cell*, **18** (12), 3370–3385.

33 Horst, I., Welham, T., Kelly, S., Kaneko, T., Sato, S., Tabata, S., Parniske, M. and Wang, T.L. (2007) TILLING mutants of *Lotus japonicus* reveal that nitrogen assimilation and fixation can occur in the absence of nodule-enhanced sucrose synthase. *Plant Physiology*, **144** (2), 806–820.

13
Transposon Tagging in Cereal Crops

Liza J. Conrad, Kazuhiro Kikuchi, and Thomas P. Brutnell

Abstract

Transposon tagging is an important tool for gene identification and characterization in many cereal crops including those with large genomes, an incomplete genome sequence, and inefficient transformation technologies. Transposon tagging resources have been most extensively developed in maize and rice but are progressing in other cereal crops such as barley and sorghum. In maize, several large collections of *Mutator* (*Mu*) transposon insertion lines have been developed for use as a reverse genetic resource. Additionally, *Activator* (*Ac*) insertions have been generated and precisely positioned on all 10 of the maize chromosomes. Using a forward genetic approach these lines can facilitate gene cloning and characterization through the creation of an allelic series, stable footprint alleles, and lineage analysis. Fundamental differences between the *Activator* and *Mutator* transposons provide unique advantages to each system. These advantages and the drawbacks of each system will be discussed along with procedures for their application in gene tagging experiments in maize. The complete genome sequence and relatively routine transformation of rice has greatly accelerated insertional mutagenesis of this important food crop. In addition to large T-DNA collections, not discussed here, extensive collections of *Ac*/*Ds*, *En*/*Spm* and *Tos17* retrotransposons have been developed. Thousands of transgenic rice insertion lines are now available to facilitate the understanding of gene function. However, the stringent regulation of transgenic maize and rice lines in plant breeding programs creates some significant hurdles in applying these technologies to the agronomic improvement of cereals. Future developments in transposon tagging utilizing newly developed sequencing technologies will help achieve the goal of near-saturation mutagenesis in rice and maize and perhaps other cereal grasses.

The Handbook of Plant Functional Genomics: Concepts and Protocols.
Edited by Günter Kahl and Khalid Meksem

ISBN: 978-3-527-31885-8

13.1 Insertional Mutagenesis in Plants

Insertional mutagenesis is an important tool for gene isolation and characterization in plants. Transposon insertion alleles are most often used to provide a molecular tag that can be used to obtain a gene sequence. Extensive transposon tagging resources have been developed for *Arabidopsis* [1–6], rice [7–12], maize [13,14], Brassica [15], tomato [16,17], and barley [18–20] using a variety of transposon families. Transposon tagging is especially critical in cereal crops that are generally recalcitrant to transformation, have incomplete genome sequences available and require a large commitment of field or greenhouse space to propagate. Exploiting endogenous transposon systems also negates the time-consuming and sometimes expensive regulatory hurdles associated with transgenic plants.

Forward genetics refers to the process of gene identification through the characterization of a mutant phenotype. In outcrossing plants such as maize, directed mutagenesis is a preferred technique for recovering insertion alleles in a gene of interest. Lines carrying an active transposon are crossed to a line carrying a reference allele and F1 progeny screened for a mutant phenotype (Figure 13.1A). Although this method requires only one generation before screening, it is limited to detection of insertions resulting in non-lethal phenotypes. This method is most effective when lines carrying highly active transposons are used as the source of mutagen. Another approach that is more feasible for many self-pollinating grasses involves developing selection schemes to identify newly transposed elements in self-pollinated progeny. DNA blot analysis is typically performed on mutant and wild-type individuals to confirm co-segregation of the mutant phenotype with a transposon insertion. Although screens of segregating F2 families require an extra generation of self-pollination before screening, it does facilitate the recovery of recessive lethal mutations that must be maintained as heterozygotes (Figure 13.1A).

Reverse genetics exploits existing sequence information to recover mutations in genes of interest. Reverse genetic screens generally do not require generations of crossing or large-scale phenotypic screens (by the end user) and thus can greatly reduce the time and expense associated with forward genetic screens. However, they

Figure 13.1 Forward and reverse genetic approaches to transposon tagging. (A) Forward genetics. (1) Directed mutagenesis of a target gene. An active transposon line is crossed by a line carrying a recessive mutant reference allele. A rare germinal insertion into the target gene will result in a mutant phenotype (yellow seedling) in the F1 generation. (2) Random mutagenesis of a target gene. An active transposon line is crossed by a line carrying a tester allele utilized to monitor transposon activity. Kernels carrying new transposition events are selected, sown and self-pollinated. The F2 generation is screened for segregation of a mutant phenotype. The triangle represents an active transposon. Upper case GENE denotes wild-type and lowercase gene is mutant. (B) Reverse genetics. Transposon insertions in a target gene are identified through either PCR screens of pooled DNA from a population of transposon lines or known gene sequences are used to BLAST transposon flanking sequence databases. After identification, researchers must carry out the appropriate regulatory steps to obtain seed for the insertion line. Typically insertions in a target gene are verified via PCR.

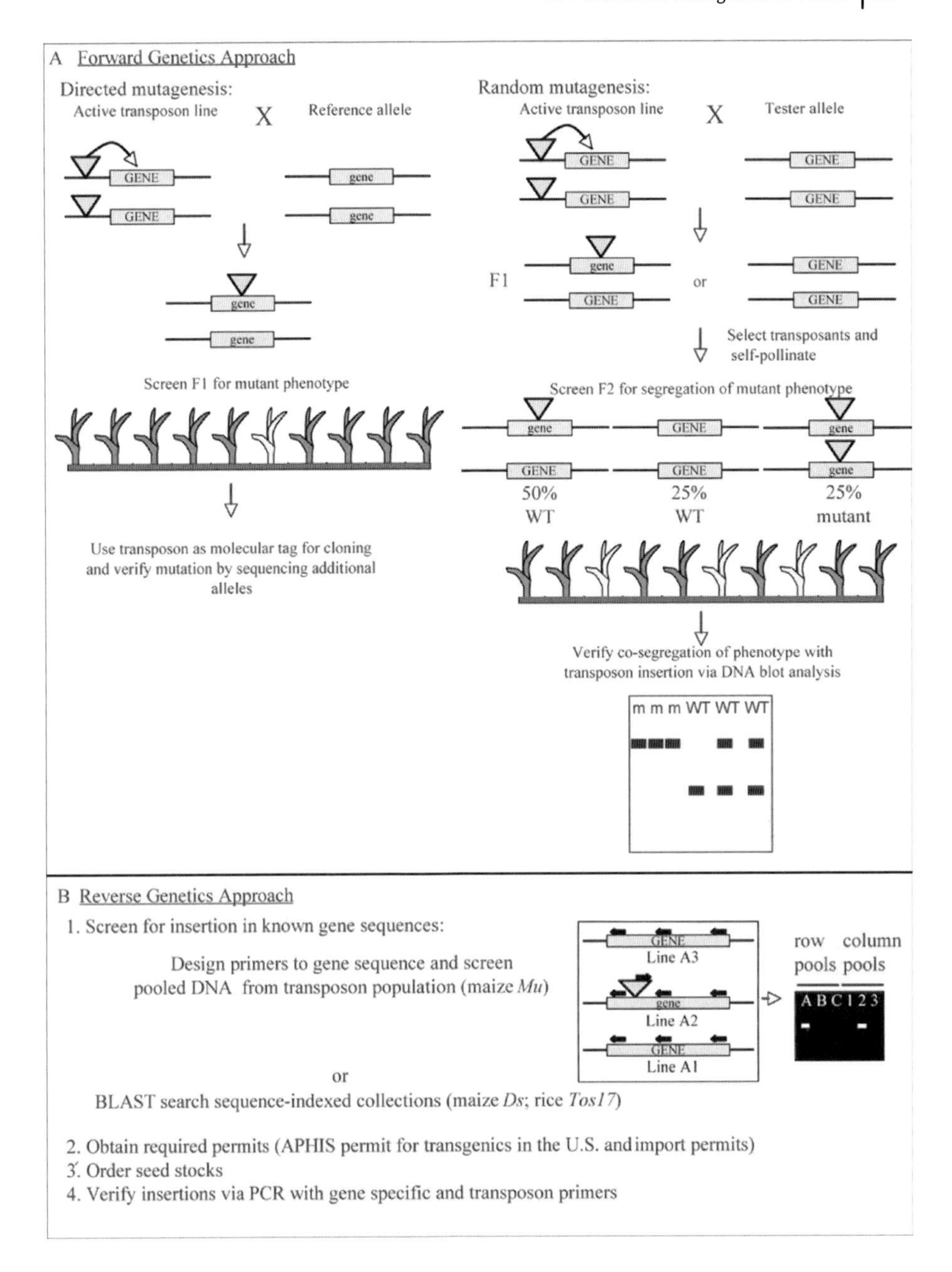

do require a candidate gene sequence and many interesting mutant phenotypes have yet to be fine-mapped in the cereals. In a general scheme, populations are established that carry multiple transposon insertions (Figure 13.1B). DNA is extracted using a pooling strategy that provides a unique plant address or an address to a small pool of individuals. Transposon insertions in a target gene can be identified through PCR screens of these DNA pools or direct sequencing of insertion sites. PCR-based screens have the obvious disadvantage of requiring an additional, often costly step of primer development and performing individualized screens on large populations with no guarantee of recovering an insertion. Sequence-indexed collections can be easily screened through database searches, but are often limited in genome coverage.

13.2 Transposon Tagging in Maize

Although map-based approaches to gene cloning are rapidly progressing [21], transposon tagging remains the principal method for gene isolation in maize [14] where numerous transposon tagging resources have been developed for use in both forward and reverse genetic screens [13,14]. These resources principally exploit two transposable element families, *Mutator* (*Mu*) and *Activator* (*Ac*)/*Dissociation* (*Ds*). Although several genes have been tagged using the *En*/*Spm* transposon family (for example [22–27]) there are no publicly available tagging resources that utilize *En*/*Spm* elements. Thus, we will focus the discussion below on the inherent differences between *Ac*/*Ds* and *Mutator* transposon families that impart unique advantages and drawbacks to the use of each in gene tagging experiments.

13.2.1 Mutator Insertional Mutagenesis

The *Mutator* (*Mu*) transposon family is the most active DNA transposon in plants with a forward mutation frequency of 10^{-3}–10^{-5} per locus per generation [28]. Active *Mu* lines can carry as many as 10 new transposition events/gamete/generation [29,30]. All *Mu* elements appear to be regulated by the autonomous *MuDR* element that contains two ORFs, *mudrA* and *mudrB* [28]. Different classes of *Mu* elements contain unrelated internal sequences, but all *Mu* elements retain conserved 220-bp terminal inverted repeats (TIR). In active *Mu* lines, the non-autonomous *Mu* elements typically outnumber the *MuDR* elements 10 : 1, thus most new *Mu* insertions are non-autonomous *Mu* elements. The behavior of *Mu* elements in maize has been studied extensively (for reviews see [28,31]). *Mu* transpositions do not show preferential reinsertion into sites linked to the donor locus [32]. However, they do insert preferentially into the 5′ region of genes [33–36]. Although several genes in maize have been cloned using *Mutator* in forward genetic approaches including *Amylose extender1* [37], *Iojap* [38], *Sugary1* [39], and *Y1* [40], *Mutator* is now more typically used in reverse genetic screens.

Mutator's high forward mutation rate and associated high copy number in the genome has facilitated the creation of large collections of *Mu*-containing lines (reviewed in [13,41]). A summary and links to these projects are located at http://www.mutransposon.org and in Table 13.1. These collections are typically surveyed by individualized PCR-based screens, though sequence-indexed collections are now being developed.

Table 13.1 *Mutator* tagging resources in maize

Mu Program	Number of lines	Cost of screening	Enrichment	URL
TUSC	45 000	Free	None	N/A – contact Bob Meeley at Pioneer Hi-Bred
MTM	43 776	$750	None	http://mtm.cshl.edu
UniformMu	31 548	Free	Seed phenotypes	http://currant.hos.ufl.edu/mutail
RescueMu	26 466	$150 per plate	None	http://www.mutransposon.org/project/RescueMu
PML	1700	Free	Photosyn-thetic mutants	http://pml.uoregon.edu

The Trait Utility System for Corn (TUSC) by Pioneer Hi-Bred Company was the first program that utilized large-scale reverse genetics using *Mutator* as the insertional mutagen [42]. This collection of 45 000 plants is likely to harbor more than 10^6 independent *Mu* insertions [14,43]. Target gene insertions are identified through PCR screens of DNA pools using *Mu*-specific and gene-specific primer pairs [42]. Positive PCR results from both column and row pools derived from independent leaf DNA samples provides a unique address for a plant carrying the insertion and reduces the number of somatic insertions recovered. Additionally, a second round of PCR is performed on five progeny from the putative insertion line to confirm heritability of the insertion prior to seed distribution. Although there is no fee for screening the TUSC collection, a Collaborative Research Agreement must be negotiated with Pioneer Hi-Bred that includes a clause for the creation of Joint Proprietary Property contract for any mutants identified from the screen. This agreement provides Pioneer with the first right of refusal to ascertain whether the mutant phenotype is of commercial relevance.

As a public sector effort, Rob Martienssen (Cold Spring Harbor Laboratory) in collaboration with Michael Freeling (University of California at Berkeley) and Danny Alexander (Syngenta) developed a collection of approximately 43 776 *Mu*-containing families known as Maize Targeted Mutagenesis (MTM). A unique aspect of this collection is that all lines have been crossed to a *MuDR* suppressor allele called *Mu*-killer in order to reduce somatic transposition of *Mu* [30]. *Mu*-killer is a novel *Mu* element that generates a double-stranded RNA targeting *Mutator* sequences for RNA-mediated silencing [44]. A multiplex pooling strategy enables screens of this collection with 1824 PCR reactions [30]. These PCR-based screens add to the expense of screening this collection and limit the detection of new events to sequenced regions of the maize genome. Nevertheless, several *Mu*-insertion alleles have been successfully retrieved from the MTM population. The mean frequency of new insertions per gene per plant has been calculated to be 2.1×10^{-5} and the probability of recovering an insertion in any given gene from this collection is estimated to be

78%, although only approximately 36% of insertions are transmitted to the next generation [30]. Several reasons have been cited for failure to recover target gene insertions, such as failure of germinal transmission through the pollen (e.g. gametophytic lethality), *Mu* target site preferences [33,35], sampling bias, and incomplete screening caused by submission of partial sequences [30]. Additionally, analysis of this collection has revealed a high frequency of *Mu*-suppressible mutations that results in the masking of the mutant phenotype in the absence of *Mutator* activity [30].

Extensive pedigree, sequence and phenotype data for this collection are available through the MTM database (MTMDB; http://mtm.cshl.edu). Currently, 43 776 ear, 8000 seedling and 2000 adult plant phenotypes have been scored. This site also processes requests for seed from insertion lines and requests for reverse genetic screening of the collection. Researchers can screen gene sequences of interest against the sequences available in the MTMDB using several search options. If the database search does not reveal an insertion in the gene of interest then the researcher can request a reverse genetic screen of the collection though the same website. The service designs primers and screens the collection. This service costs academic researchers $750 and requires a Material Transfer Agreement for seed stocks.

The Maize Gene Discovery Project has taken a unique approach to genome-wide *Mu* mutagenesis by using an engineered *Mu1* called *RescueMu*. *RescueMu* is a 4.7-kb *Mu1* element containing a pBluescript plasmid that facilitates recovery of flanking sequences by plasmid rescue in *Escherichia coli* [45]. Transposition rates from the most active lines containing both *RescueMu* and endogenous *MuDR*/*Mu* elements are from 100 to 150% (1–1.5 new germinal insertions per plant) [46]. Immortalized libraries of rescued plasmids are constructed from row and column pools and sequenced to index insertions to a subset of plants. Academic researchers can obtain library plates to identify a unique address by conducting an additional PCR-based screen. To date, 73 561 *RescueMu* genomic flanking sequences have been annotated and assembled with the maize EST collection through the MaizeGDB. However, it should be noted that the majority of these sequences represent somatic transposition events. Additionally, more than 10 000 kernel, seedling and adult mutant phenotypes have been recorded for these lines in Maize GDB (http://www.maizegdb.org/). Once an insertion in a target gene is identified seed can be obtained from the Maize Genetics Cooperative Stock Center (http://maizecoop.cropsci.uiuc.edu/) [46]. To date, one *RescueMu*-generated insertion allele has been characterized [47]. This insertion lies upstream of the *Dof1* transcription factor coding region and reduces the accumulation of *Dof1* transcript pools. The low recovery of *RescueMu* insertions characterized to date, likely reflects the fact that germinal insertions of *RescueMu* were not recovered in primary transgenic events [34]. Thus, to mobilize *RescueMu* it was necessary to maintain the construct in highly active *Mu* lines. Consequently, germinal *RescueMu* insertions represent a relatively small proportion of all *Mu* insertions in these highly active lines. Furthermore, to identify *RescueMu* insertions in genes of interest, users must first purchase a 96-well plate of DNA samples ($150) and then conduct a PCR-based screen to identify the individuals that carry the insertion allele [43]. This ‘up-front’ cost, with no guarantee of recovering a germinal insertion in the gene of interest has likely contributed to the low hit rate associated with *RescueMu*.

In an effort to reduce the mutational load and heterogeneity associated with many *Mu* populations, Don McCarty and colleagues developed the UniformMu population in the inbred, W22 [36]. By purging parental insertions from backcross populations, McCarty and colleagues selected against parental insertions and for a modest forward mutation rate of approximately 7%. The goal was to approach saturation for visible seed mutant phenotypes and over 2000 seed mutants were identified from this collection [36]. Following mobilization of the *Mu* elements, lines were selected for *Mu*-inactivation to reduce further transposition events [36]. These *Mu*-inactive lines were estimated to contain 57 germinal *Mu* elements per genome. Indeed, 89% of 106 UniformMu flanking sequences recovered were shown to be germinally inherited [36,48]. Until recently, the cost associated with generating a sequence-indexed collection of germinal insertions by sequencing libraries of MuTAIL products was prohibitive as approximately 70% of recovered sequences represented parental insertions that were shared between all libraries. However, with the advent of massively parallel sequencing technologies, sequencing cost is becoming much less of a barrier. Using 454 technologies, McCarty and colleagues have recently developed a strategy to sequence *Mu* flanking sequences from their UniformMu populations that has the potential to generate a comprehensive collection of sequence-indexed *Mu* insertions for maize (D. McCarty, personal communication).

A more specialized collection of *Mu* insertion lines was developed by Alice Barkan's group at the University of Oregon called the Photosynthetic Mutant Library (PML) (http://pml.uoregon.edu). This collection is designed to specifically identify *Mu*-induced alleles that generate non-photosynthetic phenotypes. The collection contains approximately 1700 *Mu* lines that segregate pale green, yellow, virescent, striate or high chlorophyll fluorescent leaf phenotypes. This collection can be screened by reverse genetics or by phenotype [49]. Although there is no fee associated with screening this collection, users must design primers and conduct control experiments prior to screening the population.

Although these collections of *Mu* lines have been used widely by the maize research community there are a few major drawbacks. Perhaps the greatest of which is cost. All of the publicly-funded programs require a significant financial or time investment with no guarantee of identifying an insertion in a gene of interest. This commitment often limits the number of requests that an individual investigator can place and if insertions are not recovered in initial screens, it is unlikely that additional requests will be placed. If an insertion is identified, molecular and genetic analyses of mutants are greatly complicated by the heterogeneity and high mutational load of many of these lines. Thus, several generations of backcrossing must be performed preceding any detailed phenotypic characterizations [13]. Furthermore, up to 30% of *Mu* insertions may be suppressed once *Mu* activity is lost in the line [30] preventing phenotypic characterizations of the insertion. Although it is possible to reactivate a *Mu*-suppressible insertion [50], this will require a minimum of two generations to generate lines that are homozygous for a *mop1* allele and that segregate the *Mu* insertion allele of interest.

13.2.2 Activator/Dissociation Mutagenesis

As a complement to *Mutator*-based strategies, a number of groups have exploited the well-characterized genetics of the *Ac/Ds* family of transposons to clone and characterize genes (for reviews see [13,41,51]). *Activator/Dissociation* are class II DNA transposons that belong to the *hAT* superfamily of plant transposable elements [52]. *Ac* is a 4565-bp autonomous element capable of catalyzing the transposition of itself and non-autonomous *Ds* elements [53,54]. *Ac* encodes a 3.5-kb open reading frame (ORFa) that directs the synthesis of an 807 amino acid transposase (TPase) essential for both *Ac* and *Ds* transposition [55,56]. *Ac* and *Ds* both contain 11-bp imperfect terminal repeats and approximately 240 bp of subterminal regions that are critical for TPase binding and transposition [52].

13.2.2.1 Forward Genetics

Ac offers a number of advantages for use in forward genetic approaches to gene tagging experiments in maize. Low copy number in the genome facilitates molecular and genetic characterization of *Ac/Ds*-induced alleles [55]. In addition, *Ac* copy number can be easily monitored in the genome as increasing copy number of *Ac* in the genome results in the developmental delay of *Ac* and *Ds* transposition [53]. This 'negative dosage effect' can be visualized using *Ds* insertions in genes for anthocyanin and starch biosynthesis as reporters of *Ac* activity [57,58]. These two characteristics of *Ac* provide a level of control in gene tagging experiments by avoiding lines carrying multiple *Ac* insertions while enriching selections for single transposition events.

The high frequency of somatic excision can also be exploited to confirm the identity of an *Ac*-tagged allele [59,60], and to obtain additional gene sequence from closely linked somatic transpositions [61]. It has been shown that *Ac* tends to transpose to closely linked sites [62–65]. This tendency creates the opportunity to generate an allelic series for fine mapping [61,66–70]. Furthermore, the imprecise excision of *Ac/Ds* elements can generate stable excision or 'footprint' alleles that introduce small in-frame insertions that can be used to infer functional domains, change activity or alter localization of the encoded gene product [68,71–74]. Importantly, when using endogenous transposons in such site-directed mutagenesis, the local chromatin context is maintained permitting a degree of resolution that is not possible with transgenics that alter spatial, temporal and quantitative regulation of transcription.

Two research programs have focused on developing *Ac* for use in forward genetic programs. To facilitate the use of *Ac* for tagging purposes both programs entail distributing *Ac* elements throughout the maize genome. Hugo Dooner's research group has generated over 1300 independent transpositions from *wx-m7* ⇘ *Ac* and *bz-m2* ⇘ *Ac* and sequenced DNA flanking 46 of the elements [62,65,75]. Analysis of DNA flanking the *Ac* insertions revealed that *Ac* preferentially inserts into single copy and hypomethylated DNA including a number of insertions into putative maize genes. This work suggests that *Ac* can be used as a 'gene-searching engine' to selectively disrupt coding regions of the genome [65].

The Brutnell laboratory has distributed and precisely positioned 59 *Ac* elements across all 10 of the maize chromosomes [76]. Although the majority of *Ac* insertions mapped in this study inserted into low-copy regions of the genome, many were inserted into highly repetitive DNA, contradicting previous results by Cowperthwaite *et al.* [65,76]. These differences were likely due to the use of methylation-sensitive enzymes in the Cowperthwaite study that resulted in the selective amplification of insertion sites in hypomethylated regions of the genome. These lines and others generated by members of the maize community (e.g. [77]) totaling 170 mapped *Ac* insertions are available to researchers free of charge through the Maize Genetics Cooperative Stock Center (http://maizecoop.cropsci.uiuc.edu/). Strategies have been previously described on how to best utilize *Ac* in directed and localized mutagenesis programs and are not discussed further here [78,79].

13.2.2.2 Reverse Genetics in Maize Using Ds

While the high frequency of somatic excision can be exploited in forward genetic programs, this feature of *Ac* greatly limits its use in reverse genetic strategies. Indeed, the frequency of somatic excision is so high that a 'genome walking' technique was developed to capture local sequence flanking *Ac* insertion sites [61]. To circumvent this limitation, Brutnell and colleagues Erik Vollbrecht (Iowa State University) and Volker Brendel (Iowa State University) have developed a two-component *Activator/Dissociation* reverse genetic resource. In this scheme a stabilized source of *Ac* transposase [80] is used to first mobilize *Ds* elements throughout the genome. Following a second testcross to a *Ds* reporter line, the *Ac* source is segregated away from the *Ds* and selections performed to enrich for unlinked *Ds* insertions (http://www.plantgdb.org/prj/AcDsTagging/). Families are then screened to identify novel *Ds* insertions by DNA blot analysis and flanking sequences recovered using an inverse PCR technique [76]. Importantly, all lines are maintained in a uniform W22 inbred and segregate one or two novel *Ds* insertions. The highly inbred nature of the materials and low copy number of *Ds* insertions permits near-isogenic comparisons that are often not possible with *Mutator* populations. The goal of the program is to generate 10 000 sequence-indexed *Ds* insertions distributed throughout the maize genome. To date 1037 transposed *Ds* flanking sequences have been amplified and 837 unique germinal *Ds* insertions positioned on Maize GSS assemblies. An important feature of this program is that seed is non-transgenic and available without an MTA (http://www.plantgdb.org/prj/AcDsTagging/order_instructions.php) greatly reducing the administrative hurdles which often need to be overcome to obtain rice or maize seed stocks.

13.3 Large-Scale Reverse Genetics in Rice

With the availability of the genome sequence rice, robust transformation capabilities, and support from several governmental agencies including those from the US, Japan, Korea, China and the EU, rice insertional mutagenesis has progressed rapidly in the last few years. Several insertional mutagens have been utilized in large-scale reverse genetic screens using *Ac/Ds*, *En/Spm* transposon, T-DNA insertion lines and *Tos17* retrotransposons. The most comprehensive database for rice insertion lines is available at http://orygenesdb.cines.fr/ [81]. Through this database over 140 000 insertion flanking sequences can be searched against known gene sequences to identify potential gene knock-outs. Hirochika and colleagues have estimated that anywhere from 181 000 to 460 000 insertions are required to approach saturation in rice [10]. The lower limit assumes the insertional mutagen preferentially inserts into genes and does not display an insertion site bias, whereas the upper limit assumes a random distribution of elements throughout the genome. The use of multiple insertional mutagens is one way to ensure a more uniform distribution of elements throughout the genome as each class of transposon appears to display a characteristic insertion site bias. A summary of the sequenced-indexed insertion libraries is presented in Table 13.2. Additional rice insertional databases are available through RAP-DB (http://rapdb.lab.nig.ac.jp; [82]), TIGR (http://rice.tigr.org; [83]) and RiceGE (http://signal.salk.edu/cgi-bin/RiceGE).

13.3.1 *Tos17* in Rice

One of the first programs developed for large-scale reverse genetics in rice utilized the retroelement insertion *Tos17*. *Tos17* is an endogenous copia-like retrotransposon of rice that transposes via an RNA intermediate by a copy and paste mechanism. Following the discovery that the element could be activated through tissue culture and stabilized when plants are regenerated [84], large collections of *Tos17* lines were developed [10]. There are several features of *Tos17* that are particularly attractive for mutagenesis and several genes have been tagged using this insertional mutagen [85]. *Tos17* is an endogenous retrotransposon, thus there are no regulatory restrictions associated with planting transgenic materials. Secondly, *Tos17* does not excise in regenerated rice and therefore is stably inherited. Third, the copy number of *Tos17*

Table 13.2 Summary of rice resources

Institution	Mutagen	Number of flanking sequences
CSIRO	T-DNA	787
CIRAD-INRA-IRD-CNRS, Genoplante	T-DNA	7480
National Institute of Agrobiological Sciences	*Tos17*	18 024
CerealGene Tags, European Union	*Ds*	1380
Gyeongsang National University	*Ds*	1040
Postech	T-DNA	80 259
National Center of Plant Gene Research (Wuhan)	T-DNA	15 727
National University of Singapore	*Ds*	1469
Taiwan Rice Insertional Mutant Program	T-DNA	7053
University of California at Davis	*Ds*	6878

insertions can be regulated by varying the time of tissue culture; regenerated lines typically have fewer than 10 new insertions reducing the mutational load of the lines. Moreover, the copy number of *Tos17* depends on the duration of tissue culture, making it easy to control insertion number without the need for crossing. Perhaps the biggest disadvantage of *Tos17* for large-scale mutagenesis is that it displays an insertion site preference [85] and germinal reversion events do not occur, necessitating the recovery of additional mutant alleles.

To date, nearly 50 000 *Tos17* insertion lines of rice have been generated from tissue culture-derived callus [86]. With an average copy number of 10 *Tos17* insertions/line, the population likely carries approximately 500 000 insertions [85]. Although *Tos17*-induced alleles have been identified in forward genetic screens [87], the real power of this collection is in identifying insertions in candidate genes [88–93]. Over 25 000 *Tos17* flanking sequences have been sequenced and 50 000 lines have been phenotyped [85,86], further enhancing the utility of this collection.

Like most insertional elements, *Tos17* has its limitations. An analysis of 42 000 *Tos17* flanking sequences revealed an insertion site bias into gene-dense regions of the genome such as in clusters of disease-resistance genes [85]. The element also displays a weak target site preference for ANGTT-TSD-AACNT and a slight preference for CG-rich regions of the genome. *Tos17* insertions appear to aggregate at favored sites near the distal ends of chromosomes and are refractory to insertion near pericentromeric regions. Despite these limitations, the collection of *Tos17* insertions represents a powerful tool in rice functional genomics. The *Tos17* insertion database can be searched through BLAST using known gene sequences or by mutant phenotype. Mutant seeds are available for scientific use from the Genome Resource Center at NIAS (http://www.rgrc.dna.affrc.go.jp/) at the cost of $141 or €108 per insertion line.

13.3.2
The Maize *Ac/Ds* Transposons in Rice

The maize *Activator/Dissociation* (*Ac/Ds*) transposons have been used extensively for gene tagging in rice. Several groups have developed large populations of *Ds* insertion lines [8,12,94,95]. In addition to providing gene knockouts, most of the *Ds*/T-DNA constructs were modified to serve as gene or enhancer traps [9,96,97]. A summary of *Ac/Ds* resources is shown in Table 13.3.

The largest collection of *Ds* insertion lines has been developed in Dr Venkatesan Sundaresan's laboratory at the University of California at Davis (UCD). This fluorescence-based tagging system utilizes the green fluorescent protein (GFP) and *Discosoma* sp. Red Fluorescent Protein (*DsRed*) markers to select unlinked transpositions of *Ds* in rice. *Ds* insertions in a gene of interest can be identified using BLAST searches with known gene sequences at the Sundaresan laboratory website (http://www-plb.ucdavis.edu/labs/sundar/).

CSIRO (the Commonwealth Scientific and Industrial Research Organisation) has developed a transiently-expressed transposase (TET)-mediated *Ds* insertional mutagenesis system for generating stable insertion lines in rice which will allow localized

Table 13.3 *Ac/Ds* transposon insertion collections in rice.

Group	FST	Ref.	Web site and Contact
UCD	4093	[11]	http://www-plb.ucdavis.edu/labs/sundar/ sundar@ucdavis.edu
CSIRO	611	[94]	http://www.pi.csiro.au/fgrttpub/Bug narayana.upadhyaya@csiro.au
GSNU	1072	[8]	cdhan@nongae.gsnu.ac.kr
EU-OSTID	1380	[95]	http://orygenesdb.cirad.fr/ andy.pereira@wur.nl
NUS	1469	[12]	http://www.tll.org.sg/sri.asp sri@tll.org.sg

mutagenesis of a chromosomal region [94]. In this system, *Ds* insertions are distributed throughout the genome as stably transformed T-DNA insertions. Callus tissues from single-copy *Ds*/T-DNA lines, is then transiently infected with *Agrobacterium* harboring an immobile *Ac* (*iAc*) construct. Although the frequency of linked *Ds* transposition varies widely in rice [9,12,94], it appears that many *Ds* transpositions are to linked sites. As many genes are present in tandem arrays, having a resource to mutagenize these arrays through sequential mutagenesis is particularly powerful. Indeed, this feature of *Ds* has been exploited in several organisms to mutagenize closely linked targets [66,98,99]. These lines are available for distribution under a relatively non-restrictive biological material transfer agreement with CSIRO Plant Industry (http://www.pi.csiro.au/fgrttpub/knowngene.htm).

The National University of Singapore (NUS) has also developed a two-element *Ac/Ds* gene trap system in rice by generating a collection of stable, unlinked and single-copy *Ds* insertion lines [12]. Transposition pattern data was analyzed from 4413 families carrying *Ds* elements, derived from 10 000 progeny of characterized homozygous *Ac* and *Ds* parental lines from 50 F2 families. Analysis of flanking sequences of 2057 showed that 1811 (88%) were to genomic sequences, whereas 246 (12%) were within the T-DNA. The insertions were distributed randomly throughout the genome with a bias toward gene-rich regions of the genome. As observed for *Tos17* insertions, the *Ds* insertions also displayed a tendency for clustering. Over 40 *Ds* insertions were localized to a 40-kb region on chromosome 7 suggesting an insertional preference. Interestingly, most of the transpositions to this region originated from one of the *Ds* starter lines, suggesting the use of multiple *Ds* donor platforms may compensate for these target site biases.

The Gyeongsang National University (GSNU) program utilized a regeneration procedure involving tissue culture of seed-derived calli carrying *Ac* and inactive *Ds* elements [8,9]. By analyzing 1297 *Ds*-flanking DNA sequences, a genetic map of 1072 *Ds* insertion sites was developed. The map showed that *Ds* elements transposed to each of the rice chromosomes, with preference not only near donor sites, but also to certain physically unlinked arms. To further exploit the *Ds* insertions Han and colleagues characterized the pattern of intragenic transposition of a number of stable

Ds excision alleles [100]. As observed in maize, the majority of *Ds* excision alleles generated 8- and 7-bp target site duplications [101]. However, 3-, 6-, and 9-bp insertions were also generated that resulted in in-frame insertions in coding regions. The generation of stable excision alleles is extremely important in linking a transposon insertion to a phenotype and in creating an allelic series for further genetic analysis [68].

A European consortium has also exploited *Ds* in gene tagging programs [95]. The Cereal Gene Tags, European Union (EU-OSTID) group isolated 6641 insertion sites from selected *Ac/Ds* enhancer trap lines and 250 *Ac* insertion lines using a high-throughput TAIL-PCR protocol. The collection totals 1373 unique flanking sequence tags (FSTs). These FSTs and the corresponding plant lines are publicly available through OrygenesDB database (http://orygenesdb.cirad.fr/) and from the EU consortium members.

In summary, over 8500 *Ds* elements have been distributed throughout the rice genome. These insertions serve as platforms for regional mutagenesis and as tools for defining enhancer elements and creating translational fusions through gene traps. Although the insertions display some clustering, they tend to insert into genic regions and do not display a strong insertion site bias. As *Ds* tends to move to linked sites, these lines are ideal for regional mutagenesis experiments. In general, most laboratories charge a modest fee (~$100–200) for seed shipments. However, as all *Ac/Ds* lines of rice are transgenic, an APHIS permit is required for movement through or into the US. To facilitate distribution of these materials, Sundaresan has provided step-by-step instructions on how to apply for these permits on his project website (see Figure 13.1).

13.3.3
En/Spm

The *Enhancer/Suppressor Mutator* (En/Spm) transposable element was originally identified in maize by Peterson [102] and McClintock [103] as an unstable genetic system. The autonomous 8.3-kb *En/Spm* element encodes for transposase and catalyzes the transposition of non-autonomous *I/dSpm* elements [104,105]. *En/Spm* belongs to CACTA superfamily of transposable elements and creates a 3-bp target site duplication upon insertion [52]. Although initial experiments using *En/Spm* in rice did not look promising [106], more recent studies have indicated that a two-component *Spm/dSpm* tagging system is effective in rice [11]. Sundaresan and colleagues have developed a system for insertional mutagenesis in rice using a single T-DNA construct with *Spm*-transposase and the non-autonomous *defective suppressor mutator* (*dSpm*) element [11]. Unlinked stable transpositions of *dSpm* are selected using green fluorescent protein (GFP) and *Discosoma* sp. Red Fluorescence Protein (DsRed) fluorescent markers incorporated into the constructs. Sundaresan and colleagues speculate that the size of the *dSpm* insertion in the earlier studies was likely limiting for high efficiency transposition. Over 6300 *dSpm* insertion lines and their flanking sequence are publicly available through the Sundaresan laboratory website (http://sundarlab.ucdavis.edu/rice/blast/blast.html). Insertions in a gene of interest can be identified through a BLAST search.

13.4
Ac/Ds Transposon Tagging in Barley

To demonstrate the efficacy of a two-component *Ac/Ds* tagging system in barley, Cooper *et al.* [19] developed 19 independent *Ds* insertion lines. Insertions were mapped to six of the seven barley chromosomes using the Oregon Wolfe Barley mapping population. BLAST searches and screens of barley BAC libraries revealed the majority of these *Ds* elements had inserted into predicted gene sequences. Following these preliminary experiments 100 single-copy *Ds* transposition lines were developed and elements remobilized at frequencies ranging from 11.8 to 17.1% [18]. Remobilization of *Ds* elements is critical to the success of regional mutagenesis programs. In addition, 86% of the *Ds* flanking sequences matched known or putative gene sequences [18]. These data demonstrate clearly the utility of a two-component *Ac/Ds* tagging system in barley.

Zhao *et al.* [20] have developed the most extensive collection of *Ds* insertion lines in barley to date. Single-copy *Ds* elements were reactivated by crossing to an *Ac* transposase source. A total of 101 *Ds* insertions were mapped to each of seven barley chromosomes. In order to achieve an even distribution of *tr-Ds* elements throughout the barley genome, Zhao *et al.* [20] generated transpositions from four previously positioned donor *Ds* insertions introduced to the genome through T-DNA vectors. Interestingly, this approach resulted in most *Ds* transpositions moving to sites unlinked from the donor locus. In agreement with previous studies, 72% of *tr-Ds* elements in this study reinserted in low-copy regions of the barley genome. All flanking sequences from this project are available through GenBank.

Most recently, Ayliffe *et al.* [107] has developed an activation tagging system in barley using a modified *Ds* element containing two maize *polyubiquitin* promoters termed *UbiDs*. Activation tagging involves the insertion of enhancer or promoter elements throughout the genome that can alter the expression of adjacent genes [107]. Importantly, this approach generally creates dominant, gain-of-function mutations rather than the typical loss-of-function mutations that are recessive. In this system two sets of transgenics were generated, one contained the *UbiDs* and the other with a functional transposase gene (*Ubi-transposase*). In this study, 28 independent *UbiDs* lines were generated while two *Ubi-transposase* lines were used [107]. *UbiDs* elements in these lines display insertion frequencies ranging from 0 to 52% with 36% of the new transposition going to linked sites in the genome. Approximately 9% of F2 plants contained newly transposed *UbiDs* elements with 5% of these insertions in a unique genomic location. RNA blot analysis was performed on 10 insertion lines to demonstrate transcriptional activation of the adjacent sequences. Nine of the 10 lines contained *UbiDs*-initiated transcripts confirming transcriptional activation via the *UbiDs* insertion [107]. This study demonstrates the feasibility of activation tagging in barley. This approach is especially important in large cereal genomes where gene redundancy precludes gene identification through recessive, knock-out mutations. These transposon tagging resources used in combination with the growing genomic resources in barley, including large numbers of expressed sequence tags (ESTs), single nucleotide polymorphisms (SNPs) [108], an Affymetrix Barley Genome Array, a large-insert bacterial artificial chromosome (BAC) library [109], and extensive mapping resources are making barley an attractive model system for the grasses.

13.5 Future Direction of Tagging in Cereals

13.5.1 Potential for an Endogenous Candystripe1 Tagging System in Sorghum

Although sorghum is an important grain and forage crop and has a relatively small genome that has been sequenced to 8× coverage (http://www.phytozome.net/sorghum), resources for functional genomics are lagging other grass systems. However, a potentially useful insertional mutagen has recently been characterized. *Candystripe1* (*Cs1*), the first active transposable element identified in sorghum, was cloned and shown to be a member of the CACTA family of transposable elements [110]. More recently, Carvalho *et al.* [111] demonstrated the ability of *Cs1* to transpose both somatically and germinally with a germinal excision frequency of 10%. Moreover, a screen of 800 independent germinal excisions from the *y1* locus yielded 17 mutant phenotypes. The mutant phenotypes of two out of five mutants analyzed co-segregated with a *Cs1* insertion via DNA blot analysis [111]. These studies of the endogenous *Cs1* transposon provide the groundwork for developing future tagging resources in sorghum.

13.5.2 Transposon-Mediated Deletions in Maize

In maize, certain configurations of *Ac*/*Ds* ends can undergo aberrant transposition events generating chromosome breakage and various stable chromosome rearrangements including deletions, inversions, and translocations [112]. Zhang and Peterson uncovered deletions extending >20 kb and inversions including 4.9 kb and larger that resulted from unconventional transposition reactions catalyzed by a pair of reversed *Ac* ends located 13 kb apart [113]. Furthermore, Zhang *et al.* described the creation of a novel arrangement of coding and regulatory sequences from two genes through unconventional transposition of *Ac* ends [114]. Large cereal genomes are littered with fragments of transposable element sequences offering the possibility of exploiting unconventional transposition to generate chromosomal rearrangements. Although maize transformation is still a hurdle, future engineering of *Ac*/*Ds* ends into transgenic constructs may someday be used to induce targeted deletions and rearrangements in regions of interest in the maize genome.

13.5.3
Future Tagging Resources in Rice

Transposable elements are major components of genomes and have played a significant role in natural variation and evolution. Recently, several novel active endogenous-transposons have been identified in rice that have potential for future tagging resources, such as the Long Interspersed Nuclear Element (LINE) retro-element *Karma*, the Miniature Inverted-repeat Transposable Elements (MITEs) miniature *Ping* (*mPing*), *Ping*, *Pong* and a hAT superfamily nonautonomous transposon, *nDart* [115–120]. Although most of the endogenous tranposons in rice are dormant, *Karma* and *mPing* can be activated through tissue culture conditions [84,118–120] and *mPing* elements have been mobilized following γ-irradiation [117] or through the breeding process [121–123]. Interestingly, the method of high hydrostatic pressure *in planta* has been proven to be a useful approach to mobilize *mPing* and *Pong* [122], potentially providing a rapid and cost-effective alternative to more laborious tissue-culture treatments. Given the barriers to introducing transgenic events into breeding programs, it is essential that several non-transgenic methods be developed in parallel with transgenic technologies to fully exploit the power of insertional mutagenesis in the agronomic improvement of rice and related cereals.

13.5.4
Saturation Mutagenesis

Recent advances in sequencing technology will undoubtedly have a huge impact on transposon tagging resources in maize and rice. Over the last 6 years the cost of sequencing has been reduced by 3000-fold [124]. In addition new, more efficient sequencing techniques such as 454-sequencing [125] and Solexa-sequencing [126] allow for the sequencing of hundreds of thousands to millions of fragments in a single sequencing run. Application of this inexpensive, high-throughput sequencing technology to sequencing the insertion sites of highly active transposons, such as *Mutator* in maize and *Tos17* in rice could greatly expand sequence-indexed collections of transposon insertion sites.

References

1 Kuromori, T., Hirayama, T., Kiyosue, Y., Takabe, H., Mizukado, S., Sakurai, T., Akiyama, K., Kamiya, A., Ito, T. and Shinozaki, K. (2004) A collection of 11 800 single-copy *Ds* transposon insertion lines in *Arabidopsis*. *Plant Journal*, **37**, 897–905.

2 Nishal, B., Tantikanjana, T. and Sundaresan, V. (2005) An inducible targeted tagging system for localized saturation mutagenesis in *Arabidopsis*. *Plant Physiology*, **137**, 3–12.

3 Muskett, P.R., Clissold, L., Marocco, A., Springer, P.S., Martienssen, R. and Dean, C. (2003) A resource of mapped *Dissociation* launch pads for targeted insertional mutagenesis in the *Arabidopsis* genome. *Plant Physiology*, **132**, 506–516.

4 Schneider, A., Kirch, T., Gigolashvili, T., Mock, H.P., Sonnewald, U., Simon, R., Flugge, U.I. and Werr, W. (2005) A transposon-based activation-tagging population in *Arabidopsis thaliana* (TAMARA) and its application in the identification of dominant developmental and metabolic mutations. *FEBS Letters*, **579**, 4622–4628.

5 Sundaresan, V., Springer, P., Volpe, T., Haward, S., Jones, J.D., Dean, C., Ma, H. and Martienssen, R. (1995) Patterns of gene action in plant development revealed by enhancer trap and gene trap transposable elements. *Genes & Development*, **9**, 1797–1810.

6 Ito, T., Seki, M., Hayashida, N., Shibata, D. and Shinozaki, K. (1999) Regional insertional mutagenesis of genes on *Arabidopsis thaliana* chromosome V using the *Ac/Ds* transposon in combination with a cDNA scanning method. *Plant Journal*, **17**, 433–444.

7 An, G., Jeong, D.H., Jung, K.H. and Lee, S. (2005) Reverse genetic approaches for functional genomics of rice. *Plant Molecular Biology*, **59**, 111–123.

8 Kim, C.M., Piao, H.L., Park, S.J., Chon, N.S., Je, B.I., Sun, B., Park, S.H., Park, J.Y., Lee, E.J., Kim, M.J., Chung, W.S., Lee, K.H., Lee, Y.S., Lee, J.J., Won, Y.J., Yi, G., Nam, M.H., Cha, Y.S., Yun, D.W., Eun, M.Y. and Han, C.D. (2004) Rapid, large-scale generation of *Ds* transposant lines and analysis of the *Ds* insertion sites in rice. *Plant Journal*, **39**, 252–263.

9 Chin, H.G., Choe, M.S., Lee, S.H., Park, S.H., Koo, J.C., Kim, N.Y., Lee, J.J., Oh, B.G., Yi, G.H., Kim, S.C., Choi, H.C., Cho, M.J. and Han, C.D. (1999) Molecular analysis of rice plants harboring an *Ac/Ds* transposable element-mediated gene trapping, system. *Plant Journal*, **19**, 615–623.

10 Hirochika, H., Guiderdoni, E., An, G., Hsing, Y.I., Eun, M.Y., Han, C.D., Upadhyaya, N., Ramachandran, S., Zhang, Q., Pereira, A., Sundaresan, V. and Leung, H. (2004) Rice mutant resources for gene discovery. *Plant Molecular Biology*, **54**, 325–334.

11 Kumar, C.S., Wing, R.A. and Sundaresan, V. (2005) Efficient insertional mutagenesis in rice using the maize *En/Spm* elements. *Plant Journal*, **44**, 879–892.

12 Kolesnik, T., Szeverenyi, I., Bachmann, D., Kumar, C.S., Jiang, S., Ramamoorthy, R., Cai, M., Ma, Z.G., Sundaresan, V. and Ramachandran, S. (2004) Establishing an efficient *Ac/Ds* tagging system in rice: large-scale analysis of *Ds* flanking sequences. *Plant Journal*, **37**, 301–314.

13 Brutnell, T.P. (2002) Transposon tagging in maize. *Functional & Integrative Genomics*, **2**, 4–12.

14 Walbot, V. (2000) Saturation mutagenesis using maize transposons. *Current Opinion in Plant Biology*, **3**, 103–107.

15 McKenzie, N. and Dale, P.J. (2004) Mapping of transposable element *Dissociation* inserts in *Brassica oleracea* following plant regeneration from streptomycin selection of callus. *Theoretical and Applied Genetics*, **109**, 333–341.

16 Healy, J., Corr, C., DeYoung, J. and Baker, B. (1993) Linked and unlinked transposition of a genetically marked *Dissociation* element in transgenic tomato. *Genetics*, **134**, 571–584.

17 Scofield, S.R., Harrison, K., Nurrish, S.J. and Jones, J.D. (1992) Promoter fusions to the *Activator* transposase gene cause distinct patterns of *Dissociation* excision in tobacco cotyledons. *Plant Cell*, **4**, 573–582.

18 Singh, J., Zhang, S., Chen, C., Cooper, L., Bregitzer, P., Sturbaum, A., Hayes, P.M. and Lemaux, P.G. (2006) High-frequency *Ds* remobilization over multiple generations in barley facilitates gene tagging in large genome cereals. *Plant Molecular Biology*, **62**, 937–950.

19 Cooper, L.D., Marquez-Cedillo, L., Singh, J., Sturbaum, A.K., Zhang, S., Edwards, V., Johnson, K., Kleinhofs, A., Rangel, S., Carollo, V., Bregitzer, P., Lemaux, P.G. and Hayes, P.M. (2004) Mapping *Ds* insertions in barley using a sequence-based approach. *Molecular Genetics and Genomics: MGG*, **272**, 181–193.

20 Zhao, T., Palotta, M., Langridge, P., Prasad, M., Graner, A., Schulze-Lefert, P. and Koprek, T. (2006) Mapped *Ds*/T-DNA launch pads for functional genomics in barley. *Plant Journal*, **47**, 811–826.

21 Bortiri, E., Jackson, D. and Hake, S. (2006) Advances in maize genomics: the emergence of positional cloning. *Current Opinion in Plant Biology*, **9**, 164–171.

22 Burr, F.A., Burr, B., Scheffler, B.E., Blewitt, M., Wienand, U. and Matz, E.C. (1996) The maize repressor-like gene *intensifier1* shares homology with the *r1*/*b1* multigene family of transcription factors and exhibits missplicing. *Plant Cell*, **8**, 1249–1259.

23 Schmidt, R.J., Burr, F.A. and Burr, B. (1987) Transposon tagging and molecular analysis of the maize regulatory locus *opaque-2*. *Science*, **238**, 960–963.

24 Tacke, E., Korfhage, C., Michel, D., Maddaloni, M., Motto, M., Lanzini, S., Salamini, F. and Doring, H.P. (1995) Transposon tagging of the maize *Glossy2* locus with the transposable element *En*/*Spm*. *Plant Journal*, **8**, 907–917.

25 Wienand, U., Weydemann, U., Niesback-Klosgen, U., Peterson, P.A. and Saedler, H. (1986) Molecular cloning of the *C2* locus of *Zea mays* – the gene coding for chalcone synthase. *Molecular and General Genetics*, **203**, 202–207.

26 Paz-Ares, J., Wienand, U., Peterson, P.A. and Saedler, H. (1986) Molecular cloning of the c locus of *Zea mays*: a locus regulating the anthocyanin pathway. *EMBO Journal*, **5**, 829–833.

27 Cone, K.C., Burr, F.A. and Burr, B. (1986) Molecular analysis of the maize anthocyanin regulatory locus *C1*. *Proceedings of the National Academy of Sciences of the United States of America*, **83**, 9631–9635.

28 Walbot, V. and Rudenko, G. (2002) *Mobile DNA II* (ed. N.L. Craig), ASM Press, Washington, DC, pp. 533–564.

29 Alleman, M. and Freeling, M. (1986) The *Mu* transposable elements of maize: evidence for transposition and copy number regulation during development. *Genetics*, **112**, 107–119.

30 May, B.P., Liu, H., Vollbrecht, E., Senior, L., Rabinowicz, P.D., Roh, D., Pan, X., Stein, L., Freeling, M., Alexander, D. and Martienssen, R. (2003) Maize-targeted mutagenesis: A knockout resource for maize. *Proceedings of the National Academy of Sciences of the United States of America*, **100**, 11541–11546.

31 Lisch, D. (2002) *Mutator* transposons. *Trends in Plant Science*, **7**, 498–504.

32 Lisch, D., Chomet, P. and Freeling, M. (1995) Genetic characterization of the *Mutator* system in maize: behavior and regulation of *Mu* transposons in a minimal line. *Genetics*, **139**, 1777–1796.

33 Dietrich, C.R., Cui, F., Packila, M.L., Li, J., Ashlock, D.A., Nikolau, B.J. and Schnable, P.S. (2002) Maize *Mu* transposons are targeted to the 5′ untranslated region of the *gl8* gene and sequences flanking *Mu* target-site duplications exhibit nonrandom nucleotide composition throughout the genome. *Genetics*, **160**, 697–716.

34 Fernandes, J., Dong, Q., Schneider, B., Morrow, D.J., Nan, G.L., Brendel, V. and Walbot, V. (2004) Genome-wide mutagenesis of *Zea mays* L. using *RescueMu* transposons. *Genome Biology*, **5**, R82.

35 Hardeman, K.J. and Chandler, V.L. (1989) Characterization of *bz1* mutants isolated from *Mutator* stocks with high and low numbers of *Mu1* elements. *Developmental Genetics*, **10**, 460–472.

36 McCarty, D.R., Settles, A.M., Suzuki, M., Tan, B.C., Latshaw, S., Porch, T., Robin, K., Baier, J., Avigne, W., Lai, J., Messing, J., Koch, K.E. and Hannah, L.C. (2005) Steady-state transposon mutagenesis in inbred maize. *Plant Journal*, **44**, 52–61.

37 Stinard, P.S., Robertson, D.S. and Schnable, P.S. (1993) Genetic isolation, cloning, and analysis of a *Mutator*-

induced, dominant antimorph of the maize *amylose extender1* locus. *Plant Cell*, **5**, 1555–1566.

38 Han, C.D., Coe, E.H., Jr. and Martienssen, R.A. (1992) Molecular cloning and characterization of *iojap* (*ij*), a pattern striping gene of maize. *EMBO Journal*, **11**, 4037–4046.

39 James, M.G., Robertson, D.S. and Myers, A.M. (1995) Characterization of the maize gene *sugary1*, a determinant of starch composition in kernels. *Plant Cell*, **7**, 417–429.

40 Buckner, B., Kelson, T.L. and Robertson, D.S. (1990) Cloning of the *y1* locus of maize, a gene involved in the biosynthesis of carotenoids. *Plant Cell*, **2**, 867–876.

41 Settles, A.M. (2005) Maize community resources for forward and reverse genetics. *Maydica*, **50**, 405–414.

42 Bensen, R.J., Johal, G.S., Crane, V.C., Tossberg, J.T., Schnable, P.S., Meeley, R.B. and Briggs, S.P. (1995) Cloning characterization of the maize *An1* gene. *Plant Cell*, **7**, 75–84.

43 Walbot, V. (2005) OBPC Symposium: Maize 2004 & Beyond: Regulation of the *MuDR/Mu* transposable elements of maize and their practical uses. *In Vitro: Journal of the Tissue Culture Association*, **41**, 374–377.

44 Slotkin, R.K., Freeling, M. and Lisch, D. (2005) Heritable transposon silencing initiated by a naturally occurring transposon inverted duplication. *Nature Genetics*, **37**, 641–644.

45 Raizada, M.N., Nan, G.L. and Walbot, V. (2001) Somatic and germinal mobility of the *RescueMu* transposon in transgenic maize. *Plant Cell*, **13**, 1587–1608.

46 Lunde, C.F., Morrow, D.J., Roy, L.M. and Walbot, V. (2003) Progress in maize gene discovery: a project update. *Functional & Integrative Genomics*, **3**, 25–32.

47 Cavalar, M., Phlippen, Y., Kreuzaler, F. and Peterhansel, C. (2008) A drastic reduction in DOF1 transcript levels does not affect C(4)-specific gene expression in maize. *Journal of Plant Physiology* (in press).

48 Settles, A.M., Holding, D.R., Tan, B.C., Latshaw, S.P., Liu, J., Suzuki, M., Li, L., O'Brien, B.A., Fajardo, D.S., Wroclawska, E., Tseung, C.W., Lai, J., Hunter, C.T., 3rd, Avigne, W.T., Baier, J., Messing, J., Hannah, L.C., Koch, K.E., Becraft, P.W., Larkins, B.A. and McCarty, D.R. (2007) Sequence-indexed mutations in maize using the UniformMu transposon-tagging population. *BMC Genomics*, **8**, 116.

49 Stern, D.B., Hanson, M.R. and Barkan, A. (2004) Genetics and genomics of chloroplast biogenesis: maize as a model system. *Trends in Plant Science*, **9**, 293–301.

50 Woodhouse, M.R., Freeling, M. and Lisch, D. (2006) Initiation, establishment, and maintenance of heritable MuDR transposon silencing in maize are mediated by distinct factors. *PLoS Biology*, **4**, e339.

51 Kunze, R., Saedler, H. and Lönnig, W.-E. (1997) *Advances in Botanical Research* (ed. J.A. Callow) Academic Press, London, Vol. **27**, 332–469.

52 Kunze, R. and Weil, C.F. (2002) *Mobile DNA* (ed. N.L. Craig) ASM Press, Washington, DC, Vol. **II**, 565–610.

53 McClintock, B. (1951) Chromosome organization and gene expression. *Cold Spring Harbor Symposia on Quantitative Biology*, **16**, 13–47.

54 McClintock, B. (1949) Mutable loci in maize. *Carnegie Institution of Washington Year Book*, **48**, 142–154.

55 Fedoroff, N., Wessler, S. and Shure, M. (1983) Isolation of the transposable maize controlling elements *Ac* and *Ds*. *Cell*, **35**, 235–242.

56 Kunze, R., Stochaj, U., Laufs, J. and Starlinger, P. (1987) Transcription of the transposable element *Activator* (*Ac*) of *Zea mays* L. *EMBO Journal*, **6**, 1555–1563.

57 Dooner, H.K. and Kermicle, J.L. (1971) Structure of the *R* tandem duplication in maize. *Genetics*, **67**, 427–436.

58 McClintock, B. (1955) Controlled mutation in maize. *Carnegie Institution of Washington Year Book*, **54**, 245–255.

59 Schultes, N.P., Brutnell, T.P., Allen, A., Dellaporta, S.L., Nelson, T. and Chen, J. (1996) *Leaf permease1* gene of maize is required for chloroplast development. *Plant Cell*, **8**, 463–475.

60 Schauser, L., Roussis, A., Stiller, J. and Stougaard, J. (1999) A plant regulator controlling development of symbiotic root nodules. *Nature*, **402**, 191–195.

61 Singh, M., Lewis, P.E., Hardeman, K., Bai, L., Rose, J.K., Mazourek, M., Chomet, P. and Brutnell, T.P. (2003) *Activator* mutagenesis of the *pink scutellum1/ viviparous7* locus of maize. *Plant Cell*, **15**, 874–884.

62 Dooner, H.K. and Belachew, A. (1989) Transposition pattern of the maize element *Ac* from the *bz-m2(Ac)* allele. *Genetics*, **122**, 447–457.

63 Greenblatt, I.M. (1984) A chromosome replication pattern deduced from pericarp phenotypes resulting from movements of the transposable element, *Modulator*, in maize. *Genetics*, **108**, 471–485.

64 Van Schaik, N.W. and Brink, R.A. (1959) Transposition of *Modulator*, a component of the variegated pericarp allele in maize. *Genetics*, **44**, 725–738.

65 Cowperthwaite, M., Park, W., Xu, Z., Yan, X., Maurais, S.C. and Dooner, H.K. (2002) Use of the transposon *Ac* as a gene-searching engine in the maize genome. *Plant Cell*, **14**, 713–726.

66 Alleman, M. and Kermicle, J.L. (1993) Somatic variegation and germinal mutability reflect the position of transposable element *Dissociation* within the maize *R* gene. *Genetics*, **135**, 189–203.

67 Athma, P., Grotewold, E. and Peterson, T. (1992) Insertional mutagenesis of the maize *P* gene by intragenic transposition of *Ac*. *Genetics*, **131**, 199–209.

68 Bai, L., Singh, M., Pitt, L., Sweeney, M. and Brutnell, T.P. (2007) Generating novel allelic variation through *Activator* (*Ac*) insertional mutagenesis in maize. *Genetics*, **175**, 981–992.

69 Moreno, M.A., Chen, J., Greenblatt, I. and Dellaporta, S.L. (1992) Reconstitutional mutagenesis of the maize *P* gene by short-range *Ac* transpositions. *Genetics*, **131**, 939–956.

70 Weil, C.F., Marillonnet, S., Burr, B. and Wessler, S.R. (1992) Changes in state of the *Wx-M5* allele of maize are due to intragenic transposition of *Ds*. *Genetics*, **130**, 175–185.

71 Giroux, M.J., Shaw, J., Barry, G., Cobb, B.G., Greene, T., Okita, T. and Hannah, L.C. (1996) A single mutation that increases maize seed weight. *Proceedings of the National Academy of Sciences of the United States of America*, **93**, 5824–5829.

72 Wessler, S.R., Baran, G., Varagona, M. and Dellaporta, S.L. (1986) Excision of *Ds* produces waxy proteins with a range of enzymatic activities. *EMBO Journal*, **5**, 2427–2432.

73 Liu, Y.H., Wang, L.J., Kermicle, J.L. and Wessler, S.R. (1998) Molecular consequences of *Ds* insertion into and excision from the helix-loop-helix domain of the maize *R* gene. *Genetics*, **150**, 1639–1648.

74 Liu, Y.H., Alleman, M. and Wessler, S.R. (1996) A *Ds* insertion alters the nuclear localization of the maize transcriptional activator *R*. *Proceedings of the National Academy of Sciences of the United States of America*, **93**, 7816–7820.

75 Dooner, H.K., Belachew, A., Burgess, D., Harding, S., Ralston, M. and Ralston, E. (1994) Distribution of unlinked receptor sites for transposed *Ac* elements from the *bz-m2*(*Ac*) allele in maize. *Genetics*, **136**, 261–279.

76 Kolkman, J., Conrad, L.J., Farmer, P.R., Hardeman, K., Ahern, K.R., Lewis, P.E., Sawers, R.J., Lebejko, S., Chomet, P. and Brutnell, T.P. (2005) Distribution of *Activator* (*Ac*) throughout the maize genome for use in regional mutagenesis. *Genetics*, **169**, 981–995.

77 Auger, D.L. and Sheridan, W. (1999) Maize stocks modified to enhance the recovery of *Ac*-induced mutations. *The Journal of Heredity*, **90**, 453–458.

78 Brutnell, T.P. and Conrad, L.J. (2003) Transposon tagging using *Activator* (*Ac*) in maize. *Methods in Molecular Biology*, **236**, 157–176.

79 Dellaporta, S.L. and Moreno, M.A. (1994) *The Maize Handbook* (eds V. Walbot and M. Freeling), Springer-Verlag, New York, pp. 219–233.

80 Conrad, L.J. and Brutnell, T.P. (2005) *Ac*-immobilized, a stable source of *Activator* transposase that mediates sporophytic and gametophytic excision of *Dissociation* elements in maize. *Genetics*, **171**, 1999–2012.

81 Droc, G., Ruiz, M., Larmande, P., Pereira, A., Piffanelli, P., Morel, J.B., Dievart, A., Courtois, B., Guiderdoni, E. and Perin, C. (2006) OryGenesDB: a database for rice reverse genetics. *Nucleic Acids Research*, **34**, D736–D740.

82 Ohyanagi, H., Tanaka, T., Sakai, H., Shigemoto, Y., Yamaguchi, K., Habara, T., Fujii, Y., Antonio, B.A., Nagamura, Y., Imanishi, T., Ikeo, K., Itoh, T., Gojobori, T. and Sasaki, T. (2006) The Rice Annotation Project Database (RAP-DB): hub for *Oryza sativa* ssp. *japonica* genome information. *Nucleic Acids Research*, **34**, D741–D744.

83 Ouyang, S., Zhu, W., Hamilton, J., Lin, H., Campbell, M., Childs, K., Thibaud-Nissen, F., Malek, R.L., Lee, Y., Zheng, L., Orvis, J., Haas, B., Wortman, J. and Buell, C.R. (2007) The TIGR Rice Genome Annotation Resource: improvements and new features. *Nucleic Acids Research*, **35**, D883–D887.

84 Hirochika, H., Sugimoto, K., Otsuki, Y., Tsugawa, H. and Kanda, M. (1996) Retrotransposons of rice involved in mutations induced by tissue culture. *Proceedings of the National Academy of Sciences of the United States of America*, **93**, 7783–7788.

85 Miyao, A., Tanaka, K., Murata, K., Sawaki, H., Takeda, S., Abe, K., Shinozuka, Y., Onosato, K. and Hirochika, H. (2003) Target site specificity of the *Tos17* retrotransposon shows a preference for insertion within genes and against insertion in retrotransposon-rich regions of the genome. *Plant Cell*, **15**, 1771–1780.

86 Miyao, A., Iwasaki, Y., Kitano, H., Itoh, J., Maekawa, M., Murata, K., Yatou, O., Nagato, Y. and Hirochika, H. (2007) A large-scale collection of phenotypic data describing an insertional mutant population to facilitate functional analysis of rice genes. *Plant Molecular Biology*, **63**, 625–635.

87 Agrawal, G.K., Yamazaki, M., Kobayashi, M., Hirochika, R., Miyao, A. and Hirochika, H. (2001) Screening of the rice viviparous mutants generated by endogenous retrotransposon *Tos17* insertion. Tagging of a zeaxanthin epoxidase gene and a novel ostatc gene. *Plant Physiology*, **125**, 1248–1257.

88 Kaneko, M., Inukai, Y., Ueguchi-Tanaka, M., Itoh, H., Izawa, T., Kobayashi, Y., Hattori, T., Miyao, A., Hirochika, H., Ashikari, M. and Matsuoka, M. (2004) Loss-of-function mutations of the rice GAMYB gene impair alpha-amylase expression in aleurone and flower development. *Plant Cell*,**16**, 33–44.

89 Katou, S., Kuroda, K., Seo, S., Yanagawa, Y., Tsuge, T., Yamazaki, M., Miyao, A., Hirochika, H. and Ohashi, Y. (2007) A calmodulin-binding mitogen-activated protein kinase phosphatase is induced by wounding and regulates the activities of stress-related mitogen-activated protein kinases in rice. *Plant & Cell Physiology*, **48**, 332–344.

90 Moon, S., Jung, K.H., Lee, D.E., Lee, D.Y., Lee, J., An, K., Kang, H.G. and An, G. (2006) The rice FON1 gene controls vegetative and reproductive development by regulating shoot apical meristem size. *Molecules and Cells*, **21**, 147–152.

91 Sakamoto, T., Miura, K., Itoh, H., Tatsumi, T., Ueguchi-Tanaka, M., Ishiyama, K., Kobayashi, M., Agrawal, G.K., Takeda, S., Abe, K., Miyao, A.,

Hirochika, H., Kitano, H., Ashikari, M. and Matsuoka, M. (2004) An overview of gibberellin metabolism enzyme genes and their related mutants in rice. *Plant Physiology*, **134**, 1642–1653.

92 Takano, M., Kanegae, H., Shinomura, T., Miyao, A., Hirochika, H. and Furuya, M. (2001) Isolation and characterization of rice phytochrome A mutants. *Plant Cell*, **13**, 521–534.

93 Wong, H.L., Sakamoto, T., Kawasaki, T., Umemura, K. and Shimamoto, K. (2004) Down-regulation of metallothionein, a reactive oxygen scavenger, by the small GTPase OsRac1 in rice. *Plant Physiology*, **135**, 1447–1456.

94 Upadhyaya, N.M., Zhu, Q.H., Zhou, X.R., Eamens, A.L., Hoque, M.S., Ramm, K., Shivakkumar, R., Smith, K.F., Pan, S.T., Li, S., Peng, K., Kim, S.J. and Dennis, E.S. (2006) *Dissociation* (*Ds*) constructs, mapped *Ds* launch pads and a transiently-expressed transposase system suitable for localized insertional mutagenesis in rice. *Theoretical and Applied Genetics*, **112**, 1326–1341.

95 van Enckevort, L.J., Droc, G., Piffanelli, P., Greco, R., Gagneur, C., Weber, C., Gonzalez, V.M., Cabot, P., Fornara, F., Berri, S., Miro, B., Lan, P., Rafel, M., Capell, T., Puigdomenech, P., Ouwerkerk, P.B., Meijer, A.H., Pe, E., Colombo, L., Christou, P., Guiderdoni, E. and Pereira, A. (2005) EU-OSTID: a collection of transposon insertional mutants for functional genomics in rice. *Plant Molecular Biology*, **59**, 99–110.

96 Eamens, A.L., Blanchard, C.L., Dennis, E.S. and Upadhyaya, N.M. (2004) A bidirectional gene trap construct suitable for T-DNA and *Ds*-mediated insertional mutagenesis in rice (*Oryza sativa* L.). *Plant Biotechnology Journal*, **2**, 367–380.

97 Ito, Y., Eiguchi, M. and Kurata, N. (2004) Establishment of an enhancer trap system with *Ds* and *GUS* for functional genomics in rice. *Molecular Genetics and Genomics: MGG*, **271**, 639–650.

98 Jones, D.A., Thomas, C.M., Hammond-Kosack, K.E., Balint-Kurti, P.J. and Jones, J.D. (1994) Isolation of the tomato *Cf-9* gene for resistance to *Cladosporium fulvum* by transposon tagging. *Science*, **266**, 789–793.

99 Tantikanjana, T., Mikkelsen, M.D., Hussain, M., Halkier, B.A. and Sundaresan, V. (2004) Functional analysis of the tandem-duplicated P450 genes SPS/BUS/CYP79F1 and CYP79F2 in glucosinolate biosynthesis and plant development by *Ds* transposition-generated double mutants. *Plant Physiology*, **135**, 840–848.

100 Park, S.J., Piao, H.L., Xuan, Y.H., Park, S.H., Je, B.I., Kim, C.M., Lee, E.J., Park, S.H., Ryu, B., Lee, K.H., Lee, G.H., Nam, M.H., Yeo, U.S., Lee, M.C., Yun, D.W., Eun, M.Y. and Han, C.D. (2006) Analysis of intragenic *Ds* transpositions and excision events generating novel allelic variation in rice. *Molecules and Cells*, **21**, 284–293.

101 Scott, L., LaFoe, D. and Weil, C.F. (1996) Adjacent sequences influence DNA repair accompanying transposon excision in maize. *Genetics*, **142**, 237–246.

102 Peterson, P.A. (1953) A mutable pale green locus in maize. *Genetics*, **45**, 113–115.

103 McClintock, B. (1954) Mutations in maize and chromosomal aberrations in Neurospora. *Carnegie Institution of Washington Year Book*, **56**, 254–260.

104 Masson, P., Strem, M. and Fedoroff, N. (1991) The *tnpA* and *tnpD* gene products of the *Spm* element are required for transposition in tobacco. *Plant Cell*, **3**, 73–85.

105 Pereira, A., Cuypers, H., Gierl, A., Schwarz-Sommer, Z. and Saedler, H. (1986) Molecular analysis of the *En/Spm* transposable element system of *Zea mays*. *EMBO Journal*, **5**, 835–841.

106 Greco, R., Ouwerkerk, P.B., Taal, A.J., Sallaud, C., Guiderdoni, E., Meijer, A.H., Hoge, J.H. and Pereira, A. (2004) Transcription and somatic transposition of the maize *En/Spm* transposon system in rice. *Molecular Genetics and Genomics: MGG*, **270**, 514–523.

107 Ayliffe, M.A., Pallotta, M., Langridge, P. and Pryor, A.J. (2007) A barley activation tagging system. *Plant Molecular Biology*, **64**, 329–347.

108 Kota, R., Rudd, S., Facius, A., Kolesov, G., Thiel, T., Zhang, H., Stein, N., Mayer, K. and Graner, A. (2003) Snipping polymorphisms from large EST collections in barley (*Hordeum vulgare* L). *Molecular Genetics and Genomics: MGG*, **270**, 24–33.

109 Yu, Y., Tomkins, J.P., Waugh, R., Frisch, D.A., Kudrna, D. and Kleinhofs, R.A. (2000) A bacterial artificial chromosome library for barley (*Hordeum vulgare* L). and the identification of clones containing putative resistance genes. *Theoretical and Applied Genetics*, **101**, 1093–1099.

110 Chopra, S., Brendel, V., Zhang, J., Axtell, J.D. and Peterson, T. (1999) Molecular characterization of a mutable pigmentation phenotype and isolation of the first active, transposable element from *Sorghum bicolor*. *Proceedings of the National Academy of Sciences of the United States of America*, **96**, 15330–15335.

111 Carvalho, C.H., Boddu, J., Zehr, U.B., Axtell, J.D., Pedersen, J.F. and Chopra, S. (2005) Genetic and molecular characterization of *Candystripel* transposition events in sorghum. *Genetica*, **124**, 201–212.

112 Zhang, J. and Peterson, T. (2004) Transposition of reversed *Ac* element ends generates chromosome rearrangements in maize. *Genetics*, **167**, 1929–1937.

113 Zhang, J. and Peterson, T. (2005) A segmental deletion series generated by sister-chromatid transposition of *Ac* transposable elements in maize. *Genetics*, **171**, 333–344.

114 Zhang, J., Zhang, F. and Peterson, T. (2006) Transposition of reversed *Ac* element ends generates novel chimeric genes in maize. *PLoS Genetics*, **2**, e164.

115 Tsugane, K., Maekawa, M., Takagi, K., Takahara, H., Qian, Q., Eun, C.H. and Iida, S. (2006) An active DNA transposon *nDart* causing leaf variegation and mutable dwarfism and its related, elements in rice. *Plant Journal*, **45**, 46–57.

116 Fujino, K., Sekiguchi, H. and Kiguchi, T. (2005) Identification of an active transposon in intact rice plants. *Molecular Genetics and Genomics: MGG*, **273**, 150–157.

117 Nakazaki, T., Okumoto, Y., Horibata, A., Yamahira, S., Teraishi, M., Nishida, H., Inoue, H. and Tanisaka, T. (2003) Mobilization of a transposon in the rice genome. *Nature*, **421**, 170–172.

118 Kikuchi, K., Terauchi, K., Wada, M. and Hirano, H.Y. (2003) The plant MITE *mPing* is mobilized in anther culture. *Nature*, **421**, 167–170.

119 Jiang, N., Bao, Z., Zhang, X., Hirochika, H., Eddy, S.R., McCouch, S.R. and Wessler, S.R. (2003) An active DNA transposon family in rice. *Nature*, **421**, 163–167.

120 Komatsu, M., Shimamoto, K. and Kyozuka, J. (2003) Two-step regulation and continuous retrotransposition of the rice LINE-type retrotransposon *Karma*. *Plant Cell*, **15**, 1934–1944.

121 Naito, K., Cho, E., Yang, G., Campbell, M.A., Yano, K., Okumoto, Y., Tanisaka, T. and Wessler, S.R. (2006) Dramatic amplification of a rice transposable element during recent domestication. *Proceedings of the National Academy of Sciences of the United States of America*, **103**, 17620–17625.

122 Lin, X., Long, L., Shan, X., Zhang, S., Shen, S. and Liu, B. (2006) *In planta* mobilization of *mPing* and its putative autonomous element *Pong* in rice by hydrostatic pressurization. *Journal of Experimental Botany*, **57**, 2313–2323.

123 Shan, X., Liu, Z., Dong, Z., Wang, Y., Chen, Y., Lin, X., Long, L., Han, F., Dong, Y. and Liu, B. (2005) Mobilization of the active MITE transposons *mPing* and *Pong* in rice by introgression from wild rice

(*Zizania latifolia* Griseb). *Molecular Biology and Evolution*, **22**, 976–990.

124 Service, R.F. (2006) Gene sequencing. The race for the $1000 genome. *Science*, **311**, 1544–1546.

125 Margulies, M., Egholm, M., Altman, W.E., Attiya, S., Bader, J.S., Bemben, L.A., Berka, J., Braverman, M.S., Chen, Y.J., Chen, Z., Dewell, S.B., Du, L., Fierro, J.M., Gomes, X.V., Godwin, B.C., He, W., Helgesen, S., Ho, C.H., Irzyk, G.P., Jando, S.C., Alenquer, M.L., Jarvie, T.P., Jirage, K.B., Kim, J.B., Knight, J.R., Lanza, J.R., Leamon, J.H., Lefkowitz, S.M., Lei, M., Li, J., Lohman, K.L., Lu, H., Makhijani, V.B., McDade, K.E., McKenna, M.P., Myers, E.W., Nickerson, E., Nobile, J.R., Plant, R., Puc, B.P., Ronan, M.T., Roth, G.T., Sarkis, G.J., Simons, J.F., Simpson, J.W., Srinivasan, M., Tartaro, K.R., Tomasz, A., Vogt, K.A., Volkmer, G.A., Wang, S.H., Wang, Y., Weiner, M.P., Yu, P., Begley, R.F. and Rothberg, J.M. (2005) Genome sequencing in microfabricated high-density picolitre reactors. *Nature*, **437**, 376–380.

126 Bennett, S. (2004) Solexa Ltd. *Pharmacogenomics*, **5**, 433–438.

14
Fast Neutron Mutagenesis for Functional Genomics

Christian Rogers and Giles Oldroyd

Abstract

Improvements in sequencing technologies are allowing a rapidly expanding availability of plant genomic sequence. In this knowledge environment it is critical to generate broadly applicable reverse genetic strategies to investigate gene function *in vivo* to verify the valuable but indirect evidence provided by the descriptive genomic approaches. T-DNA insertional mutagenesis provides the major platform for reverse genetics in *Arabidopsis*, but TILLING and RNAi are also heavily utilized in this species. Unfortunately, both T-DNA insertional mutagenesis and RNAi are dependent upon efficient transformation and tissue culture methods and are therefore useful in only a minority of plant species. The use of retrotransposons in rice has proved an effective alternative to T-DNA for insertional mutagenesis that requires far fewer transformation events. However, traditional chemical and radiation-induced mutagenesis followed by high throughput detection is likely to be of increasing importance for plant functional genomics, particularly in less tractable plant species. We will describe here the development of reverse genetic methods based on fast neutron mutagenesis which provide a very efficient and low cost strategy for the recovery of knockout mutants.

14.1
Introduction

14.1.1
Advantages of Fast Neutron Mutagenesis

Fast neutron is a form of ionizing radiation commonly produced by exposure to a uranium-aluminium alloy fuel source. Exposure to fast neutrons has been shown to induce a broad range of deletions and other chromosomal mutations in plants.

The Handbook of Plant Functional Genomics: Concepts and Protocols.
Edited by Günter Kahl and Khalid Meksem

ISBN: 978-3-527-31885-8

Despite having a long history of use as a mutagen in forward genetics, fast neutron mutagenesis has not been extensively exploited in the development of reverse genetic platforms. In contemporary plant genetics gene knockout is most often achieved by insertional mutagenesis (T-DNA and transposon tagging) or gene silencing (RNAi). These techniques are described in Parts II and III of this handbook. For the majority of plant species, in which existing genomic sequence information is extensive but transformation or tissue culture based methods are not practical, the TILLING (Targeting induced local lesions in Genomes) method, described in Part II Chapter 2, is the most widely applied strategy [1]. The chemical mutagen EMS may have been favored over deletion-inducing mutagens (e.g. γ-rays, X-rays and fast neutron) for reverse genetic screening because of its ability to readily induce mutant phenotypes in forward genetic screens. This is the result of the high density of point mutations which can be induced using EMS. Mutants recovered from TILLING populations will therefore possess a very large number of non-target mutations. For the *Arabidopsis* TILLING population, conservative estimates suggest the density of mutations in exons to be $\sim$3 per Mb [2]. The impact of this is around 20–25 profoundly affected genes per EMS-mutagenized genome [3]. For *Arabidopsis* lines exposed to fast neutrons at a standard dose of 60 Gy, an average of approximately 2500 lines are required to inactivate a gene once [4]. This implies that around 10 genes are randomly deleted per line. Therefore, to achieve saturation mutagenesis, fast neutron populations need to be larger than those created using EMS. The recovered mutants, however, will possess fewer non-target mutations which will simplify downstream analysis.

EMS mutagenesis gives rise to point mutations that generally have little or no effect on the activity of the mutagenized protein. Of the alleles recovered from a TILLING population, only $\sim$5% are knockout mutations, resulting from premature stop codons [3]. In contrast fast neutron mutagenesis gives rise to deletions and hence virtually every mutagenesis event identified should represent a knockout allele. Since the identification of null alleles is often the primary goal of reverse genetics, fast neutron mutagenesis offers a more efficient method for achieving this result compared with TILLING. In addition, reverse genetics platforms based on deletions have the potential to identify mutational events that remove tandemly duplicated genes, a feature unique to this method of reverse genetics.

14.1.2
Features of Fast Neutron Mutagenesis

Characterization of deletion alleles recovered in forward screens of *Arabidopsis* [4,5] has provided us with an unbiased description of the nature of fast neutron induced mutations. One of the most thorough studies of fast neutron mutagenesis was carried out by Bruggemann *et al.* [5]. A forward genetic screen of a fast neutron mutagenized *Arabidopsis* population, consisting of 300 000 M2 plants, set out to discover the full range of deletion sizes produced at the *HY4* locus. *HY4* encodes a blue light receptor, CRY1, and mutants were easily recovered from agar plates displaying elongated hypocotyl phenotypes after 5 days under 100 μE blue light. This study isolated and characterized 20 independent alleles of *hy4*. Using Southern blot analysis, deletions ranging from 300 bp to in excess of 8 kb were identified. Several lines, confirmed as *hy4* from allelism tests, were shown to have no defect detectable by this method. These were assumed to possess small deletions although the possibility of more complex rearrangements affecting *HY4* expression could not be ruled out. This study demonstrated that irradiation with fast neutron bombardment produces a range of deletion sizes from in excess of 8 kb to not detectable by Southern analysis. The phenotypes of these 20 characterized mutants indicate that they are all null alleles.

Fast neutron based forward screens in *Arabidopsis* have produced many useful mutants (Table 14.1). A review of the mutants catalogued by the NSF *Arabidopsis* Information Resource (TAIR; www.arabidopsis.org) reveals 114 fast neutron alleles. Of the 53 sufficiently characterized, 43 (81%) were deletions and 10 were other types of mutation including combinations of insertions, deletions, substitution and rearrangements. The deletions ranged from a single base pair to a 60-kb deletion

Table 14.1 *Arabidopsis* fast neutron alleles.

Allele name	Gene	Mutation type	Locus	Size (bp)
deltappd	PPD1	Deletion	AT4G14713	60 000
zip-2	AGO7	Deletion	AT1G69440	20 000
kan-12	KAN	Deletion	AT5G16560	12 000
sgr2-10	SGR2	Deletion	AT1G31480	10 000
npr4-3	NPR4	Deletion	AT4G19660	8877
era1-2	ERA1	Deletion	AT5G40280	7500
era1-3	ERA2	Deletion	AT5G40280	7500
npr3-1	NPR3	Deletion	AT5G45110	6169
ga1-3	GA1	Deletion	AT4G02780	5000
aba2-14	ABA2	Deletion	AT1G52340	951
abi3-6	ABI3	Deletion	AT3G24650	750
spr1-1	SPR1	Deletion	AT2G03680	632
zwi-9310-7	ZWI	Insertion	AT5G65930	500
flc-3	FLC	Deletion	AT5G10140	104
ddm1-5	DDM1	Insertion	AT5G66750	82
nph4-3	NPH4	Compound	AT5G20730	59
ga1-4	GA1	Deletion	AT4G02780	14
spr1-3	SPR1	Deletion	AT2G03680	10
flc-4	FLC	Deletion	AT5G10140	7
ein5-3	AIN1	Deletion	AT1G54490	5
rar1-10	PBS2	Deletion	AT5G51700	5
brt1-9	UGT84A2	Insertion	AT3G21560	4
edm2-1	EDM2	Deletion	AT5G55390	2
sex1-5	SEX1	Deletion	AT1G10760	2
sgr2-9	SGR2	Deletion	AT1G31480	2
ddm1-6	DDM1	Deletion	AT5G66750	1
itb1-7	ITB1	Deletion	AT2G38440	1

spanning 12 genes. Strikingly, deletions spanning megabases have not been identified in *Arabidopsis*, nor to our knowledge in *Medicago* or tomato. In contrast, fast neutron-induced deletions of megabases in size appear to be common in wheat, and this may be a result of the balancing nature of this hexaploid genome that can support the removal of essential genes in such large deletions through the presence of duplicate gene copies on homeologous chromosomes [6].

14.1.3 Fast Neutron Mutagenesis for Reverse Genetics

Two large scale reverse genetic methods have been developed based on fast neutron mutagenesis. The delete-a-gene method developed in *Arabidopsis* by Li *et al.* pioneered work in this area [7,8]. More recently a platform has been developed by the authors for the model legume *Medicago truncatula* known as deletion TILLING (De-TILLING). The general scheme for both these platforms is similar: an effective dose of fast neutron is determined and a large population of mutagenized plants is established. PCR is used in both methods to identify a line carrying a deletion in a target gene. These systems are designed to allow preferential amplification of the deletion mutant, even when the line carrying the mutation in that gene is diluted in pools containing an excess of lines that are wild-type at that locus. Considering that fast neutron mutagenesis populations need to be as large as 50 000 plants, this preferential amplification of the deletion fragment is essential to allow deep pooling of M2 plants. Hence, the challenge to setting up these reverse genetic systems has been the identification of procedures that provide the deleted target a competitive advantage in the PCR amplification, even when it represents a small fraction of the total available target for those primer sets.

14.2 Methods and Protocols

14.2.1 Screening Strategies

Li *et al.* [7] devised a PCR strategy for delete-a-gene using short extension times to limit amplification of the longer wild-type sequence, allowing the deletion alleles to successfully compete for amplification. A reconstruction experiment was presented using the known *Arabidopsis* deletion mutant *ga1-3* possessing a 5-kb deletion. Flanking this deletion with primers 6.4 kb apart in wild-type, a 1.4-kb deletion allele could be amplified using a 30-s extension time in pools where one plant carrying the mutation was present with 1000 wild-type plants (Figure 14.1). Amplification of the 6.4-kb wild-type fragment was suppressed under these conditions. During screening this strategy was used to amplify deletions from mega pools of 2592 lines.

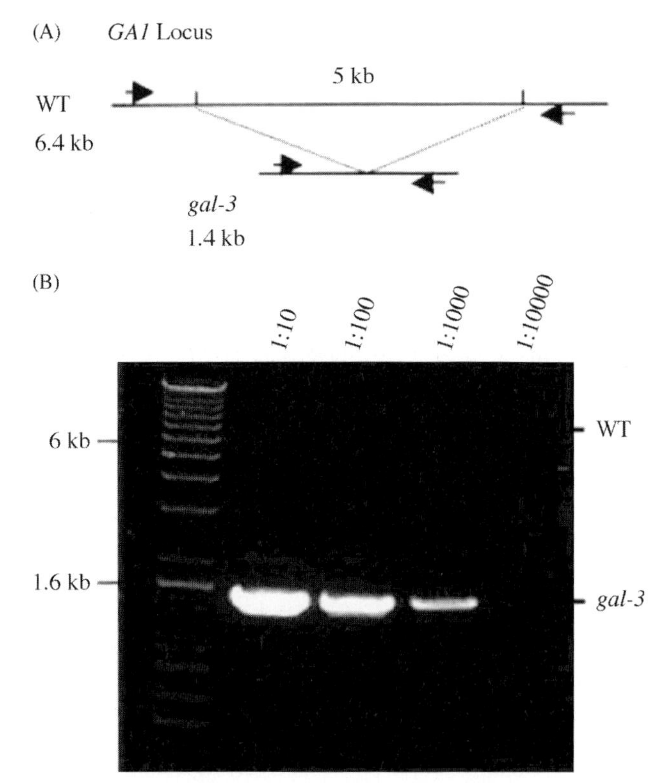

Figure 14.1 The Delete-a-gene detection strategy. PCR amplification using primers 6.4 kb apart in wild-type DNA and flanking the 5 kb *ga1-3* deletion demonstrates the detection of a mutant allele at pooling ratio of up to 1 : 1000 genomes. Amplification from the wild-type is suppressed by the 30-s extension time which only allows the amplification of the 1.4-kb deletion allele fragment.

Delete-a-gene is an effective strategy for preferentially amplifying deletion alleles in large pools but is limited in the range of detectable deletions. Small amplicons cannot be suppressed using extension time alone. Suppression of wild-type amplification is only possible using larger amplicons where the processivity of the polymerase is challenged and amplicon size becomes a limiting factor in the efficiency of the amplification. Therefore, deletions which remove only a small proportion of the targeted region are not detected by this method because they produce amplicons more similar in size to the full length product. PCR extension times alone cannot be used that allow small deletions to compete successfully for amplification with the more abundant wild-type sequences. Therefore the delete-a-gene strategy is limited to detecting large deletions and only those which remove a significant proportion of the amplified region. This presents the problem of the decreased probability of containing a large deletion within an amplicon only slightly larger in size. The reconstructed detection of the 5-kb *ga1-3* deletion by Li *et al.* [7] is based upon a deletion that removes 78% of the amplified region (Figure 14.1). The probability of randomly flanking this deletion using primers only 6.4 kb apart is low. The delete-a-gene strategy therefore relies on a broad range of large amplicon sizes and is limited in its ability to detect smaller deletions. The fact that delete-a-gene can only identify larger deletions increases the likelihood of identifying deletions that affect more than a single gene. While such an approach is advantageous for removing tandemly duplicated genes, it is limited when single gene targets are preferred, which will be the majority of cases. The ability to detect smaller deletions will increase the likelihood of discovering deletions and will generate deletions more likely to impact only a single gene.

To address these issues, an alternative detection strategy has been developed: the De-TILLING platform. High detection sensitivities for smaller deletions have been achieved through a combination of two alternative techniques for suppressing PCR amplification from wild-type sequences: restriction enzyme suppression and poison primer suppression. Restriction enzyme suppression relies upon the pre-digestion of highly complex DNA pools with a restriction enzyme which cuts once within the target sequence. This prevents a vast majority of the wild-type sequence from acting as a PCR template. This step relies on the fact that the deletion in the target gene will remove the restriction enzyme site and thus the deletion allele is protected from the restriction enzyme suppression. Wild-type target sequences escaping restriction enzyme suppression are subject to 'poison primer' suppression [9]. Poison primer suppression was first described to enhance the PCR detection of deletion mutants from pools of *Caenorhabditis elegans* treated with the chemical mutagen trimethylpsoralen (TMV) and UV light. In this strategy a third functional 'poison' primer is included in the first round of PCR. Amplification from a wild-type template leads to the production of two fragments, one full length and the other relatively short (Figure 14.2A). The shorter fragment, known as the suppressor fragment, is produced more efficiently and acts to suppress amplification of the longer fragment. Amplification from a mutant template present within the DNA pool, in which the poison primer binding site has been deleted, produces a single amplicon from the external primers. During the second round of nested PCR, the suppressor fragment, lacking one of the external primer binding sites, cannot act as a template. Only the deletion allele and wild-type allele will now be amplified. Because the production of the wild-type amplicon has been limited by competition in the first round, the mutant amplicon is able to successfully compete for amplification. The De-TILLING strategy combines restriction and poison primer suppression (Figure 14.2). A poison primer is designed adjacent to a unique restriction site within the target. A deletion has to remove both the restriction digestion site and the adjacent poison primer annealing site in order for the deletion allele to be preferentially amplified over the excess of wild-type template. A reconstruction experiment was conducted using the *nsp2* fast neutron mutant possessing a 435-bp deletion [10]. This small deletion removes only 20% of the amplified region yet is able to be detected in pools containing a 20 000-fold excess of wild-type sequences

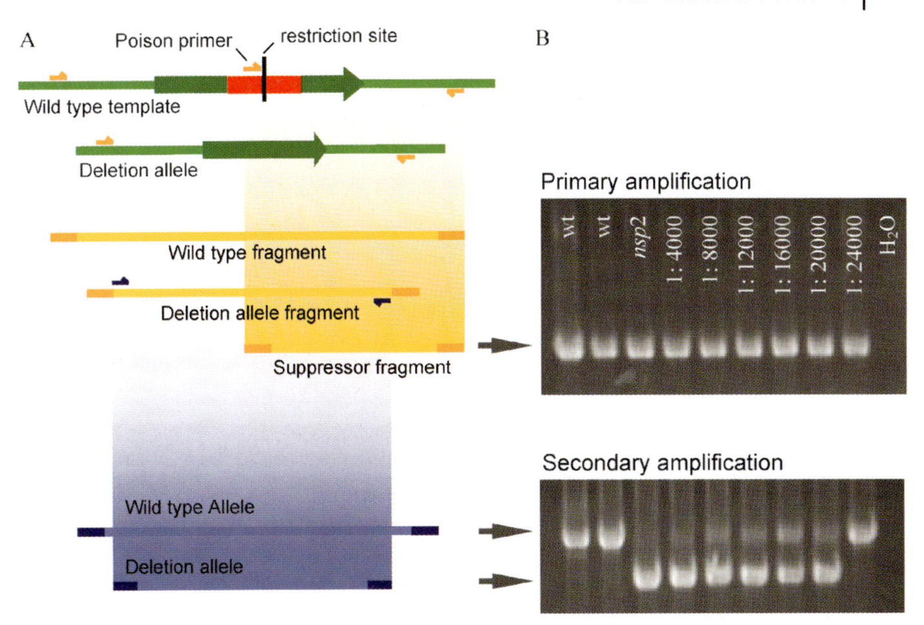

Figure 14.2 The De-TILLING deletion detection strategy. (A) Amplification from wild-type sequences in DNA pools containing a deletion allele is suppressed by template restriction and production of the suppressor fragment from the poison primer. A nested PCR reaction reveals the deletion allele. (B) A reconstruction experiment showing the amplification of the 435-bp deletion allele of *nsp2-1* in dilutions of wild-type genomic DNA. In the primary amplification only the suppressor fragment is visible. In a nested PCR the deletion allele is amplified from a pool of 20 000 wild-type genomes.

(Figure 14.2B). Hence De-TILLING has made two important improvements over delete-a-gene: it has allowed the identification of smaller deletions and it has allowed this detection at much greater pooling depths, 1:20 000 versus 1:2592 in delete-a-gene.

14.2.2
Automation

The application of automated fluorescent fragment detection strategies have been developed for De-TILLING. Because the cost of fluorescent primers is relatively high, a method has been developed using fluorescently labeled M13 primers to label all the PCR products. In this strategy, one of the first round primers is synthesized with an additional M13 sequence at the 5′ end. In the second round of the PCR, one of the nested primers is replaced with a fluorescently labeled M13 primer. The resulting fluorescently labeled PCR products can be detected and sized using the ABI3730 capillary sequencer (Figure 14.3). By combining PCR products with M13 primers variously labeled with FAM (blue), HEX (green) NED (yellow) and ROX (red), the number of capillary runs can be kept to a minimum, greatly reducing the costs and increasing the throughput of this system.

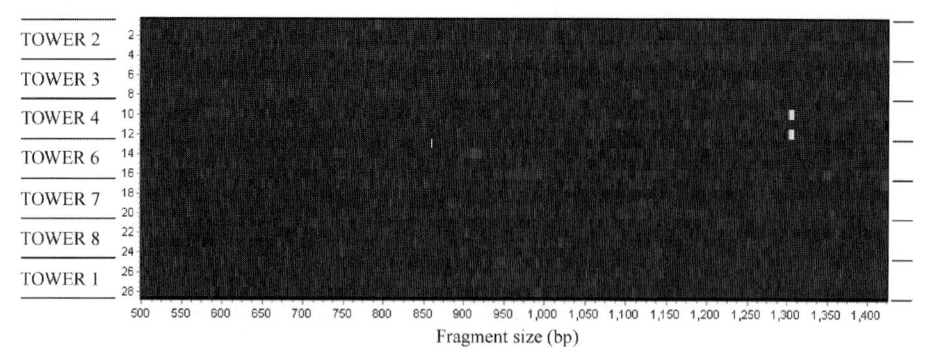

Figure 14.3 Fragment detection using capillary electrophoresis. A modification of the De-TILLING strategy uses generic fluorescently labeled primers in the second round of amplification to allow automated fragment detection. Mutants are distinguished from spurious PCR fragments as they are present as a pair of equally sized fragments within a single tower.

14.2.3
Establishing the Populations

In order to minimize the number of plants required to saturate the genome it is important to optimize the dose of fast neutrons to give the greatest number of mutations while moderating detrimental effects on the fertility of M1 plants. The gray (Gy) is a unit used to quantify fast neutron exposures and measures a quantity called absorbed dose: one gray is equal to one joule of energy deposited in one kg of a material. The dose–response to fast neutron treatment varies substantially between species. For example, 60 Gy is commonly used for *Arabidopsis*, 18–25 Gy for rice and 30–40 Gy for *Medicago* [4,5,7]. To characterize the appropriate fast neutron dose in other species, it is important to define a dose–response for each species. The measures usually used to determine mutation frequency are the M1 fertility, which should be approximately 50% and the albino frequency in the M2 [8]. Approximately 2% of M1 lines from *Arabidopsis* plants mutagenized at 60 Gy produce M2 seed segregating for albinism. For *Medicago* mutagenized at 35 Gy, the albino rate is around 2.6%. With regard to establishing an effective fast neutron dose for a new species it has previously been noted that there is no linear relationship between mutation frequency per locus and DNA content per haploid genome [4]. A clear example of this, highlighted by Koornneef *et al.* [4] is the comparable mutation frequencies per *CER* locus in barley and *Arabidopsis*. The number of plants required to achieve saturation mutagenesis using fast neutron is therefore not related to genome size and should be similar in all plants. This can be calculated from observed frequencies of deletion detection events in established reverse genetic platforms and useful populations approaching saturation are usually in excess of 50 000 lines (see Section 2.5).

When establishing these deletion-reverse genetics populations a sample of the M2 progeny of each mutagenized M1 line needs to prepared as DNA for screening.

The ideal situation is to harvest seed from individual mutagenized M1 lines and this simplifies the recovery of mutant M2 plants. The *Arabidopsis* delete-a-gene population of 51840 lines was generated in this way: seeds from individual M1 lines were collected and DNA prepared from representative M2 plants. This is different to the population structure used in TILLING, where seed and DNA is harvested from individual M2 plants. However, pooling the M1 plants can be used and this reduces the costs and effort in the establishing these populations. For the De-TILLING *Medicago* population, plants were grown as families of five M1 lines. This increases the genetic complexity of the harvested M2 seed lots but decreases the cost of generating the population by 80%. The only ramification of this strategy is that a greater number of seedlings need to be screened to recover the mutant from the M2 seed stock. The pooling strategies used for the *Medicago* population allowed this additional screening to be incorporated while still reducing the total amount of screening required.

14.2.4 **Pooling Strategies**

Pooling lines of mutants to very high depths has the obvious advantage of reducing the amount of screening required. The detection strategy employed in the TILLING method constrains the pooling depth to just eight lines. By comparison, the high detection sensitivities associated with the delete-a-gene and De-TILLING strategies allows thousands of lines to be screened simultaneously. The *Arabidopsis* delete-a-gene strategy started with DNA preparations of M2 plants from 36 M1 lines and these were combined to create sets of pools, super pools and mega pools (Figure 14.4A). Initial screening is carried out on mega pools of 2592 lines. When a mutant allele is detected screening is carried out sequentially over four additional levels of pooling to identify the individual seed lot harboring the mutation (Figure 14.4A). Individual plants from the identified line are then grown up to recover the detected deletion mutant.

A modification of this method was introduced for the De-TILLING platform which employs a three-dimensional pooling strategy. The population is segregated into tower structures consisting of five 96-well plates of DNA extractions. Each tower is pooled to create three-dimensional pools of rows, columns and plates. The pools can be screened simultaneously to identify a single seed lot within the tower in a single step. This replaces the multiple stages of screening required by the delete-a-gene strategy and makes the system more amenable to high throughput mutant detection. Initial screening of the *Medicago* De-TILLING population is carried out on half tower pools, each representing 6000 M2 plants. When a mutation is identified within a tower it can be located within a single seed lot following PCR screening of 25 pools representing the 3-D row, column and plate pools of the tower (Figure 14.4B).

A problem that arises from the use of PCR for detection is the production of spurious PCR products that can mimic genuine deletion detection events. Reciprocal half tower pools have been used in De-TILLING to provide a quality assurance step: a genuine deletion detection event requires two equal sized PCR products from two half tower

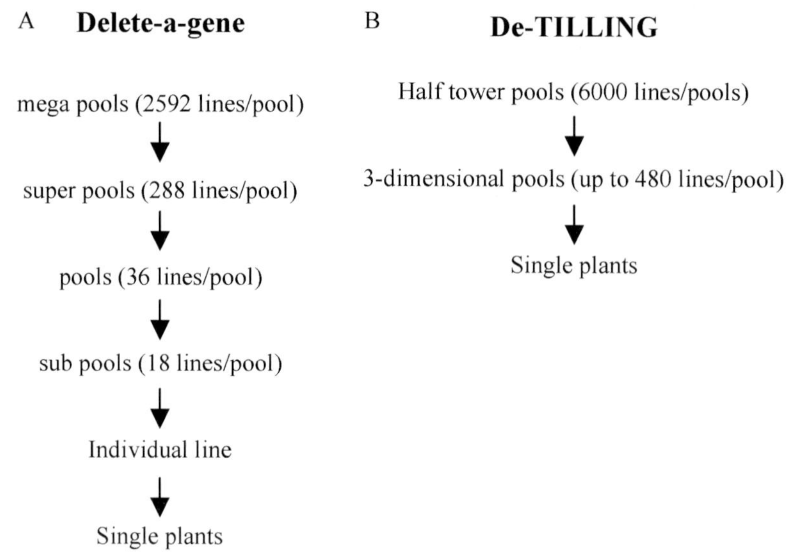

Figure 14.4 Delete-a-gene and De-TILLING pooling strategies. Flow chart of the PCR screening process for delete-a-gene (A) and De-TILLING (B). The De-TILLING strategy provides both verification of detection events using reciprocal half-tower pools and a single step dissection of positive towers facilitating high throughput screening.

pools (Figure 14.5A). Interestingly, sequencing of these spurious PCR products from the De-TILLING platform has shown that these almost invariably originate from the target sequence and are structurally identical to deletion alleles. This phenomenon was noted in *C. elegans* deletion detection platforms by Jansen *et al.* and Liu *et al.* [11,12]. Lui

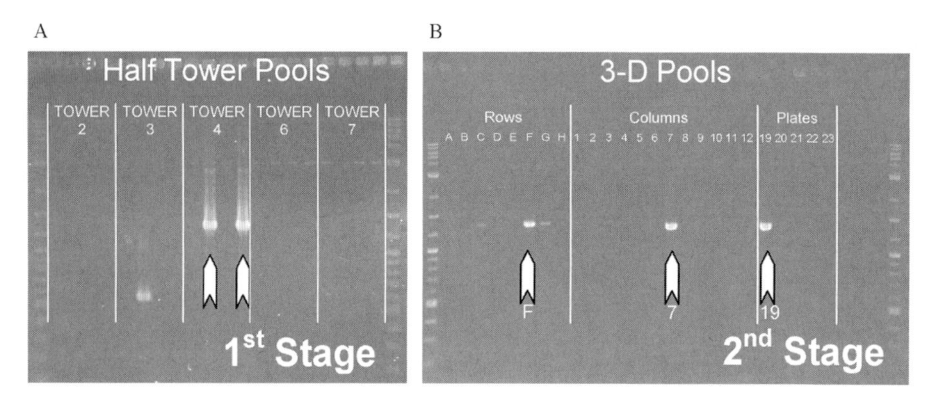

Figure 14.5 Detection of a deletion in an ERF transcription factor within the *Medicago* De-TILLING platform. (A) Two identical PCR products occurring in tower 4 indicate the presence of a 1571-bp deletion within the 2900-bp *Eco*RI target amplicon. (B) Amplification from the three-dimensional pools of tower 4 reveals the row, column and plate location of the mutant containing the M2 seed lot. This mutant was then recovered from a screen of 50 seeds (not shown).

noted a comparable number of false positive amplicons of this type using unmutagenized genomic DNA and suggested that they may arise from polymerase slippage across gaps formed by secondary loops in the DNA template.

14.2.5 Characterization of the Populations

To characterize the delete-a-gene *Arabidopsis* population, 25 loci were screened by Li *et al.* [7] for mutations in a total population of 51 840 M1 lines. Thirty-six deletion alleles were recovered from this population, representing 21 of the 25 loci screened. A broad range of deletion sizes were targeted using primer pairs spanning regions of 3–17 kb. Deletions were identified ranging from 0.8 to 12 kb. For nine loci, two deletion alleles were recovered and for three loci three alleles were recovered. Deletion alleles were therefore recovered for 84% of the targeted loci.

For the *Medicago* De-TILLING platform an initial characterization of the system targeted 10 genes in a subpopulation of 12 000 M1s (five towers). Deletions were detected for four out of 10 targeted loci ranging from 0.4 to 1.6 kb. An example of a detected mutant for an ERF transcription factor is shown in Figure 14.5. A population of 14 towers has now been established at the John Innes Centre and a further 15 towers at the Samuel Roberts Noble Foundation. Between the two institutes this represents a total *Medicago* collection of approximately 70 000 M1 families. Hence, we would predict that the discovery rate in this *Medicago* De-TILLING population should be greater than the 84% recovery rate in the *Arabidopsis* delete-a-gene population.

The number of lines (N) needed to increase the probability of recovering a mutant to any level is related to the frequency of detectable deletions (F) and the observed probability (P) of isolating a deletion through the formula:

$$N = \text{In}[1 - P]/\text{In}[1 - F]$$

Based on the screening data for the *Arabidopsis* delete-a-gene we can estimate that increasing the delete-a-gene *Arabidposis* population to 84 825 lines would enable recovery of deletion mutants for 95% of targeted loci. It would require an additional 50 000 lines to give a 99% probability of success [7]. Results from the *Medicago* De-TILLING platform suggest similarly sized populations. The 70 000 M1 population would give an 88% probability of recovering a mutant. The 99 000 M1 population would give a 95% probability and to achieve a 99% probability a population of 151 500 would be required (Figure 14.6). Given that this relationship of diminishing returns exists for any reverse genetics screening platform, a combination of approaches will always be the most effective strategy.

14.3 Applications of the Technology

These platforms are invaluable for identifying null mutations in target genes, as has been described above. However, the deletion platforms also have advantages over

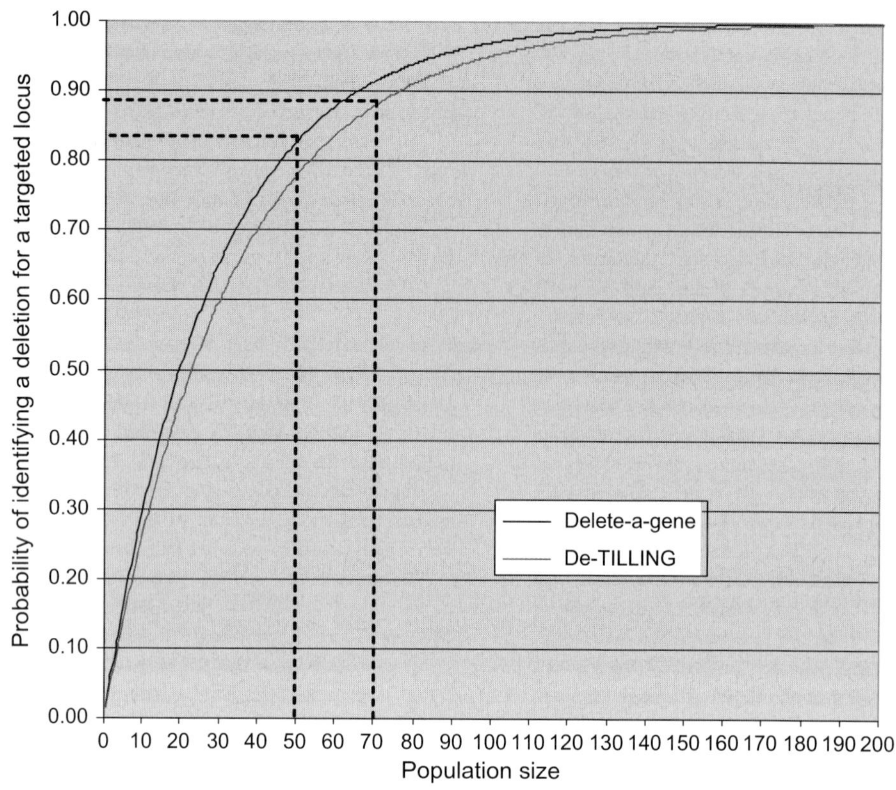

Figure 14.6 The relationship between population size and the probability of identifying a deletion allele. These data points have been extrapolated from the observed frequency of mutant identification for the Delete-a-gene and De-TILLING platforms. The size of the available populations is indicated by the broken gray lines.

alternative reverse genetic platforms, including targeting small genes and generating mutations that remove tandemly duplicated genes.

14.3.1
Targeting Small Genes

Deletion mutagenesis strategies can solve the problem of identifying mutations in small genes. The probability of identifying an insertion is dependent upon the size of the targeted gene and the structure of the gene. The probability of finding a mutant possessing an insertion in a particular gene can be calculated using the formula:

$$P = 1 - [1 - (X/125000)]^{n}$$

where P is the probability of recovering the desired mutant, X is the size of the gene in kilo base pairs, 125 000 is the approximate size of the *Arabidopsis* genome and n is the number of inserts in the mutant library [13]. Therefore, for an *Arabidopsis* collection containing 100 000 insertions there is only around a 55% probability of identifying an insertion mutant for a 1-kb target. For a similar *Medicago* collection, with a genome size of 400 Mb, the probability is approximately 22%. Using deletion mutagenesis, the probability of recovering a mutant is entirely independent of the size of the target sequence. It is far easier to hit a small target with a large deletion than an insertion or point mutation. For example, the delete-a-gene platform was used to isolate a 2.7-kb deletion that completely removed a 0.3-kb target gene [7]. The structure of the gene is also significant. Genes possessing small exons and large introns will be more difficult to mutagenize using TILLING or insertional techniques. Insertions are unlikely to be found within small exons and those which fall within introns or intergenic regions are likely to have no effect on protein function. TILLING relies on the identification of ~1-kb regions with a high probability of introducing deleterious point mutations. Where a gene is structured as small exons and large introns it is difficult to identify useful target regions for TILLING. Deletion-based reverse genetic platforms do not carry these limitations.

14.3.2 Deletions Can Span Multiple Genes

Plant genomes are highly redundant and it is estimated that fewer than 10% of the genes tagged in *Arabidopsis* are likely to generate a phenotypic change [14]. For a mutant phenotype to become apparent it is sometimes necessary for multiple members of a gene family to be inactivated. For unlinked loci 1/16 of the progeny of a cross between two homozygous mutants will be homozygous for both mutations. In this way it is possible to stack insertions or point mutations within a single line to investigate gene function. However, where homologous genes are present in tightly-linked tandem arrays, recombination becomes extremely improbable and therefore difficult to achieve. In general over 15% of the identified genes in sequenced plant genomes are members of tandem-arrayed gene families [15]. This is slightly higher in *Arabidopsis* where about 4000 genes are tandemly repeated as two or more copies [16]. In *Arabidopsis* the recombination rate is estimated to be around 200 kbp/cM. For two mutations in *Arabidopsis* separated by 5 kb a homozygous double mutant would be recovered only once in every 64 million F2 progeny [15]. Alternative strategies for recovering double mutants of tandemly homologous genes are therefore very attractive. Deletions introduced by fast neutron mutagenesis, unlike point mutations and DNA insertions, have the capacity to remove multiple adjacent genes and thus the identification of larger deletions using delete-a-gene technologies has the capacity to overcome this currently intractable problem.

14.3.3 Fast Neutron Reverse Genetics for Crop Improvement

The use of fast neutron mutagenesis is applicable to any plant species. Unlike insertional mutagenesis and RNAi, it is not limited to species with established protocols for efficient transformation and tissue culture. Mutagenesis using fast neutrons is conducted on large batches of dry seed, at very low cost and does not use toxic chemical mutagens. It is therefore ideally suited for application to crop species. The potential of EMS TILLING to crop improvement has already been demonstrate by identification of multiple alleles of loci determining the starch composition of bread wheat [17]. Fast neutron lines, with lower levels of non-target mutations may find even greater application in this area. An example is the production of soybeans with reduced concentrations of the antinutritional oligosaccharides, stachyose, raffinose and galactose, which can be generated by inactivating the biosynthetic genes through fast neutron mutagenesis [18]. Because fast neutron-generated lines do not contain any foreign DNA sequences they may be more acceptable to consumers concerned with the perceived dangers of genetic modification.

14.4 Perspectives

In comparison with the well-established TILLING method, fast neutron-based reverse genetic methods can be used to isolate mutants at a fraction of the time and cost, although the initial development costs can be higher. Fast neutron mutagenesis generates complete knockout mutants unlike many of those generated by RNAi and insertional methods and does not possess the very high number of background mutations from the TILLING platform. Fast neutron mutagenesis can also address the problems of targeting small genes as well as recovering mutations in tandemly duplicated genes, problems that are intrinsic to all methods based on insertion and point mutation. As the cost of sequencing continues to fall, the low cost, scalability and technical simplicity of fast neutron-based reverse genetics is likely to be exploited for a wide variety of plant species.

Information for accessing the De-TILLING resource can be found at www.jicgenomelab.co.uk. Services for fast neutron mutagenesis of seed are available at the Atomic Energy Research Institute in Budapest, Hungary (e-mail: palfalvi@ sunserv.kfki.hu).

References

1 McCallum, C.M., Comai, L., Greene, E.A. and Henikoff, S. (2000) Targeted screening for induced mutations. *Nature Biotechnology*, **18**, 455–457.

2 Colbert, T., Till, B.J., Tompa, R., Reynolds, S., Steine, M.N., Yeung, A.T., McCallum, C.M., Comai, L. and Henikoff, S. (2001) High-throughput screening for induced point mutations. *Plant Physiology*, **126**, 480–484.

3 Henikoff, S. and Comai, L. (2003) Single-nucleotide mutations for plant functional genomics. *Annual Review of Plant Biology*, **54**, 375–401.

4 Koornneef, M., Dellaert, L.W.M. and Vanderveen, J.H. (1982) Ems-induced and radiation-induced mutation frequencies at individual loci in *Arabidopsis thaliana* (L) Heynh.*Mutation Research*, **93**, 109–123.

5 Bruggemann, E., Handwerger, K., Essex, C. and Storz, G. (1996) Analysis of fast neutron-generated mutants at the *Arabidopsis thaliana* HY4 locus. *Plant Journal*, **10**, 755–760.

6 Roberts, M.A., Reader, S.M., Dalgliesh, C., Miller, T.E., Foote, T.N., Fish, L.J., Snape, J.W. and Moore, G. (1999) Induction and characterization of Ph1 wheat mutants. *Genetics*, **153**, 1909–1918.

7 Li, X., Song, Y.J., Century, K., Straight, S., Ronald, P., Dong, X.N., Lassner, M. and Zhang, Y.L. (2001) A fast neutron deletion mutagenesis-based reverse genetics system for plants. *Plant Journal*, **27**, 235–242.

8 Li, X., Lassner, M. and Zhang, Y.L. (2002) Deleteagene: a fast neutron deletion mutagenesis-based gene knockout system for plants. *Comparative and Functional Genomics*, **3**, 158–160.

9 Edgley, M., D'Souza, A., Moulder, G., McKay, S., Shen, B., Gilchrist, E., Moerman, D. and Barstead, R. (2002) Improved detection of small deletions in complex pools of DNA. *Nucleic Acids Research*, **30**, e52.

10 Oldroyd, G.E.D. and Long, S.R. (2003) Identification and characterization of nodulation-signaling pathway 2, a gene of *Medicago truncatula* involved in Nod factor signalling. *Plant Physiology*, **131**, 1027–1032.

11 Jansen, G., Hazendonk, E., Thijssen, K.L. and Plasterk, R.H.A. (1997) Reverse genetics by chemical mutagenesis in *Caenorhabditis elegans*. *Nature Genetics*, **17**, 119–121.

12 Liu, L.X., Spoerke, J.M., Mulligan, E.L., Chen, J., Reardon, B., Westlund, B., Sun, L., Abel, K., Armstrong, B., Hardiman, G., King, J., McCague, L., Basson, M., Clover, R. and Johnson, C.D. (1999) High-throughput isolation of *Caenorhabditis elegans* deletion mutants. *Genome Research*, **9**, 859–867.

13 Krysan, P.J., Young, J.C. and Sussman, M.R. (1999) T-DNA as an insertional mutagen in Arabidopsis. *Plant Cell*, **11**, 2283–2290.

14 Meinke, D.W., Meinke, L.K., Showalter, T.C., Schissel, A.M., Mueller, L.A. and Tzafrir, I. (2003) A sequence-based map of Arabidopsis genes with mutant phenotypes. *Plant Physiology*, **131**, 409–418.

15 Jander, G. and Barth, C. (2007) Tandem gene arrays: a challenge for functional genomics. *Trends in Plant Science*, **12**, 203–210.

16 The Arabidopsis Genome Initiative (2000) Analysis of the genome sequence of the flowering plant *Arabidopsis thaliana*. *Nature*, **408**, 796–815.

17 Slade, A.J., Fuerstenberg, S.I., Loeffler, D., Steine, M.N. and Facciotti, D. (2005) A reverse genetic, nontransgenic approach to wheat crop improvement by TILLING. *Nature Biotechnology*, **23**, 75–81.

18 Mazur, B., Krebbers, E. and Tingey, S. (1999) Gene discovery and product development for grain quality traits. *Science*, **285**, 372–375.

III
Computational Analysis

The Handbook of Plant Functional Genomics: Concepts and Protocols.
Edited by Günter Kahl and Khalid Meksem

ISBN: 978-3-527-31885-8

15
Bioinformatics Tools to Discover Co-Expressed Genes in Plants

Yoshiyuki Ogata, Nozomu Sakurai, Nicholas J. Provart, Dirk Steinhauser, and Leonard Krall

Abstract

Co-expression analysis has emerged in the past couple of years as a powerful tool for gene function and *cis*-element discovery and for hypothesis generation in the *Arabidopsis thaliana* research community. Public efforts by the AtGenExpress Consortium and by individual researchers to document the transcriptome of *Arabidopsis thaliana* have led to large numbers of data sets being made available for data mining by co-expression analysis and other methods. Given the fact that approximately 50% of the genes in *Arabidopsis* have no function ascribed to them by traditional homology-based methods, and that only around 10% of the genes have had their function validated in the laboratory, co-expression analysis can provide insight into gene function with the click of a mouse. Another emerging theme is the incorporation of co-expression networks to guide systems biological research towards the generation of a virtual plant. This chapter introduces the selected bioinformatics tools to discover co-expressed genes in plants, namely the Expression Angler tool of the Botany Array Resource, the CSB.DB tool, the KaPPA-View 2, and the KAGIANA tool.

15.1
Introduction

Co-expression analysis can be considered a generalization of more 'classical' microarray experiments in which the responses of many genes under a treatment or from a specific tissue type are compared to their responses, or lack thereof, in a reference sample. Typically, genes that are upregulated are examined for similar Gene Ontological (GO) categories and unknown genes are ascribed a function based on a similar pattern of response to known genes. The difference between co-expression analysis and these more classical microarray experiments is that in the case of a co-expression analysis it is not necessary for researchers interested in identifying co-expressed genes to actually perform microarray array experiments themselves. Rather, they may simply

The Handbook of Plant Functional Genomics: Concepts and Protocols.
Edited by Günter Kahl and Khalid Meksem

ISBN: 978-3-527-31885-8

take advantage of data sets or compendia of data sets that have been deposited in publicly-accessible databases by using web-based tools to query them. Thus co-expressed genes of interest may be identified at the click of a mouse.

As an interesting note, microarray databases can in part be thought of as databases of negative results, for while a typical researcher conducting a 'classical' microarray experiment is interested in the genes that are changing in response to a given stimulus, it is also the genes that do not respond to a given stimulus that help create a baseline across many data sets for subsequent co-expression analyses. That is, both types of gene expression responses from a microarray experiment are deposited into the database, unlike the case for most journals where negative results are not published. Thus the researcher who is actually performing a microarray experiment can take comfort in the fact that the vast majority of genes where no response is seen under his or her particular condition can actually serve some positive benefit to the larger community as the baseline level for those genes across collections of diverse data sets.

The 'guilt-by-association' paradigm for ascribing gene function to uncharacterized genes using gene expression patterns is well established in yeast, and has also been extended to higher eukaryotes, for example human and mouse [1–3]. In the case of *Arabidopsis*, many data sets have been generated in the past several years by individual researchers and also by the AtGenExpress Consortium, for example, Developmental Map data set [4] and Global Stress data set [5] and others as yet unpublished, encompassing more than 1000 sets in total, and these have been deposited in publicly-accessible microarray databases, such as GEO [6], ArrayExpress [7], TAIR [8], NASCArrays [9], Genevestigator [10], and the BAR [11]. The past 3 years has seen the development of web-based co-expression analysis tools for analyzing these data sets for genes that are co-expressed with a researcher's gene of interest. These include ATTED-II [12], the Arabidopsis Co-expression Tool – Expression Angler [11], ACT [13], AthCoR [14], KaPPA-View 2 [15], PRIMe, and the correlation tool at Loraine Lab Research [16]. Aoki and colleagues describe the various features of each of these tools and also discuss the topic of co-expression networks in plants [17] and provide the KAGIANA tool for identification of co-expressed genes based on topological properties of co-expression networks.

There have also been studies of a more computational nature that use large-scale expression data sets to infer gene function and regulatory modules, especially with regard to curated lists of genes, for example, for isoprenoid biosynthesis [18] or for cytochrome P450s [19]. These computational studies will be ignored in the present sub-chapter in favor of a discussion on the use of web-based tools for identifying co-expressed sets of genes for a gene of interest.

Finally, genes that are co-expressed often contain common *cis*-elements in their promoters. *Cis*-element enrichment analysis and *de novo* discovery aspects using publicly-available tools will be touched upon.

In the following sections, the selected bioinformatics tools for discovery of co-expressed genes are introduced, namely the Expression Angler tool of the Botany Array Resource, the CSB.DB tool, the KaPPA-View2, and the KAGIANA tool.

15.2 The Expression Angler Tool of the Botany Array Resource

15.2.1 Methods and Protocols

The Expression Angler tool of the Botany Array Resource – the BAR – offers a simple interface to the user, shown in Figure 15.1A.

To identify genes that are co-expressed with a user's gene of interest, for example with *RGL2*, a negative regulator of the response to gibberellic acid in controlling seed germination, the user simply enters its AGI identifier, At3g03450, or its corresponding Affymetrix ATH1 probe set identifier, 259042_at, into the first form field. It should be pointed out that all of the searchable gene expression data sets at the BAR are based on the ATH1 Whole Genome GeneChip from Affymetrix. The user may

(A)

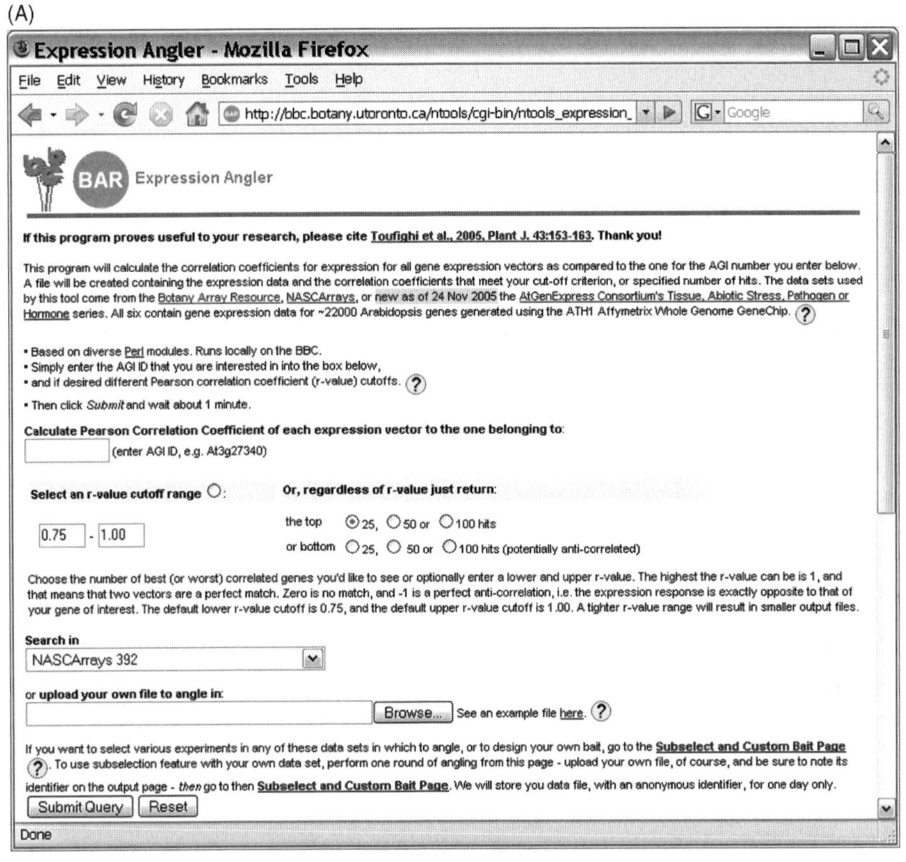

Figure 15.1 (A) Input interface of Expression Angler. (B) Output page of Expression Angler. (C) Median-centered and normalized heatmap output of Expression Angler.

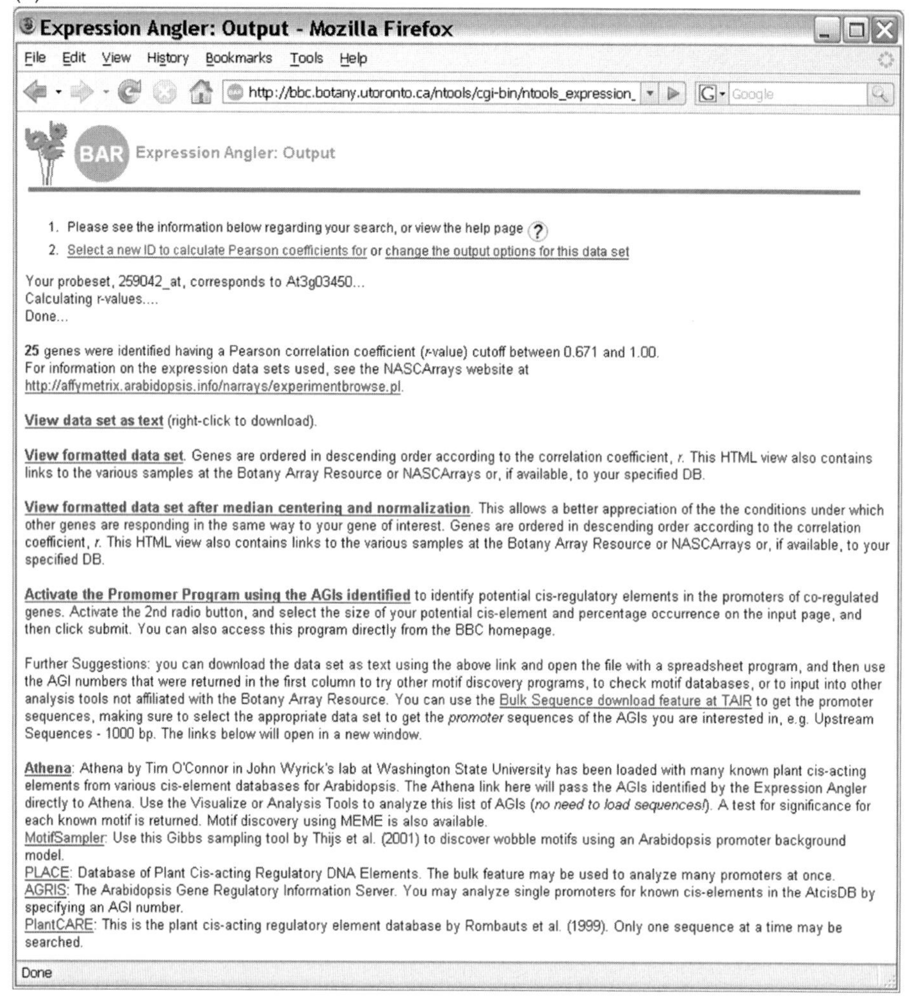

Figure 15.1 (*Continued*)

also select an *r*-value cut-off or simply request to have the top 25, 50 or 100 genes exhibiting greatest co-expression or anti-coexpression displayed. The *r*-value is calculated using the Pearson Correlation Coefficient.

The user also selects the data set in which to search. There are 392 samples in the NASCArrays data set, 93 samples in the Botany Array Database data set, 230 samples in the AtGenExpress Hormone data set, 272 samples in the AtGenExpress Stress data set, 200 samples in the AtGenExpress Pathogen data set, 250 in the AtGenExpress Tissue set, and 344 in the AtGenExpress Plus – Extended Tissue Set, for 1781 samples in total. The set in which the search is carried out is dependent on the biology of the gene in question – this will be discussed later in this sub-section.

(c)

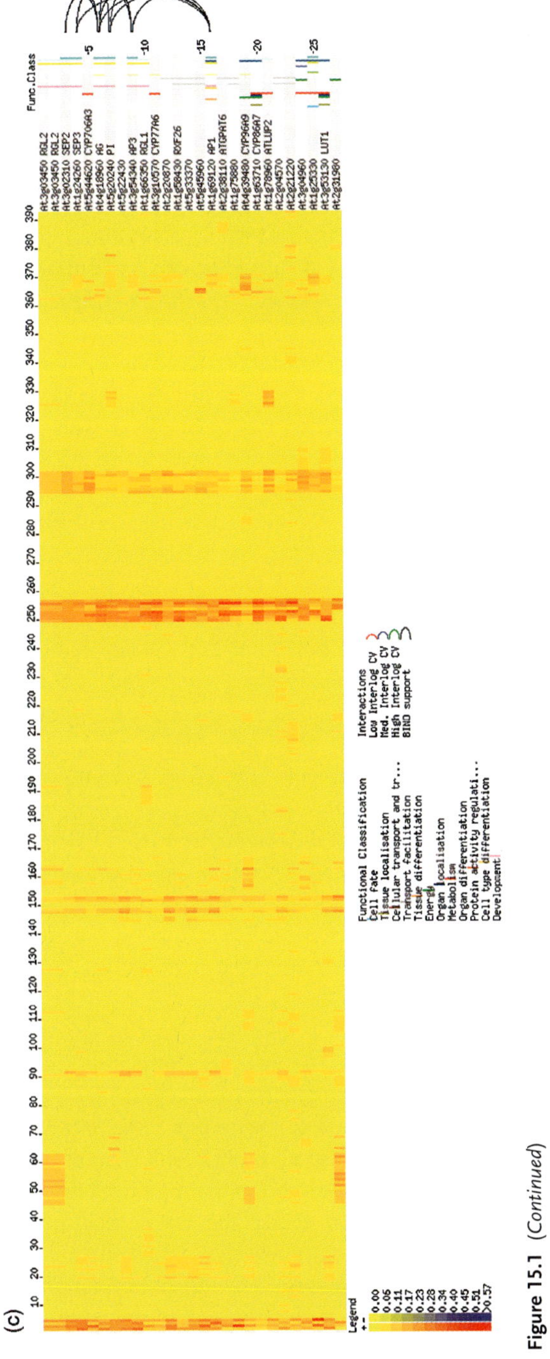

Figure 15.1 (*Continued*)

The output of Expression Angler is a series of hyperlinks, as shown in Figure 15.1A. These are to a text-based version of the ranked list that is generated by Expression Angler, to a tool that will reformat this to generate a heat map, either of the raw expression data or of median centred and normalized data, which allows better visualization of the expression features that Expression Angler was using to measure co-expression, and to various *cis*-element discovery and visualization tools, including Promomer [11], Athena [20], PLACE [21], AGRIS [22] and MotifSampler [23].

The hyperlink to the Formatted data set after median centering and normalization will activate a program called DataMetaFormatter, which will display a heatmap of the co-expressed genes that have been returned, ranked according to their *r*-value scores, see Figure 15.1C.

The lighter areas in the heatmap indicate areas of low expression, relative to the median expression level in the example above, while the darker areas indicate levels of expression above the median expression level. The DataMetaFormatter also appends multiple pieces of information, such as the MIPS functional class, whether the gene products for the genes in the list are predicted or have been documented to interact and so on. The interaction data come from the Arabidopsis Interactions Viewer, also part of the BAR, and are denoted by loops connecting the genes in question on the right side of the heatmap. Such interaction data can provide support that the genes which can be seen in the co-expressed output list are in fact true positives. Known floral homeotic gene products that have been documented to interact in the literature are connected in the output shown in Figure 15.1C.

Users may also upload their own data set in which to search, using the upload function on the input page. An example of the format for this data set is available beside the upload button. These data sets are stored temporarily on the BAR server.

The above example assumes that the researcher wishes to identify genes that are co-expressed with the gene of interest. Another question that can be asked using Expression Angler is 'are there any genes that are specific for a particular sample?' To answer this question, Expression Angler offers a 'Custom Bait' feature, which allows the identification – insofar as there is specificity – of such gene sets. An example of this feature is shown in Figure 15.2.

By designing an artificial vector – here the shape of the vector is important, rather than the absolute values, as the Pearson Correlation Coefficient effectively normalizes this vector to search the data sets – it is possible to very cleanly identify genes specific for a given stimulus or tissue type. This clearly can be useful to identify promoters for biotechnological applications. The output page format of Expression Angler in this case is identical to that produced when searching with a given gene. The heatmap output using the above example is illustrated in Figure 15.2B.

Finally, it is possible to use the 'Subselect' feature of Expression Angler to limit the correlation calculation to a subset of samples within one of the predefined sets or within an uploaded data set. Figure 15.3 shows the effect of omitting the seed samples in the BAR DB data set when angling with *RGL2*.

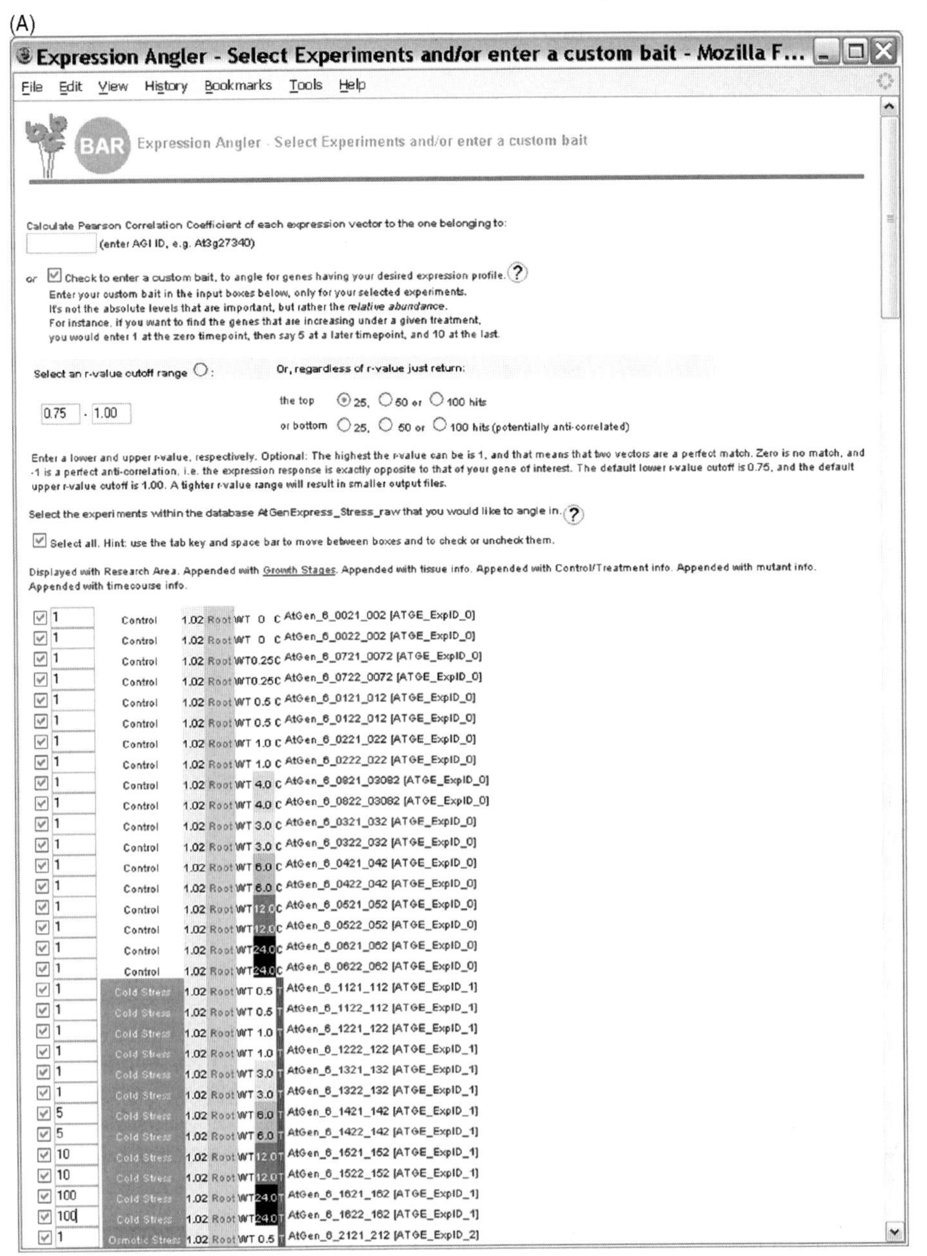

Figure 15.2 (A) Using the Custom Bait feature of Expression Angler to identify genes that are upregulated after 24 h of cold stress in the roots of *Arabidopsis* plants. The line on the right has been added to denote the 'shape' of the vector that will be used to search the AtGenExpress Stress data set. The shape of the vector is set by entering values in the fields to the left of each sample listed. A value of some arbitrarily high amount, 100 in this case as compared to the background level of 1, is sufficient to identify specific genes. (B) Heatmap output of Expression Angler using the 'Custom Bait' feature. Genes that are specifically upregulated in roots of plants subjected to cold stress for 24 h are identified, as seen by the darker stripe corresponding to samples 29 and 30. Note that there is no upregulation in the shoot samples under the same treatment, which are samples 165 and 166 in the graphic, highlighted by the arrow on the right.

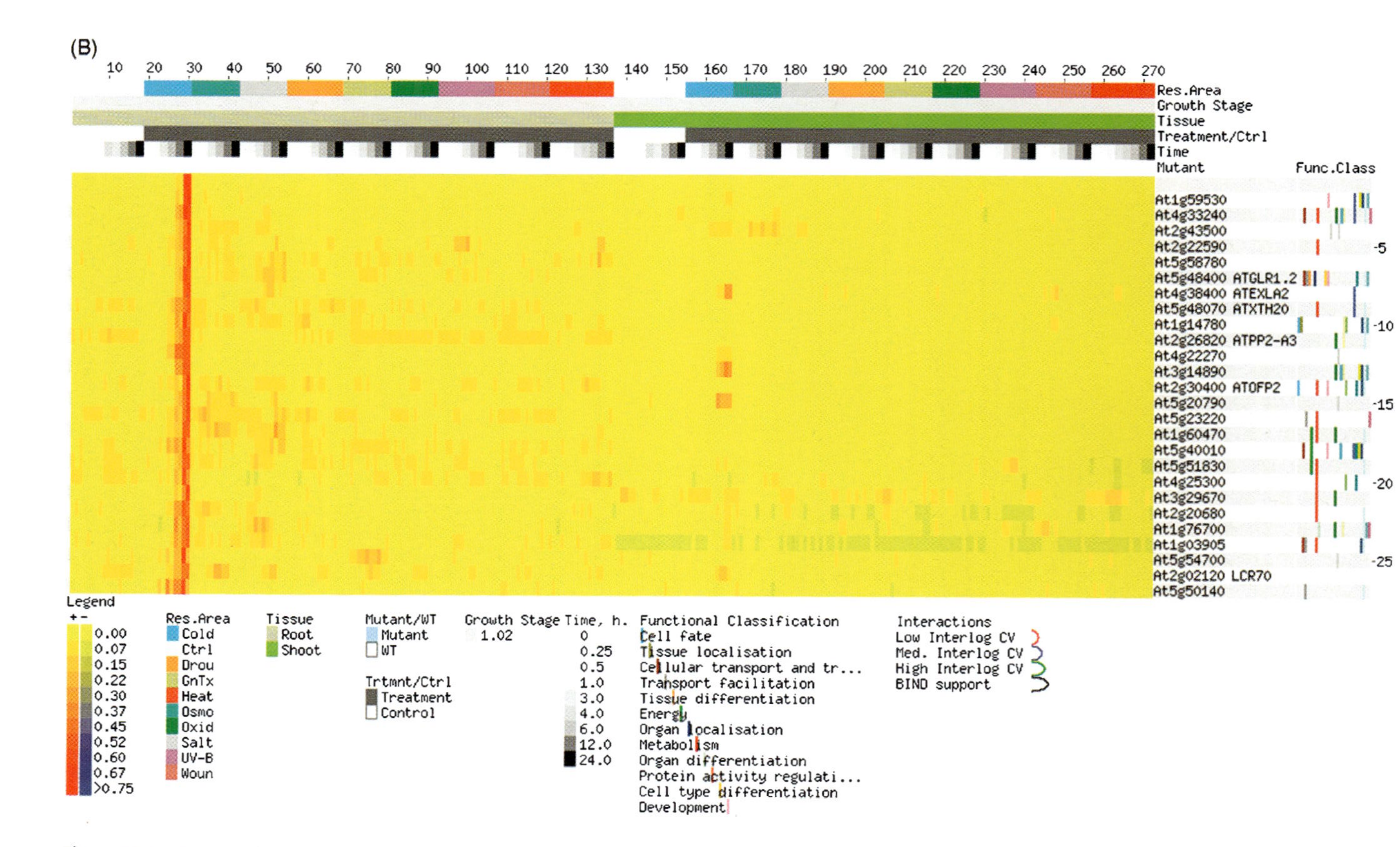

Figure 15.2 (Continued)

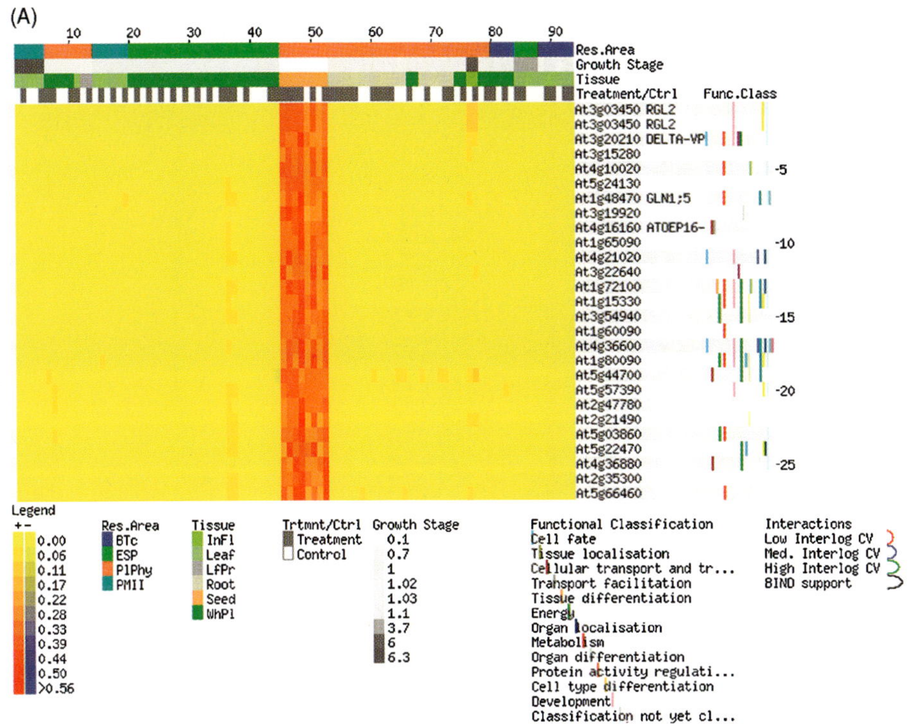

Figure 15.3 Co-expression analysis with *RGL2* in the Botany Array Resource Database with (A) and without (B) seed samples, which are samples 45–52 in the A panel. Floral homeotic genes, connected by interaction loops on the right in the bottom panel, are apparent only in the latter case as *RGL2*s expression in seeds dominates otherwise leading to many general seed-specific genes being returned.

15.2.2
Applications of the Technology

Expression Angler can provide insights into aspects of a gene's biology. For instance, in the case of *RGL2*, a number of floral homeotic genes are returned as being highly co-expressed with it. Although the role of RGL2 in seed biology is well documented [24], it was not until recently that its role in floral development was reported [25]. As seen in Figure 15.1C, the list of the top 25 *RGL2* co-expressed genes contains several floral homeotic genes, such as *SEP2, SEP3, AG, PI, AP1* and *AP3*. So in this case, while the results of Expression Angler are recapitulating laboratory work looking at the involvement of *RGL2* in flower development, it is clearly possible to imagine the converse situation where floral development would be examined because the list of genes that are returned by a co-expression analysis contains known floral homeotic genes. In addition, the list of genes that is returned contains

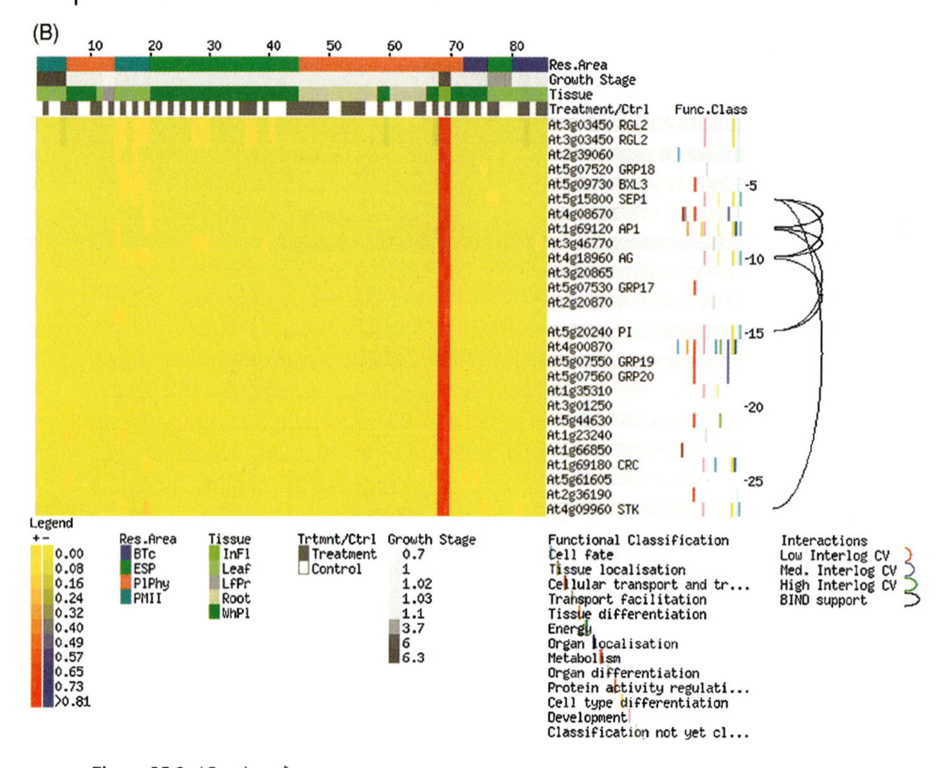

Figure 15.3 (*Continued*)

several genes annotated simply as 'expressed protein', and thus Expression Angler can provide new candidate genes for involvement in a biological process. This is also true for genes whose descriptions are quite generic, such as kinases, glycolsytransferases, and cytochrome P450s. A couple of cytochrome P450s are also returned in the *RGL2* co-expressed list. These have no specific activity ascribed to them, and it is interesting to speculate that they too may be involved in floral development.

Interestingly, depending on the data set which is being searched for co-expressed genes, different gene list are returned. There is no 'right' list of genes and the results are clearly influenced by the types of samples that are in the set being queried. For instance, in the *RGL2* example discussed in the first section, if many seed samples are present in the query database, these will mask the subtler co-expression networks of floral homeotic genes, as seen in Figure 15.3. This suggests that merging data sets may not be the best way to identify co-expression clusters, but rather that a graph network strategy is more appropriate. This will be touched upon in Section 2.3.

We have used this co-expression approach to attempt to ascribe biological roles to the large cytochrome P450 family [19]. We have also used Expression Angler to identify genes that are upregulated specifically in response to heat stress and are in the process of characterizing one of these using a reverse genetics strategy.

Finally, it is possible to use the set of promoters of co-expressed genes to search for over-representation of known *cis*-elements, or to use the promoter set for *de novo* *cis*-element prediction. We are following this strategy using all the AtGenExpress data sets and 'custom baits', that is, synthetic vectors, to generate a predicted '*cis*-ome' for *Arabidopsis thaliana*.

15.2.3
Perspectives

While the co-expression analysis described in this section can provide insight into novel aspects of biology, a couple of limitations exist. For biologists, the first limitation may not be considered limiting, as most biologists are content to extend their knowledge of a few gene candidates in a low throughput manner. While we have developed a high throughput version of Expression Angler, it is too computationally intensive to permit its use as a web-based tool. In addition, the co-expression analysis described in this section simply ranks genes according to their *r*-value score. More sophisticated computational methods involving machine learning are being developed. Here, additional metrics to the *r*-value can be used, and the expression measurements are divided into test and training sets based on a priori knowledge, such as GO categories, to be able to 'learn' the system. The results are also ranked according to the strength of prediction, rather than simply by the *r*-value score. Quaid Morris and colleagues at the Centre for Cellular and Biomolecular Research at the University of Toronto are developing a tool, called GeneMANIA, which will perform machine learning online in real time using several of the AtGenExpress data set compendia.

It is also necessary to reconcile our thinking to the idea of networks of genes – some highly connected and others not so – being involved in specific responses and in the development and manifestation of different tissues. It is far more appropriate to begin to think of which sets of genes are being expressed in each tissue and how these respond to various stress conditions, that is, to begin to think of plant biology in a systems biological context [26,27].

Another facet to current co-expression analyses is that many data sets are currently generated from tissues comprising many different cell types. Clearly, stomatal biology is very different from mesophyll cell biology but in bulk leaf samples the contribution from the stomatal component will be at most 1–2%. Thus a co-expression analysis with these sets may be misleading, as genes could be expressed in different cell types within the bulk tissue. The resolution at which expression profiling can be performed is rapidly increasing, with technologies ranging from fluorescence-activated cell sorting of GFP-tagged cell types [28], laser capture microdissection [29], or other dissection techniques [30]. My laboratory (MPI of Molecular Plant Physiology, Research Group Genes and Small Molecules) has started to make some of these available as cell-type specific data sets for searching but we have a long way to go.

An additional perspective is the concept of shifted co-expression. It is widely known that master regulators are often turned on soon after an event, for example, the plant transcription factor CBF1 in response to cold [31] or several Drosophila homeotic genes during embryogenesis [32], and these in turn serve to activate downstream suites of genes. It is likely that some sort of shifted co-expression analysis will be able to identify targets and activators, in the absence of genome-wide chIP-CHIP analyses for *Arabidopsis* transcription factors. A rudimentary form of such an analysis is possible using the 'Custom Bait' feature of Expression Angler to design 'early' and 'late' vectors.

Finally, co-expression will not tell us everything about a system – proteins may interact but may exhibit no general co-expression. A genome-wide protein interaction screen would provide additional useful information and, indeed, Michael Snyder and Mark Gerstein at Yale are developing protein microarrays to help elucidate the *Arabidopsis* interactome [33], to supplement smaller-scale yeast two-hybrid screens [34]. Such diverse data sets can be integrated to begin to model *Arabidopsis* behavior [35,36].

15.3 The CSB.DB Tool

15.3.1 Introduction

Regardless of which tool is used, for maximal application of correlation-based approaches the interpretation of the output must be made with a biological hypothesis in mind.

CSB.DB – a comprehensive systems-biology database was initiated to provide open access to the results of biostatistical analyses [14]. The basic aim of this database is to supply researchers in the field of systems biology, molecular, and applied biology with statistical tools to access transcriptional co-responses. In the first version of CSB.DB we have concentrated on the validation of gene co-response without the requirement for a-priori knowledge about statistical methods and computational algorithms. We have preferentially facilitated access for those scientists who are interested in a specific gene, or a small set of genes, in a biological pathway or process. In this sense our approach is similar to simple BLAST searches [37]. However, our approach to the generation of novel functional hypotheses is based exclusively on simultaneous changes in transcript levels and does not require structural or sequence information.

The central part of CSB.DB is a set of co-response databases which currently focus on three key model organisms, namely *Escherichia coli*, *Saccharomyces cerevisiae*, and *Arabidopsis thaliana*.

CSB.DB gives easy access to the results of large-scale co-response analyses, which are currently based exclusively on a publicly-available compendium of transcript profiles. We implicitly make the assumption that common transcriptional control of genes is reflected in co-responding, synchronous changes in transcript levels [38]. By scanning for the best co-responses among changing transcript levels, the CSB.DB tool enables users to infer hypotheses about the functional interrelation or interaction of genes of interest, as well as inferring gene functions which cannot be accessed by sequence homology.

Publicly-available expression profiles of various organisms represent a rich resource for cross-experiment co-response analysis of genes, but they need to be critically appraised. We used transcript profiles that were quality checked according to the recommendations of the respective technology platform. Furthermore, we included only accurately measured gene spots for the assembly into multi-conditional expression data matrices. For example, our data matrices comprise approximately 20–50 independent transcript profiling experiments and contain only 5% missing values per gene. Besides quality checking and reduction of missing data we chose two general strategies for combining transcript data sets prior to correlation analysis. (1) We selected representative transcript profiles of as many different experimental conditions as possible. This approach allowed the search for general, constitutive gene-to-gene correlations in each organism. (2) If available, we selected subsets of only those profiles which were generated in a single set of biological experiments or under common biological conditions. These data sets allowed investigations into conditional changes in gene-to-gene co-responses as compared to constitutive co-responses.

It must be said, however, that co-response analysis does have some associated limitations and pitfalls. If the transcript data is not carefully proofed before a correlation matrix is generated, many spurious associations can be observed. Also, in *Arabidopsis*, many transcription factors are only expressed at very low levels and are difficult to reproducibly detect through Affymetrix transcription profiling [39]. These transcripts have not been represented in the available CSB.DB correlation matrices.

15.3.2
Methods and Protocols

Rank ordered tables of pairwise gene correlations, according to the selected correlation measure, can be obtained by using the *single gene query* (sGQ) option and using a selection of pre-defined ranking strategies. Similar to typical BLAST queries, sGQ allows a gene of interest to be defined and all genes associated by co-response to be retrieved if the gene of interest is represented among the set of quality-checked genes. Moreover, the variant of sGQ made available for the *Arabidopsis* co-response databases allows the selection of filtering according to functional categories, which were reported previously together with the visualization tool MapMan [40]. The sGQ output is presented as an HTML table, which contains the rank, the gene identifier of the co-responding gene, the correlation measure, the gene description, the number of pairs (n), the covariance (cov), the probability (*P*-value), the confidence interval (CI), the power, the mutual information (d(M), converted into distance range), and the normalized Euclidean distance [d (E)]. Depending on the organism selected a few other parameters, such as the probe name, can be included in the table. The statistical parameters are dynamically calculated based on the underlying test distribution of the respective pre-selected correlation coefficient [41,42]. Graphical summaries of the set of co-responding genes are based on various external functional classification efforts [40,43,44] and/

or the text search of the returned gene annotations. This survey of gene categories present in the hit list is presented below the sGQ table.

Upon user request a detailed statistical analysis may be obtained for a selected gene pair of interest. This additional validation supports the detection of experimental outliers, which may be associated with technical errors or with the specific nature of a biological experiment. For this purpose a variety of graphical plots are offered.

The *multiple gene query option* (mGQ) allows the pre-definition of up to 60 genes of interest and returns the complete set of available correlations among these genes. This option may be used to discover interdependencies of genes which are known to contribute to a common function or pathway. To visualize this data, the interrelationship can also be displayed as a co-response network with extensive filtering and layout options in Java-enabled browsers.

Finally, an *intersection gene query tool* (isGQ) extracts those genes which exhibit common correlations with at least two pre-defined genes of interest. The threshold settings, which are available for sGQ, may also be used for isGQ. The isGQ query may be used if a few genes with a common function are already known. Using the intersection mode allows novel genes to be found, which may be involved in the function of interest, but cannot be discovered by sequence homology.

15.3.2.1 Simple Protocol for the Use of CSB.DB

For the single gene query, sGQ, any gene of interest, perhaps identified through differential display, significant up- or down-regulation on experimental transcript arrays, or interesting mutant phenotype can be detected. For this first example we will use At2g32060, a gene that encodes for the 40S ribosomal subunit S12. Because the transcripts of ribosomal proteins are know to be highly co-expressed to ensure optimal protein stochiometery [45], we assume that the majority of co-expressed genes will be components of the ribosome small and large subunits. Using matrix nasc0271 (the default matrix, consisting of 9694 genes), Spearman's non-parametric Rho rank correlation, and selecting as output all positive, significant co-responding genes we observe that 41 of the 50 highest correlated transcripts encode for ribosomal structural proteins, two encode for known ribosome associated proteins, and one encodes for a structural RNA, which is exactly the result expected. However, six of the transcripts in the top 50 are not annotated as direct or known components of the ribosome. These six may play a role in assembly of the ribosomal complex or in ensuring proper stochiometry of the complex is maintained. These genes are interesting because they allow new hypotheses to be formulated and tested, perhaps leading to novel insights about ribosome homeostasis.

The transcripts of numerous other pathways and processes in plants have been shown to be correlated to various degrees [46]. Similar sGQ searches using genes from these processes and pathways significantly identify both known factors involved in the process as well as unknown genes.

A sGQ analysis can also be performed with a gene from a relatively unknown or complex process or pathway. Here, a more detailed analysis of the correlations must be performed, or existing biological knowledge must be applied. For example, the

PIC1 gene of *Arabidopsis* (At2g15290) has been cloned and shown to encode a chloroplastic iron transporter [47].

However, if the role of PIC1 was not known, or a mutant plant was not available, correlation analysis can be a great aid in determining in what process the gene functions. If a sGQ is performed using matrix atge0100 (a developmental series using wild-type plants), Spearman's non-parametric Rho rank correlation, and selecting as output the best 5% of positive, significant co-responding genes, 611 correlating transcripts can be identified! However, by using the graphical summary provided, enrichments in two functional classes are evident: Bin 1, photosynthesis, and Bin 19, tetrapyrrole synthesis. While the large number of individual correlations would make the determination of the function of PIC1 difficult, with the largest groups of co-responding genes involved in photosynthesis, some initial hypotheses can be generated based on the protein sequence of *PIC1*. The large numbers of correlating genes in the tetrapyrrole synthesis bin shows the close link between genes for chlorophyll-producing proteins and chlorophyll-binding proteins [48]. The third largest bin contained genes coding for proteins involved in redox regulation. This result is expected, as redox regulation is involved in activating Calvin cycle enzymes [48]. By simply using correlation analysis the most likely function of PIC1 would thus be in some photosynthetic process. As we now know, it is involved in iron transport into the chloroplast, and iron is a critical cofactor for electron transport during photosynthesis.

We will not say much about the multiple gene query option, other than careful analysis of the output can elucidate potential interrelationships between groups of genes as in a cause or effect relationship. For example, gene A may strongly correlate only with gene B, while gene B may correlate strongly with all other imputed genes. The 'Network Visualization' tool shows these relationships in a graphical display which may be manipulated by the user.

The CSB.DB intersection gene query allows the user to search for the overlap of correlations between two or three genes of interest. This is an extremely useful tool when a number of genes in a common pathway are already known or have been identified. For an excellent example of the use of this tool (and CSB.DB in general) the reader is directed to [49]. Briefly, seven components of BR-signaling (*BRI1, BRL1, BRL3, BAK1, BIN2, BES1,* and *BZR1*) were tested using three data matrices, nasc0271, nasc0272, and nasc0273, both individually using sGQ, and in a variety of combinations using isGQ. The authors discovered that the *BRI1/BAK1* isGQ using data matrix nasc0271 outperformed all other queries in enrichment of known brassinolide- (BR) related genes, with a recovery rate of 34.7%. This is sensible, biologically, because BRI1 is the major BR receptor required for most BR responses while BAK1 is a co-receptor with BRI1 [50]. Other genes suspected of being involved in BR signaling or response and identified through the co-response analysis, were then tested experimentally. Interestingly, 24% of these associated genes were found to be directly BR-responsive.

Clearly, isGQs are most effective when critical components of pathways or highly correlated genes are imputed. As Lisso *et al.* [49] illustrates, sometimes numerous genes and combinations must be tested to find the best set.

15.4
The KaPPA-View 2: Co-Expression Analysis on the Plant Metabolic Pathway Maps

15.4.1
Introduction

Representation of gene-to-gene relationships as network-like figures often helps us to understand the feature of the co-expressions. Using computational software, such as Pajek [51] and BioLayout [52], we can obtain intuitive figures where genes are drawn as nodes, gene-to-gene relationships as edges. Genes densely related to each other, and less related to the other genes are considered to be candidates for functional modules [17]. However, to obtain understandable representations of gene networks, researchers are often required to rearrange and classify the gene nodes on the figure according to their functions referring to biological knowledge. It is a critical but laborious step to extract biological meanings from the calculated co-expression data.

We developed and publicly released a web based tool KaPPA-View2, which can project co-expression data onto metabolic pathway maps of *Arabidopsis*. With regard to the metabolic process, one of the most basal roles of the gene products is enzymatic reaction, and this class of knowledge is already laid out on 2D figure 'pathway maps'. Therefore the projection of the gene-to-gene relationships onto the pathway maps would facilitate the understanding of features of the calculated co-expressions. In this section, we describe co-expression analysis using KaPPA-View2 (Table 15.1).

15.4.2
Application of the Technology

We had already developed the first version of KaPPA-View, where both transcriptome data and metabolome data can be viewed simultaneously on the same *Arabidopsis* metabolic pathway maps [15,53,54]. At present, 130 leaves of *Arabidopsis* pathway maps are classified according to the metabolic flow of assimilated CO_2, and 2606 genes, 1110 compounds and 2384 enzymatic reactions are included. Users can temporarily upload their own gene expression data and/or metabolite quantified data through the Internet, and can choose two experiment sets to compare the number of transcripts and metabolites. The KaPPA-View then paints the elements for genes (squares), compounds (circles) and enzyme reactions (arrows) in color gradations according to the ratio of the two. Details of operation are described in the online manual or in other publications [15,53,54]. In the latest version, KaPPA-View2, we further implemented new functions to analyze gene co-expressions on the pathway maps, that is, drawing lines between genes to genes and compounds to compounds according to the co-expression or co-accumulation data. To improve the operation, other functions such as the selection of multiple data sets, switching of the data sets during pathway browsing, and searching deposited resources from outer systems were also added.

Table 15.1 Features of correlation coefficient data listed in KaPPA-View2.

			In whole genes detectable with the Chip (22591 genes)		In genes on the KaPPA-View maps (2606 genes)		
Category of experiments[a]	No. of original data	Filtration of correlation coefficients	Correlations	Genes involved	Correlations[b]	Genes involved[b]	Maps
Not selected	1388	>0.6	956232	13430	2405	1180	83
Not selected	1388	≤0.6	17970	3771	7	13	5
Not selected	1388	≥0.795	113945	3601	675	328	45
Hormone related	236	≥0.817	114298	3892	488	296	49
Specific tissues	237	≥0.916	114308	4990	508	324	41
Stress treatment	298	≥0.739	113870	8058	378	410	57

[a] Which were used to calculate correlation coefficients at ATTED-II.
[b] Sum of the number counted by each map.

15.4.3
Methods and Protocols

Overlaying of co-expression lines onto the pathway maps can be performed with a control panel located at the top-left of the map browsing window (Figure 15.4). By selecting co-expression data from the pull-down list, users can view representation of the co-expressions on the maps. Six gene co-expression data calculated at ATTED-II [12] are available in default. Gene-to-gene relationships in *Arabidopsis* having correlation coefficients of more than 0.6 or less than − 0.6 were calculated with 1388 of Affymetrix's GeneChip data obtained from various experimental conditions. Among them, hormone related experiments (236 Chips), tissue specific experiments (237 Chips), stress treatments (298 Chips), were selected and the correlation coefficients were calculated individually. For both selected and total data, the threshold value of the correlation coefficients was set so that the number of gene-to-gene relationships was five times greater than the gene number detectable using the GeneChip (22 591 genes). Because frequency distributions of correlation coefficient values depend on the original data sets, this setting of criteria would help to compare features of co-expression between different conditions.

Figure 15.4 shows the pathway map of the Calvin cycle (Ath00112) overlaid with positive correlation (>0.6) from 1388 Chips (ATTED-II, AthGeneCor_v3). Phosphoribulokinase gene (*PRK, At1g32060*) has many correlations to the other genes in the cycle, and appears to have a coordinated role with them. Actually PRK protein is known to form a functional complex with GAPDH [55]. Dense intra-pathway co-expressions were also observed in the pentose phosphate cycle (Ath00017) and in arginine and praline metabolism (Ath00013), which implied functional modules in the pathways. In the case of negative correlations ($\leq$0.6), only seven correlations were observed on the maps, glutamine and glutamate metabolism/nitrate assimilation (Ath00006, Glycolate pathway (Ath00118), and in the maps of very large gene families such as the Glucosyltransferase family (Ath00359), Glycoside Hydrolase (Ath00412), Peroxidase, class III (Ath00413). As the pathway maps are primarily separated according to each functional category, a few negative correlations might reflect the nature of the pathway regulation. When the correlations in the selected data sets were compared with each other, it was observed that the number of correlations was noticeably different. Despite the total number of correlation lines between genes on the Chip being adjusted to be almost the same, the line number on the KaPPA-View maps were highest on the positive correlations calculated from 1388 Chips (675 lines), and the lines were distributed on 45 maps. While the line number was lowest on the data set of stress treatment (378 lines), where lines distributed on highest number of maps (57) and the genes involved in the correlation was largest (410 genes). This implied that among the stress treated experiments, co-expressing genes were distributed to various types of biological processes. Although it is suggested that more than 100 experiments is sufficient to generate condition-independent correlations [15], application of co-expression data generated from selected samples may help to elucidate condition-dependent gene-to-gene correlations.

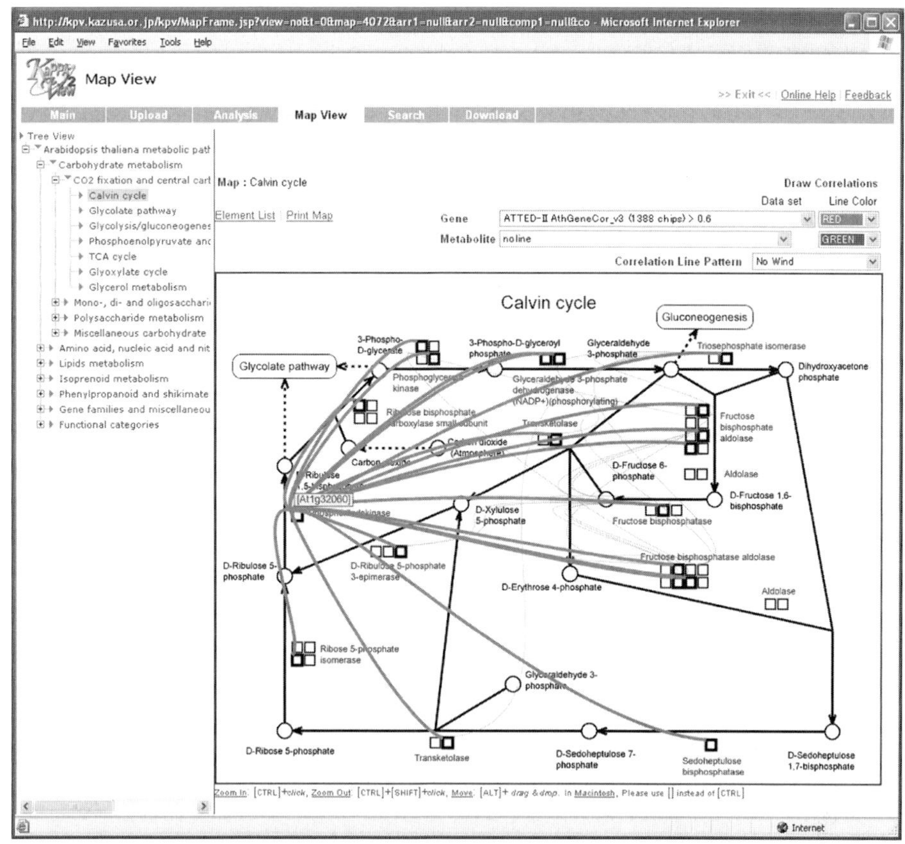

Figure 15.4 Representation of correlation lines on pathway map. The pathway map of the Calvin cycle is displayed with lines of gene-to-gene correlation coefficient calculated from 1388 GeneChip filtered with the threshold value >0.6 (curves). The mouse cursor is on the square assigned to Phosphoribulokinase gene (At1g32060), and related genes and lines are highlighted.

Other than the default data sets, users can upload their own co-expression data, such as that calculated from time series experiments after drug treatments. Taking into account that the choice of the data set may affect the extractable correlations, this function could be useful in finding novel relationships. In KaPPA-View2, correlation lines between metabolic compounds can also be represented. At present, however, there is no metabolome data that is sufficiently comprehensive and of high enough quality to calculate practical correlation coefficients to extract biological processes. Therefore, only one data calculated with 50 metabolites obtained from 16 time series of drug treated *Arabidopsis* cells is available as a sample. It is expected that a large amount of metabolome data will become available in public databases in the future.

15.4.4
Perspectives

The co-expression analysis implemented in KaPPA-View2 will facilitate the determination of functional coordination of genes and lead to a working hypothesis for further studies of the metabolic genes on the pathway maps. However we cannot obtain any information about inter-pathway relationships, such as co-expressions between a particular gene and genes on the adjacent pathway or genes that are not present on the pathway maps. In the case of secondary metabolism in particular, several types of transcription factors can affect the expression of a group of metabolic genes, resulting in considerable accumulation of the products [56,57]. To identify such regulatory correlations, it may be necessary to compare co-expressions between genes using two different maps. We are now developing the next version of the system, KaPPA-View3, which can fulfill such requirements: (1) users can make simple gene lists of any gene IDs (user maps), (2) users can lay out up to four maps including the user maps, and (3) correlation lines between the genes on the four maps can be drawn. These functions would expand the application of co-expression to plant biology. Furthermore, map drawing procedures are reviewed in KaPPA-View3 so that the pathways of any type of plant can be displayed. Lately, DNA microarray data obtained from various plant species has become available on public databases. By comparing the co-expressions in a species with those of *Arabidopsis*, species-specific genes involved in functional modules which are similar to *Arabidopsis* and/or specific to the species might be found. KaPPA-View is not only utilized as a tool to view omics data, but is also helpful in the analysis of metabolic regulatory networks.

15.5
The KAGIANA Tool for Co-Expression Network Analysis of *Arabidopsis* Genes

15.5.1
Introduction

In a gene co-expression network, complex regulatory interactions between genes (nodes) can be represented as gene-to-gene links (edges); namely, a gene can be connected to multiple genes through such links. Aoki *et al.* [17] have reviewed the recent studies of co-expression network analyses. Here, the KAGIANA tool for analyzing co-expressed genes of *Arabidopsis* is introduced. By providing AGI codes of genes, the tool provides users with the following useful functions; namely, (1) retrieval of genes that are reliably co-expressed to a query gene based on co-expression network analysis, (2) obtaining genomic information with its reliability such as Evidence Codes, (3) depiction of a gene co-expression network (preparation of a text file for Pajek visualization [51]), (4) depiction of a categorized gene expression chart for multiple genes. The KAGIANA tool is based on Microsoft Excel and thus users can utilize all the commands and functions associated with the software. In the worksheet including genomic information for example, users can extract genes including keywords of interest using the Excel 'filter' function. The system and functions of the tool are transferable to similar tools and can be used for any organism by obtaining data sets of the organism's genomic information and gene-to-gene correlation based on gene expression profiles.

15.5.2
Methods and Protocols

15.5.2.1 Initial Setting

A downloadable file of KAGIANA ('KAGIANAb309o.zip') in the common compressed ZIP format is available at the KAGIANA project home page. An Excel workbook can be produced from the ZIP-format file using any program suitable for file the extraction (the password for extraction is 'nedo'). To utilize KAGIANA with full functionality, performance of the macro programs should be validated (the 'middle' security level setting is recommended). To validate two command buttons that are added when first opening the KAGIANA workbook, the buttons should be associated with the corresponding macro programs; namely, the 'spider' button with the 'MakeNetTool' macro and the 'key' button with the 'Tools' macro. The method of association of a button to a macro is described in Excel help. To depict a gene co-expression network using the 'MakeNetFile' tool, gene-to-gene correlation data sets are available at ATTED-II or from the KAGIANA project home page; the Pajek program [50] is downloadable at the Pajek home page. To help researchers use the KAGIANA tools, four manual worksheets are provided in the KAGIANA Excel workbook.

15.5.2.2 Retrieval of Co-Expressed Genes

Users can retrieve genes co-expressed with a gene of interest, based on the 'Confeito' algorithm (unpublished) that is designed to remove 'false-positive' co-expressed genes and to extract 'false-negative' co-expressed genes which may be misclassified through the conventional clustering techniques. The method of retrieval of co-expressed genes is as follows: (1) click 'key'-shaped button or go Tools – Macro – Macros and then select Tools, (2) select 'Confeito' in the 'Analysis' frame, (3) select gene size in the 'Confeito Option' frame ('Recommended' is advised), (4) input or copy-and-paste a sing AGI code in the textbox at the left edge, (5) click 'OK' button and then click 'End' button when 'OK' changes into 'End' to create a worksheet whose name includes the query AGI code.

In the worksheet that is created through the above procedure, B and E columns should be focused; namely the AGI code and best 'network specificity' index (Figure 15.5A). The index represents reliability of identification of co-expressed genes instead of the correlation coefficient to a query gene. Genes with an index of 0.5 or higher are specifically co-expressed to the listed gene group according to the Confeito algorithm. Under the data table, 'NS' represents the network specificity of the whole gene group; the group with an index of 0.5 or higher is reliably co-expressed with each other based on the algorithm.

(A)

	A	B	C	D	E	F
1	**Query**	**At1g56650**				
2	Result of Confeito (CC : Correlation Coefficient ; ND : Network Density ; NS : Network Specificity)					
3						
4	**Class**	**AGI Code**	**NS at Fixed CC**	**CC to Initial Gene**	**Best NS**	**CC for Best NS**
5	Initial	At1g56650	1.000	1.000	1.000	0.594
6	Kernel	At5g54060	1.000	0.640	1.000	0.774
7	Kernel	At1g03495	1.000	0.607	1.000	0.767
8	Kernel	At4g14090	1.000	0.597	1.000	0.753
9	Kernel	At3g29590	1.000	0.594	1.000	0.675
10	Confeito	At5g17220	1.000	0.488	1.000	0.729
11	Confeito	At5g42800	1.000	0.474	1.000	0.767
12	Confeito	At4g22870	1.000	0.443	1.000	0.763
13						
14	**Fixed CC**	0.594				
15	**ND**	0.893				
16	**NS**	1.000				

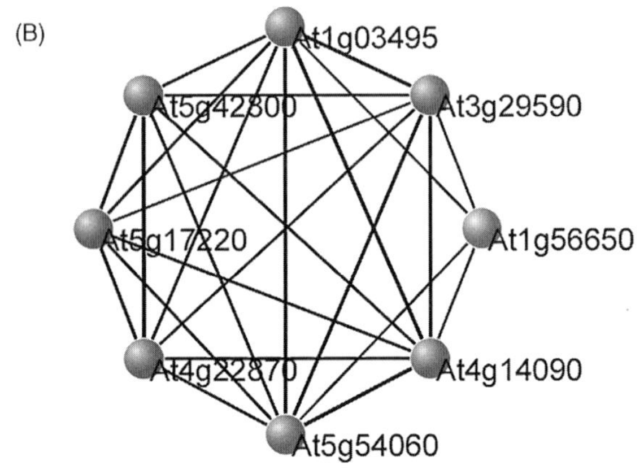

Figure 15.5 Retrieval of co-expressed genes that are correlated to *At1g56650* (A) and a co-expression network including the genes (B).

In a co-expression network originating from At1g56650 as a query gene (Figure 15.5B), circles representing genes that are co-expressed with the query gene are densely connected with gene-to-gene links representing co-expression relationships.

15.5.2.3 **The Other Tools of KAGIANA**

Users can obtain genomic information about multiple genes of interest as follows: (1) input or copy-and-paste AGI codes of genes in A4 and the lower cells on the 'Selected_Link' sheet, (2) select B4 to V4 cells, and (3) double-click at the bottom right of the selected cells, where the shape of the mouse cursor changes. Columns B to Q represent direct links to web pages for the corresponding gene in public databases providing genomic information; namely, users can access such pages with a one-click procedure. Columns R and S show genomic information obtained from TAIR database.

The KAGIANA tool also provides the other macro tools such as preparation of co-expression network and depiction of categorized gene expression chart. The uses of the tools are described in the worksheets in the KAGIANA workbook.

15.5.3 Perspective

The KAGIANA tool will be supplemented with a function that depicts the co-expression network within the tool and analyzes co-expression in other plants as well as *Arabidopsis thaliana*.

Acknowledgments

The KaPPA-View was designed by Toshiaki Tokimatsu, Takeshi Obayashi, Hideyuki Suzuki, and Daisuke Shibata. The development of the KAGIANA tool was supported by the New Energy and Industrial Technology Development (NEDO) program, which is part of the 'Development of Functional Technologies for Controlling the Material Production Process of Plants' project.

References

1 Lee, H.K., Hsu, A.K., Sajdak, J., Qin, J. and Pavlidis, P. (2004) Coexpression analysis of human genes across many microarray data sets. *Genome Research*, **14**, 1085–1094.

2 Wolfe, C., Kohane, I. and Butte, A. (2005) Systematic survey reveals general applicability of "guilt-by-association" within gene coexpression networks. *BMC Bioinformatics*, **6**, 227.

3 Zhang, W., Morris, Q., Chang, R., Shai, O., Bakowski, M., Mitsakakis, N., Mohammad, N., Robinson, M., Zirngibl, R., Somogyi, E., Laurin, N., Eftekharpour, E., Sat, E., Grigull, J., Pan, Q., Peng, W.-T., Krogan, N., Greenblatt, J., Fehlings, M., van der Kooy, D., Aubin, J., Bruneau, B., Rossant, J., Blencowe, B., Frey, B. and Hughes, T. (2004) The functional landscape of mouse gene expression. *Journal of Biology*, **3**, 21.

4 Schmid, M., Davison, T.S., Henz, S.R., Pape, U.J., Demar, M., Vingron, M., Scholkopf, B., Weigel, D. and Lohmann, J.U. (2005) A gene expression map of *Arabidopsis thaliana* development. *Nature Genetics*, **37**, 501–506.

5 Kilian, J., Whitehead, D., Horak, J., Wanke, D., Weinl, S., Batistic, O.D., Angelo, C., Bornberg-Bauer, E., Kudla, J. and Harter, K. (2007) The AtGenExpress global stress expression data set: protocols, evaluation and model data analysis of UV-B light, drought and cold stress responses. *Plant Journal*, **50**, 347–363.

6 Edgar, R., Domrachev, M. and Lash, A.E. (2002) Gene Expression Omnibus: NCBI gene expression and hybridization array data repository. *Nucleic Acids Research*, **30**, 207–210.

7 Rocca-Serra, P., Brazma, A., Parkinson, H., Sarkans, U., Shojatalab, M., Contrino, S., Vilo, J., Abeygunawardena, N., Mukherjee, G., Holloway, E., Kapushesky, M., Kemmeren, P., Lara, G.G., Oezcimen, A. and Sansone, S.A. (2003) ArrayExpress: A public database of gene expression data at EBI. *Comptes Rendus Biologies*, **326**, 1075–1078.

8 Garcia-Hernandez, M., Berardini, T.Z., Chen, G., Crist, D., Doyle, A., Huala, E., Knee, E., Lambrecht, M., Miller, N., Mueller, L.A., Mundodi, S., Reiser, L., Rhee, S.Y., Scholl, R., Tacklind, J., Weems, D.C., Wu, Y., Xu, I., Yoo, D., Yoon, J. and Zhang, P. (2002) TAIR: a resource for integrated Arabidopsis data. *Functional and Integrative Genomics*, **2**, 239–253.

9 Craigon, D.J., James, N., Okyere, J., Higgins, J., Jotham, J. and May, S. (2004) NASCArrays: a repository for microarray data generated by NASC's transcriptomics service. *Nucleic Acids Research*, **32**, D575–D577.

10 Zimmermann, P., Hirsch-Hoffmann, M., Hennig, L. and Gruissem, W. (2004) GENEVESTIGATOR. Arabidopsis microarray database and analysis toolbox. *Plant Physiology*, **136**, 2621–2632.

11 Toufighi, K., Brady, M., Austin, R., Ly, E. and Provart, N. (2005) The botany array resource: e-Northerns, expression angling, and promoter analyses. *Plant Journal*, **43**, 153–163.

12 Obayashi, T., Kinoshita, K., Nakai, K., Shibaoka, M., Hayashi, S., Saeki, M., Shibata, D., Saito, K. and Ohta, H. (2007) ATTED-II: a database of co-expressed genes and cis elements for identifying co-regulated gene groups in Arabidopsis. *Nucleic Acids Research*, **35**, D863–869.

13 Manfield, I.W., Jen, C.-H., Pinney, J.W., Michalopoulos, I., Bradford, J.R., Gilmartin, P.M. and Westhead, D.R. (2006) Arabidopsis co-expression tool (ACT): web server tools for microarray-based gene expression analysis. *Nucleic Acids Research*, **34**, W504–509.

14 Steinhauser, D., Usadel, B., Luedemann, A., Thimm, O. and Kopka, J. (2004) CSB.DB: a comprehensive systems-biology database. *Bioinformatics*, **20**, 3647–3651.

15 Sakurai, N. and Shibata, D. (2006) KaPPA-View for integrating quantitative transcriptomic and metabolomic data on plant metabolic pathway maps. *Journal of Pesticide Science*, **31**, 293–295.

16 Cui, X. and Loraine, A. (2006) Global Correlation Analysis Between Redundant Probe Sets Using a Large Collection of Arabidopsis ATH1 Expression Profiling Data. Proceedings of the LSS Computational Systems Bioinformatics, 223–226.

17 Aoki, K., Ogata, Y. and Shibata, D. (2007) Approaches for extracting practical information from gene co-expression networks in plant biology. *Plant and Cell Physiology*, **48**, 381–390.

18 Wille, A., Zimmermann, P., Vranová, E., Fürholz, A., Laule, O., Bleuler, S., Hennig, L., Prelic, A., von Rohr, P., Thiele, L., Zitzler, E., Gruissem, W. and Bühlmann, P. (2004) Sparse graphical Gaussian modeling of the isoprenoid gene network in *Arabidopsis thaliana*. *Genome Biology*, **5**, R92.

19 Ehlting, J., Provart, N.J. and Werck-Reichhart, D. (2006) Functional annotation of the Arabidopsis P450 superfamily based on large-scale co-expression analysis. *Biochemical Society Transactions*, **34**, 1192–1198.

20 O'Connor, T.R., Dyreson, C. and Wyrick, J.J. (2005) Athena: a resource for rapid visualization and systematic analysis of Arabidopsis promoter sequences. *Bioinformatics*, **21**, 4411–4413.

21 Higo, K., Ugawa, Y., Iwamoto, M. and Korenaga, T. (1999) Plant cis-acting regulatory DNA elements (PLACE) database: 1999. *Nucleic Acids Research*, **27**, 297–300.

22 Palaniswamy, S.K., James, S., Sun, H., Lamb, R.S., Davuluri, R.V. and Grotewold, E. (2006) AGRIS and AtRegNet. a platform to link cis-regulatory elements and transcription factors into regulatory networks. *Plant Physiology*, **140**, 818–829.

23 Thijs, G., Lescot, M., Marchal, K., Rombauts, S., De Moor, B., Rouze, P. and Moreau, Y. (2001) A higher-order background model improves the detection of promoter regulatory elements by Gibbs sampling. *Bioinformatics*, **17**, 1113–1122.

24 Lee, S., Cheng, H., King, K.E., Wang, W., He, Y., Hussain, A., Lo, J., Harberd, N.P. and Peng, J. (2002) Gibberellin regulates Arabidopsis seed germination via RGL2, a GAI/RGA-like gene whose expression is up-regulated following imbibition. *Genes and Development*, **16**, 646–58.

25 Yu, H., Ito, T., Zhao, Y., Peng, J., Kumar, P. and Meyerowitz, E.M. (2004) Floral homeotic genes are targets of gibberellin signaling in flower development. *Proceedings of the National Academy of Sciences of the United States of America*, **101**, 7827–7832.

26 Trewavas, A. (2006) A brief history of systems biology. "Every object that biology studies is a system of systems." Francois Jacob (1974). *Plant Cell*, **18**, 2420–2430.

27 Gutiérrez, R.A., Shasha, D.E. and Coruzzi, G.M. (2005) Systems biology for the virtual plant. *Plant Physiology*, **138**, 550–554.

28 Birnbaum, K., Shasha, D.E., Wang, J.Y., Jung, J.W., Lambert, G.M., Galbraith, D.W. and Benfey, P.N. (2003) A gene expression map of the Arabidopsis root. *Science*, **302**, 1956–1960.

29 Casson, S., Spencer, M., Walker, K. and Lindsey, K. (2005) Laser capture microdissection for the analysis of gene expression during embryogenesis of Arabidopsis. *Plant Journal*, **42**, 111–123.

30 Leonhardt, N., Kwak, J.M., Robert, N., Waner, D., Leonhardt, G. and Schroeder, J.L. (2004) Microarray expression analyses of Arabidopsis guard cells and isolation of a recessive abscisic acid hypersensitive protein phosphatase 2C mutant. *Plant Cell*, **16**, 596–615.

31 Stockinger, E.J., Gilmour, S.J. and Thomashow, M.F. (1997) *Arabidopsis thaliana* CBF1 encodes an AP2 domain-containing transcriptional activator that binds to the C-repeat/DRE, a cis-acting DNA regulatory element that stimulates transcription in response to low temperature and water deficit. *Proceedings of the National Academy of Sciences of the United States of America*, **94**, 1035–1040.

32 Harding, K., Wedeen, C., McGinnis, W. and Levine, M. (1985) Spatially regulated expression of homeotic genes in Drosophila. *Science*, **229**, 1236–1242.

33 Popescu, S.C., Popescu, G.V., Bachan, S., Zhang, Z., Seay, M., Gerstein, M., Snyder, M. and Dinesh-Kumar, S.P. (2007) Differential binding of calmodulin-related proteins to their targets revealed through high-density Arabidopsis protein microarrays. *Proceedings of the National Academy of Sciences of the United States of America*, **104**, 4730–4735.

34 de Folter, S., Immink, R.G.H., Kieffer, M., Parenicova, L., Henz, S.R., Weigel, D., Busscher, M., Kooiker, M., Colombo, L., Kater, M.M., Davies, B. and Angenent, G.C. (2005) Comprehensive interaction map of the Arabidopsis MADS Box transcription factors. *Plant Cell*, **17**, 1424–1433.

35 Jönsson, H., Heisler, M., Reddy, G.V., Agrawal, V., Gor, V., Shapiro, B.E., Mjolsness, E. and Meyerowitz, E.M. (2005) Modeling the organization of the WUSCHEL expression domain in the shoot apical meristem. *Bioinformatics*, **21**, i232–240.

36 Li, S., Assmann, S.M. and Albert, R. (2006) Predicting essential components of signal transduction networks: a dynamic model of guard cell abscisic acid signaling. *PLoS Biology*, **4**, e312.

37 Altschul, S.F., Gish, W., Miller, W., Myers, E.W. and Lipman, D.J. (1990) Basic local alignments search tool. *Journal of Molecular Biology*, **215**, 403–410.

38 Steinhauser, D., Junker, B.H., Luedemann, A., Selbig, J. and Kopka, J. (2004) Hypothesis-driven approach to predict transcriptional units from gene expression data. *Bioinformatics*, **20**, 1928–1939.

39 Czechowski, T., Bari, R.P., Stitt, M., Scheible, W.R. and Udvardi, M.K. (2004) Real-time RT-PCR profiling of over 1400 Arabidopsis transcription factors: unprecedented sensitivity reveals novel root- and shoot-specific genes. *Plant Journal*, **38**, 366–379.

40 Thimm, O., Blasing, O., Gibon, Y., Nagel, A., Meyer, S., Kruger, P., Selbig, J., Muller, L.A., Rhee, S.V. and Stitt, M. (2004) MAPMAN: a user-driven tool to display genomics datasets onto diagrams of metabolic pathways and other biological processes. *Plant Journal*, **37**, 914–939.

41 Sokal, R.R. and Rohlf, F.J. (1995) *Biometry: The Principles and Practice of Statistics in*

Biological Research, 3rd edn., W.H. Freeman and Company, New York.

42 Bonett, D.G. and Wright, T.A. (2000) Sample size requirements for estimating Pearson, Kendall and Spearman correlations. *Psychometrika*, **65**, 23–28.

43 Peterson, J.D., Umayam, L.A., Dickinson, T., Hickey, E.K. and White, O. (2001) The Comprehensive Microbial Resource. *Nucleic Acid Research*, **29**, 123–125.

44 Christie, K.R., Weng, S., Balakrishnan, R., Costanzo, M.C., Dolinski, K., Dwight, S.S., Engel, S.R., Feierbach, B., Fisk, D.G., Hirschman, J.E. *et al.* (2004) Saccharomyces genome database (SGD) provides tools to identify and analyze sequences from *Saccharomyces cerevisiae* and related sequences from other organisms. *Nucleic Acid Research*, **32**, D311–D314.

45 Baum, E.Z. and Wormington, W.M. (1985) Coordinate expression of ribosomal protein genes during Xenopus development. *Developmental Biology*, **111**, 488–498.

46 Williams, E.J.B. and Bowles, D.J. (2004) Coexpression of neighboring genes in the genome of *Arabidopsis thaliana*. *Genome Research*, **14**, 1060–1067.

47 Duy, D., Wanner, G., Meda, A.R., von Wirén, N., Soll, J. and Philippar, K. (2007) PIC1, an ancient permease in arabidopsis chloroplasts, mediates iron transport. *Plant Cell*, **19**, 986–1006.

48 Buchannan, B.B., Gruissem, W. and Jones, R.L. (2000) *Biochemistry & Molecular Biology of Plants*, American Society of Plant Physiologists (ASPP), Rockville, Maryland.

49 Lisso, J., Steinhauser, D., Altmann, T., Kopka, J. and Mussig, C. (2005) Identification of brassinosteroid-related genes by means of transcript co-response analyses. *Nucleic Acids Research*, **33**, 2685–2696.

50 Nam, K.H. and Li, J. (2002) BRI1/BAK1, a receptor kinase pair mediating brassinosteroid signaling. *Cell*, **110**, 203–212.

51 Batagelj, V. and Mrval, A. (2003) Pajek – analysis and visualization of large networks, in *Graph Drawing Software* (eds M. Jünger and P. Mutzel), Springer, Berlin.

52 Enright, A.J. and Ouzounis, C.A. (2001) BioLayout – an automatic graph layout algorithm for similarity visualization. *Bioinformatics*, **17**, 853–854.

53 Tokimatsu, T., Sakurai, N., Suzuki, H., Ohta, H., Nishitani, K., Koyama, T., Umezawa, T., Misawa, N., Saito, K. and Shibata, D. (2005) KaPPA-view: a web-based analysis tool for integration of transcript and metabolite data on plant metabolic pathway maps. *Plant Physiology*, **138**, 1289–1300.

54 Tokimatsu, T., Sakurai, N., Suzuki, H. and Shibata, D. (2006) KaPPA-view: A tool for integrating transcriptomic and metabolomic data on plant metabolic pathway maps, in *Biotechnology in Agriculture and Forestry* (eds K. Saito, R.A. Dixon and L. Willmitzer), Springer, Berlin.

55 Marri, L., Sparla, F., Pupillo, P. and Trost, P. (2005) Co-ordinated gene expression of photosynthetic glyceraldehyde-3-phosphate dehydrogenase, phosphoribulokinase, and CP12 in *Arabidopsis thaliana*. *Journal of Experimental Botany*, **56**, 73–80.

56 Tohge, T., Nishiyama, Y., Hirai, M.Y., Yano, M., Nakajima, J., Awazuhara, M., Inoue, E., Takahashi, H., Goodenowe, D.B., Kitayama, M., Noji, M., Yamazaki, M. and Saito, K. (2005) Functional genomics by integrated analysis of metabolome and transcriptome of Arabidopsis plants over-expressing an MYB transcription factor. *Plant Journal*, **42**, 218–235.

57 Hirai, M.Y., Sugiyama, K., Sawada, Y., Tohge, T., Obayashi, T., Suzuki, A., Araki, R., Sakurai, N., Suzuki, H., Aoki, K., Goda, H., Nishizawa, O.I., Shibata, D. and Saito, K. (2007) Omics-based identification of Arabidopsis Myb transcription factors regulating aliphatic glucosinolate biosynthesis. *Proceedings of the National Academy of Sciences of the United States of America*, **104**, 6478–6483.

URLs of Useful Databases or Web-Based Tools

ACT: the Arabidopsis Co-expression Tool http://www.arabidopsis.leeds.ac.uk/act/

AGRIS: Arabidopsis Gene Regulatory Information Server http://arabidopsis.med.ohio-state.edu/

ATTED-II: *Arabidopsis thaliana* trans-factor and cis-element prediction database http://www.atted.bio.titech.ac.jp/

ArrayExpress http://www.ebi.ac.uk/arrayexpress/

Athena http://www.bioinformatics2.wsu.edu/Athena/

BAR: the Botany Array Resource http://bbc.botany.utoronto.ca/or http://www.bar.utoronto.ca/

BioLayout: An automatic graph layout algorithm for similarity and network visualization http://cgg.ebi.ac.uk/services/biolayout/

CSB.DB http://csbdb.mpimp-golm.mpg.de/

GEO: Gene Expression Omnibus http://www.ncbi.nlm.nih.gov/projects/geo/

GO: Gene Ontology http://www.geneontology.org/

GeneMANIA http://morrislab.med.utoronto.ca/mania

Genevestigator https://www.genevestigator.ethz.ch/at/

KAGIANA http://pmnedo.kazusa.or.jp/kagiana/

KaPPA-View2 http://kpv.kazusa.or.jp/kappa-view/

Loraine Lab Research http://www.trnsvar.org/

Mips: munich information center for protein sequences http://mips.gsf.de/

MapMan http://gabi.rzpd.de/projects/MapMan/

MotifSampler http://homes.esat.kuleuven.be/~thijs/Work/MotifSampler.html

NASCArrays: The European Arabidopsis Stock Centre http://arabidopsis.info/

PLACE: A Database of Plant Cis-acting Regulatory DNA Elements http://www.dna.affrc.go.jp/PLACE/

PRIMe: Platform for RIKEN Metabolomics http://prime.psc.riken.jp/

Pajek http://vlado.fmf.uni-lj.si/pub/networks/pajek/

TAIR: the Arabidopsis Information Resource http://www.arabidopsis.org/

16
AthaMap, a Database for the Identification and Analysis of Transcription Factor Binding Sites in the *Arabidopsis thaliana* Genome

Reinhard Hehl

Abstract

The genome-wide Identification of Transcription Factor Binding Sites (TFBS) was created when the genome sequence of the first higher plant, *Arabidopsis thaliana*, was reported. Since then, the recent completion of the sequencing of many plant genomes permits the mapping of TFBS on a genome-wide scale. There are several approaches to mapping TFBS within genomic sequences. If a regulatory sequence has been identified experimentally, a bioinformatic approach to obtain positional information for these sequences in the genome may involve pattern recognition programs such as MatInspector, Match, Patser, or PatMatch [1–4]. This positional information can be stored in databases. For example AGRIS, AthaMap, Athena, and ATTED-II are all database resources that contain pre-calculated TFBS within genomic sequences of *A. thaliana* [5–8]. While Athena, ATTED-II, and AGRIS focus on upstream regions and use consensus sequences for the identification of putative regulatory sequences, AthaMap is the first database that generates a genome-wide map of putative TFBS mainly based on alignment matrices. AthaMap is freely available at http://www.athamap.de/. In this chapter, we will present the AthaMap database and its applications.

16.1
Introduction

AthaMap was first generated by matrix based sequence searches using alignment matrices derived from many binding sites of single transcription factors (TFs) [6]. The early version of AthaMap contained a simple search function that requires a chromosomal position or a locus identifier that would result in a sequence display window with indicated binding sites. In this early version of AthaMap the genes were simply underlined, beginning with either the transcription or the translation start site. The next version of AthaMap increased its functionality by incorporating a co-localization function [9]. This function permits the identification of chromosomal

The Handbook of Plant Functional Genomics: Concepts and Protocols.
Edited by Günter Kahl and Khalid Meksem

ISBN: 978-3-527-31885-8

positions of putative combinatorial elements. Such combinatorial elements were also pre-calculated and annotated to AthaMap based on TFs that are known to interact and on TFs that contain two DNA binding sites. Furthermore, a new function permits the restriction of displayed TFBS to those which are highly conserved. In a next step, AthaMap was extended with functionally verified single TFBS and TFBS that where predicted on the basis of these functionally verified sites [10]. The most recent update of AthaMap contains a gene analysis function that permits the identification of common or missing TFBS in a set of genes [11]. This function may be useful for the analysis of co-regulated genes and for genes that are members of the same metabolic pathway. Furthermore, the complete gene structure consisting of upstream and downstream untranslated regions, introns and exons is now displayed in the genomic sequence.

For the annotation of gene structure and the determination of TFBS in AthaMap, XML flatfiles containing sequence and gene structure information (release 5.0) were downloaded from the TIGR web site [12]. These flatfiles were parsed using a Perl script. Positional information for 5′ and 3′ UTRs, exons and introns were annotated to AthaMap. These regions are displayed in AthaMap with a color code similar to that used by TAIR, the Arabidopsis Information database [11,13].

Currently, AthaMap contains TFBS that were determined using two different methods. First, TFBS were detected with alignment matrices and second, TFBS were identified with single experimentally verified sites [6,10]. Using the pattern search program Patser [1] more than 9×10^6 putative TFBS for 49 TFs from 22 different families were detected within the *Arabidopsis thaliana* genome. With two exceptions, all detected TFBS were annotated to the database. Only in case of the CAT- and TATA-box binding factors CBF and TBP was positional information used to restrict the number of annotated TFBS to those that occur within a defined region upstream of the transcription or translation start site [9]. In some cases an alignment matrix was only available for a TF from a species other than *A. thaliana*. These alignment matrices were also used for the detection of putative *A. thaliana* TFBS because TFs and their binding site specificities are not plant species specific.

In a second approach single published TFBS were annotated to AthaMap. For many TFs no alignment matrix was available but a single binding site had been determined in a gene. This binding site may also occur at other genomic positions. Therefore, novel putative binding sites were determined within the genome that were identical to the sequence of the experimentally verified site adjacent to the core sequence of the TFBS [10] To detect TFBS based on single transcription factor binding sites, a Perl script was written for pattern-based screenings of the *Arabidopsis thaliana* genome. Both strands of the annotated genome were screened resulting in records harboring absolute positional information and orientation. In this case only sites determined with *A. thaliana* TFs were included. In total 94 191 TFBS for 55 factors from 15 TF families were identified using this method.

The third class of TFBS that was pre-calculated and annotated to AthaMap is combinatorial elements [9,11]. For this, TFBS determined with alignment matrices were used for a co-localization analysis. Combinatorial elements were identified for TFs that are either known to interact or that are known to harbor two binding domains. In total 359 867 sites for six combinatorial elements were annotated to AthaMap.

16.2 Methods and Applications

16.2.1 Using the Web Interface at http://www.athamap.de/

16.2.1.1 The Search Function

To display TFBS at any chromosomal position, the user can choose one of two options on the search page of AthaMap as shown in Figure 16.1. Either a position on a selected chromosome or a gene identifier (*Arabidopsis* genome identification number: AGI) can be submitted. Furthermore, it is possible to restrict the display to highly conserved TFBS.

A typical result screen is shown in Figure 16.2. The result window displays the nucleotide sequence 500 bp upstream and 500 bp downstream of the gene start or the chromosomal position defined by the search mode. In this case a known transcription start site (TSS) has been annotated. The gene structure is shown with a color code. The three types of TFBS are indicated on the result page with three different symbols. Matrix based TFBS (->), combinatorial element (==), and TFBS based on single sites (>>). The names of the factors or combinatorial elements for which TFBS were detected are linked to pop-up windows that show the underlying data for that particular site (Figure 16.3A–C). Furthermore, a tool tip box will open with specific positional information on the TFBS and with further parameters for matrix based TFBS (Figure 16.3D). At the bottom of the result page, two arrows allow forward and backward scrolling of the sequence window by 500 bp (not shown). Furthermore, a short description of the gene is given and links to external databases for further

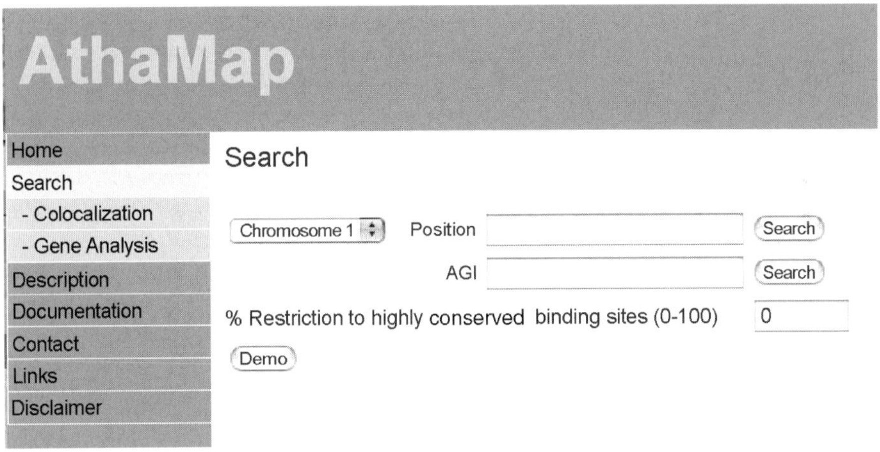

Figure 16.1 The search function of AthaMap.

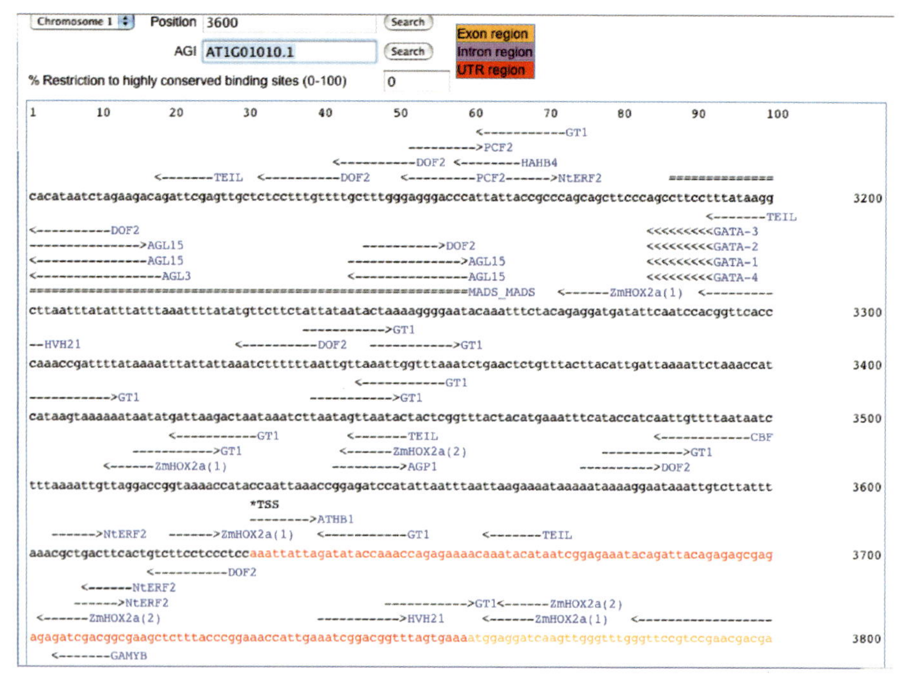

Figure 16.2 Partial screen of the search result for gene AT1G01010.1.

information on the displayed gene are implemented below the sequence display window (not shown).

Figure 16.3A and B show additional information in pop-up windows linked to every TF for which a binding site was detected in the genome. This information contains the name of the TF, the family of the TF, and the plant species. Either the matrix (Figure 16.3A) or the single sequences used for TFBS determination (Figure 16.3B) are shown. The reference from which these sequences were extracted is directly linked to the PubMed database. If the factor is annotated to the TRANSFAC database, the TRANSFAC accession number links to the factor description in the TRANSFAC professional database for licensed TRANSFAC users [14]. If the TF is from *A. thaliana*, the AGI links to the gene locus in TAIR [13]. Further specific information derived from matrix based searches is the maximum score and the threshold determined for a matrix by the search program Patser (Figure 16.3A). Each matrix-based TFBS has an individual score between threshold and maximum score which is an indication of the conservation of the binding site. A high score close to the maximum score means that this particular binding site contains nucleotides that are more frequently encountered at the corresponding position in the matrix. To obtain this information each binding site is linked to a tool tip box that opens when the cursor is moved over the site determined with a matrix. Figure 16.3D shows a tool tip box for a specific binding site. Here, in addition to the positional information, maximum score and threshold score of the matrix, the individual score of the binding site is also shown. Each matrix-based binding site has a specific score.

AthaMap

Name: AGL15

Synonyms:

Species: Arabidopsis thaliana

Family: MADS

Matrix:

A	5	4	16	0	1	17	10	17	13	19	8	19	0	9	26	19
C	4	2	2	32	15	4	3	0	0	1	1	0	0	5	1	3
G	3	1	4	0	0	1	2	1	0	2	4	12	31	4	1	4
T	20	25	10	0	16	10	17	14	19	10	19	1	1	14	4	6

Max. score: 12.09

Threshold: 4.48

Reference(s): Tang W, Perry SE. 2003 Binding site selection for the plant MADS domain protein AGL15: an in vitro and in vivo study. J Biol Chem 278: 28154-9. [PubMed:12743119]

TRANSFAC ACC: T03018

AGI: At5g13790

A

AthaMap

Name: GATA-1

Synonyms:

Species: Arabidopsis thaliana

Family: C2C2(Zn) GATA

Verified binding sequences:
TATGATAAGGCTAGCGTCGAC
TATGATAAGGCGCGCGTCGAC
GTGGATTGATGTGATATCTCC
GTGGATTCATGTAATATCTCC
AAATAGATAAATAAAAACATT
AAATAGAGATCTAAAAACATT
GAAATGGATCGCGAGCCGAGGCTCGATTTCA
GAAATGGATCGCCACGGCACCGTGGATTTCA
GCGGTATTATCGTCATAGCGTGCGGGTATCGAATATTGCCC

Additional screening sequences:
GTGGATTGA
GTGGATTCA
ATAGATAAA
AGAGATCTA
TATGATAAGG
ATGGATCGCG
CTCGATTTCA
GTGGATTTCA
TATTATCGTC
GGGTATCGAA

Reference(s): Teakle GR, Manfield IW, Graham JF, Gilmartin PM. 2002. Arabidopsis thaliana GATA factors: organisation, expression and DNA-binding characteristics. Plant Mol Biol. 50:43-57. PubMed

TRANSFAC ACC: T05705

AGI: At3g24050

B

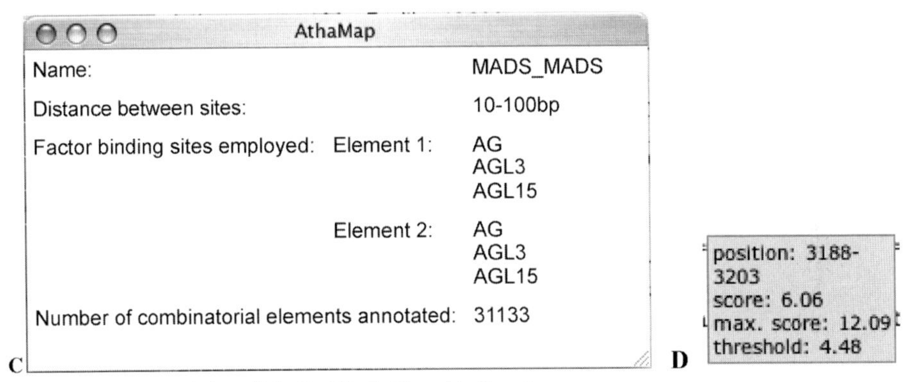

Figure 16.3 Pop-up windows linked to TFs that have binding sites determined with a matrix (A), with single experimentally verified sites (B), or that were used for identification of a combinatorial element (C). (D) A tool tip box linked to a TFBS identified with the matrix shown in A.

Figure 16.3C shows the information that is provided in a pop-up window for a combinatorial element. The distance between the first nucleotides of the two sites is indicated as well as all TFs that were used to determine the combinatorial element. Also, the total number of these combinatorial elements is shown.

It is also possible to restrict the number of displayed TFBS based on their sequence conservation. To restrict a search to those TFBS that are highly conserved a restriction value between 1 and 99 can be entered in the search window (Figure 16.1). A value of 50 means that only those TFBS are displayed that have a score that results when 50% of the difference between maximum score and threshold is added to the threshold score. If, for example, the maximum score is 6 and the threshold is 2, entering a 50 would result in the display of TFBS that have a score of at least 4. This restriction is useful, for example when multiple TFs of the same family are proposed to bind a specific TFBS. Applying a restriction will uncover those TFs that may have a higher binding affinity to this TFBS.

16.2.1.2 **Co-Localization Analysis**

Another option in AthaMap is the detection of co-localizing TFBS. This is useful for detecting TFBS of interacting TFs that occur in close proximity to their target genes. Figure 16.4 shows a composite screen of the Co-localization Analysis tool [9,11].

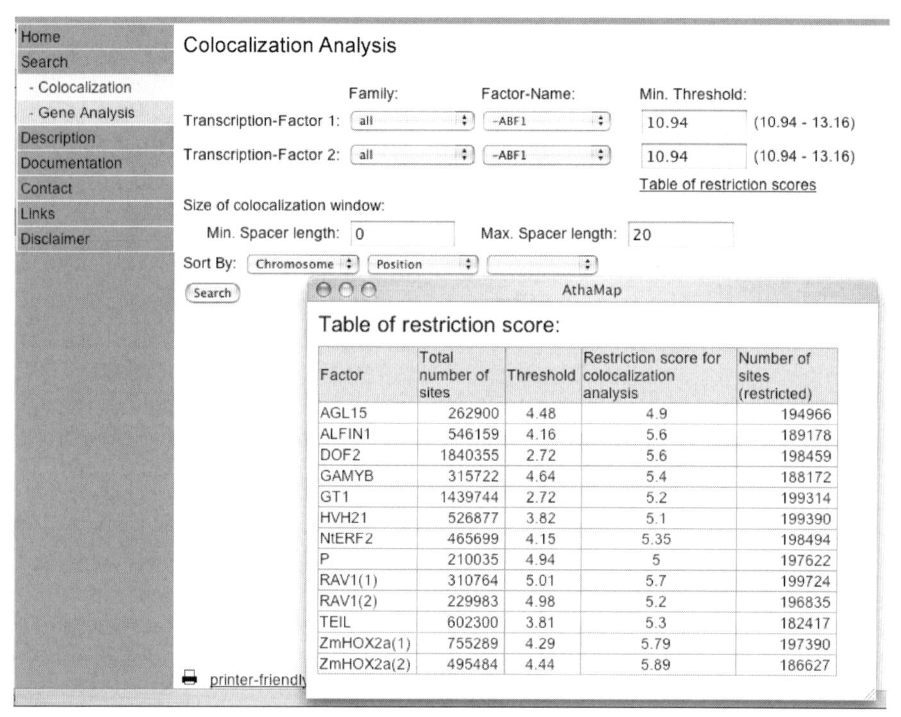

Factor	Total number of sites	Threshold	Restriction score for colocalization analysis	Number of sites (restricted)
AGL15	262900	4.48	4.9	194966
ALFIN1	546159	4.16	5.6	189178
DOF2	1840355	2.72	5.6	198459
GAMYB	315722	4.64	5.4	188172
GT1	1439744	2.72	5.2	199314
HVH21	526877	3.82	5.1	199390
NtERF2	465699	4.15	5.35	198494
P	210035	4.94	5	197622
RAV1(1)	310764	5.01	5.7	199724
RAV1(2)	229983	4.98	5.2	196835
TEIL	602300	3.81	5.3	182417
ZmHOX2a(1)	755289	4.29	5.79	197390
ZmHOX2a(2)	495484	4.44	5.89	186627

Figure 16.4 The Co-localization Analysis web tool and the 'table of restriction scores' pop-up window.

Although an unrestriced co-localization analysis with all TFBS in AthaMap is desirable, certain restrictions due to the time-performance of the web server apply. First, the difference between maximal and minimal spacer must not exceed 50 (size of co-localization window, Figure 16.4). In addition to 0 and 50 as the minimal and maximal spacer, respectively, other values can also be entered, for example, 100 and 150. Second, for 13 TFs the number of TFBS that can be entered into an online co-localization analysis should be restricted to about 200 000. This applies to all matrix-based TFBS for which more than 200 000 have been annotated. The link 'Table of restriction scores' will show a table in a pop-up window (Figure 16.4) that displays all TFs for which a restriction has been implemented. This can be achieved by selecting a restriction score higher than the threshold score required to obtain less than 200 000 TFBS.

For a co-localization analysis, the user can select two TFs from the list of all TFs for which TFBS are annotated in AthaMap. Also combinatorial elements can be selected. The three different types of TFBS that can be selected are indicated in the selection list (Factor-Name, Figure 16.4). Matrix-based TFBS are preceded by '–' as shown in Figure 16.4 for ABF1. Combinatorial elements are preceded by '=' and TFBS based on single sites by '>' in front of the TF name. Because the list of displayed TFBS is extensive, it is also possible to restrict this list to those that belong to a specific factor family (Family, Figure 16.4). For matrix-based TFBS it is also possible to increase the threshold score to restrict the search to higher conserved TFBS. For this, the threshold score and the maximum score determined by the program Patser is displayed next to the factor name (10.94–13.16 in case of ABF1, Figure 16.4). In the case shown, a user-defined threshold score has to be higher than 10.94.

The result page shows information on the TFs selected, the number of TFBS that were used for the co-localization analysis, the spacer length, and the minimum threshold. Below this information a list of all combinatorial elements is shown. This list displays the position of the two TFBS, their orientation, the spacer between both, and the nearest gene with the distance to the start codon. A minus means that the element occurs upstream of the closest translation start site. Links are implemented from the table to permit the display of the gene or combinatorial element and to show the sequence and TFBS context. To analyze the detected genes further, a link permits the export of the gene IDs to the Gene Analysis web tool of AthaMap (see below). Another link that exports the gene IDs to the PathoPlant database was also implemented [15,16]. This allows the analysis of the identified genes for co-regulation during plant pathogen interactions. The 'show overview' link on the result page is useful if a very extensive list of co-localizations is obtained. This results in a table that summarizes the number of co-localizations with the same spacer length.

16.2.1.3 Gene Analysis

The gene analysis web tool serves to determine common or missing TFBS in a set of genes [11]. This can be used, for example, to analyze a set of co-regulated genes. Figure 16.5 shows a screen of the Gene Analysis tool after activating the 'Demo' button. In this example a list of three gene IDs is submitted. The default area of these genes inspected for TFBS is −500 to +50 relative to the start codon. This region can

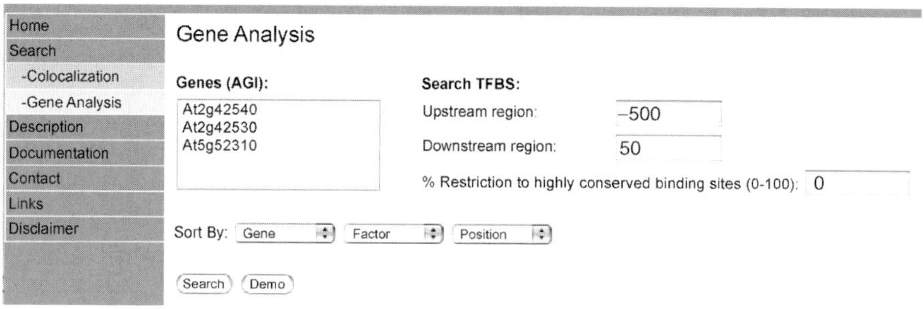

Home
Search
-Colocalization
-Gene Analysis
Description
Documentation
Contact
Links
Disclaimer

Gene Analysis

Genes (AGI):
At2g42540
At2g42530
At5g52310

Search TFBS:
Upstream region: −500
Downstream region: 50
% Restriction to highly conserved binding sites (0-100): 0

Sort By: Gene Factor Position

Search Demo

Figure 16.5 The Gene Analysis web tool (Demo).

be changed but the area inspected must not exceed 2000 bp upstream and downstream. Also, the list of gene IDs entered must not be longer than 100. It is also possible to select how the result is sorted. It is possible to sort by submitted gene, TF family, TFBS position, orientation, and distance from the identified TFBS to the start codon. When the list of genes is submitted, the result will be shown in the same window in a table. This lists the genes submitted for analysis and all factors in the corresponding factor family for which positions were detected in the selected region of the submitted genes. The positions are linked to the sequence display window. Further information on matrix-based TFBS such as maximum score and threshold score and the score of the identified TFBS is also shown. The relative distance from the start codon also indicates whether the TFBS is identified upstream or downstream of the translation start and whether the orientation of the TFBS is the same (+) or opposite (−) to the direction of the transcription of the gene. If Internet Explorer is used as a web browser, this table can be directly exported into a Microsoft Excel table for further analyses. Because these result tables are usually very long, further display options are provided. 'Show overview' will summarize the total number of TFBS detected for a specific TF. 'Show factors that are common in genes' will show all TFs for which TFBS were found in all of the genes. In this table, TFBS that occur in the submitted genes are displayed hierarchically, starting with those at the top that occur in most or all of the submitted genes and those at the bottom of the list that do not occur in the genes. This list also shows the total number of respective TFBS detected in the selected region of the submitted genes and compares this number with the theoretical number of TFBS that would be expected. These values can be subjected to a statistical analysis to obtain an indication of the significance of the observation [11].

16.2.1.4 External Links

AthaMap has been linked with other databases for further information on TFs and on all *Arabidopsis* genes. If the TF for which TFBS were determined is annotated to the TRANSFAC database, the TRANSFAC accession number (Figure 16.3A) directly leads to the factor table in the TRANSFAC database [14]. This information is only displayed for licensed users of the TRANSFAC professional database. If the factor is from *A. thaliana*, the gene ID (Figure 16.3A, AGI) links to the gene locus in TAIR [13]. Other links show up below the sequence display window on a search result (not shown). If a gene is shown in the sequence display window it is linked to the TAIR, MIPS and TIGR databases [12,13,17]. Additional external links are listed on the Links page of the website. In addition to TRANSFAC, TAIR, MIPS and TIGR, a link to the database of *Arabidopsis* transcription factors DATF is also implemented [18].

References

1 Hertz, G.Z. and Stormo, G.D. (1999) Identifying DNA and protein patterns with statistically significant alignments of multiple sequences. *Bioinformatics*, **15**, 563–577.

2 Kel, A.E., Gossling, E., Reuter, I., Cheremushkin, E., Kel-Margoulis, O.V. and Wingender, E. (2003) MATCH: a tool for searching transcription factor binding sites in DNA sequences. *Nucleic Acids Research*, **31**, 3576–3579.

3 Quandt, K., Frech, K., Karas, H., Wingender, E. and Werner, T. (1995) MatInd and MatInspector: new fast and versatile tools for detection of consensus matches in nucleotide sequence data. *Nucleic Acids*, **23**, 4878–4884.

4 Yan, T., Yoo, D., Berardini, T.Z., Mueller, L.A., Weems, D.C., Weng, S., Cherry, J.M. and Rhee, S.Y. (2005) PatMatch: a program for finding patterns in peptide and nucleotide sequences. *Nucleic Acids*, **33**, W262–266.

5 Davuluri, R.V., Sun, H., Palaniswamy, S.K., Matthews, N., Molina, C., Kurtz, M. and Grotewold, E. (2003) AGRIS: arabidopsis gene regulatory information server, an information resource of Arabidopsis cis-regulatory elements and transcription factors. *BMC Bioinformatics*, **4**, 25.

6 Steffens, N.O., Galuschka, C., Schindler, M., Bülow, L. and Hehl, R. (2004) AthaMap: an online resource for *in silico* transcription factor binding sites in the *Arabidopsis thaliana* genome. *Nucleic Acids*, **32**, D368–372.

7 O'Connor, T.R., Dyreson, C. and Wyrick, J.J. (2005) Athena: a resource for rapid visualization and systematic analysis of Arabidopsis promoter sequences. *Bioinformatics*, **21**, 4411–4413.

8 Obayashi, T., Kinoshita, K., Nakai, K., Shibaoka, M., Hayashi, S., Saeki, M., Shibata, D., Saito, K. and Ohta, H. (2007) ATTED-II: a database of co-expressed genes and cis elements for identifying co-regulated gene groups in Arabidopsis. *Nucleic Acids*, **35**, D863–D869.

9 Steffens, N.O., Galuschka, C., Schindler, M., Bülow, L. and Hehl, R. (2005) AthaMap web tools for database-assisted identification of combinatorial *cis*-regulatory elements and the display of highly conserved transcription factor binding sites in *Arabidopsis thaliana*. *Nucleic Acids*, **33**, W397–402.

10 Bülow, L., Steffens, N.O., Galuschka, C., Schindler, M. and Hehl, R. (2006) AthaMap: from *in silico* data to real transcription factor binding sites. *In Silico Biology*, **6**, 0023.

11 Galuschka, C., Schindler, M., Bülow, L. and Hehl, R. (2007) AthaMap web-tools for the analysis and identification of co-regulated genes. *Nucleic Acids*, **35**, D857–D862.

12 Haas, B.J., Wortman, J.R., Ronning, C.M., Hannick, L.I., Smith, R.K., Jr., Maiti, R., Chan, A.P., Yu, C., Farzad, M., Wu, D. *et al.* (2005) Complete reannotation of the Arabidopsis genome: methods, tools, protocols and the final release. *BMC Biology*, **3**, 7.

13 Rhee, S.Y., Beavis, W., Berardini, T.Z., Chen, G., Dixon, D., Doyle, A., Garcia-

Hernandez, M., Huala, E., Lander, G., Montoya, M. *et al.* (2003) The arabidopsis information resource (TAIR): a model organism database providing a centralized, curated gateway to Arabidopsis biology, research materials and community. *Nucleic Acids*, **31**, 224–228.

14 Matys, V., Fricke, E., Geffers, R., Gossling, E., Haubrock, M., Hehl, R., Hornischer, K., Karas, D., Kel, A.E., Kel-Margoulis, O.V. *et al.* (2003) TRANSFAC: transcriptional regulation, from patterns to profiles. *Nucleic Acids*, **31**, 374–378.

15 Bülow, L., Schindler, M., Choi, C. and Hehl, R. (2004) PathoPlant®: a database on plant–pathogen interactions. *In Silico Biology*, **4**, 529–536.

16 Bülow, L., Schindler, M. and Hehl, R. (2007) PathoPlant®: a platform for microarray expression data to analyze co-regulated genes involved in plant defense responses. *Nucleic Acids*, **35**, D841–D845.

17 Schoof, H., Ernst, R., Nazarov, V., Pfeifer, L., Mewes, H.W. and Mayer, K.F. (2004) MIPS *Arabidopsis thaliana* Database (MAtDB): an integrated biological knowledge resource for plant genomics. *Nucleic Acids*, **32**, D373–D376.

18 Guo, A., He, K., Liu, D., Bai, S., Gu, X., Wei, L. and Luo, J. (2005) DATF: a database of Arabidopsis transcription factors. *Bioinformatics*, **21**, 2568–2569.

17
Structural Phylogenomic Inference of Plant Gene Function

Nandini Krishnamurthy, Jim Leebens-Mack, and Kimmen Sjölander

Abstract

Phylogenomic inference provides a robust platform for protein function prediction, addressing the limitations of standard homology-based approaches. In annotation transfer based on homology, the function of a characterized protein is assigned to a sequence whose function is unknown if their sequence similarity is deemed significant (allowing inference of a common ancestor and a similar function). This approach has been applied widely, but is unfortunately now known to be prone to serious systematic error: two proteins can have significant similarity and yet have quite different functions due to domain rearrangements, gene duplication and mutations at key positions. Homology-based annotation transfer can thus result in a large percentage of sequences that are misannotated. Since sequence annotations are often taken at face value by biologists in designing experiments, errors in functional annotation result in overall losses in both effort and resources. Database misannotations can also be propagated by this approach (a process referred to as *transitive disaster*). In this chapter, we present *structural phylogenomic* approaches for protein function prediction. Structural phylogenomic inference of function reduces the systematic errors associated with standard homology-based protocols by integrating methods designed for protein structure prediction and phylogenetic reconstruction to optimize both the sensitivity and selectivity of functional annotation. Issues specific to functional annotation of plant proteins are discussed.

17.1
Introduction

To date, several plant genomes have been fully or partially sequenced and annotated including *Arabidopsis thaliana* [1], *Oryza sativa* [2,3], *Populus* [4], *Medicago*, the moss *Physcomitrella patens* and selected green algae (Table 17.1). Sequencing of ~40 additional plant genomes is in progress. In addition, large sets of expressed sequence

The Handbook of Plant Functional Genomics: Concepts and Protocols.
Edited by Günter Kahl and Khalid Meksem

ISBN: 978-3-527-31885-8

Table 17.1 Resources for whole plant genomes.

Organism	Resource
Arabidopsis thaliana	http://www.arabidopsis.org
Rice (*Oryza sativa*)	http://rice.tigr.org/
Poplar (*Populus trichocarpa*)	http://genome.jgi-psf.org/poplar/
Moss (*Physcomitrella patens*)	http://www.mossgenome.org/
Green algae (*Chlamydomonas reinhardtii*)	http://www.chlamy.org/
Medicago truncatula	http://www.medicago.org/

tag (EST) and transcript assemblies are available for a diverse array of plant species, well distributed across the plant phylogeny [5–9]. Taken together, these plant genome resources include millions of protein-coding sequences whose functions are largely unknown. Experimental verification of these gene structures and preliminary functional assignments is extremely costly and time consuming. To focus attention on the most desirable candidate genes and to design experiments effectively, biologists have turned to comparative computational approaches for prediction of protein function. While genome annotation includes two major steps – constructing gene models and assigning probable functions to predicted genes – our focus in this chapter is on gene function prediction. In this chapter, we present our experience and insights in the development and use of *structural phylogenomic* approaches for the prediction of protein function, and highlight issues of particular relevance to plant biologists.

The term 'protein function' is used variously, and must be interpreted in its context. In some cases, protein function means the molecular function defined by biochemical activity. In other cases, it is meant to indicate the participation of a gene in a biological process or pathway, that is, its phenotypic function based on physiological role. Function can also be defined in a hierarchical manner with varying levels of specificity. For instance, two enzymes can have the same catalytic activity but different specificities (e.g. lactate dehydrogenase (EC 1.1.1.27) versus malate dehydrogenase (EC 1.1.1.37)). Because the definition of 'protein function' is so plastic, and experimental determination of function is expensive and far from straightforward, rigorous and comprehensive benchmark datasets are still under development in the computational biology community. For excellent reviews of work in the field of automating gene function prediction, see [10–13] (Table 17.2).

Numerous annotation protocols, software tools and web servers have been developed to assist biologists in predicting the function of unknown genes. The majority of these approaches depend on homology to previously characterized genes; structural phylogenomic inference of gene function is in this class. *Non-homology* approaches use additional types of information to infer pathway or process participation based on genomic locus information in sequenced genomes, correlated expression profiles, protein–protein interaction and phylogenetic (or phylogenomic) profiles based on correlated evolutionary patterns across species [14]. Cellular localization (e.g. transmembrane or chloroplast localization, or secretion), a key aspect of protein function, is predicted based on sequence analysis for signal peptides and transmembrane segments [15,16].

Table 17.2 Selected bioinformatics resources.

Resource	URL
Interpro	www.ebi.ac.uk/interpro
PFAM	www.sanger.ac.uk/Software/Pfam/
NCBI CDD	http://www.ncbi.nlm.nih.gov/Structure/cdd/cdd.shtml
SuperFamily	http://supfam.mrc-lmb.cam.ac.uk/SUPERFAMILY/
PhyloFacts	http://phylogenomics.berkeley.edu/phylofacts
Phytome	http://www.phytome.org
PlantTribes	http://www.floralgenome.org/tribe.php
TIGR	http://www.tigr.org/plantProjects.shtml
MIPS	http://mips.gsf.de/projects/plants

Protein structure prediction or structural analysis can produce clues to molecular function when homology-based inferences fail; comparison of a newly solved structure to other solved structures is possible through numerous protein structure-comparison tools (e.g. DALI [17], VAST [18], CE [19]). As many structures have been characterized functionally, including identification of critical residues and/or active sites, it is often possible to predict an approximate molecular function for a protein based on comparison with its structural neighbors. Protein structure prediction methods can also be used in cases where a solved structure is not available for a gene or close homologs. Extensive work has been carried out in this area; see [20–22].

17.2 Challenges in Protein Function Prediction

As noted earlier, a fundamental paradigm in computational biology is function prediction by homology, or *annotation transfer*. The standard protocol for automated annotation starts by comparing an inferred protein sequence against other protein or transcript databases to identify and rank putative homologs. If a sequence can be detected whose similarity is statistically significant, the function of the unknown protein is inferred based on the known (or presumed) function of the homolog. Since sequence databases contain millions of sequences and continue to grow exponentially, computationally-efficient methods for homolog detection such as BLAST [23] are employed. The annotation transfer protocol is based on two assumptions: (a) evolution conserves function and (b) sequence similarity implies an evolutionary relationship. From this it follows that if two sequences have detectable similarity, then their functions are likely to be similar.

While annotation transfer based on top database hit is straightforward to implement and computationally efficient to apply to whole genomes, it is also prone to systematic error. The processes underlying gene family evolution – particularly gene duplication, domain shuffling, and speciation – result in modifications of function and structure between proteins with statistically significant sequence similarity [24–26].

In brief, systematic errors in functional annotation have been observed and traced to the following underlying factors.

17.2.1 Gene Duplication

Functional diversification of retained genes is expected following duplication events due to reduced selection pressure [27–30]. This can result in *neofunctionalization*, or acquisition of a completely novel function (e.g. a change in substrate specificity). Alternatively, gene duplication can result in a partition of the original function among the duplicated genes, or in specialization with respect to tissue or temporal expression, a process called *subfunctionalization*. Gene duplication is found throughout all forms of life, but is particularly dramatic in plant genomes, which include numerous multigene families that have evolved by gene or segmental duplication and polyploidization. More than 80% of *Arabidopsis* genes are present in duplicated segments [31,32]. This leads to a large number of similar sequences that may have diverged in function. A classic illustration is the receptor-like protein (RLP) family composed of an amino-terminal (extracellular) leucine-rich repeat region and a carboxy-terminal transmembrane domain (followed by a short cytoplasmic tail). RLPs have high sequence similarity and similar domain architectures [33] but span diverse functions including both defense (e.g. tomato Cf-9 [34]) and developmental pathways (e.g. CLAVATA2 [35]). Such functional divergence is not detectable from pairwise sequence similarity searches alone. In an automated annotation transfer protocol it is not uncommon for annotations to be transferred from a homolog with divergent function resulting in annotation error [36]. Differentiating between homologs that are functionally equivalent and those that have diverged functionally requires advanced methods; phylogenomic analysis is designed to handle these issues.

17.2.2 Domain Shuffling

Protein domains are independently folding structural units that often confer specific functions. Roughly 65% of eukaryotic proteins are composed of multiple domains [37,38]. Multi-domain proteins evolve by domain fission and fusion events. Some domains are termed 'promiscuous' as they are commonly found in combination with other domains resulting in protein families with very divergent functions; leucine-rich-repeat (LRR) and kinase domains are examples of this type. In automated functional inference approaches, this shared domain can be retrieved as the top hit based on 'local' (partial) homology. Since the function of a multi-domain protein is a composite of all its constituent domains, annotation transfer based on partial homology alone can lead to errors. We present one illustration of an existing misannotation that we expect is a result of transfer of annotation based on partial homology.

Oryza sativa sequence AAR00644.1 is a 508-residue protein annotated as a 'putative LRR receptor-like protein kinase' indicating membership in a well-established class of proteins termed *receptor-like kinases*, or RLKs. RLKs are composed of an amino-terminal extracellular leucine-rich repeat (LRR) region, a transmembrane domain, and a carboxy-terminal cytoplasmic kinase domain and are involved in both defense response and development [39]. PFAM analysis of this *Oryza sativa* protein finds no kinase domain; the protein appears to be composed entirely of leucine-rich repeats (Figure 17.1B). All top BLAST hits for the sequence are annotated as RLKs; we expect one of these was the probable source of the misannotation (see Figure 17.1A). The top hit to a sequence outside rice, with an E-value of 3e-62, is a 'putative receptor protein kinase (TMK1)' in *Arabidopsis thaliana* (AT1G66150.1, GenBank accession AAP04161.1). Analysis of the pairwise alignment between the rice and *Arabidopsis* sequences shows similarity is restricted to the LRR region and does not extend to the TMK1 kinase domain (Figure 17.1C). Annotation transfer protocols that require global similarity can avoid this type of error; unfortunately, standard homology-based methods typically ignore this issue.

A

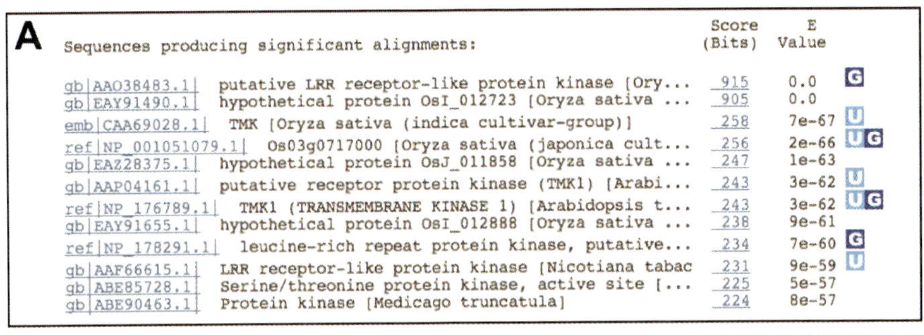

Sequences producing significant alignments:		Score (Bits)	E Value	
gb|AAO38483.1|	putative LRR receptor-like protein kinase [Ory...	915	0.0	G
gb|EAY91490.1|	hypothetical protein OsI_012723 [Oryza sativa ...	905	0.0	
emb|CAA69028.1|	TMK [Oryza sativa (indica cultivar-group)]	258	7e-67	U
ref|NP_001051079.1|	Os03g0717000 [Oryza sativa (japonica cult...	256	2e-66	U G
gb|EAZ28375.1|	hypothetical protein OsJ_011858 [Oryza sativa ...	247	1e-63	
gb|AAP04161.1|	putative receptor protein kinase (TMK1) [Arabi...	243	3e-62	U
ref|NP_176789.1|	TMK1 (TRANSMEMBRANE KINASE 1) [Arabidopsis t...	243	3e-62	U G
gb|EAY91655.1|	hypothetical protein OsI_012888 [Oryza sativa ...	238	9e-61	
ref|NP_178291.1|	leucine-rich repeat protein kinase, putative...	234	7e-60	G
gb|AAF66615.1|	LRR receptor-like protein kinase [Nicotiana tabac	231	9e-59	U
gb|ABE85728.1|	Serine/threonine protein kinase, active site [...	225	5e-57	
gb|ABE90463.1|	Protein kinase [Medicago truncatula]	224	8e-57	

B

Model	Seq-from	Seq-to	HMM-from	HMM-to	Score	E-value	Alignment	Description
!! LRRNT_2	42	67	24	54	18.1	2.4e-05		local Leucine rich repeat N-terminal domain
!! LRR_1	101	120	1	24	11.8	1.8		glocal Leucine Rich Repeat
!! LRR_1	123	145	1	24	11.6	1.9		glocal Leucine Rich Repeat
!! LRR_1	223	244	1	24	11.9	1.7		glocal Leucine Rich Repeat
!! LRR_1	247	269	1	24	13.9	0.59		glocal Leucine Rich Repeat
!! LRRNT_2	339	378	1	54	14.0	0.00053		local Leucine rich repeat N-terminal domain
!! LRR_1	413	435	1	24	9.3	5.1		glocal Leucine Rich Repeat

C

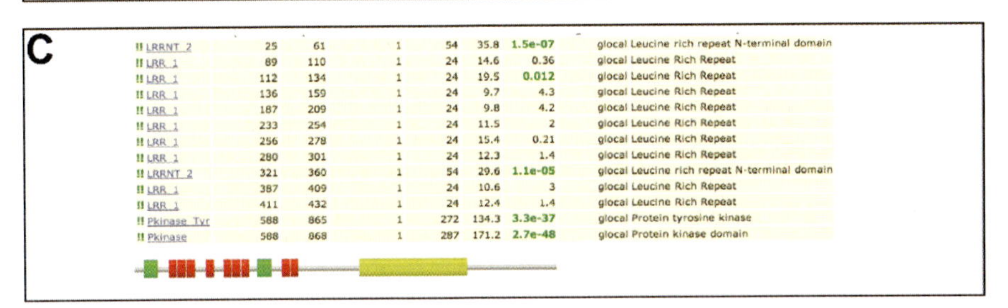

Model	Seq-from	Seq-to	HMM-from	HMM-to	Score	E-value	Description
!! LRRNT_2	25	61	1	54	35.8	1.5e-07	glocal Leucine rich repeat N-terminal domain
!! LRR_1	89	110	1	24	14.6	0.36	glocal Leucine Rich Repeat
!! LRR_1	112	134	1	24	19.5	0.012	glocal Leucine Rich Repeat
!! LRR_1	136	159	1	24	9.7	4.3	glocal Leucine Rich Repeat
!! LRR_1	187	209	1	24	9.8	4.2	glocal Leucine Rich Repeat
!! LRR_1	233	254	1	24	11.5	2	glocal Leucine Rich Repeat
!! LRR_1	256	278	1	24	15.4	0.21	glocal Leucine Rich Repeat
!! LRR_1	280	301	1	24	12.3	1.4	glocal Leucine Rich Repeat
!! LRRNT_2	321	360	1	54	29.6	1.1e-05	glocal Leucine rich repeat N-terminal domain
!! LRR_1	387	409	1	24	10.6	3	glocal Leucine Rich Repeat
!! LRR_1	411	432	1	24	12.4	1.4	glocal Leucine Rich Repeat
!! Pkinase_Tyr	588	865	1	272	134.3	3.3e-37	glocal Protein tyrosine kinase
!! Pkinase	588	868	1	287	171.2	2.7e-48	glocal Protein kinase domain

Figure 17.1 Annotation error. Analysis of *Oryza sativa* sequence AAR00644.1 protein annotated as 'putative LRR receptor-like protein kinase' suggests the annotation is in error. (A) Top BLAST hits for the sequence are annotated as 'receptor kinase'. The top hit to a sequence outside rice, with an E-value of 3e-62, is a 'putative receptor protein kinase (TMK1)' in *Arabidopsis thaliana* (AAP04161.1). (B) PFAM analysis of *O. sativa* sequence AAR00644.1 shows only LRR domains. (C) PFAM analysis of the *Arabidopsis* sequence AAP04161.1 shows that it contains both LRR and protein kinase domains.

17.2.3 Speciation

Another source of annotation error results from transfer from a top database hit that has functionally diverged due to taxonomic distance from the query sequence. Orthologs – sequences related by speciation from a common ancestor – are usually assumed to have similar functions. However, evidence of orthology alone does not prove functional similarity. Two orthologous proteins from distantly related organisms could be divergent in function due to evolutionary distance [40,41]. Moreover protein families have varying rates of evolution, which cannot be detected by simple pairwise sequence comparisons.

17.2.4 Propagation of Existing Annotation Errors

Transferring annotations based on sequence similarity can propagate existing annotation errors. The actual percentage of annotation errors is not known, but is estimated at between 8 and 40% [42,43]. This is an unfortunate combination that calls for much more rigorous methods for annotating sequences with predicted function, and for storing the source and support for these predicted functions.

17.3 The Nomenclature of Homology

Since simple sequence similarity is insufficient to assign function, the development of a standardized and informative nomenclature for different types of homology relationships is developing, along with interesting and even entertaining debate in the community [44–46]. Genes within gene families are predicted to share similar function, but the tempo and mode of divergence may differ following gene duplication or speciation events. These efforts have yielded increasingly specific terms for different types of homology, reviewed below, and illustrated in Figure 17.2.

Orthologs are sometimes referred to as the 'same gene in different species', while *paralogs* are sometimes referred to as 'duplicated genes in the same species'. These terms are useful but insufficient to describe the complexity of gene family relationships and the implications of evolutionary events on function [28–30].

Gene duplication and speciation events are differentiated in a phylogeny through the use of species tree–gene tree reconciliation [47]. Rooting the tree is a prerequisite to this analysis (reviewed later in this chapter). Within a gene family, phylogeny traces back the ancestry of orthologs to the most recent speciation events among the species under consideration (Figure 17.2). 'Super-orthologs' (also called simple orthologs) can be identified in gene trees as genes separated by speciation events with no intervening gene duplications (e.g. A^2 and R^2 in Figure 17.2; [48]). Since lineage-specific duplication events can result in new copies in some species, such a strict assignment of orthology is not always possible. Gene duplications within

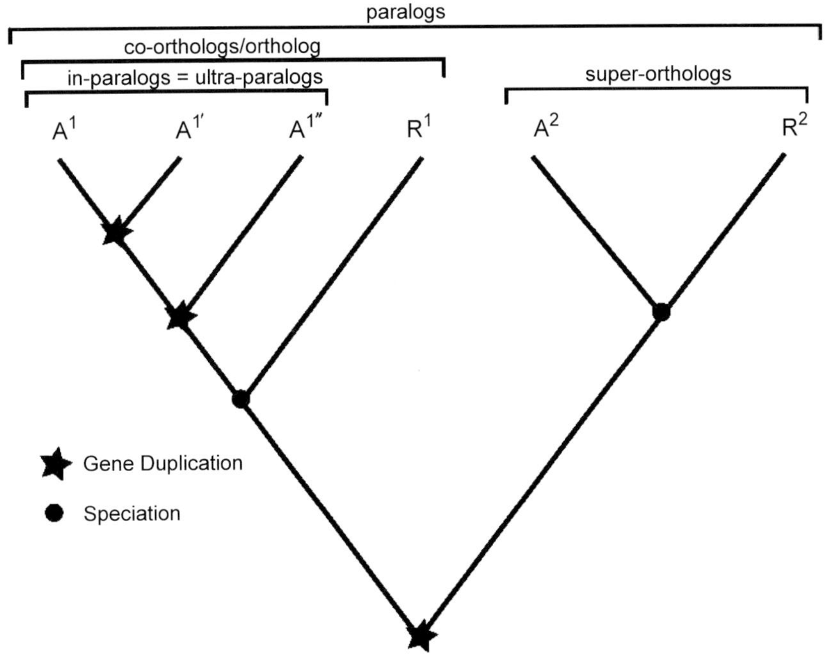

Figure 17.2 Gene tree showing various types of homology among genes sampled from two species (e.g. *Arabidopsis* and Rice). Genes in $A^{1,1',1''}/R^1$ clade are orthologous to genes in A^2/R^2 clade; $A^{1,1',1''}$ are in-paralogs (=ultra-paralogs) with respect to each other and co-orthologs with respect to R^1; A^2 and R^2 are super-orthologs. Nodes with stars represent gene duplications.

lineages can yield *co-orthologs across lineages,* sets of genes within one lineage that share a common ancestor with the same gene or gene clade in another species (e.g. A^1, $A^{1'}$ and $A^{1''}$ with respect to R^1 in Figure 17.2). From the perspective of gene family diversity within a single genome, these genes can also been classified as *in-paralogs* [49,50] (also called *ultra-paralogs* [48]) because they are paralogous within the lineage where the duplication(s) occurred (e.g. A^1, $A^{1'}$ and $A^{1''}$ in Figure 17.2). Therefore, all in-paralogs within one genome are also co-orthologs with respect to corresponding genes in another genome. Polyploidy is an important process in plant genome evolution and *homeologs* are a special class of duplicated (paralogous) genes arising from genome duplication. For excellent reviews of issues relating to this terminology, see [51,52].

Identifying ortholog/paralog relationships is particularly important in phylogenomic annotation because gene duplications can spawn functional divergence. Functional conservation is most likely in the case of super-orthologs, but there is typically some level of shared function for all genes in a gene family (e.g. all MADS box genes are transcription factors). Ortholog identification can carried out most accurately with the benefit of a rooted gene tree (Figure 17.2; see discussion of tree rooting below).

While the concepts of orthology and paralogy are explicitly phylogenetic, there are a variety of published algorithms that attempt to rapidly identify orthologs based on relative sequence similarity scores. While similarity-based methods of ortholog identification are fast, they can fail when substitution rates or rates of structural and functional divergence vary across a gene family. Formal phylogenetic analyses are not immune to all forms of rate variation, but they are typically more robust. For this reason, our protocol for phylogenomic inference includes full phylogenetic reconstruction.

Two popular methods for similarity-based approaches to ortholog identification include Inparanoid and OrthoMCL. Inparanoid [50,53] compares genes from species pairs; orthology predictions are based on reciprocal best BLAST hits found in all-against-all BLAST searches, and in-paralogs are based on genes within each species that are more similar to each other than the putative ortholog identified in the other species. OrthoMCL [54] uses Markov Clustering (MCL) [55] of all-against-all BLAST scores to identify co-orthologs from multiple species. MCL has also been used to identify more broadly defined gene families [56] from which orthology/paralogy can be estimated through formal phylogenetic analysis (e.g. [6,9]. Chen *et al.* [57] compare the relative accuracy of these approaches in terms of both false negative and false positive rates for assigning orthology.

Another class of approximate phylogenomic classification method makes use of pre-clustered clades represented by some form of statistical model (e.g. a profile or HMM); novel sequences can then be assigned to one of these clades for an automated (albeit approximate) phylogenetic classification. Main resources in this class include OrthologID [58], the PhyloFacts Phylogenomic Encyclopedias [59] and the Phytome plant gene family database [9]. We discuss these resources in detail later in this chapter.

While the nomenclature of homology may appear somewhat ponderous, it provides a framework upon which hypotheses of functional equivalence can be developed and experimentally tested. Nonetheless, inferred evolutionary relationships must be seen as just one aspect of a well-rounded functional analysis; although orthologs are often assumed to have the same function and paralogs are expected to diverge, examples to the contrary do exist [40,41]. Accurate prediction of function requires the use of a consensus approach including multiple sources of experimental data and inference protocols. We bring these approaches together in a protocol we call *structural phylogenomics.*

17.4
Structural Phylogenomic Inference of Function

Our approach to functional annotation uses *structural phylogenomics*, inferring the function of a protein in an evolutionary and structural context (Figure 17.3). The word 'phylogenomics' (combining 'phylogenetics' and 'genomics') was coined

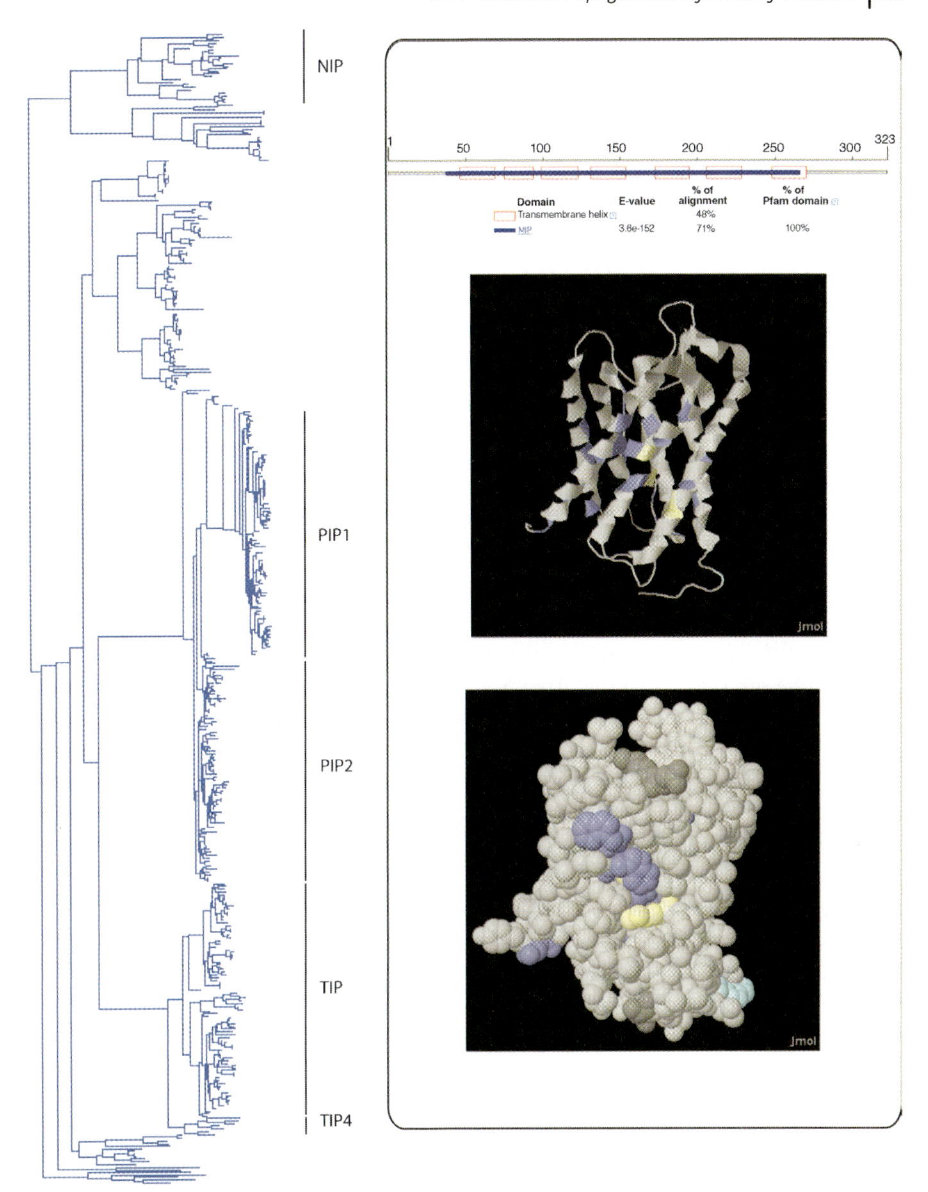

Figure 17.3 Example of phylogenomic analysis of the Major Intrinsic protein (MIP) family (http://phylogenomics.berkeley.edu/book/book_info.php?book=bpg025394). The Maximum Likelihood tree separates the various plant subtypes (NIP, PIP and TIP), which often have different substrate specificities. NIP, NOD26-like intrinsic protein; PIP, Plasma membrane intrinsic proteins; TIP, Tonoplast intrinsic protein. The NIP subfamily is known to transport both water and glycerol [106]. The TIP4 subfamily has been shown to transport glycerol and urea [107]. The panel on the right shows PFAM domain and structure prediction for this family with putative family and subfamily-specific residues mapped onto the structure. Integrating 3D structure enables identification of residues that are responsible for conferring functional specificities.

in 1998 by Jonathan Eisen to describe the use of evolutionary analysis in gene function prediction [60], and was designed to address the errors arising from annotation transfer between paralogous genes. Phylogenomics is based on the recognition that protein families are not static, but evolve novel functions through complex biological processes; integrating evolutionary analysis improves the accuracy and specificity of functional annotation [61–63]. The term 'phylogenomics' is also used to describe the integration of genomic and comparative genomic data in other types of analyses, including species tree reconstruction [64,65]. Structural phylogenomic inference of protein function integrates structural information in a phylogenomic analysis in two different ways. First homologs are restricted to those that can be inferred to share a common domain architecture, preventing annotation transfer between partial homologs related by domain shuffling. Second, three-dimensional structure prediction and analysis is employed to enable the correlation of changes in protein structure with changes in function.

Structural phylogenomics integrates phylogenomic approaches to differentiate orthologs and paralogs, structure analyses to differentiate between partial and global homologies, and distinguishes between annotations that have experimental support and those that have been derived using homology-based protocols. The integration of structural and experimental data in an evolutionary framework enables a nuanced prediction of function. Functional shifts in multi-gene families can be identified through concurrent examination of a phylogenetic tree, multiple sequence alignment, three-dimensional (3D) structure and experimental data, as illustrated in Figure 17.3.

In the remainder of this chapter we describe our recommended protocols for each step in a structural phylogenomic pipeline and discuss the challenges of this approach with particular reference to their application to plant gene families.

17.5
Recommended Protocols for a Structural Phylogenomic Pipeline

Structural phylogenomic inference of function (reviewed in [63]) starts with selection of a sequence for detailed study. Homologs are identified, and restricted to those sharing a common domain architecture based on sequence analysis. A multiple sequence alignment (MSA) is constructed and edited (a process known as 'masking'). The masked alignment is then used to construct a phylogenetic tree that is reconciled with a species tree to label internal nodes as indicating duplication or speciation events. Experimental data are overlaid on the tree and orthologous and paralogous relationships are defined based on analyzing the reconciled tree. This approach enables biologist to trace changes in molecular function or physiological role along the evolutionary tree. In cases where experimentally characterized orthologs are not available (or orthology detection is ambiguous), consistency of annotations within subtrees can be used as the basis for function prediction [48] (see Figure 17.4). Details of each of these steps follow.

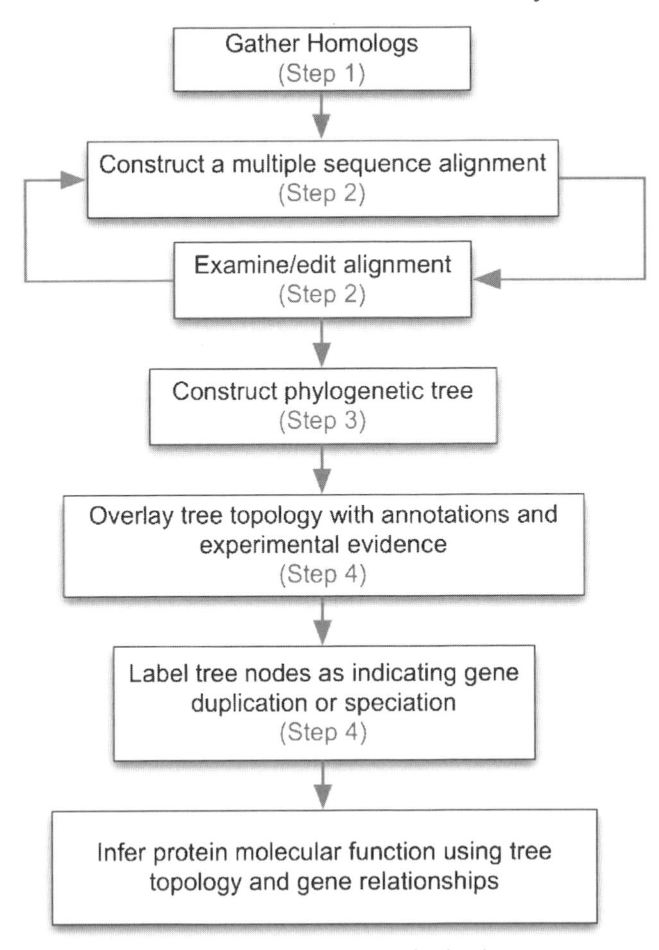

Figure 17.4 Phylogenomic pipeline. See text for details.

17.5.1
Step 1: Homolog Selection

Domain shuffling produces errors in annotation transfer based on partial homology. To overcome this problem, we recommend that homologs be restricted to those aligning over their entire lengths. An automated method for this task is FlowerPower [66], available at http://phylogenomics.berkeley.edu/flowerpower. FlowerPower was developed explicitly for use in a phylogenomic pipeline to select homologs sharing a common domain architecture [66]. FlowerPower is an iterated homolog-selection tool that uses subfamily hidden Markov models [67] to expand the cluster in each iteration, followed by alignment analysis to select sequences matching pre-specified criteria. This process helps ensure that existing subfamilies are fleshed out fully before members of new subfamilies are included, and avoids the retrieval of overly large clusters of highly variable sequences. An alternative approach would be to use similarity-based clustering (e.g. MCL [55]) of genes recovered from BLAST and PSI-BLAST [68,69], but this may not screen out all genes sharing only partial homology with the seed.

17.5.2
Step 2: Constructing and Analyzing a Multiple Sequence Alignment

Since the multiple sequence alignment is the source of phylogenetic signals, significant time and expertise should be devoted to this step. In practice, a practitioner of phylogenomic inference will alternate between homolog selection and alignment, using manual inspection of intermediate alignments to determine which sequences to include (based on agreement at apparently conserved motifs), re-aligning selected sequences, and continuing until a final set of sequences has been selected and aligned.

Available MSA (multiple sequence aligment methods include CLUSTALW [70], MUSCLE [71], MAFFT [72], T-Coffee [73] ProbCons [74] and SATCHMO [75]. The selection of an alignment method is dependent on available computational resources and the size and evolutionary divergence of the dataset, both of which affect the accuracy of alignment [76,77]. For large or divergent datasets, we recommend the use of MAFFT and MUSCLE due to their computational efficiency and excellent results. Alignment methods that employ a guide tree, such as ClustalW, include their own implicit phylogenetic bias, which may prove problematic.

A fundamental assumption of phylogenetic inference is *positional homology*, that is, that all residues in a column of an MSA descend from a common ancestral character. This assumption is problematic in multi-gene families; studies of protein 3D structures have demonstrated that homologous proteins can have limited structural superposability. Surface loop regions are particularly vulnerable to insertions and deletions across homologous proteins, and even secondary structures (beta sheet and alpha helices) can diverge across a family [20]. Moreover, protein superfamilies can include members with extremely low similarity to others (e.g. pairwise identities in the teens or lower). Alignment accuracy is affected by the degree of divergence between sequences: errors increase dramatically as sequence identities drop below 30%, and no method is successful at reconstructing the structural alignment when identities drop below 20% [20,78].

To address this issue of ambiguously aligned regions and avoid the intrusion of noise in the phylogenetic reconstruction, alignment construction is followed by *alignment masking* to restrict the alignment to regions contributing phylogenetic signals [60,63,79]. Alignment masking generally requires the removal of columns whose accuracy or positional homology appear uncertain; the editing process may also remove sequences which do not agree at apparently (or known) critical residues or motifs, or which have many gap characters. While some of these steps can be automated, the best results require manual supervision of the process including knowledge of key functional residues and some understanding of the physicochemical characteristics of amino acids. The effect of different protocols for alignment masking in application to protein superfamilies is not well understood; we have observed significant differences in tree topologies resulting from fairly modest modifications to a masking protocol (e.g. removing columns with >70% gap characters versus removing columns with >50% gap characters). Both alignment masking and the initial alignment analysis steps require the use of an alignment editor/viewer; we recommend Belvu (http://www.cgb.ki.se/cgb/groups/sonnhammer/Belvu.html) for these tasks.

17.5.3
Step 3: Constructing and Analyzing a Phylogenetic Tree

The selection of a method for phylogenetic reconstruction of a dataset involves the same considerations as the selection of an alignment method: available resources and the characteristics of the data (dataset size and evolutionary divergence). For large datasets (e.g. in the hundreds or thousands of sequences), the use of character methods such as Maximum Likelihood, Maximum Parsimony and Bayesian approaches may be infeasible due to their computational complexity. Distance-based methods such as Neighbor-Joining are generally much faster, and therefore often preferred for high-throughput studies, but may be less robust to rate variation and structural changes across family members common in protein superfamilies. Sources of phylogenetic tree construction and visualization tools are shown in Table 17.3. Due to space limitations, we have not presented guidelines for the use of these tools, and direct readers to the corresponding publications and published guidelines produced by method authors.

Phylogenetic reconstruction of protein superfamilies is inherently uncertain: tree topologies produced by different methods can vary dramatically for the same input [80]. In our experience, differences in tree topology across methods are almost always in the coarse branching order between conserved clades. Closely related sequences (e.g. those having >50% identity) are usually clustered together in a

Table 17.3 Selected phylogenetic software resources.

	Methods	URL
Tree Construction	PHYLIP	http://evolution.genetics.washington.edu/phylip.html
	PAUP	http://paup.csit.fsu.edu/
	Mr.Bayes	http://mrbayes.csit.fsu.edu/
	PHYML	http://atgc.lirmm.fr/phyml/
Tree visualization and annotation	TreeView	http://taxonomy.zoology.gla.ac.uk/rod/treeview.html
	ATV	http://www.phylosoft.org/atv/
	TreeDyn	http://www.treedyn.org/

Several methods are available for the tasks of tree construction and editing. We list a selected few here. For a comprehensive listing of phylogeny programs see http://evolution.genetics.washington.edu/phylip/software.html.

phylogenetic tree, regardless of the method used. While small variations in the branching order within such closely related subgroups can occur between phylogenetic methods, these have little impact on a phylogenomic analysis [51]. In contrast, the branching order *among* subtrees is the crux of the matter, as these can have a dramatic impact on function prediction using phylogenomic inference. In practice, a sequence whose function is being inferred using phylogenomic inference has often been selected due to having no close homologs with experimentally determined function, or for which detected homologs have multiple distinct functions. The branching order between the clade containing the unknown sequence and clades containing experimentally characterized sequences will thus have a major impact on the function predicted for this sequence.

Since tree topology is used to infer function, this inherent ambiguity in tree topology across methods presents a problem: which method should a biologist use? Simulation studies used to estimate the accuracy of phylogenetic methods do not directly and (in our opinion) adequately assess the impact of commonly occurring traits of protein superfamily evolution, including position- and lineage-specific rate variation, lack of positional homology in many alignment columns, and the unavoidable alignment errors. For instance, it has been shown that site- and lineage-specific rate variation (covarion evolution or heterotachy) can mislead phylogenetic methods [81–83] and extreme variation in rates among lineages or uneven sampling across a gene family phylogeny can cause phylogenetic reconstruction methods to wrongly infer divergent sequences as closely related due to a phenomenon called *long-branch attraction* [84,85].

Faced with these issues and the inherent ambiguity in tree topologies across methods, what should a practitioner of phylogenomic inference do? If phylogenomic predictions are to be used to guide expensive experimental investigation, we recommend a conservative approach that looks for consensus across different trees. One way to find such a consensus is the use of bootstrap analysis. Bootstrap analysis is a statistical approach to summarizing the degree of support for each node (clade) in a phylogeny [86]. Bootstrap analysis involves resampling (with replacement) the columns of the input MSA to produce a set of alignments of the same length as the original, followed by phylogenetic tree estimation for each pseudoreplicate alignment. The bootstrap value for each node in the consensus tree represents the fraction of times that the node was found in the bootstrap replicate trees (ignoring differences in tree topology below that node). Bootstrap support values >90% are generally interpreted as sufficient to infer support for an evolutionary grouping, with values below 70% seen as unreliable; intermediate values should be interpreted as simply suggestive. In Bayesian phylogenetic analyses, posterior probabilities are also interpreted in terms of nodal support (see [87]). However, Bayesian posterior probabilities are estimated on the original alignment without resampling, and are therefore not directly comparable to bootstrap support values [88]. Within a likelihood framework, an 'approximate likelihood ratio test' has recently been developed as another alternative for testing the credibility of inferred nodes within a tree [89]. While all of these approaches are certainly important, they do not completely address possible biases in the input alignment or in the phylogenetic method used. Because of this, our recommendation in phylogenetic tree construction is to hedge one's bets by supplementing bootstrap analysis, Bayesian or likelihood methods with consensus approaches across different phylogenetic methods, preferably using multiple sequence alignments constructed and masked using different protocols. It may be possible to use simulation studies to identify the root cause of conflicts in trees estimated using different methods (e.g. [90]). The PHYLIP consense program can be used to find strict or majority-rule consensus from different inputs [91].

Distinguishing between orthologs and paralogs is critical for functional inference. Once a phylogenetic tree topology has been estimated, the gene tree and an accepted species tree should be reconciled to assist in the discrimination between orthologs and paralogs. This process can be performed manually, or an automated method for tree reconciliation can be used [47]. Other automated methods for predicting orthologs based on phylogenetic tree analysis include Orthostrapper [92] and RIO [48].

In all phylogentically-based methods of ortholog identification, rooting of the gene tree is a critical step. Two approaches for rooting phylogenies are commonly applied and each can be susceptible to variation in evolutionary rates across a phylogeny. *Outgroup rooting* is commonly used in analyses of organismal relationships. For example, gymnosperms, ferns or even mosses could be used to root phylogenetic analyses of relationships among major angiosperm lineages (e.g. [90]). In gene tree estimation, where ancient duplication events can result in differences in taxonomic distributions across subfamilies in a multi-gene family, a priori assignment of outgroup genes is far from obvious. An alternative approach in this case is *midpoint rooting*. Midpoint rooting assumes a molecular clock, since it estimates the root that minimizes the average divergence distance between the root and the leaves of the tree (taxa). Multi-gene families often evolve with non-clocklike behavior, causing problems with this assumption. Given the possible pitfalls associated with both outgroup rooting and midpoint rooting, it is prudent to interpret the inferred rooting with skepticism and consider how alternative rootings would influence functional assignments.

Once a phylogenetic tree topology has been estimated, the gene tree and an accepted species tree should be reconciled to assist in the discrimination between orthologs and paralogs. This process can be performed manually, or an automated method for tree reconciliation can be used [47]. Other automated methods for predicting orthologs based on phylogenetic tree analysis include Orthostrapper [92] and RIO [48].

17.5.4 Step 4: Predicting Function using a Phylogenetic Tree

The final step in phylogenomic inference is to overlay experimental data and other annotations onto the reconciled phylogenetic tree and predict function based on an analysis of the tree topology. The Gene Ontology (GO) project established in 1998 to provide a controlled vocabulary for various types of gene function [93] is a key resource at this stage. The GO ontology has a hierarchical structure, so that functions can be described at different levels of specificity. Ontologies have been developed for describing molecular functions, biological processes and cellular localization. The ontology terms are assigned to sequences in a database based on either experimental evidence or sequence homology and different evidence codes indicate the source of the assignment. For example, the evidence code *IDA* stands for 'Inferred from Direct Assay', while *IEA* stands for 'Inferred from Electronic Annotation'. In assigning function by homology, it is critical to check the evidence code of the source, as any GO IEA assignments will also suffer from the same problems as those affecting annotation transfer as discussed earlier.

The paucity of sequences with experimental evidence cause some challenges at this point, particularly if annotation transfer is restricted to super-orthologs; our results on 370 000 UniProt [94] sequences with Gene Ontology annotation and evidence codes shows that <3% of sequences have experimental support. This limitation can be overcome by predicting function based on consistency in experimentally determined function of *subtree neighbors* [48]. In predicting function for a sequence of interest, the evolutionary persistence of different types of functional traits should be kept in mind. Some traits, such as protein 3D-fold and catalytic function, are maintained across large evolutionary distances. Other traits can diverge fairly rapidly, and may only be conserved within strict orthologs; substrate specificity, pathway participation and tissue expression fall into this class.

17.6
Web Servers and Databases useful in Phylogenomic Inference

Given the complexity of phylogenomic inference, and its dependence on expertise in many different bioinformatics analyses, a few resources provide pre-computed phylogenetic and clustering analyses for protein families, including some that specialize in protein families found in plants (see Table 17.1). PhyloFacts [59], OrthologID [58] and the Phytome plant gene family database [9] provide pre-computed phylogenies for thousands of plant gene families. *PhyloFacts* is a phylogenomic encyclopedia including protein families found across the Tree of Life. To date (May, 2007), PhyloFacts contains over 40 000 'books' representing protein families and structural domains and over 1 million hidden Markov models enabling classification of novel sequences to families and subfamilies. A specialized library for protein families involved in plant disease resistance and stress pathways is available at http://phylogenomics.berkeley.edu/PlantResistanceGene. PhyloFacts and Phytome enable biologists to submit novel sequences for classification to these phylogenies using clade-specific HMM (hidden Markov model profiles [95], which can be used to approximate the phylogenetic position. PhyloFacts includes subfamily HMMs for predicted functional subfamilies in multi-gene families using the SCI-PHY (Subfamily Classification In PHYlogenomics) tool [96], while Phytome includes HMMs for similarity-based subclusters that typically form subclades in gene trees. OrthologID takes a similar approach, but *synapomorphies* – derived character states shared by members of a clade – rather than HMM profiles are used to diagnose clade membership for genes not included in the formal parsimony-based phylogenetic analysis of rice and *Arabidopsis* gene families.

All functional annotations must be revised as new evidence becomes available for homologs, or as predictive methods improve in accuracy. A re-annotation protocol is critical to flag (and potentially correct) existing annotation errors [97]. To provide for continuous re-annotation of sequences and revised predictions of function and structure, the PhyloFacts resource is updated periodically to include new sequences and experimental data, re-estimate improved phylogenies, and check for homologous 3D structures for protein structure prediction. To facilitate the contribution of expert biologists to the annotation process, PhyloFacts includes tools for group or individual online annotation of protein families of interest in a virtual collaborative environment (http://phylogenomics.berkeley.edu/phylofacts/).

Other tools useful in phylogenomic inference include PhyloBuilder [98]. PhyloBuilder takes a user-supplied sequence, retrieves global homologs using FlowerPower, masks the alignment conservatively and constructs a Neighbor-Joining tree using the PHYLIP software. Functional subfamilies are predicted using the SCI-PHY program [96], followed by prediction of protein 3D structure and PFAM domains, retrieval of Gene Ontology annotations and evidence codes, and protein localization prediction. The alignment, tree and other data associated with the PhyloBuilder pipeline can be downloaded or viewed on the website.

17.7 Discussion

In addition to all the technical challenges inherent in a phylogenetic tree construction, automatic annotation of sequences using phylogenomic inference faces the challenge of non-standard gene and protein names and widely varying terminologies to describe the same function. For instance, a gene in *Drosophila* and its human ortholog are likely to have very different names, and their annotations may seem inconsistent. To automate the transfer of annotation and the detection of existing annotation errors, a controlled vocabulary describing protein function is required. The GO presents an advance in this area, but many gaps in specific molecular function remain. A recent method, SIFTER, tackles the problem of automatic functional assignment in a phylogenomic framework using a Bayesian approach to propagate functional annotation (such as GO annotations) along a phylogenetic tree [99].

We have focused in this chapter on the use of phylogenomic inference for function prediction. However, many apparently distinct types of bioinformatics prediction efforts can be improved through the use of structural phylogenomics to identify the functional equivalents between different species. We have focused on the phylogenomic analysis of proteins sharing a common domain architecture as the basis of function prediction. On the other hand, phylogenomic analysis is commonly applied to individual structural or functional domains [100]. Such domain-based studies can generate clues to function that may not be possible if the analysis is restricted to global homologs. Curated domain databases such as InterPro [101] are valuable resources for such analyses, and can be used as the starting point for such domain-based phylogenetic studies. Domain shuffling can result in incongruence between phylogenies estimated by the globally alignable proteins along their entire lengths and those based on individual domains; comparing these different phylogenies in the context of experimental data can yield insights into the distinct functional roles of these structural building blocks of multi-domain proteins, and assist in the evolutionary reconstruction of the larger family of proteins sharing only partial homology [102].

Another area of obvious application of phylogenomics is in the prediction of interacting partners and biological process, such as in the construction of phylogenomic profiles [14]. The assumption behind phylogenomic profiles is that functionally linked proteins will exhibit correlated evolution, that is, they will all be lost or conserved in a new species. Phylogenomic profile analysis involves construction of a matrix indicating protein presence or absence in different species. Proteins that share a similar profile have been shown to be functionally linked [103–105]. Phylogenomic profiles are sensitive to the same potential systematic errors as standard annotation transfer methods, with proteins being labeled as present or absent in a species based on simple sequence comparison. Rigorous phylogenomic approaches, such as those described here, enable far more accurate separation of orthologs and paralogs, and can be used to improve the specificity of phylogenomic profiles.

Equivalently, improving orthology identification with phylogenomics can be expected to improve the accuracy of pathway reconstruction: predicting a pathway in a novel organism based on an experimentally characterized pathway in a reference organism. This analysis can help discriminate between pathways that are essentially or completely conserved and those that may have diverged between the two species, perhaps due to duplication or domain shuffling events.

In summary, international initiatives to elucidate the function of plant genes have been extremely productive, but the pace of gene and genome sequencing exceeds the pace of experimental characterization of these genes. Further, only a small fraction of all plant species are amenable to experimental investigation. Phylogenomic analyses will play an important role in the transfer of knowledge from experimentally tractable systems to all plant species. This process will be facilitated with the development of faster and more robust computational tools and a broader diversity of whole genome sequences and experimentally tractable model systems.

References

1 Haas, B.J., Wortman, J.R., Ronning, C.M., Hannick, L.I., Smith, R.K., Jr, Maiti, R., Chan, A.P., Yu, C., Farzad, M., Wu, D., White, O. and Town, C.D. (2005) Complete reannotation of the *Arabidopsis* genome: methods, tools, protocols and the final release. *BMC Bioinformatics*, **3**, 7.

2 Itoh, T., Tanaka, T., Barrero, R.A., Yamasaki, C., Fujii, Y., Hilton, P.B., Antonio, B.A., Aono, H., Apweiler, R., Bruskiewich, R., Bureau, T., Burr, F., Costa de Oliveira, A., Fuks, G., Habara, T., Haberer, G., Han, B., Harada, E., Hiraki, A.T., Hirochika, H., Hoen, D., Hokari, H.,

Hosokawa, S., Hsing, Y.I., Ikawa, H., Ikeo, K., Imanishi, T., Ito, Y., Jaiswal, P., Kanno, M., Kawahara, Y., Kawamura, T., Kawashima, H., Khurana, J.P., Kikuchi, S., Komatsu, S., Koyanagi, K.O., Kubooka, H., Lieberherr, D., Lin, Y.C., Lonsdale, D., Matsumoto, T., Matsuya, A., McCombie, W.R., Messing, J., Miyao, A., Mulder, N., Nagamura, Y., Nam, J., Namiki, N., Numa, H., Nurimoto, S., O'Donovan, C., Ohyanagi, H., Okido, T., Oota, S., Osato, N., Palmer, L.E., Quetier, F., Raghuvanshi, S., Saichi, N., Sakai, H., Sakai, Y., Sakata, K., Sakurai, T., Sato, F., Sato, Y., Schoof, H., Seki, M., Shibata, M., Shimizu, Y., Shinozaki, K., Shinso, Y., Singh, N.K., Smith-White, B., Takeda, J., Tanino, M., Tatusova, T., Thongjuea, S., Todokoro, F., Tsugane, M., Tyagi, A.K., Vanavichit, A., Wang, A., Wing, R.A., Yamaguchi, K., Yamamoto, M., Yamamoto, N., Yu, Y., Zhang, H., Zhao, Q., Higo, K., Burr, B., Gojobori, T. and Sasaki, T. (2007) Curated genome annotation of *Oryza sativa* ssp. *japonica* and comparative genome analysis with *Arabidopsis thaliana*. *Genome Research*, **17** 175–183.

3 Ouyang, S., Zhu, W., Hamilton, J., Lin, H., Campbell, M., Childs, K., Thibaud-Nissen, F., Malek, R.L., Lee, Y., Zheng, L., Orvis, J., Haas, B., Wortman, J. and Buell, C.R. (2007) The TIGR Rice Genome Annotation Resource: improvements and new features. *Nucleic Acids Research*, **35**, D883–D887.

4 Tuskan, G.A., Difazio, S., Jansson, S., Bohlmann, J., Grigoriev, I., Hellsten, U., Putnam, N., Ralph, S., Rombauts, S., Salamov, A., Schein, J., Sterck, L., Aerts, A., Bhalerao, R.R., Bhalerao, R.P., Blaudez, D., Boerjan, W., Brun, A., Brunner, A., Busov, V., Campbell, M., Carlson, J., Chalot, M., Chapman, J., Chen, G.L., Cooper, D., Coutinho, P.M., Couturier, J., Covert, S., Cronk, Q., Cunningham, R., Davis, J., Degroeve, S., Dejardin, A., Depamphilis, C., Detter, J., Dirks, B., Dubchak, I., Duplessis, S., Ehlting, J., Ellis, B., Gendler, K., Goodstein, D., Gribskov, M., Grimwood, J., Groover, A., Gunter, L., Hamberger, B., Heinze, B., Helariutta, Y., Henrissat, B., Holligan, D., Holt, R., Huang, W., Islam-Faridi, N., Jones, S., Jones-Rhoades, M., Jorgensen, R., Joshi, C., Kangasjarvi, J., Karlsson, J., Kelleher, C., Kirkpatrick, R., Kirst, M., Kohler, A., Kalluri, U., Larimer, F., Leebens-Mack, J., Leple, J.C., Locascio, P., Lou, Y., Lucas, S., Martin, F., Montanini, B., Napoli, C., Nelson, D.R., Nelson, C., Nieminen, K., Nilsson, O., Pereda, V., Peter, G., Philippe, R., Pilate, G., Poliakov, A., Razumovskaya, J., Richardson, P., Rinaldi, C., Ritland, K., Rouze, P., Ryaboy, D., Schmutz, J., Schrader, J., Segerman, B., Shin, H., Siddiqui, A., Sterky, F., Terry, A., Tsai C.J., Uberbacher E., Unneberg P. *et al.* (2006) The genome of black cottonwood. *Populus trichocarpa* (Torr. & Gray). *Science*, **313**, 1596–1604.

5 Dong, Q., Schlueter, S.D. and Brendel, V. (2004) PlantGDB, plant genome database and analysis tools. *Nucleic Acids Research*, **32**, D354–D359.

6 Albert, V.A., Soltis, D.E., Carlson, J.E., Farmerie, W.G., Wall, P.K., Ilut, D.C., Solow, T.M., Mueller, L.A., Landherr, L.L., Hu, Y., Buzgo, M., Kim, S., Yoo, M.J., Frohlich, M.W., Perl-Treves, R., Schlarbaum, S.E., Bliss, B.J., Zhang, X., Tanksley, S.D., Oppenheimer, D.G., Soltis, P.S., Ma, H., DePamphilis, C.W. and Leebens-Mack, J.H. (2005) Floral gene resources from basal angiosperms for comparative genomics research. *BMC Plant Biology*, **5**, 5.

7 Childs, K.L., Hamilton, J.P., Zhu, W., Ly, E., Cheung, F., Wu, H., Rabinowicz, P.D., Town, C.D., Buell, C.R. and Chan, A.P. (2007) The TIGR Plant Transcript Assemblies database. *Nucleic Acids Research*, **35**, D846–D851.

8 Lee, Y., Tsai, J., Sunkara, S., Karamycheva, S., Pertea, G., Sultana, R., Antonescu, V., Chan, A., Cheung, F. and Quackenbush, J. (2005) The TIGR Gene Indices: clustering and assembling EST and known genes and integration with eukaryotic genomes. *Nucleic Acids Research*, **33**, D71–D74.

9 Hartmann, S., Lu, D., Phillips, J. and Vision, T.J. (2006) Phytome: a platform for plant comparative genomics. *Nucleic Acids Research*, **34**, D724–D730.

10 Friedberg, I. (2006) Automated protein function prediction – the genomic challenge. *Briefings in Bioinformatics*, **7**, 225–242.

11 Valencia, A. (2005) Automatic annotation of protein function. *Current Opinion in Structural Biology*, **15**, 267–274.

12 Ouzounis, C.A. and Karp, P.D. (2002) The past, present and future of genome-wide re-annotation. *Genome Biology*, **3**, COMMENT2001.

13 Rost, B., Liu, J., Nair, R., Wrzeszczynski, K.O. and Ofran, Y. (2003) Automatic prediction of protein function. *Cellular and Molecular Life Sciences: CMLS*, **60**, 2637–2650.

14 Pellegrini, M., Marcotte, E.M., Thompson, M.J., Eisenberg, D. and Yeates, T.O. (1999) Assigning protein functions by comparative genome analysis: protein phylogenetic profiles. *Proceedings of the National Academy of Sciences of the United States of America*, **96**, 4285–4288.

15 Krogh, A., Larsson, B., von Heijne, G. and Sonnhammer, E.L. (2001) Predicting transmembrane protein topology with a hidden Markov model: application to complete genomes. *Journal of Molecular Biology*, **305**, 567–580.

16 Nielsen, H., Brunak, S. and von Heijne, G. (1999) Machine learning approaches for the prediction of signal peptides and other protein sorting signals. *Protein Engineering*, **12**, 3–9.

17 Holm, L. and Sander, C. (1995) Dali: a network tool for protein structure comparison. *Trends in Biochemical Sciences*, **20**, 478–480.

18 Madej, T., Gibrat, J.F. and Bryant, S.H. (1995) Threading a database of protein cores. *Proteins*, **23**, 356–369.

19 Shindyalov, I.N. and Bourne, P.E. (1998) Protein structure alignment by incremental combinatorial extension (CE) of the optimal path. *Protein Engineering*, **11**, 739–747.

20 Baker, D. and Sali, A. (2001) Protein structure prediction and structural genomics. *Science*, **294**, 93–96.

21 Petrey, D. and Honig, B. (2005) Protein structure prediction: inroads to biology. *Molecules and Cells*, **20**, 811–819.

22 Dunbrack, R.L. Jr (2006) Sequence comparison and protein structure prediction. *Current Opinion in Structural Biology*, **16**, 374–384.

23 Altschul, S.F., Gish, W., Miller, W., Myers, E.W. and Lipman, D.J. (1990) Basic local alignment search tool. *Journal of Molecular Biology*, **215**, 403–410.

24 Bork, P. and Koonin, E.V. (1998) Predicting functions from protein sequences – where are the bottlenecks? *Nature Genetics*, **18**, 313–318.

25 Galperin, M.Y. and Koonin, E.V. (1998) Sources of systematic error in functional annotation of genomes: domain rearrangement, non-orthologous gene displacement and operon disruption. *In Silico* Biology, **1**, 55–67.

26 Gerlt, J.A. and Babbitt, P.C. (2000) Can sequence determine function? *Genome Biology*, **1**, REVIEWS0005.

27 Duarte, J.M., Cui, L., Wall, P.K., Zhang, Q., Zhang, X., Leebens-Mack, J., Ma, H., Altman, N. and dePamphilis, C.W. (2006) Expression pattern shifts following duplication indicative of subfunctionalization and neofunctionalization in regulatory genes of *Arabidopsis*. *Molecular Biology and Evolution*, **23**, 469–478.

28 Force, A., Lynch, M., Pickett, F.B., Amores, A., Yan, Y.L. and Postlethwait, J. (1999) Preservation of duplicate genes by complementary, degenerative mutations. *Genetics*, **151**, 1531–1545.

29 Lynch, M. and Conery, J.S. (2000) The evolutionary fate and consequences of duplicate genes. *Science*, **290**, 1151–1155.

30 Ohno, S. (1970) *Evolution by Gene Duplication*, Springer-Verlag, New York.

31 Simillion, C., Vandepoele, K., Van Montagu, M.C., Zabeau, M. and Van de Peer, Y. (2002) The hidden duplication past of *Arabidopsis thaliana. Proceedings of the National Academy of Sciences of the United States of America*, **99**, 13627–13632.

32 Bowers, J.E., Chapman, B.A., Rong, J. and Paterson, A.H. (2003) Unravelling angiosperm genome evolution by phylogenetic analysis of chromosomal duplication events. *Nature*, **422**, 433–438.

33 Fritz-Laylin, L.K., Krishnamurthy, N., Tor, M., Sjölander, K.V. and Jones, J.D. (2005) Phylogenomic analysis of the receptor-like proteins of rice and *Arabidopsis. Plant Physiology*, **138**, 611–623.

34 Jones, D.A., Thomas, C.M., Hammond-Kosack, K.E., Balint-Kurti, P.J. and Jones, J.D. (1994) Isolation of the tomato Cf-9 gene for resistance to *Cladosporium fulvum* by transposon tagging. *Science*, **266**, 789–793.

35 Jeong, S., Trotochaud, A.E. and Clark, S.E. (1999) The *Arabidopsis* CLAVATA2 gene encodes a receptor-like protein required for the stability of the CLAVATA1 receptor-like kinase. *Plant Cell*, **11**, 1925–1934.

36 Bork, P., Dandekar, T., Diaz-Lazcoz, Y., Eisenhaber, F., Huynen, M. and Yuan, Y. (1998) Predicting function: from genes to genomes and back. *Journal of Molecular Biology*, **283**, 707–725.

37 Apic, G., Huber, W. and Teichmann, S.A. (2003) Multi-domain protein families and domain pairs: comparison with known structures and a random model of domain recombination. *Journal of Structural and Functional Genomics*, **4**, 67–78.

38 Ekman, D., Bjorklund, A.K., Frey-Skott, J. and Elofsson, A. (2005) Multi-domain proteins in the three kingdoms of life: orphan domains and other unassigned regions. *Journal of Molecular Biology*, **348**, 231–243.

39 Shiu, S.H. and Bleecker, A.B. (2003) Expansion of the receptor-like kinase/Pelle gene family and receptor-like proteins in Arabidopsis. *Plant Physiology*, **132**, 530–543.

40 Causier, B., Castillo, R., Zhou, J., Ingram, R., Xue, Y., Schwarz-Sommer, Z. and Davies, B. (2005) Evolution in action: following function in duplicated floral homeotic genes. *Current Biology*, **15**, 1508–1512.

41 Kramer, E.M., Jaramillo, M.A. and Di Stilio, V.S. (2004) Patterns of gene duplication and functional evolution during the diversification of the AGAMOUS subfamily of MADS box genes in angiosperms. *Genetics*, **166**, 1011–1023.

42 Brenner, S.E. (1999) Errors in genome annotation. *Trends in Genetics*, **15**, 132–133.

43 Devos, D. and Valencia, A. (2001) Intrinsic errors in genome annotation. *Trends in Genetics*, **17**, 429–431.

44 Petsko, G.A. (2001) Homologuephobia. *Genome Biology*, **2**, COMMENT1002.

45 Koonin, E.V. (2001) An apology for orthologs – or brave new memes. *Genome Biology*, **2**, COMMENT1005.

46 Jensen, R.A. (2001) Orthologs and paralogs – we need to get it right. *Genome Biology*, **2**, INTERACTIONS1002.

47 Page, R.D. (1998) GeneTree: comparing gene and species phylogenies using reconciled trees. *Bioinformatics*, **14**, 819–820.

48 Zmasek, C.M. and Eddy, S.R. (2002) RIO: analyzing proteomes by automated phylogenomics using resampled inference of orthologs. *BMC Bioinformatics*, **3**, 14.

49 Sonnhammer, E.L. and Koonin, E.V. (2002) Orthology, paralogy and proposed classification for paralog subtypes. *Trends in Genetics*, **18**, 619–620.

50 Remm, M., Storm, C.E. and Sonnhammer, E.L. (2001) Automatic clustering of orthologs and in-paralogs from pairwise species comparisons.

Journal of Molecular Biology, **314**, 1041–1052.

51 Thornton, J.W. and DeSalle, R. (2000) Gene family evolution and homology: genomics meets phylogenetics. *Annual Review of Genomics and Human Genetics*, **1**, 41–73.

52 Koonin, E.V. (2005) Orthologs, paralogs, and evolutionary genomics. *Annual Review of Genetics*, **39**, 309–338.

53 O'Brien, K.P., Remm, M. and Sonnhammer, E.L. (2005) Inparanoid: a comprehensive database of eukaryotic orthologs. *Nucleic Acids Research*, **33**, D476–D480.

54 Li, L., Stoeckert, C.J., Jr and Roos, D.S. (2003) OrthoMCL: identification of ortholog groups for eukaryotic genomes. *Genome Research*, **13**, 2178–2189.

55 Van Dongen, S. (2000) Graph Clustering by Flow Simulation. Ph.D. Thesis, University of Utrecht. The Netherlands.

56 Enright, A.J., Van Dongen, S. and Ouzounis, C.A. (2002) An efficient algorithm for large-scale detection of protein families. *Nucleic Acids Research*, **30**, 1575–1584.

57 Chen, F., Mackey, A.J., Vermunt, J.K. and Roos, D.S. (2007) Assessing performance of orthology detection strategies applied to eukaryotic, genomes. *PLoS One*, **2**, e383.

58 Chiu, J.C., Lee, E.K., Egan, M.G., Sarkar, I.N., Coruzzi, G.M. and DeSalle, R. (2006) OrthologID: automation of genome-scale ortholog identification within a parsimony framework. *Bioinformatics*, **22**, 699–707.

59 Krishnamurthy, N., Brown, D.P., Kirshner, D. and Sjölander, K. (2006) PhyloFacts: an online structural phylogenomic encyclopedia for protein functional and structural classification. *Genome Biology*, **7**, R83.

60 Eisen, J.A. (1998) Phylogenomics: improving functional predictions for uncharacterized genes by evolutionary analysis. *Genome Research*, **8**, 163–167.

61 Eisen, J.A. and Hanawalt, P.C. (1999) A phylogenomic study of DNA repair genes, proteins, and processes. *Mutation Research*, **435**, 171–213.

62 Brown, D. and Sjölander, K. (2006) Functional classification using phylogenomic inference. *PLoS Computational Biology*, **2**, e77.

63 Sjölander, K. (2004) Phylogenomic inference of protein molecular function: advances and challenges. *Bioinformatics*, **20**, 170–179.

64 Philippe, H. and Blanchette, M. (2007) Overview of the first phylogenomics conference. *BMC Evolutionary Biology*, **7** (Suppl 1), S1.

65 Leebens-Mack, J., Vision, T., Brenner, E., Bowers, J.E., Cannon, S., Clement, M.J., Cunningham, C.W., dePamphilis, C., deSalle, R., Doyle, J.J., Eisen, J.A., Gu, X., Harshman, J., Jansen, R.K., Kellogg, E.A., Koonin, E.V., Mishler, B.D., Philippe, H., Pires, J.C., Qiu, Y.L., Rhee, S.Y., Sjölander, K., Soltis, D.E., Soltis, P.S., Stevenson, D.W., Wall, K., Warnow, T. and Zmasek, C. (2006) Taking the first steps towards a standard for reporting on phylogenies: Minimum Information About a Phylogenetic Analysis (MIAPA). *Omics*, **10**, 231–237.

66 Krishnamurthy, N., Brown, D., Sjölander, K. (2007) FlowerPower: clustering proteins into domain architecture classes for phylogenomic inference of protein function. *BMC Evolutionary Biology*, **7** (Suppl 1), S12.

67 Krogh, A., Brown, M., Mian, I.S., Sjölander, K. and Haussler, D. (1994) Hidden Markov models in computational biology. Applications to protein modeling. *Journal of Molecular Biology*, **235**, 1501–1531.

68 Altschul, S.F. and Koonin, E.V. (1998) Iterated profile searches with PSI-BLAST – a tool for discovery in protein databases. *Trends in Biochemical Sciences*, **23**, 444–447.

69 Altschul, S.F., Madden, T.L., Schaffer, A.A., Zhang, J., Zhang, Z., Miller, W. and Lipman, D.J. (1997) Gapped BLAST and PSI-BLAST: a new generation of protein database search programs. *Nucleic Acids Research*, **25**, 3389–3402.

70 Thompson, J.D., Higgins, D.G. and Gibson, T.J. (1994) CLUSTAL W: improving the sensitivity of progressive multiple sequence alignment through sequence weighting, position-specific gap penalties and weight matrix choice. *Nucleic Acids Research*, **22**, 4673–4680.

71 Edgar, R.C. (2004) MUSCLE: a multiple sequence alignment method with reduced time and space complexity. *BMC Bioinformatics*, **5**, 113.

72 Katoh, K., Misawa, K., Kuma, K. and Miyata, T. (2002) MAFFT: a novel method for rapid multiple sequence alignment based on fast Fourier transform. *Nucleic Acids Research*, **30**, 3059–3066.

73 Notredame, C., Higgins, D.G. and Heringa, J. (2000) T-Coffee: A novel method for fast and accurate multiple sequence alignment. *Journal of Molecular Biology*, **302**, 205–217.

74 Do, C.B., Mahabhashyam, M.S., Brudno, M. and Batzoglou, S. (2005) ProbCons: Probabilistic consistency-based multiple sequence alignment. *Genome Research*, **15**, 330–340.

75 Edgar, R.C. and Sjölander, K. (2003) SATCHMO: sequence alignment and tree construction using hidden Markov models. *Bioinformatics*, **19**, 1404–1411.

76 McClure, M.A., Vasi, T.K. and Fitch, W.M. (1994) Comparative analysis of multiple protein-sequence alignment methods. *Molecular Biology and Evolution*, **11**, 571–592.

77 Thompson, J.D., Plewniak, F. and Poch, O. (1999) BAliBASE: a benchmark alignment database for the evaluation of multiple alignment programs. *Bioinformatics*, **15**, 87–88.

78 Rost, B. (1999) Twilight zone of protein sequence alignments. *Protein Engineering*, **12**, 85–94.

79 Gatesy, J., DeSalle, R. and Wheeler, W. (1993) Alignment-ambiguous nucleotide sites and the exclusion of systematic data. *Molecular Phylogenetics and Evolution*, **2**, 152–157.

80 Citerne, H.L., Luo, D., Pennington, R.T., Coen, E. and Cronk, Q.C. (2003) A phylogenomic investigation of CYCLOIDEA-like TCP genes in the Leguminosae. *Plant Physiology*, **131**, 1042–1053.

81 Kolaczkowski, B. and Thornton, J.W. (2004) Performance of maximum parsimony and likelihood phylogenetics when evolution is heterogeneous. *Nature*, **431**, 980–984.

82 Philippe, H., Zhou, Y., Brinkmann, H., Rodrigue, N. and Delsuc, F. (2005) Heterotachy and long-branch attraction in phylogenetics. *BMC Evolutionary Biology*, **5**, 50.

83 Wang, H.C., Spencer, M., Susko, E. and Roger, A.J. (2007) Testing for covarion-like evolution in protein sequences. *Molecular Biology and Evolution*, **24**, 294–305.

84 Felsenstein, J. (1978) Cases in which parsimony and compatibility methods will be positively misleading. *Systematic Zoology*, **27**, 401–410.

85 Hendy, M.D. and Penny, D. (1989) A framework for the quantitative study of evolutionary trees. *Systematic Zoology*, **38**, 297–309.

86 Felsenstein, J. (1985) Confidence limits on phylogenies: an approach using the bootstrap. *Evolution*, **39**, 783–791.

87 Ronquist, F. and Huelsenbeck, J.P. (2003) MrBayes 3: Bayesian phylogenetic inference under mixed models. *Bioinformatics*, **19**, 1572–1574.

88 Douady, C.J., Delsuc, F., Boucher, Y., Doolittle, W.F. and Douzery, E.J. (2003) Comparison of Bayesian and maximum likelihood bootstrap measures of phylogenetic reliability. *Molecular Biology and Evolution*, **20**, 248–254.

89 Anisimova, M. and Gascuel, O. (2006) Approximate likelihood-ratio test for branches: a fast, accurate, and powerful alternative. *Systematic Biology*, **55**, 539–552.

90 Leebens-Mack, J., Raubeson, L.A., Cui, L., Kuehl, J.V., Fourcade, M.H., Chumley, T.W., Boore, J.L., Jansen, R.K. and depamphilis, C.W. (2005) Identifying the basal angiosperm node in chloroplast genome phylogenies: sampling one's way out of the Felsenstein zone. *Molecular Biology and Evolution*, **22**, 1948–1963.

91 Felsenstein, J. (2005) Free ebook, distributed by the author. Department of Genome Science, University of Washington, Seattle.

92 Storm, C.E. and Sonnhammer, E.L. (2002) Automated ortholog inference from phylogenetic trees and calculation of orthology reliability. *Bioinformatics*, **18**, 92–99.

93 Ashburner, M., Ball, C.A., Blake, J.A., Botstein, D., Butler, H., Cherry, J.M., Davis, A.P., Dolinski, K., Dwight, S.S., Eppig, J.T., Harris, M.A., Hill, D.P., Issel-Tarver, L., Kasarskis, A., Lewis, S., Matese, J.C., Richardson, J.E., Ringwald, M., Rubin, G.M. and Sherlock, G. (2000) Gene ontology: tool for the unification of biology. The Gene Ontology Consortium. *Nature Genetics*, **25**, 25–29.

94 Apweiler, R., Bairoch, A., Wu, C.H., Barker, W.C., Boeckmann, B., Ferro, S., Gasteiger, E., Huang, H., Lopez, R., Magrane, M., Martin, M.J., Natale, D.A., O'Donovan, C., Redaschi, N. and Yeh, L.S. (2004) UniProt: the Universal Protein knowledgebase. *Nucleic Acids Research*, **32**, D115–D119.

95 Eddy, S.R. (1998) Profile hidden Markov models. *Bioinformatics*, **14**, 755–763.

96 Sjölander, K. (1998) Phylogenetic inference in protein superfamilies: analysis of SH2 domains. *Proceedings of International Conference on Intelligent System for Molecular Biology*, **6**, 165–174.

97 Salzberg, S.L. (2007) Genome re-annotation: a wiki solution? *Genome Biology*, **8**, 102.

98 Glanville, J.G., Kirshner, D., Krishnamurthy, N. and Sjölander, K. (2007) Berkeley Phylogenomics Group web servers: resources for structural phylogenomic analysis. *Nucleic Acids Research*, **35**, W27–W32.

99 Engelhardt, B.E., Jordan, M.I., Muratore, K.E. and Brenner, S.E. (2005) Protein molecular function prediction by Bayesian Phylogenomics. *PLoS Computational Biology*, **1**, e45.

100 Schmidt, E.E. and Davies, C.J. (2007) The origins of polypeptide domains. *Bioessays*, **29**, 262–270.

101 Mulder, N.J., Apweiler, R., Attwood, T.K., Bairoch, A., Bateman, A., Binns, D., Bork, P., Buillard, V., Cerutti, L., Copley, R., Courcelle, E., Das, U., Daugherty, L., Dibley, M., Finn, R., Fleischmann, W., Gough, J., Haft, D., Hulo, N., Hunter, S., Kahn, D., Kanapin, A., Kejariwal, A., Labarga, A., Langendijk-Genevaux, P.S., Lonsdale, D., Lopez, R., Letunic, I., Madera, M., Maslen, J., McAnulla, C., McDowall, J., Mistry, J., Mitchell, A., Nikolskaya, A.N., Orchard, S., Orengo, C., Petryszak, R., Selengut, J.D., Sigrist, C.J., Thomas, P.D., Valentin, F., Wilson, D., Wu, C.H. and Yeats, C. (2007) New developments in the InterPro database. *Nucleic Acids Research*, **35**, D224–D228.

102 Saier, M.H. Jr (1994) Computer-aided analyses of transport protein sequences: gleaning evidence concerning function, structure, biogenesis, and evolution. *Microbiological Reviews*, **58**, 71–93.

103 Date, S.V. and Marcotte, E.M. (2003) Discovery of uncharacterized cellular systems by genome-wide analysis of functional linkages. *Nature Biotechnology*, **21**, 1055–1062.

104 Strong, M., Mallick, P., Pellegrini, M., Thompson, M.J. and Eisenberg, D. (2003) Inference of protein function and protein linkages in *Mycobacterium tuberculosis* based on prokaryotic genome organization: a combined computational approach. *Genome Biology*, **4**, R59.

105 Wu, J., Kasif, S. and DeLisi, C. (2003) Identification of functional links between genes using phylogenetic profiles. *Bioinformatics*, **19**, 1524–1530.

106 Weig, A.R. and Jakob, C. (2000) Functional identification of the glycerol permease activity of *Arabidopsis thaliana* NLM1 and NLM2 proteins by heterologous expression in *Saccharomyces cerevisiae*. *FEBS Letters*, **481**, 293–298.

107 Gerbeau, P., Guclu, J., Ripoche, P. and Maurel, C. (1999) Aquaporin Nt-TIPa can account for the high permeability of tobacco cell vacuolar membrane to small neutral solutes. *Plant Journal*, **18**, 577–587.

18
Structural, Functional, and Comparative Annotation of Plant Genomes

Françoise Thibaud-Nissen, Jennifer Wortman, C. Robin Buell, and Wei Zhu

Abstract

While genome sequencing technologies have advanced significantly in the past 5 years, advances in genome annotation have been more limited. Annotation of genomes, including plant genomes, is a challenging, imprecise, and ever-changing task. Structural annotation, identifying the genes and resolving their structure, is aided through computational algorithms and availability of transcribed sequences. However, with less than 50% of the genes within a genome having a cognate transcript available, precise structural annotation of half of the genes within a genome is reliant on algorithms or homology-based evidence. Functional annotation, determining the biochemical, physiological, and biological role of the protein or RNA encoded by a gene, is highly transitive in nature as obtaining empirical evidence for gene function on a per gene basis within a plant genome is fiscally and technically limited. Perhaps the greatest improvements to be made in the near future with respect to annotation of plant genomes will be the utilization of comparative genomics approaches to improve structure and function predictions. With the pending sequence of a number of plant genomes, this will be an area of not only active research, but also major improvements in technology.

18.1
Introduction

Genomics, the study of complete or nearly complete genomes, has revolutionized how science is performed, including plant biology. With the release of the first plant genome sequence in 2000 [1], plant biological research was brought into the genomics era which consequently led to new insights and perspectives into our understanding of plant biology. Without a genome sequence or partial genome data such as transcript sequences via Expressed Sequence Tags (ESTs), research communities are 'handicapped' in the questions they can ask and methodologies they can

The Handbook of Plant Functional Genomics: Concepts and Protocols.
Edited by Günter Kahl and Khalid Meksem

ISBN: 978-3-527-31885-8

utilize to understand key aspects of plant growth and development. A mere decade ago, attaining the genome sequences of more than a few plant genomes seemed unrealistic. However, with improvements in sequencing technologies driving rapid decreases in costs, genome sequences for a large number of plant species are available, in progress, or planned in the near future (Table 18.1). Indeed, the question of obtaining a genome sequence is no longer if but when, and what quality or depth of sequence is sufficient to meet the needs of the target community.

While it may seem that obtaining an assembled genome sequence is a major milestone for the research community, the real milestone is in the interpretation of the sequence through annotation of the functional elements of the genome. The biological features encoded in the genome sequence are the data that can be used directly by biologists to develop hypotheses and drive research. Genome annotation has focused primarily on the identification of protein coding genes, although the importance of identifying non-coding RNAs and regulatory elements has been gaining momentum in recent years. Accurately identifying gene structures remains a challenge and an active area of research. Due to the high cost and time required for manual genome annotation, most genomes are annotated via automated gene prediction pipelines. The current reality is that the annotation generated by these pipelines is often imprecise, needs to be updated constantly as new sequence data becomes available, and requires dedicated resources and significant effort for review and optimization.

Annotation is typically divided into two primary components: structural annotation, the determination of the boundaries and intron-exon structure of each gene, and functional annotation, the association of gene structures with a putative function or biological process. Structural annotation, while primarily concerned with protein coding genes, may also include the location of other functional elements in the genomic sequence, such as non-coding RNAs and regulatory elements. Functional annotation is centered on identifying gene function using homology, domain/motif information, and expression patterns.

In this chapter, we aim to provide the reader with methods and applications for both structural and functional annotation of a plant genome along with perspectives on the utility and limitation of these methods for biological applications. We will also emphasize the use of comparative genomic data within these annotation techniques, which are increasingly being leveraged to improve and interpret structural and functional annotation in a contextual manner. The reader is referred to other chapters in this book that address specific bioinformatics issues such as co-expression analyses, promoter analysis, phylogenomics, and genome-scale alignments.

18.2 Methods, Protocols, and Applications

18.2.1 Structural Annotation

What is a gene? In its modern definition, a gene consists of coding and regulatory regions with the functional gene products encoding a protein or RNA. The goal of

Table 18.1 List of genome sequences and projects for land plants within the plant kingdom.

Clade	Species	Status	Reference
Dicots			
Brassicaceae			
	Arabidopsis thaliana	Finished	Arabidopsis Genome Initiative [1]
	Brassica rapa	In progress	http://www.brassica.info/index.htm
	Arabidopsis lyrata	In progress	http://www.jgi.doe.gov/sequencing/why/CSP2006/AlyrataCrubella.html
	Capsella rubella	In progress	http://www.jgi.doe.gov/sequencing/why/CSP2006/AlyrataCrubella.html
	Thellungiella halophila	In progress	http://www.jgi.doe.gov/sequencing/why/CSP2007/thellungiella.html
	Brassica oleracea	Whole genome shotgun	http://www.tigr.org/tdb/e2k1/bog1/
Caricaceae			
	Carica papaya (papaya)	In progress	http://cgpbr.hawaii.edu/papaya/
Euphorbiaceae			
	Manihot esculenta (cassava)	In progress	http://www.jgi.doe.gov/sequencing/why/CSP2007/cassava.html
	Ricinus communis (castor bean)	Whole genome draft	http://msc.tigr.org/r_communis/index.shtml
Fabaceae			
	Glycine max (soybean)	In progress	http://www.jgi.doe.gov/sequencing/why/soybean.html
	Medicago truncatula	In progress	http://www.medicago.org/genome/
	Lotus japonicus	In progress	http://www.kazusa.or.jp/lotus/
Malvaceae			
	Gossypium (cotton)	In progress	http://www.jgi.doe.gov/sequencing/why/CSP2007/cotton.html
Myrtaceae			
	Eucalyptus tree	Planned	http://www.jgi.doe.gov/sequencing/allinoneseqplans.php
Phrymaceae			
	Mimulus guttatus (monkey flower)	In progress	http://www.jgi.doe.gov/sequencing/why/CSP2006/mimulus.html

(*Continued*)

Table 18.1 (*Continued*)

Clade	Species	Status	Reference
Ranunculaceae			
	Aquilegia formosa (columbine)	In progress	http://www.jgi.doe.gov/sequencing/why/CSP2007/aquilegia.html
Rosaceae			
	Apple	In progress	http://www.bioinfo.wsu.edu/gdr/genome/index.shtml
Saliceae			
	Populus trichocarpa (poplar)	Whole genome draft	Tuskan *et al.* (2006)[a]
Solanaceae			
	Solanum lycopersicon (tomato)	In progress	http://sgn.cornell.edu/about/tomato_sequencing.pl
	Solanum tuberosum (potato)	In progress	http://potatogenome.net/
Vitaceae			
	Vitis vinifera (grape)	In progress	http://www.cns.fr/externe/English/Projets/Projet_ML/organisme_ML.html
Monocots			
Poaceae			
	Brachypodium distachyon (*Brachypodium*)	In progress	http://www.jgi.doe.gov/sequencing/why/CSP2007/brachypodium.html
	Oryza sativa subsp *japonica* (rice)	Finished	IRGSP [81]
	Oryza sativa subsp *japonica* (rice)	Whole genome draft	[80]; Barry (2001)[b]
	Oryza sativa subsp *indica* (rice)	Whole genome draft	[82,83]
	Setaria italica (foxtail millet)	Planned	http://www.jgi.doe.gov/sequencing/why/CSP2008/foxtailmillet.html
	Sorghum bicolor (sorghum)	In progress	http://www.jgi.doe.gov/sequencing/why/CSP2006/sorghum.html
	Sorghum bicolor (sorghum)	Gene enrichment	[92]

	Zea mays (corn)	Gene enrichment	[91,89]
	Zea mays (corn)	In progress	http://www.maizesequence.org/index.html
Other Plants			
Funariaceae			
	Physcomitrella patens	In progress	http://www.jgi.doe.gov/sequencing/why/CSP2005/physcomitrella.html
Marchantiaceae			
	Marchantia polymorpha	Planned	http://www.jgi.doe.gov/sequencing/allinoneseqplans.php
Selaginellaceae			
	Selaginella moellendorffii	In progress	http://www.jgi.doe.gov/sequencing/why/CSP2005/selaginella.html

[a] Tuskan, G.A., Difazio, S., Jansson, S., Bohlmann, J., Grigoriev. I., Hellsten. U., Putnam. N., Ralph, S., Rombauts, S., Salamov, A. *et al.* (2006) The genome of black cottonwood, *Populus trichocarpa* (Torr. & Gray). *Science*, 313(5793), 1596–1604.

[b] Barry, G.F. (2001) The use of the Monsanto draft rice genome sequence in research. *Plant Physiology*, 125(3), 1164–1165.

gene prediction is to identify the components of protein-coding genes and RNAs in genomic DNA, including coding and regulatory regions. A narrower but common definition of gene prediction refers to the recognition of protein-coding regions only and excludes RNA encoding genes and regulatory elements.

Despite the historical focus on protein-coding genes, the identification of RNA genes (also called non-protein coding genes) and regulatory regions (also called promoters) are of great importance. Their locations and functional roles in plant genomes are being revealed experimentally or *in silico*. Using Massively Parallel Signature Sequencing (MPSS), Meyers *et al.* reported that small RNA genes are abundant in the Arabidopsis genome [2,3]. Similarly, promoters can also be experimentally uncovered by chromatin immunoprecipitation coupled with microarray chip technology, termed ChIP-chip analysis [4]. Computational approaches to identify RNAs and promoters have been reviewed by Meyer [5] and Rombauts *et al.* [6], respectively. Due to the scope of the chapter, we will focus on protein-coding gene prediction in eukaryotes and more specifically, in plants.

Currently, the genomes of over 100 eukaryotic organisms have been sequenced or are being sequenced, including many plant species, such as *Arabidopsis thaliana*, *Oryza sativa*, *Medicago truncatula*, sorghum, tomato, potato and corn (see Table 18.1). The development of novel sequencing technologies (e.g. 454 and Solexa) continues to foster the exponential increase of raw sequence data. Gene structure prediction is a fundamental and critical step towards understanding the functional landscape of any genomic sequence. A great number of gene prediction programs have been developed to address this issue in the last decade. In general, computational approaches to gene identification fall into three categories: intrinsic methods (*ab initio* approaches), extrinsic methods (also called homology methods or similarity-based methods), and integrated methods. A central resource for gene finding programs and resources is available at the genefinding.org web site (http://www.genefinding.org/) and a comprehensive archive of literature references concerning gene recognition is well maintained by Dr Wen-Tian Li (http://www.nslij-genetics.org/gene/). The strengths and the weaknesses of gene prediction methods have been reviewed by Mathe *et al.* [7] and more recently by Do and Choi [8]. Overall, there are a few rules of thumb to help identify the best approaches for gene structure annotation.

18.2.1.1 Cognate Transcript Sequences are the Most Reliable Data Available

Given a completed genome, the best evidence for delineating a gene model is a full-length cDNA, which can be computationally mapped to the genomic sequence and reveal the complete exon-intron structure [9]. ESTs, single-pass sequence reads, are also useful as they can reveal, if not complete, at least partial gene structure. Since most gene prediction programs do not predict alternate splice forms, cognate EST/cDNA transcripts can be leveraged to identify alternative splicing using spliced alignment [10]. Many programs exist specifically to optimally align transcript sequences to genome sequences, emphasizing splice site recognition at intron-exon boundaries. Earlier alignment tools, including AAT [11] and EST_GENOME [12], use rigorous dynamic programming algorithms. These initial algorithms proved to be too slow and compute intensive for the size and scope of most eukaryotic genome projects. Faster and more stringent tools including sim4 [13], BLAT [14], GeneSeqer [15] and GMAP [16] were developed to tackle spliced alignments more efficiently. Other tools attempt to combine spliced alignments into cohesive full or partial gene predictions including programs such as TAP [17], AIR [18], and PASA [9]. While the speed and accuracy of alignment tools has been steadily improving, challenges remain in dealing with errors in EST sequences, discriminating paralogous alignments, incorporating non-consensus splice sites, and correctly aligning small exons. Also, cognate full-length transcripts can only specify the regions that are transcribed, and it should be noted that the longest open reading frame of the full-length transcript may not always represent the correct protein coding regions.

18.2.1.2 *Ab initio* Gene Finders are Good ORF Finders

Even when EST or cDNA sequencing is completed to saturation, which is rare due to the resources involved, only a fraction of the gene complement is represented, approximately 50–60% [19]. Therefore, *ab initio* gene prediction programs are an essential part of the genome annotation process (reviewed in [8,20,21]). Gene prediction programs are based on statistical models, often hidden Markov models (HMM), that are optimized through a training routine to find features of genes, including exons, splice sites, start and stop codons, introns, and the noncoding DNA separating genes. There are a wide variety of these programs currently available, including Genscan [22], GenemarkHMM [23], GlimmerHMM [24], and Augustus [25]. Many *ab initio* gene finders achieve over 90% accuracy at the nucleotide level, but much lower accuracy at the gene level. In other words, *ab initio* approaches can predict the existence of a gene at a particular locus but not the start and end position, or the exact exon-intron boundaries. Xiao *et al.* experimentally confirmed that at least 50% of 'hypothetical genes' in Arabidopsis (the genes predicted exclusively by *ab inito* gene finders) are expressed, but that about 38% have a gene structures that is different from the predictions [26].

18.2.1.3 Integrated Approaches are Ideal Solutions for the Automated Gene Prediction

Integrated approaches combine the strengths of both intrinsic and extrinsic approaches and thus generate better gene predictions in general [7,8]. Various inputs, like gene or signal predictions, transcript/protein spliced alignments, and protein alignments, are weighted either explicitly, based on the confidence level in the input or implicitly after training against a set of known gene structures. The integration can be achieved within a single program such as EUGENE'HOM [27], AUGUSTUS+ [28], Jigsaw [29] and EvidenceModeler (http://evidencemodeler.sf.net/) or through an annotation pipeline with an ordered set of processes. With respect to the latter, we may take the TIGR rice genome annotation as an example. Initial gene models were created by the program Fgenesh (http://www.softberry.com) upon the rice pseudomolecules. Then, the gene models were refined by the program PASA [9] using rice cDNA/ESTs. The refinement operations included creating novel genes, adding novel alternative isoforms, and merging/splitting genes, extending genes, and so on. The updated gene structures were refined by PASA recursively until they were converged, that is, there was no update to be made [30]. In the end, the models with conflicting evidences can be manually curated by the TIGR rice team or via a community annotation effort.

18.2.1.4 Gene Prediction is an Iterative Process

Even though gene structures are 'perfectly' predicted with respect to all the available evidence data for the time being, some may still be found to be incorrect and need to be refined according to new genome sequences or relevant experimental data. For example, the Arabidopsis genome was annotated initially by the consortium of sequencing centers that generated the genome sequence [1], reannotated over a period of 5 years by TIGR [31,32], and is currently maintained by the Arabidopsis Information Resource [33]. The content of the annotation has changed dramatically through six major release cycles, and improvements are still being made.

18.2.1.5 Manual Curation is still an Indispensable Process in Gene Prediction

Automated gene prediction is a type of artificial intelligence, which can achieve a decent level of accuracy, but it still cannot replace the expertise of biologists in their fields. Furthermore, automated gene prediction may easily fail to address certain aberrant gene structures, for example, non-canonical introns, polycistronic [26], short genes, and so on. As a result, individual researchers should consider gene predictions or functional annotations as a reference for their genes of interest, and are recommended to browse the gene predictions together with any available evidence via an annotation viewer/editor, or even create their own gene prediction if necessary. Apollo and Artemis are two popular tools for that purpose, and both of them are written in Java and are platform-independent [34,35]. Apollo is a powerful tool designed for editing gene structure and function, but requires significant computer support. Correspondingly, Artemis is lightweight making it an ideal tool for the casual user.

18.2.1.6 Other Considerations for Gene Prediction

It is noteworthy that few gene finders are dedicated to plant species only. Eukaryotic gene finders tend to be generically written and applicable to a wide variety of organisms, although some were initially written and optimized for human/mammalian systems [22,36]. When applied to diverse genomes, these programs employ exactly the same algorithm but distinctive parameter settings, typically generated in a training process using known gene structures from a specific species. Despite compositional differences [37], the successes of many gene finders imply that substantial common properties exist among eukaryotic genomes. However, a few factors need to be considered in plant gene prediction.

One of the most significant features is the rapid evolution of plant genomes. Grasses (Poaceae), for example, diverged from their common ancestor 55 to 70 million years ago (MYA) [38], and have about 80-fold divergence in the size of their genomes [39]. By contrast, the sizes of mammalian genomes range only five-fold since the first mammal appeared on the earth about 200 MYA. In addition to polyploidization and segmental duplication, transposable elements, or more specifically, the turnover rate of the inserted transposable elements, are a major factor contributing to genome size variation [40,41]. Some transposable elements such as Helitrons [42–44] and MULEs [45] can transduplicate neighboring genes completely or partially and therefore mediate gene evolution in plant genomes, which, to some extent, blurs the distinction among genes, pseudogenes, and transposable-element related genes. Consequently, additional effort may be needed to address these distinctions.

There are also other minor differences between plant and mammalian genomes. It is estimated that approximately 70% of human genes are preceded by upstream CpG islands. The program FirstEF manages to identify the first exons in the human genome on the basis of this characteristic [46]. In plants, CpG-rich regions are hypermethylated, under-represented, and show no association with the gene promoter regions [6]. In addition, a unique feature found in the plant genomes is that intronic regions are U-rich and exonic regions are GC-rich in higher plants, and this attribute was exploited to detect splice site signals in the program SplicePredictor [47].

Finally, data from innovative technologies may stimulate the development of a new generation of gene prediction algorithms. For instance, the program ARTADE was developed to predict transcriptional structures from intensity signals obtained from the hybridization of cDNA to tiling arrays [48].

18.2.2 Functional Annotation

Once the structure of a gene is established, a variety of tools and resources may be used to infer the function of its product. The gene function is a composite of the activity of the protein, the physiological process in which it participates, the localization of the protein in the plant and in the cell, the temporal and developmental expression pattern of the gene, and more. Therefore, the depth and refinement of the annotation depends largely on the type of resources available for the species of interest.

18.2.2.1 Sequence Similarity

The function of a protein is commonly determined by sequence similarity to other proteins, a method often described as homology-based transfer. To this end, the translated gene is aligned to proteins in the database of choice using BLASTP. The quality of the results depends, in part, on the parameters used for BLASTP. Expectation value, identity and coverage cut-offs are set empirically based largely on personal experience and availability of similar sequences in the databases. As an indication, for release 5 of the Osa1 Rice Genome Annotation, BLASTP alignments of rice protein predictions were considered only if the expectation value was below e-10, the identity was above 30%, and the coverage above 50% [49].

The database against which the searches are performed also plays a role in the number and quality of the hits. For example, the UniProtKB/Swiss-Prot (http://au.expasy.org/sprot/) [50] is a database of manually annotated records, while UniProtKB/TrEMBL is a larger database of computationally analyzed records, containing all of the protein sequences translated from EMBL/GenBank/DDBJ nucleotide sequence databases, in addition to protein sequences in PDB [51]. Therefore, UniProtKB/Swiss-Prot returns highly-curated hits with low coverage and the UniProtKB/TrEMBL database is larger and therefore provides higher chances of returning a hit. In practice, there are currently several widely used large databases combining non-redundant sets of sequences of different origins. The NCBI nr (non-redundant) database is a non-redundant set of over 4 million sequences including GenBank coding sequence translations, UniProtKB/Swiss-Prot, PIR, PDB and PRF sequences. The UnitProt consortium of UniProtKB/Swiss-Prot, TrEMBL and PIR (http://www.pir.uniprot.org/) has built several non-redundant databases, UniRef50, UniRef90 and UniRef100, which combine records over 50, 90 and 100% identical in sequence, respectively, and contain over 1.3, 2.5, and 3.8 million sequences, respectively [52].

As a consequence of the diverse origins of the records in the database, the hits resulting from a protein search can be filtered and sorted, based not only on the quality of the match but also on the quality of the annotation of the hit. In order to avoid transitive annotation, preference should be given to manually curated records over proteins that were assigned a function by homology-based transfer in the context of, for example, large genome annotation projects. An additional caveat is that similarity-based function assignment assumes that similarity in sequence implies similarity in function, a principle that holds true most of the time but has exceptions (see [53]).

18.2.2.2 **Domain Searches**

Alternatively, the function of a protein can be derived from the domain information that it contains. Searches against the protein pattern database Pfam (http://www.sanger.ac.uk/Software/Pfam/search.shtml) use HMMs to predict domains in the query protein and return multiple sequence alignments of members in the family [54]. Pfam represents ~8000 families and, compared to other pattern databases, covers the largest number of proteins [55]. To accommodate the fact that proteins may contain several domains, Pfam families are further organized into clans. Pfam and other HMM-based pattern databases, such as TIGRFAM [56] and PANTHER [57], as well as other domain detection methods such as PROSITE [58] and ProDom [59] are grouped under the databases of the InterPro consortium [55]. The search tool InterProScan allows simultaneous searching of all these databases, using the method associated with each database, and returns the domains contained in the query and any additional information (protein matches, GO terms, etc.) provided by the database harboring the domain [60].

18.2.2.3 **Phylogenomics**

As mentioned above, homology-based transfer of function assumes that the more related two protein sequences are, the more similar their function is. However, evolutionary biology has shown that this principle is not always true. When genes duplicate, one paralog can preserve its function while the other becomes free to adopt a new role, that is, diversify. On the other hand, orthologous genes, resulting from speciation, are more likely to keep the function they had in the common ancestor

[61,62]. The term phylogenomics applies to the application of gene phylogeny to gene function prediction [61].

The first step requires the classification of proteins into families. Homologs of a protein of interest in the proteomes of selected species are searched iteratively using algorithms such as PSI-BLAST [63] and aligned using a multiple-sequence alignment tool (i.e. CLUSTALW [64]). Through construction of phylogenetic trees, paralogs and orthologs can be teased out and characterized proteins can be used to infer the function of neighboring genes (for details see [62]). The availability of fully sequenced genomes has led to the application of this method to the analysis of large plant proteins families [65], sometimes in combination with microsynteny information [66]. However, similar approaches can be used with large collections of ESTs. By triangulating reciprocal best matches, Wu *et al.* have identified a set of single-copy orthologs among the Euasterid clade, and by also including Arabidopsis in their analysis were able to maximize the leveraging of the functional annotation of the Arabidopsis genome [67].

18.2.2.4 Expression Data

An additional layer of functional annotation can be provided by expression information, in the form of real-time PCR data [68], microarray data, MPSS [69] or ESTs [70]. Transcription evidence not only provides support for the gene's existence or structure (as discussed above), but also indicates when and where genes are expressed and in what quantity, in the manner of an 'electronic' Northern blot. Microarray data are currently publicly available for Arabidopsis and crop species in the Gene Expression Omnibus and Array Express, and in species-specific repositories [71]. These databases can be mined for evidence of transcription at particular loci. Furthermore, large sets of data allow the identification of co-expressed genes across developmental stages, time and experimental conditions [72]. Through 'guilt by association', genes of unknown function can be presumed to participate in the same biological process as the characterized genes in the same expression cluster. Therefore, transcription information can be used for the refinement of another type of functional annotation or for a coarse delineation of function for genes of unknown function.

18.2.3 Comparative Annotation

In recent years, with the increasing availability of genome and transcriptome sequences from evolutionarily related organisms, leveraging comparative sequence data to inform the annotation process has become a major focus.

18.2.3.1 Comparative Annotation Using Transcripts

In plants, the primary type of comparative annotation performed to date is that of alignment of a genomic sequence to transcripts as there are 12 million ESTs available for plant species in Genbank dbEST (Release 061707) from 476 species. Collectively, these ESTs represent 6.54 Gb of total sequence and are derived from 122 families thereby providing a rich phylogenetic resource for comparative genomics. The redundancy, short length, and low quality inherent in single pass sequences such as ESTs make aligning individual ESTs to genomes prohibitive. Typically, ESTs are reduced in number, increased in length, and improved in quality through clustering and assembly into a set of non-redundant transcript sequences. There are multiple research groups that provide these non-redundant sets of transcripts such as the Dana Farber Gene Index group (http://compbio.dfci.harvard.edu/tgi/; [73]), Plant GDB (http://www.plantgdb.org/; [74]), TIGR Transcript Assemblies (http://plantta.tigr.org/; [75]); and the NCBI Unigene Project (http://www.ncbi.nlm.nih.gov/sites/entrez?db=unigene; [76]). To align transcripts to a genome sequence or genome assemblies, alignment programs such as GMAP [16] or AAT [11] can be used and the alignments can be viewed in a number of graphical viewers such as Artemis [35], Generic Genome Browser [77], or Apollo [34].

As described above, ESTs and full-length cDNAs derived from the same or a closely related species as the target genome are instrumental in determination of accurate gene structure, identification of alternative splice forms, and in functional annotation, that is, expression patterns. While structural alignments break down as the phylogenetic distance from the target genome is increased, alignments of ESTs from related species can be used in annotation. These cross-species alignments can be used for improvement of structural annotation and identification of alternative splice forms [78,79], however, this is challenging as the precise boundaries of the intron and exons are not as evident with divergent sequences. Thus, manual inspection of these alignments including percent identity, length of coverage, and conservation of canonical splice site should be carried out and users will need to determine the threshold for identity and length cut-off criteria to maximize information gain and minimize false alignments. Clearly, alignment across different phylogenetic groups (e.g. monocots, dicots) will require different alignment criteria.

These cross-species alignments provide three additional levels of annotation. First, where there is alignment of a heterologous transcript to a gene which lacks cognate transcript/expression support, this suggests that the gene is valid and not an artifact of the annotation process. Second, the expression metadata of the heterologous transcript could be used to infer expression of the gene in the target organism. Third, detection of a homolog, even though it is a transcript, provides information on gene conservation throughout the plant kingdom. To illustrate the power of transcript-based alignments, a rice gene in which transcripts could be detected in 19 species representing not only other Poaceae species, but also monocots, dicots, and cycads is shown in Figure 18.1.

The detection of transcripts from heterologous species is suggestive, but not empirical, evidence that the gene is expressed and should be used appropriately. Although transcript data is a rich resource and provides information on expression pattern and gene model structure, it is restricted to genes that can be captured and detected through sequence-based transcription profiling methods. These limitations can be attributed primarily to representation of transcripts within an mRNA population as well as the breadth of tissues and biological conditions surveyed as even very deep sampling such as MPSS cannot sample all of the transcripts within a transcriptome [69]. Thus, alignment of genomes or genome assemblies to a target genome provides information on gene structure and conservation and obviates the need for detection of the transcript through transcriptome profiling approaches.

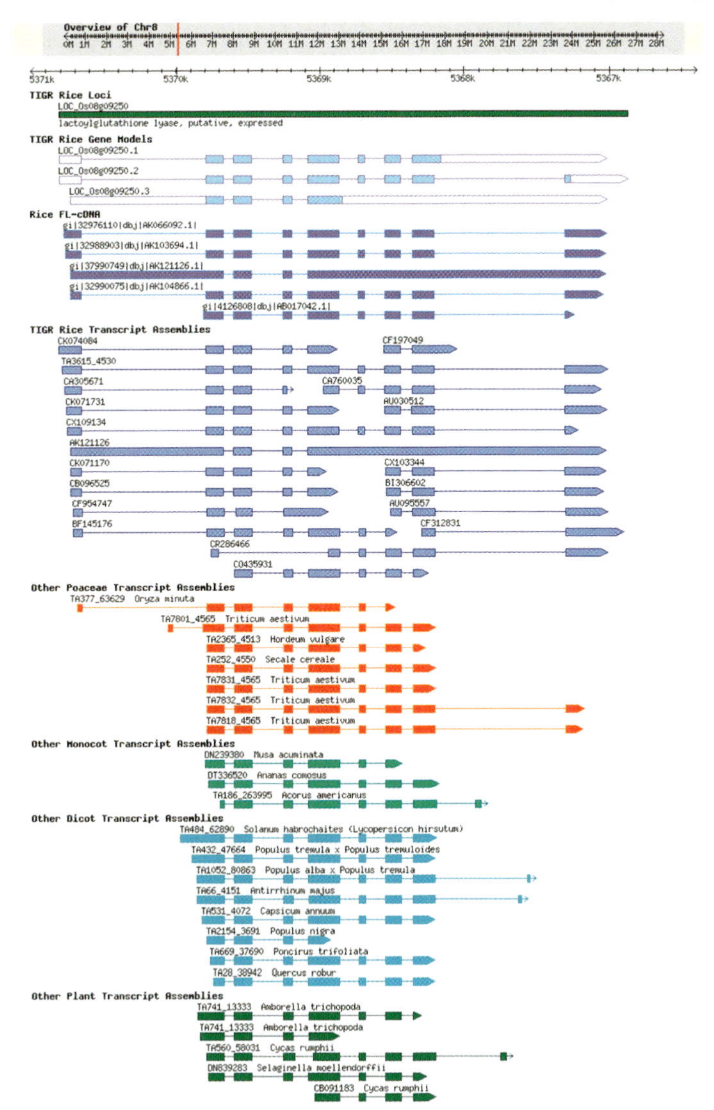

Figure 18.1 Comparative alignments between a rice gene model and sequences from 19 plant species. The rice gene model was aligned with transcript assemblies from taxa throughout the plant kingdom. Clear homology is present at the exon level across the plant kingdom.

18.2.3.2 Comparative Genomics Using Genome Sequences

Although the ultimate goal of comparative genomics is to align whole genomes, there is limited data from whole genome alignments available to date due to the lack of multiple genome sequences within a set of phylogenetically related species

(Table 18.1). Whole genome sequence is publicly available for two rice subspecies, *japonica* [80,81] and *indica* [82,83]. Not surprisingly, alignments of the *japonica* and *indica* draft genomes reveal a high degree of conservation in genic regions with divergence in the intergenic region [83]. Although the draft nature of the *indica* sequence limits to some extent interpretations on single nucleotide polymorphisms and genome re-arrangements on a micro-scale, it is clear that the two subspecies do have differences [83]. While alignment of two closely related genomes is informative for targeted researchers interested in marker development or speciation, alignment of more distantly related species can be more informative with respect to annotation. Alignment of the rice genome with a 4× draft assembly of the sorghum genome has provided new evidence for improvement of gene model structure in rice due to the ability to identify conserved exons across these two taxa (Buell, C. R., unpublished data). Although the papaya genome sequence is pending, it is apparent from alignment of papaya bacterial artificial chromosome (BAC) end sequences that there is greater synteny between papaya and poplar than papaya and Arabidopsis [84].

In the absence of a complete genome, alignment of partial or draft genome assemblies to genomes (or genome assemblies) can be informative. Indeed, alignment of small genome assemblies, such as that obtained from targeted sequencing of BAC clones, can assist in structural annotation and gaining insight into genome level features such as synteny. Whole genome shotgun reads from *Brassica oleracea* were used to improve annotation in the Arabidopsis genome, with a focus on novel gene discovery [85]. Using sequence from *Brachypodium*, a related species within the Poaceae, Bossolini *et al.* [86] were able to identify a number of rice genes that were misannotated and identify potential alternative splice forms in rice through these comparative alignments. Methylation filtration [87] and high C_0t [88], have been used to increase representation of genic sequences in a genomic DNA library in maize [89–91] and sorghum [92]. These approaches have allowed for a high rate of gene discovery prior to initiation of whole genome sequencing of these two large genomes, 2500 and 750 Mb, respectively. Even with fragmented assemblies that only represent a portion of the genome, conservation between related species can be readily detected and used in genome annotation [79].

18.2.3.3 **Algorithms for Comparative Genomics**

As more sequence data becomes available across related plant species, this data can feed algorithmic approaches, such as those developed for related mammalian and Drosophila genomes. To date, the most established and successful algorithms have exploited alignments between two genomes of an optimal evolutionary distance to indicate which nucleotides are under negative selection and therefore more likely to be coding or otherwise functional [20]. Twinscan is a stable and supported gene prediction algorithm that couples a probabilistic model of sequence conservation based on BLASTN matches between the informant and target genome to the GHMM used by a reimplemented version of GENSCAN [93]. Another class of gene prediction tools compares two genome sequences to predict gene structures in both genomes by exploiting regions of conservation using a probabilistic model called a generalized pair HMM (GPHMM). The GPHMM combines the paired HMM that describes sequence alignment with the more traditional HMM that describes gene structures. Gene prediction programs implementing the GPHMM include both SLAM and TWAIN [94,95]. One caveat in using these GPHMM-based software tools is that they emit gene structures that require identical numbers of introns and exons for the homologous gene pairs in the corresponding pair of genomes. This is a reasonable approximation for many closely related genomes, and is an attractive approach for ensuring consistency between genome annotations, but may cause problems when organisms diverge in intron-exon structure.

Individual exon prediction can be used as a complementary approach to gene prediction algorithms for highlighting missed or misannotated genes in existing annotation data sets. Exon detection is based on differentiating properties of protein-coding regions relative to those of UTRs, introns and intergenic regions. Recently published work from the Haussler laboratory introduces a new system, called shortHMM, for predicting individual exons in a pair of related genomes [96,97]. The program was shown to be effective in human–mouse data, finding short exons and AT-rich genes missed by other approaches. Another successful exon-based approach developed by the Kellis laboratory is being applied to the annotation of diverse eukaryotes [98,99]. This approach was used to revisit the genome of *S. cerevisiae*, reducing the overall gene count by 500 genes (10% of previously annotated genes) and refining the gene structure of hundreds of genes, and is currently being applied to 12 Drosophila species.

18.3 Perspectives

The identification and annotation of protein-coding genes is one of the primary goals of whole genome sequencing projects, and the accuracy of the predicted proteome is vitally important for in-depth comparative analyses and downstream functional genomic applications. Yet structural annotation of eukaryotic genomes remains a considerable challenge, despite the exponential growth in the number of sequenced genomes and improvements in eukaryotic gene prediction algorithms. Many genomes submitted to public databases, including those of major model organisms, contain significant percentages of misannotated gene structures and unvalidated gene predictions. While computational methods have been steadily improving, they have still proven inadequate for the correct identification of gene boundaries, intron-exon junctions, untranslated regions, and alternatively spliced products. These problems are especially prevalent in genomes where non-consensus splice sites, small exons, or non-standard intron size distributions are common.

This chapter highlights the importance of transcriptome sequencing, in the form of ESTs and cDNAs, and automated gene prediction, especially newer methods which leverage comparative genome data, as complementary approaches for optimally identifying correct gene structures. However, due to expression patterns and cloning biases, the coverage of EST data is limited to, on average, about half of the genes present in the genome. And even with significant advances in the technologies available for gene prediction, many predictions will contain some level of error, whether in designating the correct start site or finding all correct intron-exons boundaries. Another strategy, which is becoming more practical as sequencing costs continue to decrease, is to perform experimental validation of unsupported or weakly supported gene predictions (reviewed in [19]).

A robust automated gene prediction system will produce not only gene models well-supported by homology and experimental evidence, but also more hypothetical gene models that are ideal substrates for experimental validation. The most common experimental validation techniques are RT-PCR and direct sequencing [100–102] and recent technical innovations are making such methods more cost-efficient. Considering experimental validation as a downstream process of automated annotation underscores the importance of providing adequate evidence trails and confidence values. Selecting candidates for experimental validation based on these values can reduce the pool of possibilities and further limit costs.

It is important to note that automated annotation should occur not only prior to experimental validation, in order to identify candidates for sequencing, but also afterwards, in order to automatically incorporate the experimental results into the existing annotated gene models. While stand-alone tools exist for this purpose [9] the integration of experimental validation into annotation pipelines will be an important component in progressing automated annotation systems.

In addition, it is becoming increasingly clear that the protein-coding components of any genome sequence comprise only a limited portion of the total functional landscape. The recent publication of the results of the ENCODE pilot project, which had the aim of completely characterizing 1% of the human genome using both experimental and computational approaches, highlighted this fact for mammalian systems [103]. The major conclusions reached were that most regions of the genome are actively transcribed, that a large number of the resulting transcripts are not protein coding, and that complex mechanisms are responsible for transcriptional regulation including histone modification and chromatic structure. Surprisingly, a large number of functional elements identified experimentally in the human genome do not correspond to regions of enriched conservation and evolutionary constraint between mammalian species.

Clearly with the availability of plant genome sequences increasing, researchers as well as funding agencies are addressing the issue of interpreting this data. Large-scale functional genomics projects for Arabidopsis via the Arabidopsis 2010 project whose goal is to define the function of all of the genes within Arabidopsis by the year 2010 are critical to interpreting genome sequence data. Similar, but less extensive projects are in progress or being initiated for rice, the model monocot species. Through these efforts, in which targeted, large-scale, and public projects address the biological function of a large set of genes within a genome, we will have access to empirical data that is essential to understanding the relationship between genes, RNA, proteins, and biological function.

Acknowledgments

Work on genome annotation in the Buell group is supported by a National Science Foundation Plant Genome Research Program grant to C. R. B. (DBI-0321538).

References

1 Arabidopsis Genome Initiative. (2000) Analysis of the genome sequence of the flowering plant *Arabidopsis thaliana*. *Nature*, **408** (6814), 796–815.

2 Lu, C., Kulkarni, K., Souret, F.F., MuthuValliappan, R., Tej, S.S., Poethig, R.S., Henderson, I.R., Jacobsen, S.E., Wang, W., Green, P.J. *et al.* (2006) MicroRNAs and other small RNAs enriched in the Arabidopsis RNA-dependent RNA polymerase-2 mutant. *Genome Research*, **16** (10), 1276–1288.

3 Nakano, M., Nobuta, K., Vemaraju, K., Tej, S.S., Skogen, J.W. and Meyers, B.C. (2006) Plant MPSS databases: signature-based transcriptional resources for analyses of mRNA and small RNA. *Nucleic Acids Research*, **34** (Database issue), D731–D735.

4 Thibaud-Nissen, F., Wu, H., Richmond, T., Redman, J.C., Johnson, C., Green, R., Arias, J. and Town, C.D. (2006) Development of Arabidopsis whole-genome microarrays and their application to the discovery of binding sites for the TGA2 transcription factor in salicylic acid-treated plants. *The Plant Journal*, **47** (1), 152–162.

5 Meyer, I.M. (2007) A practical guide to the art of RNA gene prediction. *Briefings in Bioinformatics*, **8**, 47–50.

6 Rombauts, S., Florquin, K., Lescot, M., Marchal, K., Rouze, P. and van de Peer, Y. (2003) Computational approaches to identify promoters and cis-regulatory elements in plant genomes. *Plant Physiology*, **132** (3), 1162–1176.

7 Mathe, C., Sagot, M.F., Schiex, T. and Rouze, P. (2002) Current methods of gene prediction, their strengths and weaknesses. *Nucleic Acids Research*, **30** (19), 4103–4117.

8 Do, J.H. and Choi, D.K. (2006) Computational approaches to gene prediction. *Journal of Microbiology*, **44** (2), 137–144.

9 Haas, B.J., Delcher, A.L., Mount, S.M., Wortman, J.R., Smith, R.K., Jr. Hannick, L.I., Maiti, R., Ronning, C.M., Rusch, D.B., Town, C.D. *et al.* (2003) Improving the Arabidopsis genome annotation using maximal transcript alignment assemblies. *Nucleic Acids Research*, **31** (19), 5654–5666.

10 Campbell, M.A., Haas, B.J., Hamilton, J.P., Mount, S.M. and Buell, C.R. (2006) Comprehensive analysis of alternative splicing in rice and comparative analyses with Arabidopsis. *BMC Genomics*, **7**, 327.

11 Huang, X., Adams, M.D., Zhou, H. and Kerlavage, A.R. (1997) A tool for analyzing and annotating genomic sequences. *Genomics*, **46** (1), 37–45.

12 Mott, R. (1997) EST_GENOME: a program to align spliced DNA sequences to unspliced genomic DNA. *Computer Applications in the BIOSciences*, **13** (4), 477–478.

13 Florea, L., Hartzell, G., Zhang, Z., Rubin, G.M. and Miller, W. (1998) A computer program for aligning a cDNA sequence with a genomic DNA sequence. *Genome Research*, **8** (9), 967–974.

14 Kent, W.J. (2002) BLAT – the BLAST-like alignment tool. *Genome Research*, **12** (4), 656–664.

15 Usuka, J., Zhu, W. and Brendel, V. (2000) Optimal spliced alignment of homologous cDNA to a genomic DNA template. *Bioinformatics*, **16** (3), 203–211.

16 Wu, T.D. and Watanabe, C.K. (2005) GMAP: a genomic mapping and alignment program for mRNA and EST sequences. *Bioinformatics*, **21** (9), 1859–1875.

17 Kan, Z., Rouchka, E.C., Gish, W.R. and States, D.J. (2001) Gene structure prediction and alternative splicing analysis using genomically aligned ESTs. *Genome Research*, **11** (5), 889–900.

18 Florea, L., Di Francesco, V., Miller, J., Turner, R., Yao, A., Harris, M., Walenz, B., Mobarry, C., Merkulov, G.V., Charlab, R. *et al.* (2005) Gene and alternative splicing annotation with AIR. *Genome Research*, **15** (1), 54–66.

19 Brent, M.R. (2005) Genome annotation past, present, and future: how to define an ORF at each locus. *Genome Research*, **15** (12), 1777–1786.

20 Brent, M.R. and Guigo, R. (2004) Recent advances in gene structure prediction. *Current Opinion in Structural Biology*, **14** (3), 264–272.

21 Zhang, M.Q. (2002) Computational prediction of eukaryotic protein-coding genes. *Nature Reviews. Genetics*, **3** (9), 698–709.

22 Burge, C. and Karlin, S. (1997) Prediction of complete gene structures in human genomic DNA. *Journal of Molecular Biology*, **268** (1), 78–94.

23 Lukashin, A.V. and Borodovsky, M. (1998) GeneMark.hmm: new solutions for gene finding. *Nucleic Acids Research*, **26** (4), 1107–1115.

24 Majoros, W.H., Pertea, M. and Salzberg, S.L. (2004) TigrScan and GlimmerHMM: two open source *ab initio* eukaryotic gene-finders. *Bioinformatics*, **20** (16), 2878–2879.

25 Stanke, M. and Waack, S. (2003) Gene prediction with a hidden Markov model and a new intron submodel. *Bioinformatics*, **19** (Suppl 2), II215–II225.

26 Xiao, Y.L., Smith, S.R., Ishmael, N., Redman, J.C., Kumar, N., Monaghan, E.L., Ayele, M., Haas, B.J., Wu, H.C. and Town, C.D. (2005) Analysis of the cDNAs of hypothetical genes on Arabidopsis chromosome 2 reveals numerous transcript variants. *Plant Physiology*, **139** (3), 1323–1337.

27 Foissac, S., Bardou, P., Moisan, A., Cros, M.J. and Schiex, T. (2003) EUGENE'HOM: A generic similarity-based gene finder using multiple homologous sequences. *Nucleic Acids Research*, **31** (13), 3742–3745.

28 Stanke, M., Tzvetkova, A. and Morgenstern, B. (2006) AUGUSTUS at EGASP: using EST, protein and genomic alignments for improved gene prediction in the human genome. *Genome Biology*, **7 Suppl 1** (S11), 11–18.

29 Allen, J.E. and Salzberg, S.L. (2005) JIGSAW: integration of multiple sources of evidence for gene prediction. *Bioinformatics*, **21** (18), 3596–3603.

30 Yuan, Q., Ouyang, S., Wang, A., Zhu, W., Maiti, R., Lin, H., Hamilton, J., Haas, B., Sultana, R., Cheung, F. *et al.* (2005) The institute for genomic research Osa1 rice genome annotation database. *Plant Physiology*, **138** (1), 18–26.

31 Haas, B.J., Wortman, J.R., Ronning, C.M., Hannick, L.I., Smith, R.K., Jr. Maiti, R., Chan, A.P., Yu, C., Farzad, M., Wu, D. *et al.* (2005) Complete reannotation of the Arabidopsis genome: methods, tools, protocols and the final release. *BMC Biology*, **3**, 7.

32 Wortman, J.R., Haas, B.J., Hannick, L.I., Smith, R.K., Jr. Maiti, R., Ronning, C.M., Chan, A.P., Yu, C., Ayele, M., Whitelaw, C.A. *et al.* (2003) Annotation of the Arabidopsis genome. *Plant Physiology*, **132** (2), 461–468.

33 Rhee, S.Y., Beavis, W., Berardini, T.Z., Chen, G., Dixon, D., Doyle, A., Garcia-Hernandez, M., Huala, E., Lander, G., Montoya, M. *et al.* (2003) The Arabidopsis Information Resource (TAIR): a model organism database providing a centralized, curated gateway to Arabidopsis biology, research materials and community. *Nucleic Acids Research*, **31** (1), 224–228.

34 Lewis, S.E., Searle, S.M., Harris, N., Gibson, M., Lyer, V., Richter, J., Wiel, C., Bayraktaroglir, L., Birney, E., Crosby, M.A. *et al.* (2002) Apollo: a sequence annotation editor. *Genome Biology*, **3** (12), RESEARCH0082.

35 Berriman, M. and Rutherford, K. (2003) Viewing and annotating sequence data with Artemis. *Briefings in Bioinformatics*, **4** (2), 124–132.

36 Birney, E., Clamp, M. and Durbin, R. (2004) GeneWise and Genomewise. *Genome Research*, **14** (5), 988–995.

37 Karlin, S. and Mrazek, J. (1997) Compositional differences within and between eukaryotic genomes. *Proceedings of the National Academy of Sciences of the United States of America*, **94** (19), 10227–10232.

38 Kellogg, E.A. (2001) Evolutionary history of the grasses. *Plant Physiology*, **125** (3), 1198–1205.

39 Caetano-Anolles, G. (2005) Evolution of genome size in the grasses. *Crop Science*, **45**, 1809–1816.

40 Vitte, C. and Bennetzen, J.L. (2006) Analysis of retrotransposon structural diversity uncovers properties and propensities in angiosperm genome evolution. *Proceedings of the National Academy of Sciences of the United States of America*, **103** (47), 17638–17643.

41 Piegu, B., Guyot, R., Picault, N., Roulin, A., Saniyal, A., Kim, H., Collura, K., Brar, D.S., Jackson, S., Wing, R.A. *et al.* (2006) Doubling genome size without polyploidization: dynamics of retrotransposition-driven genomic expansions in *Oryza australiensis*, a wild relative of rice. *Genome Research*, **16** (10), 1262–1269.

42 Morgante, M., Brunner, S., Pea, G., Fengler, K., Zuccolo, A. and Rafalski, A. (2005) Gene duplication and exon shuffling by helitron-like transposons generate intraspecies diversity in maize. *Nature Genetics*, **37** (9), 997–1002.

43 Lai, J., Li, Y., Messing, J. and Dooner, H.K. (2005) Gene movement by Helitron transposons contributes to the haplotype variability of maize. *Proceedings of the National Academy of Sciences of the United States of America*, **102** (25), 9068–9073.

44 Gupta, S., Gallavotti, A., Stryker, G.A., Schmidt, R.J. and Lal, S.K. (2005) A novel class of Helitron-related transposable elements in maize contain portions of multiple pseudogenes. *Plant Molecular Biology*, **57** (1), 115–127.

45 Jiang, N., Bao, Z., Zhang, X., Eddy, S.R. and Wessler, S.R. (2004) Pack-MULE transposable elements mediate gene evolution in plants. *Nature*, **431** (7008), 569–573.

46 Davuluri, R.V., Grosse, I. and Zhang, M.Q. (2001) Computational identification of promoters and first exons in the human genome. *Nature Genetics*, **29** (4), 412–417.

47 Brendel, V. and Kleffe, J. (1998) Prediction of locally optimal splice sites in plant pre-mRNA with applications to gene identification in *Arabidopsis thaliana* genomic DNA. *Nucleic Acids Research*, **26** (20), 4748–4757.

48 Toyoda, T. and Shinozaki, K. (2005) Tiling array-driven elucidation of transcriptional structures based on maximum-likelihood and Markov models. *The Plant Journal*, **43** (4), 611–621.

49 Ouyang, S., Zhu, W., Hamilton, J., Lin, H., Campbell, M., Childs, K., Thibaud-Nissen, F., Malek, R.L., Lee, Y., Zheng, L. *et al.* (2007) The TIGR Rice Genome Annotation Resource: improvements and new features. *Nucleic Acids Research*, **35** (Database issue), D883–D887.

50 Boeckmann, B., Bairoch, A., Apweiler, R., Blatter, M.C., Estreicher, A., Gasteiger, E., Martin, M.J., Michoud, K., O'Donovan, C., Phan, I. *et al.* (2003) The SWISS-PROT protein knowledgebase and its supplement TrEMBL in 2003. *Nucleic Acids Research*, **31** (1), 365–370.

51 The UniProt Consortium. (2007) The Universal Protein Resource (UniProt). *Nucleic Acids Research*, **35** (Database issue), D193–D197.

52 Suzek, B.E., Huang, H., McGarvey, P., Mazumder, R. and Wu, C.H. (2007) UniRef: comprehensive and non-redundant UniProt reference clusters. *Bioinformatics*, **23** (10), 1282–1288.

53 Friedberg, I. (2006) Automated protein function prediction – the genomic challenge. *Briefings in Bioinformatics*, **7** (3), 225–242.

54 Bateman, A., Coin, L., Durbin, R., Finn, R.D., Hollich, V., Griffiths-Jones, S., Khanna, A., Marshall, M., Moxon, S., Sonnhammer, E.L. *et al.* (2004) The Pfam protein families database. *Nucleic Acids Research*, **32** (Database issue), D138–D141.

55 Mulder, N.J., Apweiler, R., Attwood, T.K., Bairoch, A., Bateman, A., Binns, D., Bork, P., Buillard, V., Cerutti, L., Copley, R. *et al.* (2007) New developments in the InterPro database. *Nucleic Acids Research*, **35** (Database issue), D224–D228.

56 Haft, D.H., Selengut, J.D. and White, O. (2003) The TIGRFAMs database of protein families. *Nucleic Acids Research*, **31** (1), 371–373.

57 Mi, H., Lazareva-Ulitsky, B., Loo, R., Kejariwal, A., Vandergriff, J., Rabkin, S., Guo, N., Muruganujan, A., Doremieux, O., Campbell, M.J. *et al.* (2005) The PANTHER database of protein families, subfamilies, functions and pathways. *Nucleic Acids Research*, **33** (Database issue), D284–D288.

58 Hulo, N., Bairoch, A., Bulliard, V., Cerutti, L., De Castro, E., Langendijk-Genevaux, P.S., Pagni, M. and Sigrist, C.J. (2006) The PROSITE database. *Nucleic Acids Research*, **34** (Database issue), D227–D230.

59 Bru, C., Courcelle, E., Carrere, S., Beausse, Y., Dalmar, S. and Kahn, D. (2005) The ProDom database of protein domain families: more emphasis on 3D. *Nucleic Acids Research*, **33** (Database issue), D212–D215.

60 Quevillon, E., Silventoinen, V., Pillai, S., Harte, N., Mulder, N., Apweiler, R. and Lopez, R. (2005) InterProScan: protein domains identifier. *Nucleic Acids Research*, **33**Web Server (issue), W116–W120.

61 Eisen, J.A. (1998) Phylogenomics: improving functional predictions for uncharacterized genes by evolutionary analysis. *Genome Research*, **8** (3), 163–167.

62 Sjolander, K. (2004) Phylogenomic inference of protein molecular function: advances and challenges. *Bioinformatics*, **20** (2), 170–179.

63 Altschul, S.F., Madden, T.L., Schaffer, A.A., Zhang, J., Zhang, Z., Miller, W. and Lipman, D.J. (1997) Gapped BLAST and PSI-BLAST: a new generation of protein database search programs. *Nucleic Acids Research*, **25** (17), 3389–3402.

64 Higgins, D.G., Thompson, J.D. and Gibson, T.J. (1996) Using CLUSTAL for multiple sequence alignments. *Methods in Enzymology*, **266**, 383–402.

65 Li, X., Duan, X., Jiang, H., Sun, Y., Tang, Y., Yuan, Z., Guo, J., Liang, W., Chen, L., Yin, J. *et al.* (2006) Genome-wide analysis of basic/helix-loop-helix transcription factor family in rice and Arabidopsis. *Plant Physiology*, **141** (4), 1167–1184.

66 Sampedro, J., Lee, Y., Carey, R.E., dePamphilis, C. and Cosgrove, D.J. (2005) Use of genomic history to improve phylogeny and understanding of births and deaths in a gene family. *The Plant Journal*, **44** (3), 409–419.

67 Wu, F., Mueller, L.A., Crouzillat, D., Petiard, V. and Tanksley, S.D. (2006) Combining bioinformatics and phylogenetics to identify large sets of single-copy orthologous genes (COSII) for comparative, evolutionary and systematic studies: a test case in the euasterid plant clade. *Genetics*, **174** (3), 1407–1420.

68 Manfield, I.W., Devlin, P.F., Jen, C.H., Westhead, D.R. and Gilmartin, P.M. (2007) Conservation, convergence, and divergence of light-responsive, circadian-regulated, and tissue-specific expression patterns during evolution of the

Arabidopsis GATA gene family. *Plant Physiology*, **143** (2), 941–958.

69 Nobuta, K., Venu, R.C., Lu, C., Belo, A., Vemaraju, K., Kulkarni, K., Wang, W., Pillay, M., Green, P.J., Wang, G.L. *et al.* (2007) An expression atlas of rice mRNAs and small RNAs. *Nature Biotechnology*, **25**, 473–477.

70 Cheung, F., Haas, B.J., Goldberg, S.M., May, G.D., Xiao, Y. and Town, C.D. (2006) Sequencing *Medicago truncatula* expressed sequenced tags using 454 Life Sciences technology. *BMC Genomics*, **7**, 272.

71 Rensink, W.A. and Buell, C.R. (2005) Microarray expression profiling resources for plant genomics. *Trends in Plant Science*, **10** (12), 603–609.

72 Persson, S., Wei, H., Milne, J., Page, G.P. and Somerville, C.R. (2005) Identification of genes required for cellulose synthesis by regression analysis of public microarray data sets. *Proceedings of the National Academy of Sciences of the United States of America*, **102** (24), 8633–8638.

73 Lee, Y., Tsai, J., Sunkara, S., Karamycheva, S., Pertea, G., Sultana, R., Antonescu, V., Chan, A., Cheung, F. and Quackenbush, J. (2005) The TIGR Gene Indices: clustering and assembling EST and known genes and integration with eukaryotic genomes. *Nucleic Acids Research*, **33** (Database issue), D71–D74.

74 Dong, Q., Lawrence, C.J., Schlueter, S.D., Wilkerson, M.D., Kurtz, S., Lushbough, C. and Brendel, V. (2005) Comparative plant genomics resources at PlantGDB. *Plant Physiology*, **139** (2), 610–618.

75 Childs, K., Hamilton, J., Zhu, W., Ly, E., Cheung, F., Wu, H., Rabinowicz, P.D., Town, C.D., Buell, C.R., Chan, A.P. The TIGR Plant Transcript Assemblies Database. *Nucleic Acids Research*, **35**, D846–D851.

76 Wheeler, D.L., Barrett, T., Benson, D.A., Bryant, S.H., Canese, K., Chetvernin, V., Church, D.M., DiCuccio, M., Edgar, R., Federhen, S. *et al.* (2007) Database resources of the National Center for Biotechnology Information. *Nucleic Acids Research*, **35** (Database issue), D5–D12.

77 Stein, L.D., Mungall, C., Shu, S., Caudy, M., Mangone, M., Day, A., Nickerson, E., Stajich, J.E., Harris, T.W., Arva, A. *et al.* (2002) The generic genome browser: a building block for a model organism system database. *Genome Research*, **12** (10), 1599–1610.

78 Chen, F.C., Wang, S.S., Chaw, S.M., Huang, Y.T. and Chuang, T.J. (2007) Plant Gene and Alternatively Spliced Variant Annotator. A plant genome annotation pipeline for rice gene and alternatively spliced variant identification with cross-species expressed sequence tag conservation from seven plant species. *Plant Physiology*, **143** (3), 1086–1095.

79 Zhu, W. and Buell, C.R. (2007) Improvement of whole-genome annotation of cereals through comparative analyses. *Genome Research*, **17** (3), 299–310.

80 Goff, S.A., Ricke, D., Lan, T.H., Presting, G., Wang, R., Dunn, M., Glazebrook, J., Sessions, A., Oeller, P., Varma, H. *et al.* (2002) A draft sequence of the rice genome (*Oryza sativa* L. ssp. *japonica*). *Science*, **296** (5565), 92–100.

81 International Rice Genome Sequencing Project. (2005) The map-based sequence of the rice genome. *Nature*, **436** (7052), 793–800.

82 Yu, J., Hu, S., Wang, J., Wong, G.K., Li, S., Liu, B., Deng, Y., Dai, L., Zhou, Y., Zhang, X. *et al.* (2002) A draft sequence of the rice genome (*Oryza sativa* L. ssp. *indica*). *Science*, **296** (5565), 79–92.

83 Yu, J., Wang, J., Lin, W., Li, S., Li, H., Zhou, J., Ni, P., Dong, W., Hu, S., Zeng, C. *et al.* (2005) The Genomes of *Oryza sativa*: a history of duplications. *PLoS Biology*, **3** (2), e38.

84 Lai, C.W., Yu, Q., Hou, S., Skelton, R.L., Jones, M.R., Lewis, K.L., Murray, J., Eustice, M., Guan, P., Agbayani, R. *et al.* (2006) Analysis of papaya BAC end sequences reveals first insights into the organization of a fruit tree genome. *Molecular Genetics and Genomics*, **276** (1), 1–12.

85 Ayele, M., Haas, B.J., Kumar, N., Wu, H., Xiao, Y., Van Aken, S., Utterback, T.R., Wortman, J.R., White, O.R. and Town, C.D. (2005) Whole genome shotgun sequencing of *Brassica oleracea* and its application to gene discovery and annotation in Arabidopsis. *Genome Research*, **15** (4), 487–495.

86 Bossolini, E., Wicker, T., Knobel, P.A. and Keller, B. (2007) Comparison of orthologous loci from small grass genomes Brachypodium and rice: implications for wheat genomics and grass genome annotation. *The Plant Journal*, **49** (4), 704–717.

87 Rabinowicz, P.D., Schutz, K., Dedhia, N., Yordan, C., Parnell, L.D., Stein, L., McCombie, W.R. and Martienssen, R.A. (1999) Differential methylation of genes and retrotransposons facilitates shotgun sequencing of the maize genome. *Nature Genetics*, **23** (3), 305–308.

88 Peterson, D.G., Schulze, S.R., Sciara, E.B., Lee, S.A., Bowers, J.E., Nagel, A., Jiang, N., Tibbitts, D.C., Wessler, S.R. and Paterson, A.H. (2002) Integration of Cot analysis, DNA cloning, and high-throughput sequencing facilitates genome characterization and gene discovery. *Genome Research*, **12** (5), 795–807.

89 Palmer, L.E., Rabinowicz, P.D., O'Shaughnessy, A.L., Balija, V.S., Nascimento, L.U., Dike, S., de la Bastide, M., Martienssen, R.A. and McCombie, W.R. (2003) Maize genome sequencing by methylation filtration. *Science*, **302** (5653), 2115–2117.

90 Yuan, Y., SanMiguel, P.J. and Bennetzen, J.L. (2003) High-Cot sequence analysis of the maize genome. *The Plant Journal*, **34** (2), 249–255.

91 Whitelaw, C.A., Barbazuk, W.B., Pertea, G., Chan, A.P., Cheung, F., Lee, Y., Zheng, L., van Heeringen, S., Karamycheva, S., Bennetzen, J.L. *et al.* (2003) Enrichment of gene-coding sequences in maize by genome filtration. *Science*, **302** (5653), 2118–2120.

92 Bedell, J.A., Budiman, M.A., Nunberg, A., Citek, R.W., Robbins, D., Jones, J., Flick, E., Rholfing, T., Fries, J., Bradford, K. *et al.* (2005) Sorghum genome sequencing by methylation filtration. *PLoS Biology*, **3** (1), e13.

93 Korf, I., Flicek, P., Duan, D. and Brent, M.R. (2001) Integrating genomic homology into gene structure prediction. *Bioinformatics*, **17** (Suppl 1), S140–S148.

94 Alexandersson, M., Cawley, S. and Pachter, L. (2003) SLAM: cross-species gene finding and alignment with a generalized pair hidden Markov model. *Genome Research*, **13** (3), 496–502.

95 Majoros, W.H., Pertea, M. and Salzberg, S.L. (2005) Efficient implementation of a generalized pair hidden Markov model for comparative gene finding. *Bioinformatics*, **21** (9), 1782–1788.

96 Siepel, A. and Haussler, D. (2004) Combining phylogenetic and hidden Markov models in biosequence analysis. *Journal of Computational Biology*, **11** (2–3), 413–428.

97 Wu, J. and Haussler, D. (2006) Coding exon detection using comparative sequences. *Journal of Computational Biology*, **13** (6), 1148–1164.

98 Kellis, M., Patterson, N., Birren, B., Berger, B. and Lander, E.S. (2004) Methods in comparative genomics: genome correspondence, gene identification and regulatory motif discovery. *Journal of Computational Biology*, **11** (2–3), 319–355.

99 Kellis, M., Patterson, N., Endrizzi, M., Birren, B. and Lander, E.S. (2003) Sequencing and comparison of yeast species to identify genes and regulatory elements. *Nature*, **423** (6937), 241–254.

100 Guigo, R., Dermitzakis, E.T., Agarwal, P., Ponting, C.P., Parra, G., Reymond, A., Abril, J.F., Keibler, E., Lyle, R., Ucla, C. *et al.* (2003) Comparison of mouse and human genomes followed by experimental verification yields an estimated 1,019 additional genes. *Proceedings of the National Academy of Sciences of the United States of America*, **100** (3), 1140–1145.

101 Wu, J.Q., Shteynberg, D., Arumugam, M., Gibbs, R.A. and Brent, M.R. (2004) Identification of rat genes by TWINSCAN gene prediction, RT-PCR, and direct sequencing. *Genome Research*, **14** (4), 665–671.

102 Yandell, M., Bailey, A.M., Misra, S., Shu, S., Wiel, C., Evans-Holm, M., Celniker, S.E. and Rubin, G.M. (2005) A computational and experimental approach to validating annotations and gene predictions in the *Drosophila melanogaster* genome. *Proceedings of the National Academy of Sciences of the United States of America*, **102** (5), 1566–1571.

103 Encode Project Consortium. (2007) Identification and analysis of functional elements in 1% of the human genome by the ENCODE pilot project. *Nature*, **447** (7146), 799–816.

19
Large-Scale Genomic Sequence Comparison and Gene Identification with ClustDB

Jürgen Kleffe

Abstract

Genome-wide sequence comparison, repeat detection, gene identification by EST or cDNA matching, as well as BAC assembly and EST clustering are important methods in genomic research. These methods rely on the efficient identification of all similar substrings in large sets of sequences and challenge software to keep up with the fast growth of data observed in the last decade. We therefore describe a new program for the simultaneous identification of similar sequences using much larger quantities of data than the currently available software can handle, given that there is only a certain amount of memory available with the current software. The algorithm for the new program is not only faster in single processor mode but also allows simple methods of parallelization. Based on count sorting, this program simultaneously finds all clusters of common substrings of a given minimal length. At the same time, the program derives and extends to maximal length all pairs of left maximal matching substrings, and extends such pairs to both sides allowing for mismatches and gaps until the number of errors exceeds a given threshold within a window of given size. Such alignments are faster to calculate than optimal alignments and more appropriate for the applications listed above. A built-in simple sequence management system quickly turns FASTA, Genbank and EMBL sequence libraries into binary format that allows fast direct access and manipulation of sequences. This program, ClustDB, is freely available for academic use under Unix/Linux operation systems. Please download from www.medizin.fu-berlin.de/molbiochem/bioinf.

19.1
Introduction

After the creation of the Genomes Online Database http://www.genomesonline.org/gold.cgi more than 126 plant genome sequencing projects now produce and annotate sequences that we could not have imagined in the past. Four plant genomes have been completed. The five chromosomes of *Arabidopsis thaliana* http://www.nature.com/

The Handbook of Plant Functional Genomics: Concepts and Protocols.
Edited by Günter Kahl and Khalid Meksem

ISBN: 978-3-527-31885-8

nature/links/001214/001214-1.html contain about 130 MB of sequence. The National Plant Genome Initiative http://www.nsf.gov.pubs/npgi2006/npgi2006.pdf and the International Rice Genome Sequencing Project http://www.nature.com/journal/v436/n7052/abs/nature03895.html reported the completion of genome sequencing for two species of *Oryza sativa* containing about 390 MB and 430 MB of sequence, respectively, and 2008 will see the completion of the maize genome with an estimated size of more than 2.4 GB. Sequencing of seven other major plant genomes is in progress including the *Avena sativa* genome with an expected size of 11 GB not to mention 95 minor projects and more than 6.2 GB of plant ESTs and cDNAs that are stored in Genbank http://ftp.ncbi.nih.gov/genbank. Highly improved sequencing technologies have made these advances possible. The biotechnology service company GATC announced a sequencing capacity of 130 GB per year, 'making it able to sequence the whole human genome in just 10 days', said its top manager in an interview to the service magazine *Laborjournal* (February 2007). The manufacturer of the new Solexa Genome Analysis System (http://www.solexa.com) claims that the cost of sequencing using their system is less than 1% of the cost of the older capillary methods and Applied Biosystems will soon be releasing the new SOLID system with a giant sequencing capacity. These systems offer an almost unlimited potential for comparative genome studies across closely related species which facilitates the elucidation of how plants manage to adapt to changing environments such as freezing temperatures and how they defend themselves against attacks by pests. Agriculture and forestry increasingly depends on genomics-based research to improve the properties of biomass for plant feed stocks and energy production.

But bioinformatics must play its part. Current programs and commonly used hardware now lag behind in their ability to compare large genomes. Treangen and Messeguer [1] describe the M-GCAT system and other software for multiple whole genome comparisons. Most of these programs were designed to compare bacterial genomes which vary from 1 to 10 MB in size. However, even the smallest plant genome is larger than this range. The first program of reasonable capacity was REPuter [2,3] and the currently most advanced string matching algorithms implemented in the programs MUMMER [4] and VMATCH [5] can compare 200 or 300 MB of sequence using 32-bit computers with 2 GB of memory [6], but this capacity is still insufficient for plant research. Table 19.1 shows that from 2002, this capacity represents less than a one year's EST production.

EST matching is believed to provide an inexpensive and direct route to the identification of the genes involved in the regulation of the life cycle of a plant. But the inability to routinely compare sufficiently large sets of ESTs with genomic sequences, results in many matches being overlooked. It also hinders quality control and the detection of sequence contamination using methods reviewed in [7] and based on comparisons of entire EST libraries. The same applies to other data such as reads, shotgun sequences, BAC-end sequences, mate pairs and CAGE tags. The large number of programs available for automated gene prediction has led to the common practice of submitting an additional gene annotation. However, the rapid growth of data and the role of alternative splicing have made computational gene prediction far more difficult than expected. The most difficult part of good annotation concerns

Table 19.1 Growth of plant ESTs submitted to Genbank over the years.

Year	New ESTs	KB	Year	New ESTs	KB
1992	618	189	2000	349 846	170 110
1993	1136	367	2001	546 223	292 794
1994	2368	781	2002	1 046 356	516 942
1995	4519	1417	2003	1 580 745	872 302
1996	3007	937	2004	1 441 989	762 410
1997	13 068	5602	2005	1 778 964	1 089 741
1998	34 359	14 423	2006	3 979 281	2 108 241
1999	58 644	27 200	2007	985 568*	345 519*

Compressed files gbest1.seq.gz to gbest589.seq.gz with ESTs were downloaded from the Genbank ftp center http://ftp.ncbi.nih.gov/genbank April 2007 release 158.0 and searched for the entries with 'viridiplantae' in the field 'ORGANISM'. The resulting 11 826 691 EST entries were distributed over 19 files, called p1 to p19, and sorted by the year of publication. All sequences add up to more than 6.2 GB. *Includes January to April 2007 only.

finding evidence for or against published gene annotations, identifying alternative gene annotations in published sequences, and distinguishing alternative splice forms from false gene predictions. Often only nearly perfect matches with full-length cDNAs are considered proof of expressed genes. Figure 19.1 shows three different published gene annotations of the same ~100-KB sequence section of *Arabidopsis thaliana* chromosome IV detected by the advanced methods described in [8]. The methods all combine gene prediction with simultaneous sequence comparison.

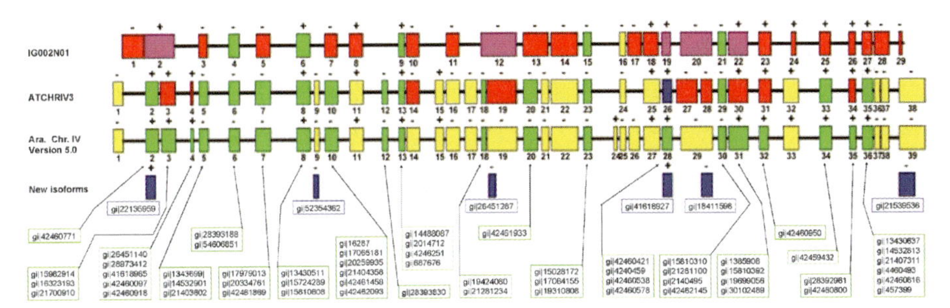

Figure 19.1 Three alternative gene annotations published in different Genbank entries for the same genomic subsequence of *Arabidopsis thaliana* chromosome IV (about 100 kB). The sequence IG002N01 shows an initial gene annotation by Green and Hillier using the program Genefinder (unpublished). The gene annotation in ATCHRIV3 was published in March 2000. We also present the current annotation of chromosome IV by TIGR. Each box denotes a complete gene. Shading of the gene models in the same color indicates that the exons and introns agree; different colors indicate that they differ. The color green is used for gene models supported by one or more full-length cDNAs listed in the corresponding text fields. In chromosome IV, genes 27 and 28 of ATCHRIV3 are summarized to a single gene, gene 29, very similar to the earlier gene prediction 20 made in IG002N01. The genes 3, 4, 14, 19, 26, 30, 31 and 34 of chromosome IV have been changed. Two new genes, 24 and 26, have been added. Only 20 out of 39 genes are supported by perfectly matching full-length cDNAs. However, the as yet unsupported gene predictions 9, 19, 29 and 39 require revision. New perfectly matching full-length cDNAs prove new isoforms (blue boxes).

Note that programs like BLAST [9] and BLAT [10] are sufficient to compare small sets of sequences with large databases as required to study single genes. The genome-wide investigation of gene annotations requires new software to answer new questions such as finding all known genes with full-length cDNA support or all genes alternatively defined in various database entries. In all these cases there is no hypothesis to begin with and irrespective of how fast we can test each candidate sequence, it takes too long to process millions of cases in turn. The known programs for simultaneous sequence comparison like MUMMER [4], VMATCH [5], QUASAR [11], MGA [12], FORREPEATS [13], PATTERNHUNTER [14] and SSAHA [15] cannot compare essentially more than 300 MB of sequence in a single run using 2 GB of memory. This limitation results from the memory consumption of the complex index structures used to efficiently generate all start positions of common substrings. But again, independently of how fast we can construct indices, it takes hours to report in turn millions of matches. There is little advantage in storing index data for generating large and complete sets of matches.

The new software ClustDB [6,16] uses a direct route to derive all start positions of common substrings for sequence data that can be close to twice as large as the computer's memory. Table 19.2 lists the performance of ClustDB for comparing growing sets of plant ESTs using a 32-bit PC with 2-GB memory and 2.6 GHz processor speed. The largest data set comes very close to the limit of the address space for 32-bit computers so that a much larger number of base pairs could not even be enumerated. The smallest data set considered is already too large for application of MUMMER, VMATCH or other existing programs.

Calling sets of common substrings clusters, ClustDB simultaneously detects all left maximal clusters of common substrings of a given minimal length. It simultaneously generates and extends all pairs of left maximal matching substrings to maximum length and extends matching pairs of substrings to alignments with a given maximum number of errors within all alignment windows of a given size. This novel heuristic alignment implies uniform local alignment quality along the entire match and is more

Table 19.2 Performance of ClustDB for finding common substrings in large sets of plant ESTs.

ESTs	Base pairs (MB)	Time	Clusters	Substrings	LM (MB)	SM (MB)
812 439	372	0:06:50	705 947	9 392 122	179.1	107.5
2 175 442	1085	0:24:35	2 330 723	40 493 389	848.6	463.4
3 606 247	1795	0:51:22	4 125 124	83 608 647	1700.0	956.8
4 984 183	2516	1:16:39	6 009 482	125 136 465	2548.0	1432.1
6 128 787	3074	1:56:33	7 416 172	200 927 613	4067.8	2068.1
6 780 291	3458	2:40:07	8 349 238	222 043 356	4505.7	2540.8
7 920 978	4153	3:08:08	9 931 631	264 392 862	5551.1	3035.7

Each line provides data for application of ClustDB to a growing number of sequence library files p1 to p12 derived as reported in the legend to Table 19.1. The time of execution includes listing all clusters of common substrings of minimal length 50 to a text file and storing them in a DNA_Stat database. The last two columns show that the DNA_Stat databases (option -SM) take less disk space than the text file output (option -LM). All computations were performed on the same PC with 2 GB of memory and 2.6 GHz processor speed.

useful for the characterization of long similar sequences than optimal linear dynamic programming alignment defined by penalties for mismatches and gaps.

19.2 Methods and Protocols

ClustDB is a command line program that performs a number of different tasks. We describe the major options and comment on the methods used to solve corresponding problems. A formal mathematical presentation of the algorithm is given in [6,16]

19.2.1 Reading Sequences

Everything begins by reading sequences which come in sequence library files. These are multiple sequence files written in different formats. The simplest format is FASTA. One line beginning with the symbol '>' provides information for sequence identification. The genetic sequence then follows and is written into one line or a number of lines. A more complex sequence format is distributed by Genbank http://ftp.ncbi.nih.gov/genbank. These files contain a number of data fields with information about the sequence and its authors. Also the nucleotide sequence is presented in a more complex form. Two forward slash characters written on a separate line terminate each sequence entry. The EMBL format is similar. ClustDB requires input written in one of these three formats. The command

```
<![ CDATA[
ClustDB file1 file2 file3 ....
]]>
```

generates from up to 20 given sequence library files, the same number of binary formatted DNA_Stat sequence libraries. This format allows rapid and direct access to individual sequences. Each library file passed to ClustDB is allowed to contain up to 2 GB of sequence. All sequences passed are numbered beginning with 1. These sequence numbers are used for communicating all results. ClustDB first lists a survey of sequence numbers to screen and to an ouptut file named using the option '-OF name'. For instance, passing sequence library file names file1, file2 and file3 generates the following log:

```
<![ CDATA[
database 1: sequences         1 to   771113 bps: 355677034 name:
file1
database 2: sequences    771114 to 1415356 bps: 329356967 name:
file2
database 3: sequences  1415357 to 2075115 bps: 355294685 name:
file3
]]>
```

The log file is also used to communicate summary information about the matches found which are generally written to very large files as shown in Table 19.2 and hence, are often impossible to examine using standard text editors. With the sequence numbers known we can limit analysis to a simple subset of sequences by defining the first sequence with command '-FS X', the last sequence with '-LS X' or the number of sequences with '-NS X'. There is no need to prepare a special input file in order to make such a selection and it is also convenient to see the results in terms of original sequence numbers. The possibility of using several sequence libraries is also important. These files generally contain different types of sequences such as ESTs and BACs or sequences from different sources. The created DNA_Stat library files receive the file extension '.SUD'. Once created, these databases can be used again by adding the extension '.SUD' to the original sequence library file names listed in the command line. This option makes sequence data instantly available compared to parsing large sequence text files.

19.2.2
Substring Clusters

The most intuitive way of comparing objects is by sorting. Finding common substrings of a long sequence is just as easy. ClustDB just looks at the set of all overlapping words of a given length, W, placed in a sequence. If the sequence length is N, it contains $N - W + 1$ overlapping words in sequence positions 1 to $N - W + 1$ also called suffix positions or suffix numbers. The suffix number j is the subsequence that starts in position j and extends to the end of the sequence. Suffixes are represented by their suffix numbers and can be sorted by the different length, W, of the words they begin with. The result is a set of disjoint clusters of suffix numbers. In small sequences some clusters are of size 1 and identify unique substrings. Each cluster of size greater than 1 is now studied in turn to find longer common substrings by using the same method but looking at the words beginning in suffix positions $j + W$. This method of finding clusters of common substring positions is illustrated in Table 19.3 and gave ClustDB its name.

The program concatenates all individual sequences to form one long sequence and calculates clusters of common substring positions. Dot characters are used for sequence separation. The following command lists and stores all clusters of common substrings found in a single database.

```
<![ CDATA[
ClustDB file1 file2 file3 -LM match -SM match -ML 50 -WL 5
]]>
```

The option '-LM match' provides the file name for listing substring clusters. Given instead or simultaneously, the option '-SM match' sets the name of a DNA_Stat database that holds all substring clusters in a more compact form. As seen in column SM of Table 19.2 the size of a DNA_Stat database is close to half of the text file output size given in column LM. All substrings of a cluster match over at least 50 characters enforced by the option '-ML 50' and word length 5 is used for computation. The following provides some detail of how ClustDB can achieve the performance listed in Table 19.2.

Table 19.3 The iterated suffix sort algorithm by example.

A			B			C		
clu	suf	suffix	clu	suf	suffix	clu	suf	suffix
0	1	aaaaa ccccc	1	1	aaaaa ccccc	1	1	aaaaa ccccc
0	2	aaaac cccca	1	11	aaaaa cccc	1	11	aaaaa ccccc
0	3	aaacc cccaa	1	21	aaaaa			
0	4	aaccc ccaaa	2	2	aaaac cccca	2	2	aaaac cccca
0	5	acccc caaaa	2	12	aaaac cccca	2	12	aaaac cccca
0	6	ccccc aaaaa	3	3	aaacc cccaa	3	3	aaacc cccaa
0	7	cccca aaaac	3	13	aaacc cccaa	3	13	aaacc cccaa
0	8	cccaa aaacc	4	4	aaccc ccaaa	4	4	aaccc ccaaa
0	9	ccaaa aaccc	4	14	aaccc ccaaa	4	14	aaccc ccaaa
0	10	caaaa acccc	5	5	acccc caaaa	5	5	acccc caaaa
0	11	aaaaa ccccc	5	15	acccc caaaa	5	15	acccc caaaa
0	12	aaaac cccca	6	6	ccccc aaaaa	6	6	ccccc aaaaa
0	13	aaacc cccaa	6	16	ccccc aaaaa	6	16	ccccc aaaaa
0	14	aaccc ccaaa	7	7	cccca aaaac	7	7	cccca aaaac
0	15	acccc caaaa	7	17	cccca aaaa			
0	16	ccccc aaaaa	8	8	cccaa aaacc	8	8	cccaa aaacc
0	17	cccca aaaa	8	18	cccaa aaa			
0	18	cccaa aaa	9	9	ccaaa aaccc	9	9	ccaaa aaccc
0	19	ccaaa aa	9	19	ccaaa aa			
0	20	caaaa a	10	10	caaaa acccc	10	10	caaaa acccc
0	21	aaaaa	10	20	caaaa a			

A: The sequence 'aaaaacccccaaaaacccccaaaaa' has length 25 and there are 21 positions for suffixes with a minimum length W = 5. They constitute the initial cluster zero shown in column 'clu' while column 'suf' provides the suffix positions. The column 'suffix' provides the initial word and the next five characters if available.B: Sorting suffix positions for the initial words of length 5 yields 10 clusters of suffix positions, one of size 3 and all others of size 2.C: Sorting each cluster obtained for the next five characters yields 10 clusters of common substrings of minimum length 10. The clusters 7 to 10 are of size 1 and represent unique substrings of length 10. The other clusters represent pairs of common substrings.A next step would prove 'aaaaacccccaaaaa' to be the only multiple substring of length 15 occurring at positions 1 and 11.

For maximum efficiency it is not only important to remove substring clusters of size 1 as soon as possible but also to avoid the consideration of clusters of common substring which are parts of longer common substrings. Such clusters have common prefixes, that is, the letters in front of all substrings are the same. Therefore ClustDB considers only substring clusters for which at least two prefixes differ. They are called left maximal.

Next, fast and memory efficient sorting is important. In our case we do not search for, but rather calculate the position of each object in the sorted list. Assume N objects have M different labels and label i occurs n(i) times. Then for any label k, $c(k) = n(1) + n(2) + \ldots + n(k)$ is the cumulative frequency of label k and tells us that all objects with label k take places $c(k - 1) + 1$ to c(k) in the list of objects sorted by labels. This method is called count sorting or distribution counting in the book by

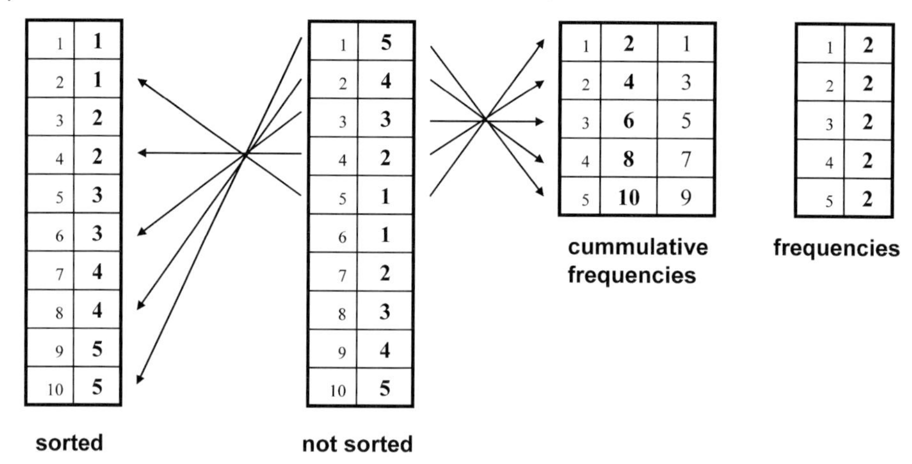

Figure 19.2 The count-sort method by example. The vector 'not sorted' contains 10 numbers in the range 1–5 with each number being represented at a frequency of 2. The cumulative frequencies are given to the left of the frequency vector. With this information, one run through the unsorted vector places each number into the correct place in the vector 'sorted'. The first number to be positioned is 5, and row 5 of the cumulative frequency table assigns it to position 10. Having positioned a number, the cumulative frequency used must be decreased by 1 to provide the correct position for another equal number that may occur. For clarity the reduced cumulative frequencies have been shown in an extra column; this is not necessary in practice.

Sedgewick [17]. It makes no comparisons, works in time proportional to the number of objects and works well for sorting of words of fixed length labeled by word numbers, that is, by the positions in the lexicographically ordered list of all possible words. Figure 19.2 illustrates the method by example. The first step is reading from the files the words at all suffix positions; this is used to calculate the cumulative frequency of each word. Then a second reading of data immediately puts each suffix position into the correct place in an array allocated to contain the sorted list of suffix positions. This sorted list is created piece by piece as large as can fit into memory and is stored on disk part by part. By saving all frequency vectors from the generated files, it is easy to collect the grand total of suffix positions which start with the same word. For instance, the first components of all frequency vectors tell us how many suffix positions to read from each file in order to obtain the complete list of all suffixes which begin with W letters 'a'. This method completely solves the initial sort shown in Table 19.3.

The second sort that produces part C in Table 19.3 is different. The advantage of this method is that the clusters of suffix positions derived by the initial sort are generally small enough to fit into memory. However, the suffix positions contained in each cluster are spread out over the entire sequence space. Efficient access to all of them requires storing the whole sequence in memory. After that there may be no space left for an extra array to store the sorted suffix positions. Count sorting on the place (rather than writing sorted numbers to an extra array) helps in such a case and is illustrated in Figure 19.3. For very large arrays count sorting on the place was observed to be even faster.

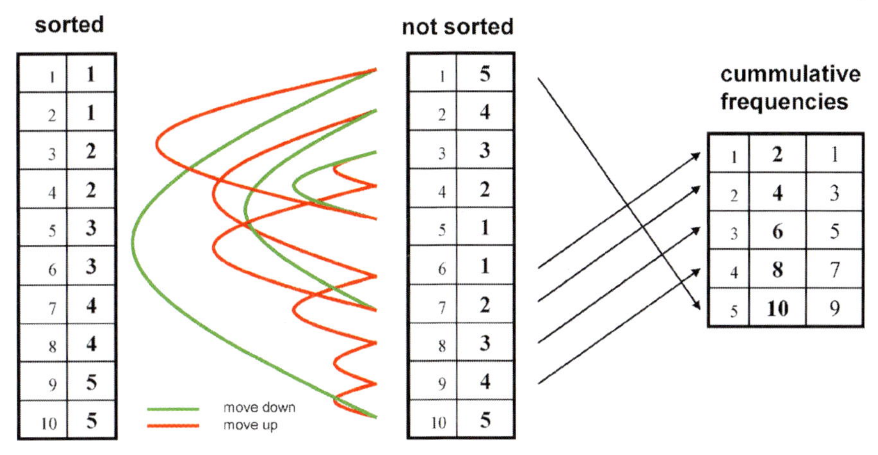

Figure 19.3 Count-sort on the position (example from Figure 19.2). Having derived cumulative frequencies we look up the correct position of each number. But before placing it correctly, the number in the target position is saved and chosen next for correct placement. The resulting cycle always returns a particular number to the place it started in. Marking correctly placed numbers allows the identification of the next as yet not correctly placed number to start the next cycle. Sorting terminates when all the numbers have been correctly positioned.

19.2.3 Maximally Extended Pairs of Common Substrings

Once left maximal clusters of common substrings of given minimum length M are derived we can ask how far each pairwise match extends to the right. It generally takes a long time to extend in turn all possible pairwise matches derived from a large cluster of common substrings. The time factor can be reduced by applying limitations to left maximal pairs of common substrings. There are only four possible prefixes and hence many if not most pairs of common substrings may have identical prefixes and need not be considered. Further, of the left maximal pairs of common substrings only those followed by different words of length W are initially considered for match extension. It takes at most W character comparisons to find the exact length of such a match. All other pairs match over at least M + W characters. This suggests splitting the considered substring cluster into sub-clusters of common substrings of length M + W as described in Section 4.2 and to study each sub-cluster in turn using the same method. All left maximal pairs of common substrings taken from different sub-clusters had been listed previously. The option '-LE pairs' sets the file name for listing all extended pairwise matches. Initially all matches are derived in unsorted order. This is the price for the high speed generation of matching pairs. Sorting large numbers of pairwise matches takes considerable time. The option '-SO 1' sets sort level 1 and initiates sorting of matches by the smaller sequence number found in each pair. Sort level 2 initiates sorting for both sequence numbers.

19.2.4
Match Extension with Errors

Pairwise exact matches are further extended on both sides until a given maximum number of mismatches occurs in a sliding window of given length. This novel window alignment method has two advantages over the methods used in VMATCH and other software which limit the total number of errors. ClustDB better detects long high quality matches and generates only one match extension of each exact match. Often different exact matches lead to the same extended match so that their number decreases. The opposite occurs in fixing the maximal number of errors. As this number is generally small each exact match becomes extended in different ways with different error numbers on both sides and generates a set of extended matches which do not all exhaust the entire regions of sequence similarity. Using ClustDB the error threshold X and window size Y are set by the options '-EM X' and '-WS Y'.

In order to cut down on unnecessary match extension it is important to identify exact matches prone to yield the same extended match. The offset helps to make such decisions. It denotes the positive difference in the start positions of both matching substrings obtained by subtracting the smaller from the larger. Sorting matches for offset brings together exact matches which could produce the same extended match. The simplest example is given by two sequences which differ by mismatches only. Then all exact matches located between mismatches have the same offset and result in the same extended match.

But at this point count sorting faces a new problem. Unlike word numbers the offsets of exact matches are distributed over a wide range. This makes it expensive to calculate and store cumulative frequencies. A bucket count sort solves this problem. ClustDB sorts all offset numbers O for the values of $O/K = X$ with an appropriately chosen integer K where '/' stands for integer division. X is the largest possible integer such that X^*K is not greater than O. The first bucket number 0 contains all offsets which are less than K and the second bucket number 1 contains all offsets which are between K and 2^*K. Denoting the maximum offset by maxO, the last bucket has the number maxO/K and the bucket sort generates at most 1 + maxO/K non-empty buckets. Hence, an extra array of size 1 + maxO/K is sufficient to undertake the bucket sort. After that bucket j is sorted for the values $O - j^*K$ which are between 0 and $K - 1$. The entire procedure takes time proportional to the number of matches and ClustDB uses $K = \text{maxO}/999$.

Next, closely neighbored extended matches, found by sorting or chaining [18], are joined to larger matches including gaps, assuming that the connecting alignment obeys the given error threshold under edit distance. A special window alignment algorithm [16] was developed in order to extend matches which find no close neighbors. These alignments are particularly useful for locating BACs and shorter sequences on chromosomes, relating single ESTs to full-length cDNA and identifying redundant and contaminated sequences.

19.2.5
Complete Matches

VMATCH calls a match complete if one sequence is found completely contained in another. More generally ClustDB names a match complete if it extends to both sides in the maximum possible manner by reaching 5′ or 3′ ends of at least one sequence. The option '-CM' limits output to those matches which play an important role in sequence assembly and full-length cDNA matching in order to confirm putative genes. In terms of the concatenated sequence, internally considered by ClustDB, a complete match reaches end of sequence symbols at both sides. It detects sequences contained in others or pairs of sequences which overlap. Selecting special types of matches is simplified by the following match type symbols printed in front of each match.

==	sequence 1 matches sequences 2
<<	sequence 1 is part of sequence 2
>>	sequence 1 contains sequence 2
->	sequence 1 extends sequence 2 downstream
<-	sequence 1 extends sequence 2 upstream
--	incomplete match

By including gaps, a complete match sometimes shows a sequence to overlap with itself or to be part of itself. The option '-DS' avoids such results by limiting the pairs of common substrings that can belong to different sequences.

19.2.6
Reference Query Problems

Many problems compare two sets of sequences called reference and query. For instance the program BLAT [10] is specialized to matching ESTs (query) with genomic sequence (reference). Usually an index for the reference is built and the query is compared with the reference using the index. The well-known program BLAST [9] functions in the reverse. ClustDB does not create a real index and hence the nature of the reference and query sequences becomes irrelevant. However, for such problems ClustDB still has a considerable advantage in speed if only pairs of common substrings in the reference and query need to be considered. During iteration, substring clusters decrease in size and are ignored as soon as they run out of reference or query substrings which are distinguished by their suffix positions being less than or greater than a cut point separating both. By default, substring clusters must include at least one suffix position that is less than the cut-off point, known as condition 1, and one suffix position that is greater than the cut-off point, known as condition 2. Optional stringency for this criterion can be weakened. The options '-EC 1' and 'EC 2' require only condition 1 or only condition 2, respectively. The options '-CS X' defines the sequences 1 to X to form the reference set. All other sequences are considered to be query. The option '-CD X' sets the cut-off point on the end of database X. Both options also imply that ClustDB splits the initial sets of sorted suffix positions into reference and query. This allows the use of already sorted reference suffix positions for different sets of queries. Using the option '-SR name', the reference suffix positions are written to the file 'name' and the option '-UR name' instructs ClustDB to read such suffix positions from file 'name' for subsequent applications.

19.2.7 Complementary Sequences

Sequence strand is a relative concept, therefore most sequence matching problems must consider both strands simultaneously, making the option '-AC X' very convenient. Beginning with database number X ClustDB automatically adds for every database numbered Y $\geq$ X a new database that contains all complementary sequences. This option satisfies the frequently employed application, where some databases contain genomic sequence and others contain EST libraries. For example, the call

```
<![ CDATA[
ClustDB genome p1 p2 -AC 2
]]>
```

generates two more databases of complementary sequences. The sequence summary output may look like this:

```
<![ CDATA[
database 1: sequences         1 to 771113  bps:   355677034 name:
genome
database 2: sequences    771114 to 1415356 bps:   329356967 name:
p1
database 3: sequences   1415357 to 2075115 bps:   355294685 name:
p2
database 4: sequences   2075116 to 2719358 bps:   329356967 name:
p1-
database 5: sequences   2717359 to 3379116 bps:   355294685 name:
p2-
]]>
```

The additional option '-CD 1' would define a reference query problem so that only the matching of ESTs to genomic sequences is considered.

19.2.8 Handling Ambiguity Letter Codes

Omitting the nucleotide U, there are only four letters A, C, G and T which completely identify the sequences which a nucleic acid should contain. In practice, however, these sequences often contain large numbers of letters N, which stand for unknown nucleotides, and other letters which code for subsets of possible nucleotides. These are known as ambiguity letter codes and make nucleotide sequence comparison considerably more difficult. Hence, and for simplicity, most programs consider all ambiguity letters to mismatch every other letter including itself. This is also the case with ClustDB when deriving substring clusters. All initially detected common substrings are free of ambiguity letters. Match extension is carried out using different approaches. Continued strict matching is invoked by the option '-RC 0'. Relaxed letter comparison is invoked by the option '-RC 1'. Ambiguity letters are then considered to mismatch if they represent disjoint sets of nucleotides. All other pairs of letters match. ClustDB uses sequence compression if the memory is insufficient to store all data in uncompressed form. ClustDB treats all ambiguity letters such as N in this manner. This may imply extended exact matches which incorrectly pair the letters W and S which stand for A or T and C or G, respectively. But this has not posed a serious problem as yet. It is much more important that ClustDB identifies as many as possible interestingly long matches. One mismatch more or less is for the most part unimportant. In [16] we successfully demonstrated advantages of this letter matching method for identifying overlapping BAC sequences in *Medicago truncatula*.

19.2.9 Sequence Clusters

The substring clusters derived in Section 4.2 allow us to group sequences into disjoint subsets, called sequence-clusters. They are defined relative to the considered match length for finding common substrings. ClustDB puts two sequences into the same cluster if both own a common substring. Sequences taken from two different clusters have no common substring. Initially each sequence forms its own cluster. Then pairs of sequence clusters are joined if they contain a pair of sequences which have a common substring. Although this task appears to be time consuming, the actual algorithm is very fast and has been published in [6]. The most important property of the derived sequence clusters is that their sequences are composed of disjoint sets of substrings. This reduces many sequence matching problems to the consideration of single sequence clusters. In most cases a few sequence clusters are very large and all others are very small. Large numbers of small clusters should be analyzed simultaneously. The sequence clusters are obtained by using the options '-LC name' or '-SL name' which list on file or store in a DNA_Stat database 'name' one or more tables of sequence numbers grouped into sequence clusters. The sequence clusters also help to derive substring clusters for large sets of sequences which cannot be compared in a single run of ClustDB. For instance the current set of plant ESTs adds up to more than 6.2 GB of sequence, impossible to simultaneously compare on a 32-bit computer. But as seen in Table 19.2, the comparison of almost half of the data, comprising 3074 MB of sequence, takes less then 2 h. Derived sequence clusters, individually or in groups, can now in turn be compared with parts of the remaining set of ESTs in order to correctly identify all substring clusters for the total set of sequences. The method fails if a single sequence cluster grows too large. Then increasing match length often helps. Another common practice is to ignore very frequent substrings for sequence clustering. The option '-MS X' sets the maximum number (X) of sequences that a common substring is allowed to belong to in order to initiate the merging of sequence clusters. Another option '-OC suff' outputs two FASTA sequences files for each library file passed to ClustDB. One is named by appending the suffix '.suff' to the original library name and contains all the sequences which were clustered. The second library file is named by appending the suffix '.NULL' and contains all sequences which were not clustered. Matching both derived parts of one database with some other database always generates disjoint sets of common substrings.

19.2.10
Memory Analysis

The memory demand of ClustDB depends on the data and cannot exactly be specified before running the program. Some data-driven flexibility of the algorithm takes into account memory restrictions. One important parameter is the number of splits SP used to perform the initial suffix sort. ClustDB sorts S/SP suffixes at the same time, where S is the total number of suffixes considered to be potential start positions of matches. Its maximum value is the total number of base pairs minus match length multiplied by the number of sequences which constitute the data. Hence the memory requirement for the initial sorting of suffix positions is essentially $4^*(Z + S/SP)$ bytes, where Z is 4^W, the number of different words of length W used for calculation. Some extra space is allocated for buffering sequence sections. It is mostly negligible in size as is Z compared to S/SP.

The next step of computation uses $(BP + NS)/3$ bytes in order to simultaneously store all sequences in compressed form (1 byte for three nucleotides). Here BP is the total number of base pairs and NS is the number of sequences. Another 8^*F bytes are required for the iterated count sort procedure where F is the maximum frequency found for a word of length W. The expected value of F is BP/Z but a great proportion of low complexity sequence may imply large deviations. Assuming there are $(BP + NS)/2$ bytes of memory, we can use approximately $(BP + NS)/6$ bytes for a table of size 8^*F, that is, F can be as large as $(BP + NS)/48$. Every 48th of all overlapping words must be the same in order to cause failure of the algorithm. We have not yet experienced such a case in practice using the word length $W = 5$ which was found to be most efficient. The option '-MB X' actively helps to avoid swapping. It limits to X MB the mount of memory ClustDB uses to allocate arrays.

19.3
Applications

Of the many possible applications of ClustDB we will describe just three, finding identical plant ESTs, finding all common substring clusters for more than 6.2 GB of plant ESTs using 2 GB of memory and calculation of all long sequence overlaps for a set of 2020 BACs from *Medicago truncatula* used for BAC assembly.

19.3.1 Deriving Clusters of Identical Plant ESTs

Following Table 19.2, a PC with 2 GB of RAM is able to calculate substring clusters for close to 4 GB of sequence; however there are more than 6 GB of plant ESTs to compare. Cross comparisons of subsets seems to be the only solution. Looking for identical sequences is a simpler course of action. It suffices to compare only substrings which begin in sequence position 1, that is, the number of initial suffix positions is equivalent to the number of sequences. ClustDB has a corresponding option. Appending ':1' to all database file names restricts valid suffixes to the first positions of each sequence. Consequently, ClustDB also stores only 50 characters of each sequence if the match length is set to this default value. Taking all these measures together the sequence matching problem becomes greatly reduced and using previously derived DNA_Stat databases (see Section 4.1) the call,

```
<![ CDATA[
ClustDB p1:1 p2:1 ... p18:1 p19:1 -SC cluster -OC x,
]]>
```

generates 19 new FASTA sequence libraries known as p1.x to p19.x in about 13 min. These libraries contain only those sequences of the input libraries p1 to p19, respectively, that match with some other sequence from the whole set. Matches necessarily begin in position 1 and extend over at least 50 characters. It is the option '-OC x' that generates the library files p1.x to p19.x based on the derived sequence clusters stored in the database 'cluster' and uses the suffix 'x' to generate names of the new and considerably smaller sequence libraries. A second call,

```
<![ CDATA[
ClustDB p1.x:1 p2.x:1 ... p18.x:1 p19.x:1 -LE pairs -CM -RO,
]]>
```

takes about 17 min to derive from the library files p1.x to p19.x as many as 16 235 566 complete pairwise matches beginning in sequence positions 1. These are all pairs of ESTs with minimum length 50 and of which one is a prefix of the other. The option '-RO' instructs ClustDB to consider the complete sequences for match extension while limiting the starts of matches.

There are 4 522 458 pairs of identical ESTs formed by 370 624 ESTs which split into 134 812 clusters. Consequently the number of redundant ESTs is 235 812, about as many as had been sequenced in the year 2000. As many as 594 194 ESTs are proper prefixes of others and may also be considered to be redundant. This number exceeds the EST production in the year 2001. Admitting only one wrong nucleotide at the 5′end we find as many as 736 035 ESTs contained in others. This number quickly grows as more errors are admitted. Moreover, the three largest clusters of identical ESTs have 1974, 850 and 751 members and careful scrutiny of the largest cluster revealed that all ESTs were submitted within 2 days, 1532 are dated 07-FEB-2007 and 442 are dated 08-FEB-2007. They all originate from the same source and all 1974 Genbank entries agree perfectly beginning with the field 'KEYWORDS'. Hence, there are 1974 identical Genbank entries just registered under different identification

Table 19.4 Cross comparisons of plant ESTs.

Runs	Options	Size (MB)	Sequences	Time	Clusters	Substrings
AB	-CD 5 -EC 1	3459	6 780 291	1:59:44	6 381 419	200 835 908
AC	-CD 5	3467	6 364 452	1:51:20	2 792 055	93 422 062
AD	-CD 5	2874	5 894 442	1:15:58	1 511 823	53 551 414
BC	-CD 5 -OP 1	3335	5 932 249	1:58:02	5 393 400	157 206 052
BD	-CD 5	2742	5 462 239	1:25:32	1 388 405	49 643 630
CD		2750	5 046 400	1:40:03	5 913 586	94 234 664

Using the option 'EC 1' ClustDB limits substring clusters to those with at least one member in the reference set of sequences while option '-EC 0' limits clusters to those with members in the reference and query sequences.

codes. The same holds for the second largest cluster of 850 identical ESTs which also originates from a single source. A total of 817 entries were submitted on 09-SEP-2005 and another 33 followed on 12-SEP-2005. Again, beginning with the field 'KEYWORDS' all the text agrees for the first 817 and the next 33 entries, respectively. Plant EST matching projects will greatly benefit from the removal of such redundant ESTs before analysis.

19.3.2
Deriving Substring Clusters for All Plant ESTs

Very large sets of sequences make it necessary to split all the data into blocks of not more than 2 GB of sequence so that ClustDB is able to process two at a time. Hence we summarize databases p1 to p19 to form four blocks A: p1 to p5, B: p6 to p10, C: p11 to p15 and D: p16 to p19 and apply the calculations reported in Table 19.4. In about 9 h six runs of ClustDB yield six files of substring clusters which must be summarized to the final result. This process includes merging of clusters which have common substrings and removal of multiple occurrences of a particular substring within the merged clusters. The first three runs are designed to cover all substring clusters with at least one member in the first set of sequences A. The option '-EC 1' used in addition to '-CD 5' ensures coverage of all clusters with members in A, only. The next two runs cover all substring clusters with at least one member in the second set of sequences B. The last run covers all substring clusters with members in set C and D.

19.3.3
Checking the TIGR Medicago BAC Assembly

Medicago truncatula represents the family of legumes as a model organism. The complete genome of about 450 Mbp is scheduled to be finished by the end of the year and a significant part is already available. As of March 2007 the website medicago.org/genome/assembly_table.php offers 9 BAC assembly files which contain 2020

Table 19.5 Two extended TIGR Medicago BAC assemblies.

Example A				Example B			
Accession	**Chromosome**	**Overlap**	**Errors**	**Accession**	**Chromosome**	**Overlap**	**Errors**
+AC166706	1	28 697	1	+CT954236	3	13 508	0
+AC087771	5	34 049	2	−CU062477		11 371	2
+CR954195	5	73 477	13	+AC157538	3	67 941	3
+AC119413	5	7577	0	−CT954233			
+CT027661	5						

A: The BAC AC166 706 is assumed to belong to chromosome 1 but perfectly overlaps with a chain of BACs from chromosome 5. B: The BAC CU062 477 has not been assigned to a chromosome by TIGR but perfectly matches two BACs of chromosome 3.

BACs and 407 chains of BACs claimed to overlap by more than 1000 nucleotides at each junction. Due to its capacity, ClustDB is able to search for complete matches of all 2020 BACs as well as their complementary sequences simultaneously. A first application clusters BACs into subsets which share long common substrings. Next ClustDB is applied to all sequence clusters in turn to find complete matches with errors and a straightforward assembly algorithm quickly written up using DNA_Stat library functions produced 856 chains, which we compared with those published by TIGR. The results include many cases where TIGR overlooked long nearly perfect matches that extend known chains. Two cases are shown in Table 19.5.

About 100 cases were reported for consideration by Nevin Young at the University of Minnesota and Steven Cannon from USDA-ARS and the Department of Agronomy Iowa State University. By using additional arguments they so far agreed that the BAC AC166 706 really seems to belong to chromosome 5 as do two other wrongly assigned BACs that are linked to it.

19.4 Perspectives

The greatest advantage of ClustDB is the simultaneous identification of multiple common substrings. This is the most important first step of all large-scale sequence comparison projects. Nucleotide BLAT [10] was pioneered by introducing the two perfect 11-mer match criterion to select those ESTs which are prone to match genomic sequence with high precision. The common substring tables derived by ClustDB are easy to scan for such results and the sequence cluster algorithm will soon be extended to give a choice of how many matches are needed to assign two sequences to the same cluster. Indeed, the sequence cluster function of ClustDB is of the greatest importance as it allows large volumes of data to be split into unrelated subsets which can be analyzed independently. These subsets are often small enough to allow the application of alternative sequence analysis software like MUMMER and VMATCH including the graphical output and easy to use links to related data. There is urgent need for easy to use follow-up software and additional options which help to restructure sequence library files according to the matches found. The option '-OC suffix' is just a beginning. An option which automatically generates non-redundant databases by removing sequences perfectly contained in others is currently in preparation. Allowing a small number of errors causes a principal problem. The non-transitivity of match criteria with errors makes it difficult to formally decide which sequence to keep and which to remove. But keeping all redundant almost perfectly matching sequences will sooner or later cause problems which are just as complex. More research is required in this field. Special applications to multiple sequence alignment, gene annotation analysis, identification of sequence contamination and rearrangements of chromosomal DNA are also in preparation.

Of next highest importance is the complete match option of ClustDB which finds many different applications. These matches were much more frequently observed than previously expected when analyzing large data sets. This suggests that a more direct approach to such results should be developed. Currently ClustDB selects complete matches from all maximally extended matches, which is a time-consuming procedure. Another urgent problem concerns the treatment of the complementary sequences described in Section 4.7. It is unnecessary to physically store complementary sequences. Virtual usage of these sequences saves considerable memory and allows much more data to be studied. Spacing suffix positions stored in reference blocks, as used in the programs BLAT [10] and Piers [19], could improve speed and memory usage at the cost of sensitivity. With regard to the problems discussed in this chapter, very little sensitivity would be lost by storing every second or third reference suffix position.

However, irrespective of future improvements to the performance of ClustDB, a single 32-bit PC will not be able to solve all the plant sequence matching problems of today and tomorrow. In Section 5.2 we reported that there is already difficultly in finding all the common substrings in all plant ESTs which have increased by 309 083 even during the last 3 months. Therefore MPI-ClustDB [20] was developed to distribute the computational load between a number of PCs using the message passing interface MPICH2 http://www-unix.mcs.anl.gov/mpi/mpich2. It has been shown in [20] that seven equally equipped PCs can solve a problem four times faster than a single PC. The described application focused on the increasing speed of computation. Large amounts of data that must be moved between computers make it a difficult task. Another version of MPI-ClustDB focuses on solving problems that cannot be solved with a single PC due to the shortage of memory. This new version will soon be available. The same holds true for ClustDB64 which was developed for 64-bit multi-processor computers and runs about three times faster when four processors are used.

Acknowledgments

This work was supported by the Bundesministerium für Bildung und Forschung (BMBF). We also thank Sven Mielordt for providing Figure 19.1.

References

1 Treangen, T.J. and Messeguer, X. (2006) M-GCAT: interactively and efficiently constructing large-scale multiple genome comparison frameworks in closely related species. *BMC Bioinformatics*, **7**, 433.

2 Kurtz, S. and Schleiermacher, C. (1999) REPuter: fast computation of maximal repeats in complete genomes. *Bioinformatics*, **15** (5), 426–427.

3 Kurtz, S., Choudhuri, J.V., Ohlebusch, E., Schleiermacher, C., Stoye, J. and Giegerich, R. (2001) REPuter: the manifold application of repeat analysis on a genomic scale. *Nucleic Acids Research*, **29**, 4633–4642.

4 Kurtz, S., Phillippy, A., Delcher, A.L., Smoot, M., Shumway, M., Antonescu, C. and Salzberg, S.L. (2004) Versantile and open software for comparing large genomes. *Genome Biology*, **5** (2), R12.

5 Abouelhoda, M.I., Kurtz, S. and Ohlebush, E. (2004) Replacing suffix trees with enhanced suffix arrays. *Journal of Discrete Algorithms*, **2**, 53–86.

6 Kleffe, J., Möller, F. and Wittig, B. (2006) ClustDB: a high-performance tool for large scale sequence matching. Proceedings DEXA 2006, Seventeenth International Workshop on Database and Expert Systems Applications, 4–8 September 2006, Krakau, Poland, 196–200.

7 Sorek, R. and Safer, M. (2003) A novel algorithm for computational identification of contaminated EST libraries. *Nucleic Acids Research*, **31** (3), 1067–1074.

8 Kleffe, J., Wessel, R., Whei, Z. and Wittig, B. (2004) Gene annotation refinement by constrained gene prediction. *Recent Developments in Nucleic Acids Research*, **1**, 289–322.

9 Altschul, S.F., Madden, T.L., Schäffer, A.A., Zhang, J., Zhang, Z., Miller, W. and Lipman, D.J. (1997) Gapped BLAST and PSI-BLAST: a new generation of protein database search programs. *Nucleic Acids Research*, **25**, 3389–3402.

10 Kent, W.J. (2002) BLAT: the BLAST-like alignment tool. *Genome Research*, **12** (4), 656–664.

11 Burkhardt, S., Crauser, A., Farragina, P., Lenhof, H.P. and Vingron, M. (1999) q-Gram based searching using a suffix array (QUASAR). Proceedings of the Third Annual International Conference on Research in Computational Molecular Biology, April 11–14, 1999, Lyon, France, 77–83.

12 Höhl, M., Kurtz, S. and Ohlebush, E. (2002) Efficient multiple genome alignment. *Bioinformatics*, **18** (Suppl. 1), 312–320.

13 Lefebvre, A., Lecroq, T., Dauchel, H. and Alexandre, J. (2003) FORRepeates: detects repeats on entire chromosoms and between genomes. *Bioinformatics*, **19** (3), 319–326.

14 Ma, B. Tromp, J. and Li, M. (2002) PatternHunter: faster and more sensitive homology search. *Bioinformatics*, **19** (3), 440–445.

15 Ning, Z., Cox, A.J. and Mullikin, J.C. (2001) SSAHA: a fast search method for large DNA databases. *Genome Research*, **11**, 1725–1729.

16 Kleffe, J., Möller, F. and Wittig, B. (2007) Simultaneous identification of long similar substrings in large sets of sequences. *BMC Bioinformatics*, **8** (Suppl 5), S7.

17 Sedgewick, R. (1992) *Algorithm in C*, Addison-Wesley, Boston, MA (USA) ISBN 3-89319-669-2.

18 Gusfield, D. (1997) *Algorithms on Strings, Trees and Graphs*, Cambridge University Press, Cambridge.

19 Cao, X., Li, S.C., Ooi, B.C. and Tung, A.K.H. (2004) Piers: an efficient model of similarity search in DNA sequence databases. *SIGMOD Record*, **33** (2), 39–44.

20 Hamborg, T. and Kleffe, J. (2006) MPI-ClustDB: a fast string matching strategy utilizing parallel computing. *Lecture Notes in Informatics (LNI)*, **83**, 33–39.

IV
Functional Genomics and Emerging Technologies

The Handbook of Plant Functional Genomics: Concepts and Protocols.
Edited by Günter Kahl and Khalid Meksem

ISBN: 978-3-527-31885-8

20
Nanotechnologies and Fluorescent Proteins for *in planta* Functional Genomics

C. Neal Stewart Jr.

Abstract

A range of bio- and nanotechnologies have been developed that can be adapted to the *in planta* monitoring of gene expression for functional genomics studies. Some technologies are amenable for transgenic monitoring such as tagging with the green fluorescent protein (GFP) or split GFP techniques, whereas endogenous gene expression and protein targeting might benefit from rapidly developing nanotechnologies utilizing aptamers, quantum dots, or molecular beacons. Other than GFP monitoring, these techniques have yet to be applied in practice to plant research but represent exciting technologies that might enhance plant functional genomics research in the future.

20.1
Introduction

Functional genomics is increasingly moving from controlled laboratory settings into the real world where plants experience biotic and abiotic stresses. It is here, in the field, where functional genomics really matters in agricultural and natural ecosystems. Therefore, researchers and, someday, practitioners, might enjoy the ability to interrogate gene expression *in planta*. Both imaging and quantification of specific transgene or endogenous gene expression would be beneficial for numerous purposes including precision agriculture and studying genetical ecology, among others.

This chapter is a critical review of various new technologies for transgene monitoring in living plants that are based on optical or fluorescent methods in various stages of development; a mix of what is and what could be – but mostly what could be. This chapter uses recent reviews [1,2] as a launch pad to speculate on how technologies that are generally more advanced in microbial and animal functional genomics can be adapted to plant functional genomics. Another very good recent review on imaging technologies for fluorescent molecules is Giepmans *et al.* [3], which is recommended to readers interested in imaging. Interestingly, there are parallel discussions ensuing about noninvasive monitoring of human gene therapy

The Handbook of Plant Functional Genomics: Concepts and Protocols.
Edited by Günter Kahl and Khalid Meksem

ISBN: 978-3-527-31885-8

using many of the same tools [4]. One relatively mature technology is based on transgenic synthesis of a marker molecule such as the green fluorescent protein (GFP). Such a platform could be used for the real-time monitoring of transgene movement and expression on a large geographic scale [1]. Essentially, most of the biological and technical components have already been developed for this application, therefore, this will be the major thrust of this chapter. Other technologies are transgene-independent and might be applied to endogenous genes, endogenous expression or monitoring the effects of silencing in an experimental system. One goal of this chapter is to identify existing technological strengths and thus, indicate where additional data and development are needed for practical implementation (Table 20.1). A macroscopic GFP monitoring protocol is included.

20.2
Green Fluorescent Protein

GFP from the jellyfish *Aequorea victoria* is extremely well characterized, has been modified for increased expression and optical properties and has been used in transgenic plants for over 10 years. The protein has the unique ability to transduce UV or blue light to green light (507 nm). Therefore, GFP fluorescence can be seen in transgenic plants merely by shining a bright ultraviolet light on leaves in an otherwise darkened location [5]. For instance, the properties unique to variant mGFP5 (discussed in [5,6]) enables it great flexibility for various plant biology applications. It has undergone site-directed mutagenesis to enable dual UV and blue light excitation as well as having better heat tolerance and folding characteristics. Either constitutive or inducible expression of GFP can be detected in intact plant organs. There is no measurable cost to plants expressing GFP [7], and GFP has been shown to be non-toxic to rats when ingested in purified form or when synthesized in transgenic plants [8].

These extraordinary properties make GFP an attractive reporter in transgenic plants that could potentially be used in commercialized products.

A *gfp* gene could be linked or fused to a gene of interest. Therefore, the presence of green fluorescence would indicate that the second transgene is present and expressed [7]; this system is also quantitative [9,10]. A priori, it would seem that a protein fusion would be better suited for using GFP to monitor the exact protein of interest. Although it has been shown that hundreds of different proteins can either be fused to GFP on the N or C termini with no loss of function [11], new fusion proteins must be thoroughly characterized to assure functional equivalency to the native protein of interest. Indeed, several projects, plants included, have been undertaken with the goal of systematic protein targeting of proteomes [12]. An alternative to producing a protein translational fusion would be a transcriptional fusion using an internal ribosomal entry site (IRES) upstream of a second transcript for bicistronic expression [13], although this technology has not proven to be robust.

GFP has become an integral tool in functional genomics and research to characterize regulatory elements on the basis of observed expression patterns in intact

Table 20.1 Intrinsic properties of various *in planta* systems for monitoring gene expression for functional genomics.

System	Description	Target(s)	Strengths	Weaknesses
GFP tagging 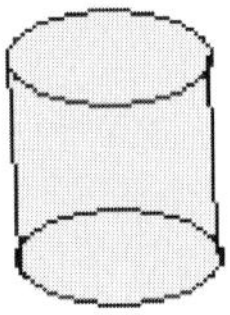	Genetically encoded fluorescent marker	Protein	• Most technologically advanced • High throughput screens • New FP colors	• Requires investment for vector construction • GFP suboptimal wavelength
Aptamer/quantum dot 	Nucleic acid aptamer tagged with a quantum dot	Protein, mRNA, metabolites	• Aptamers can be designed for many targets • Quantum dots are bright	• Requires stringent hybridization conditions to remove unbound probe • Speculative • Introduction into cells a problem
Molecular beacon 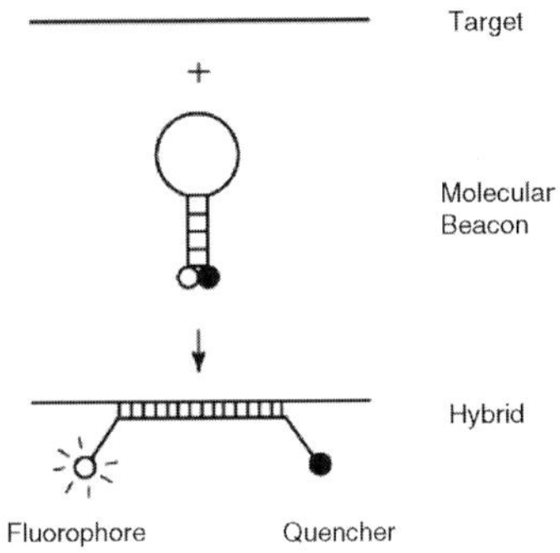 Target Molecular Beacon Hybrid Fluorophore Quencher	RNA tagged with fluorophore and quencher	mRNA	• Shown to work *in vivo* • Fluorescence quenched when not hybridized • Specific to mRNA sequences/abundant	• Sequence availability might be limited • Somewhat speculative • Introduction into cells a problem
Split GFP 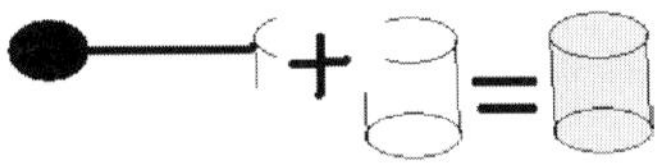	Protein fusion with small half of GFP complemented by exogenously added large half	Protein	• More appropriate tag than whole GFP	• Requires investment for vector construction • GFP suboptimal wavelength

organisms and organs. GFP has been fused or linked to numerous different proteins to monitor localization. In functional genomics, a transgenic approach utilizing FPs could well be exploited to examine promoter activity and to clone regulatory elements [14,15]. For example, the 'constitutive' CaMV 35S promoter widely used in transgenic plants is not really constitutive but has both developmental and tissue specificity as shown by differential GFP expression in tobacco [16], mustards [17], and cotton [18] in laboratory, greenhouse and field experiments. Novel promoters have been characterized by examining GFP fluorescence in living plants, for example, several promoters from the taro bacilliform virus (TaBV) have been tested in transgenic banana and tobacco plants [19].

Several groups have used FPs to tag genes *en masse* and examine fluorescence activity. Janke *et al.* [20] have described such an endeavor in yeast, and Cutler *et al.* [21] described a general approach to tagging *Arabidopsis thaliana* cDNAs with GFP to identify subcellular structures. Clearly, GFP has proven to be a powerful tool in studying various genomics phenomena. One prominent example has been RNAi experiments, in which GFP transgenic plants were infected with *Agrobacterium tumefaciens* containing various other GFP constructs to examine patterns of gene silencing [22,23]. GFP is also useful to determine whether experimental gene targeting in transgenic plants had occurred [24]. Here, with the requisite help of a yeast chromatin remodeling gene, researchers were able to determine which *A. thaliana* seeds were transgenic for the targeted sequence based on green fluorescence of primary transgenics. Population functional genomics requires information about zygosity, and for its determination dominantly or semi-dominantly expressed genes linked with GFP can be assayed by one or two allelic-labeled GFPs. Halfhill *et al.* [9] demonstrated that heterozygous (hemizygous) transgenic canola plants had approximately half the green fluorescence signal of homozygous plants for GFP. Furthermore, when GFP homozygous canola plants were hybridized with non-transgenic wild relatives, the progeny, which were necessarily heterozygous, had fluorescence measurements that were the same as heterozygous crop parent, indicating that FPs could be a tool in analyzing hybridization and introgression status. A more elegant system utilizing two FPs was applied to study the allelic conditions of mouse cells. Using gene targeting employing site-specific recombination, specific alleles of a single gene were each labeled with two different GFPs (cyan and yellow), and the mice crossed [25]. Allelic variants in mouse fibroblast or stem cells could be distinguished using cell sorting (FACS).

Although GFP has been widely used in plants and other transgenic organisms, it is not the only fluorescent protein now available. Fluorescent proteins, mainly from non-bioluminescent Anthozoa have various excitation and emission wavelengths providing researchers with a veritable rainbow of colors (reviewed in [26–28]), and these too can be expressed in plants [29]. Additional colors are necessary for multiple tagging of different traits and multiplexed imaging. Several of these 'new fluorescent' proteins from Anthozoa such as the red fluorescent protein (RFP) from the coral *Discosoma* sp. (DsRed) have the distinct disadvantage of not forming monomeric proteins when mature. The formation of tetramers or even dimers negates the possibility of the production of facile transgenic protein fusions and can also cause solubility and aggregation problems. However, DsRed has been monomerized and has also been spectrally altered [30,31]. In addition to increased brightness (once normalized for monomerization) the range of emission of DsRed-type RFP (583 nm) has been extended from yellow (537 nm) to far red (649 nm) [26,32]. RFPs are of special interest in plants because there is not much fluoresence in plants in certain red wavelengths when excited by wavelengths of light in green to red wavelengths, thus RFP should be easier to detect in intact green tissue than GFP [2]. The study by Wenck *et al.* [29] in which anthozoan FPs were expressed in plants included DsRed from the coral *Discosoma* sp., which has been the most often used Anthozoan marker in transgenic plants. Several transgenic plant tissues were examined for fluorescence and they reported that AmCyan, AsRed, DsRed, ZsGreen, and ZsYellow all displayed fluorescence. One of the most interesting results was found in rice callus. AmCyan1 transgenic material appeared to be yellow-green and AsRed appeared to be red under white light. Furthermore, DsRed1 in transgenic cells appeared reddish under white light. It is also of interest that these FPs had relatively low extinction coefficients and quantum yields compared with more recent derivatives. Thus, these results, which were much more impressive that those using *Aequorea* GFP, seem to put us at the beginning, rather than the end of the rainbow as the FP color palette seems ever-expanding (Figure 20.1).

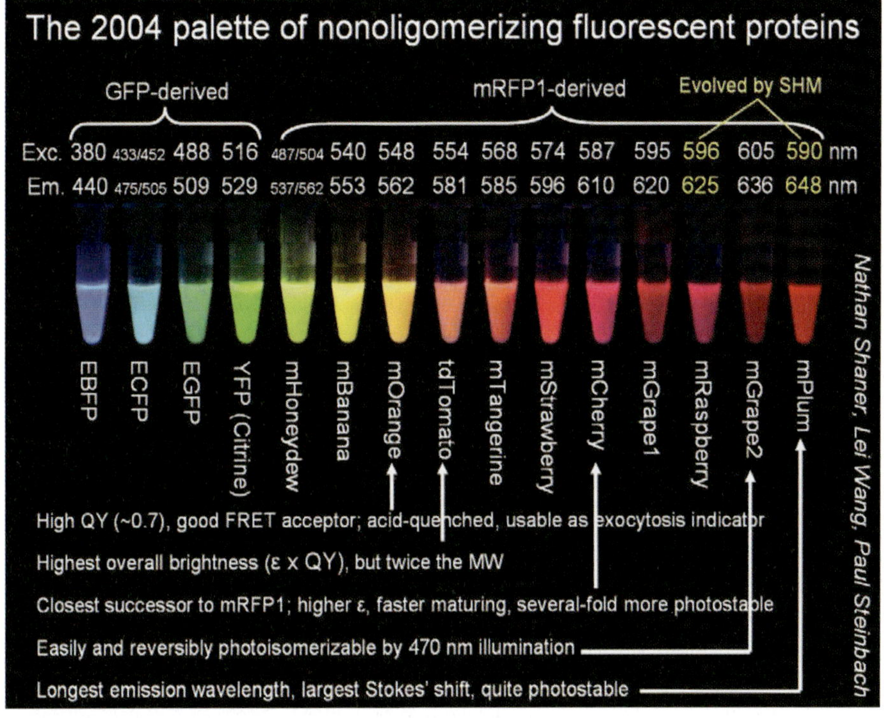

Figure 20.1 Color palette of engineered fluorescent proteins cover the full visible spectrum of emissions (Courtesy of Roger Tsien, 2005).

There are several other non-GFP-like fluorescent markers that would, apparently, not be as useful in *in planta* monitoring such as phycobiliproteins and uroporphyinogen III methyltransferase reviewed recently by Zhang *et al.* [11]. However, research is also underway toward utilizing various chromoproteins and even autocatalytic bioluminescence in transgenic plants, which could be promising (C.N. Stewart, Jr. *et al.* unpublished data).

Many reviews have focused on microscopic detection of fluorescent proteins, thus this will not be covered here, but microscopy platforms for FP detection are quite sophisticated, with parallel detection of up to eight different colors using state-of-the-art optical and computations methods [12]. As we think about expanding functional genomics to real environments, a different set of tools will be needed for macroscopic detection of FPs. While a UV spotlight [5], or 'flashlights' [33,34] in the dark can be used with certain FPs, more sophisticated techniques are likely needed for commercial applications. OptiSciences (Tyngsboro, Massachusetts, USA) produces a 'GFP Meter' that uses a fiber-optic facilitated leaf clip to sample spectra on intact leaves [35]. This portable spectrofluorometer performs similarly to laboratory-sited instruments and can take fluorescence measurements every few seconds. It can be envisaged that the investigation of gene expression studies on-the-plant under a variety of stresses will lead to a better understanding of the function of genes in relevant environments. For standoff detection, a laser-induced fluorescence imaging device has been developed and tested in GFP transgenic plants [36]. Like the GFP Meter, it can be used in the daylight, but has a broader measurement capability, and is able to interrogate entire plant canopies from standoff mode (meters).

20.3
Protocol: Seeing GFP in Transgenic Plants

In my laboratory, the majority of research is with plants transgenic for GFP – the mGFP5-ER variant [6,16] – which is excited equally by both UV (395 nm) and blue (465 nm) light. We have transformed several species of plants and have monitored GFP in all plant organs. The dual excitation profile of this GFP gives the opportunity for visualization using a standard epifluorescence microscopy setup with an appropriate filter set (e.g. FITC), but also provides for macroscopic screening using a handheld UV lamp with no requirement for an emission filter. It is important to have a non-transgenic control plant grown under the same optimal conditions for comparisons. Plants that have obtained their first true leaves can be optically assayed for GFP using a portable UV spotlight such as a UVP 100 AP (Upland, CA, USA) or the more compact Spectroline BIB-150 produced by Spectronics (Westbury, NY, USA) [5]. These lamps have a 100-W mercury bulb and a 365-nm filter. Smaller fluorescent bulb UV lamps are not as effective as these bright spotlights. Combining two or three of the Spectroline UV lamps to boost photon excitation irradiation, results in images that provide better contrast. Effective visualization of GFP in transgenic plants also requires the appropriate wavelength for excitation. While the Spectroline or UVP lamps work well for UV excitation of GFP, they would be even more effective if used with a 400-nm filter instead of the 365-nm filter, since the former better matches the GFP excitation profile. UV protective eyewear should always be used.

20.4 Nanotechnology for Monitoring Gene Expression

Nanotechnologies for molecular biology applications have greatly increased during the past few years. Although many have not been applied to gene expression analysis or functional genomics in plants (or other organisms for that matter), the prospects are clear and early work is promising. For example, it should be possible to detect any transcript and protein, perhaps *in planta*, by the rational design of complementary nucleic acids (for mRNA), antibodies, or DNA aptamers (for proteins) along with an appropriate fluorescent reporter molecule. Some of the most promising technologies will be discussed, but none has as yet been adapted for practical use in living plant cells.

20.4.1 Aptamers and Quantum Dots

DNA aptamers, which are single-stranded pieces of DNA optimized for binding to other specific molecules, are among the most intriguing potential molecules for functional genomics because of the seemingly endless variety of potential targets. They have been designed to bind ligands such as specific inorganic ions, ATP, antibiotics, and proteins using combinatorial approaches [37–40]. Aptamers can be covalently tagged with various small fluorescent molecules such as a bis-pyrenyl fluorophore [39], or quantum dots. Quantum dots, which are nanometer-sized semiconductor crystals of fluorescent metals (e.g. CdSe in the core and surrounded by a shell of CdS) are water soluble and have been used in several biological microscopic applications [41–43]. Fluorescent excitation and emission wavelengths vary as a result of materials and crystal sizing. High extinction coefficients and quantum yields (factors endowing brightness) along with a wide variety of types of quantum dots should enable increased use in sensing and monitoring applications. Instrumentation designed to measure GFP and other FPs in plants should be effective in these applications as well. One inherent difficulty with quantum dots is their continual fluorescence (bound or unbound). After binding to their target, there would need to be a stringent wash step to remove any non-bound aptamer–quantum dot probe from the detection area.

20.4.2 Molecular Beacons

Molecular beacons are nucleic acid probes that contain a fluorescent molecule at one end and a quencher molecule at the other [44]. Consisting of between 15 and 35 nucleotides, which are designed to be complementary to a specific RNA or DNA sequence, molecular beacons have quenched fluorescence when not bound to their targets because of a hairpin and self-complimentary structure enabling the quencher to come into close proximity to the fluorophore. However, when bound to its complimentary nucleic acid, fluorescence is activated. Although molecular beacons have most often been used in *in vitro* techniques such as quantitative PCR, they seem to be particularly suited to *in vivo* gene expression analysis, especially for monitoring specific transcripts in cells [45,46]. However, molecular beacons have been used in conjunction with aptamers to report the presence of specific proteins [46]. Molecular beacons have successfully been used to detect rRNA in bacterial cells [47] and mRNA in mammalian [48,49] and fly [50] cells. In each eukaryotic organism, molecular beacons were microinjected into cells which were then visualized under fluorescence or confocal microscopy. Several combinations of fluorophore and quencher pairs are available, yielding many available colors and fluorescence resonance energy transfer (FRET) pairs [46,51], in which excitation light is transduced from one fluorophore to another. Bratu *et al.* [50] provide an especially sophisticated demonstration of the power of molecular beacons in visualizing mRNAs in living cells. They utilized an altered RNA backbone by substituting an oxymethyl group for the hydrogen atom at the second position of the ribose on each nucleotide, therefore conferring nuclease resistance. RNases could otherwise digest the molecular probes prior to hybridization. These researchers also utilized two different fluorophores designed to hybridize head-to-head along an mRNA strand yielding a FRET readout. In this case FRET assures that RNA probes are interacting (if fluorescence is observed) as well as steering clear of cellular autofluorescence. As a result of target hybridization, molecular beacon fluorescence could be detected in 15 min and intracellular transport of mRNA could be visualized. Of all technologies reviewed, this one seems to have the most power for specific real-time monitoring for gene expression needed for functional genomics studies. Although it has not yet been used in plants, it has shown unequivocal results when used to probe living animal cells.

20.4.3
Split GFP Tagging and Detection

A hybrid technology combining the transgenic expression of GFP and nanotechnologies is embodied in split GFP tagging. Ghosh *et al.* [52] demonstrated that GFP could be expressed in two halves, which, as individuals do not fluoresce, but when recombined, they form a normally green-fluorescent molecule. Split GFP was recently refined to be self-associating and soluble in living systems [53] in which researchers produced a genetically encoded split-GFP fusion protein with several candidate bacteria proteins using a small (16 residues) portion of GFP. When the larger 'half' (214 residues) of GFP was synthesized in the host bacterial cells, fluorescence was recovered within hours because of the self-association of the GFP fragments. None of the target proteins used for fusions had altered solubility or functions, probably because of the diminutive size of the linker and split GFP additions [53]. It can be envisaged that specific recombinant proteins in plants could be fused with split GFP, with the larger GFP half added to cells exogenously to form a plant two-hybrid screen for interacting proteins. Similar to the other technologies discussed here, recovered GFP fluorescence could be monitored using existing GFP instrumentation, and like molecular beacons, there would be no concern over non-specific fluorescence.

20.5 Barriers to Implementation

There are a number of impediments to implementing the above technologies for functional genomics. Fluorescent protein technologies are most developed and practiced in functional genomics and they continue to evolve rapidly. Nanotechnology-based detection systems including aptamers, quantum dots, and molecular beacons simply require more basic and applied research to demonstrate that they can feasibly be used in living plants. Even though these could be used on only one plant at a time, virtually any type of transgene or metabolite could be monitored and existing fluorescence detection devices could be adapted. However, it is still not clear how nanomaterials could be efficiently introduced into plant tissues. After all, molecules such as molecular beacons are difficult to transfect through cell membranes [51], not to mention cell walls. Microinjection will not be effective at the tissue level, but perhaps microprojectile bombardment could be used as a nanoparticle delivery agent [54]. Another promising technology for introduction of nanomaterials as well as other conventional biological molecules is through the use of vertically aligned carbon nanofibers. These nanofibers, 6–10 μm in length and 20–50 nm in tip diameter, and 'grown' on silicon wafers, have been used to introduce DNA and other biochemicals into mammalian cells [55], and it appears that they can penetrate cell walls to affect plant transformation as well (Timothy McKnight *et al.* unpublished data). In any case, additional instrumentation for cellular introduction, hybridization and detection would need to be modified for this specific purpose and commercial monitoring.

20.6 Conclusions

The interface of nanotechnology and biotechnology and genomics is in its infancy. With extraordinarily large investments being made in public and private sectors in nanotechnology for various applications, accompanied by, secondarily, additional investments in biomedical research, there will be tremendous opportunities for plant science researchers to adapt these powerful enabling technologies to better understand plant gene expression in ecosystems.

References

1 Stewart, C.N., Jr. (2005) Monitoring the presence and expression of transgenes in living plants. *Trends in Plant Science*, **10**, 390–396.

2 Stewart, C.N., Jr. (2006) Go with the glow: fluorescent proteins to light transgenic organisms. *Trends in Biotechnology*, **24**, 155–162.

3 Giepmans, B.N.G. *et al.* (2006) The fluorescent toolbox for assessing protein localization and function. *Science*, **312**, 217–224.

4 Vassaux, G. and Groot-Wassink, D. (2003) *In vivo* noninvasive imaging for gene therapy. *Journal of Biomedicine and Biotechnology*, **2**, 92–101.

5 Stewart, C.N., Jr. (2001) The utility of green fluorescent protein in transgenic plants. *Plant Cell Reports*, **20**, 376–382.

6 Haseloff, J. *et al.* (1997) Removal of a cryptic intron and subcellular localization of green fluorescent protein are required to mark transgenic *Arabidopsis* plants brightly. *Proceedings of the National Academy of Sciences of the United States of America*, **94**, 2122–2127.

7 Harper, B.K. *et al.* (1999) Green fluorescent protein in transgenic plants indicates the presence and expression of a second gene. *Nature Biotechnology*, **17**, 1125–1129.

8 Richards, H.A. *et al.* (2003) Safety assessment of green fluorescent protein orally administered to weaned rats. *Journal of Nutrition*, **133**, 1909–1912.

9 Halfhill, M.D. *et al.* (2003) Additive transgene expression and genetic introgression in multiple green fluorescent protein transgenic crop × weed hybrid generations. *Theoretical and Applied Genetics*, **107**, 1533–1540.

10 Richards, H.A. *et al.* (2003) Quantitative GFP fluorescence as an indicator of recombinant protein synthesis in transgenic plants. *Plant Cell Reports*, **22**, 117–121.

11 Zhang, J. *et al.* (2002) Creating new fluorescent probes for cell biology. *Nature Reviews Molecular Cell Biology*, **3**, 906–918.

12 Sauer, S. *et al.* (2005) Miniaturization in functional genomics and proteomics. *Nature Reviews Genetics*, **6**, 465–476.

13 Urwin, P. *et al.* (2000) Functional characterization of the EMCV IRES in plants. *Plant Journal*, **24**, 583–589.

14 Jeon, J.-S. and An, G. (2001) Gene tagging in rice: a high throughput system for functional genomics. *Plant Science*, **161**, 211–219.

15 Ayalew, M. (2003) Genomics using transgenic plants, in *Transgenic Plants: Current Innovations and Future Trends*, (ed. C.N., Stewart Jr.), Horizon Scientific Press, Wymondham, UK, pp. 265–291.

16 Harper, B.K. and Stewart, C.N., Jr. (2000) Patterns of green fluorescent protein in transgenic plants. *Plant Molecular Biology Reporter*, **18**, 141a–141i.

17 Halfhill, M.D. *et al.* (2003) Spatial and temporal patterns of green fluorescent protein (GFP) fluorescence during leaf canopy development in transgenic oilseed rape, *Brassica napus* L. *Plant Cell Reports*, **22**, 338–343.

18 Sunilkumar, G. *et al.* (2002) Developmental and tissue-specific expression of CaMV 35S promoter in cotton as revealed by GFP. *Plant Molecular Biology*, **50**, 463–474.

19 Yang, M. *et al.* (2005) Facile whole-body imaging of internal fluorescent tumors in mice with an LED flashlight. *BioTechniques*, **39**, 170–172.

20 Janke, C. *et al.* (2004) A versatile toolbox for PCR-based tagging of yeast genes: new fluorescent proteins, more markers and promoter substitution cassettes. *Yeast*, **21**, 947–962.

21 Cutler, S.R. *et al.* (2000) Random GFP⇒cDNA fusions enable visualization of subcellular structures in cells of *Arabidopsis* at a high frequency. *Proceedings of the National Academy of Sciences of the United States of America*, **97**, 3718–3723.

22 Johansen, L.K. and Carrington, J.C. (2001) Silencing on the spot. Induction and suppression of RNA silencing in the *Agrobacterium*-mediated transient, expression system. *Plant Physiology*, **126**, 930–938.

23 Waterhouse, P.M. and Helliwell, C.A. (2003) Exploring plant genomes by RNA-induced gene silencing. *Nature Reviews Genetics*, **4**, 29–38.

24 Shaked, H. *et al.* (2005) High-frequency gene targeting in Arabidopsis plants expressing the yeast RAD54 gene. *Proceedings of the National Academy of Sciences of the United States of America*, **102**, 12265–12269.

25 Larson, J.S. *et al.* (2006) Expression and loss of alleles in cultured mouse embryonic fibroblasts and stem cells carrying allelic fluorescent proteins. *BMC Molecular Biology*, **7**, 36.

26 Miyawaki, A. (2002) Green fluorescent protein-like proteins in reef Anthozoa animals. *Cell Structure and Function*, **27**, 343–347.

27 Verkhusha, V.V. and Luyyanov, K.A. (2004) The molecular properties and applications of Anthozoa fluorescent proteins and chromoproteins. *Nature Biotechnology*, **22**, 289–296.

28 Carter, R.W. *et al.* (2004) Cloning of anthozoan fluorescent protein genes. *Comparative Biochemistry and Physiology. C*, **138**, 259–270.

29 Wenck, A. *et al.* (2003) Reef-coral proteins as visual, non-destructive reporters for plant transformation. *Plant Cell Reports*, **22**, 244–251.

30 Campbell, R.E. *et al.* (2002) A monomeric red fluorescent protein. *Proceedings of the National Academy of Sciences of the United States of America*, **99**, 7877–7882.

31 Shaner, N.C. *et al.* (2004) Improved monomeric red, orange and yellow fluorescent proteins derived from *Discosoma* sp. red fluorescent protein. *Nature Biotechnology*, **22**, 1567–1572.

32 Wiedenmann, J. *et al.* (2002) A far-red fluorescent protein with fast maturation and reduced oligomerization tendency from *Entacmaea quadricolor* (Anthozoa, Actinaria). *Proceedings of the National Academy of Sciences of the United States of America*, **99**, 11646–11651.

33 Tyas, D.A. *et al.* (2003) Identifying GFP-transgenic animals by flashlight. *BioTechniques*, **34**, 474–476.

34 Yang, I.C. *et al.* (2003) A promoter derived from taro bacilliform badnavirus drives strong expression in transgenic banana and tobacco plants. *Plant Cell Reports*, **21**, 1199–1206.

35 Millwood, R.J. *et al.* (2003) Instrumentation and methodology for quantifying GFP fluorescence in intact plant organs. *BioTechniques*, **34**, 638–643.

36 Stewart, C.N., Jr. *et al.* (2005) Laser-induced fluorescence imaging and spectroscopy of GFP transgenic plants. *Journal of Fluorescence*, **15**, 697–705.

37 Hermann, T. and Patel, D.J. (2000) Adaptive recognition by nucleic acid aptamers. *Science*, **287**, 820–825.

38 Jhavier, S. *et al.* (2000) *In vitro* selection of signaling aptamers. *Nature Biotechnology*, **18**, 1293–1297.

39 Yamana, K. *et al.* (2003) Bis-pyrene labeled DNA aptamer as an intelligent fluorescent biosensor. *Bioorganic and Medicinal Chemistry Letters*, **13**, 3429–3431.

40 Ho, H.A. and Lecierc, M. (2004) Optical sensors based on hybrid aptamer/conjugated polymer complexes. *Journal of the American Chemical Society*, **126**, 1384–1387.

41 Bruchez, M., Jr. *et al.* (1998) Semiconductor nanocrystals as fluorescent biological labels. *Science*, **281**, 2013–2016.

42 Chan, W.C.W. and Nie, S. (1998) Quantum dot bioconjugates for ultrasensitive nonisotopic detection. *Science*, **281**, 2016–2018.

43 Watson, A. *et al.* (2003) Lighting up cells with quantum dots. *BioTechniques*, **34**, 296–303.

44 Tyagi, S. and Kramer, F.R. (1996) Molecular beacons: probes that fluoresce upon hybridization. *Nature Biotechnology*, **14**, 303–308.

45 Tan, W. *et al.* (2004) Molecular beacons. *Current Opinion in Chemical Biology*, **8**, 547–553.

46 Drake, T.J. and Tan, W. (2004) Molecular beacon DNA probes and the bioanalytical applications. *Applied Spectroscopy*, **58**, 269A–280A.

47 Xi, C. *et al.* (2003) Use of DNA and peptide nucleic acid molecular beacons for detection and quantification of rRNA in solution and whole cells. *Applied and Environmental Microbiology*, **69**, 5673–5678.

48 Sokal, D.L. *et al.* (1998) Real time detection of DNA–RNA hybridization in living cells. *Proceedings of the National Academy of Sciences of the United States of America*, **95**, 11538–11543.

49 Peng, X.-H. *et al.* (2005) Real time detection of gene expression in cancer cells using molecular beacon imaging: new strategies for cancer research. *Cancer Research*, **65**, 1909–1917.

50 Bratu, D.P. *et al.* (2003) Visualization of the distribution and transport of mRNAs in living cells. *Proceedings of the National Academy of Sciences of the United States of America*, **100**, 13308–13313.

51 Santangelo, P., Nitin, N. and Bao, G. (2006) Nanostructured probes for RNA detection in living cells. *Annals of Biomedical Engineering*, **34**, 39–50.

52 Ghosh, I. *et al.* (2000) Antiparallel leucine zipper-directed protein reassembly: application to the green fluorescent protein. *Journal of the American Chemical Society*, **122**, 5658–5659.

53 Cabantous, S. *et al.* (2005) Protein tagging and detection with engineered self-assembling fragments of green fluorescent protein. *Nature Biotechnology*, **23**, 102–107.

54 Kummer, T.T. *et al.* (2002) Spotted substrates for focal presentation of proteins to cells. *BioTechniques*, **33**, 1018–1024.

55 McKnight, T.E. *et al.* (2003) Intracellular integration of synthetic nanostructures with viable cells for controlled biochemical manipulation. *Nanotechnology*, **14**, 551–556.

21
New Frontiers in Plant Functional Genomics Using Next Generation Sequencing Technologies

Robert C. Nutter

Abstract

The development of massively parallel sequencing systems is dramatically altering the paradigm by which scientists are thinking about the way they conduct their research. As the cost per base of DNA sequence generated continues to decrease, the number of bases sequenced has steadily increased. This is exactly the same phenomenon that is described by Moore's law in the computer industry. Not only are more bases being sequenced, but the type of sequencing projects being undertaken with these new technologies could not have been envisioned even a few years ago by groups other than a handful of large genome centers. This chapter will briefly describe the basics of several different technologies that are either commercially available or will be in the next few months. One of these new technologies, the SOLiD System, will be described in more detail. Additionally, some of the applications that are being and will be enabled by these technologies will be discussed.

21.1
Introduction

21.1.1
Advent of Massively Parallel Sequencing Systems

The technology that drove the sequencing of the human, as well as a diverse range of other organisms' genomes, was first described by Sanger *et al.* [1]. The Sanger technology was first commercialized in the early 1990s [2] and the constant improvements in the commercial processes led to the completion of the human genome in 2001 [3,4] in much less time that originally projected [5]. The amount of sequencing being carried out has continued to increase at an accelerated rate. As of April 2007, there were 534 completed, published genomes and well as an additional 1873 ongoing genome projects listed on the Genome Online Database v2.0 (www.genomesonline.org).

The Handbook of Plant Functional Genomics: Concepts and Protocols.
Edited by Günter Kahl and Khalid Meksem

ISBN: 978-3-527-31885-8

It is currently believed that Sanger-based sequencing technology can only be improved incrementally. Additionally, there are regions of genomes that cannot be sequenced with Sanger-based technology. This is assumed to be due to bias introduced during the clonal propagation of DNA templates in microbial hosts or the inability of existing sequencing chemistries to generate sequence data from problematic regions. Therefore, much effort has been put into the development of technology capable of producing a so-called $1000 genome. The first wave of new sequencing technologies, capable of producing a '$100 000 genome', has recently been introduced into the marketplace [6].

Several techniques for massively parallel DNA sequencing have recently been described in the literature [7,8]. They broadly fall into two assay categories (polymerase-based techniques, and ligation-based techniques) and two detection categories (bead detection and cluster detection). They share the requirement of clonal amplification of templates to produce sufficient target material for sequencing.

A third class of sequencing technology, single molecule sequencing, has the potential to further decrease the cost of DNA sequencing and is currently seen as the best technology to provide a $1000 genome. Different approaches are being aggressively investigated by a number of research groups with the goal of commercialization in the near future. These technologies will not be covered in this chapter.

21.1.2
Overview of the Sequencing by Synthesis System

The most familiar sequencing by synthesis system today was first developed by 454 Life Sciences and commercialized as the GS-20, and later, the GS-FLX system by Roche Applied Sciences. The key features of this system are:

1. Nucleic acid is fragmented, flanked by PCR primers and attached to large magnetic beads.
2. Emulsion PCR to provide clonal amplification of single templates.
3. Deposition of single beads into single wells in a picotiter plate.
4. Sequencing reaction carried out with DNA polymerase-based pyrosequencing chemistry designed to add one base to the template in each well.
5. The light generated each time a base is added is detected and recorded into what is called a 'flowgram'.
6. The flowgram is exported and converted to a sequence file.

This system has been in the marketplace for more than 1 year and the technology has been used to generate sequence data from a widely diverse range of organisms and has resulted in a number of publications. The system allows read lengths >100 bases and has been adapted to support a number of applications using either fragment or mate-paired libraries. There are limitations in total number of bases sequenced in one run due to the number of picotiter wells that can be machined into a plate. Additionally, the accuracy of the sequence is affected by homopolymers that are found to a high degree in all genomes. However, in general, the system has been shown to perform well.

21.1.3
Overview of Single Base Extension System

This system was developed by Solexa, and is called the 1 G Genetic Analyzer. The chemistry is characterized as DNA polymerase-based addition of one chemically modified nucleotide in a step-wise fashion. The technology has subsequently been purchased and is being distributed by Illumina, Inc. The key features of this system are:

1. Nucleic acid is fragmented to an appropriate size.
2. Fragments are ligated to specific primers arrayed on a solid surface.
3. Single templates are amplified by bridge PCR.
4. The sequencing reaction is carried out using a modified DNA polymerase adding dye-labeled nucleotides that have been modified so that only one base can be added.
5. The added dye is recorded by illumination after each round of addition.
6. The base modification is chemically removed.
7. The processes described in steps 4–6 are repeated a number of times.
8. The raw data is subsequently processed on a remote station into base calls.

This system is just being released at the time this manuscript is being written. The system currently uses a fragment library and has read lengths reported to be approximately 25–35 high quality bases. Gene resequencing, as well as tag-based sequencing, are the two applications being promoted for the system at this time. The overall system is being tested by a number of research groups and has been shown to work. However, not enough data has been released into the public domain to determine throughput capacity and robustness of the chemistry and instrument. These questions should be answered within the next few months.

21.1.4
Overview of the (SOLiD) System

The SOLiD system technology has been released by Applied Biosystems in mid-2007 and is unique in its approach to DNA sequencing in the following ways:

1. Mate-paired libraries are used to allow sequencing of DNA segments originally separated by a defined number of bases.

2. Pools of dye-labeled oligonucleotide probes are used with degenerate bases in positions 1–5 relative to the 3′-OH end of the probes. Probes with a specific sequence complimentary to the unknown target will be ligated to a universal primer in a step-wise fashion.

3. A highly efficient, chemical cleavage of ligated probes regenerating a 5′ phosphate capable of initiating the next round of ligation.

4. Removal of all labeled fragments after five rounds of ligation. This is followed by new cycles of ligation to a new universal primer that has been moved a defined distance 3′ from the previous universal primer.

5. Integral base calling error reduction resulting from proprietary algorithms and probe design.

Each of these points will be described in more detail in the following sections.

21.2
Library Generation

The SOLiD system utilizes two general types of randomly generated DNA libraries for sequencing; a 'fragment' library and a 'mate-paired' library. In the fragment library, the material to be sequenced is randomly sheared by some physical means, such as sonication, to a length of 60–90 bp, which is optimal for amplification with this system (Figure 21.1). The primers that will be used for amplification are ligated to the ends of these fragments and the fragments with unique primers on each end are selected for emulsion PCR amplification.

The methodology for construction of a mate-paired library is more involved. However, this type of library has several advantages for downstream bioinformatics analysis [9,10], such as alignment against a reference sequence. Some applications, such as detection of insertions and deletions are therefore dependent upon mate paired libraries to be successful. The nucleic acid to be sequenced is also physically sheared into smaller fragments. However, the size of the fragments used to construct the library varies depending on the application. For the purposes of this chapter, we will assume a size of 2–3 kb, though libraries ranging from 1 to 6 kb have been made. Fragments of the appropriate size are isolated from the rest of the fragments and are ligated to a synthetic primer, called a CAP primer, which has several features designed into it (Figure 21.2). The capped fragments are diluted and ligation is performed in a manner to favor recircularization of fragments over end-to-end ligation of two different fragments. This results in a random library of fragments from the starting nucleic acid, all having a pre-defined size (Figure 21.2). One of the features designed into the CAP primer is the sequence of a type III restriction enzyme, such as EcoP15I at each end of the primer. This class of restriction enzyme recognizes a specific sequence and then cleaves the DNA molecule at a defined distance from the recognition sequence (http://www.neb.com/nebecomm/

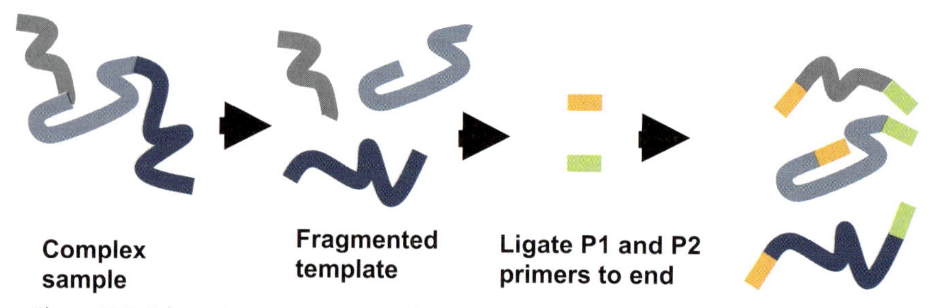

Figure 21.1 Schematic representation of the construction of a fragment library for SOLiD sequencing.

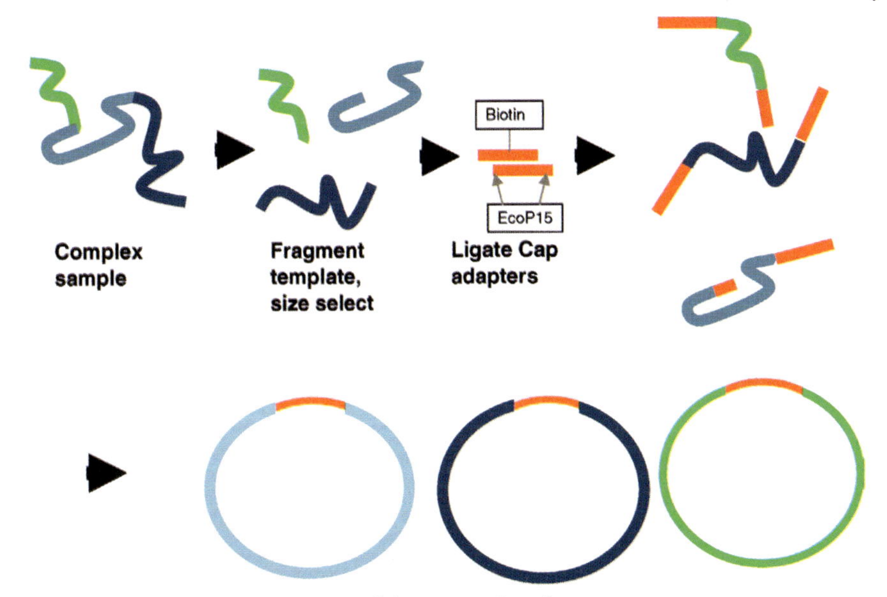

Figure 21.2 Schematic representation of the construction of a mate-paired library for SOLiD sequencing.

tech_reference/restriction_enzymes/overview.asp). In the case of EcoP15I, the enzyme cuts 25 and 27 bases from the recognition sites (Figure 21.3). This enzymatic digestion releases the intervening genomic DNA sequences from the 27 base pairs of sequence that defined the ends of the original fragment. The random, 'mate-paired' fragments can then be purified from the intervening DNA by affinity capture using a biotin tag that was designed into the CAP primer. This produces a random set of paired, 27-base DNA sequences that were originally separated in their native state by the size of the fragments used to make the library (Figure 21.3).

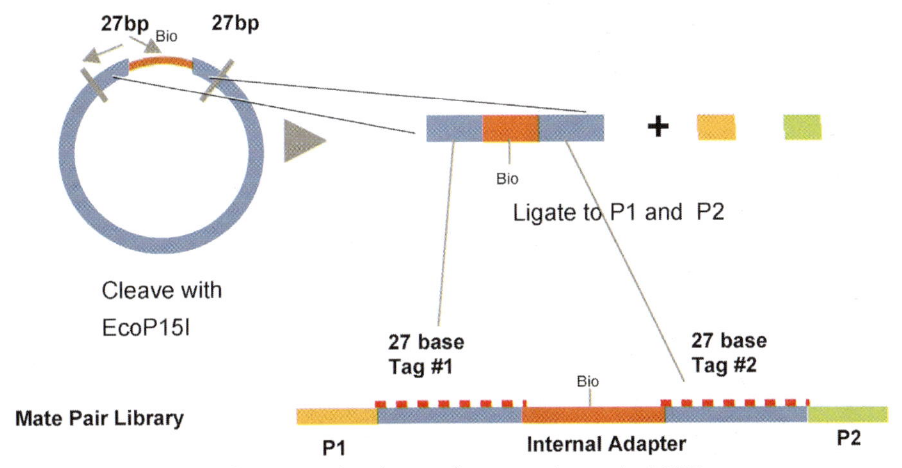

Figure 21.3 Formation of mate-paired, 27-bp tags for sequencing on the SOLiD system.

The same PCR primers used to flank the fragment library are now ligated to the mate-paired library and those fragments with unique primers on each end are selected for amplification by emulsion PCR.

21.3 Emulsion PCR

Regardless of the type of library used, the emulsion PCR process is the same and follows the protocol described by Dressman [11], with minor modifications. A reverse-phase emulsion is made using specific oils and an aqueous phase that contains the library, all of the components for the PCR reaction as well as 1-μm magnetic beads that have one of the PCR primers attached to its surface. For the purpose of clarity, the primer attached to the magnetic bead is called P1 (Figure 21.3). The PCR reaction is then carried out in the aqueous microreactors formed when the emulsion is made (Figure 21.4). Before the emulsion is made, the library in the aqueous phase has been diluted to maximize the number of microreactors that will contain one DNA template and one magnetic bead. According to a normal Poisson distribution, approximately 20% of the microreactors will contain one bead and one template from the library. The rest of the microreactors will not be suitable for further analysis for a variety of reasons (Figure 21.4). During PCR amplification, the 'productive' microreactors will drive the clonal amplification of individual templates from the library onto individual magnetic beads.

21.3.1 Bead Purification

The rest of the microreactors will not produce clonal amplification of templates for a number of reasons. It is therefore, necessary to enrich for those magnetic beads that

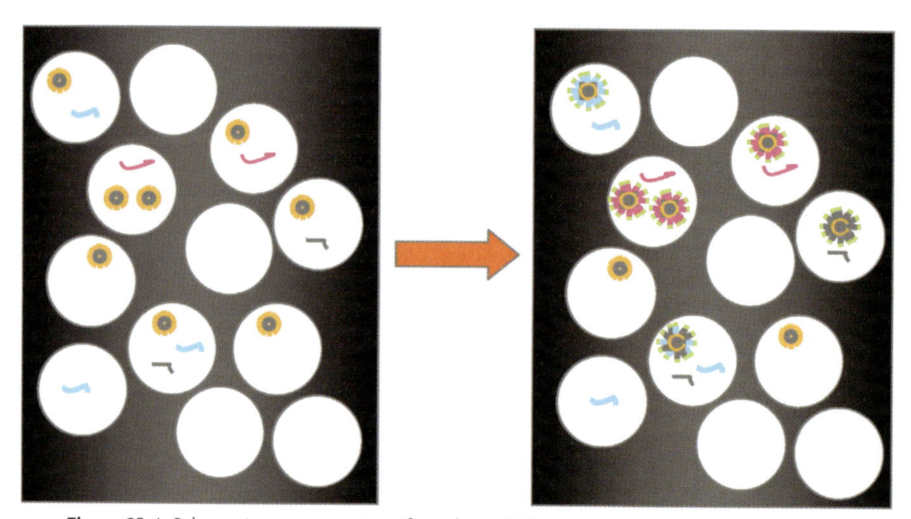

Figure 21.4 Schematic representation of emulsion PCR.

have a single, amplified template on their surface from the rest of the beads. After PCR amplification, the emulsions are broken and all magnetic beads are isolated from the oil and PCR components using standard magnetic bead-purifying racks and successive washes with the appropriate buffers. The beads are then mixed with large, polystyrene beads that have a sequence complimentary the PCR primer at the other end of the amplified template (P2) attached to their surfaces. The beads with extended PCR product will all have P2 sequences at their ends that will permit them to hybridize to the P2 complimentary sequence on the surface of the polystyrene bead. The polystyrene beads hybridized to beads with extended PCR product will float in a glycerol gradient, whereas all other beads will be driven to the bottom of the tube during centrifugation. The magnetic beads enriched for the full length PCR product are then melted off the polystyrene beads for further use. This step routinely results in total bead enrichment to >90% P2+ beads.

21.3.2 Bead Deposition

Once the beads have been purified, they are covalently attached to standard microscope slides that have had their surfaces chemically functionalized. The ends of the extended PCR-amplified templates are also chemically modified in such a manner as to allow them to be covalently linked to the surface of the slide. Unlike picotiter plates, where a defined number of wells have been machined onto a plate [6], the only limitation to the number of beads that can be deposited on the surface of the slide is the diameter of the beads (1 μm). Currently, the SOLiD system can support ~50 000 P2+ beads/0.75 mm^2 and there are approximately 1350 mm^2 on an entire slide. However, the 1-μm diameter of the beads used in this system will allow many more beads to be deposited on the slide surface. As the number of P2+ beads that are deposited on a slide increases, the number of bases that can be generated on a slide will increase accordingly.

21.4 Sequencing by Ligation

The most novel aspect of the SOLiD system is the use of DNA ligase and fluorescently labeled oligonucleotide probes to perform the sequencing reaction. The basics of the sequencing chemistry are illustrated in Figure 21.5, panels A–E. Briefly, a primer complimentary to and having its 5′ end at the P1/unknown sequence junction is used as an anchor for subsequent ligation reactions. A large pool of fluorescently labeled octamer (eight-base) probes that have all possible combinations of A, C, G and T at positions 1–5 (designated as Ns) are allowed to interrogate the sequence of the unknown template on each bead. There are a total of 1024 of these probes in the pool, though only four probes are shown in the figure for the sake of clarity (panel A). Only the probe that is exactly homologous to and therefore hybridizes to the first five bases of the unknown sequence will be in the proper position to be ligated to the anchor primer. Probes that hybridize to other regions of the unknown sequence are not

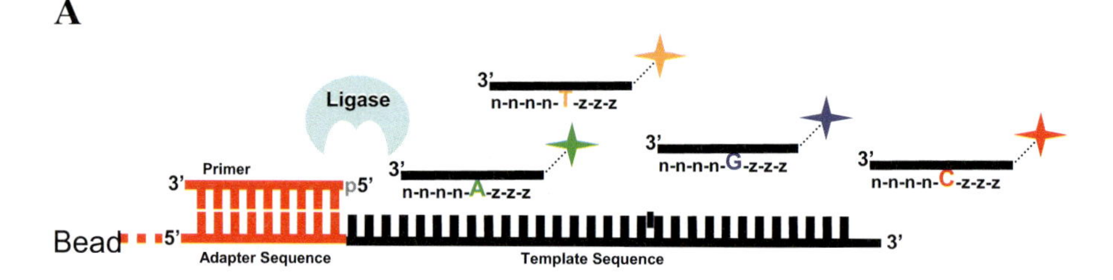

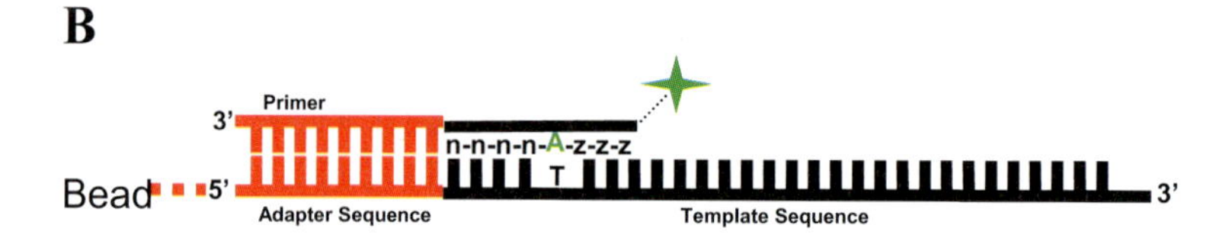

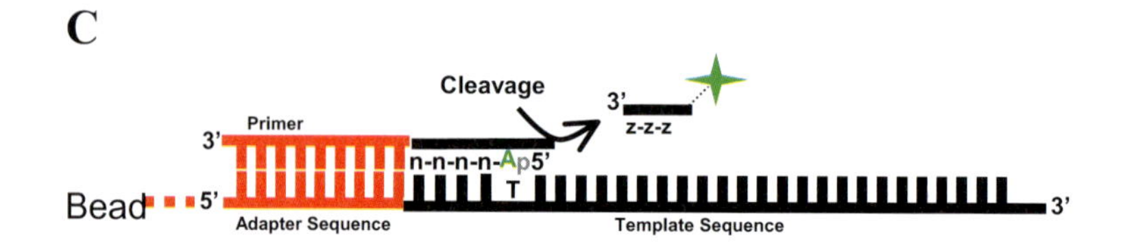

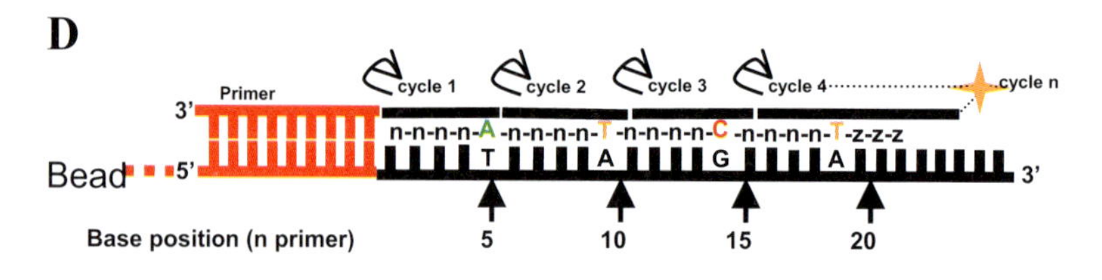

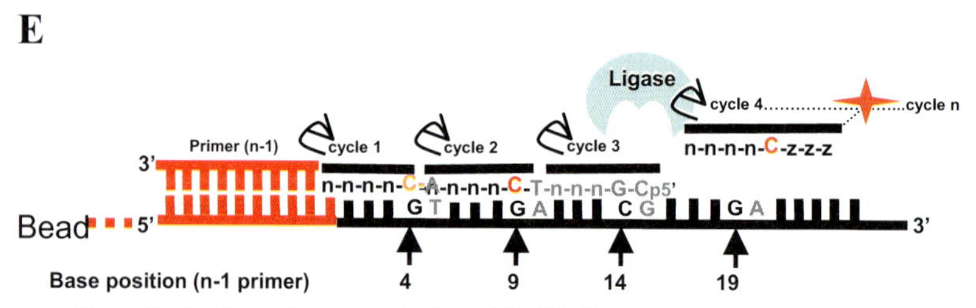

Figure 21.5 Schematic representation of SOLiD ligation chemistry. The details of each step of the process are described in more detail in the text.

substrates for ligase, since the enzyme can only establish a phosphodiester linkage between adjacent 5′ phosphate of one primer and 3′-hydroxyl of the second primer. The 3′ end of the interrogating probe will then be ligated very efficiently to the 5′ end of the anchor primer by DNA Ligase (panel B). Numerous publications have demonstrated the selectivity and specificity of ligation occurring only when complete homology exists [12–14]. The probes are labeled with specific fluorescent dyes that are associated with nucleotides at specific locations in the probe. In the example shown in Figure 21.5, these are at positions 4 and 5, but are not necessarily restricted to this example. The nature of the dye that is attached to the probe is recorded after illumination by a xenon lamp. The nucleotides at positions 6–8 are universal (represented by 'z' in the figure and do not have high discrimination for hybridization, but do help to stabilize the DNA duplex) and must be removed before the next round of ligation. This is accomplished by chemical cleavage of a modified linkage between nucleotides 5 and 6 in the probe (panel C).

Four additional rounds of ligation are conducted in exactly the same manner as the first round. At the end of the first five rounds of ligation, it is possible to recognize the nucleotides that were present in positions 5, 10, 15, 20 and 25 from the end of the unknown template (panel D). In order to determine the remaining unknown bases, it is necessary to chemically strip the extended products that resulted from the first five ligations from the template. The remaining nucleotides in the unknown template are now determined by ligation of pools of labeled probes to anchor primers that have been displaced a specific distance 3′ from the original primer (panel E). This feature is known as 'resetting' in that it resets the signal to noise of the next ligation cycle to that of the first ligation cycle. After 25 cycles of ligation, the first 25 bases of the unknown templates will have been identified. For a fragment library, these are all of the ligation reactions that will be performed.

If mate-paired libraries are being sequenced, another 25 cycles of ligation are carried out in exactly the same manner using an anchor primer complimentary to the internal CAP primer (Figure 21.3). In this manner, there will now be 25 bases of sequence produced from the two mate-paired ends of templates that were originally separated by a known distance at the beginning of library preparation (2×25 bp).

The SOLiD system is currently capable of generating in excess of 1 billion bases (1 GB) of DNA sequence with each run if the entire field of the slide is devoted to a single sample. An additional feature of the system allows multiple samples to be run on a single slide by physically segmenting the slide and depositing beads from different samples to each segment. The number of beads to be deposited can be readily calculated knowing the number of bases needed for the application being studied and the number of bases generated per bead. This will be illustrated in later sections.

21.5
Base Calling

Short sequence reads require very highly accurate base calling to unambiguously align a sequence to a reference sequence. The SOLiD system accomplishes this by the employment of a novel base-calling algorithm based on the concept of two-base encoding. As stated previously, the four dyes that can be attached to the oligonucleotide probes used in the sequencing ligation reaction represent the relationship between, for example, the nucleotides at positions 4 and 5 of the probe. In two-base encoding, the identity of a base is determined twice, once when it is in position 4 and again when it is in position 5 The color of the dye at each cycle of ligation is recorded and stored as a digital representation of that color. Dye #1 is represented by a 0, dye #2, a 1 and so on. The sequence of the 25 contiguous bases is kept as color representation, for example, 0320112332001112310021330. This concept is called 'color space'.

The benefit of two-base encoding is that by careful design of the encoding matrix, it is possible to correct for measurement error. Measurement error is the situation where an incorrect color call is made and a single color space call will be in mismatch with the reference. A real polymorphism will require two adjacent color calls to change at the same time, allowing easy discrimination between measurement errors and a real polymorphism. This is a definite advantage over single-base encoding (e.g. DNA polymerase-based systems) where there is no way to distinguish a measurement error from a polymorphism, thus requiring a higher coverage.

21.6
Potential Applications

21.6.1
Resequencing

The SOLiD system in addition to the other new sequencing systems, are ideally suited to the generation of large amounts of data for genome resequencing projects. Depending on the size and complexity of the genome, the investigator can elect to construct a simple fragment library or a mate-paired library. A fragment library has fewer manipulations and does not require enzymatic cleavage (i.e., EcoP15I), therefore it may contain certain sequences that are not represented in a mate-paired library. On the other had, a mate-paired library has considerable advantages when it comes to data assembly, especially when the genome has a more complex organization. Therefore, in some cases, it may be desirable to construct and sequence both types of libraries to maximize the sequence coverage.

The obvious advantage of a massively parallel sequencing platform such as the SOLiD system is the ability to achieve extremely deep coverage of most organisms with one or a limited number of runs. For instance, a 2-MB organism has been sequenced to a depth in excess of 200-fold with a single run on the SOLiD system using a pre-commercial protocol (unpublished data). Other systems with the ability to generate fewer bases/run, such as the GS-FLX, can achieve the same coverage with more runs and therefore, more cost.

21.6.2
De novo Sequencing

It is not yet clear how well short sequences can be used for *de novo* sequencing of most organisms. This question should be addressed quickly once data from the SOLiD and other systems have been tested in this application and the bioinformatic tools for assembly have been refined. However, it is currently possible to take advantage of a combination of traditional Sanger-based sequencing with the SOLiD system for cost-effective *de novo* sequencing projects. This is due to the very long, highly accurate, read-lengths that are routinely generated with the Sanger system. These sequences can readily be used to assemble a genome 'scaffold', or backbone. Goldberg [15] showed that sequences generated by the GS-20 sequencing system were able to fill in a number of assembly gaps that remained after 5× coverage of a bacterial genome using Sanger-based sequencing. Gaps in the scaffold can just as easily be filled in by using short reads from the SOLiD system.

Assembly of *de novo* sequences using a hybrid Sanger sequencing/SOLiD approach is also expected to benefit from the fact that massively parallel sequencing systems, such as SOLiD, clonally amplify single molecules using PCR. This type of amplification eliminates bias in the regions of a genome that are sequenced due to the inability to clone these regions in bacteria, which is known to occur with traditional Sanger sequencing. The so-called 'unclonable' regions presumably contain genes or sequences that are either directly toxic to the host cell or somehow interfere with normal cellular function. Additionally, sequences, such as GC-rich regions, homopolymers and other simple repeats that are very difficult to sequence with polymerase-based sequencing chemistries, are more likely to be represented in assemblies using sequences derived from the SOLiD ligase-based sequencing chemistry. This is due to the fact that ligation-based sequencing reactions require all five nucleotides of the correct probe from the sequencing probe pool to recognize the unknown sequence adjacent to the end of the sequencing primer before ligation will occur (Figure 21.5). This effectively eliminates the possibility of out-of-phase extensions seen when homopolymers are sequenced using pyrosequencing chemistry [6].

21.6.3
Gene Expression via Sequence Tags

In addition to traditional sequencing applications, such as *de novo* and resequencing, massively parallel sequencing systems now make tag-based sequencing applications practical on a genome-wide basis. A number of sequence tag-based gene expression applications have been developed over the past few years. These applications are known by a number of acronyms, such as SAGE [16], SuperSAGE [17], CAGE [18] and 5′-SAGE [19]. A number of these applications are described in other chapters of this book and the reader is encouraged to review those chapters for more details. However, they all take advantage of the presence of short sequences that are unique for a specific species of RNA. Methodology exists to isolate these 'tags', manipulate them and determine the sequence of the tags by any sequencing method. The number of tags sequenced has been shown to be proportional to the number of mRNA molecules in the population. The advantage of these applications is that they do not require a priori knowledge of the sequence of the genome being studied. This is of extreme value for plant-based applications due to the extensive up-front DNA sequence information needed for array-based gene expression applications. Additionally, tag-based sequencing allows for more precise measurement of the differences in gene expression levels than array technology simply by increasing the number of tags sequenced (Figure 21.6). While existing sequence tag methodologies

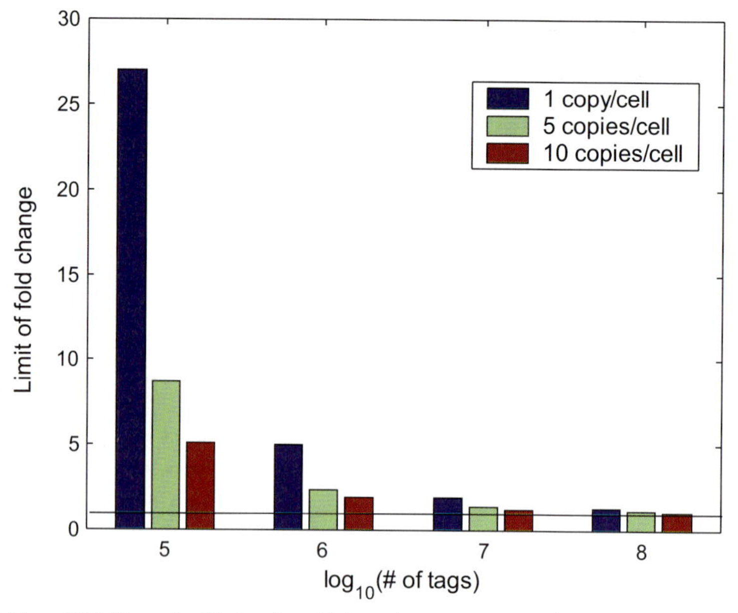

Figure 21.6 Theoretical limits of sensitivity to detect transcripts of different copy numbers using sequence tags. The model shows that even two-fold changes in expression levels of single copy transcripts can be detected using 10^7 sequence tags.

have been validated using other platforms, it will be necessary to also validate the data generated using SOLiD, or any other new sequencing technology, as a part of their adoption by the scientific community.

The number of sequence tags needed for a 5′ SAGE gene expression experiment is normally in the range of 1×10^6 to 5×10^7, depending on the specific application (I. Hashomoto, personal communication). The SOLiD system takes advantage of random deposition of beads containing clonally amplified templates. Therefore, hundreds of thousands of 1-μm beads can be deposited on each square millimeter of the surface of the slide. The slide surface can also be physically separated into a number of different sectors when the beads are deposited. If, for instance, it has been determined that 2×10^6 sequence tags are needed for a sample, and 12 000 mappable (sequences will unambiguously map to the reference with 0 or 1 mismatch) beads (or tags) can be deposited on each 0.75 mm^2, it is quite straightforward to calculate the area of a slide that will be needed to achieve the desired number of tags ($2\,000\,000/12\,000 \times 0.75$).

Since only a portion of the entire slide is likely to be needed to provide the number of tags for a specific experiment, a number of different samples or biological controls can be run on a slide. By configuring the slide appropriately, it is possible to control for the run-to-run variation that may exist. As the number of usable beads (tags) that can be packed onto the surface of the slide increases, it will become possible to use more beads in the same area, giving greater sensitivity to differences in gene expression. Alternatively, more samples could be run on each slide.

Whole transcriptome analysis of complex genomes is a second application that is amenable to the massively parallel sequencing output possible with the SOLiD system. For examples, total, non-polysomal RNA can be isolated, fragmented and converted to cDNA flanked by P1 and P2 primers. After sequencing, the resulting sequence is compared to the appropriate reference sequence to identify unambiguous transcribed regions. This work is the subject of an ongoing collaboration. Preliminary analysis of sequence tags generated on the SOLiD system from mouse embryonic stem cell RNA has been shown to uniquely identify over 20 000 transcripts. The same RNA run on the Illumina gene expression bead array, on the other hand, only identifies almost 9000 unique transcripts. Further analysis of the novel transcripts identified from SOLiD sequence tags shows the majority of them are low abundance transcripts that are present below the level of detection of standard arrays. A further benefit from the SOLiD sequencing data also permits identification and quantification of splicing events. (S. Grimmond, personnel communication). This work is being extended and validated prior to submission for publication. This data is the first demonstration of the ability of sequence-tag transcriptome applications to identify more genes than arrays in a cost-effective manner.

The sequence data generated from each experiment must be stored and processed by software that has been specifically designed for each application. For instance, the sequence tags, mRNAs they correspond to and number of times they are found in the sample must be tracked and tabulated. Serviceable analysis packages for each type of application have been developed by members of the scientific community and have been made freely available to others interested in developing the application in their laboratories. The amount of data generated with the new massively parallel sequencing systems present significant new challenges to the bioinformatics community. These systems generate hundreds of times more data in the same time as current systems. Downstream analysis software to manage the sequence data generated with the SOLiD system will need to be developed and distributed in a similar manner.

21.6.4
Other Tag-Based Applications

As the cost of generating sequence data has decreased, it is becoming possible to move essentially all genetic analysis applications to sequenced-based. Even before the new sequencing systems became available, forward-looking scientists were already describing genome-wide genetic analysis applications that are readily converted to sequence-based. While gene expression was the first to take advantage of sequence-specific tags, other applications such as digital karyotyping [20–22], chromosomal immunoprecipitation to locate genetic regions specific for recognition by transcriptional binding proteins [23], small RNA discovery [24], methylation patterns [25] and biomarker detection of microbial organisms [26] are now possible. Interested readers can obtain the details of each application from the original publications. They are similar in that they have developed ways to enrich regions of interest from the rest of the genome using a number of different approaches. The regions of interest have been represented by either size or sequence using existing platforms. All of these applications can readily be converted to run on one of the massively parallel sequencing platforms by using system-specific primers, clonal amplification and sequencing. Application-specific data analysis tools must also be developed that will allow enormous amounts of raw sequence data to be organized in an appropriate manner. It is very likely that the lack of appropriate data analysis tools capable of handling the flood of sequence data being generated by the sequencing platforms described in this chapter will be the rate-limiting factor for genetic analysis over the next few years.

21.7 Conclusions

The development of massively parallel sequencing systems is revolutionizing the way in which the scientific community is conducting genetic analysis. This revolution is due to lowering the cost of an experiment and increasing the amount of data generated per experiment. Applications that take advantage of using large numbers of specific sequence tags, such as gene expression, digital karyotyping, chromosome immunoprecipitation and miRNA discovery will now be possible on a genome-wide scale. Additionally, systems such as the SOLiD system, will enable essentially any research group to develop and execute genome-wide genetic applications which will significantly increase the amount of data generated and accelerate the pace at which new discoveries are made and used to improve agriculture and other scientific endeavors.

References

1 Sanger, F., Nicklen, S. and Coulson, A.R. (1977) DNA sequencing with chain-terminating inhibitors. *Proceedings of the National Academy of Sciences of the United States of America*, **74**, 5463–5467.

2 Hunkapiller, T., Kaiser, R.J., Koop, B.F. and Hood, L. (1991) Large-scale and automated DNA sequence determination. *Science*, **254**, 59–67.

3 Venter, C., Adams, M., Myers, E., Li, P., Mural, R. *et al.* (2001) The Sequence of the Human Genome. *Science*, **291**, 1304–1351.

4 International Human Genome Sequencing Consortium. (2001) Initial sequencing and analysis of the human genome. *Nature*, **409**, 860–921.

5 Marshall, E. (1995) A strategy for sequencing the genome 5 years early. *Science*, **267**, 783–784.

6 Margulies, M., Eghold, M. *et al.* (2005) Genome sequencing in microfabricated high-density picolitre reactors. *Nature*, **437**, 326–327.

7 Brenner, S., Johnson, M., Bridgham, J., Golda, G., Lloyd, D.H., Johnson, D., Luo, S., McCurdy, S., Foy, M., Ewan, M., Roth, R., George, D., Eletr, S., Albrecht, G., Vermaas, E., Williams, S.R., Moon, K., Burcham, T., Pallas, M., DuBridge, R.B., Kirchner, J., Fearon, K., Mao, J. and Corcoran, K. (2000) Gene expression analysis by massively parallel signature sequencing (MPSS) on microbead

arrays. *Nature Biotechnology*, **18**, 630–634.

8 Shendure, J., Porreca, G.J., Reppas, N.B., Lin, X., McCutcheon, J.P., Rosenbaum, A.M., Wang, M.D., Zhang, K., Mitra, R.D. and Church, G.M. (2005) Accurate multiplex polony sequencing of an evolved bacterial genome. *Science*, **309**, 1728–1732.

9 Raphael, B., Volik, S., Collins, C. and Pevzner, P. (2003) Reconstructing tumor genome architectures. *Bioinformatics*, **19**, 162–171.

10 Whiteford, N., Haslam, N., Weber, G., Prügel-Bennett, A., Essex, J.W., Roach, P.L., Bradley, M. and Neylon, C. (2005) An analysis of the feasibility of short read sequencing. *Nucleic Acids Research*, **33** (19), e171.

11 Dressman, D., Yan, H., Traverso, G., Kinzler, K. and Vogelstein, B. (2003) Transforming single DNA molecules into fluorescent magnetic particles for detection and enumeration of genetic variations. *Proceedings of the National Academy of Sciences of the United States of America*, **100**, 8817–8822.

12 Luo, J. and Barany, F. (1996) Identification of essential residues in *Thermus thermophilus* DNA ligase. *Nucleic Acids Research*, **24**, 3079–3085.

13 Liu, P., Burdzy, A. and Sowers, L. (2004) DNA ligases ensure fidelity by interrogating minor groove contacts. *Nucleic Acids Research*, **32** (15), 4503–4511.

14 Bhagwat, A., Sanderson, R. and Lindahl, T. (1999) Delayed DNA joining at 3′ mismatches by human DNA ligases. *Nucleic Acids Research*, **27** (20), 4028–4033.

15 Goldberg, S., Johnson, J., Busam, D., Feldblyum, T., Ferriera, S. *et al.* (2006) A Sanger/pyrosequencing hybrid approach for the generation of high-quality draft assemblies of marine microbial genomes. *Proceedings of the National Academy of Sciences of the United States of America*, **103**, 11240–11245.

16 Velculescu, V.E., Zhang, L., Vogelstein, B. and Kinzler, K.W. (1995) Serial analysis of gene expression. *Science*, **270**, 368–369, 371.

17 Matsumura, H., Reich, S., Ito, A., Saitoh, H., Winter, P., Kahl, G., Reuter, M., Krueger, D. and Terauchi, R. (2003) SuperSAGE: A universal functional genomics tool for eukaryotes. *Proceedings of the National Academy of Sciences of the United States of America*, **100**, 15718–15723.

18 Shiraki, T. Kondo, S. *et al.* (2003) Cap analysis gene expression for high-throughput analysis of transcriptional starting point and identification of promoter usage. *Proceedings of the National Academy of Sciences of the United States of America*, **100**, 15776–15781.

19 Hashimoto, S.I., Suzuki, Y., Kasai, Y., Morohoshi, K., Yamada, T., Sese, J., Morishita, S., Sugano, S. and Matsushima, K. (2004) 5′-end SAGE for the analysis of transcriptional start sites. *Nature Biotechnology*, **22**, 1146–1149.

20 Wang, T.L., Maierhofer, C., Speicher, M.R., Lengauer, C., Vogelstein, B., Kinzler, K.W. and Velculescu, V.E. (2002) Digital karyotyping. *Proceedings of the National Academy of Sciences of the United States of America*, **99**, 16156–16161.

21 Volk, S., Zheo, S., Chin, K., Brekner, J.H., Herndon, D.R., Tao, Q., Kowbel, D., Huang, G., Lapuk, A., Kuo, W.-L., Magrane, G., de Jong, P., Gray, J.W. and Collins, C. (2002) End-sequence profiling: Sequence-based analysis of aberrant genomes. *Proceedings of the National Academy of Sciences of the United States of America*, **100**, 7696–7701.

22 Tengs, T., LaFramboise, T., Den, R., Hayes, D., Zhang, J., DebRoy, S., Gentleman, R., O'Neill, K., Birren, B. and Meyerson, M. (2004) Genomic representations using concatenates of Type IIB restriction endonuclease, digestion fragments. *Nucleic Acids Research*, **32**, e121–e129.

23 Wei, C.L., Wu, Q., Vega, V.B., Chiu, K.P., Ng, P., Zhang, T., Shahab, A., Ridwan, A., Fu, Y.T., Weng, Z., Lee, Y.L., Liu, J.J., Kuznetsov, V.A., Sung, K., Lim, B., Liu,

E.T., Yu, Q., Ng, H.H. and Ruan, Y. (2006) A global mapping of p53 transcription factor binding sites in the human genome. *Cell*, **124**, 207–219.

24 Berezikov, E., Cuppen, E. and Plasterk, R.H.A. (2006) Approaches to microRNA discovery. *Nature Genetics*, **38**, S2–S7.

25 Rollins, R.A., Haghighi, F., Edwards, J.R., Das, R., Zhang, M.Q., Ju, J. and Bestor, T.H. (2006) Large-scale structure of genomic methylation patterns. *Genome Research*, **16**, 157–163.

26 Tengs, T., LaFramboise, T., Den, R.B., Hayes, D.N., Zhang, J., DebRoy, S., Gentleman, R.C., O'Neill, K., Birren, B. and Meyerson, M. (2004) Genomic representations using concatenates of Type IIB restriction endonuclease digestion fragments. *Nucleic Acids Research*, **32** (15), e121.

22
454 Sequencing: The Next Generation Tool for Functional Genomics

Lei Du, Jan Frederik Simons, Maithreyan Srinivasan, Thomas Jarvie, Bruce Taillon, and Michael Egholm

Abstract

DNA sequencing has been a powerful technique in genetics and molecular biology, allowing analysis of genes, operons and whole genomes at the nucleotide level. As a key component of modern day genomics studies, sequencing technology has gone through several generations of rapid improvement and cost reduction. For more than 30 years, a large proportion of DNA sequencing has been carried out using the chain-termination method developed by Frederick Sanger and coworkers in 1975, and optimized with the development and commercialization of automated sequencers. The current process for decoding a whole genome involves shearing of DNA into small fragments, shotgun cloning into plasmid vectors, amplification and purification in bacteria, and sequencing on a 96- or 384-lane automated capillary sequencer. The entire process takes about 3 weeks in a fully automated large-scale facility, with significant investment in robotic hardware, disposables and human labor.

In 2004, 454 Life Sciences commercialized the first next-generation sequencing instrument, the Genome Sequencer 20 [1]. It combined a novel technique of emulsion-based cloning and amplification (in lieu of the traditional cloning step which introduces coverage bias and areas of non-sequenceability), with ultra-fast and parallelized pyrophosphate-based sequencing using beads on a PicoTiterPlate (PTP). The PTP is composed of fused optical fibers with chemically etched picoliter volume wells. Light generated during the course of pyrosequencing is efficiently conducted by way of total internal reflection from the well through the remaining fiber for detection by a juxtaposed astronomical grade CCD camera. The massively parallel nature of this technology allows simultaneous decoding of more than 200 000 DNA fragments totaling 20–40 million base pairs. Since then, more than 70 peer-reviewed articles have been published, demonstrating the utility of 454 Sequencing in a broad range of research applications, such as microbial genomics and drug resistance [2–7], plant genetics [8–10], human genetics [38–40], small RNA [11–26] and gene

The Handbook of Plant Functional Genomics: Concepts and Protocols.
Edited by Günter Kahl and Khalid Meksem

ISBN: 978-3-527-31885-8

regulation [27–30], transcriptome analysis [31–37], metagenomics and environmental diversity [41–47], and ancient DNA analysis [48–51]. In December 2006, 454 Life Sciences introduced the second generation Genome Sequencer system, the GS FLX, with a throughput of 100 million base pairs and average read length of more than 240 bases. The increased read length and throughput will expand the utility of this technology and enable scientists to conduct experiments never before possible due to the prohibitive requirements in time and cost of traditional sequencing technologies.

22.1 Introduction

The GS FLX technology has three main components: DNA library preparation, emulsion PCR, and PicoTiterPlate sequencing. The library preparation step (Figure 22.1) generates a pool of single-stranded DNA fragments, each carrying distinct universal adapters, A and B, on each end of the molecule. The universal adapters are 44 bases in length and are composed of a 20-base PCR primer, a 20-base sequencing primer and a 4-base key sequence for read identification and signal normalization. The DNA fragments can be derived from mechanical shearing of a long stretch of DNA such as whole genome or long PCR products and subsequently the library is generated by ligation of the universal adapters to each fragment. Alternatively, sequencing templates can be generated by targeted PCR amplification of genes and loci of interest using site-specific primers with 5′ adapter overhangs. For emulsion PCR, the library material is mixed at limited dilution with beads carrying one of the adapter primers on its surface in a water-in-oil emulsion set-up (Figure 22.2). The aqueous phase of

The Genome Seqencer FLX System provides a complete solution – from sample preparation through digital data analysis.

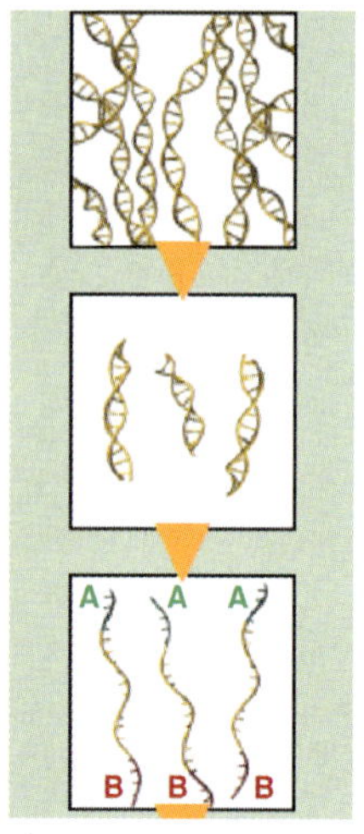

(1) Sample Input: The Genome Sequencer FLX System supports the sequencing o samples from a variety of starting materials, including genomic DNA, PCR products, BACs, and cDNA.

(1) Sample Fragmentation: Samples such as genomic DNA and BACs are fractionated into small 300 to 800 base-pair fragments. For some samples, such as small non-coding RNA. fragnentation is not required. Short PCR products can be amplified using Genome Sequencer fusion primers to g directly to Step 4, shown below.

(1) Adaptor Ligation: Using a series of standard molecular biology techniques, short adptors (A and B) – specific for both the 3′ and 5′ ends – are added to ech fragment. The adaptors will also be used purification. amplification. and sequencing steps. Single-stranded fragnent. shown here. are used in subsequent steps in the workflow.

Figure 22.1 DNA sample preparation.

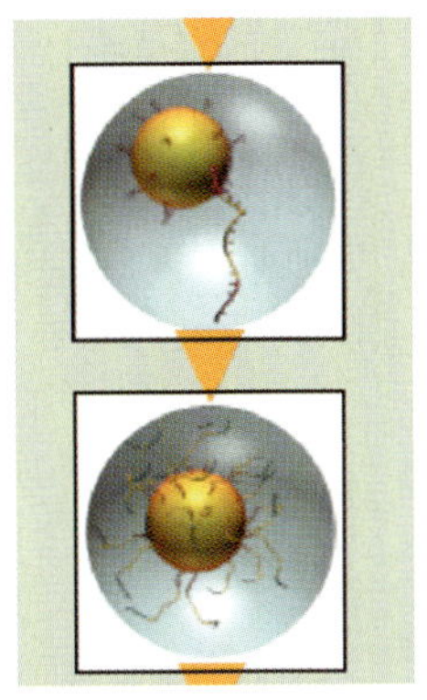

Figure 22.2 Emulsion PCR.

(4) One Fragment = One Bead: The first step in emPCR (emulsion PCR) is shown. The adaptors enable hundreds of thousands of single-stranded fragments to bind to their own unique beads. The beads are then encapsulated into individual droplets formed by a water-in-oil emulsion. creating a microreactor containing one bead with one unique fragment. Eacg unique fragnent is amplified without the introduction of competing or contaminating nating sequences. The entire fragment collection is amplified in parallel.

(5) One Bead = One Read: The emPCR process amplifies each fragment to a copy number of several million per beab. Subsequently. the emulsion is broken while the fragments remain bound to their specific beabs. After enrichment. the clonally amplified bead is ready to load onto the PicoTiterPlate device for sequencing.

the emulsion encloses the bead, PCR primers (with the same sequence as the universal adapters and one of the primers having a biotin at the 5′ end), nucleotides and polymerase, and thus forms hundreds of thousands of microreactors. Through limited dilution, each bead will encounter mostly zero or one DNA molecule prior to PCR. The emulsion mix undergoes thermal cycling, leading to clonal amplification of the DNA template in each vesicle and population of that bead. At the end of the PCR process, emulsions are solubilized and template-carrying beads are enriched using magnetic streptavidin beads. The bead-bound double-stranded DNA fragments are rendered single-strand by denaturing and sequencing primer is annealed to the free 3′ end of the millions of clonal fragments on each bead. The beads are deposited into the wells of the PicoTiterPlate (PTP) and overlaid with beads carrying sulfurylase and luciferase, enzymes needed for the pyrosequencing reaction. The PTP is then loaded into a flow cell on the Genome Sequencer and placed directly in front of the CCD camera (Figure 22.3). Nucleotides are flown across the open surface of the PTP via the fluidics subsystem inside the sequencer, and DNA synthesis is carried out in real time in each well. Positive incorporation of one or more nucleotides at a given flow will generate pyrophosphate, which is converted to ATP by sulfurylase. ATP

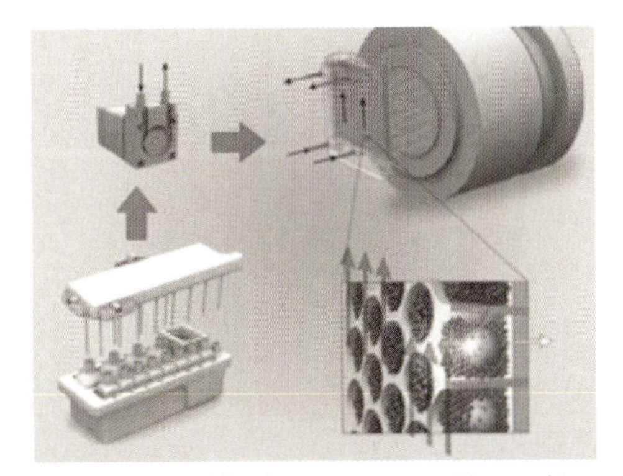

Figure 22.3 Bead loading into PicoTitrePlate and instrument.

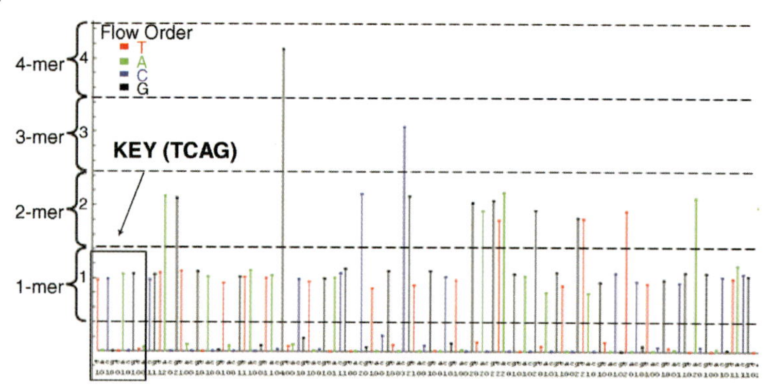

Figure 22.4 Signal flowgram and base calling.

subsequently drives the oxidation of luciferin by luciferase and light is emitted in a stochiometric fashion. The emitted photons are then captured from the bottom of the PTP well by a CCD camera. The intensity of light is proportional to the number of bases incorporated as well as the signal-transduction enzymatic activity in each well (a stretch of four Gs will generate four times the light than that from a single G). Since the amount of enzyme activity in each well is dictated by the number of enzyme beads deposited in each well, and that the four-base key sequence at the beginning of each polymerase reaction is a fixed stretch of monomer nucleotides, the overall signal is normalized by the light intensity generated during the sequencing of the key for each well (Figure 22.4). The data processing workflow includes image processing, signal processing and normalization, noise reduction and phase correction, and base-calling. The GS FLX system also comes with graphical user interface software to allow browsing of run data, quality monitoring as well as tools for *de novo* assembly and remapping against references to identify mutations.The original GS20 system was published with a detailed protocol available in the supplemental materials section [1]. The GS FLX system has the same core components as GS20, with a series of improvements in sample preparation, sequencing, instrument run and bioinformatics. The detailed protocol for DNA sequencing on the GS FLX is described in the following section, and references can also be found on-line (http://www.roche-applied-science.com/sis/sequencing/flx/index.jsp)

22.2
Methods and Protocols

22.2.1
DNA Library Preparation

There are two main routes by which DNA can be prepared for emulsion PCR. Option A is to mechanically shear long DNA molecules, hereby called 'nebulized library'. Option B is to use PCR primers to amplify specific regions of interest, and is referred to as 'amplicon library'.

22.2.1.1 **Option A: Nebulized Library Procedure**

DNA Fragmentation (Nebulization)

DNA Sample Dilution

1. Obtain 3–5 μg of sample DNA (in TE) and pipette it to the bottom (cup) of a nebulizer.
2. Add TE Buffer to a final volume of 100 μl. Add 500 μl of Nebulization Buffer, and mix thoroughly.

Nebulizer Assembly

1. Affix the cap and the condenser tube to the nebulizer top part. Press firmly. Transfer to the externally vented hood. Insert the nebulizer into the holder. Connect to the nitrogen tank.

DNA Nebulization and Collection/Purification of the Fragmented DNA

1. Apply 45 psi of nitrogen for 1 min. Allow the pressure to normalize, and disconnect the tubing. Measure the volume of nebulized material. Total recovery should be greater than 300 μl.
2. Add 2.5 ml of Buffer PB; swirl.
3. Purify the nebulized DNA using two columns from a MinElute PCR Purification Kit (Qiagen), with the following exceptions:
4. Each column should be loaded and spun in two aliquots (~750 μl each).
5. After the PE dry spin, rotate the column 180° and spin for an additional 30 s.
6. Elute with 25 μl of Buffer EB (room temperature; supplied in the Qiagen kit).
7. Pool the eluates of the two columns, for a total volume of ~50 μl.

Small Fragment Removal

1. Measure the volume of the pooled eluates, using a pipetter. Add Buffer EB (Qiagen) to a final volume of 50 μl. Add 35 μl of AMPure SPRI beads. Vortex.
2. Incubate for 5 min at room temperature (22 °C). Using a Magnetic Particle Collector (MPC), pellet the beads against the wall of the tube.
3. Remove the supernatant and wash the beads twice with 500 μl of 70% ethanol. Remove all the supernatant and allow the SPRI beads to air dry completely (can use a 37 °C heating block).
4. Remove the tube from the MPC, add 24 μl of 10 mM Tris-HCl, pH 8.0 (or Qiagen's Buffer EB), and vortex to resuspend the beads. This elutes the nebulized DNA from the SPRI beads.
5. Using the MPC, pellet the beads against the wall of the tube once more, and transfer the supernatant containing the purified nebulized DNA to a fresh microcentrifuge tube.

DNA Sample Quality Assessment (Nebulized or LMW DNA Sample)

1. Run 1 μl of the pooled nebulized material on a BioAnalyzer DNA 7500 LabChip (LMW sample: BioAnalyzer DNA 1000 LabChip). The mean size should be between 400 and 800 bp (70–500 bp for LMW DNA).

Fragment End Polishing

1. In a microcentrifuge tube, add the following reagents, in the order indicated: ~23 μl nebulized DNA (or 1 μg of a LMW DNA sample, in TE), 5 μl 10× Polishing Buffer, 5 μl BSA, 5 μl ATP, 2 μl dNTPs, 5 μl T4 PNK, 5 μl T4 DNA polymerase. 50 μl final volume.
2. Mix and incubate for 15 min at 12 °C. Continue incubation at 25 °C for an additional 15 min.
3. Purify the polished fragments using one column from a MinElute PCR Purification Kit (Qiagen). Elute with 15 μl of buffer EB (room temperature).

Adapter Ligation

1. Mix the following reagents in a microcentrifuge tube: ~15 μl Polished DNA, 20 μl 2× Ligase BuffeR, 1 μl adapters, 4 μl ligase, 40 μl total.
2. Mix, spin briefly, incubate the ligation reaction at 25 °C for 15 min.
3. Purify the ligation products using one column from a MinElute PCR Purification Kit (Qiagen), with the following exception: elute with 25 μl of Buffer EB (room temperature) directly into the washed Library Immobilization Beads (see point 4 in "Library Immobilization" below).

Library Immobilization

1. Transfer 50 μl of Library Immobilization Beads to a fresh 1.5-ml tube. Using a Magnetic Particle Collector (MPC), pellet the beads and remove the buffer. Wash the beads twice with 100 μl of 2× Library Binding Buffer, using the MPC.
2. Resuspend in 25 μl of 2× Library Binding Buffer. Elute DNA from the MinElute column (25 μl; see see "Adaptor Ligation" above) directly into the tube of washed Library Immobilization Beads.
3. Mix well and place on a tube rotator at ambient temperature (22 °C) for 20 min.
4. Using the MPC, wash the immobilized Library twice with 100 μl of Library Wash Buffer.

Fill-In Reaction

1. In a 1.5-ml tube, add the following reagents, in the order indicated, and mix:
   - 40 μl Molecular Biology Grade water, 5 μl 10× Fill-in Polymerase Buffer, 2 μl dNTPs, 3 μl Fill-in Polymerase, 50 μl total.
2. Using the MPC, remove the 100 μl of Wash Buffer from the Library-carrying beads from the "Library Immobilization" step above.
3. Add the 50 μl of fill-in reaction mix prepared in step 1. Mix well and incubate at 37 °C for 20 min.

4. Using the MPC, wash the immobilized Library twice with 100 µl of Library Wash Buffer.

Single-stranded Template DNA (ssDNA) Library Isolation

1. In a 1.5-ml tube, prepare the neutralization solution by mixing 500 µl of Qiagen's PB buffer and 3.8 µl of 20% acetic acid.
2. Using the MPC, remove the 100 µl of Library Wash Buffer from the Library-carrying beads from from the "Fill-in Reaction" step above. Add 50 µl of Melt Solution to the washed Library-carrying beads.
3. Vortex well and using the MPC, pellet the beads away from the 50 µl supernatant.
4. Carefully remove and transfer the supernatant to the freshly-prepared neutralization solution.
5. Repeat steps 2–4 for a total of two 50-µl Melt Solution washes of the beads (pooled together in the same tube of neutralization solution).
6. Purify the neutralized sstDNA library using one column from a MinElute PCR Purification Kit. Follow the manufacturer's instructions for spin columns using a microcentrifuge, with the following exception, elute in 15 µl of TE Buffer (from the GS DNA Library Preparation Kit; room temperature).

sstDNA Library Quality Assessment and Quantitation

Library Quality Assessment and Physical Quantitation

1. Run 1 µl of library on an RNA Pico 6000 LabChip.
2. Quantitate the sstDNA library (1 µl, in triplicate) by fluorometry using the RiboGreen method (Molecular Probes), following the manufacturer's instructions.
3. Assess the quality of the sstDNA library: the average fragment size should be between 400 and 800 bp, with <10% below 300 nt (150–500 bp for LMW DNA, no lower size cut-off); the total DNA yield should be $\geq$10 ng; and there should be no visible dimer peak.

Library Primary Dilution and Storage

1. Using the RiboGreen quantitation results (ng/µl), calculate the concentration equivalence in molecules/µl.

$$\text{Molecules}/\mu\text{l} = \frac{(\text{Sample conc.;ng}/\mu\text{l}) \times (6.022 \times 10^{23})}{(328.3 \times 10^{9}) \times (\text{avg.fragment length;nt})}$$

2. Make a primary dilution of 100×10^6 molecules/µl in TE Buffer, using 1 µl of the library.
3. Store the concentrated library and, if not used immediately, the 100×10^6 molecules/µl library stock, should be stored at -15 to $-25\,^{\circ}$C.

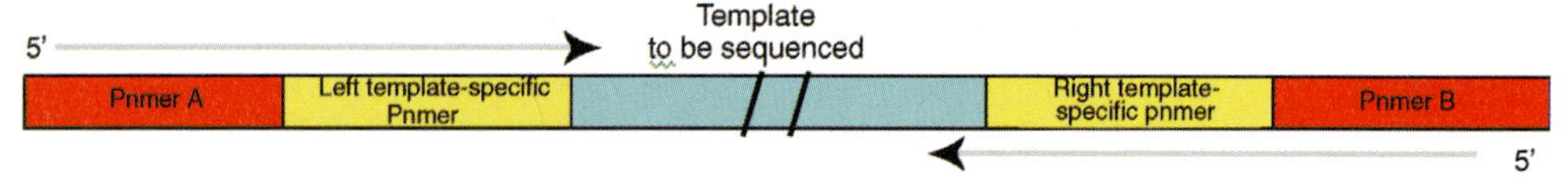

Figure 22.5 Schematic representation of an amplification product generated by the Amplicon library preparation procedure described in this chapter. The composite primers each comprise a 20–25-bp target-specific sequence region at their 3′-end; and a 19-bp region (Primer A or Primer B) that will be used in subsequent clonal amplification and sequencing reactions, at their 5′-end.

22.2.1.2 **Option B: Amplicon Library Procedure**

1. In a 0.2-ml microcentrifuge tube, add the following reagents, in the order indicated:
   - 41 μl Molecular Biology Grade Water
   - 5 μl 10× FastStart High Fidelity Reaction Buffer with 18 mM $MgCl_2$
   - 1 μl dNTPs (10 mM each)
   - 1 μl Fusion Primer A (10 μM)
   - 1 μl Fusion Primer B (10 μM)
   - 1 μl FastStart High Fidelity Enzyme Blend (5 U/μl)
   - 50 μl final volume (Figure 22.5)
2. Mix by vortexing, spin down briefly, and add the sample DNA. The initial amount of DNA required depends on the nature of the sample. In all cases, however, the volume should not exceed 2 μl.
   - For a complex DNA sample (e.g. genomic DNA), use 10–50 ng of DNA, in no more than 2 μl.
   - For a template cloned in a plasmid or for PCR-generated template DNA, use 1–5 ng of DNA, in no more than 2 μl.
3. Place the tube in a thermocycler and launch an amplification program appropriate for the particular sample such that the total amount of product does not exceed 10^{12} molecules (e.g. ~200 ng of a 200-mer product).
4. For amplified fragments larger than 100 bp, purify the amplified DNA using SPRI size exclusion beads, as follows (for smaller amplicons, see 'Note' below):
   - After the amplification program completes, vortex and spin down the amplification reaction.
   - Transfer 45 μl of the amplified DNA to a fresh 1.7-ml microcentrifuge tube.
   - Add exactly 72 μl of Ampure SPRI beads. Vortex to mix. Incubate 3–5 min at room temperature (22 °C).
   - Using a Magnetic Particle Collector (MPC), pellet the beads against the wall of the tube (this may take several minutes due to the high viscosity of the solution).
   - *Note*: Leave the tube of beads in the MPC during all wash steps.
   - Remove the supernatant and wash the beads twice with 200 μl of 70% ethanol.

- Remove all the supernatant and allow the SPRI beads to air-dry completely. The drying time can vary due to environmental conditions and the amount of residual fluid left in the tube. The tube may be placed in a heating block set to 37 °C to help speed up the drying process; the beads are dry when visible cracks form in the pellet.
- Remove the tube from the MPC, add 15 µl of 1× TE Buffer, and vortex to resuspend the beads. This elutes the amplified DNA from the SPRI beads.
- Using the MPC, pellet the beads against the wall of the tube once more, and transfer the supernatant containing the Amplicon library to a fresh microcentrifuge tube.
- *Note*: The SPRI beads are not appropriate for amplicons smaller than 100 bp. If the target(s) is (are) smaller, use an alternative method such as gel electrophoresis to purify them.

5. Run a 1-µl aliquot of the amplicon library on a BioAnalyzer DNA 1000 LabChip to assess the quality of the amplification product.
   - Ensure that the amplification product is of the expected size; the size displayed on the BioAnalyzer typically does not diverge from the calculated size by more than a few bp for amplicons in the 100–200 bp range.
   - Examine the trace closely for extraneous products, such as primer dimers, as contaminants could seriously reduce the number of useful reads that will be obtained from sequencing the Amplicon library. Repeat the SPRI bead purification if primer dimers are present.

6. Quantitate the library by fluorometry, using the Quant-iT PicoGreen dsDNA Assay Kit (Invitrogen), following the manufacturer's instructions.

7. *Note*: Whereas various quantitation methods are possible, fluorometry has been found to be highly reproducible and to provide consistent emPCR and sequencing results. It is especially important to accurately determine the concentration of individual Amplicon libraries when they are to be pooled prior to emPCR, for example, to monitor multiple targets together; this will ensure an even representation of all the targets in the sequencing reaction.

8. Given the library concentration (in ng/µl) measured by fluorometry, calculate the equivalence in molecules/µl, using the following equation:

$$\text{Molecules}/\mu\text{l} = \frac{(\text{Sample conc.};\text{ng}/\mu\text{l}) \times (6.022 \times 10^{23})}{(656.6 \times 10^{9}) \times (\text{amplicon length};\text{bp})}$$

   6.022×10^{23} is Avogadro's number (molecules/mole), and 656.6 is the average molecular weight of nucleotide pairs, in g/mole.

9. Dilute an aliquot of the concentrated library stock to 2×10^{5} molecules/µl, in TE Buffer.

10. Distribute this working stock into 50-µl aliquots, and store them and the concentrated stock at −15 to −25 °C.

Note: There is usually no need to optimize the amount of an Amplicon library to use for emPCR. (If more information is required on the optimization of the quantity of a library to use in emPCR, as is recommended for the shotgun sequencing of a sstDNA library, see the *GS DNA Library Preparation Kit User's Manual*.)

22.2.2 Emulsion PCR

Preparation of the Live Amplification Mix

1. Allow the frozen kit components to fully thaw (except enzymes). Vortex the reagents for 5 s.
2. Prepare the Live Amplification Mix (table below).

Reagent	Volumes for one emulsion	Volumes for four emulsions	Volumes for 16 emulsions
Amplification mix	181.62 μl	726.48 μl	2905.92 μl
$MgSO_4$	10.00 μl	40.00 μl	160.00 μl
Amplification Primer Mix (I, A, or B)	2.08 μl	8.32 μl	33.28 μl
Platinum HiFi *Taq* Polymerase	6.00 μl	24.00 μl	96.00 μl
PPiase	0.30 μl	1.20 μl	4.80 μl
Total:	**200.00 μl**	**800.00 μl**	**3200.00 μl**

3. Vortex the Live Amplification Mix for 5 s. Store it at + 2 to + 8 °C until ready for use.

DNA Library Capture

Washing the Capture Beads

1. Transfer 600 000 (450 000 for Amplicon libraries) DNA Capture Beads *per reaction* from the stock tube to a 1.5-ml tube (stock is 10 000 beads/μl; use 60/45 μl *per reaction, for up to 16 reactions, that is, to a maximum of* 960/720 μl). Wash the beads with 500 μl of 1× Capture Bead Wash buffer, vortex for 5 s. Remove and discard the supernatant.

2. Resuspend the bead pellet in 50 μl of 1× Capture Bead Wash buffer *per reaction*. Remove and discard 30 μl of the supernatant. The Capture Beads are now in ~20 μl of buffer and ready for binding the DNA library.

3. Addition (and Annealing) of the DNA Library Fragments to the Capture Beads.
   - Obtain a sufficient amount of the quantitated DNA library to be amplified.
   - To the *(each)* tube of washed Capture Beads, add the correct amount of library to provide optimal amplification (as per the table below; between 1 and 10 μl).

Library type	Library concentration	# Beads per emulsion	Target copies per bead	Volume of library to use per emulsion
sstDNA	Per titration	600 000	Per titration	Per titration
Paired end	2×10^5 mol/μl	600 000	1.5 cpb	4.5 μl
Amplicon	2×10^5 mol/μl	450 000	1.0 cpb	2.25 μl

4. Vortex the ***(each)*** tube for 5 s to mix its contents.

Emulsification

1. In the Emulsion enclosure, vortex one tube of Emulsion Oil for 10 s. Add 240 μl of Mock Amplification Mix to the Emulsion Oil. Set the TissueLyser for 25/s for 5 min, and press the start button to begin shaking.
2. While the TissueLyser is running, add 160 μl of Live Amplification Mix to the DNA library beads (≤40 μl). Remove the TissueLyser tube rack. Pipette the bead mixture up and down three times, and add it to the emulsion tube. Place the tube back into the TissueLyser tube rack (outer row), and insert the rack into the TissueLyser. Set the TissueLyser to 15/s for 5 min and start shaking.

Amplification

1. Dispensing the Emulsions ('Controlled Room') After emulsification, remove the rack from the TissueLyser, and carefully open the emulsion tube.
   Use an Eppendorf Repeater Plus to draw and dispense a reaction into eight 0.2-ml tubes. Place a 1.0-ml combitip plus tip onto the pipetter and set it to dispense 100 μl. Slowly draw the emulsion from the tube and the cap, into the tip. Dispense the emulsion into the tubes.
2. Amplification Reaction ('Amplicon Room')
   Place the emulsified amplification reactions into a thermocycler. Check to ensure that the lid is set to track within 5 °C of the block temperature. Set up and launch the amplification program.

Emulsion Breaking

Bead Resuspension/Pooling

1. Prepare a 16-gauge needle on a 10-ml syringe. Assemble the Swinlock filter unit with the nylon filter.
2. Add 100 μl of isopropanol to each tube containing the emulsion of amplified material.
3. Draw the emulsion–isopropanol mix from each of the tubes into the syringe. Add another 100 μl of isopropanol to each tube, and cap and vortex them for 10 s.
4. Draw the emulsion–isopropanol mix into the syringe. Invert the syringe and expel all air.

Emulsion Breaking and Bead Washing

1. Draw additional, fresh isopropanol into the syringe to 9 ml. Invert the syringe and draw in 1 ml of air to facilitate the mixing of emulsions with isopropanol in the following steps. Remove the blunt needle, attach the Swinlok filter, and attach the blunt needle to the Swinlok filter.

2. Mix the isopropanol with the emulsified beads by vigorously shaking the syringe for 5 s.

3. Squirt the contents of the syringe through the Swinlok filter unit into a waste jar containing bleach. Gently squirt out the contents of the syringe into the waste jar containing bleach.

4. Syringe wash once more with 9 ml of fresh isopropanol (vigorous shaking). Syringe wash with 9 ml of 1× DNA Bead Wash Buffer (vigorous shaking). Syringe wash with 9 ml of 1× Enhancing Fluid (vigorous shaking).

5. Draw in 0.5 ml of 1× Enhancing Fluid to resuspend the amplified library beads.

Bead Recovery

1. Remove the Swinlok filter unit and expel the contents of the syringe into a 1.5-ml tube. Remove as much of the supernatant as possible, without disturbing the beads.
2. Add 100 μl *(per reaction)* of 1× Enhancing Fluid to the beads. Vortex to mix.

DNA Library Bead Enrichment

Enrichment of the DNA-Carrying Beads

1. Add 100 μl *(per reaction)* of washed Enrichment Beads to the amplified DNA beads. Mix.
2. Rotate on a LabQuake tube roller at 15–25 °C for 5 min. *Do not vortex.*
3. After 5 min, bring the bead suspension volume up to 1 ml with 1× Enhancing Fluid.
4. Place the tube in the MPC, and wait 2 min to pellet the paramagnetic Enrichment Beads.
5. Carefully remove all the supernatant, taking care not to draw off any pelleted Enrichment Beads.
6. Remove the tube from the MPC and gently add 1 ml of 1× Enhancing Fluid to the beads.

Collection of the Enriched DNA Beads

1. Remove the tube from the MPC and resuspend the bead pellet in 700 μl (80 μl for Amplicon libraries) of Melt Solution. Vortex for 5 s, and put the tube back into the MPC to pellet the Enrichment Beads.

2. Transfer the supernatant, containing enriched DNA beads, to a separate 1.5-ml microfuge tube (0.2-ml tube for Amplicon libraries).
3. Pellet the enriched DNA beads by centrifugation as before.
4. Remove and discard the supernatant, and wash the enriched DNA beads with 1 ml (100 μl for Amplicon libraries) of 1× Annealing Buffer.

Sequencing Primer Annealing

1. Pellet the enriched (or non-enriched) DNA beads. Remove the supernatant without disturbing the bead pellet.
2. Add 15 μl of 1× Annealing Buffer *(per reaction)* to the DNA Bead pellet. Add 3 μl of Sequencing Primer *(per reaction)*. Vortex for 5 s.
3. Place the 0.2-ml tube into the thermocycler and run the Sequencing Primer annealing program.
4. Wash the beads once with 200 μl of 1× Annealing Buffer, resuspend in 100 μl of 1× Annealing Buffer.
5. Count a 3-μl aliquot (10 μl for Amplicon libraries) of the beads in the Coulter Counter.
6. Store the beads (immobilized, clonally amplified DNA library) at +2 to +8 °C.

Second-strand Removal (for Non-Enriched DNA Beads)

1. Pellet the beads in a bench top minifuge (the minifuge has only one speed), as follows:
   - spin for 10 s,
   - rotate the tube 180°, and
   - spin again for 10 s.
2. Remove as much of the supernatant as possible, without disturbing the beads, and discard it.
3. Add 1 ml of Melt Solution to the beads *(regardless of the number of pooled reactions)*.
4. Vortex for 2 s to resuspend the beads.
5. Rotate on a LabQuake tube roller at ambient temperature (+15 to +25 °C) for 3 min.
6. Pellet the sstDNA beads by centrifugation as before.
7. Remove and discard the supernatant (which contains the melted strands), and wash the (non-enriched) sstDNA beads with 1 ml of 1× Annealing Buffer. Remove all the supernatant without disturbing the pellet.
8. Resuspend once again in 1 ml of 1× Annealing Buffer to completely neutralize the Melt Solution, centrifuge as above, and remove and discard 900 μl of the supernatant without disturbing the pellet.
9. Vortex and transfer the remaining (non-enriched) sstDNA bead suspension to a 0.2-ml tube.
10. Rinse the tube that contained the beads with 100 μl of 1× Annealing Buffer, and add to the 0.2-ml tube.
11. Anneal the Sequencing Primer as described above.

22.2.3
Loading of PTP and Instrument Run

Initial Procedures

1. Bring the Sequencing Reagents Insert of the GS LR70 Sequencing Kit out of frozen storage. (Keep the Sequencing Enzymes at -15 to $-25\,^{\circ}$C.)
2. Open the barrier bag and allow the concentrated reagents to thaw for 2.5–3 h at room temperature ($+15$ to $+25\,^{\circ}$C), with the kit's Sequencing Reagents Insert kept upright and protected from bright light.
3. Place the bottle of 'Buffer CB for Bead Buffer 1' on ice.
4. When the contents of the Sequencing Reagents Insert are thawed, transfer the Insert to $+2$ to $+8\,^{\circ}$C to keep the reagents chilled until the run (not more than 8 h).

The Pre-Wash Run

Discard the Spent Reagents and Clean the Reagents Cassette

1. Open the exterior fluidics door and raise the sipper manifold completely. Slide out the Reagents Cassette. Carry the Reagents Cassette to a sink and pour in the fluids remaining in the reagent bottles.
2. Tip the Reagents Cassette into the sink to drain out the waste. Rinse the empty Reagents Cassette with warm tap water. Dry all the outside surfaces of the Reagents Cassette with a paper towel.
3. Replace all the Sipper Tubes.

Prepare the Pre-Wash Cassette

1. Prepare the pre-wash cassette by placing the Pre-wash Tube Insert (tube holder) into the cassette.
2. Place the 11 small Pre-wash Tubes on the right-hand side of the cassette and the four large tubes on the left-hand side of the cassette. Fill all the tubes with Pre-wash Buffer.
3. Slide the Reagents Cassette into the fluidics area, lower the sipper carefully, and close the exterior fluidics door.

Launch the Pre-Wash Run

1. If the Instrument Run window is not open, double-click on the *Run* icon to launch the application. At the top of the Instrument Run window are two menus: *File* and *Help*.
2. Select *Pre-wash Run*, and the Instrument Run – Pre-wash Run window will open.
3. Click Start and the pre-wash Run will be initiated and will run to completion without further intervention. (Start the preparation of the PicoTiterPlate device straight away).

PicoTiterPlate Device Preparation

Preparation of the Bead Buffers

1. Prepare Bead Buffer 2 by adding 34 μl of Apyrase solution (keep the remainder on ice) to the 200 ml of 'Buffer CB for Bead Buffer 1' (pre-chilled). Label the bottle 'Bead Buffer 2'. Swirl to mix and keep on ice.
2. Prepare two tubes of Bead Buffer 3 by mixing 930 μl of Bead Buffer 2 and 20 μl of the Bead Buffer Additive (each tube). Mix and keep on ice.

Preparation of the PicoTiterPlate and Bead Deposition Devices

1. PicoTiterPlate Buffer Equilibration
   - Pour Bead Buffer 2 into the tray until the PTP is completely submerged.
   - Leave the PTP in Bead Buffer 2 at room temperature for at least 10 min, until ready to assemble the Bead Deposition Device (BDD).

2. Assembly of the Bead Deposition Device with the PicoTiterPlate and Gasket
   - Remove the PTP from the shipping tray where it has been soaking. Wipe the back side of the PTP with a Kimwipe. Place the PTP onto the BDD Base. Make sure that the notched corners of the PTP and the BDD Base are aligned. Secure the washed and dried Bead Loading Gasket to the BDD Base.

   - Place the BDD Top over the assembled BDD Base/PTP/Gasket. Rotate the two latches from the BDD Base into their grooves in the BDD Top to firmly secure the assembly.

   - Fill each loading region with the volume of Bead Buffer 2 appropriate for the type of loading gasket which is being used:

Loading region size	PicoTiterPlate size	Volume to load (μl)
Large (30 × 60 mm)	70 × 75 mm	1860 (×2)
Medium (14 × 43 mm)	70 × 75 mm	660 (×4)
Small (2 × 53 mm)	70 × 75 mm	110 (×16)

   - Place both the assembled BDD and the BDD counterweight into centrifuge swinging baskets and place the baskets onto the rotor, opposite each other. Check that the microplate carriers are correctly positioned.
   - Centrifuge the PTP in the Bead Deposition Device, for 5 min at 1430 × g RCF (2640 rpm for the Beckman Coulter X-12 or X-15 centrifuges). Leave the Bead Buffer 2 on the PTP.

Preparation of the DNA Beads (Sample)

1. Find the number of DNA Beads required for the PicoTiterPlate size and gasket type that is being used, in the table matching the library type (below, third column).

sstDNA and Paired End Libraries

Loading region size	PicoTiter Plate size	# DNA beads per region	DNA beads (example; μl)	Control DNA beads (μl)
Large (30 × 60 mm)	70 × 75 mm	900 000 (×2)	450 (×2)	18 (×2)
Medium (14 × 43 mm)	70 × 75 mm	300 000 (×4)	150 (×4)	6 (×4)
Small (2 × 53 mm)	70 × 75 mm	50 000 (×16)	25 (×16)	1 (×16)
Small (Titration)	70 × 75 mm	24 000 (×4)	12 (×4)	1 (×4)

Amplicon Libraries

Loading region size	PicoTiter Plate size	# DNA beads per region	DNA beads (example; μl)	Control DNA beads (μl)
Large (30 × 60 mm)	70 × 75 mm	750 000 (×2)	375 (×2)	18 (×2)
Medium (14 × 43 mm)	70 × 75 mm	250 000 (×4)	125 (×4)	6 (×4)
Small (2 × 53 mm)	70 × 75 mm	40 000 (×16)	20 (×16)	1 (×16)
Small (Titration)	70 × 75 mm	24 000 (×4)	12 (×4)	1 (×4)

2. Vortex the DNA Library Beads to resuspend them, and transfer the appropriate amount of beads into 2-ml (for large or medium regions) or 0.2-ml tubes (for the small 2 × 53 mm regions of a multi-lane gasket).
3. Add the appropriate amount of Control DNA Beads to each DNA Bead tube (above, fifth column).
4. Centrifuge for 1 min at 10 000 rpm (9300 × g RCF). Rotate the tube 180° and centrifuge again for 1 min.
5. Calculate the volume of supernatant to remove: 30 μl for the large and medium size loading regions, and 10 μl for the small sizes. Draw off the appropriate volume of supernatant and discard.
6. Separately, prepare the DNA Bead Incubation Mix by combining the reagents listed below into each of *two* 1.7-ml microfuge tubes. Vortex gently. Spin briefly in a microcentrifuge.

Loading region size	PicoTiterPlate size	Bead buffer 3 (μl)	Polymerase cofactor (μl)	DNA pol (μl)	Total volume (μl)
All sizes	70 × 75 mm	785 (×2)	75 (×2)	150 (×2)	1010 (×2)

7. Transfer the appropriate volume of DNA Bead Incubation Mix as listed below, to the tubes containing the DNA Beads. Vortex well. Save the leftover DNA Bead Incubation Mix on ice; it will be used for the Packing Bead incubation in the next section.

Loading region size	PicoTiterPlate size	DNA beads (μl)	DNA bead incubation mix (μl)	Total volume (μl)
Large (30 × 60 mm)	70 × 75 mm	30 (×2)	870 (×2)	900 (×2)
Medium (14 × 43 mm)	70 × 75 mm	30 (×4)	290 (×4)	320 (×4)
Small (2 × 53 mm)	70 × 75 mm	10 (×16)	50 (×16)	60 (×16)

8. Place the samples on the laboratory rotator and incubate at room temperature (+15 to +25 °C) for 30 min.

Preparation of the Packing Beads

1. Transfer the appropriate volumes of washed Packing Beads and of DNA Bead Incubation Mix to 2.0-ml tubes as listed below:

Loading region size	PicoTiterPlate size	Packing beads (μl)	DNA bead incubation mix (μl)	Total volume (μl)
Large (30 × 60 mm)	70 × 75 mm	360 (×2)	80 (×2)	440 (×2)
Medium (14 × 43 mm)	70 × 75 mm	132 (×4)	160 (×4)	292 (×4)
Small (2 × 53 mm)	70 × 75 mm	360	860	1220

2. Place the Packing Beads in Bead Incubation Mix on the laboratory rotator and incubate them at room temperature (+15 to +25 °C).

Preparation of the Enzyme Beads

1. Begin by pelleting the Enzyme Beads using a magnetic particle collector (MPC):
   - Place the tubes of Enzyme Beads in the MPC and wait 30 s for the beads to form a pellet. Invert the MPC several times to wash off any beads that may be lodged inside the cap. Wait another 30 s for the beads to settle.
   - Remove the supernatants (including any liquid in the tube caps), being careful not to bring the pipette tip into contact with the beads. Remove the tubes from the MPC.
   - Add 1 ml of Bead Buffer 2 to each tube of Enzyme Beads. Vortex the beads to resuspend them.
2. Place the tubes in the MPC and wait 30 s for the beads to form a pellet.
3. In each of two 2.0-ml tubes, combine the amounts of Bead Buffer 2 and Enzyme Beads indicated below. Vortex the beads to produce a uniform suspension before transferring.
4. Keep the final Enzyme Bead suspension on ice, as well as the leftover Bead Buffer 2.

Deposition of the First Bead Layer (DNA Beads)

1. Preparation of the First Layer Beads
   - When the 30-min incubation of the DNA Beads is finished, remove the tubes from the laboratory rotator.
   - To the tubes of DNA Beads, add the volumes of Bead Buffer 2 indicated below. Vortex for 20 s and keep on ice. This is the first layer bead suspension.

Loading region size	PicoTiterPlate size	DNA beads (µl)	Bead buffer 2 (µl)	Total volume (µl)
Large (30 × 60 mm)	70 × 75 mm	900 (×2)	960 (×2)	1860 (×2)
Medium (14 × 43 mm)	70 × 75 mm	320 (×4)	340 (×4)	660 (×4)
Small (2 × 53 mm)	70 × 75 mm	60 (×16)	50 (×16)	110 (×16)

2. Deposition of the First Layer
   - Return to the BDD, with the wetted PTP. Using a pipettor, draw as much of the Bead Buffer 2 as possible back out (through the loading ports) and discard it, with the pipette tip.
   - Vortex the first layer bead suspension (DNA Beads), for 5 s.
   - Draw the amount of first layer beads corresponding to the loading region gasket (last column, table above) and promptly load them onto the first region of the PTP in a single, smooth dispensing action.
   - Repeat for all loading regions of the PTP.
   - Cover the loading ports and the vent holes with the BDD Port Seals provided in the kit (or MicroSeal A strips).
   - Leave the PTP on the bench top (level) for at least 10 min to allow the DNA beads to settle into the wells by gravity. *Do not centrifuge the PTP.*

Deposition of the Second Bead Layer (Packing Beads)

1. Dilution of the Packing Beads for the Second Bead Layer
   - Remove the tubes of Packing Beads from the laboratory rotator
   - When the 10-min gravity deposition of the DNA Beads is finished, gently and slowly draw out the Bead Incubation Mix supernatants from each PTP region and transfer them to a set of fresh tubes:
   - Two 2.0-ml tubes for the large regions gasket, and four 1.7-ml tubes for the medium regions gasket; skip this recovery step for the small region gasket (lanes), and discard the supernatants instead.
   - Centrifuge the collected supernatants for 10 s at 10 000 rpm (9300 × g RCF). Without disturbing the pellet, carefully remove the amount of the 'recovered supernatant' listed in the table below and add it to the Packing Beads. *Do not draw any of the bead pellet* from the bottom of the tubes containing the recovered supernatants. (For small regions, use Bead Buffer 2 instead.)

Loading region size	PicoTiterPlate size	Packing beads (µl)	Recovered supernatant (µl)	Bead buffer 2 (µl)	Total volume (µl)
Large (30 × 60 mm)	70 × 75 mm	440 (×2)	1460 (×2)	0	1900 (×2)
Medium (14 × 43 mm)	70 × 75 mm	292 (×4)	408 (×4)	0	700 (×4)
Small (2 × 53 mm)	70 × 75 mm	1220	0	680	1900

2. Deposition of the Second Layer
   - Vortex the diluted Packing Beads for 5 s to obtain a homogeneous suspension.
   - Draw the appropriate amount of diluted Packing Beads for the size of loading region being used.

Loading region size	PicoTiterPlate Size	Volume to load (μl)
Large (30 × 60 mm)	70 × 75 mm	1860 (×2)
Medium (14 × 43 mm)	70 × 75 mm	660 (×4)
Small (2 × 53 mm)	70 × 75 mm	110 (×16)

- Centrifuge the loaded PTP in the BDD for 10 min at 1430 × g RCF (2640 rpm for the X-12 or the X-15 centrifuges).

Deposition of the Third Bead Layer (Enzyme Beads)

1. After the centrifugation of the second bead layer is complete, remove the BDD from the centrifuge, and remove and discard the BDD Port Seals (or MicroSeal A strips) covering the loading ports and air vents.
2. Return to the Enzyme Bead preparation (page 463), and vortex it thoroughly to obtain a uniform suspension.
3. Draw the appropriate amount of diluted Enzyme Bead suspension for the size of loading region being used, per the table *below* (same volumes as for the first and second layers).

Loading region size	PicoTiterPlate size	Volume to load (μl)
Large (30 × 60 mm)	70 × 75 mm	1860 (×2)
Medium (14 × 43 mm)	70 × 75 mm	660 (×4)
Small (2 × 53 mm)	70 × 75 mm	110 (×16)

4. Centrifuge the BDD for 10 min at 1430 × g RCF (2640 rpm for the X-12 or the X-15 centrifuges).

The Sequencing Run

Load the Sequencing Reagents into the Instrument

1. Thaw the reagents from the Kit's 'Sequencing Enzymes' tray, and keep on ice.
2. Add 1 ml of 1 M DTT to each of the four bottles of Buffer CB. Re-cap and gently swirl the bottles to mix. Place the four bottles of supplemented buffer CB in the left-hand side of the Reagents Cassette.
3. Place the Sequencing Reagents Insert (thawed, but cold; see 'Preliminary Procedures') in the right-hand side of the Reagents Cassette.
4. Add the reagent supplements one tube at a time, as follows:

- Unscrew the cap of the 'Buffer for Apyrase' tube (yellow, position 11), and add 164 μl of Apyrase reagent (yellow cap, from the 'Sequencing Enzymes' part of the Kit). Screw the cap back tightly.
- Unscrew the cap of the 'Buffer for dATP (A)' tube (lavender, position 10), and add 1.5 ml of dATP reagent (lavender caps; two tubes from the 'Sequencing Enzymes' part of the Kit). Screw the cap back tightly. Make sure to change gloves after handling the concentrated dATP.

5. Load the Reagents Cassette into the instrument.
6. Lower the sipper carefully, and close the exterior fluidics door.

Load and Set the Run Script and Other Run Parameters

1. Return to the instrument computer. At the top of the Instrument Run window are two menus: *File* and *Help*. The *File* menu contains the following commands:
   - Select *Sequencing Run*, and the Run Wizard's 'Settings' window will open.
   - (or select *LIMS Lookup* and enter the PTP Barcode in the 'Instrument Run – Sequencing Run' window; in this case, skip the rest of the present section.
   - Do *not* click 'Start' until the PTP has been inserted in the cartridge).
2. To set up a new sequencing Run manually, complete the following:
   - In the *Scripts* field at the top of the Settings window, select the sequencing Run script for this Run. The current release of the Genome Sequencing FLX System has two sequencing Run scripts; scripts for the sequencing Runs are located inside folders as follows:
     70 × 75\TACG\100×_TACG_70 × 75 (the standard script; read-length >200 nt)
     70 × 75\TACG\42×_TACG_70 × 75 (for a quicker Run; read-length about 100 nt).
   - In the *Run Name* field, type in a specific name for this Run.
   - Under *PicoTiterPlate Barcode*, enter the barcode of the PicoTiterPlate device used in this Run (type in the field or use the barcode scanner).
3. Click on the [Next] button at the bottom of the Settings window to proceed. This opens the Run Wizard's 'Set Up Data Analysis' window.
   - In the *Choose a Layout* drop-down menu, select the number of regions in the Bead Loading Gasket that were used when preparing the PTP.
   - In the *Choose Configuration* drop-down menu, select the Data Analysis Config uration file to be used for this Run.
4. Click on the [Next] button at the bottom of the Set Up Data Analysis window to proceed.
5. The Run Name Confirmation window will open.
   - If the name is specified correctly (as designated by the User), click [Yes].
   - If the name is not correct, select [No]; the window will close and return the user to the Settings window.
6. After you click [Yes], the Requirements window will appear. If there is *not* sufficient disk space available, some files must be transferred or deleted from the disk before proceeding with the Run. Do *not* click 'Finish' yet.

Insert the PicoTiterPlate Device and Launch the Sequencing Run

1. Move to the camera door. If the PicoTiterPlate frame is open, close it (*not* the camera door).
2. Install the Cartridge Seal as described below:
   - Verify that the square ridge on the seal is facing up, and drop the seal in the cartridge groove.
   - Gently tap the seal into place with a gloved hand. *Do not* wipe the seal with anything.
3. Press the PicoTiterPlate Cartridge spring latch to lift the PicoTiterPlate frame from the cartridge.
4. Remove the PTP from the BDD, as follows:
5. Slide the PTP into the frame, making sure that the notch is on the lower right hand corner.
6. Close the PTP frame, making sure it is caught by the latch.
7. Wipe the backside of the PTP with a Kimwipe. Close the camera door.
8. Click on the [Finish] button at the bottom of the Requirements window to exit it (or [Start] in the LIMS 'Instrument Run – Sequencing Run' window). The sequencing Run will start immediately and proceed to completion without any further user intervention.
9. When the sequencing Run is complete, a small window will open on screen to inform the user. The system can be configured by a *System Administrator* to send an e-mail to the user when the Run is complete.

22.2.4
Data Analysis

Depending on the research application, the samples used and the amount of sequence data generated, there are a variety of data analysis methods that will help manage and interpret results from the GS FLX system. Three main methods are described here which include whole genome assembly, whole genome mapping and mutation detection, and amplicon based ultra-deep sequencing.

22.2.4.1 Whole Genome Assembly

A bacterial whole genome assembly typically requires sequencing the organism to 20× depth, followed by running Newbler. Newbler constructs *de novo* assemblies of the reads from one or more sequencing runs, using as input the signal flowgrams (as SFF files), and generates a set of contigs and a consensus sequence for each contig. An option allows the inclusion of paired end sequencing data into the analysis to orient the assembled contigs into scaffolds. SFF stands for Standard Flowgram File format, developed in collaboration with the Whitehead Institute and the Sanger center (http://www.ncbi.nlm.nih.gov/Traces/trace.cgi?cmd=show&f=formats&m=doc&s=formats).

As an example, the following files are generated after one run of *E. coli* sequencing:

```
<![CDATA[
1.ATG.454Reads.fna
2.ATG.454Reads.fna
1.TCA.454Reads.fna
2.TCA.454Reads.fna
1.ATG.454Reads.qual
2.ATG.454Reads.qual
1.TCA.454Reads.qual
2.TCA.454Reads.qual
454BaseCallerMetrics.txt
454QualityFilterMetrics.txt
454RuntimeMetricsAll.txt
sff/ (folder contains EC6P7LH01.sff and EC6P7LH02.sff)
]]>
```

The .fna files are sequence reads in FASTA format, and .qual files are corresponding quality scores (phred equivalent). The first numbers 1 and 2 correspond to the two loading regions on the PTP. ATG is the key sequence header for control DNA and TCA is the key for sample library. The .txt files record various run-related metrics such as nucleotide flow order, quality of the control fragment and overall signal distribution. The sff/directory contains the SFF files for each region which can be used for assembly and mapping.

To perform *de novo* assembly, the following command in the Linux environment should be used:

```
<![CDATA[
runAssembly -o Ecoli.assembly sff/*.sff
]]>
```

This will generate a result directory called `Ecoli.assembly` using all the SFF files contained in the sff/directory as input. The assembly takes about 15 min on a single 3.4-GHz Intel dual core CPU with 4 Gb RAM, and results will typically contain the following files:

```
<![CDATA[
 File name                 Description
454NewblerMetrics.txt    summary of assembly outcome
454ReadStatus.txt record of how each read was used
454LargeContigs.fna      FASTA file of contigs above 500 bp
454LargeContigs.qual     quality scores of contigs above 500 bp

454AllContigs.fna      FASTA file of all contigs
454AllContigs.qual     quality scores of all contigs
454Contigs.ace    .ace file for assembly, viewable by consed,
clview, DNAStar, Sequencher
]]>
```

22.2.4.2 **Resequencing and Mutation Detection**

Data from a whole genome shotgun sequencing can be used to map against a reference genome with highly homologous sequence, and individual mutations can be detected from comparing the consensus base-call with the reference (both homozygous and heterozygous, SNPs and indels). The tool for this analysis is the Mapper, and can be executed by the following simple command.

```
<![CDATA[
runMapping -o Ecoli.map.ref Escherichia_coli_K12 sff/*.sff
]]>
```

Here the file `Escherichia_coli_K12` is the reference genome sequence, and the result directory `Ecoli.map.ref` will contain the following list of files. Note that many of these share the same names as those in the *de novo* assembly output, but the mapper does not perform *de novo* assembly, rather it uses the reference as a guide.

```
<![CDATA[
 File name                 Description
454AlignmentInfo.tsv     tab delimited file showing base by base
coverage information for the mapping consensus
454HCDiffs.txt            list of high confidence variations
454AllDiffs.txt           list of all variations
454MappingQC.xls   summary of mapping outcome (coverage, error,
etc.)
454NewblerMetrics.txt summary of mapping outcome
454LargeContigs.fna "mapped" reference contigs above 500 bp
454LargeContigs.qual quality scores of mapped contigs
454RefStatus.txt      coverage statistics of reference genome
454AllContigs.fna all "mapped" reference contigs
454AllContigs.qual quality scores of all mapped contigs
454Contigs.ace ace file for mapping assembly
454ReadStatus.txt record of how each read was used
]]>
```

22.2.4.3 **Ultra-deep Sequencing**

Using the amplicon library procedure, targeted gene and chromosomal regions can be selectively amplified and sequenced to a very high depth. This will allow the detection of variants with an allele frequency of as low as 0.5%. Applications include sequencing tumor DNA to detect rare somatic mutations, sequencing disease-associated regions from individuals or pools for gene discovery and genotyping, and viral sequencing for drug resistance characterization. The Amplicon Variant Analyzer software is a comprehensive and user friendly package that was developed to facilitate such data analysis. Figure 22.6A shows an example of a 15-base deletion occurring at 3% frequency from about 300 aligned reads, and in Figure 22.6B, a single base substitution of A to G is detected at 6% frequency with 60 reads.

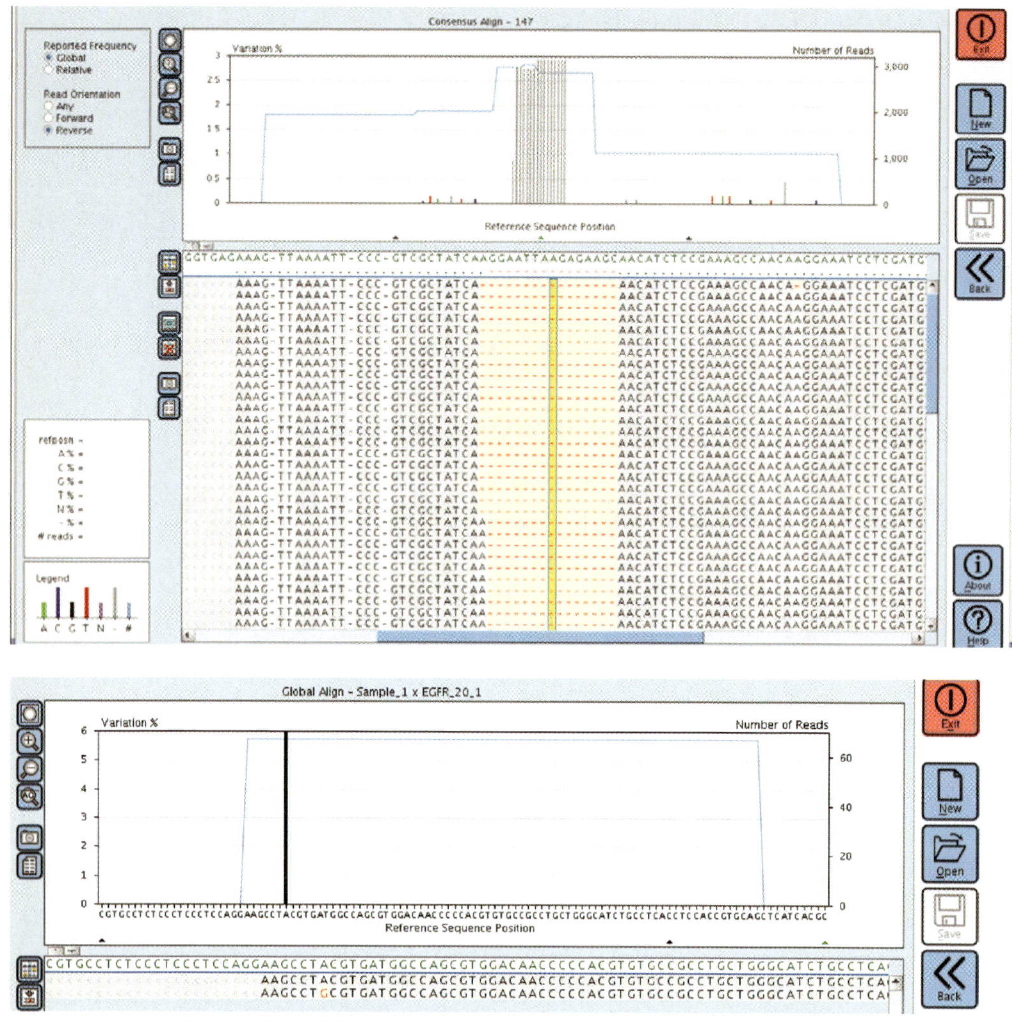

Figure 22.6 Read alignment showing variation in the Amplicon Variant Analyzer tool.

22.3
Applications of the Technology

Since the original publication of the GS20 technology [1], more than 70 peer-reviewed articles have appeared in the scientific literature that used the GS sequencing platform in a variety of research applications. These are listed in the Reference section below, grouped by research category. Here are some highlights.

Andries and colleagues published the first paper [2] describing the discovery and validation of a point mutation in the ATP synthase gene in *Mycobacterium tuberculosis*. The power of whole genome shotgun sequencing and mutation detection using

GS20 was elegantly demonstrated in this study where multiple strains of a model organism *M. smegmatis* harboring the same drug resistance phenotype were sequenced. After sequencing one sensitive and two mutant strains to a depth of 15×, four mutations were identified which were shared between the resistant strains and absent from the sensitive strain. One of them, found in the coding region of ATP synthase, was subsequently verified to confer resistance to a potent *M. tuberculosis* inhibitor R207910.

The GS system has since been rapidly adopted as a robust and cost effective method for analyzing whole bacterial genomes [3–7]. Plant geneticists took advantage of their existing tiled BAC libraries and sequenced them in pools [8–10]. Paired end sequencing is already possible on the GS system, which generates 21-bp tags that can build genome scaffolds. In the near future long paired end sequencing (>100 bp pairs) and universal ID-tag labeling will become available, which will enable filling of repetitive regions and construction of long range genomic scaffolds, as well as multiplexing of samples.

One research area that has witnessed rapid advancement in the last few years is the discovery and characterization of small RNA molecules in model organisms [11–26]. Small RNAs are known to play important roles in regulating RNA stability, protein synthesis, chromatin structure and genome organization. Henderson *et al.* [11] studied micro RNA patterns from four dicer gene mutants in *Arabidopsis thaliana* and characterized their enzymatic function in small RNA production. Girard and colleagues [12] identified a new class of small RNA called piRNA from mouse, and Lau *et al.* did the same for rat [13]. Hannon and colleagues have since published a series of articles on the study of piRNA in Drosophila and Zebrafish [22,25], as did Bartel and colleagues in the study of Arabidopsis and *C. elegans* [16,19].

In the area of environmental sequencing and molecular diversity studies, the GS system has been shown to be very effective [41–47]. Sogin and colleagues sequenced the V6 hypervariable region of ribosomal RNAs and demonstrated that bacterial diversity estimates for the diffuse flow vents of Axial Seamount and the deep water masses of the North Atlantic are much greater than any published description of marine microbial diversity at the time [42]. Other metagenomic efforts include assessing microbial diversity in deep mines [41], soil [44], the virome in multiple ocean locations [45], and the relationship between microbial communities in animal gut and obesity [46].

The sequencing of large eukaryotic genomes such as human and mouse will continue to challenge the technology as the requirement for whole genome coverage exceeds tens of gigabases of raw data. As an intermediate step toward the ultimate whole human/animal sequencing, genes and disease association regions can be selectively amplified and sequenced in the GS FLX system using the amplicon protocol [38–40]. This protocol can facilitate two types of genetic interrogation on large genomes: (a) ultra-deep sequencing and detection of rare alleles such as those found in tumors, and (b) parallel sequencing of a large pool of amplicons covering many genes of interest. As a demonstration, Thomas and colleagues published their work in the profiling of EGFR mutations in human non-small-cell lung carcinoma, and were able to detect low abundance mutations that are invisible to traditional Sanger technology [38].

Another powerful method of studying large eukaryotic genomes is to analyze the total transcriptome [31–37]. This has been shown to be very effective either by a modified form of short tags [31,34] or direct sequencing of full length cDNA [33,35,37] for the discovery of novel transcripts, splice variants, and expression profiles.

The promise of routine human sequencing cannot be underestimated as the technology continues to undergo rapid improvement. Individuals will be able to obtain their own genetic blueprint and assess risks relative to disease, behavior and interaction potential with the environment and prescription drugs. A new generation of functional genomic studies can be carried out where an organism or group of organisms can be analyzed in parallel by studying their genome, transcriptome, expression profiles, chromosomal modification and somatic mutations. The era of personalized medicine will become reality when sequencing and interpreting individual genomes becomes affordable and routine.

References

Sequencing Technology

1 Margulies, M., Egholm, M., Altman, W.E., Attiya, S., Bader, J.S., Bemben, L.A., Berka, J., Braverman, M.S., Chen, Y.J., Chen, Z., Dewell, S.B., Du, L., Fierro, J.M., Gomes, X.V., Godwin, B.C., He, W., Helgesen, S., Ho, C.H., Irzyk, G.P., Jando, S.C., Alenquer, M.L., Jarvie, T.P., Jirage, K.B., Kim, J.B., Knight, J.R., Lanza, J.R., Leamon, J.H., Lefkowitz, S.M., Lei, M., Li, J., Lohman, K.L., Lu, H., Makhijani, V.B., McDade, K.E., McKenna, M.P., Myers, E.W., Nickerson, E., Nobile, J.R., Plant, R., Puc, B.P., Ronan, M.T., Roth, G.T., Sarkis, G.J., Simons, J.F., Simpson, J.W., Srinivasan, M., Tartaro, K.R., Tomasz, A., Vogt, K.A., Volkmer, G.A., Wang, S.H., Wang, Y., Weiner, M.P., Yu, P., Begley, R.F., Rothberg, J.M. (2005) Genome sequencing in microfabricated high-density picolitre reactors. *Nature*, **437**, 376–380.

Whole Genome Sequencing

2 Andries, K., Verhasselt, P., Guillemont, J., Gohlmann, H.W., Neefs, J.M., Winkler, H., Van Gestel, J., Timmerman, P., Zhu, M., Lee, E., Williams, P., de Chaffoy, D., Huitric, E., Hoffner, S., Cambau, E., Truffot-Pernot, C., Lounis, N. and Jarlier, V. (2005) A diarylquinoline drug active on the ATP synthase of *Mycobacterium tuberculosis*. *Science*, **307**, 223–227.

3 Velicer, G.J., Raddatz, G., Keller, H., Deis, S., Lanz, C., Dinkelacker, I. and Schuster, S.C. (2006) Comprehensive mutation identification in an evolved bacterial cooperator and its cheating ancestor. *Proceedings of the National Academy of Sciences of the United States of America*, **103**, 8107–8112.

4 Goldberg, S.M., Johnson, J., Busam, D., Feldblyum, T., Ferriera, S., Friedman, R., Halpern, A., Khouri, H., Kravitz, S.A., Lauro, F.M., Li, K., Rogers, Y.H., Strausberg, R., Sutton, G., Tallon, L., Thomas, T., Venter, E., Frazier, M. and Venter, J.C. (2006) A Sanger/pyrosequencing hybrid approach for the generation of high-quality draft assemblies of marine microbial genomes. *Proceedings of the National Academy of Sciences of the United States of America*, **103**, 11240–11245.

5 Hofreuter, D., Tsai, J., Watson, R.O., Novik, V., Altman, B., Benitez, M., Clark, C., Perbost, C., Jarvie, T., Du, L. and Galán, J.E. (2006) Unique features of a highly pathogenic *Campylobacter jejuni* strain. *Infection and Immunity*, **74**, 4694–4707.

6 Jung, D.O., Kling-Backhed, H., Giannakis, M., Xu, J., Fulton, R.S., Fulton, L.A., Cordum, H.S., Wang, C., Elliott, G., Edwards, J., Mardis, E.R., Engstrand, L.G. and Gordon, J.I. (2006) The complete genome sequence of a chronic atrophic gastritis *Helicobacter pylori* strain: Evolution during disease progression. *Proceedings of the National Academy of Sciences of the United States of America*, **103**, 9999–10004.

7 Smith, M.G., Gianoulis, T.A., Pukatzki, S., Mekalanos, J.J., Ornston, L.N., Gerstein, M. and Snyder, M. (2007) New insights into *Acinetobacter baumannii* pathogenesis revealed by high-density pyrosequencing and transposon mutagenesis. *Genes & Development*, **21**, 601–614.

BACs/Plastids/Mitochondria

8 Wicker, T., Schlagenhauf, E., Graner, A., Close, T.J., Keller, B. and Stein, N. (2006) 454 sequencing put to the test using the complex genome of barley. *BMC Genomics*, **7**, 275–295.

9 Moore, M.J., Dhingra, A., Soltis, P.S., Shaw, R., Farmerie, W.G., Folta, K.M. and Soltis, D.E. (2006) Rapid and accurate pyrosequencing of angiosperm plastid genomes. *BMC Plant Biology*, **6**, 17–29.

10 Cai, Z., Penaflor, C., Kuehl, J.V., Leebens-Mack, J., Carlson, J.E., dePamphilis, C.W., Boore, J.L. and Jansen, R.K. (2006) Complete plastid genome sequences of Drimys, Liriodendron, and Piper: implications for the phylogenetic relationships of magnoliids. *BMC Evolutionary Biology*, **6**, 77–96.

Small RNA

11 Henderson, I.R., Zhang, X., Lu, C., Johnson, L., Meyers, B.C., Green, P.J. and Jacobsen, S.E. (2006) Dissecting *Arabidopsis thaliana* DICER function in small RNA processing, gene silencing and DNA methylation patterning. *Nature Genetics*, **38**, 721–725.

12 Girard, A., Sachidanandam, R., Hannon, G.J. and Carmell, M.A. (2006) A germline-specific class of small RNAs binds mammalian Piwi proteins. *Nature*, **442**, 199–202.

13 Lau, N.C., Seto, A.G., Kim, J., Kuramochi-Miyagawa, S., Nakano, T., Bartel, D.P. and Kingston, R.E. (2006) Characterization of the piRNA complex from rat testes. *Science*, **313**, 363–367.

14 Lu, C., Kulkarni, K., Souret, F.F., MuthuValliappan, R., Tej, S.S., Poethig, R.S., Henderson, I.R., Jacobsen, S.E., Wang, W., Green, P.J. and Meyers, B.C. (2006) MicroRNAs and other small RNAs enriched in the Arabidopsis RNA-dependent RNA polymerase-2 mutant. *Genome Research*, **16**, 1276–1288.

15 Qi, Y., He, X., Wang, X.J., Kohany, O., Jurka, J. and Hannon, G.J. (2006) Distinct catalytic and non-catalytic roles of ARGONAUTE4 in RNA-directed DNA methylation. *Nature*, **443**, 1008–1012.

16 Axtell, M.J., Jan, C., Rajagopalan, R. and Bartel, D.P. (2006) A two-hit trigger for siRNA biogenesis in plants. *Cell*, **127**, 565–577.

17 Berezikov, E., Thuemmler, F., van Laake, L.W., Kondova, I., Bontrop, R., Cuppen, E. and Plasterk, R.H.A. (2006) Diversity of microRNAs in human and chimpanzee brain. *Nature Genetics*, **38**, 1375–1377.

18 Pak, J. and Fire, A. (2007) Distinct populations of primary and secondary effectors during RNAi in *C. elegans*. *Science*, **315**, 241–244.

19 Ruby, J.G., Jan, C., Player, C., Axtell, M.J., Lee, W., Nusbaum, C., Ge, H. and Bartel, D.P. (2006) Large-scale sequencing reveals

21U-RNAs and additional MicroRNAs and endogenous siRNAs in *C. elegans*. *Cell*, **127**, 1193–1207.

20 Rajagopalan, R., Vaucheret, H., Trejo, J. and Bartel, D.P. (2006) A diverse and evolutionarily fluid set of microRNAs in *Arabidopsis thaliana*. *Genes & Development*, **20**, 3407–3425.

21 Fahlgren, N., Howell, M.D., Kasschau, K.D., Chapman, E.J., Sullivan, C.M., Cumbie, J.S., Givan, S.A., Law, T.F., Grant, S.R., Dangl, J.L. and Carrington, J.C. (2007) High-throughput sequencing of arabidopsis microRNAs: evidence for frequent birth and death of MIRNA genes. *PLoS One*, **2**, e219–e232.

22 Brennecke, J., Aravin, A.A., Stark, A., Dus, M., Kellis, M., Sachidanandam, R. and Hannon, G.J. (2007) Discrete small RNA-generating loci as master regulators of transposon activity in Drosophila. *Cell*, **128**, 1089–1103.

23 Zhang, X., Henderson, I.R., Lu, C., Green, P.J. and Jacobsen, S.E. (2007) Role of RNA polymerase IV in plant small RNA metabolism. *Proceedings of the National Academy of Sciences of the United States of America*, **104**, 4536–4541.

24 Howell, M.D., Fahlgren, N., Chapman, E.J., Cumbie, J.S., Sullivan, C.M., Givan, S.A., Kasschau, K.D. and Carrington, J.C. (2007) Genome-wide analysis of the RNA-DEPENDENT RNA POLYMERASE6/DICER-LIKE4 pathway in arabidopsis reveals dependency on miRNA- and tasiRNA-directed targeting. *Plant Cell*, **19**, 926–942.

25 Houwing, S., Kamminga, L.M., Berezikov, E., Cronembold, D., Girard, A., van den Elst, H., Filippov, D.V., Blaser, H., Raz, E., Moens, C.B., Plasterk, R.H.A., Hannon, G.J., Draper, B.W. and Ketting, R.F. (2007) A role for Piwi and piRNAs in germ cell maintenance and transposon silencing in Zebrafish. *Cell*, **129**, 69–82.

26 Aravin, A.A., Sachidanandam, R., Girard, A., Fejes-Toth, K. and Hannon, G.J. (2007) Developmentally regulated piRNA clusters implicate MILI in transposon control. *Science*, **316**, 744–747.

Chromosome Structure

27 Dostie, J., Richmond, T.A., Arnaout, R.A., Selzer, R.R., Lee, W.L., Honan, T.A., Rubio, E.D., Krumm, A., Lamb, J., Nusbaum, C., Green, R.D. and Dekker, J. (2006) Chromosome Conformation Capture Carbon Copy (5C): a massively parallel solution for mapping interactions between genomic elements. *Genome Research*, **16**, 1299–1309.

28 Johnson, S.M., Tan, F.J., McCullough, H.L., Riordan, D.P. and Fire, A.Z. (2006) Flexibility and constraint in the nucleosome core landscape of *Caenorhabditis elegans* chromatin. *Genome Research*, **16**, 1505–1516.

29 Albert, I., Mavrich, T.N., Tomsho, L.P., Qi, J., Zanton, S.J., Schuster, S.C. and Pugh, B.F. (2007) Translational and rotational settings of H2A.Z nucleosomes across the *Saccharomyces cerevisiae* genome. *Nature*, **446**, 572–576.

30 Nagel, S., Scherr, M., Kel, A., Hornischer, K., Crawford, G.E., Kaufmann, M., Meyer, C., Drexler, H.G. and MacLeod, R.A.F. (2007) Activation of TLX3 and NKX 2-5in t (5;14)(q35;q32) T-cell acute lymphoblastic leukemia by remote 3′-BCL11B enhancers and coregulation PU.1 and HMGA1. *Cancer Research*, **67**, 1461–1471.

Transcriptomes

31 Ng, P., Tan, J.J., Ooi, H.S., Lee, Y.L., Chiu, K.P., Fullwood, M.J., Srinivasan, K.G., Perbost, C., Du, L., Sung, W.K., Wei, C.L. and Ruan, Y. (2006) Multiplex sequencing of paired-end ditags (MS-PET): a strategy for the ultra-high-throughput analysis of transcriptomes and genomes. *Nucleic Acids Research*, **34**, e84–e103.

32 Gowda, M., Li, H., Alessi, J., Chen, F., Pratt, R. and Wang, G.L. (2006) Robust analysis of 5′-transcript ends (5′-RATE): a novel technique for transcriptome analysis and genome annotation. *Nucleic Acids Research*, **34**, e126.

33 Bainbridge, M.N., Warren, R.L., Hirst, M., Romanuik, T., Zeng, T., Go, A., Delaney, A., Griffith, M., Hickenbotham, M., Magrini, V., Mardis, E.R., Sadar, M.D., Siddiqui, A.S., Marra, M.A. and Jones, S.J. (2006) Analysis of the prostate cancer cell line LNCaP transcriptome using a sequencing-by-synthesis approach. *BMC Genomics*, **7**, 246–256.

34 Nielsen, K.L. Høgh, A.L. and Emmersen, J. (2006) DeepSAGE – digital transcriptomics with high sensitivity, simple experimental protocol and multiplexing of samples. *Nucleic Acids Research*, **34**, e133.

35 Cheung, F., Haas, B.J., Goldberg, S.M., May, G.D., Xiao, Y. and Town, C.D. (2006) Sequencing *Medicago truncatula* expressed sequenced tags using 454 Life Sciences technology. *BMC Genomics*, **7**, 272–281.

36 Emrich, S.J., Barbazuk, W.B., Li, L. and Schnable, P.S. (2006) Gene discovery and annotation using LCM-454 transcriptome sequencing. *Genome Research*, **17**, 69–73.

37 Weber, A.P., Weber, K.L., Carr, K., Wilkerson, C. and Ohlrogge, J.B. (2007) Sampling the arabidopsis transcriptome with massively-parallel pyrosequencing. *Plant Physiology*, **144**, 32–42.

Amplicons

38 Thomas, R.K., Nickerson, E., Simons, J.F., Jänne, P.A., Tengs, T., Yuza, Y., Garraway, L.A., LaFramboise, T., Lee, J.C., Shah, K., O'Neill, K., Sasaki, H., Lindeman, N., Wong, K.K., Borras, A.M., Gutmann, E.J., Dragnev, K.H., DeBiasi, R., Chen, T.H., Glatt, K.A., Greulich, H., Desany, B., Lubeski, C.K., Brockman, W., Alvarez, P., Hutchison, S.K., Leamon, J.H., Ronan, M.T., Turenchalk, G.S., Egholm, M., Sellers, W.R., Rothberg, J.M. and Meyerson, M. (2006) Sensitive mutation detection in heterogeneous cancer specimens by massively parallel picoliter reactor sequencing. *Nature Medicine*, **12**, 852–855.

39 Binladen, J., Gilbert, M.T., Bollback, J.P., Panitz, F., Bendixen, C., Nielsen, R. and Willerslev, E. (2007) The use of coded PCR primers enables high-throughput sequencing of multiple homolog amplification products by 454 parallel sequencing. *PLoS One*, **2**, e197.

40 Dahl, F., Stenberg, J., Fredriksson, S., Welch, K., Zhang, M., Nilsson, M., Bicknell, D., Bodmer, W.F., Davis, R.W. and Ji, H. (2007) Multigene amplification and massively parallel sequencing for cancer mutation discovery. *Proceedings of the National Academy of Sciences of the United States of America*, **104**, 9387–9392.

Metagenomics and Microbial Diversity

41 Edwards, R.A., Rodriguez-Brito, B., Wegley, L., Haynes, M., Breitbart, M., Peterson, D.M., Saar, M.O., Alexander, S., Alexander, E.C. and Rohwer, F. (2006) Using pyrosequencing to shed light on deep mine microbial ecology. *BMC Genomics*, **7**, 57.

42 Sogin, M.L., Morrison, H.G., Huber, J.A., Welch, D.M., Huse, S.M., Neal, P.R., Arrieta, J.M. and Herndl, G.J. (2006) Microbial diversity in the deep sea and the underexplored "rare biosphere". *Proceedings of the National Academy of Sciences of the United States of America*, **103**, 12115–12120.

43 Krause, L., Diaz, N.N., Bartels, D., Edwards, R.A., Puhler, A., Rohwer, F., Meyer, F. and Stoye, J. (2006) Finding novel genes in bacterial communities isolated from the environment. *Bioinformatics*, **22**, e281–e289.

44 Leininger, S., Urich, T., Schloter, M., Schwark, L., Qi, J., Nicol, G.W., Prosser, J.I., Schuster, S.C. and Schleper, C. (2006) Archaea predominate among ammonia-oxidizing prokaryotes in soils. *Nature*, **442**, 806–809.

45 Angly, F.E., Felts, B., Breitbart, M., Salamon, P., Edwards, R.A., Carlson, C., Chan, A.M., Haynes, M., Kelley, S., Liu, H., Mahaffy, J.M., Mueller, J.E., Nulton, J., Olson, R., Parsons, R., Rayhawk, S., Suttle, C.A. and Rohwer, F. (2006) The marine viromes of four oceanic regions. *PLoS Biology*, **4**, e368.

46 Turnbaugh, P.J., Ley, R.E., Mahowald, M.A., Magrini, V., Mardis, E.R. and Gordon, J.I. (2006) An obesity-associated gut microbiome with increased capacity for energy harvest. *Nature*, **444**, 1027–1031.

47 Huson, D.H., Auch, A.F., Qi, J. and Schuster, S.C. (2007) MEGAN analysis of metagenomic data. *Genome Research*, **17**, 377–386.

Ancient DNA

48 Poinar, H.N., Schwarz, C., Qi, J., Shapiro, B., Macphee, R.D., Buigues, B., Tikhonov, A., Huson, D.H., Tomsho, L.P., Auch, A., Rampp, M., Miller, W. and Schuster, S.C. (2006) Metagenomics to paleogenomics: large-scale sequencing of mammoth DNA. *Science*, **311**, 392–394.

49 Gilbert, M.T., Binladen, J., Miller, W., Wiuf, C., Willerslev, E., Poinar, H., Carlson, J.E., Leebens-Mack, J.H. and Schuster, S.C. (2006) Recharacterization of ancient DNA miscoding lesions: insights in the era of sequencing-by-synthesis. *Nucleic Acids Research*, **35**, 1–10.

50 Stiller, M., Green, R.E., Ronan, M., Simons, J.F., Du, L., He, W., Egholm, M., Rothberg, J.M., Keates, S.G., Ovodov, N.D., Antipina, E.E., Baryshnikov, G.F., Kuzmin, Y.V., Vasilevski, A.A., Wuenschell, G.E., Termini, J., Hofreiter, M., Jaenicke-Despres, V. and Pääbo, S. (2006) Patterns of nucleotide misincorporations during enzymatic amplification and direct large-scale sequencing of ancient DNA. *Proceedings of the National Academy of Sciences of the United States of America*, **103**, 13578–13584.

51 Green, R.E., Krause, J., Ptak, S.E., Briggs, A.W., Ronan, M.T., Simons, J.F., Du, L., Egholm, M., Rothberg, J.M., Paunovic, M. and Pääbo, S. (2006) Analysis of one million base pairs of Neanderthal DNA. *Nature*, **444**, 330–336.

Glossary

This glossary lists a number of relevant terms for the reader, who is not so familiar with the topics of the present *Handbook of Plant Functional Genomics* (*Concepts and Protocols*)

Aberrant RNA (abRNA) A hypothetical RNA molecule, produced directly from a transgene, being double-stranded ('aberrant') and serving as template for the synthesis of short complementary RNA molecules (cRNA) by specialized cellular *R*NA-*d*ependent *R*NA *p*olymerases (RdRPs). These cRNAs in turn could pair with transgene messenger RNAs to form double-stranded RNAs (e.g. catalyzed by cellular RNA-dependent RNA polymerases), the substrates for mRNA degradation.

Abundance The average number of molecules of a specific mRNA or a specific protein in a given cell at a given time.

Accession number A unique identification code for each sequence deposited in the databanks (e.g. GenBank). This number can be used to search for example, GenBank records for a specific sequence.

Activation domain (AD; activating domain; C-terminal activation domain, CTAD, transcriptional activation domain) A specific 30–100 amino acid domain of transcription factors, located at the C-terminus and rich in acidic amino acids, that can form amphipathic α-helical structures and is necessary for the transcriptional activation of the target gene. For example, the yeast transcription factor GAL4 harbors such an activation domain, which can be discriminated into two regions (I, residues 148–236; II, residues 768–881), either of which activates transcription when fused to the DNA-binding domain (residues 1–147). The activity of region I is directly proportional to its content of acidic residues. Principally, three different features of ADs can be discriminated: an acidic, negatively charged domain (e.g. in GAL4, GCN4), a glutamine-rich domain (e.g. in HAP1, HAP2, GAL11, OCT-1, OCT-2, Jun, AP-2, SRF, Sp1), and a proline-rich domain (e.g. in CTF/NF-1, AP-2, Jun, OCT-2, SRF). All these regions establish contacts with other proteins.

Active chromatin Any, mostly euchromatic region of the nucleus, that supports transcription of the underlying genes.

The Handbook of Plant Functional Genomics: Concepts and Protocols.
Edited by Günter Kahl and Khalid Meksem

ISBN: 978-3-527-31885-8

Agarose gel An inert, macroporous and nontoxic polysaccharide matrix for the electrophoretic separation of RNA or DNA molecules according to their size and conformation.

Allele-specific expression The transcription of only one allele, or the transcription of both alleles of a genetic locus to different extents. Allele-specific expression may be detected by for example, allele-specific amplification or allele-specific polymerase chain reaction.

Allele-specific oligonucleotide (ASO) probe A synthetic, approximately 20 nucleotides long oligodeoxynucleotide designed to locate single base mismatches in complex genomes, and to discriminate between two alleles. Such probes are long enough to detect unique sequences in the genome, but sufficiently short to be destablilized by a single internal mismatch during their hybridization to a target sequence. The technique involves the immobilization of target DNA, hybridization with oligonucleotide probes, and finally washing under carefully controlled conditions, which allows sequences with one single nucleotide mismatch to be discriminated from their wild-type genomic counterparts on the basis of different hybridization behavior.

All-exon array A microarray, onto which synthetic oligonucleotides are immobilized, that span all exons of a genome. Usually the number of probes for each exon is reduced to four 70mers, and no mismatch oligonucleotides are spotted as controls. The individual spot architecture is reduced to 8 μm diameter, which allows several millions of exons to be accommodated on a single chip. All-exon arrays are used for the detection of all individual exons of a genome independently of one another and for the identification of alternatively spliced transcripts.

Alternative polyadenylation A variant of the conventional post-transcriptional polyadenylation process of eukaryotic messenger RNAs (mRNAs), in which the 3′-end processing of the message either starts earlier in the transcript, or poly(A) tails of various lengths are added to the 3′terminus of heterogeneous nuclear RNA or messenger RNA transcribed from the same gene (caused by the use of different poly(A) addition signals by poly(A)-polymerase. The former leads to a shortening of the original mRNA, and to a protein with a lower molecular weight than its wild-type double.

Alternative splicing (AS, alternative RNA splicing, alternate splicing, differential splicing, alternative pre-messenger RNA splicing) The unconventional ligation of exons of a particular pre-messenger RNA to form a functional messenger RNA (mRNA), that differs in information content from the normal message. Alternative splicing produces either longer or shorter mRNA variants, if compared to the wild-type mRNA, and consequently the encoded protein contains more or less functional domains. For example, through exon skipping an in-frame deletion is introduced into the resulting protein, intron retention produces an alternative C-terminus of the protein, and alternative 3′- or 5′- sites lead to an out-of-frame deletion or an alternative initiation, respectively, in the protein. Alternative splicing is regulated by the balance between splicing factors. For example, the SR protein SF2/ASF defines the 5′splice site by an interaction with the SR domain of U1 small nuclear ribonucleoprotein. SF2/ASF preferentially induces the use of a proximal splice site. Inversely, *h*eterogeneous *n*uclear *r*ibo*n*ucleo*p*rotein (hnRNP) A1 promotes use of a distal splice site. Therefore, the relative concentrations of both SF2/ASF and hnRNP A1 determine whether a proximal or distal splice site is cut. Alternative splicing expands the information content of a single gene, so that different domains can be shuffled to create novel proteins.

Anchored oligo(dT) primer A synthetic homopolymeric oligodeoxynucleotide consisting of a string of deoxythymidylic acid residues followed by dV (dG, dA, or dC), and then dN (dA, dT, dG, or dC), that can be annealed to the 5′-end of the poly(A) tail of polyadenylated mRNA to prevent priming from within the poly(A)-tail. Anchored oligo(dT) primers are used as primers in cDNA labeling and *r*everse *t*ranscriptase *p*olymerase *c*hain *r*eaction (RT-PCR) protocols.

Antisense RNA (aRNA; asRNA; complementary RNA, cRNA) An RNA whose sequence is antiparallel to the corresponding sense RNA. If such antisense RNA is produced in the nucleus (e.g. by the transfer and expression of an antisense gene), it will interfere with the normal expression of the sense gene (i.e. binds to the sense mRNA and blocks its translation). Different mechanisms of interference, either solely or in combination, are possible.

(1) Antisense RNA binds to sequences in the major groove of duplex DNA to form triple helix structures that interfere with the binding of DNA-affine proteins (e.g. transcription factors).
(2) Antisense RNA can form double-stranded RNA molecules with its sense messenger RNA, which cannot be processed and/or exported to the cytoplasm (at least in mammals), will be preferentially degraded (as in *Drosophila*), or arrest translation by blocking ribosomal binding sites. Usually, antisense RNAs contain one to three stem-loop structures. The corresponding sense RNAs are frequently longer, and possess the complementary stem-loops and additional structures. The loops determine the specificity of pairing between antisense and sense RNAs, and the stems are responsible for the stability of the antisense RNAs.
(3) Antisense RNA can enter transcription bubbles (where single-stranded DNA is available for RNA polymerase), bind to their cognate sequences and reduce processivity of the transcriptional complex.

Autofluorescent protein (AFP) Any protein carrying an autofluorescent domain which can be excited by light and emits fluorescent light of longer wavelengths. Autofluorescent proteins are widely distributed among reef corals (e.g. anthozoa species as *Anemone majano, Zoanthus* sp., *Discosoma striata* and other *Discosoma* species, and *Clavularia* sp.). Many of them share protein sequence homology with the green fluorescent protein. Such proteins or their genes are increasingly being used as reporter molecules.

BAC DNA microarray (BAC microarray, BAC clone array, 'BAC array') The ordered alignment of different bacterial artificial chromosome (BAC) clones, immobilized on supports of minute dimensions (e.g. nylon membranes, silicon, glass or quartz chips). Each colony harbors DNA fragments of 100–150 kb. Such microarrays are used to isolate genomic DNA that contains a gene or genes of interest, detected by hybridization of radiolabeled or fluorescent gene probes to the microarray. BAC DNA microarrays may contain clones that represent the whole genome of an organism, or each of its chromosomes for comparative genomic hybridization. Such microarrays facilitate the detection and characterization of chromosomal abnormalities.

Bacteriophage promoter Any promoter located on a bacteriophage (e.g. phage SP6, T7 or T3) genome, that consists of only 23 base pairs, numbered −17 to +6 (+1: transcription start site) and is used in *in vitro* transcription of specific target genes and the generation of large amounts of RNA (amplified RNA). The +1 base is guanine and is the first base incorporated into RNA during transcription.

Bacteriophage promoters:

−17 +1
SP6 promoter: 5′-ATTTAGGTGACACTATA**G**AAGNG-3′ (N: any nucleotide)

−17 +1
T7 promoter: 5′-TAATACGACTCACTATA**G**GGAGA-3′

−17 +1
T3 promoter: 5′-AATTAACCCTCACTAAA**G**GGAGA-3′

Basal expression The (usually) very low level of transcription of a gene and the translation of the resulting messenger RNA into a protein, that occurs constantly in a cell nucleus in the absence of exogenous or intrinsic activators.

Biallelic expression The unequal transcription of both alleles of a single genetic locus. For example, many genes of genetically improved modern hybrids of corn (*Zea mays* L.) express both alleles, in contrast to the preferentially monoallelic expression of several genes in old maize varieties. Frequently the various alleles respond differently to for example, abiotic stresses, or a response to different environments may be either mono- or biallelic. Examples are the *l*ipid *t*ransfer *p*rotein (LTP)-encoding gene and the *a*uxin *r*epressed *d*ormancy *a*ssociated protein (ARDA) gene of *Zea mays*, corn. Biallelic expression may be detected by for example, allele-specific amplification or allele-specific polymerase chain reaction.

Bidirectional transcription The simultaneous (or also asynchronous) transcription of the sense and antisense strand of a particular gene, resulting in both a sense and an antisense transcript. Two slightly different bidirectional transcription modes exist in the human genome: the so-called single bidirectional transcription leads to a single transcript each from the sense and the antisense strand, whereas multiple bidirectional transcription produces one sense transcript and two smaller antisense transcripts.

Bimorphic transcript Any transcript, that is initially polyadenylated, but processed to reduce or totally remove the 3′-poly(A) tail under specific environmental conditions.

Box A laboratory slang term for a DNA consensus sequence, or element. The AGGA-, CAAT-, G-, GC-, homeo-, I-, Pribnow-, TACTAAC- and TATA-box are examples of such boxes.

Cap analysis of gene expression (CAGE) A technique for the high-throughput identification of sequence tags representing 5′-ends of messenger RNAs at the cap sites, the identification of *t*ranscription *s*tart *s*ites (TSSs) and the isolation of promoters. More details are presented in this book.

Capillary array electrophoresis (CAE) A technique for the sequencing of many DNA samples in parallel, which uses arrays of capillaries filled with polyacrylamide (or other carrier material) for the electrophoresis and separation of the sequencing products. Normally 96 such capillaries are arranged in an array, but the number can be increased up to 384 capillaries. CAE is therefore a high-throughput technique that circumvents the production of polyacrylamide gels, allows automated probe application, reduces electrophoresis time, and has the capacity to sequence about 700 bases per sample in less than 2 h.

Capping The post-transcriptional addition of a cap at the 5′-terminus of eukaryotic mRNA molecules, that is, a post-transcriptional modification reaction.

Cap trapper technique (biotinylated cap trapper technique, cap trapping) A method for the isolation of full-length messenger RNAs (mRNAs), that capitalizes on the covalent coupling of a biotin moiety to the cap at the 5′-end of the mRNA. In short, in a first step the diol groups of the ribose in the cap structure are oxidized by $NaIO_4$, which opens the pentose ring. The second step aims at derivatizing the oxidized diol groups by biotin hydrazide, a reaction which is completed in an overnight run and at room temperature. As a result, the biotin molecule is now covalently linked to the cap of the mRNA, and can in turn be bound by streptavidin-coated beads (e.g. porous glass beads). The cap trapper technique therefore captures mRNAs with an intact 5′-end, that can be released from the beads and completely synthesized to a full-length cDNA using a 3′-oligo(dT) primer (capturing the poly[A] tail of the message) and reverse transcriptase.

cDNA (complementary DNA, copy DNA) A single- or double-stranded DNA molecule that is complementary to an RNA (usually mRNA) template from which it has been copied by RNA-dependent DNA polymerase (reverse transcriptase, RTase). The synthesis of a single-stranded cDNA is the first step in cDNA cloning procedures. A second strand may be synthesized using a DNA polymerase after removal of the RNA either through RNase H or alkaline hydrolysis (*d*ouble-*s*tranded cDNA, dscDNA). cDNAs can be used as hybridization probes for the isolation of full-length genes from genomic libraries, or as probes to be spotted onto chips.

cDNA-AFLP (cDNA amplification fragment length polymorphism; restriction-mediated differential display, RMDD) A technique to monitor the steady-state levels of a large number of messenger RNAs in a cell, tissue, organ, or organism. In short, mRNAs are isolated, reverse-transcribed into double-stranded cDNAs, the cDNA duplexes are first restricted with a rare cutter restriction enzyme (e.g. *Bst* Y 1), then with a frequent cutter (e.g. *Mse* I), and the restriction fragments ligated to *Bst* Y 1 and *Mse* I adaptors for selective amplification in an AFLP procedure. This technique produces patterns resembling the complex AFLP patterns produced with genomic DNA. More details are presented in this book.

cDNA clone A DNA duplex molecule complementary to an mRNA molecule, generated by the reverse transcription of the message using retroviral *r*everse *t*ranscript*ase* (RTase), and cloned into an appropriate cloning vector (e.g. a plasmid).

cDNA expression array (cDNA array; gene expression array; gene *e*xpression *m*icro-array, GEM, transcript array, mRNA expression array, RNA expression microarray, REM) The ordered alignment of different complementary DNAs (cDNAs), or fragments of cDNAs, or cDNA-complementary oligonucleotides immobilized on a support (e.g. a nylon-based membrane). Such arrays may contain tens of thousands of different cDNAs on a small space (e.g. 1 Ã –1 cm, or less), and are used to determine differential gene expression patterns. cDNA arrays can be produced by different techniques. One particular method uses PCR-amplified partial sequences of cDNAs. In short, reverse transcriptase PCR primers are designed from known cDNA sequences and used to amplify the corresponding cDNAs such that the amplification products are 200–600 bp in length (optimal for hybridization). These amplicons are cloned and (partially) sequenced. The cloned cDNA fragments are again amplified, normalized (adjusted to the same concentration, e.g. 10 ng) and immobilized on positively-charged nylon membranes. The sequence homologies among the different cDNA amplicons are kept at a minimum. Usually two identical cDNA fragments are spotted side by side, and cDNA classes are arranged according to functional relationships (e.g. cDNAs of genes encoding glycolytic enzymes, or genes involved in tumorigenesis as apoptosis genes, oncogenes, tumor suppressor and cell cycle regulator genes).

The hybridization probes are derived from total RNA or polyadenylated RNA of different specimen (e.g. different organisms, tissues, or cells), reverse transcribed and labeled using oligo(dT), random or gene-specific primers, and hybridized to the arrays. The hybridization patterns can then be detected by autoradiography and/or phosphorimaging. The visual expression profiles allow large-scale up- or down-regulation of functionally related genes or gene classes to be detected.

cDNA library (cDNA bank) A collection of cloned DNA sequences derived from reverse transcription of all mRNAs of a cell and thus representing the active genes in that cell. Vectors used for cDNA libraries are for example lambda ORF 8 and lambda ZAP.

Cell-specific gene expression The transcription of specific genes and the translation of the resulting messenger RNAs (mRNAs) into proteins in a particular cell at a particular time, producing specific patterns of mRNAs and proteins. Cell-specific gene expression is best explored by laser-capture microdissection of single cells, extracting their mRNAs on a microscale, reverse transcribing them into cDNAs, labeling the cDNAs and probing them on for example, microarrays.

Cell-specific splicing The splicing of a specific set of exons of a particular gene in one cell type, and the splicing of another set of exons of the same gene in another cell type. For example, the gene encoding the peptide hormone calcitonin (inhibits the mobilization of calcium from the bones at high Ca^{2+} ion concentrations in the serum) consists of six exons. In the parathyroid gland, exons 1–4 are spliced, and exons 5–6 are skipped. The product is calcitonin. However, in nerve cells, exon 4 is skipped and the remaining exons are spliced (exons 1-2-3-5-6). The product is the neuropeptide *c*alcitonin *g*ene *r*elated *p*rotein (CGRP). Cell-specific splicing is therefore a mechanism to increase the information content of a particular gene and to exploit new information to identify new functions.

Chimeric transcript Any messenger RNA that contains sequences from two different chromosomal loci, and most probably represents an artifact of cDNA library construction, or results from chromosomal rearrangements (e.g. translocation).

ChIP chip technique ('ChIP on chip' technique) A technique for the identification of chromosomal DNA binding sites for proteins (e.g. transcription factors), for the determination of areas of active transcription in chromatin, and the study of histone modifications (histone code). This technique is based on (1) the chromatin immunoprecipitation technology and (2) the hybridization of identified protein-binding sequences to a DNA microarray containing genomic DNA. In short, DNA-binding proteins are first chemically cross-linked to their binding sites within the nuclear DNA *in situ* (by e.g. formaldehyde), the cells subsequently lysed, the chromatin fragmented into smaller DNA–protein complexes by ultrasound, and the protein–DNA complexes selectively enriched by immunoprecipitation using a monoclonal antibody raised against the protein of interest (e.g. a transcription factor). The crosslinks are then destroyed, and the enriched pools of DNA fragments amplified by ligation-mediated polymerase chain reaction, labeled with for example, a fluorochrome, and hybridized to a DNA microarray, onto which a series of for example, chromosomal fragments or a series of oligonucleotides (50- or 60-mers) spanning these fragments (or a whole genome) are immobilized. Hybridization allows the DNA-binding site to localize to a chromosomal region.

Chromatin immunoprecipitation (CHIP, ChIP) A technique for the localization of specific proteins or their modified forms in chromatin. In short, chromatin is isolated, fragmented by micrococcal nuclease (fragments sizes: mononucleosomes) or sonication (fragment sizes: $\sim$500 bp), and the resulting nucleosomes (input chromatin) mixed with an antibody raised against the protein in question (e.g. an acetylated histone). Immuno-conjugates are then immobilized on protein A agarose beads. The non-bound (=non-acetylated) nucleosomes are washed off, and the proteins and DNA from the antibody-bound, unbound, and input chromatins comparatively analyzed. Alternatively, formaldehyde can be used to covalently cross-link proteins to DNA *in vivo* (formaldehyde reacts with lysine and arginine side chains of proteins and the purine and pyrimidine moieties of DNA). The DNA is then sheared into small fragments, antibodies against target proteins used to purify cross-linked DNA, the target sequences amplified by PCR and sequenced.

Chromatin immunoprecipitation paired-end ditag (ChIP-PET) technique A method for scanning whole genomes for *cis*-regulatory elements, especially transcription factor binding sites, and histone modifications, that combines *ch*romatin *i*mmuno*p*recipitation (ChIP) with the *p*aired-*e*nd di*t*ag (PET) technologies. In short, a monoclonal antibody raised against a specific transcription factor (or modified histone) is used to precipitate the corresponding chromatin fragment in a conventional *ch*romatin *i*mmuno*p*recipitation experiment. End-polished ChIP DNA fragments are then ligated to a special cloning vector containing two *Mme*I recognition sites. The ligation mixture is transformed into target cells ('ChIP DNA library'). Plasmids from this library are digested with *Mme*I and end-polished with T4 DNA polymerase. The resulting vector containing a signature tag from each end of the ChIP DNA insert is self-ligated and then transformed into target cells ('single-PET library'). The plasmids from this library in turn are restricted with *Bam*HI to release 50-bp paired-end tags that are concatenated into 1–2-kb fragments, cloned into an appropriate vector ('final ChIP-PET library') and sequenced (10–15 PETs per sequence read). PET sequences from the raw sequence reads from the ChIP-PET library can be mapped to an already sequenced genome, and will define the boundaries of the cloned ChIP fragments and thereby mark the positions of specific transcription factor binding sites.

Chromosome expression map A graphical description of the location of expressed genes together with their relative transcription frequencies along linearized chromosomal DNA. Abundance levels of transcripts are displayed on a vertical axis with transcription from the (+)-strand above and transcription from the (−)-strand below the chromosomal DNA.

***cis* natural antisense transcript (cis-NAT)** Any *n*atural *a*ntisense *t*ranscript (NAT) with a sequence completely or only partially complementary to another endogenous RNA (e.g. messenger RNA, mRNA), that forms a sense–antisense complex with this RNA, and is transcribed from the same genomic locus (*in cis*). Most of the *cis*-NAT pairs overlap at their 3′ termini (tail-to-tail arrangement), others pair at their 5′-ends (head-to-head arrangement), while still others are composed of one transcript starting within an intron of the second transcript, or overlap completely. Many *cis*-NAT mRNAs encode proteins for DNA repair. *cis*-NATs are identified by searching for SA gene pairs represented by transcripts (mRNAs or ESTs) in opposite directions at the same genomic locus with a 20-nt overlap in exonic regions.

Coding strand (sense strand; +strand) The strand of a DNA duplex molecule whose nucleotide sequence is identical to that of the RNA (except that U is exchanged for T), which is transcribed from the corresponding antisense strand. In some cases, both strands may be transcribed, but from opposite directions. Then the given strand is a sense strand for one and an antisense strand for the other RNA.

Codons optimized to deliver deleterious lesions (CODDLe) A web-based program for the design of gene-specific primers for TILLING, available at http://www.proweb.org/input/. The program generates a gene model with defined intron/exon positions from information relating to an entry sequence, and also provides a protein conservation model using the Blocks Databases.

Concatemer A DNA molecule consisting of linearly repeated, identical monomeric DNA units that are linked to each other in the same relative orientation (e.g. lambda phage multimers).

Consensus sequence (canonical sequence, consensus motif, conserved sequence box)

(a) The sequence of nucleotides which – in a set of DNA sequences – is the most frequent at a defined position.

(b) A particular nucleotide sequence characteristic for a specific functional part of a gene (e.g. the promoter region) which occurs in the same context in other genes, also of other organisms (e.g. the TATA box, CAAT box, Shine-Dalgarno sequence). A consensus sequence is also often referred to as a box or element.

Conserved *a*lternative *s*plicing (conserved AS) Any specific alternative splicing of a specific pre-messenger RNA (as e.g. skipping of a particular exon, or retention of a particular intron) that is conserved over evolutionary periods. Since conservation indicates function, conserved AS events are considered real, and the splicing products therefore are not substrates for example, for nonsense-mediated mRNA decay (or other mRNA surveillance mechanisms).

Constitutive expression (constitutive activity, ca) The permanent transcription of a gene which is directed by a constitutive promoter. The term 'constitutive expression' is misleading, since it implies, that it always occurs. However, it is not clear whether this 'constitutive expression' is only observed under the prevailing experimental conditions. The term 'normal expression' is sometimes used instead.

Constitutive promoter Any promoter that permanently drives the expression of a linked gene.

Copy RNA (cRNA, amplified antisense RNA) Any RNA, that is *in vitro* transcribed from cloned genes driven by either T7 RNA polymerase or SP6 RNA polymerase promoters and catalyzed by the corresponding RNA polymerases.

Core promoter (promoter core) The minimal sequence requirements, usually ~40 bp, within a promoter (core sequence) that are necessary to allow the correct initiation of transcription of the adjacent gene by DNA-dependent RNA polymerase I, II, or III, consisting of the TATA box, the TFII*B* *r*ecognition *e*lement (BRE), the *d*ownstream *p*romoter *e*lement (DPE) and the cap site. The core promoter directs the assembly of the pre-initiation complex.

Correlated messenger RNA *e*xpression (CE) The prediction of a functional interaction of various proteins from the expression pattern of their genes under different experimental conditions. If this expression pattern is similar or identical in a series of different environments, the proteins are inferred to work in a close functional relationship (i.e. in a metabolic pathway or in a protein machine).

Cotranscriptional splicing The working hypothesis, that splicing of pre-messenger RNA (pre-mRNA) occurs simultaneously with its synthesis during transcription of a gene. Alternatively, splicing of most of the introns of a pre-mRNA may also take place post-transcriptionally.

Countertranscript (countertranscript RNA, ctRNA; antisense messenger RNA, anti-mRNA) Any RNA transcript that is synthesized on a gene in the opposite orientation to a messenger RNA transcript. This type of antisense RNA may bind to mRNA and impair its function(s). Also known as *m*essenger-RNA-*i*nterfering *c*omplementary (mic) RNA.

Cyanine 5 (Cy 5) The fluorochrome indodicarbo*cy*anine, that is used as a marker for fluorescent primers in for example, automated sequencing procedures or for labeling in DNA chip technology. The molecule can be excited by light of 643 nm wavelength, and emits red fluorescence light at 667 nm. Since the wavelength of the excitation and emission maxima is pH-dependent, the exact values vary. A series of bathochromically shifted variants of Cy5 are available, as for example, Cy5.5 (excitation at 675 nm, emission at 694 nm) or Cy7 (excitation at 743 nm, emission at 767 nm).

Cyanine 3 (Cy 3) The fluorochrome indodicarbo*cy*anine, that is used as a marker for fluorescent primers in for example, automated sequencing procedures or for labeling in DNA chip technology. The molecule can be excited by light of 552 nm wavelength, and emits green fluorescence light at 570 nm. Since the wavelength of the excitation and emission maxima is pH-dependent, the exact values vary. A series of bathochromically shifted variants of Cy3 are available, as for example, Cy3.5 (excitation at 588 nm, emission at 604 nm).

Database of expressed sequence tags (dbEST) A database containing end sequences of random, arrayed cDNA clones from a large number and variety of tissues of an organism (e.g. fetal and adult, healthy and diseased, inactive and activated tissues). The cDNA libraries from each tissue are oligo(dT)-primed, directionally cloned, have average insert sizes of 1–2 kb, and are usually arrayed in microtiter plates for single-run sequencing, yielding about 300 nucleotides of sequence information from the 3′-end (the so-called 3′ sequence), and from a region 1–2 kb upstream of the poly(A)-tail (which is known as the 5′ sequence, though it is not identical with the real 5′-end of the original message). dbESTs are used to fish for novel genes in a target cell, tissue, organ, or organism.

Deadenylation The removal of poly(A) tracts at the 3′-terminus of eukaryotic messenger RNA, catalyzed by the so-called poly(A) removing nuclease. Deadenylation is the initial step in the degradation of mRNAs.

Deadenylation-dependent mRNA degradation The regulated destruction of mammalian messenger RNAs (mRNAs) by an attack on the 3′-poly(A) tail. First, the poly (A)tail at the 3′-end of the mRNA is shortened by 3′ 5′-ribonucleases (e.g. poly(A) nuclease [PAN] in yeast, deadenylating nuclease [DAN] in mammals) in the cytoplasm ('deadenylation'). The 7-methylguanosine cap at the 5′-end of the mRNA is then removed by cap-specific pyrophosphatases ('decapping'). The resulting decapped mRNA is susceptible to cytoplasmic 5′ → 3′exonucleases that complete the degradation process ('5′ → 3′exonucleolytic decay').

Deadenylation-independent mRNA degradation The regulated destruction of messenger RNAs (mRNAs) by an initial attack on the interior sequences of the message. In contrast to deadenylation-dependent mRNA degradation, which presupposes shortening of the poly(A) tail, the whole process starts with the endonucleolytic cleavage and the resulting exposure of free ends that are then completely degraded by cytoplasmic 5′ → 3′exonucleases.

Decoy promoter A DNA sequence, fully or only partly identical to a promoter that contains consensus sequences for the binding of transcription factors. Such decoy promoter sequences can be injected into nuclei of target cells, where they compete with endogenous promoters for common transcription factors. Competition results in reduced availability of these factors for the endogenous promoter, and consequently a reduced transcription of the adjacent gene. Decoy promoters may be used in gene therapy.

ΔC_t The difference between the C_t values of two samples in a conventional quantitative polymerase chain reaction. For example, the ΔC_t allows the up- or down-regulation of a specific gene of interest to be calculated quantitatively as compared to a non-regulated so-called house-keeping gene.

Designer microarray Any solid support (e.g. glass, nylon, nitrocellulose, polypropylene, silicon), onto which a defined group of genes, cDNAs, or oligonucleotides representing these genes are spotted in an ordered array. The spotted sequences may be related functionally (e.g. code for similar enzymes, such as protein kinases), may encode co-regulated proteins, or proteins working in the same metabolic pathway. Designer microarrays are usually low-density formats designed by a customer.

Dicer (Dicer nuclease, Dicer-1, dimeric RNaseIII RNase) A complex eukaryotic protein, encoded by a single gene, and consisting of an N-terminal DEXH box (DEAD box) ATP-dependent RNA helicase domain, an ATP-binding PAZ ('*P*iwi/*A*rgonaute/*Z*wille') domain (recognizing the end of RNAs), tandemly arranged ribonuclease III (RNase III) domains and a C-terminal double-stranded RNA-binding domain that cleaves *d*ouble-*s*tranded (ds) RNA precursor molecules into 21–22-bp microRNAs, more specifically small interfering RNAs or short hairpin RNAs in an ATP-dependent mechanism. The RNase III specifically recognizes the termini of dsRNA molecules, binds to them, and cleaves the dsRNA successively into 21-nucleotide long dsRNA fragments with 3′overhangs of two to three nucleotides and 5′-phosphate and 3′-hydroxyl termini, as it moves along the RNA. This process either occurs in the nucleus or the cytoplasm. The small dsRNAs bound to the Dicer-ribonucleoprotein complex then become denatured and guide the complex to target RNAs with complementary sequences in the cytoplasm. As a consequence, the target RNAs (usually messenger RNAs) are endonucleolytically cleaved in the center of the recognized 21-nucleotide sequence, which incites the decay of the message. Dicer requires cooperation of ALG-1 and ALG-2 proteins of the RDE-1(*R*NAi *de*ficient-1)/Argonaute protein family that associate with Dicer. Both the RNA interference and small temporal RNA pathways require Dicer as the key enzyme complex. Dicer proteins are preferentially concentrated in epithelia of animals, where they probably defend the tissues against viral attack.

Differential cDNA polymerase chain reaction (differential cDNA PCR) A variant of the conventional polymerase chain reaction that allows the detection and amplif cation of messenger RNA subsets of a cell. In short, cDNA is synthesized from all mRNAs of a sample using a 3′ primer consisting of a stretch of dT residues and additionally two bases. This primer anchors at the 3′-end of polyadenylated mRNA and primes cDNA synthesis by reverse transcriptase. The two additional bases allow a subpopulation of total mRNA to be selected. For example, a primer with the sequence 5′-TTTTTTTTTTTTCA-3′ will only anneal to mRNAs containing TG just upstream of their poly(A) tail. Then a second 6–10 bp long arbitrary primer is used to amplify the selected cDNAs in a polymerase chain reaction. The amplification products are then resolved on a sequencing gel.

Differential display reverse transcription polymerase chain reaction (DDRT-PCR; differential display reverse transcription PCR; differential display, DD; DD-PCR; RNA fingerprinting) A technique for estimating the number of expressed genes in different cell types, and detecting differences in expression by a differential RNA display. In short, the complete set of messenger RNAs of a particular cell type is used as the template for reverse transcriptase to synthesize cDNAs, employing either oligo dTVN (V = A,C,G; N = any deoxynucleotide triphosphate), or simply oligod$T_{(12-18)}$ primers. The cDNAs are then amplified in a conventional polymerase chain reaction, using either primers of arbitrary sequence or specially designed amplimers as reverse primers. The amplified fragments are then separated in denaturing or native agarose or polyacrylamide sequencing gels. The native gels are used to reduce the band pattern complexity. The bands are then detected by autoradiography, if for example, ^{32}P- or ^{33}P-dATP is used to label the amplified fragments during their synthesis, or by simple staining with ethidium bromide. The highly resolved banding patterns from different cell types allow the visualization of cDNAs that are specific for one, but not another cell type. More details are presented in this book.

Differential gene expression technology (DGE technology) Any one of several, usually high-throughput and automatable platforms for the genome-wide detection and analysis of all genes expressed in a cell, tissue, organ or organism at any given time and all the changes occurring during a particular period of time or after various natural or experimental challenges. Basically two concepts can be discriminated: the so-called closed DGE technologies (e.g. macroarray or microarray techniques, in which only those genes that are spotted on an array can be probed) and the open DGE technologies (e.g. cDNA-AFLP, differential display, serial analysis of gene expression, total gene expression analysis, to name but a few, in which all differentially expressed genes can be profiled).

Directional cDNA library A collection of DNA sequences derived from reverse transcription of all mRNAs in a cell. These cDNAs are cloned in a specific orientation relative to the transcriptional polarity of the original mRNAs, or relative to an inducible promoter in the vector. Directional cDNA libraries are established by the forced cloning of cDNAs into appropriate directional vectors, and can be used to drive the expression of the cloned cDNAs, or for the production of subtractive libraries.

Dispersed gene family Any group of homologous or mostly homologous genes, that arose by gene duplication of a common ancestral gene and spread throughout the genome. Such families may comprise few (actin genes, 5–30; globin genes, five; myosin heavy chains, 5–10; ovalbumins, three; tubulins, 3–15) or many genes (insect eggshell protein genes, 50; histone genes, 100–1000; immunoglobin variable region genes, 500). Sequence divergence may lead to new functions for these genes.

DNA chip (DNA array, DNA microarray) A combinatorial array of DNA sequences (e.g. oligonucleotides, cDNAs, genes or part of genes, also PNAs) on a solid support (e.g. nylon membrane, glass or quartz, polypropylene) of minute dimensions. The DNA is either synthesized directly on the chip, or first synthesized *in vitro* and then cross-linked to the chip surface. DNA chips are used to detect for example, mutations (e.g. single nucleotide polymorphisms) or to monitor gene expression profiles. More details are presented in this book.

DNA chip technology The whole repertoire of techniques to generate, maintain and use solid supports ('chips', such as e.g. nylon membranes, glass or quartz slides, polypropylene chips) onto which DNA fragments (such as e.g. oligodeoxynucleotides, genes, gene fragments, cDNAs) have been fixed. Further details are presented in this book.

Double-stranded DNA microarray Any microarray onto which double-stranded DNA is spotted (in contrast to most of the arrays used for hybridization experiments, which require single-stranded target DNA to be immobilized on the surface of the chip). Double-stranded DNA arrays are used to characterize the binding sites of fluorescently labeled transcription factors in a massively parallel experiment. Transcription factors usually bind only to double-stranded target DNA sequences.

Downstream A term used to describe sequences in a linear DNA, RNA or protein molecule proceeding in the direction of gene expression, translation, or protein synthesis, respectively, compared to a point of reference (e.g. on the 3′ side of a given site in DNA or RNA, and on the free carboxyl terminus side of a given site in a protein). For example, downstream sequences in a gene lie in the 3′ direction from the transcription initiation site (designated as +1). Conventionally, nucleotides downstream of this site are marked + (plus), nucleotides upstream of this site – (minus).

Downstream promoter One of a pair of promoters which both drive the expression of a particular gene that is located 3′ downstream of the gene. The other promoter (upstream promoter) lies at the 5′-end of the gene. For example, transcription of the human *RCC 1* gene is initiated at two different promoters about 9 kb apart. Initiation at the downstream promoter produces a pre-mRNA, in which a 5′-terminal single noncoding exon is spliced to downstream exons encoding the RCC 1 protein. Initiation at the upstream promoter leads to the synthesis of a transcript containing four short noncoding exons spliced to the coding part of the mRNA.

Dual chip (dual chip microarray) Any glass slide which carries two identical microarrays that are physically separated from each other. These two microarrays allow a parallel gene expression study to be carried out on one single slide and with only one single fluorochrome.

Dual promoter (bidirectional promoter; twin promoter) A couple of promoters that are part of a dual promoter vector, separated from each other by a polylinker, and driving the transcription of the inserted DNA in opposite directions. Such systems frequently consist of a T7 and an SP6 promoter and allow the *in vitro* synthesis of both a sense and an antisense RNA. Many human dual promoters have 66% GC content (unidirectional or non-bidirectional promoters: 53%). About 80% of all bi-directional promoters in the human genome are located within a CpG island (unidirectional promoters: 38%). Only 8% of human bidirectional promoters possess a TATA box.

Dynamic array Any microarray that consists of a dense network of fluid-handling chemically-inert elements (channels, valves, pumps, collectively called *i*ntegrated *f*luidic *c*ircuits, IFCs) patterned into ultrathin layers of elastomers. IFCs manage to partition extremely small sample and reagent volumes in a fraction of the time that is necessary for traditional microarrays.

Ectopic expression (ectopic gene expression) The expression of a gene outside of its normal location (domain) in a genome. For example, all transgenes underly an ectopic expression at their insertion site in the transformant's genome.

Electronic expression The exploitation of the huge amount of *e*xpressed *s*equence *t*ag (EST) data deposited in gene data banks (e.g. GenBank) for a comparison with sequences of cDNAs isolated from a target cell, tissue, organ or organism. If a sequence is found in the cDNA collection, that matches a sequence in the database, the potential function of the encoded protein can immediately be inferred.

Electrophoretic mobility shift assay (EMSA) A rapid and simple method for the detection of sequence-specific DNA-binding proteins. In short, an end-labeled DNA fragment containing the binding site for the protein is electrophoresed through a non-denaturing polyacrylamide gel together with a nuclear protein extract. Proteins that bind to the DNA fragment decrease its electrophoretic mobility which allows discrimination from the non-bound fragment.

Emulsion polymerase chain reaction (emPCR) A variant of the conventional *p*olymerase *c*hain *r*eaction (PCR), in which the PCR reagents (buffer, salts, primers, *Taq* DNA polymerase, deoxynucleotide triphosphates and template DNA) are enclosed in an aqueous compartment of some 5–15 μm in diameter surrounded by mineral oil. The aqueous compartment ideally contains only one single template DNA molecule such that any amplification proceeds quasi-clonally. Microemulsions are produced by stirring the PCR reagents into the oil phase, composed of a mixture of the detergents Span 80, Tween 80, and Triton X-100 in mineral oil. The emulsions are stable at or above 90 °C, and therefore can be temperature-cycled in a traditional PCR. emPCR then allows the simultaneous amplification of multiple template DNAs in completely separate compartments with a concomitant reduction in the amount and volume of reagents and template required per reaction.

Epigenetic code The specific distribution of methylated cytosines along the DNA of a chromosome, and/or the specific side chain modifications of histones in the chromatin of this chromosome. Since both the cytosine methylation patterns as well as histone side chain modifications (e.g. acetylation, methylation, phosphorylation) in a specific region of the genome varies with time, so does the epigenetic code.

Epigenetic signature The characteristic pattern of cytosine methylation in a specific region of a promoter (or a gene) at a given time. The epigenetic signatures vary with the state of a cell, and changes in response to environmental, and also to intrinsic factors. Methylation of strategic cytosines in promoters recruits proteins which bind to the methylated sites, prevents the binding of activating transcription factors and silences the adjacent gene.

Epigenotype The normally stable and heritable genotype (based on the four-base genetic code), upon which the so-called epigenetic code is superimposed (i.e. the specific pattern of methylated or otherwise modified bases in the genome). For example, the genotype in all cells of a multicellular organism is identical, but the different types of cells have different distribution patterns of 5-methylcytosine (i.e. have different epigenotypes).

Equalized cDNA library Any cDNA library that contains fewer clones derived from redundant mRNAs than conventional cDNA libraries. The construction of an equalized cDNA library starts with the ligation of a DNA adaptor to both ends of the double-stranded cDNA. This adaptor contains as the template a polymerase chain reaction (PCR) primer. The cDNA is then amplified via PCR using a primer complementary to the template primer. After this it is denatured and allowed to reanneal under specific conditions which preferentially allow the reannealing of abundant cDNAs, while less abundant cDNAs remain single-stranded. The double-stranded cDNA is then separated from the single-stranded cDNA by hydroxyapatite chromatography, and the single-stranded cDNA re-amplified by PCR. Repeated cycles (equalization cycles) yield double-stranded cDNA originating from rare mRNAs.

Exon The sequences of a eukaryotic gene that are conserved during processing of the pre-mRNA, and make up the mature message. Exons principally code for three different functions:

(a) Leader function: the first exon usually contains signals for transcription initiation and sequences that function as a guide to direct the message to the ribosomes. This exon is not translated into protein.
(b) Message function: the core exons contain the information that directs the sequence of amino acids in a protein.
(c) Termination function: the last exon usually contains sequences which appear in the message and signal the termination of translation and the addition of a homopolymeric adenyl tail (polyA tail) to the mRNA.

Exon array Any microarray (e.g. a glass slide) onto which up to 25 000 different 60mer oligonucleotides complementary to exons (whose sequences can be retrieved from data banks) are spotted by for example, ink-jet printing. The best candidate probes for a given exon are selected on the basis of their base composition, sequence complexity, binding- and cross-hybridization energies and secondary structure. Usually two (or more) such probes per exon are spotted to ensure exon coverage. The array is hybridized with fluorochrome-labeled cDNAs from mRNA preparations, and fluorescence intensities scanned.

Exonic single nucleotide polymorphism (exon SNP) Any single nucleotide polymorphism, that is present in an exon of a gene. Synonymous with expressed single nucleotide polymorphism.

Exon skipping (exonS) The elimination of one (or more) exons from a transcript during splicing such that the combination of residual exons results in a new messenger RNA and consequently a protein with a new arrangement of domains. For example, the deletion ('skipping') of an exon B, that was originally linking two other exons A and C, allows exons A and C to recombine, creating a new exon combination (exon shuffling). Exon skipping is a route to the generation of new genes.

Exon tiling array Any microarray, onto which 25mer synthetic oligonucleotides are spotted, that are complementary to various regions of (preferably all) exons of a genome. Usually four different oligonucleotides represent different parts of the underlying exon. Exon tiling arrays are hybridized to fluorescently labeled test cDNAs (or cRNAs) and used to monitor the expression of virtually all exons of a genome and to detect splice variants.

Expressed sequence tag (EST) A short synthetic oligonucleotide of 300–500 bp, complementary to the 5′- or 3′-end of a specific messenger RNA and usually derived from a cDNA library by random sequencing. ESTs represent tags for the state of expression of genes at a given time and for the cell or tissue type. Many thousands of ESTs have been sequenced and deposited in databases for gene discovery.

Expressed sequence tag array (EST array) The ordered alignment of different expressed sequence tags on supports of minute dimensions (e.g. nylon membranes, glass or quartz slides, silicon chips). EST arrays allow the simultaneous detection of thousands of expressed genes in a particular cell, tissue, organ, or organism at a given time by hybridization of fluorochrome-labeled cDNA preparations to the array. Any hybridization event between an EST and cDNA is then detected by fluorescence.

Expressed single nucleotide polymorphism (eSNP) Any single nucleotide polymorphism that is present in exons, that is, expressed sequences.

Expression marker Any expressed sequence, as for example, a cDNA, or a tag derived from *s*erial *a*nalysis of *g*ene *e*xpression (SAGE) or an *e*xpressed *s*equence *t*ag (EST), that has been identified by high-throughput expression profiling (as e.g. massively parallel signature sequencing, any of the microarray platforms, serial analysis of gene expression) and serves as a diagnostic (or even prognostic) marker for a disease.

Expression profile (transcript profile, 'RNA fingerprint', expression fingerprint) A complex, context-dependent and genome-wide pattern of (preferably all) expressed genes at a given time. The expression profile is characteristic for a certain cell, tissue, organ or organisms (e.g. a bacterial cell), but changes continuously, dependent on the developmental stage and the environment. Transcript profiles can be established by high-throughput techniques such as cDNA microarrays or methods such as massively parallel signature sequencing, or serial analysis of gene expression and its variant SuperSAGE. More details are presented in this book.

Fiber bead array A collection of 96 (or more) optical fibers, glued to a microtiter plate, each containing 2000 (or more) unique latex beads with a diameter varying from a few microns to a few hundred nanometers (50 000 in total). Every bead in turn hosts between 500 000 and one million molecules of DNA. Though still in an experimental stage, fiber bead arrays have the potential for high-through-put genotyping, gene expression analysis and proteome analysis.

Fiber-optic reactor sequencing (picoliter reactor sequencing, massively parallel picoliter reactor sequencing) The rapid and highly parallel estimation of the sequence of base pairs in multiple samples of DNA by combining the capture of fragments of genomic DNA on beads with a variant of the *em*ulsion *p*olymerase *c*hain *r*eaction (emPCR) for the isolation and *in vitro* amplification of these DNA fragments and their simultaneous sequencing by a variant of the pyrosequencing procedure. In short, entire genomes (up to now limited to genome sizes of maximally 50 Mb) are first isolated, randomly fragmented by nebulization ('shotgun fragmentation') into pieces of 300–500 bp, which are polished (blunt-ended). These blunted fragments are then ligated to short, specially designed common adaptors A and B, that contain sequences complementary to primers for subsequent amplification and sequencing steps. Adaptor B harbors a 5′-biotin tag that allows the immobilization of the library onto streptavidin-coated beads. After nick repair, the non-biotinylated strand is released to form a single-stranded template DNA library (sstDNA library). Following a purification and quantitation step, individual single-stranded DNA fragments are bound to beads by limiting dilution (favoring the binding of only one single fragment per bead), and the individual DNA fragments clonally amplified within droplets of an oil emulsion by emPCR. These droplets contain all amplification reagents (including DNA polymerase) and act as closed microreactors. This step results in millions of beads each carrying 10–20 millions of copies of a unique DNA template. Subsequently, the emulsion is broken, and the released beads with the single-stranded DNA 'clones' deposited into the wells ('picoliter reactors') of a 6 × 6 cm fiber-optic slide by centrifugation. The diameter of the 1.6 million wells of a single slide each permits accomodation of only one single bead of 28 μm diameter per well. About 60% of the beads do not carry DNA. Therefore, an enrichment step for DNA-bound beads follows. Then smaller beads with immobilized enzymes for the pyrosequencing protocol (i.e. ATP sulphurylase and luciferase) are added to the wells, and the sequencing reaction started by pumping deoxynucleotides across the reactor array. Extension reactions can then simultaneously occur on all the bead-bound single strands in the open wells by convective and diffusive transport of the nucleotides. After the flow of one nucleotide, the panel is washed with a solution containing apyrase (removing residual nucleotides) prior to the addition of the next nucleotide, and the waste collected in a receptacle. The nucleotides are added in the series T → A → C → G. Nucleotide incorporation occurs when the template strand carries the complementary base, and results in the release of inorganic pyrophosphate and the generation of photons. These photons emitted from the bottom of each well are captured by fiber-optic imaging bundles bonded to a large format *c*harge-*c*oupled *d*evice (CCD) camera. The images are processed into sequence information simultaneously for all wells with beads carrying template DNA. Normally, sequence reads comprise 80–120 bases with at least 99% accuracy in a single run, generating over 25 million bases with a Phred quality score of 20 or more. The resulting sequences are then assembled by the powerful software of an onboard computer. For example, the so-called assembler consists of various modules, of which the Overlapper identifies overlaps between different reads, the Unitigger constructs larger contigs of overlapping sequences, the Multialigner generate consensus calls and quality scores for each base in each contig, and the FlowMapper maps individual reads to a reference genome. Fiber-optic reactor sequencing avoids robotics for colony picking and for handling of microtiter plates, cloning into bacterial vectors and subcloning, and processing of individual clones. Using this technique bacterial and lower eukaryotic genomes can be sequenced within a few days, open reading frames can be identified, sequenced genomes can be compared with other sequenced genomes of the same species, and conserved sequence elements, mutational hotspots and rare mutations can be identified.

5′-end (five prime end, 5′ carbon end, 5′ carbon atom end, 5′-terminus) The end of a linear DNA or RNA molecule that carries the free phosphate group at the 5′ carbon of the pentose. Conventionally this terminus is written to the left when depicting a nucleic acid molecule.

Fluorescent differential display (FDD; fluorescent differential display reverse transcription polymerase chain reaction, FDDRT-PCR) A variant of the conventional differential display reverse transcription polymerase chain reaction, which uses an oligo dTVN (V = A, C, G; N = any deoxynucleotide triphosphate) or simply oligo dT $_{(12–18)}$ as upstream primer, and a primer of arbitrary sequence labeled with a fluorochrome (e.g. rhodamine) as downstream primer to amplify specific messenger RNAs and display differentially expressed cDNAs. The use of a fluorescent primer avoids radioactivity, increases the sensitivity and allows high throughput.

Fluorescent nucleotide (fluorescently labeled nucleotide) Any ribonucleotide or deoxyribonucleotide, that is covalently bound to a fluorochrome (e.g. fluorescein) via a linker. Such fluorescent nucleotides can be incorporated into target nucleic acids (RNA or DNA) and used to detect these labeled nucleic acids after for example, laser excitation and fluorescence measurement.

Fluorescent primer Any oligonucleotide that has been labeled by one (or more) fluorochromes and is used as primer in polymerase chain reaction-based amplification of DNA sequences.

Format I microarray Any microarray, onto which cDNAs of a length between 500 and 5000 bases are immobilized.

Format II microarray Any microarray, onto which oligonucleotide probes of length 20–25 bases are immobilized.

Full-length cDNA (flcDNA) Any cDNA that contains a complete reading frame (from the ATG start codon to the stop codon), or, more precisely, the 5′-untranslated region as well. Full-length cDNA is important for several aspects of functional genomics. For example, the prediction of transcription units from genomic sequence data can only be validated with full-length cDNA, the occurrence of splice sites are correctly identified with full-length cDNA, alternative splicing events are detected by the identification of a cDNA containing the alternatively spliced region, and full-length cDNAs can be exploited to produce large amounts of specific proteins in homo- or heterologous expression systems (such as e.g. *E. coli* or yeast). Additionally, the sequencing of full-length transcripts identifies RNAs from different members of a gene family.

Functional genomics The whole repertoire of large-scale and high-throughput techniques and subsequent computational analysis used for deciphering the roles of DNA and RNA in the progression from information (DNA) to function (protein). For example, specific gene disruption (gene knock-down, gene knock-in, gene knock-out), allows the function of a gene to be revealed, gene expression patterns to be determined (i.e. expressed genes in a given cell, tissue, organ, or organism at a time by e.g. micorarrays), gene function(s) to be related to developmental processes, and foreign genes to be transferred and integrated so that their influence(s) on the activity of other resident genes can be studied. One area of functional genomics focuses on posttranscriptional events such as messenger RNA stability, frequency of translation of a specific mRNA, and the stability of the protein product, but also the protein–protein interactions of all cellular proteins (as e.g. detected by two-hybrid analysis). This book is dedicated to the functional genomics of plants.

Fusion transcript (intergenically spliced transcript) Any messenger RNA composed of exons from different genes, that is generated by trans-splicing.

Gene-based single nucleotide polymorphism (gene-based SNP) Any single nucleotide polymorphism, that is located in either an exon, an intron, or a promoter of a gene.

Gene co-expression The simultaneous and coordinated transcription of two (or more) genes and the translation of the resulting messenger RNAs into proteins. Gene co-expression suggests a functional relationship between the encoded proteins (as e.g. in membrane or multiprotein complexes, or signal transduction cascades).

Gene content The absolute number of genes per genome or chromosome. The gene content varies between genomes of different yet related organisms, and between different, yet equally sized chromosomes within the same genome. For example, human chromosome 9 (145 Mbp) carries 1248, chromosome 10 (144 Mbp) 1371, chromosome 11 (144 Mbp) 1755, and chromosome 12 (143 Mbp) 1585 genes.

Gene density The number of genes per unit length of DNA. For example, in the gene space the gene density is much higher than extrapolated from a uniform distribution in the genome, whereas in the intergenic space it is equal to or lower than expected from randomness. The gene density varies from organism to organism (e.g. in *Arabidopsis thaliana* chromosome 1 it is around one gene per 4–5 kb), and from chromosome to chromosome in one organism (e.g. human chromosomes 4, 5, 8, 13, 18 and X have considerably lower gene density than chromosomes 1, 11, 17, 19 and 22).

Gene duplication A process by which an ancestral gene is copied so that the corresponding genome contains two identical gene sequences. One of these genes subsequently undergoes mutation(s) which may convert it to a pseudogene or retain its functions in spite of changed sequence composition

Gene expression

(a) The appearance of a phenotypic trait as a consequence of the transcription of a specific gene (or specific genes).
(b) The transcription of a gene (or genes) into structural RNA (rRNA, tRNA) or messenger RNA (mRNA) with subsequent translation of the latter into a protein. Experimentally, expression can be detected by for example, Northern or Western blotting, or various tag- or chip-based techniques.

Gene expression fingerprint The specific pattern of expressed genes (or their transcripts, or the encoded proteins) in a specific cell, tissue, organ, or organism at a specific time.

Gene expression profiling The determination of the pattern of expressed genes in a cell, tissue, or organ at a given time.

Gene expression signature The specific pattern of gene expression of a cell, or a population of identical cells, that differs from the patterns of other cells. Such gene expression signatures can be established by high-throughput transcript profiling techniques such as the various types of expression microarrays, and 'open architecture methods' such as for example, massively parallel signature sequencing (MPSS) and the different serial analysis of gene expression (SAGE) variants, to name but a few.

Gene *e*xpression *q*uantitative *l*ocus (*e*xpression QTL, eQTL) Any genetic locus, whose gene expression level is determined, treated as a quantitative trait and linked to a genetic map by linkage analysis.

Gene *i*dentification *s*ignature (GIS) technique A variant of the conventional *s*erial *a*nalysis of gene *e*xpression (SAGE) method, that combines the generation of 3′- and 5′-tags (i.e., short sequences encompassing both the transcription start and poly(A) sites) of full-length transcripts on a genome-wide scale. In short, poly(A)$^{+}$-messenger RNA (mRNA) is first isolated, reverse-transcribed into cDNA, using an oligo$(dT)_{16}$ primer harboring a *Gsu*I recognition site and methylated deoxycytidines to be incorporated instead of the normal deoxycytidines. Double-stranded full-length cDNAs are then selected by the so-called cap trapper procedure, and a so-called linker I harboring a *Mme*I recognition site is ligated to the 5′-end of the cDNA. Then the cDNA is digested with *Gsu*I to remove the poly(A)-tail (with the exception of an AA dinucleotide, that facilitates orientation of the tags produced later in the procedure). This enzyme is methylation-sensitive and does not cleave hemimethylated cDNA. Subsequently another linker II with a second *Mme*I site is ligated to the 3′-end of the cDNA, and the linkered cDNAs cloned into a special plasmid vector that does not contain a *Mme*I site. The resulting full-length cDNA library (flcDNA library) is then restricted with *Mme*I to release the central part of the cDNA, yet leaves the tags from both the 5′- and the 3′-ends in the plasmid vector. These tags are blunt-ended (the original cohesive ends in the majority of tags do not match), producing two tags of 18 bp originating from the 5′- and 3′-terminus, respectively, of the corresponding cDNA. The subsequent ligation of two tags each from an end of the original cDNA forms *p*aired-*e*nd di*t*ags (PETs), that are amplified, digested with the original enzymes, concatenated, cloned into an appropriate vector (GIS library, GIS PET library), and sequenced. Sequencing of such a concatemer identifies >15 PETs per reaction, and each PET can be mapped to the genome, thereby defining the boundaries of the underlying gene. Additionally, primers complementary to PET sequences are used to amplify the transcript sequences between the 5′- and 3′-tags in a conventional *p*olymerase *c*hain *r*eaction (PCR), which facilitates annotation of the resulting transcript sequence to databanks. Abnormal messenger RNAs originating from chromosomal aberrations or rearrangements can be analyzed, expressed genes can be localized on a physical genome map, full-length transcripts can be isolated, and the transcription boundaries of genes can be demarcated using GIS.

Gene island Any cluster of genes, that is separated from neighboring clusters by regions of repetitive DNA. Such gene islands are characteristic of eukaryotic genomes.

Gene knock-out A laboratory slang term for the disruption of a gene by the insertion of a DNA sequence or mutation(s) that abolishes gene function.

Gene map (map) A graph depicting the arrangement, that is the relative position of genes, on a chromosome or plasmid; the product of gene mapping.

Gene mapping The estimation of the linear arrangement of genes, the determination of the relative location of specific genes on specific chromosomes or plasmids, and their relative distance from one another. Gene maps may be based on classical genetic recombination analysis or on direct DNA data obtained by DNA sequencing.

Gene number paradox The discrepancy between morphological complexity of an organism and its number of genes. For example, the morphological complexity of a human being (*Homo sapiens*) is by far greater than that of a worm (e.g. *Caenorhabditis elegans*), yet the number of genes is similar in both organisms.

Gene Ontology (GO) The term 'gene ontology' encompasses both a collaborative project to develop structured, controlled vocabularies ('ontologies'), that relate the molecular functions of gene products (any protein or RNA encoded by a gene) to their role in multi-step biological processes and their localization to cellular components in a species-independent manner, using different databases (http://www.geneontology.org/), and the process of linking gene functions to biological processes and cellular localization. The ontology terms are assigned to sequences in a database based on either experimental evidence or sequence homology.

Gene pool The total sum of genes in a specific population of reproductively active organisms at a given time.

Gene product The product of the transcription of a gene (e.g. rRNA or tRNA in case of ribosomal RNA genes or transfer RNA genes, also messenger RNA in case of structural genes). The term also refers to proteins as the products of structural genes.

General transcription factor (GTF) Any one of a series of transcription factors (proteins) that are necessary for the formation of a transcription initiation complex with RNA polymerase II (B).

Gene re-expression The repeated expression of a gene at two (or more) times during a physiological or developmental process with an intermittent period of no expression. For example, during *Drosophila melanogaster* development, many genes expressed in the embryo are activated in a second wave in pupae and larvae (i.e. are re-expressed). This re-expressed class of genes encompasses endopeptidase-, chaperone-, cytoskeleton-, signaling-, cell adhesion- and transcription-protein encoding genes, to name but a few.

Gene-rich region Any segment of a chromosome or part of a physical map, which harbors genes at a higher frequency than the expected average.

Gene silencing The inactivation of a previously active gene.

Gene size The length of a gene from the cap site to the poly(A) addition signal (eukaryotes), expressed in number of base pairs (bp). Gene sizes vary tremendously, from 21 bp (Enterobacterial gene *mcc*A, encoding the antibiotic heptapeptide microcin C7) to 2.34 Ã–10^6 bp (dystrophin gene of *Homo sapiens*).

Gene surfing The identification of genic DNA sequences in an anonymous DNA by comparing it against sequences in genome and/or protein databases, that have already had functions assigned (and in some cases, proven) (e.g. coding sequences). Also, computational programs such as Genescan or GeneWise can predict the occurrence of genes in raw sequence data.

Genomics A term describing the whole repertoire of technologies used to describe the organization of genomes and the functions of their constituents (e.g. genes).

Global gene expression (genome expression, genome-wide expression profiling, global transcription profiling, genomic profiling) The comprehensive search for all genes of a genome that are expressed, and their isolation, characterization, and sequencing to establish a genome-wide expression profile. Global gene expression can be determined by for example, massively parallel signature sequencing, or serial analysis of gene expression.

Hairpin RNA vector (hpRNA vector) Any cloning and transformation vector encoding two self-complementary hairpin RNA (hpRNA) sequences in an inverted-repeat orientation that are expressed in transgenic organisms to produce double-stranded RNA, which in turn post-transcriptionally silences the gene from which its complementary mRNA is derived.

Helix-loop-helix (HLH) A specific three-dimensional structure adopted by DNA hyphen;binding proteins within their respective DNA-binding or protein–-protein interacting domain. It consists of an HLH domain formed by two amphipathic α-helices of 12–15 amino acids connected by a nonconserved looped region of varying dimensions, and an adjacent basic domain. The latter is located at the N-terminal end of the HLH domain, and comprises some 10–20, mostly basic amino acid residues. Such helix-loop-helix configurations are characteristic of a number of nuclear proteins (e.g. Myo D) and probably mediate protein–protein contacts (e.g. for homo or heterodimerization). In some cases dimers are known to bind to target sequences in DNA (e.g. Myo D-E 12 heterodimers bind to a specific sequence of the muscle creatine kinase and the κ light chain enhancer whereas the corresponding monomers show only a slight affinity).

Helix-turn-helix A specific three-dimensional structure adopted by DNA-binding proteins within their respective DNA-binding domains. It consists of two α helices ('recognition helix') bridged by a sharp β-turn. One of the two α-helices forms a close specific contact with the major groove of the DNA, which is thought to be mediated by hydrogen bonds and van der Waals forces between the side chains of the protein and the edges of the base pairs exposed within the grooves of the DNA.

Hemimethylation The presence of methylated nucleotides (e.g. 5-methylcytosine, N^6-methyl-adenine) in only one strand of a DNA duplex molecule, resulting from semiconservative replication. Usually the newly synthesized strand is methylated by methyltransferases, so that both strands of a DNA duplex are normally methylated at comparable sites.

Heterochromatin The part of chromatin that is maximally condensed in interphase nuclei, replicates late in the S phase, and contains DNA that is mostly transcriptionally inactive (e.g. satellite DNA). Heterochromatin stains maximally in the interphase nucleus and can be broadly categorized into (1) constitutive heterochromatin that is permanently and densely packaged (e.g. around the centromere) and (2) facultative heterochromatin whose staining properties vary at different developmental stages. The assembly of heterochromatin requires an orchestrated array of chromatin modifications. For example, in fission yeast the deacetylation of N-termini of histone H3 by class I and II histone deacetylases Clr3 and Clr6 as well as the class III NAD-dependent deacetylase Sir2, is followed by methylation of histone H3 at lysine 9 (K9) by the methyltransferase Clr4 to create a binding site for Swi6 and Chp1 chromodomain proteins. Histone H3K9 methylation is a conserved hallmark of heterochromatin.

High abundancy messenger RNA (high abundance mRNA, high abundance message, abundant RNA, superprevalent mRNA) A sub-family of eukaryotic messenger RNAs, comprising messages encoded by some 100 tissue-specific genes and present in about 1000–20 000 copies per cell (for example globin mRNA in erythrocytes or actin mRNA in muscle cells).

High density chip A laboratory slang term for a DNA chip, onto which from 10 000–200 000 (or more) probes are spotted.

High density oligonucleotide array A general term for any high density chip, onto which hundreds of thousands or even millions of oligonucleotides are synthesized by a photolithographic method.

High mobility group protein (HMG protein) A member of an abundant class of non-histone proteins of higher eukaryotes consisting of three structurally unrelated subgroups (HMG-1/−2; HMG-14/−17, and HMG I/Y). The conserved proteins of the HMG 1/2 family contain a so-called HMG-box functioning as a DNA-binding domain, and a highly acidic carboxy-terminal region. Both HMG-14 and HMG-17 bind to the nucleosome core, possibly by replacing histones H2A and H2B, and thereby altering the interaction between the nucleosomal DNA and the histones. As a consequence, the DNA helix is locally bent and unwound. Both proteins probably maintain active or potentially active genes in a special chromatin conformation.

High-resolution microarray Any microarray, onto which DNA fragments (e.g. PCR products), cDNAs, or oligonucleotides are immobilized in spots that are smaller than 250 μm.

Histone Any one of a group of low-molecular weight, basic nuclear proteins of eukaryotic organisms, which are highly conserved throughout evolution and serve to package nuclear DNA into the nucleosomes of eukaryotic chromatin. Histones fall into three main categories: the lysine-rich (e.g. H1), the slightly lysine-rich (e.g. H2A and H2B), and the arginine-rich histones (e.g. H3 and H4). Histones interact with the negatively charged phosphate backbone of DNA via salt bridges, and can be posttranslationally modified by acetylation, methylation, phosphorylation, poly (ADP) ribose polymerization, or reduction.

Histone acetylation The enzymatic transfer of acetyl groups from acetyl-CoA to some amino acids of certain histone molecules. Acetylation, especially of serine residues at the N-terminus of for example, histones H1, H2A and H4 may occur during histone synthesis and is irreversible. Other acetylations, especially of N-terminal lysine residues of histones H2A, H2B, H3 and H4, may facilitate repulsion of histones from the phosphate backbone of DNA in nucleosomes, because of the introduced negative charges. This induces conformational changes (nucleosome to lexosome), a prerequisite for gene activation.

Histone code A somewhat misleading term for the various posttranslational modifications of histone proteins at a given time that are recognized by other proteins involved in chromatin modeling, chromatin remodeling and transcriptional regulation. The acetylation of 13 different lysine residues in all core histones, methylation of lysine and arginine residues in histone H3 and H4, phosphorylation and ubiquitinylation of all histones are examples of such 'codes'. For example, lysine residues 4 and 9 (K9) in histone H3 and lysine 20 in histone H4 are methylated by histone methyltransferase $SU(VAR)_{39}$ (in mammals) or Clr4 (in yeast). This methylated lysine is the only binding site for heterochromatin protein HP_1 that is associated with silent heterochromatic regions of a genome. Phosphorylation of the adjacent S10 residue by Aurora kinase B loosens this association. Acetylation at H3K14, catalyzed by *h*istone *a*cetyl*t*ransferases (HATs) prevents HP1 binding.

Homologous gene Any gene in species A, that has an identical counterpart in species B. Both genes have a common origin and encode identical or similar proteins.

Homology (sequence homology) The extent of identity between two nucleotide or amino acid sequences, as a measure of a common evolutionary origin. The term is frequently and incorrectly used as synonym for 'similarity'.

Hybrid promoter (chimeric promoter) An artificial promoter which has been engineered to contain a consensus sequence (e.g. the Pribnow- or generally TATA-box) from one and a second consensus sequence (e.g. the −35 region TTGACA in bacteria or the CAAT-box in eukaryotes) from another promoter. Such hybrid promoters are designed to direct maximal expression of linked genes.

Hybrid transcription factor (chimeric transcription factor, hybrid TF) Any transcription factor, that is composed of two (or more) parts from different transcription factors, and therefore combines different specificities or activities. For example, a specific hybrid TF consists of the GAL4 activation domain and the LexA DNA-binding domain which specifically binds to its cognate sequence and activates genes with LexA operators.

Initiator element (INR element; Inr; initiator) A short sequence (consensus 5′-CTCA-3′) of RNA polymerase II promoters, located at +1 to +11. This 'CTCA box' is necessary for efficient transcription of the adjacent gene.

***In silico* transcriptomics** A computational screen of for example, cDNA and/or expressed sequence tag data banks for genes specific for a particular cellular state (e.g. disease, stress, inflammation, injury, tumor).

***In situ* hybridization (ISH)** A technique for the identification of specific DNA sequences on intact chromosomes (or also RNA sequences in a cell) by hybridization of radioactively labeled or fluorescent complementary nucleic acid probes (frequently synthetic oligonucleotides) to denatured metaphase or interphase chromosomes. The hybridizing loci are then detected either by autoradiography or laser excitation of the fluorochrome and emission light capture with CCD cameras for example.

Interaction transcriptome An infelicitous term for the transcriptional responses of two (or more) interacting organisms (as e.g. a host and a parasite, a host and a pathogen).

Intergenic DNA Any DNA sequence, that is located between two adjacent genes. Mostly synonymous with intergenic region.

Intergenic microRNA gene Any one of a series of genes that encode microRNAs and are located in intergenic regions of eukaryotic genomes. Most of the human microRNA genes are intronic genes (55%), the rest are intergenic (28%) and exonic (17%).

Intergenic transcript Any messenger RNA ('transcript') encoded by genomic sequences outside genes

Intergenic transcription The RNA polymerase II-catalyzed transcription of genomic sequences that does not encode proteins.

Intermediate abundancy messenger RNA (intermediate abundance mRNA, intermediate abundance message) A sub-family of eukaryotic messenger RNAs, comprising messages encoded by some 500–1000 house-keeping genes and present at about 100–500 copies per cell.

Internal exon Any exon embedded between two introns within the main body of a mosaic gene, as opposed to an initial exon, or terminal exon.

Internal poly(A) priming (internal priming) The binding of oligo(dT) primers to poly(A) tracts within a messenger RNA ('internal poly[A] site'), additionally to the 3′-poly(A) tract as a prerequisite for reverse transcription of this mRNA into a complementary DNA (cDNA). Such internal A-stretches, if consisting of at least eight adenosyl residues, compete favorably with the common poly(A)tail at the 3′-end of an mRNA, such that full-length and truncated cDNAs from the same transcript are generated by oligo(dT) priming. Internal priming therefore leads to the occurrence of truncated cDNAs in the databases, which is in the range of 12% of cDNAs from human genes. The effect of internal priming can be minimized by replacing the traditional oligo(dT) primer by a set of oligo(dT) primers with different anchoring bases ('anchored primers').

Internal promoter Any promoter that is located within a gene. For example, the retrotransposon *jockey* transposes via a poly(A)$^+$-RNA intermediate, and would not be able to take an external promoter to a new insertion site. *Jockey* has solved this problem: it harbors an internal promoter.

Intragenic DNA Any DNA sequence located within a gene (e.g. exons, introns, 5′- and 3′-untranslated regions).

Intron (intragenic region;*intervening* *s*equence, IVS) A sequence of nucleotides within eukaryotic genes that is transcribed into pre-mRNA but subsequently excised (splicing) and degraded within the nucleus. The residual sequences of the transcript (exons) are joined to produce the translatable message, so that intron sequences are not normally represented in a protein. Introns vary in number per gene (one in some rRNA genes, more than 30 in *Xenopus* yolk protein genes), in size (from less than 50 to more than 12 000 nucleotides) and in sequence. Only the borders between exon and intron (splice junctions) are identical in most introns. These boundaries direct the correct excision of the intron and the splicing of the exons.

Intron-exon mapping The localization of introns and exons within the coding region of a eukaryotic gene with the aid of S1-mapping or heteroduplex mapping procedures.

Intronic microRNA gene Any one of a series of genes that encode microRNAs and are located in introns of eukaryotic mosaic genes. Most of the human microRNA genes are intronic genes (55%); the remainder is intergenic (28%) and exonic (17%).

Intronic *s*ingle *n*ucleotide *p*olymorphism (intronic SNP, intron SNP) Any single nucleotide polymorphism, that occurs in introns of eukaryotic genes. Intron SNPs are more frequent than SNPs in coding regions.

Intronic transcript (IT) Any mature messenger RNA, that does not only contain exons, but also one (or more) introns.

Intron retention (IntronR) The inclusion of an intron in a final messenger RNA. Normally, the introns are spliced out of the pre-mRNA, but in certain cases one or more introns can be left unspliced, with drastic consequences. For example, intron 3 retention in the P-element mRNA of *Drosophila melanogaster* generates a repressor protein of transposition, whereas splicing of intron 3 allows the expression of the transposase in the germline.

Intron shuffling The recombination of intron sequences such that new combinations of introns are generated. For example, in some cases, all the functional sequences are contained within introns, not exons. The *s*mall *n*ucleolar RNA (snoRNA) genes encode stable and low molecular weight RNAs that are necessary for proper rRNA processing. Some of these snoRNAs are encoded by introns and transcribed as part of the parent pre-mRNA. After intron excision, exonucleases trim back the surrounding intron to produce the mature snoRNA. In extreme cases, as for example, the *U22 host* gene (UHG), eight of its nine introns harbor snoRNA-encoding sequences (U22, U25–U31), whereas the exons have no coding function, are spliced and destroyed. Therefore, the terms intron and exon should simply signify RNA sequences that become physically separated during RNA splicing.

***In vitro* polyadenylation** The selective attachment of about 30–120 adenosine residues to the 3′ termini of poly(A)$^-$-RNAs (e.g. messenger RNAs of bacteria) *in vitro*, catalyzed by poly(A)polymerase. Ribosomal and transfer RNAs remain unmodified. This technique is used to adenylate the 3′-ends of for example, bacterial messenger RNAs, that do not carry poly(A) tails. After *in vitro* adenylation, these mRNAs can then easily be isolated by oligo (dT) cellulose chromatography.

***In vitro* transcription (*in vitro* RNA synthesis; cell-free transcription)** A method of transcribing cloned genes into their corresponding transcripts *in vitro*, using specially prepared cell extracts for example, from HeLa cells or *Drosophila* embryos, and specific transcription vectors (expression vectors). Such vectors contain promoters for RNA polymerases (e.g. SP6 RNA polymerase or T7 RNA polymerase promoters, or promoters of eukaryotic class II genes) flanking polylinkers. Any foreign DNA, inserted into one of the polylinker cloning sites will be transcribed under the control of the promoter. The transcripts accumulate to high concentrations (e.g. per mg plasmid DNA the SP6 promoter/RNA polymerase system produces up to 25 mg RNA that can also be labeled with ^{32}P-ribonucleoside triphosphates) and can be used as specific probes in Southern blotting, Northern blotting and *in situ* hybridizations. *In vitro* transcribed RNA may also be used in *in vitro* translation systems, for studies of RNA splicing and S1 mapping procedures. *In vitro* transcription systems are also ideal for the study of promoter sequences (e.g. the mapping of transcriptional control sequences in promoter DNA).

Knock-down mutation Any mutation that reduces the expression of a gene, but does not abolish it.

Laser microdissection (LMD) A technique for the isolation of specific chromosomes or cells from tissue sections or also culture dishes. The mounted tissue is moved either manually or robotically around a stationary laser which cuts the target cell from its surrounding neighbors. The excised cell is then trapped in various ways (laser-capture microdissection, laser pressure catapulting). LMD can also be used to destroy a particular cell amidst the surrounding tissue ('negative selection').

Linked marker Any molecular marker(s) located closely to a target gene on the same chromosome such that the recombination frequency between them approaches zero. Linked markers are exploited for the isolation of the linked gene(s) via positional cloning.

Locked nucleic acid (LNA, L-DNA, 'bridged nucleic acid') A nucleic acid derivative, that contains one or more 2′-C,4′-oxy-methylene-linked bicyclic ribonucleotide monomers (furanose rings locked in a 3′-endo conformation) embedded among DNA nucleotides as constituents of an antisense oligonucleotide. This conformation allows the formation of extremely stable → Watson-Crick base-pairing between the LNA and complementary DNA or RNA ($T_m = +3$ to $+10$Â° C per LNA monomer introduced). Therefore, LNA–DNA mixmers are potent duplex stabilizers, but biologically inert (non-toxic). Moreover, they are resistant to 3′-exonucleolytic degradation, soluble in aqueous media, and can be cut by restriction endonucleases. LNA primers are recognized by various DNA polymerases and reverse transcriptases, and show excellent mismatch discrimination in for example, SNPing. Moreover, the high binding affinity of LNA oligonucleotides allows the probe length to be reduced without hampering its hybridization to the target DNA.

Locus (plural: loci) The position of a gene (generally DNA sequence) on a chromosome or a genetic map.

Long form (LF) A laboratory slang term for the longer, normally spliced wild-type messenger RNA (or its cDNA) transcribed from a particular gene, as compared to the short form(s) arising from the transcript of the same gene undergoing alternative splicing. A longer form can also arise from alternative splicing with intron retention. The shorter form is then the wild-type form.

Long serial analysis of gene expression (LongSAGE, LS) A variant of the conventional *s*erial *a*nalysis of *g*ene *e*xpression (SAGE) technique for the quantification of transcript abundance in the RNA population of a cell, tissue, organ, or organism that generates 21-bp tags derived from the 3′-ends of *m*essenger RNAs (mRNAs) rather than the 14 bp in the original SAGE protocol. In short, RNA is first extracted from the target cells, and mRNA isolated. This mRNA preparation is then treated (e.g. converted to cDNA) according to the conventional SAGE procedure with the following changes. After digestion of the cDNAs with *Nla*III, linkers containing a *Mme*I recognition site are ligated to the 3′-ends of the cDNAs. Linker-tag molecules are then released from the cDNA using the type IIS restriction enzyme *Mme*I. The resulting tags are then directly ligated with DNA ligase. Tag concatemers are sequenced, and the longer tags analyzed and matched to genomic sequence data. Matching of tags to genomic sequences allows the precise localization of genes, from which the tags ultimately are derived.

Loss-of-expression mutation Any mutation in a gene that silences the gene (i.e. leads to the disappearance of its transcript). A loss-of-expression mutation represents a loss-of-function mutation.

Loss-of-function mutation (lf) Any mutation that completely abolishes the function of the encoded protein.

Low abundancy messenger RNA (low abundance mRNA, low abundancy message) A subfamily of eukaryotic messenger RNAs, comprising messages encoded by unique genes and present in some 5–10 copies per cell.

Molecular marker (DNA marker) Any specific DNA segment whose base sequence is different (polymorphic) in different organisms and is therefore diagnostic for each of them. Molecular markers can be visualized by either hybridization (as e.g. in DNA fingerprinting or restriction fragment length polymorphism, RFLP) or polymerase chain reaction (PCR) techniques. Ideal molecular markers are highly polymorphic between two organisms, inherited codominantly, distributed evenly throughout the genome and visualized easily.

Massively parallel signature sequencing (MPSS) A high-throughput technique for the sequencing of millions of cDNAs conjugated to oligonucleotide tags on the surface of 5-mm diameter microbeads that avoids separate cDNA isolation, template processing and robotic procedures. In short, 32mer capture oligonucleotides are attached to the surface of separate microbeads (diameter: 5 mm) by combinatorial synthesis, such that each microbead has a unique tag for its complementary cDNA. The messenger RNA is then reverse transcribed into cDNA using oligo(dT) primers, restricted at both ends with for example, *Dpn* I, complements of the capture oligonucleotides are attached to the poly(A) tail of each cDNA molecule and the construct cloned into an appropriate vector containing PCR handles which serve as primer-binding sites for polymerase chain reaction-based amplification of the tagged cDNA. The cDNA is now amplified with a fluorochrome-labeled primer, denatured, and the single-stranded address tag-containing fragments annealed ('cloned') to the surface of microbeads containing address tag sequences as hybridization anchors, and then ligated ('*in vitro* cloning'). Each microbead displays about 100 000 identical copies of a particular cDNA ('microbead library'). The fluorescent microbeads (all containing a cDNA) are then separated from the non-fluorescent beads (not containing a cDNA) by a *f*luorescence-*a*ctivated *c*ell *s*orter (FACS). Each single microbead in the library harbors multiple copies of a cDNA derived from different mRNA molecules. If a particular mRNA is highly abundant in the original sample, its sequence is represented on a large number of microbeads, and vice versa. In the original version of MPSS, 16–20 bases at the free ends of the cloned templates on each microbead are sequenced ('signature sequences'). First, millions of template-containing microbeads are assembled in a densely packed planar array at the bottom of a flow cell such that they remain fixed as sequencing reagents are pumped through the cell, and their fluorescence can be monitored by imaging. The fluorophore at the end of the cDNA is then removed, and the sequence at the end of the cDNA determined in repetitive cycles of ligation of a short adaptor carrying a restriction recognition site for a class IIS restriction endonuclease (binding within the adaptor and cutting the cDNA remotely, producing a four-nucleotide overhang; e.g. *Bbv*I). Next, a collection of 1024 specially encoded adaptors are ligated to the overhangs, and the coded tails interrogated by the successive hybridization of 16 different fluorescent decoder oligonucleotides. This process is repeated several times to determine the signature of the cDNA on the surface of each bead in the flow cell. The abundance of each mRNA in the original sample is estimated by counting the number of clones with identical signatures.

Megagene Any unusually large gene whose length exceeds 10–20 kb (e.g. the X-linked Duchenne muscular dystrophy (DMD) gene of about 1000 kb, or the dystrophin gene with a total length of 2300 kb and 100 introns).

Messenger RNA (mRNA) A single-stranded RNA molecule synthesized by RNA polymerase (RNA polymerase II or B in eukaryotic organisms) from a protein-encoding gene template (structural gene) or several adjacent genes (polycistronic mRNA). An mRNA specifies the sequence of amino acids in a protein during the process of translation.

Messenger RNA isoform Any one of a series of messenger RNAs which all originate from one single gene but differ in the combination of their exons. Isoforms are generated by alternative splicing.

Messenger RNA profiling (mRNA profiling) The simultaneous detection of thousands of messenger RNAs (indicative for the transcription of thousands of genes) involved in developmental, physiological, environmentally influenced or pathological processes. Profiling can be achieved by cDNA expression arrays, massively parallel signature sequencing, or serial analysis of gene expression, to name only a few techniques.

Microarray Any microscale solid support (e.g. nylon membrane, nitrocellulose, glass, quartz, silicon, or other synthetic material) onto which either DNA fragments, cDNAs, oligonucleotides, genes, open reading frames, peptides or proteins (e.g. antibodies) are spotted in an ordered pattern ('array') at extremely high density. Such microarrays (laboratory jargon: 'chips') are increasingly used for high-throughput expression profiling.

MicroRNA (miRNA, also tiny RNA) Any one of a class of hundreds (vertebrates: more than 1000) of ubiquitous, usually single-stranded, evolutionarily conserved, 16–24 nucleotide long non-coding, regulatory, eukaryotic RNAs that are processed *in nucleo* by the double-stranded RNA-specific ribonuclease III Drosha from longer and normally polyadenylated transcripts (pri-miRNAs, usually 70–171 nucleotides, in extreme cases up to 1 kb long) carrying a stem-loop structure. Drosha, in concert with its cofactor DGCR8 that binds the junction between the double-stranded stem and the flanking single-stranded regions of the pri-miRNA, cuts the stem-loop at an 11-bp distance from the junction. The resulting hairpin RNAs (precursor miRNAs, 'pre-miRNAs') are then transported to the cytoplasm by a transportin-5- ('exportin') dependent mechanism, where they are again trimmed by a second, double-stranded RNA-specific ribonuclease called Dicer. One of the two strands (active strand) of the resulting 19–23 nt long RNA is bound by a complex similar or identical to the *R*NA-*i*nduced *s*ilencing *c*omplex (RISC) involved in RNA interference (RNAi). The complex-bound single-stranded miRNA is targeted to and binds specific messenger RNAs (mRNA) with complete or only partial sequence complementarity. The bound mRNA remains untranslated, resulting in reduced expression of the corresponding gene without degradation of the mRNA. MicroRNAs associate with proteins to form so-called micro-ribonucleoprotein (microRNP) complexes. One of the proteins in this RNA–protein complex is the *e*ukaryotic translation *i*nitiation *f*actor eIF2C2, others are Argonaut, Gemin3 and 4 (components of the *s*urvival of *m*otor *n*eurons (SMN) complex). Some of the miRNAs (e.g. *Lin*-4 and *Let*-7) are also called small temporal RNAs because their mutational inactivation affects developmental timing in *Caenorhabditis elegans*. MicroRNAs inhibit the translation of target mRNAs containing 3′-untranslated region (3′-UTR) sequences with partial complementarity, and are probably involved in the development of spinal muscular atrophy, a hereditary neurodegenerative disease of (predominantly) children. The SMN complex is involved in the assembly and restructuring of diverse ribonucleoprotein machines, as for example, the spliceosomal small nuclear RNPs (snRNPs), the small nucleolar RNPs (snoRNPs), the heterogenous nuclear RNPs (snRNPs), and the transcriptosomes. MicroRNAs should not be confused with short interfering RNAs, though the two RNA species are both generated by Dicer from longer precursors. However, siRNAs are not encoded by discrete genes, while microRNAs are. Numerous miRNAs are encoded by introns, and these miRNAs are different from the intergenic miRNAs, because they are transcribed by RNA polymerase II and use specific spliceosomal components for their processing.

MicroSAGE A variant of the original serial analysis of gene expression (SAGE) technique for the global analysis of gene expression patterns that requires only minute quantities of starting material (e.g. bioptic material or microdissections). MicroSAGE is run in a single streptavidin-coated PCR tube (in which the RNA or cDNA remains immobilized) from RNA isolation to the release of tags, thus avoiding losses at each step. Also, re-amplification of excised ditags is reduced to only 8–15 cycles. In between different steps, enzymes from the previous reactions are removed by heat inactivation and disposal, so that after washing the reaction buffer and all ingredients for the next step can easily be added. MicroSAGE also uses total RNA rather than polyadenylated RNA, because the poly(A)$^+$ -fraction is directly bound to the strepavidin-coated wall of the tube via a biotinylated oligo(dT) primer that also serves as primer in subsequent cDNA synthesis.

Molecular beacon A single-stranded oligonucleotide that contains a fluorochrome (e.g. fluorescein, TAMRA, Cy3, Cy5, Texas red) at its 5′-terminus and a non-fluorescent quencher dye (e.g. [4(4-(*d*imethyl*a*mino)phenyl)azo] *b*enzoi*c* acid; DABCYL) at its 3′-terminus. The sequence of such a molecular beacon is designed such that it forms a hairpin structure intramolecularly, with a 15–30-bp probe region (complementary to the target DNA), and 5–7-bp long stem region (self-complementary). In this folded state the fluorochrome is quenched (i.e. any photon emitted by the fluorophore through excitation light is absorbed by the quencher (e.g. TAMRA) and emitted in the non-visible spectrum). After binding to a homologous target sequence, the beacon undergoes a conformational change forcing the stem of the hairpin apart, displacing the fluorochrome from the quencher, and abolishing the quenching (i.e. fluorescence occurs). Such molecular beacons are used for quantitation of the number of amplicons synthesized during conventional polymerase chain reactions, for the discrimination of homozygotes from heterozygotes, the detection of single nucleotide polymorphisms, *in situ* visualization of messenger RNA within living cells, and the simultaneous detection of different target sequences in one sample, if various fluorochromes with differing emission spectra are used.

Multigene analysis The simultaneous determination of the expression patterns of hundreds, thousands or even hundreds of thousands of genes in a particular cell, tissue, or organ at a given time, as opposed to the analysis of the expression of only a single gene or a few genes. Multigene analysis can be conducted with microarrays (cDNA array, cDNA expression array, expression microarray, transcript array) and high-throughput profiling techniques such as massively parallel signature sequencing or serial analysis of gene expression.

Multigene family (gene family) A set of closely related genes originating from the same ancestral gene by duplication and mutation processes. They may either be clustered on the same chromosome (e.g. genes coding for ribosomal RNAs, rDNA) or be dispersed throughout the genome (e.g. heat shock protein genes). Most of the members of such multigene families retain a far-reaching homology in the coding region, but are divergent in the intron and promoter regions.

Multiple sequence alignment (MSA) The iterative search for homologs of a protein of interest in the proteomes of selected species with algorithms such as PSI-BLAST, and their sequence alignment using a multiple-sequence alignment tool as for example, CLUSTALW, MUSCLE, MAFFT, T-Coffee, ProbCons and SATCH-MO. The selection of an alignment method is dependent on available compute resources and the size and evolutionary divergence of the dataset, both of which affect alignment accuracy. For large or divergent datasets, MAFFT and MUSCLE are recommended because they are computationally efficient.

Natural antisense transcript (NAT) Any one of a series of naturally occuring antisense messenger RNAs in pro- and eukaryotic organisms. NATs are able to form double-stranded RNAs with sense transcripts and therefore function in the regulation of pre-mRNA splicing, alternative splicing, control of translation, the degradation of target RNA ('turnover'), RNA stability and trafficking (the transport of mRNA from the nucleus into the cytoplasm), RNA interference, genomic imprinting, X chromosome inactivation, or RNA editing. At least 2500 human genes are also transcribed into the corresponding antisense variants. Changes in antisense transcription have been implicated in pathogenesis, such as cancer or neurological diseases. In maize, as a representative of the plant kingdom, more than 70% of all genes are transcribed in both sense and antisense transcripts that tend to be inversely expressed. Frequently, NATs anneal to 3′-UTRs.

Neofunctionalization The acquisition of a novel, beneficial function by a gene duplicated at some time during evolution, which is preserved by natural selection. The gene copy with the original function is retained.

Nested primer Any primer whose sequence is complementary to an internal site of a DNA molecule that has been amplified with other primers in a conventional polymerase chain reaction (PCR). Such nested primers are used to re-amplify the target sequence at sites different from the original primer sites and thereby increase the specificity of the amplification reaction.

Next generation sequencing A generic term for novel DNA and RNA sequencing technologies with the potential to sequence a human genome for 100 000, or even only 1000 US$, that are not based on the conventional Sanger sequencing procedure.

Non-annotated expressed gene (NAE) Any one of a class of genes, for which either a tag (e.g. SAGE or SuperSAGE tag), an *e*xpressed *s*equence *t*ag (EST) or a cDNA is present in an organism's transcriptome, but whose sequence has not been identified as coding in a sequenced genome of an organism. Most frequently NAEs reside in intergenic regions.

Non-coding DNA Any DNA that does not encode either a polypeptide or an RNA. Non-coding DNA is a major constituent of most eukaryotic genomes, and includes introns, spacers, pseudogenes, centromeres, and most repetitive DNA.

Non-coding RNA (ncRNA, non-protein-encoding RNA, non-protein-coding RNA, npcRNA) Any ribonucleic acid that does not encode a protein and can therefore not be annotated by a search for open reading frames. MicroRNAs, ribosomal RNAs,

7SL-RNAs, small nuclear RNAs, small nucleolar RNAs, small interfering RNAs, small temporal RNAs, telomerase RNAs, transfer RNAs, and Xist-RNAs are examples of such ncRNAs.

Non-contact spotting (non-contact printing) The deposition of target oligonucleotides, cDNAs, DNAs, peptides or proteins onto solid supports ('chips') of glass, quartz, silicon or nitrocellulose by an electrically induced discharge of the solution from the pin onto the surface of the chip. The pin does not come into physical contact with the solid support.

Non-exon probe (NEP) Any one of tens of thousands of 36-nucleotide long oligonucleotide probes on a microarray which is complementary to intronic or intergenic regions. Such oligonucleotides are synthesized on a glass substrate by for example, *m*askless *a*rray *s*ynthesis (MAS) and hybridized to cDNA labeled with a fluorochrome to determine the expression status of the underlying sequences.

Nonprocessive transcription Any gene transcription whose initiation occurs normally but with inefficient elongation. The transcription complex pauses and is rapidly released from the template, leading to an accumulation of short, non-polyadenylated RNAs, and only rarely full-length messenger RNAs.

Nonsense-mediated mRNA decay (Nonsense-mediated decay, NMD) The destruction of eukaryotic messenger RNAs (mRNAs) containing frameshift or nonsense mutations, that would otherwise lead to the synthesis of truncated and thus non-functional proteins. All mRNAs are first monitored for errors that would encode potentially deleterious proteins ('RNA surveillance'). During their exit from the nucleus to the cytoplasm, they are recruited for NMD by the shuttle protein Upf3p (in yeast) if they cannot be translated along their full length. In this case they will remain in a transition complex (i.e. associated with mRNP proteins and Upf3p) which triggers their decay. First, Upf3p forms a binary Upf3p–Upf2p complex ('recruitment complex'), and then a transient bridge between recruitment and termination complexes (mediated by Upf1p in yeast). Finally, Upf1p-associated ATP-dependent $5' \rightarrow 3'$ RNA/DNA helicase unwinds the faulty RNA in the $5' \rightarrow 3'$ direction and induces a topology change that exposes the $5'$ cap making it accessible to the decapping enzyme Dcp1p. Once decapped, the mRNA is fully degraded by Xrn1p from the $5'$-end. NMD requires active translation. Without NMD or similar processes, the eukaryotic cell would produce truncated and most probably non-functional proteins.

Nonstop messenger RNA decay (non-stop decay) A process that eliminates eukaryotic messenger RNAs that do not possess termination codons. Such mRNAs ('non-stop mRNAs') are degraded by the exosome, a highly conserved complex of $3'$ $5'$-exonucleases.

Northern blotting (Northern transfer, RNA blotting) A gel blotting technique in which RNA molecules, separated according to size by agarose or polyacrylamide gel electrophoresis, are transferred directly to a nitrocellulose filter or other matrices by electric or capillary forces (Northern transfer). Single-stranded nucleic acids may be fixed to the nitrocellulose filter by baking and are thus immobilized. Hybridization of specific, radioactively or non-radioactively labeled, single-stranded probes to the immobilized RNA molecules (Northern hybridization) allows the detection of individual RNAs in complex RNA populations.

Nuclear RNA (nRNA) Any RNA that either remains within the nucleus after its synthesis, or is exported into the cytoplasm only after processing. For example, heterogeneous nuclear RNA (hnRNA), including the primary transcripts of many genes (e.g. pre-mRNA, pre-tRNA, pre rRNA), occurs only in the nucleus. The processed transcripts (e.g. mRNA, tRNA, rRNA) are associated with specific proteins and transported into the cytoplasm.

Nucleosome (nu particle, nu body) A disk-shaped structure of eukaryotic chromosomes consisting of a core of eight histone molecules (two each of H2A, H2B, H3 and H4) complexed with 146 bp of DNA and spaced at roughly 100 Å intervals by 'linker' DNA of variable length (8–114 bp) to which histone H1 attaches. Nucleosomes mainly serve to package DNA within the nuclei of eukaryotic cells, but also play important roles in gene activation/inactivation. *In vitro* reconstitution of nucleosomes is possible.

Nucleosome occupancy The density of nucleosome positioning along a stretch of DNA, as revealed for example, by chromatin immunoprecipitation–chip (ChIP-chip) assays employing protein-specific antibodies, mostly directed against specific histones. For example, ChIP-chip experiments in yeast using an antibody against histone H3 or epitope-tagged histone H2B or H4 revealed that promoters and coding regions of transcribed genes generally have fewer (more widely spaced) nucleosomes than non-transcribed genes.

Off-target silencing The undesirable silencing of a gene (or genes), that has a similar sequence to a gene targeted by RNA interference. In the normal RNA interference process, the antisense strand of the siRNA binds to the cognate messenger RNA (mRNA) within the so-called RISC complex. If, by chance, the sense strand of a different mRNA with far-reaching sequence identity is also identified by the siRNA, then this mRNA is also destroyed, although not by intention.

Oligo-capping A technique for the *in vitro* capping of eukaryotic *m*essenger RNA (mRNA) to define the 5′- cap site accurately. In short, isolated mRNA is first treated with alkaline phosphatase to remove the 3′-terminal phosphate, and then with *t*obacco nucleotide *a*cid *p*yrophosphatase (TAP) to remove the 5′cap of the message. Subsequently, a T4 RNA ligase is used to ligate a specific 38-mer oligoribonucleotide to the 5′-end of the de-capped message ('re-capping'). The sequence of the 38-mer oligo cap is only rarely represented in mRNA databases. The oligo-capped mRNA is then converted to a stable cDNA by reverse transcriptase employing either a random hexamer or an oligo(dT) primer. The double-stranded cDNA.is then purified and used to determine the exact sequence around the original cap site.

Oligonucleotide array (oligonucleotide chip, oligonucleotide microarray) A two-dimensional arrangement of thousands, hundreds of thousands, or even millions of short oligonucleotides, immobilized on a membrane, silicon, or glass support, and used to screen for complementary sequences by hybridization. For example, in sequencing by hybridization (SBH), the immobilized oligonucleotides have overlapping sequences and are used to reconstruct the sequence of a target molecule by computer analysis of the resulting hybridization signals.

Open promoter complex A promoter configuration in which the DNA double helix is locally unwound to facilitate the binding of various transcription factors and RNA polymerase to form a pre-initiation complex.

ORFeome The complete set of open reading frames (ORFs) in a particular genome. Specific ORFeomes are designated according to their organism of origin (e.g. hORFeome for *h*uman ORFeome).

Orphan gene (orphan) Any one of a series of open reading frames discovered in genome sequencing projects, whose function is unknown and whose sequence does not reveal any homology with entries in the sequence databanks.

Ortholog (orthologous gene) One of two or more genes (generally, DNA sequences) with similar sequence and identical function(s) in two different genomes that are direct descendants of a sequence in a common ancestor (i.e. without having undergone a gene duplication event). Also called 'homology by descent'.

Overexpression The transcription of a gene at an extremely high rate so that its mRNA is more abundant than under normal conditions. Such overexpression usually occurs in host cells that have been transformed with a cloning vector containing a gene driven by a very strong promoter, allowing the accumulation of its protein product (in some cases this will form up to 40% of the total cellular protein of the host cell). Over-expression may also be due to the presence of a runaway plasmid in a bacterial cell. In eukaryotes it can be responsible for the transforming activity of oncogenes.

Overlapping transcript Any transcript, that overlaps for at least 20 nucleotides with another transcript.

Padlock probe A linear single-stranded oligodeoxynucleotide with target-complementary sequences of 20 bp located at both termini, which are separated by a central spacer element of about 50 bp. Upon hybridization of such a padlock probe to a target sequence, the two ends of the probe are brought into juxtaposition and can then be joined by enzymatic ligation (i.e. by DNA ligase). This leads to a circularization of the oligonucleotide. This intramolecular reaction is highly specific, and discriminates among very similar sequences from two genomes (that differ by only one or a few nucleotides). The circles can then be amplified and identified by for example, hybridization to a microarray. Padlock probes are used for the detection of gene variants and mutations (e.g. determination of copy numbers of specific genomic sequences).

Paired end ditagging (PETting) The ligation of 18-bp long sequence signatures from the 5′- and 3′-ends, respectively, of a cDNA molecule to form a ditag, that can be concatenated with other ditags from other transcripts, be sequenced and mapped to a physical map of a genome to localize the corresponding gene, and simultaneously determine the boundaries of the corresponding transcript. PETs are the basis for the so-called *g*ene *i*dentification *s*ignature (GIS) technique that enables the isolation of tags from both ends of virtually all full-length transcripts of a cell at a given time. The sequence of the PETs can further be exploited to design primers for the amplification of the intervening transcripts by conventional polymerase chain reaction (PCR) techniques.

Passenger strand One of the two RNA strands in double-stranded *s*mall *i*nterfering RNA (siRNA) that is not recognized by the double-strand RNA-binding protein R2D2 and therefore not incorporated into the RISC-loading complex (RLC). However, its complementary strand, the guide strand, is recognized by R2D2 and finally incorporated into the *R*NA-*i*nduced *s*ilencing *c*omplex (RISC), and guides the destruction of complementary messenger RNA. The passenger strand is excluded and destroyed.

Photolithography A technique for the light-dependent engraving of a specific pattern on a solid support, as used in printing processes. The solid support ('plate') is coated with a light-sensitive emulsion and overlaid with a photographic film. The coated plate is then illuminated, and the image of the film is reproduced on the plate. Photolithography is employed in DNA chip technology, where modifications of the usual phosphoramidite reagents are used (i.e. the *d*i*m*ethox-*y*t*r*ityl (DMT) group that protects the 5′hydroxyl is replaced by a photolabile protective group). The synthesis of the oligonucleotides on the chip proceeds by photolithographically deprotecting all the areas that will receive a common nucleoside, and coupling this nucleoside by exposing the entire chip to the appropriate phosphoramidite. This is achieved by using so-called masks made from chromium/glass that contain holes at positions where deprotection is desired. A more advanced procedure exploits a so-called virtual mask. Up to 480 000 (or even more) digitally controlled micro-mirrors allow the illumination of only defined spots on a DNA chip depending on their precise angular position ('mask-less photolithography'). After the oxidation and washing steps the procedure has to be repeated for the next nucleoside.

Phylogenomics (phylogenetics/genomics) A branch of genomics that exploits existing sequence information from various organisms ('evolutionary information') in the databases to assign a specific function to a particular sequence, and that links genome analysis to phylogenetics. Integrating evolutionary analysis improves the accuracy and specificity of functional annotation. Functional predictions are improved by concentrating on questions such as for example, *how* genes became similar in sequence during evolution rather than focusing on sequence similarity *itself*. The term 'phylogenomics' also describes the integration of genomic and comparative genomic data in for example, species tree reconstruction.

PNA array (PNA microarray, PNA chip) Any microarray, onto which *p*eptide *n*ucleic *a*cids (PNAs) instead of conventional nucleic acids with deoxyribose-phosphate backbones are bound via N-terminal groups. PNA arrays do not require any labeling of hybridization probes with radioisotopes, stable isotopes or fluorochromes. Another advantage of PNA arrays is the neutral backbone of PNAs and the increased strength of PNA–DNA pairing. The lack of inter-strand charge repulsion improves the hybridization properties in DNA–PNA duplexes as compared to DNA–DNA duplexes (e.g. the higher binding strength leads to better sequence discrimination in PNA–DNA hybrids than in DNA–DNA duplexes). PNA arrays are used for genome diagnostics, sequencing of DNA or RNA, detection of sequence polymorphisms and identification of expressed genes.

Poly(A) addition signal (poly[A] signal; poly[A] site; poly[A] addition site; poly[A] signal sequence; polyadenylation site)

(a) A hexanucleotide consensus sequence (animals: 5′-AATAAA-3′, 5′-ATTAAA-3′, 5′-AATTAA-3′, 5′-AATAAT-3′, 5′-CATAAA-3′ or 5′-AGTAAA-3′; plants: 5′-AA-TAAN-3′, generally 5′-AATAA-3′ sequence) close (within the last 50 bp) to the 3′-end of most eukaryotic genes transcribed by RNA polymerase II.

(b) The consensus sequence 5′-AAUAAA-3′ in an mRNA molecule that directs the cleavage of the message 10–30 bases 3′ of the element. The cleaved mRNA then serves as a substrate for processive poly- adenylation. First, the so-called *p*olyadenylation *s*pecificity *f*actor (CPSF), a tetrameric protein with subunits of 33, 73, 100, and 160 kDa, binds to the 5′-AAUAAA-3′ signal, the trimeric *c*leavage-*st*imulating *f*actor (CstF; 50, 64, and 77 kDa) then binds to a GU-rich sequence element further downstream of the RNA. The *p*oly(*A*)*p*olymerase (PAP) binds in between the two elements. This complex is joined by two (or more) other proteins, of which the *c*leavage *f*actors CF I and CF II are positioned upstream of the GU-rich box and terminate the mRNA.

Polyadenylated RNA An RNA molecule, that contains a homopolymeric tail of adenyl residues at its 3′-terminus (e.g. poly[A]$^+$-mRNA).

Polyadenylation The post-transcriptional addition of poly(A) tails of up to 200 adenine residues to the 3′-termini of heterogeneous nuclear RNA and messenger RNA in eukaryotes.

Polymerase chain reaction (PCR) An *in vitro* amplification procedure by which DNA fragments of up to 15 kb in length can be amplified about 108-fold. In brief, two 10–30-nucleotides long oligonucleotides complementary to nucleotide sequences at the two ends of the target DNA and designed to hybridize to opposite strands, are synthesized. Excessive amounts of these two oligonucleotide primers (amplimers) are mixed with genomic DNA, and the mixture is heated to denature the duplexes. During the subsequent decrease in temperature the primers will anneal to their genomic homologs and can be extended by DNA polymerase. This sequence of denaturation, annealing of primers and extension is repeated 20–40 times. During the second cycle, the target DNA fragment bracketed by the two primers is among the reaction products, and serves as template for subsequent reactions. Thus repeated cycles of heat denaturation, annealing, and elongation result in an exponential increase in copy number of the target DNA. The use of thermostable DNA polymerases (e.g. *Thermus aquaticus* DNA polymerase; *Pfu* DNA polymerase; Vent DNA polymerase) obviates the necessity of adding new polymerase for each cycle. About 25 amplification cycles increase the amount of the target sequence selectively and exponentially by approximately 106-fold. In later phases of the amplification cycle undesirable, incompletely elongated products may accumulate.

Precursor RNA Any ribonucleic acid synthesized from a gene as a long precursor that is not yet mature but still contains many different regions cut out or modified in later processing steps. Such modifications include capping, polyadenylation, and splicing, which together lead to its final functional form.

Pre-messenger RNA (pre-mRNA) Any complete primary transcript from a structural gene before its post-transcriptional modification. Pre-mRNA is packaged with proteins into *m*essenger *r*ibo*n*ucleo*p*rotein complexes (mRNPs), also called *h*eterogeneous *n*uclear *r*ibo*n*ucleo*p*rotein complexes (hnRNPs), that contain for example, proteins of the hnRNP A family and specific splicing/mRNA export-associated factors such as THO/TREX complexes.

Primary microRNA (pri-miRNA) Any long (up to 1 kb) primary transcript containing a hairpin of 60–120 nucleotides, that encodes a mature microRNA in one of the two strands. The hairpin is cleaved from the pri-miRNA molecule *in nucleo* by the double-strand-specific ribonuclease Drosha. The resulting precursor miRNA ('pre-miRNA') is transported to the cytoplasm by exportin-5, and then further processed by Dicer to generate a short, partially double-stranded RNA, in which one strand represents the mature microRNA. The latter associates with a protein complex similar or identical to the *R*NA-*i*nduced *S*ilencing *C*omplex (RISC).

Primary transcript An RNA molecule immediately after its transcription from DNA (i.e. before any post-transcriptional modifications take place). The primary transcript corresponds to a transcription unit.

Promoter (promotor) A *cis*-acting DNA sequence, 80–120 bp long and located 5′ upstream of the initiation site of a gene to which RNA polymerase may bind and initiate correct transcription. Prokaryotic promoters contain the sequences 5′-TATAATG-3′ (Pribnow box) approximately at position −10, and 5′-TTGACA-3′ at position −35. Eukaryotic promoters differ for the different DNA-dependent RNA polymerases. RNA polymerase I recognizes one single promoter for rDNA transcription, RNA polymerase II transcribes a multitude of genes from very different promoters, which have specific sequences in common (e.g. theTATA box at about position −25 and the CAAT box at about position −90. The so-called house-keeping genes contain promoters with multiple GC-rich stretches with a consensus core sequence, 5′-GGGCGG-3′. RNA polymerase III recognizes either single elements (e.g. in 5S RNA genes) or two blocks of elements (e.g. in all transfer tRNA genes) within the gene. All these consensus sequences function as address sites for DNA-affine proteins (transcription factors) that promote or reduce transcription.

Promoter strength The frequency with which an RNA polymerase molecule can bind to specific consensus sequences within a promoter and express the linked gene. It depends on specific sequences (e.g. TATA box, CAAT-box) and their exact spacing within the promoter region.

Promoter *s*ingle *n*ucleotide *p*olymorphism (promoter SNP, pSNP) Any single nucleotide polymorphism, that occurs in the promoter sequence of a gene. If a pSNP prevents the binding of a transcription factor to its recognition sequence in the promoter, the promoter becomes partly dysfunctional.

Quantitative polymerase chain reaction (Q-PCR; kinetic PCR; real-time PCR; TaqMan technique) The detection of the accumulation of amplification products during conventional polymerase chain reactions and their quantification. Basically, the various techniques of Q-PCR fall into two broad categories. First, the intercalator-based methods include intercalating dyes (such as e.g. ethidium bromide) in each amplification reaction, irradiation of the sample with UV-light in a specialized thermocycler, and detection of the resulting fluorescence light with a computer-controlled, cooled, *c*harge-*c*oupled *d*evice (CCD) camera. By plotting the increase in fluorescence versus cycle number, amplification plots are generated, allowing the products to be quantified. However, this type of Q-PCR suffers from the disadvantage, that both specific and non-specific products generate fluorescence signals, which makes quantitation obsolete. Second, the so-called 5′ nuclease PCR and similar probe-based quantification protocols allow detection of only specific amplification products in real-time. The 5′ nuclease PCR assay exploits the 5′ nuclease activity of *Taq* DNA polymerase to cleave probe–target hybrids during amplification when the enzyme extends from an upstream primer into the region of the probe. This cleavage can be visualized by increased fluorescence, if the oligonucleotide probe contains both a reporter fluorochrome at its 5′-end and a quencher dye at its 3′-end. The close proximity of both fluorochromes (1.5–6.0 nm) results in a Förster-type fluorescence energy transfer, leading to the suppression of the reporter ('quenching') which is relaxed when the probe is hydrolyzed.

Probe-based Q-PCR has been refined to be reproducible. First, the endpoint measurement of the amount of accumulated PCR products is skipped in favor of the more reliable threshold cycle (T_c), which is defined as the fractional cycle number at which the reporter fluorescence generated by cleavage of the probe passes a fixed threshold above baseline. T_c is inversely proportional to the number of target copies in the sample. Quantification is achieved by calculating the unknown target concentration relative to an absolute standard (e.g. a known copy number of plasmid DNAs, or a house-keeping gene as internal control). In contrast to the endpoint approach, T_c is measured when PCR amplification is still in the exponential phase (i.e. the amplicons accumulate at a constant rate, the amplification efficiency is not influenced by variations and limitations of the reaction components, and the enzymes and reactants are still stable). Also, primer–primer artifacts are low in number.

In another version of Q-PCR, the competitive PCR, a synthetic DNA or RNA is used as internal standard (competitor amplicon) which contains the same primer binding sites and (optimally) has the same amplification efficiency as the target, but is of a different size to discriminate it from the target. A known amount of this competitor is co-amplified with the target nucleic acid in the same tube. If the amplification efficiency of target and competitor is identical, then the ratio target/competitor will be constant throughout the PCR process. By determining the target/competitor ratio at the end of the process, and accounting for the starting quantity of the spiked-in competitor, the initial amount of target can be calculated. As opposed to the superior real-time Q-PCR, competitive PCR is tedious, as it requires finding the most suitable ratio of target to competitor by dilution series, and moreover necessitates construction and characterization of a different competitor for every target to be quantified. Also, a series of experiments need to be undertaken to insure that the amplification efficiencies of target and competitor are in fact identical.

Rapid amplification of cDNA ends (RACE) A variant of the conventional polymerase chain reaction technique that uses gene-specific oligodeoxynucleotide primers to amplify cDNAs reverse-transcribed from low-abundance messenger RNAs. Basically, the 3′- or the 5′-end of a cDNA can be amplified. Accordingly, the somewhat different techniques are called 3′-RACE or 5′-RACE, respectively. In short, **3′-RACE** works with an oligo(dT)-containing adaptor primer, partly complementary to the poly(A) tail of mRNAs. This primer allows first strand synthesis with reverse transcriptase. After destruction of the mRNA with RNase H a gene-specific primer complementary to a region at the 5′-end of the original mRNA and a universal adaptor-primer complementary to its 3′-end allow the amplification of the cDNA with an intact 3′-end. The **5′-RACE** technique starts with the annealing of a gene-specific antisense primer complementary to the 3′-region of the mRNA, first strand synthesis with reverse transcriptase, degradation of the mRNA with RNase H, purification of the cDNA, its homopolymer tailing with dCTP, the anchoring of an oligo(dG)-sequence, and the amplification of the cDNA using the anchored primer and a nested gene primer, and PCR.

Rapid analysis of gene expression (RAGE) A technique for the expression analysis of tens to hundreds of genes in multiple samples. In short, RNA is isolated from the target tissue, converted to cDNA using a biotinylated oligo(dT) primer, the cDNAs digested with *Dpn*II, the 3′-most *Dpn*II fragment of each cDNA adsorbed to streptavidin-coated magnetic beads and thereby non-biotinylated fragments removed. A linker with a *Dpn*II-generated overhang (B-linker) is then annealed to the cDNA fragments on the beads and ligated using T4 DNA ligase. The preparation is restricted by *Nla*III, the fragments released from the beads recovered and ligated to a linker with an *Nla*III-generated overhang (A-linker). The cDNA fragments containing the gene-specific targets ligated to A- and B-linkers (or B- and A-linkers) are referred to as A/B or B/A ditags ('bitags'). The templates are then amplified in a polymerase chain reaction using linker–complementary RAGE primers, containing 3–4-nucleotide long specificity regions at the 3′-end. After electrophoresis on 8% polyacrylamide gels, the fragments are simply stained with a fluorescent dye and fluorescent signals digitized in a fluorescence imager.

Read through The transcription of a gene beyond a termination signal sequence through an occasional failure of RNA polymerase to recognize the stop codon. This leads to the synthesis of a so-called read-through messenger RNA.

Recombinant RNA Any RNA molecule composed of two or more heterologous RNAs ligated *in vitro* by T4 RNA ligase.

Region of increased gene expression (RIDGE) Any region of a genome, where the transcription of genes per unit DNA length is at least five times higher than the average and about 20- to 200-fold higher than in the weakly expressed regions. For example, the *m*ajor *h*isto*c*ompatibility locus (MHC) on human chromosome 6 represents such a RIDGE. In yeast, a common RIDGE comprises 2–30 genes, in

Drosophila it covers from 10 to 900 kb. Most of the RIDGEs are characterized by high gene densities, probably harbour house-keeping genes, and locate to subtelomeric regions.

Regulated promoter Any promoter whose activity is limited to those occasions when an inducer (a transcription factor recognizing a consensus sequence in the promoter region) is present.

Regulatory single nucleotide polymorphism (regulatory SNP, rSNP) A relatively rare single nucleotide polymorphism, that affects the expression of a gene (or several genes). Usually this SNP is located in the promoter of the gene.

Repeated exon Any one of a series of exons of similar length and with a high level of sequence homology in genes encoding highly specialized proteins. For example, the *flag*elliform silk (*Flag*) gene of spiders (e.g. *Nephila clavipes*, Araneae, Tetragnathidae) is composed of numerous iterations encoding the three amino acid motifs GPGG(X)n, GGX (G = glycine; P = proline; X = any other amino acid) and a 28-residue spacer. These motifs are organized into complex arrangements each of ~440 amino acids in the encoded protein. The *Flag* gene spans 30 kb and contains 13 exons that are evenly distributed and separated by introns. The first two exons encode the nonrepetitive NH_2-terminus, the last exon contains repetitive sequences and the non-repetitive COOH-terminus. Exons 3–13 each encode an individual repeat, all are about 1.32 kb in length and differ only in the number of tandem GPGG(X)n and GGX motifs ('variable number of tandem motifs'). This exceptionally high sequence similarity of the exons (73%) is indicative of a concerted evolution. The introns separating the repeated exons also share high sequence similarity (87%). Introns 3–12 are each 1.42 kb long and are even more identical to each other than the exons ('repeated introns'). The encoded protein represents an elastic filament that forms the capture spiral of an orb-web and has extreme extensibilities (more than 200%).

Repetitive exon Any one of a series of exons in silk genes of spiders that share extensive sequence similarity and overall architecture. For example, the 30-kb *Flag* gene of *Nephila clavipes* (Araneae), encoding flagelliform silk (the elastic filament forming the capture spiral of an orb-web) is evenly divided into exonic and intronic regions. The first two exons encode the NH_2-terminus, the final exon a repetitive sequence and the carboxy terminus of the corresponding protein. Exons 3 to 13 are of similar lengths (1.32 kb) and identical organization, and each exon encodes a so-called ensemble. An ensemble consists of sequences encoding $GPGG(X)_n$, GGX, and a 28-bp spacer, where the centrally located spacer is flanked by $GPGG(X)_n$ and GGX repeats (G = glycine; P = proline; X = any other amino acid), whose numbers vary between spider species. The introns separating these repeated exons also share high sequence similarity and overall length (1.42 kb) within, but less so, between species.

Repetitive intron Any one of a series of introns in silk genes of spiders that share extensive sequence similarity and separate the so-called repetitive exons from each other. For example, the repetitive introns of the *Flag* gene of *Nephila clavipes* (Araneae) each comprise 1.42 kb and are conserved within species but are divergent between species.

Reporter gene Any gene that is well characterized both genetically and biochemically, may easily be fused to regulatory regions of other genes, and whose activity is normally not detectable in the target organism into which it is transferred. Most reporter gene activities can be easily tested by simple assays (for example the enzymatic activity of the protein product, as for β-galactosidase, β-glucuronidase, chloramphenicol acetyl transferase, luciferase, neomycin phosphotransferase II, nopaline synthase, or octopine synthase). Also, a series of reporters are autofluorescent proteins, as for example, green fluorescent protein (GFP) and the various analogs.

Repression RNA Any RNA, that represses the transcription of a gene by for example, transcription interference, promoter competition, or RNA interference. For example, the *SRG1* RNA transcribed by RNA polymerase II from sequences within the promoter of the *SER3* gene of *Saccharomyces cerevisiae* is such a repression RNA that interferes with the transcription of the gene by binding of activating factors ('activators').

Response element (RE) Any one of a series of short consensus sequences in DNA occurring in the promoters or enhancers of a number of genes that are controlled by the same external stimulus (e.g. temperature: heat-shock element; hormones: glucocorticoid response element, GRE; heavy metals: metal regulatory element).

Restriction The exclusion of foreign DNA from bacterial cells by restriction endonuclease-catalyzed recognition and degradation.

Restriction endonuclease (restriction enzyme) Any bacterial enzyme, that recognizes specific target nucleotide sequences ('recognition site') in double-stranded DNA and catalyzes the breakage of internal bonds between specific nucleotides within these targets, or within a specific distance from there. Restriction generates double-stranded breaks with either cohesive or blunt ends. Restriction endonucleases are part of bacterial restriction-modification systems that protect against foreign DNA. The cell's own DNA is protected by methylation of cytosyl residues in the recognition sites.

Retained intron Any intron that is not removed from the pre-messenger RNA during the splicing process, but retained such that it appears in the mature messenger RNA and encodes part of the corresponding protein. Retained introns are frequently considered to be bona fide exons, and therefore may have biological functions.

Reverse transcription polymerase chain reaction (RT-PCR, cDNA-PCR, RNA-PCR) An *in vitro* RNA amplification procedure that uses retroviral reverse transcriptase or thermostable *Thermus thermophilus* (*Tth*) DNA polymerase to produce a cDNA on the RNA template. This cDNA is then amplified using conventional polymerase chain reaction techniques. The reverse transcriptase catalyzes both the reverse transcription in the presence of $MnCl_2$, and the amplification of the resulting cDNA in the presence of $MgCl_2$. Furthermore its catalytic activity is unimpaired by elevation of the reaction temperature to destabilize complex secondary structures of the RNA for stringent primer annealing. Thus all reactions can be carried out in one test tube.

RNA-PCR allows the amplification of cDNA derived from small amounts of purified mRNA, tRNA, rRNA and viral RNAs and the detection of specific RNAs at a very low copy number, and is therefore used in the analysis of gene expression at the RNA level (e.g. the study of posttranscriptional modifications such as alternative splicing; also for the analysis of mRNA populations of very small cell populations, ideally even of single cells). RNA-PCR may also be combined with *in vitro* translation.

Ribonome (RNome)

(1) The complete set of ribonucleic acid molecules in a cell at a given time.
(2) The complete set of non-coding RNAs in a cell at a given time.

Riboregulation The regulation of the expression of a gene by an RNA molecule, acting either in *cis* or in *trans*. Various small RNAs riboregulate gene activity in for example, *E. coli*. For example, the 10Sa RNA (=tm RNA) binds and inactivates repressor proteins, or the small regulatory RNA, DsrA, functions both as a gene anti-silencer and translation modifier. It overcomes nucleoid-associated H-NS-protein mediated transcriptional silencing of genes, and stimulates the translation of the stationary phase stress s factor, RpoS. Riboregulation is one of the gene-controlling processes in bacteria, plants, and mammals.

Ribotype A formal intermediate between genotype and phenotype, comprising all ribonucleic acids in a cell at a given time (i.e. thousands of messenger RNAs, transfer RNAs, ribosomal RNAs, small nuclear RNAs).

RNA (ribonucleic acid) A mostly single-stranded polynucleotide characterized by its sugar component (ribose) and by the presence of the pyrimidine uracil (instead of thymine in DNA). Single-stranded or also double-stranded RNA are constituents of many viral genomes. In pro- and eukaryotic organisms, RNA serves very different functions. It mediates information flow (e.g. transfer RNA, messenger RNA), has enzymatic functions (e.g. ribozyme), and serves as structural backbone for subcellular particles (e.g. ribosomal RNA). The cellular RNAs can be classified into three main groups: ribosomal RNA (rRNA; about 80–90% of the total cellular RNA), transfer RNA (tRNA; about 6–8%) and messenger RNA (mRNA; usually less than 2%). In cells RNA probably never occurs free, but is complexed with proteins forming ribonucleoprotein particles. Such RNPs also contain a great number of specific small RNAs which do not belong to any of the main groups described above (e.g. adjacent hairpin RNA, ambisense RNA, amplified RNA, antisense RNA, antisense siRNA, catalytic RNA, chromosomal RNA, *cisR*, complementary RNA, degradation-resistant signal RNA, guide RNA, hairpin RNA, heterogenous nuclear RNA, intron-containing hairpin RNA, micro RNA, non-coding RNA, precursor RNA, pre-messenger RNA, ribosomal RNA, scan RNA, sense RNA, sense siRNA, 7SL RNA, short hairpin RNA, short interfering RNA, short stop RNA, SLRNA, small auxin up RNA, small cytoplasmic RNA, small endogenous RNA, small interfering RNA, small interfering stable RNA, small non-coding RNA, small non-mRNA, small nuclear RNA, small nucleolar RNA, small regulatory RNA, small RNA, small temporal RNA, spatial development RNA, spliced leader RNA, stress-response RNA, subgenomic RNA, TAR RNA, telomerase RNA, tiny RNA, tiny expressed RNA, trans-acting RNA, trans-acting small interfering RNA, trans-activation response region RNA, transfer RNA, U-RNA, Xist RNA).

RNA display (RNA fingerprint) The visualization of all, or a subset of all messenger RNA molecules in a given cell at a given time by techniques such as for example, differential display reverse transcription polymerase chain reaction.

RNA gene A sloppy laboratory slang term for any gene encoding an RNA that is not translated into a protein. Genes encoding microRNA, non-coding RNA, ribosomal RNA, short hairpin RNA, short interfering RNA, small RNA, small endogenous RNA, small non-messenger RNA, small regulatory RNA, small temporal RNA, tiny RNA and tRNA are examples of such RNA genes. However, siRNAs are not encoded by discrete genes.

RNA interference (RNAi, RNA-mediated interference, double-stranded RNA-mediated messenger RNA degradation, 'gene silencing') A process of sequence-specific, posttranslational gene silencing in all eukaryotic organisms, that is initiated by double-stranded (ds) RNA homologous to the silenced gene ('RNAi pathway'). In short, RNAi can be divided into two phases. In the so-called initiation phase, dsRNA is processed by the RNase III family nuclease Dicer to produce 21–23-nucleotide long double-stranded so-called *s*mall *i*nterfering RNAs (siRNAs) with symmetric two-nucleotide 3′-overhangs for local interference (and 24–26-nucleotide long siRNAs for systemic interference). In the subsequent so-called effector phase, these siRNAs are incorporated into the multiprotein complex RNA-induced silencing complex (RISC), that targets transcripts by base-pairing between one of the siRNA strands and the endogenous RNA (generally messenger RNA). A nuclease associated with the RISC complex ('slicer') then cleaves the mRNA--siRNA duplex and thus targets cognate mRNA for destruction. Therefore, the RNAi pathway silences specific genes and interferes with gene expression. In *Caenorhabditis elegans*, the dsRNA is amplified and moves from cell to cell, causing a systemic response and ensuring a robust RNAi. RNAi represents a protection mechanism against viruses, retrotransposons, transposons, also transgenes and aberrant single-stranded RNAs. It is also involved in heterochromatin stability (e.g. regulation of histone H3 lysine-9 methylation) of fission yeast, or genome rearrangements in *Tetrahymena*.

RNAi has the potential to engineer the specific control of gene expression and to serve as potent tool for functional genomics. For these purposes, 21 nucleotide siRNAs with two-nucleotide 3′overhangs are designed for the inhibition of specific genes (i.e. for the degradation of the messenger RNAs encoded by these genes). These siRNAs can either be prepared by chemical synthesis, *in vitro* transcription by for example, SP6 *in vitro* transcription system, or the digestion of long double-stranded RNA by RNase III or Dicer. The synthetic siRNAs are then introduced into target cells by electroporation, lentiviral vectors, microinjection, retroviral vectors, transfection, or other techniques, without inducing antiviral response. Also, animals can be fed with bacteria that contain plasmids with cloned siRNA-expressing genes. The siRNAs are then liberated in the digestive tract and extracted. Or siRNA-producing cassettes can be stably integrated into embryonic stem cells and transmitted in the germ-line. The design of a distinct siRNA includes selection of a region located 50–100 nucleotides downstream of the AUG → start codon of the corresponding mRNA. In this region, the sequence AA (N19)TT or AA(N21) is searched, and its G/C percentage calculated (should be 50%, but must be less than 70% and more than 30%). Then a BLAST (using e.g. the NCBI EST database) for the nucleotide sequence fitting the above criteria is performed to ensure that only one single gene is silenced. More than one siRNA for any given target mRNA can be designed to be more effective. Also, siRNAs consisting of negatively charged → peptide nucleic acids ('gripNA') can be employed for gene silencing, since they are more resistant to nucleases and display better sequence specificity than conventional siRNAs.

RNA machine A generic name for any intracellular complex of several to many RNAs that interact physically and cooperate synergistically. Ribosomes are examples of such RNA machines, in which different ribosomal RNAs act in concert with transfer RNAs and the RNA core of peptidyltransferase to synthesize proteins.

RNA-only gene A laboratory slang term for any gene encoding RNA that is not translated into protein. For example, microRNAs are encoded by such RNA-only genes.

RNA polymerase (DNA-dependent RNA polymerase, nucleoside triphosphate: RNA nucleotidyltransferase, transcriptase; EC 2.7.7.6; RNAP) An enzyme catalyzing the formation of RNA using the antisense strand of a DNA duplex as template. In prokaryotes two types of RNA polymerases exist, one synthesizing the RNA primer necessary for DNA replication, the other transcribing structural, ribosomal and transfer RNA genes. In eukaryotes, three distinct nuclear RNA polymerases with different template specificities transcribe rDNA (polymerase I, A), tDNA, 7S-DNA, snDNA and 5S-DNA (polymerase III, C) and the protein-encoding ('structural') genes (polymerase II,B), and can be discriminated by their different sensitivity towards a-amanitin. In plants (e.g. *Arabidopsis thaliana, Oryza sativa*), a fourth RNA polymerase IV is present and silences certain transposons and repetitive DNA in a short interfering pathway involving RNA-dependent RNA polymerase 2 and Dicer-like 3.

RNA profiling The isolation, separation and visualization of (preferably) all ribonucleic acids (RNAs) in a cell, tissue, organ or organism.

RNA–RNA interaction Any interaction between two (or more) identical or different RNA molecules, usually initiated by loops and/or single-stranded stretches of the RNAs. The base-pairing between the codon of a messenger RNA and the anticodon of a transfer RNA represents an example of such an RNA–RNA interaction.

RNA topology The three-dimensional arrangement of a single-stranded RNA chain by the formation of internal fold-backs (hairpin loops) and stem- and -loop structures, its folding into tertiary structures and the changes of these structures in response to physical (e.g. temperature) or chemical parameters (e.g. intercalating agents, or proteins).

RNA world A pre-biotic era in which RNA was the genetic template (not DNA), and able to replicate itself (autocatalytically) and to modify other RNAs (heterocatalytically; analogous to ribozymes).

SAGE adaptation for downsized extracts (SADE) A variant of the conventional serial analysis of gene expression (SAGE) technique for minute amounts of starting tissue (e.g. 0.5 mg tissue, or 50 000 cells, or less from e.g. microdissected specimens) that is run in a single tube from tissue lysis to cDNA tag recovery.

SAGE-Lite A variant of the serial analysis of gene expression (SAGE) technique for global analysis of gene expression patterns that is characterized by a very large reduction in starting material and thus the amount of total RNA required (less than 50 ng). SAGE-Lite is therefore used for expression analysis in rare specimens, bioptic probes and microdissection material.

Scan RNA A specific type of small interfering RNA that is involved in genomic rearrangements (including DNA deletions and chromosomal breakages) during conjugation of the protozoon *Tetrahymena thermophila*. In particular, the scan RNA-mediated DNA deletion requires histone methylation at the recombinational regions. The scan RNAs are expressed prior to the chromosome rearrangements.

Scrambled transcript (scrambled exon) A synonym for an alternatively spliced transcript, in which pairs of exons are joined accurately at consensus splice sites, but in a different order from that present in the genomic DNA and the primary transcript (pre-messenger RNA).

Secondary small interfering RNA (secondary siRNA) Any small interfering RNA (siRNA) that is derived from regions in a messenger RNA located upstream of the original 'trigger' double-stranded (ds) RNA, and exhibits a distinct $5' \rightarrow 3'$ polarity (on the antisense strand). The abundance of secondary siRNAs decreases with increasing distance from the primary trigger RNA region.

Seed region (seed) A laboratory slang term for the six nucleotides at positions 2–7 of the 5′-end of a microRNA, that bind to the 3′-untranslated region (3′-UTR) of a target messenger RNA by Watson–Crick base pairing ('seed pairing'), forming a seed duplex.

Sense–antisense gene (SA gene) Any gene that is transcribed into both the sense and the antisense orientation and therefore encodes two transcripts with opposing polarity. Hundreds of SA pairs are conserved across different species, even maintaining the same overlapping patterns (e.g. complete overlap, head-to-head and tail-to-tail arrangements). Chromosome X in human and mouse, but not fly or worm contains only a few SA genes, probably a consequence of X-inactivation in mammals. SA genes predominantly encode proteins with catalytic activities and basic metabolic functions. At least 25% of the human transcripts are transcribed from SA genes. The abundance of SA genes is low in the worm *Caenorhabditis elegans* (2.8% of all genes), but higher in simpler eukaryotes (yeast: 11%; *Plasmodium falciparum*: 12%).

Sense–antisense pair (SA pair) Any two RNAs (frequently messenger RNAs), that are simultaneously transcribed from the different strands (sense and antisense strand) of the same gene. SA pairs appear in the nucleus (and also cytoplasm) for about 30% (or more) of all human genes. Frequently the transcription of both strands is asynchronous, differentially regulated (e.g. sense up-regulated, and antisense down-regulated), and in many cases the antisense form only partially covers the sense strand.

Sense RNA (sense mRNA) The transcript of a structural gene. Sense RNAs possess the same sequence as the coding strand (sense strand), but a complementary sequence to the template strand (antisense strand).

Sense siRNA (s-siRNA) One of the two strands of double-stranded small interfering RNAs (siRNAs) that is complementary to the antisense siRNA, and therefore does not target any messenger RNA for degradation (i.e. does not contribute to the gene silencing effect).

Sequence alignment The computational juxtaposition of two (or more) linear sequences of nucleotides (in DNA or RNA) or amino acids (in proteins or peptides) for the identification of the extent of homology, sequence variants (e.g. single nucleotide polymorphisms) and stretches of unique and conserved target sequences for example, for the design of primers.

Sequence tag analysis of genomic enrichment (STAGE) A technique for the identification of genomic sequences to which proteins (e.g. transcription factors) bind *in vivo* that is based on the isolation and sequencing of concatemerized short sequence tags derived from target DNA enriched by chromatin immunoprecipitation. In short, proteins (in this case transcription factors) are *in vivo* cross-linked to their target sites in chromatin with formaldehyde, and the chromatin isolated and sheared. After immunoprecipitation of the cross-linked protein–DNA complexes with a specific antibody raised against a given transcription factor, the recovered DNA fragments are dissociated from the antibody and amplified by conventional *p*olymerase *c*hain *r*eaction (PCR) with biotinylated degenerate primers. The amplified DNA fragments are digested with the four-base cutter restriction endonuclease *Nla*III (recognition sequence: 5′CAGT-3′). The biotinylated fragments are captured on streptavidin-coated magnetic beads and ligated to linkers containing a recognition site for the type IIS restriction enzyme *Mme*I. This enzyme cleaves 21 bp away from its recognition site and releases 21-bp tags containing *Nla*III sites from DNA fragments enriched by immunoprecipitation. Two tags each are now ligated to form so-called ditags which are amplified with nested primers, gel-purified, and trimmed by *Nla*III. The ditags are then concatemerized by ligation with T4 DNA ligase, cloned into an appropriate plasmid vector, and the inserts sequenced. Subsequent mapping to the genome identifies genomic loci harboring transcription factor-binding site sequences.

Sequence-tagged site (STS) Any short track of about 200–500 base pairs that is unique to a given genomic DNA fragment and serves to identify that fragment among thousands of other fragments used to construct a genetic and physical map of a eukaryotic genome. STSs, if known for all genomic DNA fragments used for the mapping procedure, eliminate the need to store and exchange clones. If a clone from a specific part of the genome is needed, a database search for an STS mapped to the region of interest will facilitate the design of primers for the polymerase chain reaction amplification of the STS. The amplification product is then labeled and used as a probe to fish the corresponding DNA fragment from a gene library. About 30 000 to 50 000 STSs distributed throughout for example, the human genome at about 100 kb-intervals are sufficient to construct a physical map of the entire human genome (STS map).

Serial analysis of gene expression (SAGE) A high-throughput technique for the simultaneous detection, identification and quantitation of virtually all genes expressed in a given cell at a given time which in addition enables the identification of unknown genes, novel genes, up- or down-regulated genes, to monitor patterns of gene expression at various developmental stages and define disease marker transcripts. SAGE is based on the isolation of a short, 9–14 bp so-called SAGE tag from a defined location within a transcript that contains unique and sufficient information to identify this transcript specifically ('diagnostic tag'). Such tags from various transcripts are then concatenated serially into a single long DNA molecule for efficient sequencing and for identification of the multiple tags simultaneously. The expression pattern of any transcript population can be quantitated by determining the abundance of individual tags and identifying the gene corresponding to each tag. The sequence data is analyzed by special software to identify each gene expressed in the cell, and to determine its expression level. In short, total polyadenylated messenger RNAs are first prepared from the target cell or tissue, reverse transcribed into cDNAs in the presence of a biotinylated oligo (dT) primer (biotin-5′T_{18}-3′) such that they all carry biotin at their 3′-termini. Then the cDNAs are cleaved with the restriction endonuclease *Nla*III ('anchoring enzyme'), and the 3′-terminal cDNA fragment captured with streptavidin-coated magnetic beads. After ligation of an oligodeoxynucleotide linker containing the recognition site for *Bsm* FI ('tagging enzyme', that cleaves 14–20 bp away from its asymmetric recognition site), the linkered cDNA is released from the beads by digestion with *Bsm* FI. The resulting overhang of the released tag is filled in with the Klenow fragment (or DNA polymerase I), the tags are ligated to one another, concatemerized, and amplified in a conventional polymerase chain reaction to create hundreds of copies of each tag. From 30 to 50 such tags are serially ligated in a single DNA molecule, which is cloned and sequenced. The number of times each tag is represented correlates with the number of mRNAs originally present in the cell or tissue (i.e. is an index for the expression of the corresponding gene). However, SAGE detects neither transcripts that lack an *Nla*III site, nor very low abundance messenger RNAs.

Short hairpin RNA (shRNA, small hairpin RNA) Any one of a series of artificial small RNAs, either synthesized exogenously by a T7 RNA polymerase system and transfected into a target cell, or endogenously transcribed from corresponding genes incorporated into the target cell genome and controlled by RNA polymerase III promoters. The shRNAs consist of short, usually 19–30-bp stems and a loop of unpaired bases and variable length, and suppress the expression of target genes through a mechanism resembling RNA interference (i.e. the shRNAs are recognized by Dicer and subsequently cut into the sense and antisense strands of the resulting siRNAs). If expressed constitutively in target cells, hsRNAs can silence specific genes permanently and therefore allow continuous cell lines or transgenic organisms to be established. The presence of a spliceable intron in the shRNA transgene enhances its silencing efficiency.

Short RNA (sRNA) Any one of a series of cytoplasmic and nuclear poly-adenylated RNAs shorter than 200 nucleotides. Such sRNAs map to intronic, intergenic and annotated regions of the human genome. About 20% of sRNAs are evolutionarily conserved and probably have cellular functions. sRNA-encoding sequences cluster at the 5′- or 3′-end of genes.

Single cell analysis of gene expression (SCAGE) A technique for the analysis of (preferably all) transcripts in a single cell, that starts with the lysis of this cell, isolation of its messenger RNA through binding to oligo(dT)-coated solid supports (chips) and the chip-based reverse transcription of the bound RNA into cDNA, using random primers with a 5′-oligo(dC)-tail and a 3′-tailing reaction with dGTP, employing terminal deoxynucleotidyltransferase, generating a 3′-oligo(dG) flanking region. Subsequently the cDNAs are amplified with a single poly(dC) primer (e. g. 5′-TCA GAA TTC ATG CCC CCC CCC CCC CCC-3′) in a conventional polymerase chain reaction. An aliquot is then reamplified in the presence of labeled nucleotides. Specific transcripts can then be detected by PCR using gene-specific primers, or by hybridization of the labeled transcripts to specific microarrays (e.g. a medium-density array loaded with cancer-specific gene sequences). The mRNA profiling of a single cell requires about 60 pg of mRNA for a representative expression analysis (amount in a single cell is 3–6 pg). SCAGE is used to detect and characterize rare cells, such as for example, occult systemically spread tumor cells, or stem cells.

Small interfering RNA (siRNA, also called small inhibitory RNA) Any 21–22-nucleotide long double-stranded RNA (dsRNA) molecule with a 3′ overhang of two nucleotides, that is generated by ribonuclease III from longer double-stranded RNAs, and mediates sequence-specific messenger RNA degradation in eukaryotic cells. SiRNAs efficiently recruit cellular proteins to form an endonuclease complex that specifically recognizes the homologous target RNA and destroys it. Since siRNAs are stable over several cell generations, they have potential in gene-specific therapies, especially since they are effective at concentrations that are several orders of magnitude below those of conventional antisense or ribozyme gene-targeting approaches. Synthetic siRNAs are used to knock-down cognate genes. However, the uptake of siRNAs by target cells presents problems, the stability of the siRNAs *in vivo* has to be secured (by e.g. phosphorylation, methylation or fluorylation of the 2′carbon to prevent endonucleolytic degradation), at the same time the protecting groups should not interfere with the function of the siRNAs, and off-target effects have to be avoided, to name only a few obstacles in siRNA treatment of genetic disorders for example.

Small non-coding RNA (sncRNA) Any one of a multitude of relatively small RNAs, that are not translated into proteins (i.e. are 'non-coding'), but influence or regulate multiple cellular functions. Cell cycle RNAs, *cisR*, microRNAs, non-coding RNA, short hairpin RNA, short interfering RNA, small RNA, small endogenous RNAs, small interfering RNAs, small non-messenger RNAs, small nucleolar RNAs, small regulatory RNAs, small temporal RNAs, spatial development RNAs, stress response RNAs, tiny RNAs, and others all belong to this group of RNAs.

Spatial expression pattern The differential expression of multiple genes of an organism in different organs, tissues, or cells at a given time. This spatial expression pattern changes during cell, tissue and organ development.

Splice junction (splicing junction, splice junction signal, RNA splice site, splice site, ss, plural: sss) Consensus sequences at the ends of introns which are involved in excision and splicing reactions during the posttranscriptional modification of primary transcripts from eukaryotic split genes. The junction signal at the 5′-end of an intron transcript is the donor splice junction, the signal at the 3′-end the acceptor splice junction.

Splicing (RNA splicing, RNA processing, nuclear processing of RNA, pre-messenger RNA splicing) The small nuclear RNA catalyzed excision of introns from a pre-mRNA molecule (primary transcript) and the ligation of exons to create translatable mRNA molecules. This process is part of the posttranscriptional modification of RNA. RNA splicing is essentially a two-step process. First, cleavage occurs at the 5′ splice site (the junction between exon 1 and the intron) to generate intermediates with a free 5′ exon and a lariat form of the intron plus the 3′ exon. Second, the 5′ and 3′ exons are ligated, releasing the mRNA and the fully excised intron lariat. Both reactions are catalyzed by the spliceosome.

Strong promoter ('high level promoter') Any promoter that allows the frequent attachment of DNA-dependent RNA polymerase with high affinity and concomitant increase in the rate of transcriptional initiation of the adjacent gene.

SuperSAGE A variant of the conventional serial analysis of gene expression (SAGE) technique, that allows the genome-wide and quantitative gene expression profiling of cells, tissues, organs and organisms. SuperSAGE basically follows the original SAGE protocol, but involves the type III restriction endonuclease *Eco*P15I that cleaves the cDNA template most distantly from its recognition site. Therefore, the resulting tags are 26 bp long, and much longer than the tags from traditional SAGE (13 bp) or LongSAGE (19–21 bp). The advantages of SuperSAGE are two-fold. First, the information content of a SuperSAGE tag of 26 bp is higher than the conventional tags and enables the identification of a gene directly from the Genbank databases. Second, the ends of linker-tag fragments generated by SuperSAGE are blunt-ended to insure random association of the tags to form ditags. SuperSAGE has the additional benefit of discovering host and pathogen messages simultaneously from the same infected material.

Symmetrical transcription The complete transcription of both strands of a double hyphen;stranded DNA molecule such that two RNAs each of the length of the corresponding strand are produced. This type of transcription is rare, but typical for mitochondria, where the D-loop region contains two promoters, one for the transcription of the H strand, and the other for the L strand of mitochondrial DNA.

Synteny-based positional cloning Any positional cloning of a gene (or genes) in the genome of organism A for which mapping information from the genome of organism B is used. For example, if molecular markers bracket the gene of interest in the genetic map, can be located on a physical map (e.g. constructed with bacterial artificial chromosomes) and facilitate the isolation of the target gene of organism B, then the same molecular markers can be exploited to tag the orthologous gene in organism A such that it can be isolated by the same techniques. This approach is based on synteny.

Synthetic promoter Any promoter that contains regulatory sequences that have been synthesized *in vitro* (e.g. TATA boxes, CAAT boxes enhancer cores, negative elements). The term 'synthetic promoter' is also frequently used synonymously with hybrid promoter.

Tag-based transcription profiling A comprehensive term for all techniques, that allow the isolation and enumeration of so-called tags, short sequences representing (preferably all) the transcripts in a cell at a given time. For example, 5′-SAGE, LongSAGE, *m*assively *p*arallel *s*ignature *s*equencing (MPSS), serial analysis of gene expression (SAGE), SuperSAGE, 3′-SAGE, and other techniques allow transcripts to be profiled on a genome-wide scale.

Tag sequencing techniques A series of high-throughput RNA profiling techniques that are based on the sequencing of short nucleotide stretches ('tags') identifying a specific RNA (e.g. messenger RNAs which differ from each other). For example, tag sequencing is a prerequisite for *m*assively *p*arallel *s*ignature *s*equencing (MPSS) or *s*erial *a*nalysis of *ge*ne *e*xpression (SAGE).

Tandem promoters A special sequence arrangement where a promoter is duplicated and the two promoters are localized in series. Such tandem promoters are characteristic for rDNA genes and serve to accumulate RNA polymerase I molecules for efficient transcription of the linked genes (RNA polymerase I trap, RNA polymerase I trapping center). Promoters in tandem array can also be found in histone genes, and are used in gene technology to insure high expression of cloned genes.

Targeted display A variant of the conventional differential display technique for the identification and isolation of differentially expressed genes. In short, total RNA is first isolated, and then reverse transcribed into cDNA using oligo(dT) primers. The resulting cDNAs are then amplified in a conventional polymerase chain reaction with specially designed so-called targeted display primers, the fragments produced are separated by agarose gel electrophoresis and stained with ethidium bromide. Differentially expressed cDNAs are isolated from the multiple banding pattern, cloned and sequenced (and thereby characterized) or used as probes in Northern blotting or RT-PCR analyses to verify the results of the targeted display.

Targeting induced local lesions in genomes (TILLING, 'targeted knockout') A reverse genetics technique, that introduces a high density of point mutations into a genome by conventional chemical mutagenesis and a subsequent mutational screening to rapidly detect induced lesions. In short, the target organism is first mutagenized by ethylmethane sulfonate (EMS), which primarily induces CG $\rightarrow$ TA transitions (of which about 50% are silent, and most of the rest are missense mutations). DNA is then isolated, and the region of interest (e.g. a gene) is amplified by conventional polymerase chain reaction techniques with region-specific primers. The amplified products are denatured and allowed to reanneal to form heteroduplexes, which are analyzed by *d*enaturing *h*igh *p*ressure *l*iquid *c*hromatography (DHPLC). DHPLC detects mismatches in heteroduplex molecules that appear as extra peaks in the chromatogram. The mutant allele can then be sequenced. As an alternative for mutation detection, the plant endonuclease CEL I from *cel*ery can be employed. This enzyme recognizes a mismatch and cleaves exactly at the 3′ side of the mismatch. Therefore, cutting the mutated DNA with CEL I and subsequent DHPLC analysis or high-resolution polyacrylamide gel electrophoresis pinpoints the precise position of the mismatch. TILLING is used for functional genomics (e.g. the proof of the function of a gene of interest by introducing EMS mutations into it).

T-DNA tagging (T-DNA gene tagging) A method to isolate a gene that has been mutated by the insertion of a T-DNA sequence. In short, T-DNA is integrated into the genomes of plant protoplasts, the transformants regenerated to complete plants, and these plants screened for mutant phenotypes (e.g. a change in growth behavior due to loss-of-function or gain-of-function of a gene of interest). Then a genomic library is constructed from a T-DNA-induced mutant, and screened with a radiolabeled T-DNA as the probe. The T-DNA-containing clones are sequenced, and the gene into which the T-DNA has been inserted, can be isolated directly.

Temporal gene expression Any transcription of a gene and the translation of the resulting messenger RNA into a protein, that is restricted to only a limited time period (e.g. in the development of an organism).

Tentative consensus sequence (TC) A unique virtual transcript, derived from comprehensive *e*xpressed *s*equence *t*ag (EST) databanks by the clustering of the sequences and assembly of cluster elements at high stringency (i.e. removal of low quality, misclustered or chimeric sequences). TCs are generally longer than the individual ESTs that comprise them, so that TCs can be used more efficiently for functional annotation.

Tentative human consensus (THC) A consensus sequence for each putative protein derived from potential protein-coding regions deposited in databases. Sequences are first grouped together if they contain at least 40 bases with greater than 95% identity. Then the groups are assembled to generate a THC, and discordant sequences are eliminated.

Tentative unique contig (TUC) Any group of expressed sequence tags (ESTs), whose members share significant sequence similarity.

Tentative unique gene (TUG) Any gene, that is represented in a comprehensive cDNA or EST library only once.

Tentative unique singlet (TUS) Any cDNA or expressed sequence tag (EST) that lacks significant sequence similarity to other cDNAs or EST sequences from the same organism.

Tentatively unique transcript (TUT) Any messenger RNA, cDNA, or expressed sequence tag, that is represented only once in a cDNA or EST library. Its uniqueness must be proven experimentally.

Thermal asymmetric interlaced (TAIL) polymerase chain reaction (TAIL-PCR) A variant of the conventional polymerase chain reaction that uses two primers of different lengths with different thermal stabilities. First, PCR reactions are carried out at relatively high annealing temperatures that favor priming by the longer primer. Then lower temperatures allow both primers to anneal. By switching the amplification cycles from high to low stringency, target sequences detected by the long and sequence-specific primer are amplified preferentially. TAIL-PCR can be used to isolate promoters of specific genes fairly easily: gene-specific primers are designed and used in concert with arbitrary primers to walk upstream of the gene. Any amplification product will contain promoter sequences. Full promoters can be isolated by repeating this step, i.e. designing promoter-specific primers and using another arbitrary primer to walk still further upstream. Characterization of the promoter can be accomplished by functional analysis (e.g. by *Bal*31 deletion and transient expression of reporter genes with truncated promoter fragments) and sequencing.

***Thermus aquaticus* (Taq) DNA polymerase (*Taq* polymerase, Taquenase^{TM}; EC 2.7.7.7)** A 94-kDa enzyme from the thermophilic eubacterium *Thermus aquaticus*, strain YT 1 or BM, polymerizing deoxynucleotides with little or no $3' \rightarrow 5'$ or $5' \rightarrow 3'$ exonuclease activity, which is highly thermostable (optimum temperature: 70–75 °C) and allows the selective amplification of any cloned DNA about 10 million-fold with very high specificity and fidelity in the so-called polymerase chain reaction. *Taq* polymerase can also be used to label DNA fragments either with radioactive nucleotides, or non-radioactively with biotin or digoxygenin. Furthermore it is ideal for Sanger sequencing of templates with a high degree of secondary structure, since high temperatures will destroy such secondary structures. DNA sequencing with *Taq* DNA polymerase produces uniform band intensities and low background on sequencing gels. The enzyme is also available as recombinant *Taq* polymerase (Ampli Taq, Taquenase).

3′-promoter A laboratory slang term for any promoter sequence that is located at the 3′-end of a gene (rather than at the 5′-end, which is normally the case). For example, in the insulin-like growth factor 2 receptor (Igf2r) gene, a second promoter sequence close to the 3′-end directs the synthesis of an antisense RNA. If this occurs, the antisense RNA forms a duplex RNA with the corresponding sense RNA, leading to the translational incompetence, or destruction of both by RNA interference. As a consequence, both alleles of the *Igf2r* gene are silenced.

Total gene expression analysis (TOGA) A technique for the automated high-throughput analysis of the expression of nearly all genes in a given cell, tissue, or organ. The method is based on the fact that almost all messenger RNAs can be identified by an eight-nucleotide sequence and the distance of this sequence from the poly(A)-tail. In short, poly(A)$^{+}$-mRNA is first isolated and double-stranded cDNA synthesized by reverse transcription, using a pool (e.g. 48) of equimolar *Not*I-containing 5′-biotinylated anchor primers, degenerate in their 3′ ultimate three positions (e.g. 5′-T_{18}VNN [V = A,C or G; N = A,C,G or T]). One primer of this primer mixture initiates synthesis at a fixed position at the 3′-end of all copies of each mRNA species in the sample (defining a 3′ endpoint for each species). Then the cDNAs are cleaved with *Msp*I (recognition site: 5′-CCGG-3′), and the 3′-fragments isolated by streptavidin bead capture and released from the beads by *Not*I digestion. *Not*I cleaves at an eight-nucleotide sequence within the anchor primers (but rarely within the mRNA-derived part of the cDNAs). The resulting *Not*I-*Msp*I fragments are then directionally cloned into a *Cla*I-*Not*I-cleaved expression vector in an antisense orientation to its T3 RNA polymerase promoter, and the constructs transformed into an *E. coli* host. The plasmids are then isolated, the insert-containing vectors linearized with *Msp*I which cleaves at several sites within the vector but not in the cDNA inserts or the T3 promoter (insert-less plasmids are concomitantly inactivated), and antisense cRNA transcripts of the cloned inserts produced with T3 RNA polymerase. These transcripts contain known vector sequences ('tags') abutting the *Msp*I and *Not*I sites. These cRNAs, after removal of the plasmid DNA template with RNase-free DNase, serve as substrates for reverse transcriptase using a primer complementary to the vector sequences. The resulting cDNA is then amplified with a primer extending across the non-reconstituted *Msp*I/*Cla*I site (with either A, C, G or T) and a universal 3′-primer in a conventional polymerase chain reaction. A subsequent PCR with a fluorescent 3′-primer and each of the 256 possible 5′-primers extending four bases into the inserts (each one in a separate reaction), generates products that are separated on denaturing sequencing gels, and the bands detected by laser-induced fluorescence. Each final PCR product carries an identity tag, a combination of an eight-nucleotide sequence (in the case of *Msp*I: $CCGGN_1N_2N_3N_4$) and its distance from the 3′-end of the mRNA (also a known vector-derived sequence added during TOGA processing).

Trans-acting small interfering RNA (tasiRNA) Any one of a series of endogenous low-abundance *s*mall *i*nterfering RNAs (siRNAs), that share a 21- to 22-nucleotide long region of sequence similarity with members of their target gene family. tasiRNAs therefore bind to messenger RNAs derived from these genes and direct their cleavage, leading to the silencing of the genes. Since these siRNAs act *in trans* (i.e. the genes encoding tasiRNAs are located at genomic loci different from those of the target genes) to cleave endogenous mRNA targets, they are coined trans-acting siRNAs. For example, a tasiRNA from *Arabidopsis thaliana*, transcribed from a non-coding locus located in an intergenic region, targets mRNAs from three different *a*uxin *r*esponse *f*actor (*ARF*) genes, *ARF2*, *ARF3/ETT*, and *ARF4*. Both the tasiRNA and its target genes are conserved in rice and maize. Accumulation of this tasiRNA depends on both *R*NA-*d*ependent *R*NA polymerase6, RDR6 (producing bimolecular RNA duplexes from single-stranded RNA molecules) and *Dic*er-*l*ike1 (DCL1), that processes double-stranded microRNA (miRNA) precursors into 21- and 22-nucleotide mature single-stranded miRNAs. During tasiRNA biogenesis, polyadenylated RNAs transcribed from non-protein-coding *TAS* genes are cleaved by a microRNA (miRNA)-programmed RNA-induced silencing complex. In contrast to classical miRNA targets, RDR6 and SGS3 convert one of the *TAS* RNA cleavage products into double-stranded RNA, which is subsequently processed in a phase determined by the initial miRNA cleavage site, by Dicer-like 4 to generate a 21-nucleotide tasiRNA population. tasiRNAs guide endogenous mRNA cleavage through the action of AGO1 or, in some cases, AGO7. Some of the tasiRNA targets regulate the juvenile-to-adult phase transition.

Transcript The single-stranded RNA molecule produced by RNA polymerase I (A) on ribosomal genes (transcript: ribosomal RNA), by RNA polymerase II (B) on structural genes (transcript: messenger RNA), and by RNA polymerase III (C) on transfer RNA genes (transcript: transfer RNA).

Transcript-derived fragment (TDF) Any sequence derived from a transcript (or its cDNA) that is generated by restriction of the corresponding cDNA. Usually, messenger RNA is first isolated, reverse-transcribed into a double-stranded cDNA, the cDNA restricted with appropriate restriction endonucleases (e.g. *Eco* RI and *Mse* I), *Eco* RI- and *Mse* I-complementary adaptors ligated to the restriction fragment, and the fragment amplified by conventional polymerase chain reaction, using adaptor-specific primers. By necessity, TDFs represent only parts of mRNAs.

Transcription The synthesis of an RNA molecule on a DNA or RNA template, catalyzed by DNA-dependent or RNA-dependent RNA polymerases, respectively.

Transcriptional control The regulation of the expression of a particular gene by controlling the number of transcripts produced per unit time, as opposed to translational control.

Transcriptional desert Any genomic region that contains relatively few transcribed genes.

Transcriptional interference The negative *in cis* effect of one promoter on a second adjacent promoter. For example, in tandem promoters, the elongating RNA polymerase II emanating from an upstream promoter negatively interferes with the binding of an activator protein (or proteins) at the downstream promoter.

Transcriptionally active region (TAR) Any region of a genome in which active genes are located. Such TARs can be detected by genome tiling arrays, onto which millions of oligonucleotides are immobilized such that on average every 50 (or less) nucleotides are interrogated. Sense and antisense strands of the target genome are both represented. Fluorescence-labeled cDNA, reverse-transcribed from poly(A)$^{+}$-messenger RNA, is hybridized to the tiling array, and the hybridization pattern detected by fluorescence using a laser scanner.

Transcriptional network The concerted activation (or silencing) of whole batteries of genes (frequently not clustered on the same chromosome, but distributed all over the genome) as a consequence of receiving an internal (e.g. a hormone) or external (e.g. a light flash) signal.

Transcriptional silencing The repression of the activity of a distinct protein-encoding gene, that depends on its location on the chromosome and not on its nucleotide sequence. Three possible mechanisms of gene silencing prevail: (1) the steric hindrance of the binding of upstream activator proteins or DNA-dependent RNA polymerase II by silenced, compacted or sequestered chromatin, or (2) the blockage of the transcription process by an obstacle downstream of the transcription pre-initiation complex (i.e. *T*ATA *b*ox-binding *p*rotein (TBP) and polymerase II have access to their cognate sites), which is coined the 'downstream inhibition model', or (3) the reduced probability of promoter occupancy by RNA polymerase II (i.e. RNA polymerase II is virtually absent from the promoter because transcription factor IIB cannot bind (or is unavailable).

Transcription factor (TF; trans-acting factor, trans-acting protein, nuclear factor, transcriptional activator) Any one of a class of nuclear DNA-binding proteins that interacts with its recognition sequence (binding site), and facilitates the initiation of transcription by eukaryotic DNA-dependent RNA polymerase. Transcription factors may bind to *u*pstream *r*egulatory *s*equences (so-called upstream binding factors, UBFs), to the TATA box or also to sequences within the coding region (e.g. in the case of class III genes). The so-called general transcription factors are highly conserved and interchangeable between mammals, *Drosophila*, yeast and plants.

Transcription factor family A group of regulatory proteins involved in transcription, that either recognize similar or identical target sequences as in the case for example, of transcription factors NF1, CTF, NFY, or CBF, which all recognize the motif 5′-CCAAT-3′, or contain similar protein domains (e.g. helix-turn-helix, leucine zipper, or zinc fingers).

Transcription initiation The start of the transcription of a gene into the corresponding messenger RNA, which presupposes the formation of the RNA polymerase holoenzyme (in prokaryotes) or a transcription initiation complex, consisting of various transcription factors and DNA-dependent RNA polymerase (in eukaryotes). The transcription initiation site is located downstream of the TATA box and upstream of the translation initiation site in eukaryotes.

Transcription profiling The determination of all expressed (transcribed) genes in a cell, tissue, organ or organism at a given time. The process of transcription profiling produces an expression profile.

Transcription start site (TSS; initiator box; initiator, Inr; transcription initiation site, TIS; mRNA initiation site; transcription start point, TSP; cap site) The transcription initiation site of a eukaryotic gene (its location on linear gene maps is indicated by +1) with the consensus sequence 5′-Py-Py-C-**A**-$(Py)_5$-3′ (in animals), 5′-A $(A_{rich})_5$NPy**A**(A/T)-NN$(A_{rich})_6$-3′ (in *Saccharomyces cerevisiae*), or 5′-Py-Py-C-**A**-$(Py)_n$-3′ (in plants). This sequence element also contains the start codon ATG. The TSS in higher eukaryotes is typically located 25 bp downstream of the TATA element, but may also be positioned from 25 to 40 bp (*Schizosaccharomyces pombe*) to 45–125 nucleotides from the TATA box (*S. cerevisiae*). The majority of yeast genes have multiple transcription start sites. The Inr acts as a major transcription promoter in TATA box-less genes. The term 'cap site' refers to the addition of a 7-methyl-guanosine cap to the first nucleotide (mostly an A) of the primary transcript during its processing.

Transcript level (expression level) The relative abundance of a transcript encoded by a specific gene at a specific time.

Transcript mapping The localization of specific cDNAs, cDNA fragments (e.g. expressed sequence tags) or open reading frame sequences on a physical map (i.e. on bacterial artificial chromosome or yeast artificial chromosome clones). The result of transcript mapping is a complete transcript map of a genome.

Transcriptome The entirety of all expressed genes of a genome. Also called 'expressed genome'.

Transcriptome atlas A comprehensive collection of data on the patterns of expressed genes, their expression levels, and their tissue-specific regulation at a particular time point in the life of an organism.

Transcriptome mapping The procedure to establish a genetic map with cDNA-AFLP fragments that differ in size in the respective parents and segregate in the progeny. Usually the resulting map reveals maternal or paternal inheritance of the fragments. Since the cDNA-AFLP fragments are displayed on Northern-type gels, interesting fragments can directly be isolated, sequenced and the sequence annotated such that the underlying gene can be identified.

Transcriptomics The whole repertoire of techniques to analyze and characterize the transcriptome of an organelle or a cell, including RNA isolation, messenger RNA isolation, reverse transcription into complementary DNA, agarose or polyacrylamide gel electrophoresis, Northern blotting, cDNA array techniques, isolation of specific transcripts, their sequence analysis, and use in transgenics.

Transfrag (transcribed fragment) Any one of multiple contiguous genomic fragments that are transcribed under a certain condition. Such transfrags are deduced from hybridization of regularly spaced neighboring oligonucleotides (physically covering a distinct genomic region or a whole genome at 35–70-bp intervals) to cDNA derived from polyadenylated messenger RNA. Neighboring transfrags on the genome are aligned to form transfrag maps.

Transient expression (transient gene expression) The expression of foreign genes that have been introduced into cells, spheroplasts or protoplasts by direct gene transfer, but are not covalently integrated into cellular DNA. These genes are nevertheless transcribed until they are degraded by cytoplasmic and/or nuclear nucleases so that their expression is only transient. Transient expression assays are preferentially used to test the functionality of gene constructs in host cells, especially their promoter strength, and their compatibility with transcription factors. They also serve to optimize DNA delivery into the host cell.

Trans-splicing (trans-RNA splicing, pre-mRNA trans-splicing) The ligation of exons from two (or more) different messenger RNA (mRNA) molecules to form one mature message with a new combination of coding sequences. *Trans*-splicing falls into two categories. First, the spliced leader type of *trans*-splicing, characteristic for protozoa (e.g. trypanosomes) and lower invertebrates (e.g. nematodes) results in the addition of short, capped 5′-noncoding sequences to the mRNA. Second, the discontinuous group II intron types in chloroplasts of algae and higher plants (and also plant mitochondria) involves the joining of independently transcribed coding sequences through unusual interactions between intronic RNA stretches. Both types of *trans*-splicing processes probably accelerate the evolution of novel proteins.

T7 *in vitro* transcription system An *in vitro* system to generate large amounts of specific, homogeneous, biologically active and labeled RNA. In short, total RNA is first isolated from the target cell, reverse transcribed with reverse transcriptase using oligo(dT)-primers, and the template RNA removed by RNase H. The resulting single-stranded antisense cDNA is then exposed to a mixture of random primers that carry a T7 promoter sequence at their 5′-terminus. The resulting double-stranded cDNA then harbors a functional T7 promoter at one end. After denaturation, another T7 random primer is added and amplification again leads to a double-stranded cDNA with a T7 promoter sequence. After purification, T7 RNA polymerase amplifies the sense strand *in vitro* to high quantities. The resulting sense mRNA therefore possesses defined sequences at both ends that can be used as primer binding sites for further amplification. The T7 *in vitro* transcription system is a linear isothermal amplification procedure that increases otherwise limiting amounts of mRNAs for microarray experiments, for example.

Tuschl rules A set of empirically developed rules for the design of effective *s*mall *i*nterfering RNAs (siRNAs). For example, mismatches in the central part of the siRNA duplex prevent the destruction of the target messenger RNA (mRNA) and should be avoided, the G/C content should range between 30 and 52%, at position 15–19 of the sense strand there should be three (or more) A/U base pairs, no internal repeats or palindromes should be possible, and the sense strand should carry an A (at positions 3 and 19), and a T (position 10), but no G at position 13, and no G or C at position 19.

Two-dimensional gene expression fingerprinting (2D-GEF) A variant of the gene *e*xpression *f*ingerprinting (GEF) technique for the visualization and characterization of (preferably) all messenger RNAs in a cell at a given time, that combines the original GEF procedure with the resolving power of two-dimensional polyacrylamide gel electrophoresis. In short, a first step leads to a set of cDNA fragments generated by the same procedure as used in conventional GEF. This primary cDNA fragment population is then resolved according to size in the first dimension by denaturing polyacrylamide gel electrophoresis. For transfer to the second dimension, the resulting gel is subdivided into 96 fractions, and each fraction eluted into a well of a 96-well microtiter plate. Eluted single-stranded fragments are captured by streptavidin-coated beads, a second strand synthesized with *Taq* DNA polymerase and a ^{32}P-labeled adaptor-primer, and the double-stranded fragments sequentially restricted with several restriction endonucleases (selected so as to minimize variations in the number of cDNA fragments liberated after each round of restriction). The cDNA fragments are then resolved by two-dimensional polyacrylamide gel electrophoresis and subsequent autoradiography. The complex patterns can be analyzed using appropriate software, or interesting bands can be isolated, cloned, sequenced, and compared to database entries.

Unassigned reading-frame (URF, unidentified reading-frame) A gene-like nucleotide sequence with proper start and stop codons but without any known function, usually detected by inspection of DNA sequences.

Uncharacterized open reading frame (uncharacterized ORF) Any open reading frame (ORF), detected by sequencing a genome from one species, that is assumed to be real since orthologs exist in one or more other species, which have not however, been verified experimentally (i.e. no gene product has been found as yet).

Unique open reading frame (unique ORF) Any open reading frame that occurs only once in a sequenced genome.

Universal cDNA library (UCL) A comprehensive cDNA library that contains the messenger RNA transcripts of many different cells, tissues, or organs of an organism. Normally these universal superlibraries are extensively normalized to reduce the over-representation of abundant messages. The clones can be arrayed on supports (e.g. nitrocellulose filters, microchips) and serve to detect specific cDNAs of a sample by hybridization procedures.

Universal reference RNA A mixture of DNA-free total RNAs from several tissue types of an organism (e.g. 10 different human tissues or cell lines such as e.g.

B-lymphocytes, brain, cervix, glia, kidney, liver, macrophages, mammary glands, skin and testis) in equal quantities, representing preferably all expressed genes in these tissues/organs. These pooled RNAs are used as internal standards in expression array experiments. For example, universal reference RNA enables the comparison of multiple expression arrays for homogenous spotting and hybridization.

Untranslated RNA A generic term for all ribonucleic acids that are not translated into proteins (as e.g. messenger RNAs are). Non-coding RNA, ribosomal RNA, short interfering RNA, small endogenous RNA, small non-messenger RNA, small regulatory RNA, tiny RNA, transfer RNA (among others) are examples of such untranslated RNAs.

Upstream promoter One of a pair of promoters which both drive the expression of a particular gene that is located 5′ upstream of the gene. The other promoter (downstream promoter) lies at the 3′-end of the gene. For example, transcription of the human *RCC 1* gene is initiated at two different promoters about 9 kb apart. Initiation at the downstream promoter produces a pre-mRNA, in which a 5′-terminal single noncoding exon is spliced to downstream exons encoding the RCC 1 protein. Initiation at the upstream promoter leads to the synthesis of a transcript containing four short noncoding exons spliced to the coding part of the mRNA.

Weak promoter (low level promoter) Any promoter that does not allow the frequent attachment of DNA-dependent RNA polymerase so that the adjacent gene can only be transcribed at a low frequency.

Weak splice site (weak splice junction) Any splice junction, that is only used if a splicing enhancer is present. Weak splice sites are underlying several alternative splicing events.

Whole cell transcription system A concentrated and dialyzed whole-cell extract prepared from eukaryotic cells (e.g. HeLa cells) that contains endogenous RNA polymerase II and is used to initiate messenger RNA synthesis from exogenous templates *in vitro*.

Zinc finger protein transcription factor (ZFP TF, TF_{ZF}) Any one of a series of transcription factors that contains one or more zinc finger protein motifs. Most natural ZFP TFs have three fingers, but some possess as many as 37 such motifs, arranged one after the other such that they can contact multiple adjacent base triplets along the DNA double helix. ZFP TFs can also be assembled *in vitro*, producing novel combinations of the basic motifs that can bind to virtually any gene or promoter sequence in the genome and thereby either activate or inactivate the corresponding gene.

Index

The Handbook of Plant Functional Genomics: Concepts and Protocols.
Edited by Günter Kahl and Khalid Meksem

ISBN: 978-3-527-31885-8

h

i

n